Lecture Notes in Artificial Intelligence 10246

Subseries of Lecture Notes in Computer Science

More information about this series at http://www.springer.com/series/1244

Leszek Rutkowski · Marcin Korytkowski
Rafał Scherer · Ryszard Tadeusiewicz
Lotfi A. Zadeh · Jacek M. Zurada (Eds.)

Artificial Intelligence and Soft Computing

16th International Conference, ICAISC 2017
Zakopane, Poland, June 11–15, 2017
Proceedings, Part II

 Springer

Editors

Leszek Rutkowski
Częstochowa University of Technology
Częstochowa
Poland

Marcin Korytkowski
Częstochowa University of Technology
Częstochowa
Poland

Rafał Scherer
Częstochowa University of Technology
Częstochowa
Poland

Ryszard Tadeusiewicz
AGH University of Science and Technology
Kraków
Poland

Lotfi A. Zadeh
University of California
Berkeley, CA
USA

Jacek M. Zurada
University of Louisville
Louisville, KY
USA

ISSN 0302-9743   ISSN 1611-3349   (electronic)
Lecture Notes in Artificial Intelligence
ISBN 978-3-319-59059-2   ISBN 978-3-319-59060-8   (eBook)
DOI 10.1007/978-3-319-59060-8

Library of Congress Control Number: 2017941502

LNCS Sublibrary: SL7 – Artificial Intelligence

Printed on acid-free paper

This Springer imprint is published by Springer Nature
The registered company is Springer International Publishing AG
The registered company address is: Gewerbestrasse 11, 6330 Cham, Switzerland

Preface

This volume constitutes the proceedings of 16th International Conference on Artificial Intelligence and Soft Computing ICAISC 2017, held in Zakopane, Poland, during June 11–15, 2017. The conference was organized by the Polish Neural Network Society in cooperation with the University of Social Sciences in Łódź, the Institute of Computational Intelligence at the Częstochowa University of Technology, and the IEEE Computational Intelligence Society, Poland Chapter. Previous conferences took place in Kule (1994), Szczyrk (1996), Kule (1997), and Zakopane (1999, 2000, 2002, 2004, 2006, 2008, 2010, 2012, 2013, 2014, 2015, and 2016) and attracted a large number of papers and internationally recognized speakers: Lotfi A. Zadeh, Hojjat Adeli, Rafal Angryk, Igor Aizenberg, Shun-ichi Amari, Daniel Amit, Piero P. Bonissone, Jim Bezdek, Zdzisław Bubnicki, Andrzej Cichocki, Swagatam Das, Ewa Dudek-Dyduch, Włodzisław Duch, Pablo A. Estévez, Erol Gelenbe, Jerzy Grzymala-Busse, Martin Hagan, Yoichi Hayashi, Akira Hirose, Kaoru Hirota, Adrian Horzyk, Eyke Hüllermeier, Hisao Ishibuchi, Er Meng Joo, Janusz Kacprzyk, Jim Keller, Laszlo T. Koczy, Tomasz Kopacz, Zdzisław Kowalczuk, Adam Krzyzak, James Tin-Yau Kwok, Soo-Young Lee, Derong Liu, Robert Marks, Evangelia Micheli-Tzanakou, Kaisa Miettinen, Krystian Mikołajczyk, Henning Müller, Ngoc Thanh Nguyen, Andrzej Obuchowicz, Erkki Oja, Witold Pedrycz, Marios M. Polycarpou, José C. Príncipe, Jagath C. Rajapakse, Šarunas Raudys, Enrique Ruspini, Jörg Siekmann, Roman Słowiński, Igor Spiridonov, Boris Stilman, Ponnuthurai Nagaratnam Suganthan, Ryszard Tadeusiewicz, Ah-Hwee Tan, Shiro Usui, Thomas Villmann, Fei-Yue Wang, Jun Wang, Bogdan M. Wilamowski, Ronald Y. Yager, Xin Yao, Syozo Yasui, Gary Yen, and Jacek Zurada. The aim of this conference is to build a bridge between traditional artificial intelligence techniques and so-called soft computing techniques. It was pointed out by Lotfi A. Zadeh that "soft computing (SC) is a coalition of methodologies which are oriented toward the conception and design of information/intelligent systems. The principal members of the coalition are: fuzzy logic (FL), neurocomputing (NC), evolutionary computing (EC), probabilistic computing (PC), chaotic computing (CC), and machine learning (ML). The constituent methodologies of SC are, for the most part, complementary and synergistic rather than competitive." These proceedings present both traditional artificial intelligence methods and soft computing techniques. Our goal is to bring together scientists representing both areas of research. This volume is divided into five parts:

- Data Mining
- Artificial Intelligence in Modeling, Simulation and Control
- Various Problems of Artificial Intelligence
- Special Session: Advances in Single-Objective Continuous Parameter Optimization with Nature-inspired Algorithms
- Special Session: Stream Data Mining

The conference attracted a total of 274 submissions from 33 countries and after the review process, 133 papers were accepted for publication.

The ICAISC 2017 scientific program included the following special sessions:

1. "Granular and Human-Centric Approaches in Data Mining and Analytics" – Special session devoted to the 70th anniversary of Prof. Janusz Kacprzyk, organized by:

   - Witold Pedrycz, University of Alberta, Edmonton, Canada and Systems Research Institute, Polish Academy of Sciences, Warsaw, Poland
   - Rudolf Kruse, Otto von Guericke University, Magdeburg, Germany
   - Leszek Rutkowski, Czestochowa University of Technology, Poland
   - Jacek Żurada, University of Louisville, USA
   - Ronald R. Yager, Hagan School of Business, Iona College, New Rochelle, NY, USA
   - Sławomir Zadrożny, Systems Research Institute, Polish Academy of Sciences, Warsaw, Poland

2. "Biology as a Source of Technical Inspirations" — Special session devoted to the 70th anniversary of Prof. Ryszard Tadeusiewicz, organized by:

   - Krzysztof Cios, Virginia Commonwealth University, USA
   - Leszek Rutkowski, Czestochowa University of Technology, Poland
   - Jacek Żurada, University of Louisville, USA
   - Bogdan M. Wilamowski, Auburn University, USA

3. "Advances in Single-Objective Continuous Parameter Optimization with Nature-inspired Algorithms" organized by:

   - Swagatam Das, Indian Statistical Institute, India
   - P.N. Suganthan, Nanyang Technological University, Singapore
   - Janez Brest, University of Maribor, Slovenia
   - Roman Senkerik, Tomas Bata University in Zlin, Czech Republic
   - Rammohan Mallipeddi, Kyungpook National University, Republic of Korea

4. "Stream Data Mining" organized by:

   - Piotr Duda, Czestochowa University of Technology, Poland
   - Maciej Jaworski, Czestochowa University of Technology, Poland

I would like to thank our participants, invited speakers, and reviewers of the papers for their scientific and personal contribution to the conference.

Finally, I thank my co-workers Łukasz Bartczuk, Piotr Dziwiński, Marcin Gabryel, Marcin Korytkowski, Agnieszka Piersiak-Puchała, and the conference secretary Rafał Scherer for their enormous efforts to make the conference a very successful event. Moreover, I would like to acknowledge the work of Marcin Korytkowski, who designed the Internet submission system.

June 2017   Leszek Rutkowski

Organization

ICAISC 2017 was organized by the Polish Neural Network Society in cooperation with the University of Social Sciences in Łódź, the Institute of Computational Intelligence at Częstochowa University of Technology.

ICAISC Chairs

Honorary Chairs

Lotfi Zadeh (USA)
Hojjat Adeli (USA)
Jacek Żurada (USA)

General Chairs

Leszek Rutkowski (Poland)

Co-chairs

Włodzisław Duch (Poland)
Janusz Kacprzyk (Poland)
Józef Korbicz (Poland)
Ryszard Tadeusiewicz (Poland)

ICAISC Program Committee

Rafał Adamczak, Poland
Cesare Alippi, Italy
Shun-ichi Amari, Japan
Rafal A. Angryk, USA
Jarosław Arabas, Poland
Robert Babuska, The Netherlands
Ildar Z. Batyrshin, Russia
James C. Bezdek, Australia
Marco Block-Berlitz, Germany
Leon Bobrowski, Poland
Piero P. Bonissone, USA
Bernadette Bouchon-Meunier, France
Tadeusz Burczynski, Poland
Andrzej Cader, Poland
Juan Luis Castro, Spain
Yen-Wei Chen, Japan

Wojciech Cholewa, Poland
Kazimierz Choroś, Poland
Fahmida N. Chowdhury, USA
Andrzej Cichocki, Japan
Paweł Cichosz, Poland
Krzysztof Cios, USA
Ian Cloete, Germany
Oscar Cordón, Spain
Bernard De Baets, Belgium
Nabil Derbel, Tunisia
Ewa Dudek-Dyduch, Poland
Ludmiła Dymowa, Poland
Andrzej Dzieliński, Poland
David Elizondo, UK
Meng Joo Er, Singapore
Pablo Estevez, Chile

Danil Prokhorov, USA
Anna Radzikowska, Poland
Ewaryst Rafajłowicz, Poland
Sarunas Raudys, Lithuania
Olga Rebrova, Russia
Vladimir Red'ko, Russia
Raúl Rojas, Germany
Imre J. Rudas, Hungary
Enrique H. Ruspini, USA
Khalid Saeed, Poland
Dominik Sankowski, Poland
Norihide Sano, Japan
Robert Schaefer, Poland
Rudy Setiono, Singapore
Paweł Sewastianow, Poland
Jennie Si, USA
Peter Sincak, Slovakia
Andrzej Skowron, Poland
Ewa Skubalska-Rafajłowicz, Poland
Roman Słowiński, Poland
Tomasz G. Smolinski, USA
Czesław Smutnicki, Poland
Pilar Sobrevilla, Spain
Janusz Starzyk, USA
Jerzy Stefanowski, Poland
Vitomir Štruc, Slovenia
Pawel Strumillo, Poland
Ron Sun, USA
Johan Suykens, Belgium
Piotr Szczepaniak, Poland

Eulalia J. Szmidt, Poland
Przemysław Śliwiński, Poland
Adam Słowik, Poland
Jerzy Świątek, Poland
Hideyuki Takagi, Japan
Yury Tiumentsev, Russia
Vicenç Torra, Spain
Burhan Turksen, Canada
Shiro Usui, Japan
Michael Wagenknecht, Germany
Tomasz Walkowiak, Poland
Deliang Wang, USA
Jun Wang, Hong Kong, SAR China
Lipo Wang, Singapore
Zenon Waszczyszyn, Poland
Paul Werbos, USA
Slawo Wesolkowski, Canada
Sławomir Wiak, Poland
Bernard Widrow, USA
Kay C. Wiese, Canada
Bogdan M. Wilamowski, USA
Donald C. Wunsch, USA
Maciej Wygralak, Poland
Roman Wyrzykowski, Poland
Ronald R. Yager, USA
Xin-She Yang, UK
Gary Yen, USA
John Yen, USA
Sławomir Zadrożny, Poland
Ali M. S. Zalzala, United Arab Emirates

ICAISC Organizing Committee

Rafał Scherer, Secretary
Łukasz Bartczuk, Organizing Committee Member
Piotr Dziwiński, Organizing Committee Member
Marcin Gabryel, Finance Chair
Marcin Korytkowski, Databases and Internet Submissions

Additional Reviewers

R. Adamczak
M. Al-Dhelaan
T. Babczyński
M. Baczyński
M. Blachnik
L. Bobrowski
P. Boguś
G. Boracchi
B. Boskovic
J. Botzheim
J. Brest
T. Burczyński
R. Burduk
Y. Cheng
W. Cholewa
K. Choros
P. Cichosz
C. Coello Coello
B. Cyganek
J. Cytowski
R. Czabański
I. Czarnowski
F. Deravi
N. Derbel
A. Doi
W. Duch
L. Dymowa
A. Dzieliński
S. Ehteram
B. Filipic
I. Fister
D. Fogel
M. Fraś
A. Galuszka
E. Gelenbe
P. Głomb
F. Gomide
Z. Gomółka
M. Gorawski
M. Gorzałczany
D. Grabowski
E. Grabska
K. Grąbczewski

J. Grzymala-Busse
B. Hammer
Y. Hayashi
Z. Hendzel
F. Hermann
H. Hikawa
Z. Hippe
A. Horzyk
E. Hrynkiewicz
M. Hwang
J. Ishikawa
D. Jakóbczak
E. Jamro
A. Janczak
T. Jiralerspong
R. Jong-Hei
W. Kamiński
Y. Kaneda
W. Kazimierski
V. Kecman
E. Kerre
J. Kitazono
F. Klawonn
J. Kluska
F. Kobayashi
L. Koczy
Z. Kokosinski
A. Kołakowska
J. Konopacki
J. Korbicz
M. Kordos
P. Korohoda
J. Koronacki
M. Korytkowski
M. Korzeń
L. Kotulski
Z. Kowalczuk
M. Kraft
R. Kruse
A. Krzyzak
A. Kubiak
E. Kucharska
J. Kulikowski

O. Kurasova
V. Kurkova
M. Kurzyński
H. Kwaśnicka
J. Kwiecień
A. Ligęza
M. Ławryńczuk
J. Łęski
K. Madani
W. Malina
R. Mallipeddi
J. Mańdziuk
U. Markowska-Kaczmar
A. Martin
A. Materka
J. Mazurkiewicz
V. Medvedev
J. Mendel
D. Meyer
J. Michalkiewicz
Z. Mikrut
S. Misina
W. Mitkowski
M. Morzy
O. Mosalov
G. Nalepa
S. Nasuto
F. Neri
M. Nieniewski
R. Nowicki
A. Obuchowicz
S. Osowski
E. Ozcan
M. Pacholczyk
W. Palacz
K. Patan
A. Pieczyński
A. Piegat
V. Piuri
P. Prokopowicz
A. Przybył
R. Ptak
A. Radzikowska

Contents – Part II

Artificial Intelligence in Modeling, Simulation and Control

Various Problems of Artificial Intelligence

Special Session: Stream Data Mining

Contents – Part I

Evolutionary Algorithms and Their Applications

Computer Vision, Image and Speech Analysis

Bioinformatics, Biometrics and Medical Applications

Data Mining

Computer Based Stylometric Analysis of Texts in Polish Language

Maciej Baj and Tomasz Walkowiak[✉]

Faculty of Electronics, Wroclaw University of Science and Technology,
Wybrzeze Wyspianskiego 27, 50-370 Wroclaw, Poland
tomasz.walkowiak@pwr.edu.pl

Abstract. The aim of the paper is to compare stylometric methods in a task of authorship, author gender and literacy period recognition for texts in Polish language. Different feature selection and classification methods were analyzed. Features sets include common words (the most common, the rarest and all words) and grammatical classes frequencies, as well as simple statistics of selected characters, words and sentences. Due to the fact that Polish is a highly inflected language common words features are calculated as the frequencies of the lexemes obtained by morpho-syntactic tagger for Polish. Nine different classifiers were analysed. Authors tested proposed methods on a set of Polish novels. Recognition was done on whole novels and chunked texts. Performed experiments showed that the best results are obtained for features based on all words. For ill defined problems (with small recognition accuracy) the random forest classifier gave the best results. In other cases (for tasks with medium or high recognition accuracy) the multilayer perceptron and the linear regression learned by stochastic gradient descent gave the best results. Moreover, the paper includes an analysis of statistical importance of used features.

Keywords: Stylometric · Polish · Text analysis · Classification · Machine learning

1   Introduction

Stylometric methods aimed to provide automatic answers to questions about the text class based on the statistical analysis of writing style. It was originally developed to verify authorship of anonymous literary works. Later stylometry goals were extended to assess style differentiation, chronology, genre or author's gender. Stylometry draws on the assumption that there is a similarity of style between the texts in the same group. It relies mostly on machine analysis of some features generated from texts and attempts to define feature characteristics that can identify the class of the text [7].

Stylometry uses machine learning techniques [10]. Representing text as a vector of features can be then analyzed by mathematical tools, eg. classification. The effectiveness of analysis may be affected by various parameters such as the number of classes to define whether the number of corrupted data.

© Springer International Publishing AG 2017
L. Rutkowski et al. (Eds.): ICAISC 2017, Part II, LNAI 10246, pp. 3–12, 2017.
DOI: 10.1007/978-3-319-59060-8_1

Most of stylometric analysis concerns texts in English and the authorship attribution. Studies proved effectiveness of designing specific stylometric analysis for different languages. Frequencies of the most frequent words is the most effective features in English texts authorship attribution [16]. However Polish (compared to English) is a language with a fairly rich inflection and weakly constrained word order [7]. Moreover [6] shows that using the same stylometric methods to analyze the same texts in different languages brings different results.

The aim of this paper is to compare stylometric methods for texts in Polish language. Different methods of feature generation and classification in terms of their effectiveness in identifying three different classification tasks (recognition of gender, author and literacy period) are analyzed. Multiple stylometric methods were implemented and benchmarked.

The paper is structured as follows. It starts with a short review of types of features used in stylometry. It is followed by a description of used data sets, performed classification tasks, feature generation methods and used classifiers. The next section presents achieved recognition results. It also includes statistical feature importance analysis.

2   Features Used in Stylometry

A wide range of used in stylometry features includes characteristics [3] such as:

- **lexical** - these features are based on statistical metrics that can be computed based on the segments used in the text, the most common feature is the frequency of specific words within the text;
- **syntactic** - features based on a study of the relationship between segments of text; sample features:
  - the frequency of nouns, verbs or adjectives,
  - the frequency of grammatical classes;
- **structural** - features specific for the processed forms of writing texts; they are particularly useful in the analysis of documents containing not only text, but also other structures (eg. HTML tags);
- **idiosyncratic** - these features are used in rules established for the specific problem; this can be a defining keywords for texts-establishing rules [23].

For stylometric analysis of text lexical features based on common segments are widely used. Analyzing the common text segments, it is possible to consider the most [6] and rarest [19] frequent words (MFW and RFW) in texts [6].

Changing the number of MFW to be taken into consideration during the analysis affect the efficacy of the method. Different numbers of MFW had been using analyzed: 50 not always most frequent words [14], 150 MFW [2], 300 MFW [21], 500 MFW [4], more than 1000 MFW [11], common words in group of text [14] and all words from all texts [13]. The reported results show that with the increase of a number of used MFW, effectiveness of recognition increases.

3   Performed Stylometric Analysis

Methods vary depending on the data on which the analysis is carried out (harvesting), the type of text to identify the class, characteristics by which the text is described and used classifier. The amount of tested stylometric methods can be calculated by the formula:

$$N_{methods} = |S| \cdot |C| \cdot |F| \cdot |K| \tag{1}$$

where: S-data sets, C-classification tasks, F-feature generation methods, K-classifiers.

It gives following number of performed experiments for different combinations of data sets, classification tasks, types of text features and types of classifiers:

$$N_{methods} = 2 \cdot 3 \cdot 5 \cdot 9 = 270 \tag{2}$$

3.1   Data Sets and Classes

The study was conducted on a collection of Polish authors novels. The firsts set includes each novel whereas the second set was obtained by dividing texts from first set into ten equal parts. Set differs then with number of texts and the size of each of them (see Table 1) Three different classification tasks were analyzed, classifications by authors, literary periods and author genders (see Table 2).

Table 1. Text sets comparison.

	Full texts set	Chunked texts set
Set size	105	1058
Text average size [in characters]	~103 926	~19 652

3.2   Feature Generation Methods

Five different feature generation methods were tested:

- grammatical classes frequency,
- based on statistical features,
- based on common words appearance, three methods:
  - top used,
  - the rarest,
  - all words.

Table 2. Classes representatives.

Class	Representatives number	Representatives
Author	33	Adolf Dygasiński, Kornel Makuszyński, Aleksander Świętochowski, Józef Ignacy Kraszewski, Maria Kuncewiczowa, Irena Zarzycka, Pola Gojawczyńska, Maria Rodziewiczówna, Waleria Marenne, Maria Dąbrowska, Ludwika Godlewska, Gabriela Zapolska, Magdalena Samozwaniec, Antoni Sygietyński, Władysław Reymont, Zofia Nałkowska, Tadeusz Mostowicz, Wacław Berent, Emma Dmochowska, Jarosław Iwaszkiewicz, Józef Korzeniowski, Bolesław Prus, Helena Mniszek, Jadwiga Łuszczewska, Wanda Bęczkowska, Hanna Krzemieniecka, Antonina Domańska, Eliza Orzeszkowa, Zofia Kossak, Henryk Sienkiewicz, Stefan Żeromski, Zygmunt Kaczkowski, Jerzy Żuławski, Michał Bałucki
Period	3	Positivism, Young Poland, Contemporary
Author gender	2	Female, Male

Grammatical Classes Frequency. To gather syntactic features from text, prior morpho-syntactic analysis needs to be made. For these purposes WCRFT [18] tagger was used. Tagger describes the grammatical class of each segment (it also produces many different data about each of text segments). Based on taggers result, for each document a list of grammatical classes frequencies is created. In the case of short texts there is a high probability that there not all grammatical classes will be present. To handle such cases during the subsequent classification, it is necessary to normalize the instances so that set of grammatical classes was the same.

Statistical Features. Features based on statistics are simple to obtain and - due to their low number- fastest in the further classification process. Following suggestions from [17] features were used:

1. average words length - segments classified as special characters are not taken into account during the counting;
2. average sentences length - sentences were recognized based on occurrences of characters *? ! . ;*
3. special characters frequency - the ratio of special characters *? ! . : ; - # * ( )*, to other segments;
4. segments diversity - the ratio of segments occurrences to their number
5. Hapax legomenon - the ratio of segments appearing in the text just once to the amount of all segments.

Statistics were calculated based on altered lexeme of each word.

Features Based on Common Words. According [8] a few assumptions had been made to describe texts with common words used:

- words sets should contain at least 1000 positions [11],
- to compare texts basing top [2], rarest [19] and all words [13].

The method of representation of text using common words is the map, where keys are words present in all positions of the set, and the values defines the occurrence of specific lexeme (obtained by morpho-syntactic tagger) in the analyzed text. Presence of each segment is expressed with TF-IDF (term frequency-inverse document frequency) [20]:

$$N = 1/n \sum_{i=1}^{n} n_i \cdot \log \frac{n}{d} \tag{3}$$

where:
n-number of elements in the whole set
n_i-number of segment occurrences in a text i
d-number of texts, where segment appears at least once.

Words frequency will be the smaller, the more likely to occur in all documents. This solution favors words that are unique in the context of a document, a feature desirable for in stylometry.

3.3   Classification

Following classification algorithms were used in performed experiments:

- ridge regression (Ridge) [10],
- multilayer perceptron (Perceptron) [10],
- k-nearest neighbor (kNN) [10],
- random forest (RF) [1],
- passive-aggresive (PA) [5],
- Elastic net learned by stochastic gradient descent (SGD) [22],
- Rocchio classifier (RC) [12],
- Naive Bayes classifier for multinomial models (MNB) [10],
- Naive Bayes classifier for multivariate Bernoulli models (BNB) [10].

Effectiveness of stylometric methods was tested by splitting texts into training and test sets. The study was performed according to the stratified k-fold cross-validation [10].

4   Results

4.1   Stylometric Features

Stylometric methods based on the similarity of the segments have proven to be much better than those based on classes of grammatical or statistical characteristics. Detailed ranking of stylometric features is presented in the Table 3.

Table 3. Summary of the best classifiers for specific stylometric methods within tasks, distinguishing between sets.

Sets	Tasks	Feature set	Best classifier	Accuracy [in %]
Chunked texts	Literary period	Statistical features	RF	77
		Grammatical classes	kNN	83
		Top used words	PA	98
		Rarest words	MNB	97
		All words	Perceptron	100
	Author	Statistical features	RF	56
		Grammatical classes	RF	87
		Top used words	MNB	100
		Rarest words	Ridge	100
		All words	Ridge	100
	Gender	Statistical features	RF	83
		Grammatical classes	kNN	85
		Top used words	SGD	98
		Rarest words	MNB	100
		All words	PA	100
Full texts	Literary period	Statistical features	SGD	65
		Grammatical classes	Random forest	71
		Top used words	SGD	82
		Rarest words	Perceptron	76
		All words	Perceptron	94
	Author	Statistical features	RF	21
		Grammatical classes	RF	36
		Top used words	BNB	74
		Rarest words	Ridge	42
		All words	SGD	79
	Gender	Statistical features	RF	83
		Grammatical classes	RF	76
		Top used words	RF	88
		Rarest words	Perceptron	82
		All words	SGD	94

Cell values represents the accuracy of the best classifier while recognizing the particular class in a given set.

Averaging analysis (Table 4) of accuracy of both sets and all tasks, ranking of stylometric feature sets is as follows:

1. all words,
2. top 1000 used words,
3. top 1000 rarest words,
4. grammatical classes,
5. statistical features.

Table 4. Summary of average accuracy [%] in the attribution of all classes with each of stylometric methods.

Feature set	Chunked texts	Full texts	Average
Statistical features	72	56	**64**
Grammatical classes	85	61	**73**
Top used words	99	81	**90**
Rarest words	99	67	**83**
All words	100	89	**95**

Similar studies were carried out on the texts of English literature [14]. Test details, such as the size of the data set used for training or recognition phase has not been described. In the research, various methods have been tested and the results of attribution differ. Accuracy in recognition of American authors (Cooper, Crane, Hawthorne, Irving, Twain) of novels of the 19th century (truncated up to 100000 characters) ranged from 50% to 100% depending on the methods used. The method that proved to be the most accurate was using the mix of stylometric features such as "unstable" words (words with commonly used substitutes [15]), function words, 500 MFW (in training), and part-of-speech tags and used SVM as the classifier.

4.2 Training Set Size

The study clearly shows, that bigger the set size is (number of files of chunked texts is ten times greater then in case of full novels), better results can be achieved in attributes recognition. This is true even at expense of the smaller (in experiments ten times smaller) size of a single document.

4.3 Classifiers Accuracy

Results presented in Table 5 show that the random forest classifier was the most accurate. It was especially useful, when other classifiers turned out to be inaccurate. Good quality goes hand in hand with the longest training and learning times. Random Forest would be a good choice where the effectiveness of the stylometric methods are far from maximum.

Where appropriate stylometric characteristics are used for a given problem, accuracy of all classifiers are a way higher. The multilayer perceptron and the elastic net learned by stochastic gradient descent recognize a class of equally high or higher accuracy compared to the random forest. Also times of training and learning of these methods are several times less.

4.4 Statistical Features Importance

Statistical features proved to be the least accurate method from all of stylometric methods tested. However, this method is very fast and requires low memory.

Table 5. Classifiers rank based on how many times they gave the best results on different stylometric methods in different text sets.

Classifier	For both	Chunked texts	Full texts
RF	12	7	5
SGD	6	4	2
MNB	5	2	3
BNB	3	2	1
Perceptron	2	0	2

The RFECV [9] analysis showed that the optimal number of characteristics that should be involved in the classification is less than the number of characteristics used. That proves dependency of statistical characteristics from each other.

Choosing the random forest [1] as a classification method has another advantage that results from the specificity of the analysis training set-build decision trees. The decision about belonging to a particular attribute is taken as a result of a series of decisions - response to another, closely related questions. One question is usually connected with a single feature of the set training facility. The constructed decision tree relevance of questions, and thus also features, is all the greater, the sooner it will be the higher of the desired is a decision tree. This value is called Gini Importance [1].

Valuing the characteristic features with Gini Importance value, comparison due to feature importance is as follows:

1. average sentence length,
2. segments diversity,
3. hapax legomenon,
4. special characters frequency,
5. average segments length.

RFECV [9] analysis confirmed the results obtained from the analysis of the Gini importance calculated while classifying with the random forest. Most often exclude non-optimal statistical feature was the average segments length. The optimal amount of statistical characteristics selected on the basis RFECV can usually be limited to four: the average length of sentences, special characters frequency, segments diversity and hapax legomenon.

4.5   Grammatical Classes Importance

Similar analysis of importance of grammatical classes didn't give similar, easy to define results. Importance of each grammatical classes was unclear. It proves that all grammatical classes are independent and have influence on the final result.

5   Summary

The study examined various feature generation methods, data sets, tasks (recognition of gender, author and literacy period), and methods of classification. Methods examining the similarity on the basis of all the words from texts turned out to be the most accurate. However, the choice of the appropriate method should depend on the specifics of the problem. Classification of long texts may prove to be too costly in case of used memory and computational power, because of too large features vector that have to accommodate data of all words used. Approaches to comparing the texts on the basis of common segments, however, they do not guarantee a fixed size representation of these vectors as analysis of texts based on the frequency of grammatical classes or statistical characteristics.

In a case of classification method the Random Forest is recommended as the best for ill defined problems (with small classification accuracy). In other cases multilayer perceptron and linear regression learned by stochastic gradient descent performs the best.

Future plans includes usage of features based on bi-grams of grammatical classes and mixture of bi-grams and common words. Moreover, we plan to investigate open set classification problem for literary period and author recognition.

References

1. Breiman, L.: Random forests. Mach. Learn. **45**(1), 5–32 (2001). http://dx.doi.org/10.1023/A:1010933404324
2. Burrows, J.F.: Delta: a measure of stylistic difference and a guide to likely. Lit. Linguist Comput. **17**(3), 267–287 (2002)
3. Canales, O., Monaco, V., Murphy, T., Edyta Zych, J.S., Tappert, C., Castro, A., Sotoye, O., Torres, L., Truley, G.: A stylometry system for authenticating students taking online tests. In: Proceedings of Student-Faculty Research Day, CSIS. Pace University (2011)
4. Craig, H., Kinney, A.: Shakespeare, Computers, and the Mystery of Authorship. Cambridge University Press, Cambridge (2009)
5. Crammer, K., Dekel, O., Keshet, J., Shalev-Shwartz, S., Singer, Y.: Online passive-aggressive algorithms. J. Mach. Learn. Res. **7**, 551–585 (2006). http://dl.acm.org/citation.cfm?id=1248547.1248566
6. Eder, M.: Style-markers in authorship attribution: a cross-language study of the authorial fingerprint. Stud. Pol. Linguist. **6**, 99–114 (2011)
7. Eder, M., Piasecki, M., Walkowiak, T.: Open stylometric system based on multi-level text analysis. Cogn. Stud. **17** (2017, to appear)
8. Fomenko, A.T., Fomenko, V.P., Fomenko, T.G.: The authorial invariant in Russian literary texts. Its application: who was the real author of the "quiet don"? In: Fomenko, A.T., Nosovskiy, G.V. (eds.) History: Fiction or Science?, pp. 425–444 (2005)
9. Guyon, I., Weston, J., Barnhill, S., Vapnik, V.: Gene selection for cancer classification using support vector machines. Mach. Learn. **46**(1–3), 389–422 (2002)
10. Hastie, T.J., Tibshirani, R.J., Friedman, J.H.: The Elements of Statistical Learning: Data Mining, Inference, and Prediction. Springer Series in Statistics. Springer, New York (2009). autres impressions: 2011 (corr.), 2013 (7e corr.)

11. Hoover, D.L.: Testing burrows's delta. Liter. Linguist. Comput. **19**(4), 453–475 (2004)
12. Joachims, T.: A probabilistic analysis of the rocchio algorithm with TFIDF for text categorization. In: Fisher, D.H. (ed.) ICML, pp. 143–151. Morgan Kaufmann (1997). http://dblp.uni-trier.de/db/conf/icml/icml1997.html#Joachims97
13. Jockers, M.L., Witten, D.M.: A comparative study of machine learning methods for authorship attribution. Lit. Linguist Comput. **25**(2), 215–223 (2010)
14. Juola, P.: Authorship attribution. Found. Trends Inf. Retr. **1**(3), 233–334 (2006). http://dx.doi.org/10.1561/1500000005
15. Koppel, M., Akiva, N., Dagan, I.: Feature instability as a criterion for selecting potential style markers. J. Am. Soc. Inf. Sci. Technol. **57**(11), 1519–1525 (2006)
16. Koppel, M., Schler, J., Argamon, S.: Computational methods in authorship attribution. J. Am. Soc. Inf. Sci. Technol. **60**(1), 9–26 (2009)
17. Peng, R.D.: Hengartner: quantitative analysis of literary style. Am. Stat. **56**(3), 175–185 (2002)
18. Piasecki, M., Radziszewski, A.: Morphological prediction for polish by a statistical a tergo index. Syst. Sci. **34**(4), 7–17 (2008)
19. Riloff, E.: Little words can make a big difference for text classification. In: Proceedings of the 18th Annual International ACM SIGIR Conference on Research and Development in Information Retrieval, SIGIR 1995, NY, USA, pp. 130–136. ACM, New York (1995). http://doi.acm.org/10.1145/215206.215349
20. Salton, G., McGill, M.J.: Introduction to Modern Information Retrieval. McGraw-Hill Inc., New York (1986)
21. Smith, P., Aldridge, W.: Improving authorship attribution: optimizing burrows' delta method. J. Quant. Linguist. **18**(1), 63–88 (2011)
22. Tsuruoka, Y., Tsujii, J., Ananiadou, S.: Stochastic gradient descent training for l1-regularized log-linear models with cumulative penalty. In: Proceedings of the Joint Conference of the 47th Annual Meeting of the ACL and the 4th International Joint Conference on Natural Language Processing of the AFNLP, ACL 2009, pp. 477–485. Association for Computational Linguistics, Stroudsburg (2009)
23. de Vel, O., Anderson, A., Corney, M., Mohay, G.: Mining e-mail content for author identification forensics. SIGMOD Rec. **30**(4), 55–64 (2001)

Integration Base Classifiers Based on Their Decision Boundary

Robert Burduk$^{(\boxtimes)}$

Department of Systems and Computer Networks,
Wroclaw University of Science and Technology,
Wybrzeze Wyspianskiego 27, 50-370 Wroclaw, Poland
robert.burduk@pwr.edu.pl

Abstract. Multiple classifier systems are used to improve the performance of base classifiers. One of the most important steps in the formation of multiple classifier systems is the integration process in which the base classifiers outputs are combined. The most commonly used classifiers outputs are class labels, the ranking list of possible classes or confidence levels. In this paper, we propose an integration process which takes place in the "geometry space". It means that we use the decision boundary in the integration process. The results of the experiment based on several data sets show that the proposed integration algorithm is a promising method for the development of multiple classifiers systems.

Keywords: Ensemble selection · Multiple classifier system · Decision boundary

1 Introduction

An ensemble of classifiers (EoC) or multiple classifiers systems (MCSs) [7,10] have been a very popular research topics during the last two decades. The main idea of EoC is to employ multiple classifier methods and combine their predictions in order to improve the prediction accuracy. Creating EoC is expected to enable better classification accuracy than in the case of the use of single classifiers (also known as base classifiers).

The task of constructing MCSs can be generally divided into three steps: generation, selection and integration [1]. In the first step a set of base classifiers is trained. There are two ways, in which base classifiers can be learned. The classifiers, which are called homogeneous are of the same type. However, randomness is introduced to the learning algorithms by initializing training objects with different weights, manipulating the training objects or using different features subspaces. The classifiers, which are called heterogeneous, belong to different machine learning algorithms, but they are trained on the same data set. In this paper, we will focus on homogeneous classifiers which are obtained by applying the same classification algorithm to different learning sets.

The second phase of building MCSs is related to the choice of a set of classifiers or one classifier from the whole available pool of base classifiers. If we

© Springer International Publishing AG 2017
L. Rutkowski et al. (Eds.): ICAISC 2017, Part II, LNAI 10246, pp. 13–20, 2017.
DOI: 10.1007/978-3-319-59060-8_2

choose one classifier, this process will be called the classifier selection. But if we choose a subset of base classifiers from the pool, it will be called the ensemble selection. Generally, in the ensemble selection, there are two approaches: the static ensemble selection and the dynamic ensemble selection [1]. In the static classifier selection one set of classifiers is selected to create EoC during the training phase. This EoC is used in the classification of all the objects from the test set. The main problem in this case is to find a pertinent objective function for selecting the classifiers. Usually, the feature space in this selection method is divided into different disjunctive regions of competence and for each of them a different classifier selected from the pool is determined. In the dynamic classifier selection, also called instance-based, a specific subset of classifiers is selected for each unknown sample [2]. It means that we are selecting different EoCs for different objects from the testing set. In this type of the classifier selection, the classifier is chosen and assigned to the sample based on different features or different decision regions [4]. The existing methods of the ensemble selection use the validation data set to create the so-called competence region or level of competence. These competencies can be computed by K nearest neighbours from the validation data set. In this paper, we will use the static classifier selection and regions of competence will be designated by the decision boundary of the base classifiers.

The integration process is widely discussed in the pattern recognition literature [13,18]. One of the existing way to categorize the integration process is by the outputs of the base classifiers selected in the previous step. Generally, the output of a base classifier can be divided into three types [11].

- The abstract level – the classifier ψ assigns the unique label j to a given input x.
- The rank level – in this case for each input (object) x, each classifier produces an integer rank array. Each element within this array corresponds to one of the defined class labels. The array is usually sorted and the label at the top being the first choice.
- The measurement level – the output of a classifier is represented by a confidence value (CV) that addresses the degree of assigning the class label to the given input x. An example of such a representation of the output is a posteriori probability returned by Bayes classifier. Generally, this level can provide richer information than the abstract and rank levels.

For example, when considering the abstract level, voting techniques [16] are most popular. As majority voting usually works well for classifiers with a similar accuracy, we will use this method as a baseline.

In this paper we propose the concept of the classifier integration process which takes place in the "geometry space". It means that we use the decision boundary in the integration process. The decision boundary is another type of information obtained from the base classifiers. In our approach, the decision boundary from the selected base classifiers is averaged in each region of competence separately.

Geometry reasoning is used in the formation of so-called "geometry-based ensemble". The method proposed in [12,14] used characteristic boundary points to define the decision boundary. Based on the characteristic boundary points, there is create a set of hyperplanes that are locally optimal from the point of view of the margin.

The remainder of this paper is organized as follows. Section 2 presents the basic concept of the classification problem and EoC. Section 3 describes the proposed method for the integration base classifiers in the "geometry space" which have been selected earlier (in particular we use Fisher linear discriminant method as a base classifier). The experimental evaluation is presented in Sect. 4. The discussion and conclusions from the experiments are presented in Sect. 5.

2   Basic Concept

Let us consider the binary classification task. It means that we have two class labels $\Omega = \{0, 1\}$. Each pattern is characterized by the feature vector x. The recognition algorithm Ψ maps the feature space x to the set of class labels Ω according to the general formula:

$$\Psi(x) \in \Omega. \tag{1}$$

Let us assume that $k \in \{1, 2, ..., K\}$ different classifiers $\Psi_1, \Psi_2, \ldots, \Psi_K$ are available to solve the classification task. In MCSs these classifiers are called base classifiers. In the binary classification task, K is assumed to be an odd number. As a result of all the classifiers' actions, their K responses are obtained. Usually all K base classifiers are applied to make the final decision of MCSs. Some methods select just one base classifier from the ensemble. The output of only this base classifier is used in the class label prediction for all objects. Another option is to select a subset of the base classifiers. Then, the combining method is needed to make the final decision of EoC.

The majority vote is a combining method that works at the abstract level. This voting method allows counting the base classifiers outputs as a vote for a class and assigns the input pattern to the class with the majority vote. The majority voting algorithm is as follows:

$$\Psi_{MV}(x) = \arg\max_{\omega} \sum_{k=1}^{K} I(\Psi_k(x), \omega), \tag{2}$$

where $I(\cdot)$ is the indicator function with the value 1 in the case of the correct classification of the object described by the feature vector x, i.e. when $\Psi_k(x) = \omega$. In the majority vote method each of the individual classifiers takes an equal part in building EoC. This is the simplest situation in which we do not need additional information on the testing process of the base classifiers except for the models of these classifiers.

3 Proposed Method

The proposed method is based on the observation that the large majority of the integration process used the output of a base classifiers. In addition, the method called "geometry-based ensemble" used characteristic boundary points not the decision boundary [12,14]. Therefore, we propose the method of integrating (fusion) base classifiers based on their decision boundary. Since the proposed algorithm also uses the selection process, it is called *decision-boundary fusion with selection* and labelled Ψ_{DBFS}. The proposed method can be generally divided into five steps.

Step 1: Train each of base classifiers $\Psi_1, \Psi_2, \ldots, \Psi_K$ using different training sets by splitting according to the cross-validation rule.
Step 2: Divide the feature space in different separable decision regions. The regions can be found using points in which the decision boundaries of base classifiers are equal.
Step 3: Evaluate the base classifiers competence in each decision region based on the accuracy. The classification accuracy is computed taking into account the learning set of each base classifier separately.
Step 4: Select l best classifiers from all base classifiers for each decision regions, where $1 < l < K$.
Step 5: Define the decision boundary of the proposed EoC classifier Ψ_{DBFS} as an average decision boundary of the selected in the previous step base classifiers in the geometry space. The decision boundary of Ψ_{DBFS} is defined in each decision region separately. In this step we make the integration process of the selected base classifiers.

The decision boundary obtained in step 5 is applied to make the final decision of the proposed EoC. Graphical interpretation of the proposed method for two-dimensional data set and three base classifiers is shown in Fig. 1.

The method proposed above may be modified at various stages. For example, another division of the training set can be made using different subspaces of the feature space for different base classifiers or by using the bagging method. Another modification relates to step 2, when the competence regions can be found using a clustering method [9]. It should also take into account the fact that the method proposed in step 5 is suitable for linear classifiers.

4 Experimental Studies

In the experiential research 6, benchmark data sets were used. Four of them come from the KEEL Project and two are synthetic data sets – Fig. 2. The details of the data sets are included in Table 1. All the data sets constitute two class problems. In the case of data sets with more than 2 features, the feature selection process [8,15] was performed to indicate two most informative features.

In the experiment 3 Fisher linear discriminant classifiers are used as base classifiers. This means that in the experiment we use an ensemble of the homogenous

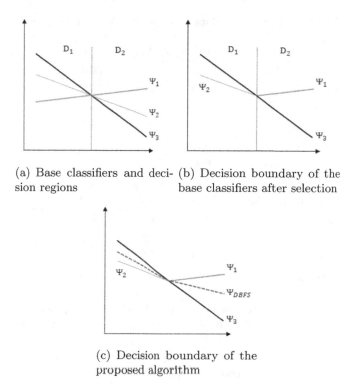

(a) Base classifiers and decision regions

(b) Decision boundary of the base classifiers after selection

(c) Decision boundary of the proposed algorithm

Fig. 1. Example with two-dimensional data set and three base classifiers of the proposed method

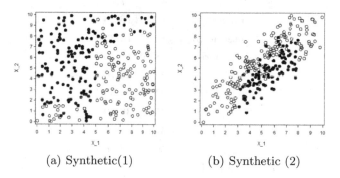

(a) Synthetic(1)

(b) Synthetic (2)

Fig. 2. Syntectic data sets

base classifiers. Their diversity is created by learning either from subsets of the training patterns according to 3-cross-validation method. The learning process was repeated ten times. In each decision region two base classifiers are selected to perform "Step 5" from the algorithm proposed.

Table 2 shows the results of the classification error and the mean ranks obtained by the Friedman test for the proposed method Ψ_{DBFS} and the

Table 1. Description of data sets selected for the experiments

Data set	Example	Attribute	Ration (0/1)
Syntectic$(1)_{150}$	150	2	0.5
Syntectic$(1)_{300}$	300	2	0.5
Syntectic$(1)_{600}$	600	2	0.5
Syntectic$(2)_{150}$	150	2	0.5
Syntectic$(2)_{300}$	300	2	0.5
Syntectic$(2)_{600}$	600	2	0.5
Ionosphere	351	34	1.8
Pima Indians diabetes	768	8	1.9
Sonar	208	60	0.87
Ring$_{7400}$	7400	20	0.5
Ring$_{3700}$	3700	20	0.5
Ring$_{1850}$	1850	20	0.5

Table 2. Classification error and mean rank positions for the proposed method Ψ_{DBFS} and the majority voting method without selection Ψ_{MV} produced by the Friedman test

Data set	Ψ_{MV}	Ψ_{DBFS}
Syntectic$(1)_{150}$	20.36	19.40
Syntectic$(1)_{300}$	27.02	25.34
Syntectic$(1)_{600}$	23.82	22.89
Syntectic$(2)_{150}$	17.00	17.17
Syntectic$(2)_{300}$	17.29	16.36
Syntectic$(2)_{600}$	16.67	15.33
Ionosphere	18.70	19.20
Pima	25.02	25.10
Sonar	24.99	24.52
Ring$_{7400}$	35.03	33.83
Ring$_{3700}$	37.15	36.67
Ring$_{1850}$	39.53	39.36
Mean rank	1.25	1.75

majority voting method without selection Ψ_{MV}. The results were compared with the use of the post-hoc test [17]. This test is useful for pairwise comparisons of the methods considered. The critical difference (CD) for this test at $p = 0.05$, $p = 0.1$, equals $CD = 0.56$ and $CD = 0.47$ respectively. We can conclude that the post-hoc Nemenyi test detects significant differences between the proposed algorithm Ψ_{DBFS} and Ψ_{MV} method at $p = 0.10$. Additionally, at $p = 0.05$ the post-hoc test is not powerful enough to detect any significant differences between those algorithms, but the obtained difference between the mean ranks (0.5) is very close to $CD = 0.56$. This observation confirms that the algorithm Ψ_{DBFS} proposed in the paper can improve the quality of the classification as compared to the method without the selection. It should also be noted that the size of the data set does not allow formulating requests for increasing the data set size and the difference between considered algorithms.

5   Conclusion

In this paper we have proposed a concept of a classifier integration process taking place in the "geometry space". It means that we use the decision boundary in the integration process but we do not consider information produced by the base classifiers such as class labels, a ranking list of possible classes or confidence levels. In the proposed approach the selection process is carried out additionally, while the decision boundary from the selected base classifiers is averaged in each region of competence separately.

The experiments have been carried out on six benchmark data sets. The aim of the experiments was to compare the proposed algorithm Ψ_{DBFS} and the majority voting method without selection Ψ_{MV}. The results obtained show an improvement in the quality of the proposed method with respect to the majority voting method.

Future work might include another division of a training set using different subspaces of the feature space for different base classifiers, using the clustering method to partition the feature space in decision regions or application of the proposed methods for various practical tasks [3,5,6] in which base classifiers are used.

Acknowledgments. This work was supported by the statutory funds of the Department of Systems and Computer Networks, Wroclaw University of Science and Technology.

References

1. Britto, A.S., Sabourin, R., Oliveira, L.E.: Dynamic selection of classifiers-a comprehensive review. Pattern Recogn. **47**(11), 3665–3680 (2014)
2. Cavalin, P.R., Sabourin, R., Suen, C.Y.: Dynamic selection approaches for multiple classifier systems. Neural Comput. Appl. **22**(3–4), 673–688 (2013)
3. Cyganek, B., Woźniak, M.: Vehicle logo recognition with an ensemble of classifiers. In: Nguyen, N.T., Attachoo, B., Trawiński, B., Somboonviwat, K. (eds.) ACIIDS 2014. LNCS, vol. 8398, pp. 117–126. Springer, Cham (2014). doi:10.1007/978-3-319-05458-2_13
4. Didaci, L., Giacinto, G., Roli, F., Marcialis, G.L.: A study on the performances of dynamic classifier selection based on local accuracy estimation. Pattern Recogn. **38**, 2188–2191 (2005)

5. Forczmański, P., Łabędź, P.: Recognition of occluded faces based on multi-subspace classification. In: Saeed, K., Chaki, R., Cortesi, A., Wierzchoń, S. (eds.) CISIM 2013. LNCS, vol. 8104, pp. 148–157. Springer, Heidelberg (2013). doi:10.1007/978-3-642-40925-7_15

6. Frejlichowski, D.: An algorithm for the automatic analysis of characters located on car license plates. In: Kamel, M., Campilho, A. (eds.) ICIAR 2013. LNCS, vol. 7950, pp. 774–781. Springer, Heidelberg (2013). doi:10.1007/978-3-642-39094-4_89

7. Giacinto, G., Roli, F.: An approach to the automatic design of multiple classifier systems. Pattern Recogn. Lett. **22**, 25–33 (2001)

8. Guyon, I., Elisseeff, A.: An introduction to variable and feature selection. J. Mach. Learn. Res. **3**, 1157–1182 (2003)

9. Jackowski, K., Krawczyk, B., Woźniak, M.: Improved adaptive splitting and selection: the hybrid training method of a classifier based on a feature space partitioning. Int. J. Neural Syst. **24**(03), 1430007 (2014)

10. Korytkowski, M., Rutkowski, L., Scherer, R.: From ensemble of fuzzy classifiers to single fuzzy rule base classifier. In: Rutkowski, L., Tadeusiewicz, R., Zadeh, L.A., Zurada, J.M. (eds.) ICAISC 2008. LNCS (LNAI), vol. 5097, pp. 265–272. Springer, Heidelberg (2008). doi:10.1007/978-3-540-69731-2_26

11. Kuncheva, L.I.: Combining Pattern Classifiers: Methods and Algorithms. Wiley Inc., Hoboken (2004)

12. Li, Y., Meng, D., Gui, Z.: Random optimized geometric ensembles. Neurocomputing **94**, 159–163 (2012)

13. Ponti, Jr., M.P.: Combining classifiers: from the creation of ensembles to the decision fusion. In: 2011 24th SIBGRAPI Conference on Graphics, Patterns and Images Tutorials (SIBGRAPI-T), pp. 1–10. IEEE (2011)

14. Pujol, O., Masip, D.: Geometry-based ensembles: toward a structural characterization of the classification boundary. IEEE Trans. Pattern Anal. Mach. Intell. **31**(6), 1140–1146 (2009)

15. Rejer, I.: Genetic algorithms for feature selection for brain computer interface. Int. J. Pattern Recogn. Artif. Intell. **29**(5), 1559008 (2015)

16. Ruta, D., Gabrys, B.: Classifier selection for majority voting. Inf. Fusion **6**(1), 63–81 (2005)

17. Trawiński, B., Smętek, M., Telec, Z., Lasota, T.: Nonparametric statistical analysis for multiple comparison of machine learning regression algorithms. Int. J. Appl. Math. Comput. Sci. **22**(4), 867–881 (2012)

18. Tulyakov, S., Jaeger, S., Govindaraju, V., Doermann, D.: Review of classifier combination methods. In: Marinai, S., Fujisawa, H. (eds.) Machine Learning in Document Analysis and Recognition. SCI, vol. 90, pp. 361–386. Springer, Heidelberg (2008)

Complexity of Rule Sets Induced by Two Versions of the MLEM2 Rule Induction Algorithm

Patrick G. Clark[1], Cheng Gao[1], and Jerzy W. Grzymala-Busse[1,2(✉)]

[1] Department of Electrical Engineering and Computer Science,
University of Kansas, Lawrence, KS 66045, USA
patrick.g.clark@gmail.com, {cheng.gao,jerzy}@ku.edu
[2] Department of Expert Systems and Artificial Intelligence,
University of Information Technology and Management,
35-225 Rzeszow, Poland

Abstract. We compare two versions of the MLEM2 rule induction algorithm in terms of complexity of rule sets, measured by the number of rules and total number of conditions. All data sets used for our experiments are incomplete, with many missing attribute values, interpreted as lost values, attribute-concept values and "do not care" conditions. In our previous research we compared the same two versions of MLEM2, called true and emulated, with regard to an error rate computed by ten-fold cross validation. Our conclusion was that the two versions of MLEM2 do not differ much, and there exists some evidence that lost values are the best. In this research our main objective is to compare both versions of MLEM2 in terms of complexity of rule sets. The smaller rule sets the better. Our conclusion is again that both versions do not differ much. Our secondary objective is to compare three interpretations of missing attribute values. From the complexity point of view, lost values are the worst.

Keywords: Incomplete data · Lost values · Attribute-concept values · "Do not care" conditions · MLEM2 rule induction algorithm · Probabilistic approximations

1   Introduction

In this paper we report results of experiments conducted on incomplete data. In data mining, missing attribute values are usually handled using imputation. Our approach is based on rough set theory, we rather modify the process of rule induction, taking into account an interpretation of missing attribute values, than replace missing attribute values by existing values.

We distinguish between three interpretations of missing attribute values: lost values, attribute-concept values and "do not care" conditions [1]. A lost value is denoted by "?", an attribute-concept value is denoted by "−", and a "do not

© Springer International Publishing AG 2017
L. Rutkowski et al. (Eds.): ICAISC 2017, Part II, LNAI 10246, pp. 21–30, 2017.
DOI: 10.1007/978-3-319-59060-8_3

care" condition is denoted by "*". Lost values are interpreted as values that are not accessible, e.g., erased. Attribute-concept values are typical attribute values for a given concept. "Do not care" conditions, are interpreted as any possible attribute value. Incomplete data sets need special kind of approximations, called singleton, subset and concept [2]. In our experiments we used probabilistic approximations, a generalization of lower and upper approximations. A probabilistic approximation is associated with some parameter α, interpreted as a probability. If $\alpha = 1$, the probabilistic approximation is identical with the lower approximation, if α is a positive number slightly larger than 0, the probabilistic approximation is the upper approximation. Probabilistic approximations are usually applied to completely specified data [3–11], however when applied to incomplete data must be generalized [12]. Some experimental research on probabilistic approximations applied to incomplete data was originated in [13,14].

In this paper we compare two different approaches to rule induction from incomplete data, both based on the MLEM2 algorithm (Modified Learning from Examples Module, version 2). Experiments comparing the two versions MLEM2 from the view point of error rate were reported in [15]. When experiments were conducted on incomplete data sets with many missing attribute values, also on error rate, results were more conclusive [16], this paper presents the results on the complexity of the induced rule sets.

2   Incomplete Data Sets

An example of incomplete data set is presented in Table 1. A *concept* is a set of all cases with the same decision value. In Table 1 there are two concepts, e.g., the set of all cases with flu is the set {1, 2, 3, 4}.

Table 1. An incomplete data set

	Attributes			Decision
Case	Temperature	Headache	Cough	Flu
1	normal	*	yes	yes
2	high	yes	no	yes
3	–	no	yes	yes
4	high	?	?	yes
5	high	no	*	no
6	?	no	*	no
7	high	–	no	no
8	–	no	no	no

We use notation $a(x) = v$ if an attribute a has the value v for the case x. The set of all cases will be denoted by U. In Table 1, $U = \{1, 2, 3, 4, 5, 6, 7, 8\}$.

For complete data sets, for an attribute-value pair (a, v), a *block* of (a, v), denoted by $[(a, v)]$, is the following set

$$[(a, v)] = \{x | x \in U, a(x) = v\}.$$

For incomplete decision tables the definition of a block of an attribute-value pair must be modified in the following way [1,2]:

- If for an attribute a and a case x, if $a(x) = ?$, the case x should not be included in any blocks $[(a, v)]$ for all values v of attribute a,
- If for an attribute a and a case x, if $a(x) = -$, the case x should be included in blocks $[(a, v)]$ for all specified values $v \in V(x, a)$ of attribute a, where

$$V(x, a) = \{a(y) \mid a(y) \text{ is specified}, y \in U, d(y) = d(x)\},$$

and d is the decision.
- If for an attribute a and a case x, if $a(x) = *$, the case x should be included in blocks $[(a, v)]$ for all specified values v of attribute a.

For a case $x \in U$ the *characteristic set* $K_B(x)$ is defined as the intersection of the sets $K(x, a)$, for all $a \in B$, where B is a subset of the set A of all attributes and the set $K(x, a)$ is defined in the following way:

- If $a(x)$ is specified, then $K(x, a)$ is the block $[(a, a(x))]$ of attribute a and its value $a(x)$,
- If $a(x) = ?$ or $a(x) = *$ then the set $K(x, a) = U$,
- If $a(x) = -$, then the corresponding set $K(x, a)$ is equal to the union of all blocks of attribute-value pairs (a, v), where $v \in V(x, a)$ if $V(x, a)$ is nonempty. If $V(x, a)$ is empty, $K(x, a) = U$.

For the data set from Table 1, the set of blocks of attribute-value pairs is

$[(Temperature, normal)] = \{1, 3\}$,
$[(Temperature, high)] = \{2, 3, 4, 5, 7, 8\}$,
$[(Headache, no)] = \{1, 3, 5, 6, 7, 8\}$,
$[(Headache, yes)] = \{1, 2\}$,
$[(Cough, no)] = \{2, 5, 6, 7, 8\}$,
$[(Cough, yes)] = \{1, 3, 5, 6\}$.

For Table 1, $V(3, Temperature) = \{normal, high\}$, $V(7, Headache) = \{no\}$, and $V(8, Temperature) = \{high\}$. The corresponding characteristic sets are

$K_A(1) = \{1, 3\}$,
$K_A(2) = \{2\}$,
$K_A(3) = \{1, 3, 5\}$,
$K_A(4) = \{2, 3, 4, 5, 7, 8\}$,
$K_A(5) = \{3, 5, 7, 8\}$,
$K_A(6) = \{1, 3, 5, 6, 7, 8\}$,
$K_A(7) = \{5, 7, 8\}$,
$K_A(8) = \{5, 7, 8\}$.

3   Probabilistic Approximations

For incomplete data sets there exist a number of different definitions of approximations. In this paper we will use only *concept* approximations, skipping the word *concept*.

The B-*lower approximation* of X, denoted by $\underline{appr}(X)$, is defined as follows

$$\cup\ \{K_B(x) \mid x \in X, K_B(x) \subseteq X\}. \tag{1}$$

Such lower approximations were introduced in [2,17].

The B-*upper approximation* of X, denoted by $\overline{appr}(X)$, is defined as follows

$$\cup\ \{K_B(x) \mid x \in X, K_B(x) \cap X \neq \emptyset\} = \cup\ \{K_B(x) \mid x \in X\}.$$

These approximations were studied in [2,17].

A B-probabilistic approximation of the set X with the threshold α, $0 < \alpha \leq 1$, denoted by $B\text{-}appr_\alpha(X)$, is defined as follows

$$\cup\{K_B(x) \mid x \in X,\ Pr(X \mid K_B(x)) \geq \alpha\},$$

where $Pr(X \mid K_B(x)) = \frac{|X \cap K_B(x)|}{|K_B(x)|}$ is the conditional probability of X given $K_B(x)$. A-probabilistic approximations of X with the threshold α will be denoted by $appr_\alpha(X)$.

4   Rule Induction

In our experiments we used two versions of the MLEM2 rule induction algorithm, called true MLEM2 and emulated MLEM2.

4.1   True MLEM2

In this approach, the MLEM2 rule induction algorithm, modified to accept parameter α, is implemented from scratch. For a given concept X and parameter α, we compute the probabilistic approximation $appr_\alpha(X)$. The set $appr_\alpha(X)$ is globally definable [18]. Thus, we may use the MLEM2 strategy to induce rule sets directly from the set $appr_\alpha(X)$ [19,20]. For example, for Table 1, for the concept $[(Flu, yes)] = \{1, 2, 3, 4\}$ and for the probabilistic approximation $appr_{0.5}(\{1, 2, 3, 4\}) = \{1, 2, 3, 4, 5, 7, 8\}$, using the true MLEM2 approach, the following two rules are induced

1, 3, 6
(Temperature, high) -> (Flu, yes)
1, 2, 2
(Temperature, normal) -> (Flu, yes)

Every rule is preceded by three numbers: the total number of attribute-value pairs on the left-hand side of the rule, the total number of cases correctly classified by the rule during training, and the total number of training cases matching the left-hand side of the rule, i.e., the rule domain size.

4.2   Emulated MLEM2

It is much more convenient to use the standard MLEM2 rule induction algorithm. The standard MLEM2 algorithm computes lower and upper approximations for all concepts. The possible rule set, induced from an upper approximation, should be used. The usual strategy of MLEM2 must be slightly modified. We will illustrate it by inducing rules for the concept [(*Flu*, *yes*)] and for the probabilistic approximation $appr_{0.5}(\{1,2,3,4\})$. A new data set must be created in which for all cases from the set $appr_1(\{1,2,3,4\})$ the decision values are copied from the original data set (Table 1). For cases that are not in the set $appr_1(\{1,2,3,4\})$, a new decision value is added. In our experiments this new decision value was called SPECIAL. Such a new data set is presented in Table 2.

Table 2. A preliminary modified data set

	Attributes			Decision
Case	Temperature	Headache	Cough	Flu
1	normal	*	yes	yes
2	high	yes	no	yes
3	–	no	yes	yes
4	high	?	?	yes
5	high	no	*	no
6	?	no	*	SPECIAL
7	high	–	no	no
8	–	no	no	no

This data set is an input to the standard MLEM2 algorithm. The MLEM2 algorithm computes the upper approximation of the set {1, 2, 3, 4} to be {1, 2, 3, 4, 5, 7, 8}, and induces the following rule set

1, 3, 6
(Temperature, high) -> (Flu, yes)
1, 2, 2
(Temperature, normal) -> (Flu, yes)
2, 3, 4
(Temperature, high) & (Headache, no) -> (Flu, no)
1, 1, 6
(Headache, no) -> (Flu, SPECIAL)

where the three numbers that precede every rule are computed from Table 2. Because we are inducing rules for the concept [(Flu, yes)], only the first two rules should be saved and the remaining two rules should be deleted in computing the final rule set. In our example, the three numbers preceding every rule do not need any modification, but in general, such numbers should be adjusted.

5   Experiments

In our experiments, we used six data sets available from the University of California at Irvine *Machine Learning Repository*. For every data set, an incomplete data set was created. First, we gradually used "?"s (lost values) for a random replacement of specified values, until an entire row of a data set was full of "?"s. If so, the last replacement was called void. Thus, the data set was saturated with "?"s, and in any row there exists at least one specified value. Then two additional incomplete data sets were created by global editing, all "?"s were replaced by "−"s and by "*"s, respectively.

Our main objective was to compare two versions of MLEM2, true and emulated, in terms of complexity of the induced rules. The smaller rule sets the better. Results of our experiments, presented in Figs. 1, 2, 3, 4, 5, 6, 7, 8, 9, 10, 11 and 12, show that there is no significant difference between the two versions.

For every data set and three interpretations of missing attribute values, we compare the size of rule sets. Since the experiments were conducted on six data sets, with three interpretations of missing attribute values, the total number of combinations is 18. The number of rules in induced rule sets was smaller for the emulated version of MLEM2 than for the true version of MLEM2 in five out of 18 combinations (for *Breast cancer* with "−"s and "*"s, *Echocardiogram* with "*"s, *Lymphography* with "*"s and *Wine recognition* with "*"s). For four combinations, the true version of MLEM2 was better than the emulated version of MLEM2 (for *Image segmentation* with "?"s and "−"s and for *Wine recognition* with "−"s and "*"s. For the remaining nine combinations, the difference between emulated and true versions of MLEM2 is not statistically significant (5% significance level, two-tailed Wilcoxon matched-pairs signed-ranks test). Note that the number of rules induced by both versions of MLEM2 is identical for *Hepatitis* with "?"s.

The total number of conditions in induced rule sets was smaller for the emulated version of MLEM2 than for the true version of MLEM2 in seven out of 18 combinations (for *Breast cancer* with "−"s and "*"s, *Echocardiogram* with "?"s, "−"s and "*"s, *Lymphography* with "*"s and *Wine recognition* with "*"s). For three combinations, the true version of MLEM2 was better than the emulated version of MLEM2 (for *Image segmentation* with "−"s, for *Lymphography* with "−"s and for *Wine recognition* with "?"s. For the remaining eight combinations, the difference between emulated and true versions of MLEM2 was not statistically significant. The total number of conditions in rules induced by both versions of MLEM2 was identical for *Hepatitis* with "?"s. Although statistically there is no significant difference between both versions of MLEM2, there is some evidence that the emulated version of MLEM2 may be better than the true version of MLEM2.

Our secondary objective was to compare three interpretations of missing attribute values in terms of complexity of induced rule sets. The total number of combinations is 36 (since we have two versions of MLEM2, three interpretations of missing attribute values and six data sets). Here results of our experiments are quite decisive: for the number of rules and for the total number of conditions

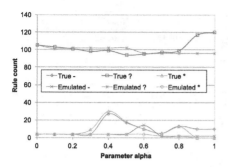

Fig. 1. Number of rules for the *breast cancer* data set

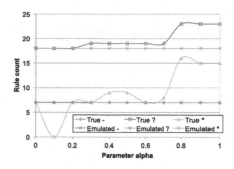

Fig. 2. Number of rules for the *echocardiogram* data set

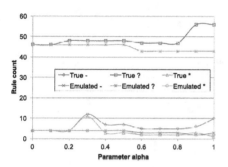

Fig. 3. Number of rules for the *hepatitis* data set

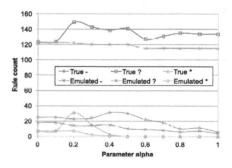

Fig. 4. Number of rules for the *image segmentation* data set

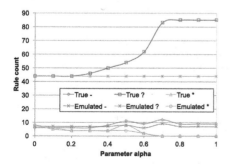

Fig. 5. Number of rules for the *lymphography* data set

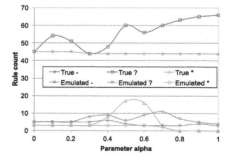

Fig. 6. Number of rules for the *wine recognition* data set

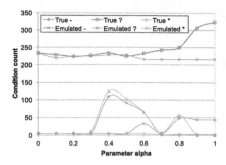

Fig. 7. Total number of conditions for the *breast cancer* data set

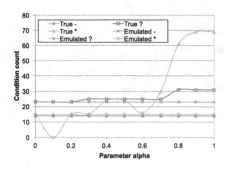

Fig. 8. Total number of conditions for the *echocardiogram* data set

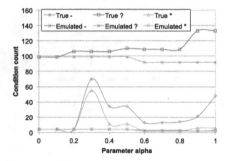

Fig. 9. Total number of conditions for the *hepatitis* data set

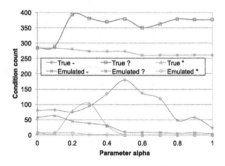

Fig. 10. Total number of conditions for the *image segmentation* data set

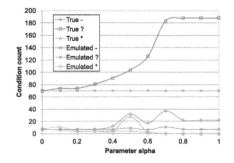

Fig. 11. Total number of conditions for the *lymphography* data set

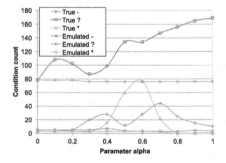

Fig. 12. Total number of conditions for the *wine recognition* data set

in induced rule sets, the lost value is always associated with more complex rule sets, in all possible 24 combinations. On the other hand, the difference between the attribute-concept values and "do not care" conditions is not that clear. The number of rules in induced rule sets was smaller for attribute-concept values than for "do not care" conditions for three combinations (out of 12), for *Echocardiogram* with both emulated and true versions of MLEM2 and for *Lymphography* with true version of MLEM2. The total number of conditions in induced rule sets was smaller for attribute-concept values than for "do not care" conditions for two combinations (out of 12), for *Echocardiogram* with true version of MLEM2 and for *Lymphography* with true version of MLEM2, for one combination (*Wine recognition* with emulated version of MLEM2) "do not care" conditions were better than attribute-concept values.

6   Conclusions

In experiments reported in this paper, we compared true and emulated versions of the MLEM2 algorithm using complexity of induced rule sets as the main criterion of quality. Results of our experiments show that both versions of the MLEM2 rule induction algorithm do not differ much. However, the worst interpretation of missing attribute values, from the view point of complexity of induced rule sets, is associated with incomplete data sets with lost values. It is surprising since as follows from our previous research [16], experiments conducted on the same data sets with lost values provided the smallest error rate.

References

1. Grzymala-Busse, J.W.: Three approaches to missing attribute values–a rough set perspective. In: Proceedings of the Workshop on Foundation of Data Mining, in Conjunction with the Fourth IEEE International Conference on Data Mining, pp. 55–62 (2004)
2. Grzymala-Busse, J.W.: Rough set strategies to data with missing attribute values. In: Notes of the Workshop on Foundations and New Directions of Data Mining, in Conjunction with the Third International Conference on Data Mining, pp. 56–63 (2003)
3. Grzymala-Busse, J.W., Ziarko, W.: Data mining based on rough sets. In: Wang, J. (ed.) Data Mining: Opportunities and Challenges, pp. 142–173. Idea Group Publ., Hershey (2003)
4. Pawlak, Z., Skowron, A.: Rough sets: some extensions. Inf. Sci. **177**, 28–40 (2007)
5. Pawlak, Z., Wong, S.K.M., Ziarko, W.: Rough sets: probabilistic versus deterministic approach. Int. J. Man Mach. Stud. **29**, 81–95 (1988)
6. Ślęzak, D., Ziarko, W.: The investigation of the Bayesian rough set model. Int. J. Approximate Reasoning **40**, 81–91 (2005)
7. Wong, S.K.M., Ziarko, W.: INFER–an adaptive decision support system based on the probabilistic approximate classification. In: Proceedings of the 6-th International Workshop on Expert Systems and their Applications, pp. 713–726 (1986)
8. Yao, Y.Y.: Probabilistic rough set approximations. Int. J. Approximate Reasoning **49**, 255–271 (2008)

9. Yao, Y.Y., Wong, S.K.M.: A decision theoretic framework for approximate concepts. Int. J. Man Mach. Stud. **37**, 793–809 (1992)
10. Ziarko, W.: Variable precision rough set model. J. Comput. Syst. Sci. **46**(1), 39–59 (1993)
11. Ziarko, W.: Probabilistic approach to rough sets. Int. J. Approximate Reasoning **49**, 272–284 (2008)
12. Grzymała-Busse, J.W.: Generalized parameterized approximations. In: Yao, J.T., Ramanna, S., Wang, G., Suraj, Z. (eds.) RSKT 2011. LNCS, vol. 6954, pp. 136–145. Springer, Heidelberg (2011). doi:10.1007/978-3-642-24425-4_20
13. Clark, P.G., Grzymala-Busse, J.W.: Experiments on probabilistic approximations. In: Proceedings of the 2011 IEEE International Conference on Granular Computing, pp. 144–149 (2011)
14. Clark, P.G., Grzymala-Busse, J.W.: Rule induction using probabilistic approximations and data with missing attribute values. In: Proceedings of the 15-th IASTED International Conference on Artificial Intelligence and Soft Computing ASC 2012, pp. 235–242 (2012)
15. Clark, P.G., Grzymala-Busse, J.W.: A comparison of two versions of the MLEM2 rule induction algorithm extended to probabilistic approximations. In: Cornelis, C., Kryszkiewicz, M., Ślęzak, D., Ruiz, E.M., Bello, R., Shang, L. (eds.) RSCTC 2014. LNCS, vol. 8536, pp. 109–119. Springer, Cham (2014). doi:10.1007/978-3-319-08644-6_11
16. Clark, P.G., Gao, C., Grzymala-Busse, J.W.: A comparison of two MLEM2 rule induction algorithms applied to data with many missing attribute values. In: Proceedings of DBKDA 2016 International Conference on Advances in Databases, Knowledge, and Data Applications, pp. 60–65 (2016)
17. Grzymala-Busse, J.W.: Data with missing attribute values: generalization of indiscernibility relation and rule induction. Trans. Rough Sets **1**, 78–95 (2004)
18. Grzymala-Busse, J.W., Rzasa, W.: Local and global approximations for incomplete data. In: Greco, S., Hata, Y., Hirano, S., Inuiguchi, M., Miyamoto, S., Nguyen, H.S., Słowiński, R. (eds.) RSCTC 2006. LNCS, vol. 4259, pp. 244–253. Springer, Heidelberg (2006). doi:10.1007/11908029_27
19. Grzymala-Busse, J.W.: A new version of the rule induction system LERS. Fundamenta Informaticae **31**, 27–39 (1997)
20. Grzymala-Busse, J.W.: MLEM2: a new algorithm for rule induction from imperfect data. In: Proceedings of the 9th International Conference on Information Processing and Management of Uncertainty in Knowledge-Based Systems, pp. 243–250 (2002)

Spark-Based Cluster Implementation of a Bug Report Assignment Recommender System

Adrian-Cătălin Florea[1]([✉]), John Anvik[2], and Răzvan Andonie[1,3]

[1] Electronics and Computers Department, Transilvania University of Braşov,
Braşov, Romania
acflorea@unitbv.ro
[2] Department of Mathematics and Computer Science, University of Lethbridge,
Lethbridge, AB, Canada
john.anvik@uleth.ca
[3] Computer Science Department, Central Washington University,
Ellensburg, WA, USA
andonie@cwu.edu

Abstract. The use of recommenders for bug report triage decisions is especially important in the context of large software development projects, where both the frequency of reported problems and a large number of active developers can pose problems in selecting the most appropriate developer to work on a certain issue. From a machine learning perspective, the triage problem of bug report assignment in software projects may be regarded as a classification problem which can be solved by a recommender system. We describe a highly scalable SVM-based bug report assignment recommender that is able to run on massive datasets. Unlike previous desktop-based implementations of bug report triage assignment recommenders, our recommender is implemented on a cloud platform. The system uses a novel sequence of machine learning processing steps and compares favorably with other SVM-based bug report assignment recommender systems with respect to prediction performance. We validate our approach on real-world datasets from the Netbeans, Eclipse and Mozilla projects.

1 Introduction

Most software projects use an issue tracking system (ITS) to organize the bug fixing process. The consistent use of an ITS offers a number of advantages, including a shared, central location for all issues, accountability, permissions management, workflow and notifications, and integration with various other software development tools such as version control systems.

Reports entered into an ITS follow a standard workflow. After a report is submitted to a project's ITS, it is *triaged* by a project member. The triage process involves several steps including determining if the report is a valid report or duplicates an existing report, if the report has been correctly classified by product, component, or subcomponent, and to whom to assign responsibility for

© Springer International Publishing AG 2017
L. Rutkowski et al. (Eds.): ICAISC 2017, Part II, LNAI 10246, pp. 31–42, 2017.
DOI: 10.1007/978-3-319-59060-8_4

addressing the report. Triage of the reports is a mandatory activity in the use of an ITS [19], and can be time-consuming due to a large number of new reports, especially for large projects [3]. As a means of reducing the overhead to software development projects caused by the triage process, the use of recommenders for various triage decisions has been proposed.

Most bug report triage research efforts have focused on the assignment decision, with the goal of making suggestions for optimal assignment. The process of bug report assignment is complex as it can be difficult to identify the right developer for fixing a new bug, and other considerations such as workload and availability need to be addressed. As projects add more components, developers and testers, the number of bug reports submitted daily increases, and manually recommending developers based on their expertise becomes more challenging.

From a Machine Learning (ML) perspective, the bug report assignment problem can be regarded as a supervised classification problem. Specifically, the input is textual and categorical information extracted from bug reports, and the output classes are developer names. Several classifiers have been used in this context, including Naïve Bayes, Bayesian Networks, C4.5, Support Vector Machines (SVM), and k-Nearest Neighbors (kNN). Also, feature selection/extraction techniques have been used to reduce the input space [7].

Since all these ML tools are presently available on many platforms, the challenge is how to choose, for a specific application, the right combination of them and how to optimize their hyperparameters. In addition, since bug datasets are increasing in size (more than 14,000 new entries and almost 2,000 developer profiles were added for Mozilla in the first two months of 2016, for a total of over 500,000 entries and more than 50,000 profiles created in the last five years[1]), the ability to scale the system to an increasingly large amount of data becomes critical, especially during the training phase.

As the training of an ML bug report assignment recommender can be computationally expensive, our novel approach is to implement a highly scalable recommender system on a cloud platform able to run on massive datasets. Compared to previous efforts, including the results in Anvik et al. [2–4] which will be our references for comparisons, our contributions are as follows:

- *A cloud-based approach to assignment recommender creation.* Our implementation focuses primarily on the recommender scalability and makes use of a highly scalable cloud platform, specifically Apache Spark hosted on a Google Cloud DataProc cluster.
- *A unique approach to processing the issue tracking data.* Compared to prior works, we use a different sequence of processing steps. These steps demonstrate a significant improvement in the prediction performance of the recommender system. Specifically, we use the Stanford parts-of-speech (POS) tagger to tokenize the data, Term Frequency/Inverse Document Frequency (TF/IDF) for term weighting, Latent Dirichlet Allocation (LDA) for dimensionality reduction, and an SVM classifier for generating the recommender.

---

[1] Data extracted from a Mozilla's Bugzilla database dump as of March 4, 2016.

The paper proceeds as follows. First, an overview of previous efforts in creating bug report assignment recommenders using ML is presented. Next, our recommender system is described, followed by the results of the approach applied to three real-world datasets and by discussions about the recommender performance and scalability. The paper is then concluded.

2 Related Work

Many approaches for creating bug report assignment recommenders have been investigated, with the majority of approaches focusing on the analysis of categorical and textual information using information retrieval and ML techniques [9]. We focus on previous efforts for bug report assignment that use ML.

The first attempt at using ML specific techniques for bugs assignment was by Cubranic and Murphy [12]. The authors reported a 30% classification accuracy on 15,859 bug reports from the Eclipse project. Filtering out noisy data based on bug status [3,4] improved the classifier's accuracy to more than 50%. Unlike the original attempts, results from several ML algorithms (Naïve Bayes, SVM, and C4.5) were compared in Anvik et al. [3], with SVM providing the best results.

Shokripour et al. [21] found that further filtering of the input data by parts-of-speech proved to be useful. Their approach extracted unigram nouns exclusively in order to predict source files that would be changed for the fix of a new bug and used the predicted location to decide which developer to recommend, based on past interventions with these particular files. Similar to Shokripour et al., our approach uses only noun terms but additionally explores the use of χ^2 or LDA for dimensionality reduction.

Ahsan et al. [1] examined the use of dimensionality reduction techniques in creating a bug report assignment recommender, such as feature extraction (LSI) and feature selection. Combining these techniques with various ML algorithms, they found that SVM outperforms the other ML methods they considered.

Nasim et al. [16] investigated the use of an Alphabet Frequency Matrix for assignment recommendation, in combination with different ML algorithms. Simple logistic, Sequential Minimal Optimization (a variant of SVM) and Complement Naïve Bayes algorithms were found to perform better than the other investigated ML algorithms.

Xia et al. [25] used a multi-label kNN in their bug report assignment approach to determine the similarity of previously fixed bugs to new bug reports and the expertise of the developers. They used four features of each bug (term, topic, component, and product) to determine the similarity.

Banitaan and Alenezi [5] proposed an automatic bug report assignment approach that combined the use of ML with the χ^2 method for dimensionality reduction. They constructed a vector space model using 1% of the most discriminating terms in the bug report descriptions, as well as the reporter and component features as metadata.

Nguyen et al. [17] developed a topic-based automatic bug report assignment method wherein the bug reports used for training are classified based on the topics in the descriptions using the LDA. The time to resolve the bugs for each topic was used as metadata for developer recommendation.

Wu *et al.* [24] suggested the use of a kNN algorithm for classifying bug reports. Their method combined the use of a Vector Space Model with TF/IDF [15] term weighting, and developer ranking by frequency and social network metrics.

3   The Recommender System

This section introduces our recommender system, describing the input datasets, the data cleansing procedure, the preprocessing steps, the training algorithm, and the recommender implementation.

3.1   Datasets

In selecting the datasets for our work, we examined previous bug report assignment research efforts to determine the most commonly used software projects. Based on this criteria, we selected three projects. Specifically, we use the following datasets:

- Eclipse Bugzilla[2] dataset available at the MSR 2011 website.[3]
- Netbeans Bugzilla[4] dataset, also from the MSR 2011 website.
- Mozilla Bugzilla[5] made available to us, on request, by Mozilla Foundation.

For all three datasets, we use a MySQL dump of their respective Bugzilla databases.[6] Each dump contains all of the tables used by Bugzilla, with the exception of the `profiles` table,[7] which was either omitted from the dump or anonymized.

For our recommender we use data from the `bugs`, `duplicates`, `longdescs` and `bugs_activity` tables. The `bugs` table is the core of Bugzilla system and stores most of the current information about a bug. `Bugs_activity` stores information regarding what changes are made to bugs and when, providing a complete history for each bug. The `duplicates` table is where the information about duplicate bug reports is maintained. The `longdescs` table stores all of the user comments and is considered to be "the meat of Bugzilla"[8].

We extract categorical bug report information from the `bugs` table, including the id of the bug, its creation date, status, and the product and component ids.

---

[2] https://bugs.eclipse.org/bugs/.
[3] http://2011.msrconf.org/msr-challenge.html.
[4] https://netbeans.org/bugzilla/.
[5] https://bugzilla.mozilla.org/.
[6] https://www.bugzilla.org/docs/2.18/html/dbdoc.html.
[7] The `profiles` table contains personal information, such as names and email addresses of the project members.
[8] https://www.bugzilla.org/docs//2.18/html/dbdoc.html.

The textual information is extracted from the longdescs table. We query the bugs_activity table in order to determine which developer marked the bug as FIXED. We also use information from the duplicates table to identify bugs marked as duplicates of existing bug reports, and remove them from the training dataset as they did not improve the recommender's predictions.

3.2   Cleansing

To be useful, a bug report assignment recommender should suggest only developers that are currently active on the project. Therefore, we filter the original input data, selecting only the developers with an average of at least three fixed issues per month for the last three months. A similar threshold was used in Anvik et al. [4]. This filtering helped us to eliminate people with only occasional interventions in the project. From the prefiltered data in the form of $(developerId, fixedBugs)$ we compute the mean and standard deviation for $fixedBugs$ and disregard the developers that marked as FIXED more than $mean + 2 * stddev$, in an attempt to eliminate the project members we believe to fill a management like role.

We consider only the issues marked as FIXED. As each issue can be assigned to several developers during its life cycle, for issues that are still in progress the current assignee is not necessarily the most appropriate one. Also, there are cases where a new bug report is automatically assigned to a default user id; this information is not useful in our context. Bug reports can also be closed with a resolution other than FIXED, such as WONTFIX, WORKSFORME or INVALID. These other resolutions are usually set by a bug report triager, who is also not appropriate for assignment recommendation.[9] Also, bug reports with resolutions other than FIXED do not trigger code changes, do not require a developer to work on them, and do not contribute to improving the quality of the project.

For each FIXED issue, the developer who fixed it can be considered either the one appearing in the assigned_to field from the bugs table, or the one who last marked the issue as FIXED according to the bugs_activity table. Based on project heuristics, as detailed in Anvik et al. [4], we chose the second approach.

Besides the textual data from each issue's title, description and comments, we also use the component_id and product_id information from the bugs table. Instead of training a separate recommender for each component (as described in Anvik et al. [4]), we transform the values into binary vectors using one-hot encoding [14]. A unique feature of our approach is applying a weight to the binary vectors. The optimal values for the weights are determined through grid search [6], using 100 runs with a feature space between one and 100, followed by another round of tests in an interval centered on the results of the first optimization round.

---

[9] In cases where a bug report triager is also an active developer, the person will have been assigned to issues that are marked as FIXED.

The combination of `component_id` and `product_id` offers a solid selection criteria. For each pair of values, we look at the number of developers fixing bugs that fall under a particular combination of values. We discovered that for more than half of these distinct `component_id`, `product_id` pairs, at most two developers were assigned.

3.3 Preprocessing and Feature Reduction

We apply the Stanford parts-of-speech tagger [23] to the filtered textual information and extract and preserve exclusively the nouns,[10] similar to the approach described in Shokripour et al. [21]. The preprocessing phase converts the textual input into numerical data using TF/IDF [18]. We decided not to explicitly remove English stop words (i.e. commonly used English words), as using IDF already greatly lower their weights. We did, although, filter the terms based on their frequency as described in Yang et al. [26], removing any term which appears in less than two documents and more than 15% of the corpus.

For training, we chose the data from the last 240 days, as we believe the most recent data is the most relevant. Table 1 shows for each project the number of reports in the training and test datasets, the number of vocabulary terms, and the number of developers for potential recommendation using the described filtering.

Table 1. Details on used datasets

Project	Training reports	Test reports	Vocabulary terms	Developers #
Eclipse	8,379	934	41,652	76
Netbeans	5,283	592	25,022	27
Mozilla	26,118	2,908	225,473	70

For training/testing, we use either the entire feature space or one of the following two feature reduction techniques: χ^2 [26] or Latent Dirichlet Allocation (LDA) [8]. When applying feature reduction, both for χ^2 and LDA, we build the models based only on the training dataset and use them to map the test dataset to the smaller feature space.

3.4 Training the Recommender

As SVMs were found to outperform other classifiers in the context of bug report assignment triage [1,4], we chose to use an SVM classifier in our recommender system. We kept the most recent 10% of the data for testing and use the other 90% for training.

---

[10] Those terms tagged as either Noun {NN}, Noun Plural {NNS}, Proper Noun {NNP}, or Proper Noun Plural {NNPS}.

We train a linear SVM [11] model for each developer class, applying a one-versus-all strategy, and predict a single class - the recommended developer. For prediction performance metrics, we compute per-class precision, recall, and F1-measure [22] and we report their average values, weighted by class frequencies.

We train our model using Stochastic Gradient Descent with L2 regularization, step size of 1, at most 100 iterations, 0.01 for the regularization (i.e. cost) parameter, and a convergence tolerance of 0.001. These optimal hyperparameters were found by a grid search.

3.5 Implementation

Our implementation performs on the Apache Spark cloud platform engine.[11] Spark favors a direct, in-memory data processing strategy which results in impressive speed gains over classical MapReduce implementations [20].

We use the LIBLINEAR [13] SVM implementation in MLLib,[12] the Apache Spark's scalable ML library. Compared to the more popular LIBSVM [10] implementation, LIBLINEAR is proven to achieve at least similar performance when the number of features is large, especially in the case of document classification, and it does so by maintaining a far better scalability [13].

We develop our recommender system in Scala 2.10.6[13] and the code is publicly available on GitHub.[14] The tests were performed on a Google DataProc cluster[15] consisting of one master node and three worker nodes, with all four VMs of type n1-highmem-2 (2 vCPU, 13.0 GB memory) with 100 GB disk size each.

4   Results and Discussion

To evaluate our approach for creating a cloud-based bug report assignment recommender, for each dataset we create three recommenders where the variation is either no feature reduction, use χ^2, or use LDA. On the test data, our recommender achieved a best precision of 0.79 for Eclipse (using χ^2), 0.77 for Mozilla (no feature reduction), and 0.89 for Netbeans (no feature reduction). For recall, the best achieved values were 0.77 for Eclipse (no feature reduction), 0.75 for Mozilla (no feature reduction), and 0.88 for Netbeans (no feature reduction), resulting in an F1-measure of 0.76 for Eclipse, 0.73 for Mozilla, and 0.88 for Netbeans. We also found using either feature reduction technique allowed the recommender to be trained in a non-clustered environment in a reasonable amount of time without a significant drop in prediction performance.

---

[11] http://spark.apache.org/.

[12] http://spark.apache.org/mllib/.

[13] http://www.scala-lang.org/.

[14] https://github.com/acflorea/columbugus.

[15] https://cloud.google.com/dataproc/.

4.1   Results of Dimensionality Reduction Techniques

When using χ^2, selecting less than 1,000 input features had a negative impact on the classifier's precision. With 1,000 features, the system remains stable, showing less than a 0.01 drop in precision for Netbeans and Mozilla, and even a small precision increase for Eclipse when compared to the case with no feature reduction applied.

When using LDA, the precision dropped, regardless of the number of LDA topics used. The optimal number of LDA topics was around 1,000. With 750 topics, the precision dropped by less than 0.03 for Eclipse and Mozilla, and remained almost unchanged for Netbeans. For less than 500 topics, the prediction accuracy of the recommender dropped significantly in the case of Eclipse and Netbeans, and more slowly for Mozilla. Figures 1, 2 and 3 show the values for precision, recall and F1-measure using only the textual information (i.e. title, description, and comments) of the bug reports.

When examining the categorical data from the bug reports (i.e. `product_id` and `component_id`), we observed that the Netbeans dataset is the most homogeneous of the three datasets. This explains why using the component and product information together with any of the dimensionality reduction techniques led to the smallest increase in the precision and the recall. At the other extreme, the most scattered dataset is Mozilla, and using component and product information together with LDA almost doubled both the precision and the recall.

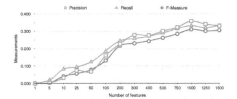

Fig. 1. Precision, recall and F1-measure values vs. the number of LDA topics (Eclipse).

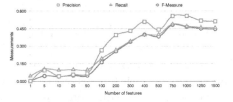

Fig. 2. Precision, recall and F1-measure values vs. the number of LDA topics (Netbeans).

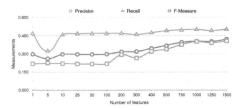

Fig. 3. Precision, recall and F1-measure values vs. the number of LDA topics (Mozilla).

4.2   Choice of SVM Kernel

A legitimate question is whether the linear SVM kernel is the best choice for creating a bug report assignment recommender. Perhaps using a Gaussian kernel would provide better results. After reducing the feature space with LDA,[16] we tested recommenders where the distributed LIBLINEAR implementation was replaced with the sequential LIBSVM implementation provided by WEKA, and compared using both the RBF and linear kernels. For both kernels, we used the same value of 0.01 for the cost parameter as was used in our Spark implementation. For RBF we used the default value for gamma (1/number of features). Even in the context of this reduced feature space, the precision and recall values obtained with our Spark-based distributed implementation were superior for all datasets (Tables 2, 3, and 4). Based on these results, LIBLINEAR is the optimal choice for our application in terms of prediction performance.

Table 2. Precision, recall, F1-measure for different SVM algorithms (Eclipse).

	LIBLINEAR			LIBSVM	
	No feature reduction	χ^2 1,000 features	LDA 1,000 features	Weka RBF	Weka linear
Precision	0.78	**0.79**	0.71	0.70	0.70
Recall	**0.77**	0.75	0.67	0.66	0.67
F1-measure	**0.76**	0.74	0.65	0.63	0.64

Table 3. Precision, recall, F1-measure for different SVM algorithms (Netbeans).

	LIBLINEAR			LIBSVM	
	No feature reduction	χ^2 1,000 features	LDA 1,000 features	Weka RBF	Weka linear
Precision	**0.89**	0.88	0.85	0.78	0.78
Recall	**0.88**	0.87	0.82	0.78	0.78
F1-measure	**0.88**	0.87	0.82	0.76	0.78

4.3   Comparison with Previous Approaches and Scalability

We compared our results with previous SVM-based recommenders. Table 5 shows the results of this comparison, with the best-obtained values are marked in bold. Our implementation achieved similar values for precision but had improved recall values.

---

[16] Without this reduction, the dataset was found to be too large for WEKA to process.

Table 4. Precision, recall, F1-measure for different SVM algorithms (Mozilla).

	LIBLINEAR			LIBSVM	
	No feature reduction	χ^2 1,000 features	LDA 1,000 features	Weka RBF	Weka linear
Precision	**0.77**	0.75	0.72	0.69	0.70
Recall	**0.75**	0.74	0.74	0.72	0.72
F1-measure	**0.73**	**0.73**	0.71	0.68	0.69

Table 5. Precision and Recall comparison with previous SVM implementations

	Eclipse	Mozilla
	Precision/Recall	Precision/Recall
Our Spark implementation	0.78/**0.77**	**0.76/0.75**
WEKA RBF Kernel	0.72/0.74	0.71/0.67
Recommender One prediction [3]	0.86/0.12	0.64/0.02
Recommender Three predictions [3]	0.77/0.32	0.53/0.06
Recommender One prediction [4]	**0.97**/0.18	0.70/0.01
Recommender Three predictions [4]	0.79/0.41	0.64/0.03

The creation time for a recommender using our approach is less than 10 min on the test infrastructure described in Sect. 3.5, and is achieved with no compromise in the prediction performance of the system. As the execution time was not reported for the other bug report assignment recommender approaches, there is no available baseline for comparison. However, we believe the creation time values for our approach are highly promising.

We tested the scalability of our implementation using up to six worker cores on increasing amounts of data. Table 6 shows the values for the training step speedup (i.e., $\frac{sequential\ algorithm\ execution\ time}{parallel\ algorithm\ execution\ time}$).

Table 6. Speedup on different training data sizes.

Training samples	2 cores	3 cores	4 cores	5 cores	6 cores
2,000,000	1.97	2.86	3.85	4.41	4.89
900,000	1.97	2.58	3.32	3.99	4.12
90,000	1.95	2.59	3.41	3.91	4.12
30,000	2.11	2.67	3.16	3.33	2.86

5   Conclusion

To the best of our knowledge, this is the first bug report assignment recommender system to be implemented on a cloud platform (Apache Spark hosted on a Google Cloud DataProc platform). Our recommender system provides a fast and highly scalable alternative for the existing desktop implementations and achieves comparable performance in terms of precision and recall.

Acknowledgment. The authors are grateful to the Mozilla Foundation for providing a dump of their Bugzilla database.

References

1. Ahsan, S.N., Ferzund, J., Wotawa, F.: Automatic software bug triage system (BTS) based on latent semantic indexing and support vector machine. In: Fourth International Conference on Software Engineering Advances, ICSEA 2009, pp. 216–221, September 2009
2. Anvik, J.: Automating bug report assignment. In: Proceedings of the 28th International Conference on Software Engineering, ICSE 2006, NY, USA, pp. 937–940 (2006). http://doi.acm.org/10.1145/1134285.1134457
3. Anvik, J., Hiew, L., Murphy, G.C.: Who should fix this bug? In: Proceedings of the 28th International Conference on Software Engineering, ICSE 2006, NY, USA, pp. 361–370 (2006). http://doi.acm.org/10.1145/1134285.1134336
4. Anvik, J., Murphy, G.C.: Reducing the effort of bug report triage: recommenders for development-oriented decisions. ACM Trans. Softw. Eng. Methodol. **20**(3), 10:1–10:35 (2011). http://doi.acm.org/10.1145/2000791.2000794
5. Banitaan, S., Alenezi, M.: Tram: an approach for assigning bug reports using their metadata. In: 2013 Third International Conference on Communications and Information Technology, pp. 215–219, June 2013
6. Bergstra, J., Bengio, Y.: Random search for hyper-parameter optimization. J. Mach. Learn. Res. **13**, 281–305 (2012). http://dl.acm.org/citation.cfm?id=2188385.2188395
7. Bhattacharya, P., Neamtiu, I., Shelton, C.R.: Automated, highly-accurate, bug assignment using machine learning and tossing graphs. J. Syst. Softw. **85**(10), 2275–2292 (2012)
8. Blei, D.M., Ng, A.Y., Jordan, M.I., Lafferty, J.: Latent Dirichlet allocation. J. Mach. Learn. Res. **3**, 993–1022 (2003)
9. Cavalcanti, Y.C., da Mota Silveira Neto, P.A., do Carmo Machado, I., Vale, T.F., de Almeida, E.S., de Lemos Meira, S.R.: Challenges and opportunities for software change request repositories: a systematic mapping study. J. Softw. Evol. Process **26**(7), 620–653 (2014). http://dx.doi.org/10.1002/smr.1639
10. Chang, C.C., Lin, C.J.: LIBSVM: a library for support vector machines. ACM Trans. Intell. Syst. Technol. **2**, 27:1–27:27 (2011). http://www.csie.ntu.edu.tw/~cjlin/libsvm
11. Cortes, C., Vapnik, V.: Support-vector networks. Mach. Learn. **20**(3), 273–297 (1995). http://dx.doi.org/10.1023/A: 1022627411411
12. Cubranic, D., Murphy, G.C.: Automatic bug triage using text categorization. In: Proceedings of the Sixteenth International Conference on Software Engineering & Knowledge Engineering (SEKE 2004), Banff, Alberta, Canada, 20–24 June 2004, pp. 92–97 (2004)
13. Fan, R.E., Chang, K.W., Hsieh, C.J., Wang, X.R., Lin, C.J.: LIBLINEAR: a library for large linear classification. J. Mach. Learn. Res. **9**, 1871–1874 (2008)

14. Harris, D., Harris, S.: Digital Design and Computer Architecture, 2nd edn. Morgan Kaufmann Publishers Inc., San Francisco (2012)
15. Jones, K.S.: A statistical interpretation of term specificity and its application in retrieval. J. Documentation **28**, 11–21 (1972)
16. Nasim, S., Razzaq, S., Ferzund, J.: Automated change request triage using alpha frequency matrix. In: Frontiers of Information Technology (FIT), pp. 298–302, December 2011
17. Nguyen, T.T., Nguyen, A.T., Nguyen, T.N.: Topic-based, time-aware bug assignment. SIGSOFT Softw. Eng. Notes **39**(1), 1–4 (2014). http://doi.acm.org/10.1145/2557833.2560585
18. Rajaraman, A., Ullman, J.D.: Mining of Massive Datasets. Cambridge University Press, New York (2011)
19. Reis, C.R., de Mattos Fortes, R.P., Pontin, R., Fortes, M.: An overview of the software engineering process and tools in the mozilla project (2002)
20. Shinnar, A., Cunningham, D., Saraswat, V., Herta, B.: M3r: Increased performance for in-memory Hadoop jobs. Proc. VLDB Endow. **5**(12), 1736–1747 (2012). http://dx.doi.org/10.14778/2367502.2367513
21. Shokripour, R., Anvik, J., Kasirun, Z.M., Zamani, S.: Why so complicated? simple term filtering and weighting for location-based bug report assignment recommendation. In: Proceedings of the 10th Working Conference on Mining Software Repositories, MSR 2013, pp. 2–11. IEEE Press, Piscataway (2013). http://dl.acm.org/citation.cfm?id=2487085.2487089
22. Sokolova, M., Lapalme, G.: A systematic analysis of performance measures for classification tasks. Inf. Process. Manage. **45**(4), 427–437 (2009). http://dx.doi.org/10.1016/j.ipm.2009.03.002
23. Toutanova, K., Klein, D., Manning, C.D., Singer, Y.: Feature-rich part-of-speech tagging with a cyclic dependency network. In: Proceedings of the 2003 Conference of the North American Chapter of the Association for Computational Linguistics on Human Language Technology, NAACL 2003, vol. 1, pp. 173–180. Association for Computational Linguistics, Stroudsburg (2003). http://dx.doi.org/10.3115/1073445.1073478
24. Wu, W., Zhang, W., Yang, Y., Wang, Q.: Drex: developer recommendation with k-nearest-neighbor search and expertise ranking. In: 2011 18th Asia Pacific Software Engineering Conference (APSEC), pp. 389–396, December 2011
25. Xia, X., Lo, D., Wang, X., Zhou, B.: Accurate developer recommendation for bug resolution. In: Proceedings of the 20th Working Conference on Reverse Engineering, pp. 72–81, October 2013
26. Yang, Y., Pedersen, J.O.: A comparative study on feature selection in text categorization. In: Proceedings of the Fourteenth International Conference on Machine Learning, ICML 1997, pp. 412–420. Morgan Kaufmann Publishers Inc., San Francisco (1997). http://dl.acm.org/citation.cfm?id=645526.657137

The Bag-of-Words Method with Dictionary Analysis by Evolutionary Algorithm

Marcin Gabryel[1]([✉]) and Giacomo Capizzi[2]

[1] Institute of Computational Intelligence, Częstochowa University of Technology,
Al. Armii Krajowej 36, 42-200 Częstochowa, Poland
marcin.gabryel@iisi.pcz.pl
[2] Department of Electric, Electronic and Informatics Engineering,
University of Catania, Catania, Italy
gcapizzi@diees.unict.it
http://iisi.pcz.pl

Abstract. In this paper we present innovative solutions improving general operational efficiency of the Bag-of-Words algorithm (BoW). The first innovation which we put forward is creating a visual words' dictionary using the clustering algorithm which in itself is responsible for selecting the appropriate number of clusters. This solution results in significant automation of image database creation. Another innovation is adding to the BoW model an analytical module whose task is to analyse the visual words' dictionary and to modify histogram values before storing them in a database. This algorithm is operated with the use of the evolutionary algorithm. The modifications of the BoW algorithm significantly improve the efficiency of image search and classification, which has been presented in a variety of experiments.

1   Introduction

Digital image processing is a very complex issue. Image analysis is so difficult and complicated for computers and a very serious challenge for researchers. For the time being there is no method which would always be effective. In solving image processing and retrieval problems algorithms from different fields of computational intelligence [11,12,20] are used, in particular fuzzy systems [1,13,21], neural networks [4,23,26], evolutionary algorithms [27,29,30], mathematics [28], neuro-fuzzy systems [31,32], clustering [22] and data mining [24].

One of the most popular and widely spread algorithms used for indexation and image retrieval is the Bag-of-Words (BoW) model [6]. The BoW model is an algorithm used in natural language processing and information retrieval. In this model a test document is represented as a sparse vector of occurrence of words presented as a histogram of dictionary.

The new image searching and classifying algorithm presented in this paper is based on the idea of the BoW algorithm. The classical BoW model consists of a few various algorithms, which can be selected according to current needs [18]. The authors of this paper use this flexibility of the BoW model and introduce

© Springer International Publishing AG 2017
L. Rutkowski et al. (Eds.): ICAISC 2017, Part II, LNAI 10246, pp. 43–51, 2017.
DOI: 10.1007/978-3-319-59060-8_5

so-far unused algorithms for image clustering, comparison and representation. At present in the literature on the subject, the most frequently used clustering algorithm which is supposed to create a visual words dictionary is k-means [16,33]. However, with this algorithm it is necessary to specify the number of clusters at the very beginning. If the cluster happens to be too small, then visual words will not represent all patterns, and if the cluster is too large, it will result in overfitting and overlearning. In our case using the Growing Self-Organizing Map (GSOM) algorithm [8] makes it possible to automatize this process due to the very fact that the GSOM automatically chooses a suitable number of clusters in the unsupervised learning process. Visual words are created via clustering of local image features. The literature on the subject shows that local features are most frequently obtained by way of using the Scale-invariant feature transform (SIFT) algorithm [10,17] or by dividing an image into segments [14,15]. In our method local features are detected with the use of the Speeded Up Robust Features (SURF) algorithm [3], which is a fast algorithm allowing for creating short vectors describing the region of the points of interest. Those points remain invariant in spite of changes in image scale or rotation.

This paper is divided into several sections. Section 2 provides the description of the presented retrieval and classification algorithm. Section 3 presents the research results and effectiveness of the new algorithm. The paper ends with the conclusions.

2   Description of Proposed Methods

The image search and classification algorithm developed under the presented approach consists of three modules: (i) the initiating module, which is supposed to prepare the images stored in a database, (ii) the analytical module, responsible for the visual words' dictionary analysis and histogram modification, and (iii) the retrieval module, whose task is to retrieve and classify similar images.

The initiating module is meant to save images in a database in the way the Bag-of-Words algorithm does. Local characteristic features are retrieved from an image, and next they are clustered in order to create a visual words' dictionary. Each image $\mathbf{I}_i$ has histogram $\mathbf{h}_i^p$ which mapping characteristic points to group centers.

We are considering herein a set of given images $\mathbf{I}_i$, where $i = 1, ..., I_L$, I_L is the number of all images. Each image $\mathbf{I}_i$ has a class $c(\mathbf{I}_i)$ assigned to it, where $c(\mathbf{I}_i) \in \Omega$, $\Omega = \{\omega_i, ..., \omega_C\}$ is a set of all classes and C is the number of all classes.

1. For each class $\omega_i \in \Omega$ images L_i are selected randomly, and J is the number of all randomly selected images:

$$J = \sum_{i=1}^{C} L_i \qquad (1)$$

2. Find the characteristic points (for example with the SURF algorithm [3]) for all images J, $\mathbf{x}_i = [x_{i1}, x_{i2}, ..., x_{iK}]$, $i = 1, ..., L$, L – the total number of all characteristic points, K – the dimension of the vector describing a characteristic point (for SURF $K = 64$).

3. Group the points $\mathbf{x}_i$ with the use of the GSOM algorithm [9]. Obtain cluster centres $\mathbf{w}_j$ of neurons N_j, $j = 1, ..., N_c$, N_c – the number of clusters (number of neurons created as the result of applying the GSOM algorithm) or, in other words, the size of the visual words' dictionary.

4. Create a database:
   (a) Create histograms $\mathbf{h}_i^p = [h_{i1}^p, ..., h_{iN_c}^p,]$ for an image i, $i = 1, ..., I_L$, where

$$h_{ik}^p = \sum_{n=1}^{L} \delta_{nk}(i), \ k = 1, ..., N_c, \tag{2}$$

$$\delta_{nk}(i) = \begin{cases} 1 \text{ if } \|\mathbf{w}_k - \mathbf{x}_n\| \leq \|\mathbf{w}_j - \mathbf{x}_n\| \text{ for } \mathbf{x}_n \in \mathbf{I}_i, j = 1, ..., N_c, j \neq k \\ 0 \hspace{5.5cm} \text{otherwise} \end{cases}. \tag{3}$$

   Variable $\delta_{nk}(i)$ is an indicator if a group $\mathbf{w}_k$ is the closest vector (a winner) for any sample $\mathbf{x}_n$ from an image $\mathbf{I}_i$.
   (b) Store histogram $\mathbf{h}_i^p$ for image $\mathbf{I}_i$ in the database along with the image identifier and the label $c(\mathbf{I}_i)$ of the class to which it belongs.

The task of the next module of the presented algorithm is to analyse the dictionary of all the images in order to improve the quality of retrieved images and to specify more precisely the class to which a retrieved image belongs. We have applied here the new algorithm called Nonactive Visual Words Thresholding (NVWT). Its task is (i) to analyse occurrence of the visual words in images in a given group (all the images or in a given class of images), (ii) to remove from the dictionary those visual words whose occurrence number is below the set threshold, and (iii) to filter histogram $\mathbf{h}_i^p$ and to reset those histogram elements which contain information on removed visual words. After this operation the histograms are normalized and the remaining histogram element values which have not been reset increase their values automatically. This is why the histogram has information on the most significant visual words occurring in a given group of images. Our experiments clearly confirm that removing inactive visual words boosts the classification efficiency. The algorithm comprises a few steps. The first step in this method is to calculate neuron activity for each image class. For every class ω_k we calculate the activity of visual words α_{jk}

$$\alpha_{jk} = \sum_{\substack{i=1 \\ c(\mathbf{I}_i)=\omega_k}}^{I_L} \sum_{n=1}^{L} \delta_{nj}(i) \tag{4}$$

for $c(\mathbf{I}_i) = \omega_k$, $j = 1, ..., N_c$ and $k = 1, ..., C$. If there occurs inequality:

$$\alpha_{jk} \leq \Theta \tag{5}$$

then
$$h_{ij}^p = 0 \qquad (6)$$

where Θ is the threshold value of visual words' activity, below which the histogram value h_{ij}^p is reset. The algorithm version presented above concerns setting threshold Θ for all images of all classes in a database. The study results and the algorithm itself have been partly presented in [9]. The numerous conducted experiments have produced far better results when the threshold values Θ_k have been set for each of the classes. In such case formula (5) is replaced with the following one:

$$\alpha_{jk} \leq \Theta_k \qquad (7)$$

On completion of this operation histograms of images of a particular class only contain information on the most significant visual words.

The analytical part also makes use of operation of the Differential Evolution algorithm [25], which is meant to choose threshold values Θ_k for the NVWT algorithm. For the sake of our DE algorithm a chromosome consists of a natural number vector. Each chromosome makes a prospective solution, and that is the reason why number values correspond to threshold values of each of the classes. Fitness function $f(\cdot)$ calculates the value of recall index, which is one of the indexes of performance metrics in classification problems [19]. The recall rate of a classifier is the ratio of the number of relevant records retrieved to the total number of relevant records in the database

$$Recall = \frac{TruePositiveCount}{TruePositiveCount + FalseNegativeCount} \cdot 100\% \qquad (8)$$

It is usually expressed as a percentage. Another frequently-used index is precision, which is the ratio of the number of relevant records retrieved to the total number of irrelevant and relevant records retrieved

$$Precision = \frac{TruePositiveCount}{TruePositiveCount + FalsePositiveCount} \cdot 100\% \qquad (9)$$

It is also usually expressed as a percentage.

The Differential Evolution method has been slightly modified for the purpose of the algorithm developed under our approach. All the modifications relating to the operation of the original algorithm are as follows:

1. The recall value is taken as the value of the fitness function $f(\cdot)$ (8), i.e. the value calculated as the percentage of the correctly classified images out of all the selected ones.
2. The total number of algorithm generations is 100.
3. The population comprises 100 individuals.
4. Chromosome $\mathbf{x}_i^G$, (G is the number of the generation, $i = 1, ..., NP$, NP – population size) is a natural number vector, in which natural numbers take threshold values of Θ_k. The vector length corresponds to the number of image classes. Vector element values are initiated by random numbers from the range of $[0; 25]$.

5. Recall value is calculated by random selection of 30 images for each of the classes in the whole database.
6. An additional condition has been provided in the algorithm, which improves efficiency of its operation. If the fitness function value does not improve throughout 10 algorithm iterations, there is another random selection of 30 images for each of the classes.

Operation of this algorithm allows us to automatically obtain threshold values Θ_k for the operational needs of the NVWT algorithm. The experimental research experiments presented in the following section show its improved efficiency after the analytical part has been added to the retrieval and classification process.

The last module of the presented algorithm is responsible for classification of the query image marked as $\mathbf{Q}$ and also for retrieving its similar images from the database. The process starts with initiation of the SURF algorithm and retrieval of the image local features from $\mathbf{Q}$. Next, similar to formula (3) a histogram of visual words' frequency occurrence $\mathbf{h}_Q^p$ is created. The next stage involves calculating distances between all the histograms in the database using the L1 metric; yet, for the histogram $\mathbf{h}_Q^p$ we also use the NVWT algorithm depending on the class to which a given image i belongs:

$$d_i^p = \sum_{k=1}^{N_c} |h_{Qk}^p \cdot m_{c(\mathbf{I}_i),k} - h_{ik}^p| \tag{10}$$

for $i = 1, ..., L$, where $m_{j,k}$ is a mask, which accounts for non-active visual words removed in the analysis process completed according to formula (5) in the version for one threshold for all images:

$$m_{j,k} = \begin{cases} 1 \text{ if } \alpha_{jk} \geq \Theta \\ 0 \text{ if } \alpha_{jk} < \Theta \end{cases} \tag{11}$$

For the version with a threshold set for each class individually (7), the mask $m_{j,k}$ takes the form of:

$$m_{j,k} = \begin{cases} 1 \text{ if } \alpha_{jk} > \Theta_k \\ 0 \text{ if } \alpha_{jk} < \Theta_k \end{cases} \tag{12}$$

The value α_{jk} is calculated according to formula (4). The image which has minimum of d_i^p is marked as $\mathbf{I}_p$ and is treated as images similar to the query image $\mathbf{Q}$.

The next section presents the results obtained with the use of the described algorithm.

3   Experimental Research

In this section we present the research experiments which were conducted with the use of various configuration of the algorithm operation. The algorithm was implemented in the Java language as well as JavaCV [2] library function. JavaCV is a library which adopts functions available in OpenCV [5] for the Java language needs. The research was performed on the Caltech 101 image database (collected by L. Fei-Fei et al. [7]). Six sample categories comprising motorbikes, car sides, revolvers, airplanes, leopards and wrenches were selected. Each start of the initiating part of our algorithm involves development of an initial database of images by random selection of 30 images out of the whole repository for each of the categories ($L_i = 30$). Out of the remaining group of images 20% are randomly selected and marked as a set of testing images. They are used to test efficiency of the final classification. The images left in the database are used during the operation of the analytical algorithm. The main value which is used to compute classification efficiency is the recall value expressed as a percentage (8) (percentage of correct images which are selected). Our algorithm requires that a few parameters be initiated. Identification of their optimum values is also the object of our research.

The experiment was set to test the effect of the analytical model operation on classification efficiency. A number of tests were carried out in order to compare operation of the following algorithms: one without an analytical module, one with operating analytical module with one threshold value for all the classes (according to formulas (5) and (11)), and also one with an optimization module with individual thresholding for each of the classes according to formulas (7) and (12)). In the last case particular values Θ_k for each class were selected using the DE algorithm. In Table 1 there are recall values for the three cases under

Table 1. Effect of the analytical process on classification efficiency for different values τ_{max}. In the Table there are recall classification efficiency values [%] for three different cases of the operation of our algorithm, i.e. without the analysis module, with the analysis module applying thresholding with one value Θ, and with the analysis module for the threshold value Θ_k for each of the classes individually. The results are given for both: the images from the database and the testing images.

τ_{max}	N_c	algorithm without optimization		Θ	thresholding with one value		threshold values Θ_k for particular classes						individual thresholding for each of the classes	
		train recall	test recall		train recall	test recall	revolver	motorbike	car side	airplane	leopard	wrench	train recall	test recall
125	1176	73.33	**74.13**	10	77.77	**76.92**	3	23	11	14	10	10	82.22	**77.62**
250	300	72.77	**63.63**	20	77.77	**69.93**	18	50	14	27	26	21	84.44	**79.72**
500	255	77.22	**75.52**	20	76.66	**77.32**	22	50	4	37	34	15	82.77	**77.62**
750	104	63.88	**65.73**	35	63.88	**69.93**	41	10	15	32	40	22	70.55	**69.23**
1000	208	70.00	**68.53**	15	69.44	**68.53**	22	28	44	45	25	25	70.00	**69.93**

consideration while the results for the images in the database are also accounted for. The research experiments were also conducted for different values τ_{max}. It can be noticed that introducing new stages of the analysis algorithm improves the result significantly both for the images used for testing as well as for the images stored in the database. The Table also presents threshold values Θ_k with the class division, which were also obtained as a result of the operation of the DE algorithm. It is worth noting that the threshold values differ from one another as well as from the common value Θ for thresholding with the use of one value. Thus, manual selection of those numbers would be problematic, so application of an optimization algorithm such as the DE one considerably automates this process. One of the columns contains the number of neurons which was generated during the operation of the GSOM clustering algorithm. As it can be noticed the value τ_{max} has a great influence on the number of clusters. Its smaller value results in generating a greater number of groups during the clustering process.

4 Conclusions

In this paper we have presented a new analytical module which is to extract from the dictionary those visual words which are most characteristic of all images or of a set of images of a given class. The experiments which we conducted clearly show that each of the new algorithm elements, which we propose, considerably affects its efficiency. It is rather difficult to compare the results obtained within the scope of our research with the latest image recognition results obtained by deep-learning networks; however, an obvious advantage of our algorithm is the very fact that it copes well with small databases (in our case the system learnt on 30 images from each class), and most importantly, it does not require great computing power of the present-day computers or a long learning period.

References

1. Almohammadi, K., Hagras, H., Alghazzawi, D., Aldabbagh, G.: Users-centric adaptive learning system based on interval type-2 fuzzy logic for massively crowded e-learning platforms. J. Artif. Intell. Soft Comput. Res. **6**(2), 81–101 (2016)
2. Audet, S.: JavaCV (2017). http://bytedeco.org/. Online; Accessed 1 Feb 2017
3. Bay, H., Tuytelaars, T., Gool, L.: SURF: speeded up robust features. In: Leonardis, A., Bischof, H., Pinz, A. (eds.) ECCV 2006. LNCS, vol. 3951, pp. 404–417. Springer, Heidelberg (2006). doi:10.1007/11744023_32
4. Bilski, J., Smolag, J.: Parallel architectures for learning the RTRN and elman dynamic neural networks. IEEE Trans. Parallel Distrib. Syst. **26**(9), 2561–2570 (2015)
5. Bradski, G.: The OpenCV library. Dr. Dobb's J. Softw. Tools **25**, 120–126 (2000)
6. Csurka, G., Dance, C.R., Fan, L., Willamowski, J., Bray, C.: Visual categorization with bags of keypoints. In: Workshop on Statistical Learning in Computer Vision, ECCV, pp. 1–22 (2004)

7. Fei-Fei, L., Fergus, R., Perona, P.: Learning generative visual models from few training examples: an incremental bayesian approach tested on 101 object categories. In: Conference on Computer Vision and Pattern Recognition Workshop, CVPRW 2004, pp. 178–178 (2004)

8. Fritzke, B.: Growing grid – a self-organizing network with constant neighborhood range and adaptation strength. Neural Process. Lett. **2**(5), 9–13 (1995)

9. Gabryel, M., Grycuk, R., Korytkowski, M., Holotyak, T.: Image indexing and retrieval using GSOM algorithm. In: Rutkowski, L., Korytkowski, M., Scherer, R., Tadeusiewicz, R., Zadeh, L.A., Zurada, J.M. (eds.) ICAISC 2015. LNCS, vol. 9119, pp. 706–714. Springer, Cham (2015). doi:10.1007/978-3-319-19324-3_63

10. Gao, H., Dou, L., Chen, W., Sun, J.: Image classification with bag-of-words model based on improved sift algorithm. In: 2013 9th Asian Control Conference (ASCC), pp. 1–6 (2013)

11. Korytkowski, M.: Novel visual information indexing in relational databases. Integr. Comput.-Aided Eng. **24**(2), 119–128 (2017)

12. Lan, K., Sekiyama, K.: Autonomous viewpoint selection of robot based on aesthetic evaluation of a scene. J. Artif. Intell. Soft Comput. Res. **6**(4), 255–265 (2016)

13. Łapa, K., Cpałka, K., Wang, L.: New method for design of fuzzy systems for nonlinear modelling using different criteria of interpretability. In: Rutkowski, L., Korytkowski, M., Scherer, R., Tadeusiewicz, R., Zadeh, L.A., Zurada, J.M. (eds.) ICAISC 2014. LNCS, vol. 8467, pp. 217–232. Springer, Cham (2014). doi:10.1007/978-3-319-07173-2_20

14. Lazebnik, S., Schmid, C., Ponce, J.: Beyond bags of features: spatial pyramid matching for recognizing natural scene categories. In: 2006 IEEE Computer Society Conference on Computer Vision and Pattern Recognition, vol. 2, pp. 2169–2178 (2006)

15. Li, F.F., Perona, P.: A bayesian hierarchical model for learning natural scene categories. In: Proceedings of the 2005 IEEE Computer Society Conference on Computer Vision and Pattern Recognition (CVPR 2005), vol. 2, pp. 524–531. IEEE Computer Society (2005)

16. Li, W., Dong, P., Xiao, B., Zhou, L.: Object recognition based on the region of interest and optimal bag of words model. Neurocomputing **172**, 271–280 (2016)

17. Lin, W.C., Tsai, C.F., Chen, Z.Y., Ke, S.W.: Keypoint selection for efficient bag-of-words feature generation and effective image classification. Inf. Sci. **329**, 33–51 (2016)

18. Liu, J.: Image retrieval based on bag-of-words model. CoRR abs/1304.5168 (2013). http://arxiv.org/abs/1304.5168

19. Olson, D.L., Delen, D.: Advanced Data Mining Techniques, 1st edn. Springer Publishing Company Incorporated, Heidelberg (2008)

20. Pabiasz, S., Starczewski, J.T., Marvuglia, A.: SOM vs FCM vs PCA in 3D face recognition. In: Rutkowski, L., Korytkowski, M., Scherer, R., Tadeusiewicz, R., Zadeh, L.A., Zurada, J.M. (eds.) ICAISC 2015. LNCS, vol. 9120, pp. 120–129. Springer, Cham (2015). doi:10.1007/978-3-319-19369-4_12

21. Prasad, M., Liu, Y.T., Li, D.L., Lin, C.T., Shah, R.R., Kaiwartya, O.P.: A new mechanism for data visualization with tsk-type preprocessed collaborative fuzzy rule based system. J. Artif. Intell. Soft Comput. Res. **7**(1), 33–46 (2017)

22. Starczewski, A.: A clustering method based on the modified RS validity index. In: Rutkowski, L., Korytkowski, M., Scherer, R., Tadeusiewicz, R., Zadeh, L.A., Zurada, J.M. (eds.) ICAISC 2013. LNCS, vol. 7895, pp. 242–250. Springer, Heidelberg (2013). doi:10.1007/978-3-642-38610-7_23

23. Starczewski, J.T., Pabiasz, S., Vladymyrska, N., Marvuglia, A., Napoli, C., Woźniak, M.: Self organizing maps for 3D face understanding. In: Rutkowski, L., Korytkowski, M., Scherer, R., Tadeusiewicz, R., Zadeh, L.A., Zurada, J.M. (eds.) ICAISC 2016. LNCS, vol. 9693, pp. 210–217. Springer, Cham (2016). doi:10.1007/978-3-319-39384-1_19

24. Staszewski, P., Woldan, P., Korytkowski, M., Scherer, R., Wang, L.: Query-by-example image retrieval in microsoft SQL server. In: Rutkowski, L., Korytkowski, M., Scherer, R., Tadeusiewicz, R., Zadeh, L.A., Zurada, J.M. (eds.) ICAISC 2016. LNCS, vol. 9693, pp. 746–754. Springer, Cham (2016). doi:10.1007/978-3-319-39384-1_66

25. Storn, R., Price, K.: Differential evolution - a simple and efficient heuristic for global optimization over continuous spaces. J. Global Optim. 11(4), 341–359 (1997)

26. Szarek, A., Korytkowski, M., Rutkowski, L., Scherer, R., Szyprowski, J.: Application of neural networks in assessing changes around implant after total hip arthroplasty. In: Rutkowski, L., Korytkowski, M., Scherer, R., Tadeusiewicz, R., Zadeh, L.A., Zurada, J.M. (eds.) ICAISC 2012. LNCS, vol. 7268, pp. 335–340. Springer, Heidelberg (2012). doi:10.1007/978-3-642-29350-4_40

27. Woźniak, M.: Novel image correction method based on swarm intelligence approach. In: Dregvaite, G., Damasevicius, R. (eds.) ICIST 2016. CCIS, vol. 639, pp. 404–413. Springer, Cham (2016). doi:10.1007/978-3-319-46254-7_32

28. Wozniak, M., Polap, D.: On manipulation of initial population search space in heuristic algorithm through the use of parallel processing approach. In: 2016 IEEE Symposium Series on Computational Intelligence, pp. 1–6. IEEE (2016)

29. Wozniak, M., Polap, D., Napoli, C., Tramontana, E.: Graphic object feature extraction system based on cuckoo search algorithm. Expert Syst. Appl. 66, 20 31 (2016)

30. Yin, Z., O'Sullivan, C., Brabazon, A.: An analysis of the performance of genetic programming for realised volatility forecasting. J. Artif. Intell. Soft Comput. Res. 6(3), 155 172 (2016)

31. Zalasiński, M., Cpałka, K.: New algorithm for on-line signature verification using characteristic hybrid partitions. In: Wilimowska, Z., Borzemski, L., Grzech, A., Świątek, J. (eds.) Information Systems Architecture and Technology: Proceedings of 36th International Conference on Information Systems Architecture and Technology – ISAT 2015 – Part IV. AISC, vol. 432, pp. 147–157. Springer, Cham (2016). doi:10.1007/978-3-319-28567-2_13

32. Zalasiński, M., Cpałka, K., Er, M.J.: New method for dynamic signature verification using hybrid partitioning. In: Rutkowski, L., Korytkowski, M., Scherer, R., Tadeusiewicz, R., Zadeh, L.A., Zurada, J.M. (eds.) ICAISC 2014. LNCS, vol. 8468, pp. 216–230. Springer, Cham (2014). doi:10.1007/978-3-319-07176-3_20

33. Zhao, C., Li, X., Cang, Y.: Bisecting k-means clustering based face recognition using block-based bag of words model. Optik - Int. J. Light Electron Opt. 126(19), 1761–1766 (2015)

The Novel Method of the Estimation of the Fourier Transform Based on Noisy Measurements

Tomasz Galkowski[1]([✉]) and Miroslaw Pawlak[2,3]

[1] Institute of Computational Intelligence, Czestochowa University of Technology,
Czestochowa, Poland
tomasz.galkowski@iisi.pcz.pl
[2] Information Technology Institute, University of Social Sciences, Lodz, Poland
[3] Department of Electrical and Computer Engineering, University of Manitoba,
Winnipeg, Canada
pawlak@ee.umanitoba.ca

Abstract. This article refers to the problem of the analysis of spectrum of signals observed in the presence of noise. We propose a new concept of estimation of the frequency content in the signal. The method is derived from the nonparametric methodology of function estimation. We refer to the model of the system $y_i = R(x_i) + \epsilon_i$, $i = 1, 2, \ldots n$, where x_i is assumed to be the set of deterministic inputs, $x_i \in D$, y_i is the set of probabilistic outputs, and ϵ_i is a measurement noise with zero mean and bounded variance. $R(.)$ is a completely unknown function. In this paper we are interested in a question about frequency spectrum of unknown function. Finding of unknown function in the model could be realized using algorithms based on the Parzen kernel. The alternative approach is based on the orthogonal series expansions. Nonparametric methodology could also be used in the task of implicit estimation of its spectrum. The main aim of this paper is to propose an original integral version of nonparametric estimation of spectrum based on trigonometric series - referring to the classic Fourier transform. The results of numerical experiments are presented.

Keywords: Nonparametric estimation · Frequency spectrum · Noisy signals · Orthogonal series

1 Introduction

Several methods using the artificial intelligence methods like neural networks, fuzzy sets, genetic algorithms were adopted to classification and modelling tasks, see e.g. [1–6,14–16,18–21,23,25,30,38–41,45,49]. Especially the nonparametric methodology have been proposed in literature for modelling and

M. Pawlak—Carried out this research at ASS during his sabbatical leave from University of Manitoba.

L. Rutkowski et al. (Eds.): ICAISC 2017, Part II, LNAI 10246, pp. 52–61, 2017.
DOI: 10.1007/978-3-319-59060-8_6

classification problems in stationary conditions [7,9–12,28,29,31,33,34,43,44], quasi-stationary and/or time-varying environment [13,32,35–37].

Current work is about the data analysis basing on Fourier transform. The Fourier transform is used in several areas by engineers and scientists, like electronic circuit designers, signal processing and telecommunication, chemists and physicist interested in spectroscopy and crystallography, vision and imaging engineers, and many others. In fundamental mathematics the Fourier analysis is used for periodic phenomena, via Fourier series. In literature of the subject one may find the extension of those insights to nonperiodic phenomena, via the Fourier transform, see e.g. [26].

The preliminary problem in our investigation is to estimate regression function $R(.)$ in the model of type:

$$y_i = R\left(x_i\right) + \epsilon_i, \ i = 1, 2, \dots n \tag{1}$$

where x_i is assumed to be the set of deterministic inputs, $x_i \in D$, y_i is the set of probabilistic outputs, and ϵ_i is a measurement noise with zero mean and bounded variance. $R(.)$ is a completely unknown function. We mean "completely unknown" as: no assumption neither on its shape (like e.g. in the spline methods) nor on any mathematical formula with certain set of parameters to be found (so-called parametric approach). The possible solutions of finding unknown function in nonparametric approaches are based on Parzen kernel [8] or methods derived from orthogonal series [35]. Note that the Parzen kernel methods are much more often applied and analysed for estimation of probability density functions and/or regressions with probabilistic input than in a deterministic case.

Proposition submitted in this paper bases on the second approach, particularly nonparametric orthogonal series expansion.

In many situations we do not need to approximate the original function $R(.)$. The essential information on observed system may be obtain by the analysing of frequency content of the function. For instance, the possible damage of mechanical device may be detected in the graph of its vibrations as a unexpected frequency in the spectrum. The second example concerns of the telecommunication signals. Signal modulation is the fundamental in transmission of information for a long distances, particularly by wireless media. For other possible applications see e.g. [17,22,24,27,42,47,48]. Of course in the transmission channel the signal is affected by external noises. A recipient needs to extract sent information from imperfect observations of received signals. The right detection also leads by the analysis of the spectrum and appropriate filtering algorithm.

2   Orthogonal Series Estimation of Regression Function and Its Spectrum

The nonparametric algorithm of the estimation of the unknown function $R(.)$ derived from the orthogonal series expansion is in the form

$$\hat{R}_n\left(x\right) = \sum_{k=0}^{N} \hat{a}_k \cdot g_k(x) \tag{2}$$

where

$$\hat{a}_k = \sum_{i=1}^{n} y_i \int_{D_i} g_k(u)du \tag{3}$$

Functions $g_k(x)$, $k = 0, 1, 2, ...$, form the orthogonal system on the interval $D = [a, b]$, without loss of generality we assume $a = 0$, $b = 2\pi$. The deterministic inputs x_i are selected so that $x_i \in D_i$, where D_i are subintervals of D such that $D_i \wedge D_l = \emptyset$ for $i \neq l$ and $\cup D_i = D$.

Equations (2) and (3) could be written jointly:

$$\hat{R}_n(x) = \sum_{k=0}^{N} g_k(x) \sum_{i=1}^{n} y_i \int_{D_i} g_k(u)du \tag{4}$$

When we choose the trigonometric orthogonal system it is easily seen that the sequence of expressions (3) forms the frequency components in the Fourier series of unknown function. The parameter $N(n)$ depends on n and it determine on how many Fourier components we take into considerations. The complete collection of these components is the frequency spectrum of analysed function. Note that the frequencies in this instance are the multiplication of the base frequency (i.e. for $k = 1$) and sequence of natural numbers. So they have the discrete character. Let us to remind that continuous Fourier transform of function $x(t)$ is defined by

$$F(s) = \int_{-\infty}^{\infty} x(t)e^{-j2\pi st}dt \tag{5}$$

where $e^{j\theta} = \cos\theta + j\sin\theta$.

Note the analogue character of function $x(t)$. In real situations we often dispose of a finite set of data x_i, $i = 0, 1, ..., n$. The way of getting from Fourier series to the Fourier transform is to consider nonperiodic phenomena (and thus just about any general function) as a limiting case of periodic phenomena as the period tends to infinity (see [26]). We commonly apply the discrete Fourier transform (DFT) X_k in the form:

$$X_k = \sum_{i=0}^{n-1} x_i e^{-j2\pi \frac{i}{n}k} \tag{6}$$

where $k = 0, 1, ..., n - 1$, and n is the number of measured values x_i, $i = 0, 1, ..., n - 1$.

3   Nonparametric Orthogonal Series Estimation of Fourier Transform

In this paper we propose the method of calculation of Fourier transform in semi-continuous case. It is applicable when we observe outputs of analogue systems when the observer can decide at what moments of time to take measurements y_i. In nonparametric terminology we say that the system has deterministic inputs.

Comparing Eqs. (3) and (5) we can propose the novel modification applied to the Eq. (5) obtaining the original nonparametric estimator:

$$\hat{F}_n(s) = \sum_{i=0}^{n-1} y_i \int_{D_i} e^{-j2\pi st} dt \tag{7}$$

The time period $D = [t_0, t_{n-1}]$ is divided into n intervals D_i and the points t_i, $i = 0, 1, ..., n$ should be placed inside D_i, i.e. $t_i \in D_i$. In papers one can find the theorems of convergence of the algorithm defined by (2) and (3). The standard conditions for the mean square convergence [8] are:

$$
\begin{align}
(i) \quad & R(t) \subset L^1 & for\ t \in D \\
(ii) \quad & \max|D_i| = O(n^{-1}) & for\ i = 0, ..., n - 1 \\
(iii) \quad & N(n) \to \infty, \tfrac{N(n)}{n} \to 0\ if \quad n \to \infty
\end{align} \tag{8}
$$

We assume that conditions (8) are fulfilled for the new proposed estimator (7). In the following section we present the simulation example of application of algorithm (7) in telecommunication system using amplitude modulation method in the presence of noise in the communication channel.

4   Simulation Example

Let us assume that the investigated function is in the form:

$$y(t) = p(t) + jq(t) \tag{9}$$

where $p(t)$ is an even function of t, and $q(t)$ is an odd function of t. This assumption is justified in telecommunication systems in which we commonly use the trigonometric (harmonic) functions. Then the Fourier transform can be rewritten as [26]

$$F(s) = 2 \int_0^\infty p(t) cos(2\pi st) dt - 2j \int_0^\infty q(t) \sin(2\pi st) dt \tag{10}$$

Functions $p(t)$ and $q(t)$ are so-called in-phase and quadrature components of signal, respectively [46]. For instance such signals are useful in modern digital telecommunications systems based on multiple orthogonal carriers.

In the simple case of amplitude modulation two signals are multiplied. First of them is the message (or information) modulating signal, the second one is the carrier (the modulated signal).

The nonparametric estimator of spectrum now takes the form:

$$\hat{F}_n(s) = 2 \sum_{i=0}^{n-1} \text{Re}[y_n] \int_{D_i} \cos(2\pi st) dt - 2j \sum_{i=0}^{n-1} \text{Im}[y_n] \int_{D_i} \sin(2\pi st) dt \tag{11}$$

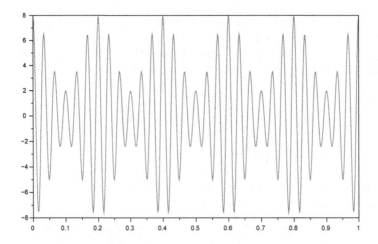

Fig. 1. The generated amplitude modulated signal without noise

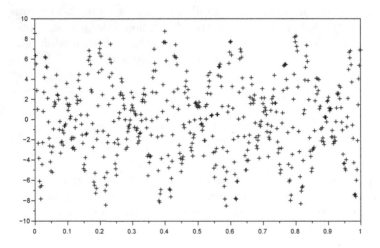

Fig. 2. Sampled amplitude modulated signal with additive white noise

This formula significantly simplifies the subsequent calculations.

The algorithm (11) was tested on one of the classical telecommunication rules which is the amplitude modulation of signals. Without loss of generality the message signal is in the form

$$m(t) = A_m \sin(2\pi f_m t) \tag{12}$$

and the carrier signal is in the form

$$c(t) = A_c \sin(2\pi f_c t) \tag{13}$$

Then the resulting modulated signal is:

$$X_{AM}(t) = A_c \left(1 + \mu \cos(2\pi f_m t)\right) \cdot \cos(2\pi f_c t). \tag{14}$$

where $\mu = A_m / A_c$.

After some algebra we obtain the expression (14) in the form presenting three components clearly showing its spectrum components:

$$X_{AM}(t) = A_c \cos(2\pi f_c t) + \frac{1}{2} A_m \cos(2\pi(f_c + f_m)t) + \frac{1}{2} A_m \cos(2\pi(f_c - f_m)t) \tag{15}$$

It consists of three parts: main - representing carrier at frequency f_c, and two side-chunks representing the information signal - at frequencies $(f_c - f_m)$ and $(f_c + f_m)$. For simulation we choose the parameters: $A_n = 1$, $A_c = 3$, $f_m = 5$, $f_c = 30$.

Figure 1 presents the generated amplitude modulated signal without noise - such signal usually occurs on transmitter side. In the transmission channel an additive noise affects sent signal, so on the receiver side we may observe and register the noised signals - exemplary samples are presented in Fig. 2.

The problem is to estimate Fourier spectrum of $X(.)$ basing on set of the measurements in that model:

$$y_n = X_{AM}(t_n) + Z_n \tag{16}$$

Sequence Z_n represents the measurement noise, i.e. realization of white stochastic process of zero mean $EZ_i = 0$, and finite variation $EZ_i^2 = \sigma_i^2 < \sigma^2$, $i = 1, ..., n$. Pictures show the simulation performed for $n = 512$ - number of samples, $f_s = 128$ - sampling frequency, and $t_\vartheta = 0,0078125$ - the sampling interval.

The graph in Fig. 3. shows the result of simulation of spectrum estimation applying procedure (7). We can see three peaks: the highest in the middle corresponding to the frequency $f_c = 30$ and the pair of sidebands frequencies $(f_c - f_m) = 25$ and $(f_c + f_m) = 35$. This clearly corresponds to our expectation and Eq. (15). For comparison we present the calculation of DFT using standard module of Scilab environment in Fig. 4. Note that the classical DFT has more restricted method of sampling (like uniform distance between samples, and not-continuous resulting spectrum). Our new method yields to similar conclusions on frequency content of spectrum. This spectrum is continuous which means that it may be calculated for any frequency (under obvious limitations). The choice of placement of points t_i in subintervals D_i seems to be important and it will be investigated in future works.

5 Remarks and Extensions

The new application area of using the nonparametric algorithm, derived from orthogonal series expansion, is proposed. The presented idea relays on the harmonic system of functions for estimation of Fourier transform in deterministic

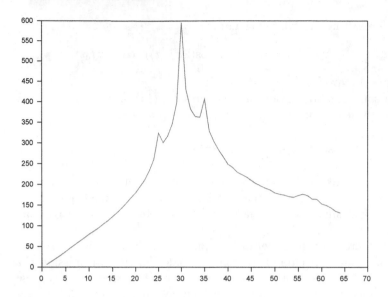

Fig. 3. Nonparametric estimation of Fourier transform based on noised samples

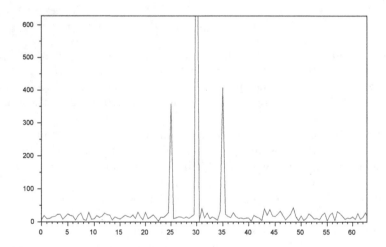

Fig. 4. Discrete Fourier transform obtained in $Scilab^{TM}$

case of input signals. Our algorithm is an alternative way for obtaining frequency spectrum of signal in the presence of the additive noise. The results of numerical simulations show that the conclusions of its analysis are similar to classical discrete Fourier transform. The future works should be continued in such directions as multivariate case of analysed signals, optimization of convergence of algorithm, and applications in solving problems of real systems.

References

1. Bas, E.: The training of multiplicative neuron model artificial neural networks with differential evolution algorithm for forecasting. J. Artif. Intell. Soft Comput. Res. **6**(1), 5–11 (2016)
2. Bertini Jr., J.R., Carmo, N.M.: Enhancing constructive neural network performance using functionally expanded input data. J. Artif. Intell. Soft Comput. Res. **6**(2), 119–131 (2016)
3. Chang, O., Constante, P., Gordon, A., Singana, M.: A novel deep neural network that uses space-time features for tracking and recognizing a moving object. J. Artif. Intell. Soft Comput. Res. **7**(2), 125–136 (2017)
4. Cierniak, R., Rutkowski, L.: On image compression by competitive neural networks and optimal linear predictors. Sig. Process.-Image Commun. **15**(6), 559–565 (2000)
5. Cpałka, K., Zalasiński, M., Rutkowski, L.: A new algorithm for identity verification based on the analysis of a handwritten dynamic signature. Appl. Soft Comput. **43**, 47–56 (2016)
6. Duch, W., Korbicz, J., Rutkowski, L., Tadeusiewicz, R. (eds.): Biocybernetics and Biomedical Engineering 2000. Neural Networks, vol. 6. Akademicka Oficyna Wydawnicza, EXIT, Warsaw (2000) (in Polish)
7. Gałkowski, T., Rutkowski, L.: Nonparametric recovery of multivariate functions with applications to system identification. In: Proceedings of the IEEE, vol. 73, pp. 942–943, New York (1985)
8. Gałkowski, T., Rutkowski, L.: Nonparametric fitting of multivariate functions. IEEE Trans. Autom. Control **AC-31**, 785–787 (1986)
9. Gałkowski, T.: Nonparametric estimation of boundary values of functions. Arch. Control Sci. **3**(1–2), 85–93 (1994)
10. Gałkowski, T.: Kernel estimation of regression functions in the boundary regions. In: Rutkowski, L., Korytkowski, M., Scherer, R., Tadeusiewicz, R., Zadeh, L.A., Zurada, J.M. (eds.) ICAISC 2013. LNCS (LNAI), vol. 7895, pp. 158–166. Springer, Heidelberg (2013). doi:10.1007/978-3-642-38610-7_15
11. Galkowski, T., Pawlak, M.: Nonparametric extension of regression functions outside domain. In: Rutkowski, L., Korytkowski, M., Scherer, R., Tadeusiewicz, R., Zadeh, L.A., Zurada, J.M. (eds.) ICAISC 2014. LNCS (LNAI), vol. 8467, pp. 518–530. Springer, Cham (2014). doi:10.1007/978-3-319-07173-2_44
12. Greblicki, W., Rutkowski, L.: Density-free Bayes risk consistency of nonparametric pattern recognition procedures. Proc. IEEE **69**(4), 482–483 (1981)
13. Greblicki, W., Rutkowska, D., Rutkowski, L.: An orthogonal series estimate of time-varying regression. Ann. Inst. Stat. Math. **35**(1), 215–228 (1983)
14. Knop, M., Kapuscinski, T., Mleczko, W.K.: Video key frame detection based on the restricted Boltzmann machine. J. Appl. Math. Comput. Mech. **14**(3), 49–58 (2015)
15. Korytkowski, M., Nowicki, R., Scherer, R.: Neuro-fuzzy rough classifier ensemble. In: Alippi, C., Polycarpou, M., Panayiotou, C., Ellinas, G. (eds.) ICANN 2009. LNCS, vol. 5768, pp. 817–823. Springer, Heidelberg (2009). doi:10.1007/978-3-642-04274-4_84
16. Korytkowski, M., Rutkowski, L., Scherer, R.: Fast image classification by boosting fuzzy classifiers. Inf. Sci. **327**, 175–182 (2016)
17. Kozieł, G.: Steganographic algorithm of hiding information in sound based on fourier transform and masking. Control Cybern. **40**(4), 1231–1247 (2011)

18. Najeeb, A.K., Amber, S.: A smart amalgamation of spectral neural algorithm for nonlinear Lane-Emden equations with simulated annealing. J. Artif. Intell. Soft Comput. Res. **7**(3), 215–224 (2017)
19. Laskowski, L.: A novel hybrid-maximum neural network in stereo-matching process. Neural Comput. Appl. **23**(7–8), 2435–2450 (2013)
20. Laskowski, L., Jelonkiewicz, J.: Self-correcting neural network for stereo-matching problem solving. Fundamenta Informaticae **138**, 1–26 (2015)
21. Laskowski, L., Laskowska, M., Jelonkiewicz, J., Boullanger, A.: Molecular approach to hopfield neural network. In: Rutkowski, L., Korytkowski, M., Scherer, R., Tadeusiewicz, R., Zadeh, L.A., Zurada, J.M. (eds.) ICAISC 2015. LNCS (LNAI), vol. 9119, pp. 72–78. Springer, Cham (2015). doi:10.1007/978-3-319-19324-3_7
22. Liflyand, E.: Integrability spaces for the fourier transform of a function of bounded variation. J. Math. Anal. Appl. **436**, 1082–1101 (2016)
23. Łapa, K., Przybył, A., Cpałka, K.: A new approach to designing interpretable models of dynamic systems. In: Rutkowski, L., Korytkowski, M., Scherer, R., Tadeusiewicz, R., Zadeh, L.A., Zurada, J.M. (eds.) ICAISC 2013. LNCS (LNAI), vol. 7895, pp. 523–534. Springer, Heidelberg (2013). doi:10.1007/978-3-642-38610-7_48
24. Nguyen, C.T., Havlicek, J.P.: On the amplitude and phase computation of the AM-FM image model. In: 2014 IEEE International Conference on Image Processing (ICIP) (2015)
25. Nowak, B., Nowicki, R., Starczewski, J., Marvuglia, A.: The learning of neuro-fuzzy classifier with fuzzy rough sets for imprecise datasets. In: Artificial Intelligence and Soft Computing, pp. 256–266 (2014)
26. Osgood, B.: Lecture notes for EE 261 the fourier transform and its applications. Electrical Engineering Department, Stanford University, CreateSpace Independent Publishing Platform, 18 December 2014
27. Paarmann, L.D., Najar, M.D.: Analysis of the Wigner-Ville transform of periodic signals. In: Proceedings of the IEEE-SP International Symposium on Time-frequency and Time-scale Analysis (1994)
28. Rafajłowicz, E.: Nonparametric least squares estimation of a regression function Statistics. J. Theor. Appl. Stat. **19**(3), 349–358 (1988)
29. Rafajłowicz, E., Schwabe, R.: Halton and Hammersley sequences in multivariate nonparametric regression. Stat. Prob. Lett. **76**(8), 803–812. Elsevier (2006)
30. Rutkowska, A.: Influence of membership function's shape on portfolio optimization results. J. Artif. Intell. Soft Comput. Res. **6**(1), 45–54 (2016)
31. Rutkowski, L.: Sequential estimates of probability densities by orthogonal series and their application in pattern classification. IEEE Trans. Syst. Man Cybern. **SMC-10**(12), 918–920 (1980)
32. Rutkowski, L.: On bayes risk consistent pattern recognition procedures in a quasi-stationary environment. IEEE Trans. Pattern Anal. Mach. Intell. **PAMI-4**(1), 84–87 (1982)
33. Rutkowski, L.: A general approach for nonparametric fitting of functions and their derivatives with applications to linear circuits identification. IEEE Trans. Circ. Syst. **CAS-33**, 812–818 (1986)
34. Rutkowski, L.: Sequential pattern recognition procedures derived from multiple fourier series. Pattern Recogn. Lett. **8**, 213–216 (1988)
35. Rutkowski, L.: Application of multiple fourier series to identification of multivariable nonstationary systems. Int. J. Syst. Sci. **20**(10), 1993–2002 (1989)
36. Rutkowski, L.: Non-parametric learning algorithms in the time-varying environments. Sig. Process. **18**(2), 129–137 (1989)

37. Rutkowski, L.: Adaptive probabilistic neural networks for pattern classification in time-varying environment. IEEE Trans. Neural Netw. **15**(4), 811–827 (2004)
38. Rutkowski, L., Pietruczuk, L., Duda, P., Jaworski, M.: Decision trees for mining data streams based on the McDiarmid's bound. IEEE Trans. Knowl. Data Eng. **25**(6), 1272–1279 (2013)
39. Rutkowski, L., Jaworski, M., Duda, P., Pietruczuk, L.: Decision trees for mining data streams based on the gaussian approximation. IEEE Trans. Knowl. Data Eng. **26**(1), 108–119 (2014)
40. Rutkowski, L., Jaworski, M., Pietruczuk, L., Duda, P.: The CART decision trees for mining data streams. Inf. Sci. **266**, 1–15 (2014)
41. Rutkowski, L., Jaworski, M., Pietruczuk, L., Duda, P.: A new method for data stream mining based on the misclassification error. IEEE Trans. Neural Netw. Learn. Syst. **26**(5), 1048–1059 (2015)
42. Singh, P., Joshi, S.D.: Some studies on multidimensional fourier theory for hilbert transform, analytic signal and space-time series analysis. Comput. Sci. Inf. Theor. (2015). arXiv: 1507.08117
43. Skubalska-Rafajłowicz, E.: Pattern recognition algorithms based on space-filling curves and orthogonal expansions. IEEE Trans. Inf. Theor. **47**(5), 1915–1927 (2001)
44. Skubalska-Rafajłowicz, E.: Random projection RBF nets for multidimensional density estimation. Int. J. Appl. Math. Comput. Sci. **18**(4), 455–464 (2008)
45. Szarek, A., Korytkowski, M., Rutkowski, L., Scherer, R., Szyprowski, J.: Application of neural networks in assessing changes around implant after total hip arthroplasty. In: Rutkowski, L., Korytkowski, M., Scherer, R., Tadeusiewicz, R., Zadeh, L.A., Zurada, J.M. (eds.) ICAISC 2012. LNCS (LNAI), vol. 7268, pp. 335–340. Springer, Heidelberg (2012). doi:10.1007/978-3-642-29350-4_40
46. Wade, G.: Signal Coding and Processing, vol. 1, 2nd edn. Cambridge University Press, Cambridge (1994). ISBN 0521412307
47. Wielgus, M.: Amplitude demodulation of interferometric signals with a 2D Hilbert transform. Challenges Mod. Technol. Found. Young Sci. **2**(1), 8–11 (2011)
48. Wu, Y., Ma, J., Yang, Y., Sun, P.: Improvements of measuring the width of Fraunhofer diffraction fringes using fourier transform. Optik **126**, 4142–4145 (2015)
49. Zalasiński, M., Łapa, K., Cpałka, K.: New algorithm for evolutionary selection of the dynamic signature global features. In: Rutkowski, L., Korytkowski, M., Scherer, R., Tadeusiewicz, R., Zadeh, L.A., Zurada, J.M. (eds.) ICAISC 2013. LNCS (LNAI), vol. 7895, pp. 113–121. Springer, Heidelberg (2013). doi:10.1007/978-3-642-38610-7_11

A Complete Efficient FFT-Based Algorithm for Nonparametric Kernel Density Estimation

Jarosław Gramacki[1] and Artur Gramacki[2(✉)]

[1] Computer Center, University of Zielona Góra, Zielona Góra, Poland
j.gramacki@ck.uz.zgora.pl
[2] Institute of Control and Computation Engineering, University of Zielona Góra,
Zielona Góra, Poland
a.gramacki@issi.uz.zgora.pl

Abstract. Multivariate kernel density estimation (KDE) is a very important statistical technique in exploratory data analysis. Research on high performance KDE is still an open research problem. One of the most elegant and efficient approach utilizes the Fast Fourier Transform. Unfortunately, the existing FFT-based solution suffers from a serious limitation, as it can accurately operate only with the constrained (i.e., diagonal) multivariate bandwidth matrices. In the paper we propose a crucial improvement to this algorithm which results in relaxing the above mentioned limitation. Numerical simulation study demonstrates good properties of the new solution.

Keywords: Multivariate kernel density estimation · Unconstrained bandwidth matrix · Fast Fourier Transform · Nonparametric estimation

1 Introduction

Kernel density estimation (KDE) is a very important statistical technique with many practical applications. It has been applied successfully to both univariate and multivariate problems. There exists extensive literature on this issue, including several classical monographs, see [8,10,12,13].

A general form of the d-dimensional multivariate kernel density estimator is

$$\hat{f}(\boldsymbol{x}, \boldsymbol{H}) = n^{-1} \sum_{i=1}^{n} K_{\boldsymbol{H}}\left(\boldsymbol{x} - \boldsymbol{X}_i\right),$$

$$K_{\boldsymbol{H}}(u) = |\boldsymbol{H}|^{-1/2} K\left(\boldsymbol{H}^{-1/2} u\right), \tag{1}$$

where $\boldsymbol{H}$ is the $d \times d$ *bandwidth* or *smoothing* matrix, d is the problem dimensionality, $\boldsymbol{x} = (x_1, x_2, \ldots, x_d)^T$ and $\boldsymbol{X}_i = (X_{i1}, X_{i2}, \ldots, X_{id})^T$, $i = 1, 2, \ldots, n$ is a sequence of independent identically distributed (iid) d-variate random variables drawn from a (usually unknown) density function f. Here K and $K_{\boldsymbol{H}}$ are the unscaled and scaled kernels, respectively. In most cases the kernel has the form of a standard multivariate normal density.

© Springer International Publishing AG 2017
L. Rutkowski et al. (Eds.): ICAISC 2017, Part II, LNAI 10246, pp. 62–73, 2017.
DOI: 10.1007/978-3-319-59060-8_7

It seems that both uni- and multivariate KDE techniques have reached maturity and recent developments in this field are primarily focused on computational improvements. There are two main computational problems related to KDE: (a) the fast evaluation of the kernel density estimates $\hat{f}$, and (b) the fast estimation of the optimal bandwidth matrix H (or scalar h in the univariate case). In the paper the first problem is reported. The second one is shortly mentioned in the conclusions section.

It is obvious from (1) that the naive direct evaluation of the KDE at m evaluation points x for n data points X_i requires $O(mn)$ kernel evaluations. Evaluation points can be of course the same as data points and then the computational complexity is $O(n^2)$, making it very expensive, especially for large datasets and higher dimensions.

A number of methods have been proposed to accelerate the computations. See for example [9] for an interesting review of the methods. Other techniques, like for example usage of Graphics Processing Units (GPUs) and Field-Programmable Gate Array (FPGA) are also used [1,6]. One of the most elegant and effective methods is based on using the Fast Fourier Transform (FFT). A preliminary work on using FFT to kernel density estimation was given in [11] (but only for univariate case).

In the paper we are concerned with an FFT-based method that was originally described in [14]. The original method works very well for univariate case but, unfortunately, its multivariate extension does not support unconstrained bandwidth matrices. From now on this method will be called *Wand's algorithm*.

The paper is organized as follows: In Sect. 2, based on a simple example, we demonstrate the problem. In Sect. 3 we give details of our improved FFT-based algorithm for density estimation. In Sect. 4 we demonstrate the main idea of a kind of data discretization known as *binning* (binning is the required data preprocessing step in our algorithm). In Sect. 5 we give results from some numerical experiments based on both synthetic and real data sets. In Sect. 6 we conclude the paper.

2   Problem Demonstration

Wand's algorithm is based on rewriting (1) in such a way that it might be efficiently calculated using FFT. This makes possible to reduce the computational complexity from $O(n^2)$ to $O(M_1 \log M_1 \ldots M_d \log M_d)$, where $M_1 \ldots M_d$ are constant values and they will be defined precisely in Sect. 3. A preliminary step required to construct the FFT-based algorithm is a kind of data discretization, known as *binning*. This concept is briefly described in Sect. 4.

Although Wand's algorithm is very fast and accurate it suffers from a serious limitation. It supports only a small subset of all possible multivariate kernels of interest. Two commonly used kernel types are *product* and *radial* (also known as *spherically symmetric*) ones [13]. The problem reveals if the radial kernel is used and the bandwidth matrix H is *unconstrained*, that is $H \in \mathcal{F}$, where $\mathcal{F}$ denotes the class of symmetric, positive definite $d \times d$ matrices. If, however, the bandwidth matrix belongs to a more restricted *constrained* (*diagonal*) form (that is $\boldsymbol{H} \in \mathcal{D}$), the problem doesn't manifest itself.

To the best of our knowledge the above mentioned problem is not clearly presented and solved in literature, except a few short mentions in [13,14] and in the kde{ks} R function [2]. Moreover, many authors cite the FFT-based algorithm for KDE mechanically, without any qualification or mentioning its greatest limitation.

Note that Wand's algorithm is implemented in the **ks** R package [2], as well as in the **KernSmooth** R package [15]. However, the **KernSmooth** implementation supports only product kernels. The standard density{stats} R function uses FFT to compute only univariate kernel density estimates.

For simplicity in the paper only 2D examples are presented, but extension for higher dimensions is rather immediate. In Fig. 1 we demonstrate the problem. A sample *unicef* dataset from the **ks** R package was used. The density estimation depicted in Fig. 1a can be treated as a reference. It was calculated directly from (1). The unconstrained bandwidth $\boldsymbol{H}$ was calculated using the Hpi{ks} R function. In Fig. 1b the density estimation was calculated using Wand's algorithm. The bandwidth $\boldsymbol{H}$ was also unconstrained, exactly the same as in Fig. 1a. It is easy to notice that the result is obviously inaccurate, as the results in Fig. 1a and b should be the same. Figure 1c and d are equivalents to Fig. 1a and b, respectively, except that the bandwidth $\boldsymbol{H}$ is now constrained (calculated using the Hpi.diag{ks} R function). Both figures are identical and, moreover, very similar to Fig. 1b!

This similarity suggests that Wand's algorithm lose in some way most (or even all) the information carried by off-diagonal entries of the bandwidth matrix $\boldsymbol{H}$. In other words, Wand's algorithm (in it's original form) is adequate only for constrained bandwidths. Additionally, in Fig. 1e and f we show the results where the *product* kernel is used. Two individual scalar bandwidths were calculated using hpi{ks} R function (note that the hpi{ks} function implements the univariate plug-in selector, while the Hpi{ks} function is its multivariate equivalent). Both figures are identical as the problem presented in the paper does not affect product kernels.

3   The Improved FFT-Based Algorithm for Density Estimation

In this section we briefly present original Wand's algorithm. Then we propose a crucial modification to this algorithm (see (7) and (8) and compare them with the ones given in [14]). This modification results in $\boldsymbol{H} \in \mathcal{F}$ support.

The algorithm consists of 3 basic steps. In the **first step** the multivariate *linear binning* (a kind of data discretization, see [14] and Sect. 4) of the input data points $\boldsymbol{X}_i$ is required. The binning approximation of (1) is

$$\tilde{f}(\boldsymbol{g}_i, \boldsymbol{H}, \boldsymbol{M}) = n^{-1} \sum_{j_1=1}^{M_1} \cdots \sum_{j_d=1}^{M_d} K_{\boldsymbol{H}}\left(\boldsymbol{g}_i - \boldsymbol{g}_j\right) c_j, \tag{2}$$

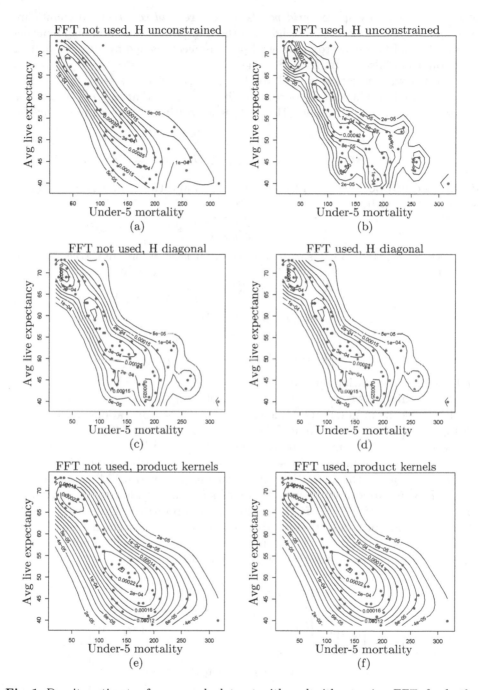

Fig. 1. Density estimates for a sample dataset with and without using FFT, for both unconstrained and constrained bandwidth matrices. Description of each plot is given in the text

where g are equally spaced *grid points* and c are *grid counts*. Grid counts are obtained by assigning certain weights to the grid points, based on neighbouring observations. In other words, each grid point is accompanied by a corresponding grid count (see also examples in Sect. 4).

The following notation is used: for $k = 1, \ldots, d$, let $g_{k1} < \cdots < g_{kM_K}$ be an equally spaced grid in the kth coordinate directions such that $[g_{k1}, g_{kM_k}]$ contains the kth coordinate grid points. Here M_k is a positive integer representing the *grid size* in direction k. Let

$$g_i = (g_{1i_1}, \ldots, g_{di_d}), \quad 1 \le i_k \le M_k, \quad k = 1, \ldots, d, \tag{3}$$

denote the grid point indexed by $i = (i_1, \ldots, i_d)$ and the kth *binwidth* (or *mesh size*) be denoted by

$$\delta_k = \frac{g_{kM_k} - g_{k1}}{M_k - 1}. \tag{4}$$

In the **second step** (2) is rewritten so that it takes a form of the *convolution*

$$\tilde{f}_i = \sum_{j_1=-(M_1-1)}^{M_1-1} \cdots \sum_{j_d=-(M_d-1)}^{M_d-1} c_{i-j} k_j = c * k, \tag{5}$$

where

$$k_j = n^{-1} K_H(\delta_1 j_1, \cdots, \delta_d j_d), \tag{6}$$

and $*$ is the convolution operator.

In the **third step** we compute the convolution between c_{i-j} and k_j using the FFT algorithm in only $O(M_1 \log M_1 \ldots M_d \log M_d)$ operations compared to $O(M_1^2 \ldots M_d^2)$ operations required for direct computation of (2).

To compute the convolution between c and k they must first be reshaped (*zero-padded*) according to precise rules which are described in detail in [4]. It is worth to note in this place that the reshaping proposed in [14] supports only $H \in \mathcal{D}$. Our proposition, however, supports the most wide class of bandwidth matrices, that is $H \in \mathcal{F}$ (the class of symmetric, positive definite matrices). In the further part of this section we give details of our improvements.

Here, for simplicity, only two-dimensional variant is presented but extension to higher dimensions is straightforward. We have

$$k_{zp} = \begin{bmatrix} k & 0 \\ 0 & 0 \end{bmatrix} = \begin{bmatrix} k_{-M_1,-M_2} & \cdots & k_{-M_1,0} & \cdots & k_{-M_1,M_2} & \\ \vdots & \ddots & \vdots & \ddots & \vdots & \\ k_{0,-M_2} & \cdots & k_{0,0} & \cdots & k_{0,M_2} & 0 \\ \vdots & \ddots & \vdots & \ddots & \vdots & \\ k_{M_1,-M_2} & \cdots & k_{M_1,0} & \cdots & k_{M_1,M_2} & \cdots \\ & & 0 & & \vdots & 0 \end{bmatrix}, \tag{7}$$

and

$$
c_{zp} = \begin{bmatrix} 0 & 0 & 0 \\ 0 & c & 0 \\ 0 & 0 & 0 \end{bmatrix} = \begin{bmatrix} 0 & \vdots & 0 & \vdots & 0 \\ \cdots & c_{1,1} & \cdots & c_{1,M_2} & \cdots \\ 0 & \vdots & \ddots & \vdots & 0 \\ \cdots & c_{M_1,1} & \cdots & c_{M_1,M_2} & \cdots \\ 0 & \vdots & 0 & \vdots & 0 \end{bmatrix}, \tag{8}
$$

where the entry $c_{1,1}$ in (8) is placed in row M_1 and column M_2. The sizes of the zero matrices are chosen so that after reshaping of c and k, they both have the same dimension $P_1 \times P_2, \times, \ldots, \times P_d$ (highly composite integers; typically, a power of 2). P_k $(k = 1, \ldots, d)$ are computed according to the following equation

$$
P_k = 2^{\lceil \log_2(3M_k - 1) \rceil}. \tag{9}
$$

Now, to evaluate (5), we can apply the discrete convolution theorem, that is, we must do the following operations

$$
C = \mathcal{F}(c_{zp}), \quad K = \mathcal{F}(k_{zp}), \quad S = CK, \quad s = \mathcal{F}^{-1}(S), \tag{10}
$$

where $\mathcal{F}$ stands for the Fourier transform and $\mathcal{F}^{-1}$ is its inverse. The sought convolution $(c * k)$ corresponds to a subset of s in (10) divided by the product of $P_1, P_2, \ldots, P_d$ (the so-called normalization), that is

$$
(c * k) = \frac{1}{(P_1 P_2 \ldots P_d)} s[(2M_1 - 1) : (3M_1 - 2), \ldots, (2M_d - 1) : (3M_d - 2)], \tag{11}
$$

where, for the two-dimensional case, $s[a : b, c : d]$ means a subset of rows from a to b and a subset of columns from c to d of the matrix s.

In practical implementations of Wand's algorithm the limits $\{M_1, \cdots, M_d\}$ can be additionally shrunk to some smaller values $\{L_1, \cdots, L_d\}$, which significantly reduces the computational burden. In most cases, the kernel K is the multivariate normal density distribution and an *effective support* can be defined, i.e., the region outside which the values of K are practically negligible. Now (5) can be finally rewritten as

$$
\tilde{f}_i = \sum_{i_1 = -L_1}^{L_1} \cdots \sum_{i_d = -L_d}^{L_d} c_{i-j} k_j. \tag{12}
$$

We propose to calculate L_k using the following formula $(k = 1, \ldots, d)$

$$
L_k = \min \left(M_k - 1, \left\lceil \frac{\tau \sqrt{|\lambda|}}{\delta_k} \right\rceil \right), \tag{13}
$$

where λ is the largest eigenvalue of H and δ_k is the mesh size computed from (4). After some empirical tests we found that τ can be set to around 3.7 for a standard two-dimensional normal kernel. Finally, we can calculate sizes P_k of matrices (7) and (8) according to the following formula

$$
P_k = 2^{\lceil (\log_2(M_k + 2L_k - 1) \rceil}. \tag{14}
$$

4   Binning

In this section we present some details on the binning procedure required in
the FFT-based algorithm. The idea of binning is presented graphically in Fig. 2.
Mathematical background is given in [14], so here we don't reproduce it, giving
only an illustrative example.

A sample 2D dataset (*unicef*, available in the **ks** R package) of size $n = 73$
shown as small red filled squares is replaced by a grid of *equally spaced grid
points* (black filled circles) of sizes 5×8, 10×10 and 20×20 – see Fig. 2a, b
and c, respectively. In the figure the actual values of *grid counts* (c_j, see (2)) are
proportional to the circle diameters. Grid counts are bigger for corresponding
grid points which are located in the area of bigger concentration of X_i. Grid
points with the corresponding grid counts equal zero are presented as small blue
open squares in the plots. Below we show the numerical results of binning from
Fig. 2a. Here $M_1 = 5$, $M_2 = 8$ (see (3)) and:

- evaluation points $g_{11}, \cdots, g_{15}$ in x direction: $\{19.0, 93.2, 167.5, 241.8, 316.0\}$,
- evaluation points $g_{21}, \cdots, g_{28}$ in y direction: $\{39.0, 43.9, 48.7, 53.6, 58.4, 63.3,$
  $68.1, 73.0\}$,
- matrix of grid counts c_j ($\sum c_j = 73 = n$):

	[,1]	[,2]	[,3]	[,4]	[,5]	[,6]	[,7]	[,8]
[1,]	0.00	0.00	0.0	0.0	0.35	1.9	6.470	4.87
[2,]	0.23	1.81	2.1	5.2	5.71	5.8	4.233	1.48
[3,]	2.80	5.54	6.0	6.2	0.95	0.2	0.092	0.13
[4,]	1.22	3.30	2.5	2.0	0.00	0.0	0.000	0.00
[5,]	0.83	0.74	0.2	0.0	0.00	0.0	0.000	0.00

In Fig. 2b the relationship between grid size and percent of non-zero grid
points is presented for our sample dataset (*unicef*). For simplicity we assume
that $M_1 = M_2$ but in general these values do not need to be the same. It can be
easily observed that the percentage of non-zero grid points declines very rapidly,
as the grid size increases. This phenomena is also easily observed if we compare
Fig. 2a, b and c. Such behaviour can obviously be used in practical software
implementations reducing computational burden significantly. In Fig. 2e and f
we show the final density estimates calculated for relatively small grid sizes
(5×8 and 10×10). Actually, in practical applications such small grid sizes are
not acceptable.

We think that for bivariate densities grid sizes around 100×100 should be
appropriate in most applications. For example in Fig. 1 the grid sizes of 30×30
were used and in this particular example is seems to be enough for visualization
purposes. For univariate data grid size around several hundred should guarantee
a very accurate results for the majority of situations. For more then two dimensions the problem with selecting roughly appropriate grid sizes is more challenging and in fact it is not well studied in literature. We should also remember that
in general, KDE is not a good choice for dimensions higher that five/six.

A more accurate analysis of the accuracy of binned KDE can be found in [3,7]. It is also worth to stress that it is not possible to derive a precise formula for calculating the optimal grid sizes for a particular dataset. Grid sizes depend on many data characteristics like dimensionality, data size and probably some other not easy to define factors. The two above cited papers give only a coarse estimates of grid sizes (only for univariate and bivariate case).

5 Experiments

In original Wand's paper [14] (published in 1994) some speed comparisons have been performed. The maximal speedups (the ratio between time required for direct and FFT-based computations) don't exceed the value of 5.50. Nowadays, we live in totally different computer's age and that's why we decided to perform a similar numerical experiments and see the differences.

All the computations were performed in the R environment. The computations were based on samples of sizes $n = 100$, 500, 1000 and 2000 of a trimodal normal mixture density of the following form

$$f(x) = \sum_{k=1}^{3} w_k \mathcal{N}(\mu_{k1}, \mu_{k2}, \sigma_{k1}^2, \sigma_{k2}^2, \delta_k)$$

$$= 3/7\,\mathcal{N}\left(-2, -1, (3/5)^2, (7/10)^2, 1/4\right)$$

$$+ 3/7\,\mathcal{N}\left(1, 2/\sqrt{3}, (3/5)^2, (7/10)^2, 0\right)$$

$$+ 1/7\,\mathcal{N}\left(1, -2/\sqrt{3}, (3/5)^2, (7/10)^2, 0\right). \tag{15}$$

Additionally, a real dataset was used (*unicef* dataset from the **ks** R package). The kernel K was normal and the bandwidths were calculated using the `Hpi{ks}` R function. Grids $M = M_1 = M_2 = \{10, 20, 30, 40, 50\}$ were considered (for the purpose of making 2D plots, grid sizes greater than 50 are usually not necessary in practical applications). The results are presented in Fig. 1. The following versions of R functions were implemented:

1. The *FFT-based* implementation (in Table 1 abbreviated to F). It implements Eq. (12). L_i are calculated according to (13). Grid counts c and kernels k need to be pre calculated first before the FFT convolution is performed. Computation of k is fully vectorized using typical techniques known in R.
2. The pure *sequential* non-FFT implementation, based on *for* loops (in Table 1 abbreviated to S). This version is just a literal implementation of (12). It takes advantage of the fact that a high proportion of grid counts c are zeros (see also Fig. 2d).
3. The maximally *vectorized* non-FFT implementation, with no *for* loops (in Table 1 abbreviated to V). It implements Eq. (2). Grid counts c and grid points g need to be pre calculated first. This is possibly the fastest non-FFT version, as all the computations are carried out in one compact R command.

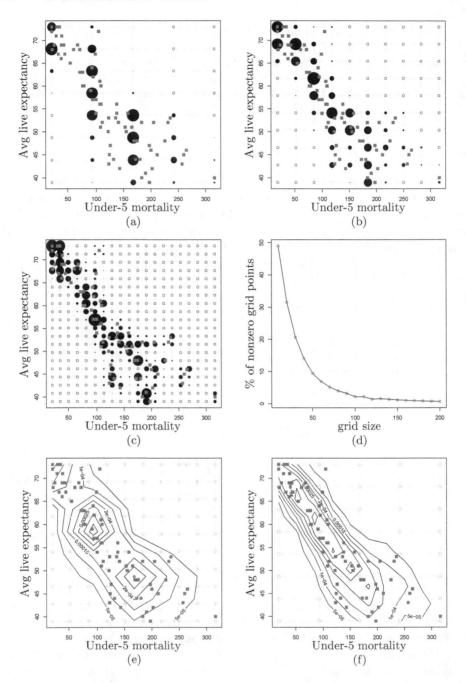

Fig. 2. An illustrative example of data binning. A sample 2D dataset *unicef* is binned and the grid sizes are: (a) – $M_1 = 5$, $M_2 = 8$, (b) – $M_1 = 10$, $M_2 = 10$, (c) – $M_1 = 20$, $M_2 = 20$, (d) – the relationship between grid size (for simplicity $M_1 = M_2$) and the percent of non-zero grid points, (e) – final density estimate for grid sizes $M_1 = 10$, $M_2 = 10$, (f) – final density estimate for grid sizes $M_1 = 5$, $M_2 = 8$. Red filled squares – data points, black filled circles – grid points where corresponding grid counts are not equal zero, blue open squares – grid points where corresponding grid counts are equal zero (Color figure online)

It is worth to note here that extensive code vectorization results in a huge growth of resources (RAM memory) needed to be allocated (if M_i is big, as 100 or more). This may be a source of rapid performance degradation, depending on actual operation system used, total RAM memory available, processor type and many other factors.

For each setting of grids ($M = M_1 = M_2$) and for each sample size n, 100 replications were performed and execution times were recorded. The means of the 100 replications constitute the final results depicted in Table 1. It is

Table 1. Speed comparisons of the FFT-based, sequential non-FFT and vectorized non-FFT versions. The abbreviations mean as follows: F – FFT-based method, S – sequential non-FFT method, V – vectorized non-FFT method. Speedups were rounded to the nearest integer value. More detailed description is given in the text

n	$M = M_1 = M_2$	F (sec)	V (sec)	S (sec)	Speedup (V/F)	Speedup (S/F)	Speedup (S/V)	L_1	L_2
100	10	0.005	0.008	0.49	2	98	61	4	4
	20	0.006	0.064	4.88	11	813	76	7	8
	30	0.011	0.383	17.92	35	1629	47	11	12
	40	0.007	0.934	37.01	133	5287	40	14	15
	50	0.007	2.081	67.66	297	9666	33	18	19
500	10	0.018	0.016	0.32	1	18	20	2	3
	20	0.018	0.071	3.315	4	184	47	5	5
	30	0.019	0.411	12.715	22	669	31	7	7
	40	0.014	0.976	36.765	70	2626	38	9	10
	50	0.016	2.089	71.09	131	4443	34	11	12
1000	10	0.036	0.035	0.19	1	5	5	2	2
	20	0.039	0.106	2.105	3	54	20	4	4
	30	0.031	0.405	8.285	13	267	20	5	6
	40	0.034	0.953	25.81	28	759	27	7	8
	50	0.031	2.169	59.8	70	1929	28	9	10
2000	10	0.059	0.058	0.22	1	4	4	2	2
	20	0.059	0.119	1.285	2	22	11	3	3
	30	0.066	0.442	7.375	7	112	17	5	5
	40	0.059	1.049	17.785	18	301	17	6	6
	50	0.061	2.175	41.195	36	675	19	8	7
73 (unicef dataset)	10	0.006	0.003	0.605	0	101	202	4	9
	20	0.004	0.06	6.615	15	1654	110	8	19
	30	0.009	0.398	21.685	44	2409	54	11	29
	40	0.008	0.919	51.665	115	6458	56	15	39
	50	0.014	2.063	96.11	147	6865	47	19	49

easy to notice that our FFT-based implementation is absolutely unbeatable, even comparing with highly vectorized codes. Vectorization gives roughly similar speedups (comparing to the FFT-based method) only for very small, usually useless in practice, grid sizes. Sequential implementation, as was foretold, is very slow and completely impractical for M_i greater than a dozen or so.

6   Conclusions

In the paper we have described a very serious problem related to using FFT for calculation of multivariate kernel estimators when unconstrained bandwidth matrices are used. Next, we have discovered a satisfactory solution which rectifies the problem. As a consequence, the results given by direct evaluation of (5) (or (12)) and by our proposed FFT-based algorithm based on (7) and (8) are identical for any form of the $\boldsymbol{H}$ bandwidth matrix.

Our results can be used not only for direct KDE calculations (as presented in the paper), but also for calculation of integrated density derivative functionals involving an arbitrary derivative order. This is extremely important in implementing almost all modern bandwidth selection algorithms, like those based on *cross-validation* and *plug-in* ideas. An appropriate authors' research paper is [5]. It is worth to note that in [5] we have also reported some numerical-like problems which remain a challenging open problem.

References

1. Andrzejewski, W., Gramacki, A., Gramacki, J.: Graphics processing units in acceleration of bandwidth selection for kernel density estimation. Int. J. Appl. Math. Comput. Sci. **23**(4), 869–885 (2013)
2. Duong, T.: Kernel smoothing (2016). http://CRAN.R-project.org/package=ks, **R** package version 1.10.4
3. González-Manteigaa, W., Sánchez-Selleroa, C., Wand, M.: Accuracy of binned kernel functional approximations. Comput. Stat. Data Anal. **22**, 1–16 (1996)
4. Gramacki, A., Gramacki, J.: FFT-based fast computation of multivariate kernel estimators with unconstrained bandwidth matrices. J. Comput. Graph. Stat. **26**(2), 459–462 (2016). doi:10.1080/10618600.2016.1182918
5. Gramacki, A., Gramacki, J.: FFT-based fast bandwidth selector for multivariate kernel density estimation. Comput. Stat. Data Anal. **106**, 27–45 (2017). doi:10.1016/j.csda.2016.09.001
6. Gramacki, A., Sawerwain, M., Gramacki, J.: FPGA-based bandwidth selection for kernel density estimation using high level synthesis approach. Bull. Pol. Acad. Sci.: Tech. Sci. **64**(4), 821–829 (2016). doi:10.1515/bpasts-2016-0091
7. Hall, P., Wand, M.P.: On the accuracy of binned kernel density estimators. J. Multivar. Anal. **56**(2), 165–184 (1996)
8. Härdle, W.: Smoothing Techniques: With Implementation in S. Springer Series in Statistics. Springer, New York (1991)
9. Raykar, V., Duraiswami, R., Zhao, L.: Fast computation of kernel estimators. J. Comput. Graph. Stat. **19**(1), 205–220 (2010)

10. Scott, D.W.: Multivariate Density Estimation: Theory, Practice, and Visualization. Wiley Inc., New York (1992)
11. Silverman, B.W.: Algorithm AS 176: kernel density estimation using the fast fourier transform. Appl. Stat. **31**, 93–99 (1982)
12. Silverman, B.W.: Density Estimation for Statistics and Data Analysis. Chapman & Hall/CRC, London (1986)
13. Wand, M.P., Jones, M.C.: Kernel Smoothing. Chapman & Hall, London (1995)
14. Wand, M.: Fast computation of multivariate kernel estimators. J. Comput. Graph. Stat. **3**(4), 433–445 (1994)
15. Wand, M., Ripley, B.: Functions for kernel smoothing supporting Wand & Jones (1995) (2016). http://CRAN.R-project.org/package=KernSmooth, **R** package version 2.23-15

A Framework for Business Failure Prediction

Irem Islek[1(✉)], Idris Murat Atakli[1], and Sule Gunduz Oguducu[2]

[1] Idea Teknoloji Cozumleri, Istanbul, Turkey
{irem.islek,idris.atakli}@ideateknoloji.com.tr
[2] Department of Computer Engineering, Istanbul Technical University,
Istanbul, Turkey
sgunduz@itu.edu.tr

Abstract. Business failure prediction systems help predict financial failures before they actually happen and provide an early warning for enterprises. Using machine learning techniques, instead of traditional statistical models, has brought a considerable increase in performance into the area of business failure prediction. This paper presents a framework for predicting business failures by using different machine learning techniques. We, also, implemented a novel model for business failure prediction based on NARX (nonlinear autoregressive network with exogenous inputs) feedback neural network to be included into this framework which is a recurrent dynamic network with feedback connections. Detailed experiments are conducted to compare the performance of these approaches. Especially, for the long-term business failure predictions, there are no other papers investigating the performance of NARX. To the best of our knowledge, this is the first time NARX algorithm is applied for long-term business failure prediction.

Keywords: Business failure prediction · Financial distress prediction · Machine learning · NARX

1 Introduction

From 1960's to present, researchers have paid a great deal of attention to finding a successful way for predicting business failures. It can be described as developing a methodology to predict financial distress using several existing financial features of an enterprise. Business failure prediction, which is also known as financial distress prediction or firm failure prediction has a considerable importance to shareholders, investors, credit managers, etc. Business failure prediction models as such alert a stakeholder or a manager to take timely precautions to prevent failures before they occur. For investors, this model provides vital information which helps them deciding whether to invest in a firm or not. In other words, this model reduces the risk of false investment decisions and prevents financial loss. Also, this model can be used by credit managers to evaluate the level of risk and credit limit for an enterprise.

There exist many studies in the literature about business failure prediction. The first study about this topic was done by Beaver in 1966 [1]. Beaver used

© Springer International Publishing AG 2017
L. Rutkowski et al. (Eds.): ICAISC 2017, Part II, LNAI 10246, pp. 74–83, 2017.
DOI: 10.1007/978-3-319-59060-8_8

univariate analysis to forecast bankruptcy. After that, Altman proposed multivariate discriminant analysis to solve this problem [2]. Most of the subsequent studies were based on Altman's study. After 1980, different types of regression models, such as logit and probit, were proposed to develop a model which can predict business failures accurately. Afterwards, machine learning algorithms were introduced as alternatives to the statistical models. Most of the recent studies compare traditional statistical models with machine learning models or combine several models in one methodology [3–8]. In general, obtained results show that machine learning algorithms overcome statistical models in predicting business failure.

In this study, we proposed a framework for successfully predicting business failures. This framework contains nine different prediction models, namely, Logistic Regression, Multilayer Perceptron (MLP), Sequential Minimal Optimization (SMO), Bayesian Network, Naive Bayes, J48, Random Forest, Random Tree and NARX (nonlinear autoregressive network with exogenous inputs) feedback neural network. To the best of our knowledge, NARX has never been used for business failure prediction before this study. In addition to that, this framework gives chance of making multistep ahead prediction with NARX model. For the evaluation purposes, nine different models were applied to same datasets on the same framework and obtained results are given in detail.

The paper organized as follows: In Sect. 2, we reviewed the related work. Details of constructed datasets are given in Sect. 3. In addition to that, proposed methodology is explained in Sect. 3. The performances of applied methodologies are evaluated in Sect. 4. Comparisons of these performances are also given in this section. The paper is concluded by summarizing achievements and giving future directions in Sect. 5.

2   Related Work

Financial distress prediction has remained highly popular since 1960's. After Altman's multivariate discriminant analysis, Ohlson proposed logit analysis for bankruptcy prediction for the first time [9].

After that, machine learning algorithms came into use as an alternative to statistical models. For instance, neural networks were used in numerous studies in order to predict business failure [3–5, 10, 11]. In these studies, neural networks were compared with traditional statistical models such as multivariate discriminant analysis. Most of these studies claim that neural networks gave better performance than discriminant analysis. In several studies, SVM has been also used for predicting business failures. It has been found that SVM outperformed the classical methods [12, 13]. Another popular machine learning approach which is used for firm failure prediction is tree algorithms such as ID3 and decision trees [14, 15]. In these studies, tree algorithms were compared with discriminant analysis and provided better results than statistical models. According to the literature review, we can say that machine learning models generally outperform traditional statistical models such as multivariate discriminant analysis.

Combining a model with other models to strengthen the weak points of the model is a common approach in machine learning studies. In this direction, researchers compared neural networks to decision trees, SVM, majority voting and concluded that neural networks was the best method for forecasting financial distress in comparison to other methods [7]. Azayite and Achchab composed a hybrid model based on discriminant analysis, back propagation neural network and self-organizing maps [8]. They applied the hybrid model to Moroccan firms and claimed that the hybrid model outperformed discriminant analysis. Wu et al. proposed a genetic based SVM to predict bankruptcy [16]. This methodology tested on Taiwan dataset to compare with discriminant analysis, logit, probit, neural networks and traditional SVM. Proposed hybrid methodology gave the best predictive accuracy according to the experimental results. Another hybrid study brought together SVM and logistic regression [17]. The methodology modified the outputs of the SVM classifiers according to the result of logistic regression analysis. In [18], single classifiers were trained by SVM algorithms with different kernel functions on different feature subsets of one initial dataset. This ensemble SVM provided better performance than individual SVM classifier. Lin et al. proposed another hybrid method which combines locally linear embedding (LLE) and SVM to predict firm failures [19].

Even though big data approach is extremely popular, it is not used for predicting business failures. In literature, there is only one study which uses big data approach for business failure prediction [20]. The reason for that may be that it is quite difficult to obtain huge amounts of data for business failure prediction.

In this study, we propose a framework for business failure prediction by making following contributions:

- Our framework contains NARX network algorithm which has never been used for business failure prediction before.
- Thanks to NARX network, multistep ahead prediction can be done in addition to one-step ahead prediction.
- Proposed framework can be used for not only business failure prediction, but also other suitable prediction problems in some areas such as finance, biomedical etc., due to its flexible structure.

3   The Dataset and the Proposed Framework

3.1   Details of Dataset

Financial statements of enterprises, which are registered to IMKB BIST [21], are published on Public Disclosure Platform, periodically. In addition to that, deteriorated firms are published on Public Disclosure Platform, as well. Datasets for our study are derived from these resources. 10 different financial ratios are defined as input variables from these datasets. These variables are selected according to Aktan's study which detects 10 best financial ratios for bankruptcy prediction within 53 financial ratios [22]. Selected financial ratios can be seen in Table 1.

Table 1. Selected financial ratios

Ratio name	Calculation formula	Ratio name	Calculation formula
u1	Cash/Total assets	u6	Total debts/Total assets
u2	Quick assets/Total assets	u7	Short term debts/Total assets
u3	Financial debts/Total assets	u8	Return on assets
u4	Inventory/Net sales	u9	Operating income/Total assets
u5	Current assets/Total assets	u10	Cash flow/Total assets

Class values, which correspond to financial status of firms are defined as *good*, *bad* and *very bad* in constructed datasets.

In the first dataset, input variables and class values are calculated for quarterly periods. Apart from that, a second dataset is constructed using yearly values of selected variables.

3.2   Proposed Framework

The proposed framework contains three main steps, Data Preparation, Prediction and Evaluation as seen in Fig. 1.

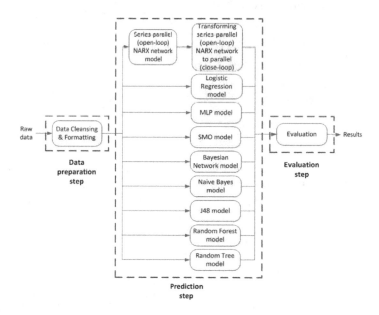

Fig. 1. Proposed framework.

Data Preparation Step. In data preparation step, data rows, which include null values for some financial ratios, are removed from the dataset. After cleaning, 10 financial ratios are calculated using several financial variables. Lastly, a matrix data structure is composed from calculated financial ratios and class values.

Prediction Step. This step is responsible for producing business failure prediction results. For this purpose, we constructed Logistic Regression [23], Multilayer Perceptron [24], Sequential Minimal Optimization [25], Bayesian Network [26], Naive Bayes [27], J48 [28], Random Forest [29], Random Tree [30] and NARX models in this step. A prediction model should be selected within these nine models to continue this step of the framework. Afterwards, the selected model is trained using given data and the prediction results are produced according to the trained model.

Due to page limitations, we, very briefly, explain NARX model, which has not been employed for business failure prediction purposes before.

NARX, which is a dynamic network, is useful for time series modeling. As can be seen in Eq. 1, the previous output value of the network and previous values of input parameters are used for producing next step value of the output.

$$y(t) = f(y(t-1), y(t-2), ..., y(t-n_y), u(t-1), u(t-2), ..., u(t-n_u)) \quad (1)$$

In this equation, u represents the training inputs while y represents the target variables to be predicted. t means the discrete time step in this equation. For predicting next values of $y(t)$, previous values of the exogenous input and previous values of the output regress together using f function. A general NARX network architecture can be seen in Fig. 2.

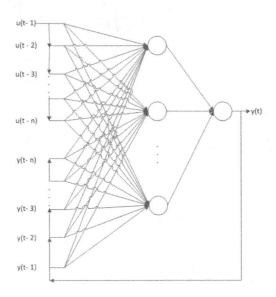

Fig. 2. NARX network architecture.

There are two types of NARX network: series-parallel architecture and parallel architecture. Series-parallel architecture which is also called open-loop, uses existing output as one of network inputs. Parallel network (close-loop) uses the output produced by previous iteration as one of network inputs.

Firstly, the series-parallel architecture is constructed in order to train network. In this network, inputs of the network are selected financial ratios ($u1(t)$, $u2(t)$, .., $u10(t)$) and existing outputs ($y(t)$). Series-parallel NARX completes training phase in a shorter time than parallel NARX because series-parallel one uses existing output values.

Afterwards, the architecture of the realized network is transformed to parallel architecture in prediction step. Reason of using parallel NARX network is that parallel architecture provides opportunity to make multi-step ahead prediction.

Evaluation Step. Accuracy, Type I error and Type II error are measured for the performance review of applied algorithms. Brief descriptions and formulas of them are given below:

Accuracy, calculates the ratio of total number of correct predictions to total number of predictions.

Type I error (false positive), means predicting a firm's financial status as good when it is actually bad or very bad. Also, predicting a firm's financial status as bad when it is actually very bad is Type I error, as well.

Type II error (false negative), means predicting a firm's financial status as bad when it is actually good. In addition to that, predicting a firm's status as very bad when it is actually bad or good is also Type II error.

If we compare Type I and Type II, we can easily say that Type I error is more significant than Type II error for our problem. If a firm's financial status is bad or very bad but our methodology says that it is good, firm's managers will not take necessary precautions and possibly, end up with bankruptcy.

4 Performance Evaluation Results

As we mentioned before, two separate datasets are constructed from raw data. First one contains data in quarters and second one contains annual data. In both datasets, 2015 data is used for testing. Test data sample counts of quarter-period dataset and annual dataset are 222 and 66, respectively. Class values for datasets are defined as good, bad and very bad. The optimal parameters are defined using validation set which includes 2014 data.

Constructed NARX network contains 10 neurons in the hidden layer and Levenberg - Marquardt [31] algorithm is used as the training step for the network. The applied NARX network contains one hidden layer. In our NARX model, 10 financial ratio values which are given in Table 1, are used as input. There is one value as output of the network which corresponds financial status of firm. Transfer function of the NARX model is sigmoid function. In Eq. 1, n_y and n_u are the lags of the input and output of our NARX model. $n = 1$ means one-step ahead, while any larger value of n means multi-step ahead prediction (If $n = 2$, model predicts 2 step ahead value).

Besides NARX, other prediction algorithms are applied using Weka. NARX algorithm is implemented using MATLAB. Evaluated results of applied methods for the first dataset (quarter-period dataset) are given in Table 2.

Table 2. Comparison results for quarter-period dataset

Method name	Accuracy (%)	Type I error (%)	Type II error (%)
Logistic Regression	93.42	2.9	3
Multilayer Perceptron	89.18	4	6
Sequential Minimal Opt.	89.18	3.5	6.5
Bayesian Network	93.69	2.5	2.5
Naive Bayes	70.27	27	2
J48 Tree	95.04	3	1.8
Random Forest	96.39	1	1
Random Tree	91.44	4	3
One step ahead NARX	95.81	2.3	1.8

As can be seen in Table 2, Random Forest gives the best accuracy for quarter-period dataset. In addition to that, lowest Type I and Type II error rates are obtained with Random Forest for quarter-period dataset. One step ahead NARX provides second best results for accuracy, Type I and Type II error rates.

As shown in Table 3, one step ahead NARX gives the best accuracy for the annual dataset. For Type I error, Random Forest outperforms one step ahead NARX. For Type II error, NARX gives lowest error rate.

Table 3. Comparison results for annual dataset

Method name	Accuracy (%)	Type I error (%)	Type II error (%)
Logistic Regression	94.45	1.5	3
Multilayer Perceptron	93.93	1.5	4.5
Sequential Minimal Opt.	92.42	2	5
Bayesian Network	96.48	1.5	2
Naive Bayes	81	15	4
J48 Tree	96.96	1	2
Random Forest	96.96	0	3
Random Tree	95.45	1.5	3
One step ahead NARX	97.20	1.4	1.2

The reason Random Forest gives satisfying results is that it is actually an ensemble learning methodology. It contains multitude of decision trees and prediction results are chosen according to the voting mechanism. Ensemble learning approach is based on obtaining highly accurate classifiers by combining less accurate ones.

In addition to Random Forest, one step ahead NARX, also, gives better results than other prediction models of the framework for our datasets as NARX is commonly used for modelling time series based prediction and our datasets also have a temporal ordering for several different financial ratios.

Since, this framework also provides a multi-step ahead business failure prediction, we did some extra experiments for multi-step ahead prediction using parallel NARX network. Detailed results of these experiments are given in Tables 4 and 5.

In Table 4, 5 step ahead prediction gives result for one year later in quarter-period dataset. As you can see from Table 4, accuracy value of 3 step ahead NARX is lower than expected. We guess that this decrease causes from imbalanced dataset of 3 step ahead test.

Table 4. Comparison results for one step and multistep ahead NARX for quarter-period dataset

Method name	Accuracy (%)	Type I error (%)	Type II error (%)
1 step ahead NARX	95.8	2.3	1.8
2 step ahead NARX	95.3	2.7	1.8
3 step ahead NARX	69.7	5.5	24
4 step ahead NARX	80	2.7	16.6
5 step ahead NARX	71	5.9	23

Table 5. Comparison results for one step and multistep ahead NARX for annual dataset

Method name	Accuracy (%)	Type I error (%)	Type II error (%)
1 step ahead NARX	97.2	1.4	1.2
2 step ahead NARX	94.4	3	1.8
3 step ahead NARX	66	5	28.3
4 step ahead NARX	62	8	29
5 step ahead NARX	58	11	30

In Table 5, each step indicates one year, thus 5 step ahead prediction gives results for five years later. Not surprisingly, prediction accuracy, Type I and Type II error rates drop year after year. It is obvious that the long-term business failure prediction is challenging since political and societal changes also play a role in business failure. However, it is difficult to predict political and societal changes.

5   Conclusions

In this study, we presented a framework for business failure prediction. To achieve that, Logistic Regression, Multilayer Perceptron, Sequential Minimal Optimization, Bayesian Network, Naive Bayes, J48 Tree, Random Forest, Random Tree and NARX models are constructed in this framework. We also want to emphasize that this is the first study, which uses NARX for business failure prediction. All prediction models of framework are tested separately using two different datasets which contain firms from Turkey. The first dataset uses quarterly period data but second one uses annual data for financial ratios and class values estimations.

In conclusion, we can confidently say that proposed framework is very useful for business failure prediction. Using this framework, suitable business failure prediction model for a dataset can be chosen easily. Moreover, NARX model gives a chance of predicting multi-step ahead business failure.

Acknowledgments. This research was partially supported by The Scientific and Technological Research Council of Turkey (TUBITAK) under TEYDEB grant 3150156.

References

1. Beaver, W.H.: Financial ratios as predictors of failure, empricial research in accounting: selected studies. J. Acc. Res. **5**, 179–199 (1966)
2. Altman, E.I.: Financial ratios, discriminant analysis and the prediction of corporate bankruptcy. J. Financ. **23**, 589–609 (1968)
3. Tam, K.Y., Kiang, M.Y.: Managerial applications of neural networks: the case of bank failure predictions. Manage. Sci. **38**, 926–947 (1992)
4. Coats, P.K., Fant, L.F.: Recognizing financial distress patterns using neural network tool. Financ. Manage. **22**, 142–155 (1993)
5. Wilson, R.L., Sharda, R.: Bankruptcy prediction using neural networks. Decis. Sci. **11**, 545–557 (1994)
6. Ahn, B.S., Cho, S.S., Kim, C.Y.: The integrated methodology of rough set theory and artificial neural network for business failure prediction. Expert Syst. Appl. **18**(2), 65–74 (2000)
7. Geng, R., Bose, I., Chen, X.: Prediction of financial distress: an empirical study of listed Chinese companies using data mining. Eur. J. Oper. Res. **241**(1), 236–247 (2015)
8. Azayite, F.Z., Achchab, S.: Hybrid discriminant neural networks for bankruptcy prediction and risk scoring. Procedia Comput. Sci. **83**, 670–674 (2016)
9. Ohlson, J.: Financial ratios and the probabilistic prediction of bankruptcy. J. Account. Res. **18**, 109–131 (1980)
10. Odom, M.D., Sharda, R.A.: A neural networks model for bankruptcy prediction. In: The 2nd IEEE International Joint Conference on Neural Network, pp. 163–168 (1990)
11. Altman, E.I., Marco, G., Varetto, F.: Corporate distress diagnosis: comparisons using linear discriminate analysis and neural networks. J. Bank. Financ. **18**, 505–529 (1994)
12. Gestel, T.V., Baesens, B., Suykens, J., Espinoza, M., Baestaens, D.E., Vanthienen, J., De Moor, B.: Bankruptcy prediction with least squares support vector machine classifiers. In: Computational Intelligence for Financial Engineering, pp. 1–8. IEEE Press (2003)
13. Shin, K.S., Lee, T.S., Kim, H.: An application of support vector machines in bankruptcy prediction model. Expert Syst. Appl. **28**(1), 127–135 (2005)
14. Messier, W.F., Hansen, J.V.: Inducing rules for expert system development: an example using default and bankruptcy data. Manage. Sci. **34**(12), 1403–1415 (1988)

15. Gepp, A., Kumar, K., Bhattacharya, S.: Business failure prediction using decision trees. J. Forecast. **29**(6), 536–555 (2010)
16. Wu, C.H., Tzeng, G.H., Goo, Y.J., Fang, W.C.: A real-valued genetic algorithm to optimize the parameters of support vector machine for predicting bankruptcy. Expert Syst. Appl. **32**(2), 397–408 (2007)
17. Hua, Z., Wang, Y., Xu, X., Zhang, B., Liang, L.: Predicting corporate financial distress based on integration of support vector machine and logistic regression. Expert Syst. Appl. **33**(2), 434–440 (2007)
18. Sun, J., Li, H.: Financial distress prediction using support vector machines: ensemble vs. individual. Appl. Soft Comput. **12**(8), 2254–2265 (2012)
19. Lin, F., Yeh, C.C., Lee, M.Y.: A hybrid business failure prediction model using locally linear embedding and support vector machines. Romanian J. Econ. Forecast. **16**(1), 82–97 (2013)
20. Hafiz, A., Lukumon, O., Muhammad, B., Olugbenga, A., Hakeem, O., Saheed, A.: Bankruptcy prediction of construction businesses: towards a big data analytics approach. In: Big Data Computing Service and Applications (BigDataService), pp. 347–352. IEEE Press (2015)
21. IMKB BIST. http://www.borsaistanbul.com
22. Aktan, S.: Application of machine learning algorithms for business failure prediction. Investment Manage. Financ. Innov. **8**(2), 52–65 (2011)
23. Hosmer Jr., D.W., Lemeshow, S.: Applied Logistic Regression. Wiley, New York (2004)
24. Alpaydin, E.: Introduction to Machine Learning. The MIT Press, Cambridge (2004)
25. Platt, J.: Sequential Minimal Optimization: A Fast Algorithm for Training Support Vector Machines (1998)
26. Friedman, N., Geiger, D., Goldszmidt, M.: Bayesian Network Classifers. Mach. Learn. **29**(2-3), 131–163 (1997)
27. Mitchell, T.M.: Machine Learning. McGraw-Hill, Maidenhead (1997)
28. Quinlan, J.R.: C4.5: Programs for Machine Learning. Elsevier, New York (2014)
29. Breiman, L.: Random forests. Mach. Learn. **45**(1), 5–32 (2001)
30. Hall, M., Frank, E., Holmes, G., Pfahringer, B., Reutemann, P., Witten, I.H.: The WEKA data mining software: an update. SIGKDD Explor. **11**(1), 10–18 (2009)
31. Weisstein, E.W.: Levenberg-Marquardt Method. From MathWorld-A Wolfram Web Resource. http://mathworld.wolfram.com/Levenberg-MarquardtMethod.html

Fuzzy Clustering with ε-Hyperballs and Its Application to Data Classification

Michal Jezewski$^{(\boxtimes)}$, Robert Czabanski, and Jacek Leski

Faculty of Automatic Control, Electronics and Computer Science,
Institute of Electronics, Silesian University of Technology,
16 Akademicka Str., 44-100 Gliwice, Poland
{michal.jezewski,robert.czabanski,jacek.leski}@polsl.pl

Abstract. In the presented paper the Fuzzy Clustering with ε-Hyperballs being the prototypes is proposed. It is based on the idea of regions of insensitivity – described by the hyperballs of radius ε, in which the distances of objects from the centers of the hyperballs are considered as equal to zero. The proposed clustering was applied to determine the parameters of fuzzy sets in antecedents of the classifier based on fuzzy if-then rules. The classification quality obtained for six benchmark datasets was compared with the reference classifiers. The results show the improvement of the classification accuracy using the proposed method.

Keywords: Fuzzy clustering · Fuzzy if-then rules · Data classification

1 Introduction

The main task of clustering methods [1,2,4,9,17] is partition of objects into groups (clusters) containing the objects that are somehow similar to each other. The similarity is determined on the basis of the object characteristics, which are represented by the feature vector. Clustering based on minimizing (or maximizing) the criterion function (usually defined as a scalar index) forms one of the basic classes of the partitioning methods. As a subclass, we can distinguish the fuzzy clustering, which uses the idea of partial membership of an object to a group, that originates from the fuzzy sets theory [18].

In the presented paper the Fuzzy Clustering with ε-Hyperballs (FCεH) being the prototypes is proposed. The FCεH is based on the idea of regions of insensitivity described by the hyperballs of radius ε, in which the distances of objects from the hyperballs centers are considered as equal to zero. The idea of the ε-insensitive distance between prototypes and objects has been previously presented in [10]. However, in contrast to [10] where the ℓ_1 distance was assumed, the FCεH uses quadratic loss function leading to simplification of the clustering criterion minimization procedure. The typical goal of clustering is to attain high intra-cluster similarity and low inter-cluster similarity. To evaluate the clustering results validity indices are used [2]. However, good scores concerning validity indices do not necessarily translate into good performance in a target application. Hence, an alternative approach is indirect evaluation of the clustering

L. Rutkowski et al. (Eds.): ICAISC 2017, Part II, LNAI 10246, pp. 84–93, 2017.
DOI: 10.1007/978-3-319-59060-8_9

in the application. In the presented approach the FCεH was applied to determine antecedents of the fuzzy rule-based classifier and hence its efficiency was evaluated based on classification results.

Fuzzy rule-based classifiers can be considered as a mixture of experts, each represented by a single fuzzy if-then rule. The fuzzy rules operate in regions that are described by antecedents fuzzy sets. To determine these sets the fuzzy clustering can be used [7,8,11]. In our study we applied the FCεH to find the parameters of fuzzy sets in antecedents of the fuzzy rule-based classifier presented in [11]. The main reason behind the idea of application of the FCεH is the improvement of the generalization ability of the classifier based on fuzzy rules. We assume that estimation of the antecedents on the basis of clustering with the areas of insensitivity will improve the 'smoothness' of the resulting decision curve, and thus increase the classification accuracy of test data. Similarly to [11], to determine the consequents parameters we used the Iteratively Reweighted Least Squares (IRLS) procedure. The IRLS-based classifier may be interpreted as an extension of the Ho-Kashyap [6] algorithm, which is considered as one of the most efficient methods of classifier design. The IRLS enables various approximations of the misclassification error and achieves high efficiency of data classification [13]. The resulting fuzzy rules provide compact and easy to understand decision explanation, however there is an accuracy-interpretability trade-off to be considered [5]. In the proposed approach we focused mainly on the classification efficiency in terms of low misclassification error. As reference, we applied IRLS combined with Fuzzy c-Means (FCM) [2] and Fuzzy $(c+p)$-Means (FCPM) [11] clustering. We used also the well-known Support Vector Machine (SVM) [3]. As the quadratic programming in classic SVM solution is characterized by large computational complexity we applied the Lagrangian SVM [14].

2 Fuzzy Clustering with ε-Hyperballs

In a fuzzy clustering the partition of N objects into c groups is defined by the partition matrix $\mathbf{U} \in \mathbb{R}^{c \times N}$ that satisfies the following conditions [2]:

$$\mathop{\forall}_{\substack{1 \leq i \leq c \\ 1 \leq k \leq N}} u_{ik} \in [0,1], \quad \mathop{\forall}_{1 \leq k \leq N} \sum_{i=1}^{c} u_{ik} = 1, \quad \mathop{\forall}_{1 \leq i \leq c} \sum_{k=1}^{N} u_{ik} \in (0, N), \qquad (1)$$

where u_{ik} is the membership degree of the k-th object $\mathbf{x}_k \in \mathbb{R}^p$ in the i-th group. Zero value of u_{ik} indicates that the object $\mathbf{x}_k$ is not the member of the i-th group, while $u_{ik} = 1$ represents the full membership in the group. Each i-th group is represented by the prototype $\mathbf{v}_i$. Hence, the clustering can be described as the process of finding the partition matrix $\mathbf{U}$ (fulfilling the criteria (1)) and the prototypes matrix $\mathbf{V} = [\mathbf{v}_1, \mathbf{v}_2, \cdots, \mathbf{v}_c] \in \mathbb{R}^{p \times c}$, which minimize the assumed criterion. Thus, the optimal solution $(\mathbf{U}_{\mathrm{opt}}, \mathbf{V}_{\mathrm{opt}})$ can be obtained as

$$\min J(\mathbf{U}, \mathbf{V}) = J(\mathbf{U}_{\mathrm{opt}}, \mathbf{V}_{\mathrm{opt}}), \qquad (2)$$

where J is a criterion function.

One of the simplest and most commonly used criteria is a weighted sum of the squares of distances d_{ik} of the object $\mathbf{x}_k$ from the prototype $\mathbf{v}_i$, with weights determined by the m-th power of the membership degree u_{ik}

$$J\left(\mathbf{U}, \mathbf{V}\right) = \sum_{i=1}^{c} \sum_{k=1}^{N} J_{ik} = \sum_{i=1}^{c} \sum_{k=1}^{N} \left(u_{ik}\right)^m d_{ik}^2, \tag{3}$$

where $m \in (1, +\infty)$ is the parameter determining the groups fuzziness and d_{ik} is the Euclidean norm

$$d_{ik} = \|\mathbf{x}_k - \mathbf{v}_i\| = \sqrt{\left(\mathbf{x}_k - \mathbf{v}_i\right)^\top \left(\mathbf{x}_k - \mathbf{v}_i\right)}. \tag{4}$$

Among the methods of fuzzy clustering based on the above criterion there is the well-known Fuzzy c-Means [2].

In the classical approach to fuzzy clustering, we get the zero addend J_{ik} of the criterion (3) if:

1. the object $\mathbf{x}_k$ is not a member of a group represented by the prototype $\mathbf{v}_i$, i.e. $u_{ik} = 0$, or
2. the distance d_{ik} of the object $\mathbf{x}_k$ from the group prototype $\mathbf{v}_i$ is equal to zero (the object covers the prototype), i.e. $d_{ik} = 0 \iff \mathbf{x}_k = \mathbf{v}_i$.

Thus, the criterion calculated for the i-th group

$$J_i = \sum_{k=1}^{N} J_{ik}, \tag{5}$$

is zero only if it consists entirely of the objects that are equal to the group prototype.

In practice, objects located near the group prototype are considered as strictly belonging to the group, and consequently, as objects for which the addend J_{ik} should equal to zero. This leads to the idea of application of the limit distance ε, for which $J_{ik} = 0$, i.e. a zero J_{ik} is defined for objects $\mathbf{x}_k$ located inside a hyperball of radius ε and center $\mathbf{v}_i$. Therefore the new clustering criterion function can be defined as

$$J^{(\varepsilon)}\left(\mathbf{U}, \mathbf{V}\right) = \sum_{i=1}^{c} \sum_{k=1}^{N} \left(u_{ik}\right)^m g_{ik}^2, \tag{6}$$

where

$$\underset{\substack{1 \leq i \leq c \\ 1 \leq k \leq N}}{\forall} g_{ik} = \begin{cases} d_{ik} - \varepsilon, & d_{ik} > \varepsilon, \\ 0, & d_{ik} \leq \varepsilon. \end{cases} \tag{7}$$

On the basis of (6) the new method of Fuzzy Clustering with ε-Hyperballs (FCεH) is introduced. The optimal solution $(\mathbf{U}_{\mathrm{opt}}, \mathbf{V}_{\mathrm{opt}})$ can be determined using the method of Lagrange multipliers subject to equality constraints

$$\underset{1 \leq k \leq N}{\forall} \sum_{i=1}^{c} u_{ik} = 1. \tag{8}$$

The derivation of the $\mathbf{U}_{opt}$ is very similar to the FCM method and can be obtained by replacing the distances d_{ik} (FCM) with distances g_{ik} (FCεH). Hence, we skip it and refer to [2] instead. The optimal position of the ε-hyperball center is calculated from the gradient of (6) with respect to $\mathbf{v}_s$. Denoting $\Omega_k = \{k \,|\, d_{sk} > \varepsilon\}$, after transformations we obtain

$$\mathop{\forall}_{1 \leq s \leq c} \ \mathbf{v}_s = \sum_{\Omega_k} (u_{sk})^m \left(\frac{\varepsilon}{d_{sk}} - 2 \right) \mathbf{x}_k \bigg/ \sum_{\Omega_k} (u_{sk})^m \left(\frac{\varepsilon}{d_{sk}} - 2 \right). \qquad (9)$$

By introducing additional notation

$$\mathop{\bigvee}_{1 \leq k \leq N} \begin{cases} I_k = \{i \,|\, 1 \leq i \leq c; \ g_{ik} = 0\}, \\ \tilde{I}_k = \{1, 2, \cdots, c\} \setminus I_k, \end{cases} \qquad (10)$$

we get $\mathbf{U}$, which may be the global minimum of (6) if:

$$\mathop{\forall}_{1 \leq i \leq c} \mathop{\forall}_{1 \leq k \leq N} \ u_{ik} = \begin{cases} (g_{ik})^{\frac{2}{1-m}} \bigg/ \sum\limits_{j=1}^{c} (g_{jk})^{\frac{2}{1-m}}, & I_k = \emptyset, \\[2mm] \begin{cases} \mathop{\forall}_{i \in \tilde{I}_k} 0, \\ \sum\limits_{i \in I_k} u_{ik} = 1, \end{cases} & I_k \neq \emptyset. \end{cases} \qquad (11)$$

If $I_k \neq \emptyset$, then selection of u_{ik} according to (11) results in minimal value of the criterion (6) as the elements of the partition matrix for the non-zero distances are zeros, and non-zero for zero distances. Optimal $\mathbf{U}_{opt}$ and $\mathbf{V}_{opt}$ are a fixed point of (11) and (9), which is determined using the Picard algorithm. From (9) one can notice, that the current position of the center of a given ε-hyperball (for a fixed $\mathbf{U}$) depends on the distances that were calculated in relation to its previous position. Hence, each center is a fixed point of (9) and (4) determined using the Picard algorithm. Finally, the algorithm of the FCcH can be written as:

1. Fix $\varepsilon \geq 0$, c ($1 < c < N$) and $m \in (1, +\infty)$. Set the iteration index $t = 0$. Initialize $\mathbf{V}^{(t)}$.
2. Calculate the partition matrix $\mathbf{U}^{(t)}$ using (11).
3. Set the iteration index of the Picard algorithm of the ε-hyperball center estimation $j = 0$. Using $\mathbf{U}^{(t)}$ calculate $\mathbf{V}^{(t+1)} = \left[\mathbf{v}_1^{(t+1)}, \mathbf{v}_2^{(t+1)}, \cdots, \mathbf{v}_c^{(t+1)} \right]$ as a fixed point of (9) and (4). Stop the iterations for the i-th group if $\left\| \mathbf{v}_i^{(j+1)} - \mathbf{v}_i^{(j)} \right\|_F < \delta$, where $\| \bullet \|_F$ is the Frobenius norm, and δ is a preset parameter.
4. If $\left\| \mathbf{V}^{(t+1)} - \mathbf{V}^{(t)} \right\|_F \geq \Delta$, then $t \leftarrow t + 1$ and go to 2, where Δ is a preset parameter.

3   FCεH Based Classification

The classification based on FCεH was performed using the fuzzy rule-based classifier [11] with the L following fuzzy if-then rules

$$\bigvee_{1 \leq \ell \leq L} \mathcal{R}^{(\ell)} \triangleq \mathbf{if} \left(\operatorname*{and}_{n=1}^{p} \ x_{kn} \text{ is } A_{\ell n} \right), \mathbf{then} \ y_{\ell} = \delta_{y,y_{\ell}}, \tag{12}$$

where: x_{kn} stands for the n-th component (feature) of the k-th object, $A_{\ell n}$ denotes a fuzzy set for the n-th component in antecedent of the ℓ-th rule, and $\delta_{y,y_{\ell}}$ is location of the ℓ-th consequent singleton in the output space. The fuzzy sets in antecedents are defined by Gaussian membership functions, which ensure the classifier decision for any input values combination.

The final output of the classifier is calculated as

$$y_{0k} = \frac{\sum_{\ell=1}^{L} \mu_{\mathbf{A}_{\ell}} (\mathbf{x}_k) \, y_{\ell}}{\sum_{\ell=1}^{L} \mu_{\mathbf{A}_{\ell}} (\mathbf{x}_k)}, \tag{13}$$

where

$$\bigvee_{1 \leq \ell \leq L} \mu_{\mathbf{A}_{\ell}} (\mathbf{x}_k) = \exp \left[-\frac{1}{2} \sum_{n=1}^{p} \left(\frac{x_{kn} - v_{\ell n}}{\gamma s_{\ell n}} \right)^2 \right] \tag{14}$$

denotes the membership function of the ℓ-th antecedent, where $v_{\ell n}$ and $s_{\ell n}$ are center and dispersion (scaled by the parameter γ) determined using the proposed clustering.

Each of two classes (ω_1 and ω_2) of objects in a given dataset is separated and clustered using the FCεH into the initial number of clusters per class (c). As a result centers $\mathbf{v}_i^{(1)}$ of the ε-hyperballs in class ω_1 and $\mathbf{v}_i^{(2)}$ in class ω_2 ($i = 1, 2, \cdots, c$) are obtained, together with corresponding memberships $u_{ik}^{(1)}$ and $u_{ik}^{(2)}$. When the same centers in a given class are found, only one is chosen and the corresponding partition matrix is updated. Also, all centers representing the "empty" clusters (with all membership degrees equal to 0) are rejected. After possible rejections $c^{(1)}$ and $c^{(2)}$ centers in classes ω_1 and ω_2 are obtained (if $c^{(1)} = 1$ or $c^{(2)} = 1$ then such a case is excluded from further analysis). Centers of the ε-hyperballs are centers of the Gaussian membership functions, the corresponding dispersions are calculated as

$$\bigvee_{\substack{1 \leq i \leq c^{(j)} \\ j \in \{1,2\}}} \left[\mathbf{s}_i^{(j)} \right]^{(\bullet 2)} = \sum_{k | \mathbf{x}_k \in \omega_j} u_{ik}^{(j)} \left(\mathbf{x}_k - \mathbf{v}_i^{(j)} \right)^{(\bullet 2)} \bigg/ \sum_{k | \mathbf{x}_k \in \omega_j} u_{ik}^{(j)}, \tag{15}$$

where $(\bullet 2)$ denotes the component-by-component squaring of a given vector $\mathbf{a}$, i.e. $\mathbf{a}^{(\bullet 2)} = [a_1^2, a_2^2, \cdots, a_p^2]$. Finally, centers and dispersions from both classes are denoted by $\mathbf{v}_{\ell}$ and $\mathbf{s}_{\ell}$ (see (14)), $\ell = 1, 2, \cdots, L = c^{(1)} + c^{(2)}$.

By introducing the notation $\mathbf{g}(\mathbf{x}_k)^{\mathsf{T}} = [\overline{\mu_{\mathbf{A}_1}}(\mathbf{x}_k), \overline{\mu_{\mathbf{A}_2}}(\mathbf{x}_k), \ldots, \overline{\mu_{\mathbf{A}_L}}(\mathbf{x}_k)]$, where $\overline{\mu_{\mathbf{A}_{\ell}}}(\mathbf{x}_k) = \mu_{\mathbf{A}_{\ell}}(\mathbf{x}_k) / \sum_{j=1}^{L} \mu_{\mathbf{A}_j}(\mathbf{x}_k)$, and $\mathbf{y}^{\mathsf{T}} = [y_1, y_2, \ldots, y_L]$, the final output of the classifier (13) can be written in the form $y_{0k} = \mathbf{g}\left(\mathbf{x}_k\right)^{\mathsf{T}}\mathbf{y}$. To find the consequents $(\mathbf{y})$ such that $\mathbf{g}\left(\mathbf{x}_k\right)^{\mathsf{T}}\mathbf{y} > 0$ if $\mathbf{x}_k \in \omega_1$, and $\mathbf{g}\left(\mathbf{x}_k\right)^{\mathsf{T}}\mathbf{y} \leq 0$ if $\mathbf{x}_k \in \omega_2$, we minimize the following criterion function [11]

$$J^{(r)}\left(\mathbf{y}^{(r)}\right) \triangleq \frac{1}{2}\left(\mathbf{G}\mathbf{y}^{(r)} - \mathbf{1}\right)^{\mathsf{T}}\mathbf{H}^{(r)}\left(\mathbf{G}\mathbf{y}^{(r)} - \mathbf{1}\right), \tag{16}$$

where: $\mathbf{G}^{\mathsf{T}} = [\Phi_1\mathbf{g}(\mathbf{x}_1), \Phi_2\mathbf{g}(\mathbf{x}_2), \cdots, \Phi_N\mathbf{g}(\mathbf{x}_N)]$ is the $N \times L$ matrix, Φ_k denotes the class label (equals $+1$ for class ω_1 and -1 for class ω_2) of the object $\mathbf{x}_k$ from the train set of size N, $\mathbf{1}$ denotes a vector with all entries equal to one, $\mathbf{H}^{(r)} = \operatorname{diag}\left(h_1^{(r)}, h_2^{(r)}, \cdots, h_N^{(r)}\right)$, $h_k^{(r)} = 0$ if $e_k^{(r-1)} \geq 0$, and $h_k^{(r)} = 1$ if $e_k^{(r-1)} < 0$, $\mathbf{e}^{(r-1)} = \mathbf{G}\mathbf{y}^{(r-1)} - \mathbf{1}$. The r is the iteration index in the IRLS error minimization procedure with the conjugate gradient approach [13], that was applied to minimize the criterion (16). The above definition of the $h_k^{(r)}$ provides the asymmetric square loss function (approximation of the misclassification error). Changing definition of the $h_k^{(r)}$ other loss functions can be obtained [13].

4   Results and Discussion

To evaluate the classification quality using the FCεH clustering five following benchmark datasets described in [15,16] were applied: Breast Cancer (BRE), Diabetis (DIA), Heart (HEA), Thyroid (THY) and Titanic (TIT). The sixth benchmark dataset was the Ripley synthetic two-class problem (SYN). Each dataset was represented by 100 splits into train and test sets. For the SYN we used divisions as in [11], for the remaining datasets the divisions were as in [15,16]. The numbers of features (p) and sizes of train (N_{trn}) and test (N_{tst}) sets are presented in Table 1. We used three methods as a reference to the proposed algorithm: the LSVM and the same fuzzy rule-based classifier, but with the parameters of antecedents calculated on the basis of FCM and FCPM. The class labels were assigned according to the sign of classifiers output value: class ω_1 for value > 0 and ω_2 for value ≤ 0. The FCεH clustering requires data to be scaled into the range $[0, 1]$, therefore clustering methods processed the scaled data. The LSVM classification was performed using data scaled to the range $[-1, 1]$.

The ε is set arbitrarily, however we investigated two preliminary approaches to automatic determination of its value using the results of FCM clustering. Finally we studied three procedures:

1. Data Independent ε (DIε), where the following values of ε were analyzed: 0, 0.01, 0.02, 0.05, 0.1, 0.2, 0.5.
2. Multiple Data Dependent ε (DDε_M), where separate value of $\varepsilon_i^{(j)}$ for each center $\mathbf{v}_i^{(j)}$ was calculated as a mean of p components of the dispersion $\mathbf{s}_i^{(j)}$, for $1 \leq i \leq c$ and $j \in \{1, 2\}$. The dispersions were calculated using (15) and the results of FCM clustering of class ω_j into c clusters.
3. Single Data Dependent ε (DDε_S), where single value of $\varepsilon^{(j)}$ for all centers in class ω_j was calculated as a mean of $c \times p$ components of all dispersions $\mathbf{s}_i^{(j)}$, for $1 \leq i \leq c$ and $j \in \{1, 2\}$.

To find centers and dispersions in antecedents using the clustering procedures, ten first train sets were merged into a single dataset and each of its classes of objects was clustered into considered initial number of clusters per class (c, changed from 2 to 20). The consequents were determined using single train set. The number of clusters (rules) together with the value of the scale parameter γ (searched in the range [0.2, 1.6] with the step of 0.1) providing the lowest mean classification error for ten first test sets were chosen to calculate the final result – mean classification error and its standard deviation for all 100 test sets. The classification error was defined as the percentage of incorrectly classified cases. Such a procedure was conducted for three values of m (1.1, 1.5, 2) and for three approaches to setting the value of ε. In the case of LSVM, the Gaussian kernel $K(\mathbf{x}, \mathbf{x}_i) = \exp\left(-\chi \|\mathbf{x} - \mathbf{x}_i\|^2\right)$, $\chi \in \mathbb{R}_+$ was assumed, and the parameters (ν and χ) providing the lowest mean classification error for ten first test sets were selected from the set $\{0.00001, 0.00004, 0.00007, 0.0001, 0.0004, \cdots, 70000, 100000\}$. The remaining parameters of LSVM were set to the default values [14].

Clustering procedures were started from the prototypes located in the boundary of the convex hull [11] of the clustered class. Maximum number of iterations was set to 100 and the following values of parameters in stop conditions were assumed: $\delta = \Delta = 10^{-5}$ (FCεH), $\xi = \zeta = 10^{-5}$ (FCPM). In the FCM clustering the iterations were stopped when the Frobenius norm of the difference between successive prototype matrices was less than 10^{-5}. In the IRLS procedure with the conjugate gradient approach we set $\xi = \zeta = 10^{-5}$.

If the calculation of a prototype (center of the ε-hyperball in the case of FCεH) was not possible since the denominator was equal to 0 (in special cases), then its location from the previous iteration was preserved. Analyzing the formula (11) it can be noticed, that for $m = 1.1$ the quotient $1/(g_{ik})^{20}$ is obtained, which for small g_{ik} goes toward infinity. Therefore, in all clustering procedures distances between prototypes and objects (g_{ik} in the FCεH) less than 10^{-10} were treated as zero and the relevant calculation of the partition matrix was done. Also, any components of dispersions equal to 0 were set to 10^{-6}. If the summary membership function of all antecedents ($\sum_{\ell=1}^{L} \mu_{\mathbf{A}_\ell}(\mathbf{x}_k)$) was equal to 0, then $\mu_{\mathbf{A}_\ell}(\mathbf{x}_k) = 10^{-6}$ for each antecedent was assumed.

In case of the approach with the fixed values of ε (DIε), for all datasets except TIT the lowest mean classification errors were obtained for $\varepsilon > 0$. However, we did not notice monotonous relation between the value of ε and the classification accuracy. For all datasets except SYN the best results were obtained applying FCεH with $m = 1.1$. The results of the highest accuracy for the DIε approach are shown in the fourth column of Table 1. Considering the best results for the automatically determined ε (DDε_M and DDε_S), in the case of BRE, HEA and THY we obtained higher accuracy for DDε_M, while for DIA and TIT it was better to apply DDε_S. For SYN the mean classification errors were the same. However, comparing the best results of three approaches DIε, DDε_M and DDε_S (columns 4–6 in Table 1) it can be noticed, that the automatically determined ε improves the classification quality only in the case of BRE and HEA datasets.

Table 1 compares the best results (mean classification errors and its standard deviations calculated for all 100 test sets) achieved when applying the FCεH clustering with the outcome of the reference methods, the values of classifiers parameters are given at the bottom of each cell. All clustering-based classification schemes outperformed the LSVM. For all datasets the proposed clustering provided lower classification error than FCM. Comparing FCεH with FCPM it can be seen, that only for SYN and THY the higher classification quality was obtained applying the FCPM. Taking the above into consideration we can conclude, that the FCεH allows for an efficient estimation of the antecedents parameters of the applied fuzzy rule-based classifier.

Table 1. The comparison of the results with the applied reference methods

LSVM	FCM	FCPM	FCεH		
			DIε	DDε_M	DDε_S
$\nu \mid \chi$	$m \mid L \mid \gamma$	$m \mid L \mid \gamma$	$m \mid L \mid \gamma \mid \varepsilon$	$m \mid L \mid \gamma$	$m \mid L \mid \gamma$
BRE ($p = 9, N_{trn} = 200, N_{tst} = 77$)					
24.27 (3.95)	17.19 (3.41)	20.66 (3.89)	16.51 (3.44)	**16.18** (3.58)	17.66 (4.22)
0.04 | 0.4	1.1 | 34 | 1.4	1.1 | 40 | 0.6	1.1 | 40 | 0.2 | 0.01	1.1 | 38 | 1.4	1.5 | 34 | 0.7
DIA ($p = 8, N_{trn} = 468, N_{tst} = 300$)					
23.00 (1.78)	20.50 (2.07)	22.08 (1.67)	**18.17** (1.82)	19.56 (1.86)	19.26 (1.56)
1000 | 0.004	1.1 | 38 | 1.4	1.1 | 30 | 1.4	1.1 | 40 | 1.2 | 0.1	1.1 | 40 | 1.3	1.1 | 38 | 1.0
HEA ($p = 13, N_{trn} = 170, N_{tst} = 100$)					
16.33 (2.67)	9.36 (2.83)	10.94 (2.72)	8.82 (2.58)	**8.39** (2.29)	8.65 (2.40)
10000 | 0.0001	1.1 | 28 | 0.5	1.1 | 28 | 0.3	1.1 | 34 | 0.6 | 0.1	1.1 | 34 | 0.4	1.1 | 34 | 0.5
SYN ($p = 2, N_{trn} = 250, N_{tst} = 1000$)					
9.54 (0.60)	8.92 (0.91)	**8.31** (0.63)	8.87 (0.76)	9.07 (0.43)	9.07 (0.47)
0.7 | 7	1.5 | 12 | 0.5	1.1 | 16 | 0.2	2 | 12 | 0.2 | 0.02	1.5 | 4 | 1.0	1.5 | 4 | 1.1
THY ($p = 5, N_{trn} = 140, N_{tst} = 75$)					
4.21 (2.11)	3.77 (4.52)	**1.65** (1.46)	2.35 (1.84)	2.41 (2.07)	2.64 (2.43)
10 | 4	1.1 | 28 | 1.6	1.1 | 20 | 0.6	1.1 | 37 | 0.4 | 0.1	1.5 | 38 | 0.5	1.1 | 28 | 1.6
TIT ($p = 3, N_{trn} = 150, N_{tst} = 2051$)					
22.90 (1.33)	22.05 (0.86)	22.37 (1.07)	**22.04** (0.87)	22.28 (1.15)	22.13 (1.37)
0.01 | 4000	1.1 | 8 | 0.6	2 | 12 | 0.7	1.1 | 8 | 0.6 | 0	1.5 | 24 | 0.6	1.1 | 6 | 1.1

The classification efficiency of the considered benchmark data was also studied in [15, 16] and the best results are: BRE (25.8 (4.6)), DIA (23.2 (1.6)), HEA (16.0 (3.3)), THY (4.2 (2.1)) and TIT (22.4 (1.0)). For BRE, DIA and THY the Kernel Fisher Discriminant classifier was used, while for HEA and TIT the Support Vector Machine. It can be noticed, that all proposed FCεH clustering-based methods provided lower mean classification errors for the above datasets.

To estimate the computational complexity we measured the computing time (t_1) of the FCεH clustering ($m = 1.1$, $\varepsilon = 0.1$) and time (t_2) needed for the classification (FCεH + IRLS). The results are normalized to the computing times of FCM clustering ($m = 1.1$) and LSVM classification ($\nu = 1, \chi = 1$) respectively (Table 2). The FCεH is characterized by the higher computing time than the FCM, especially for SYN, THY and TIT ($L = 40$) datasets. The reason are two nested loops realizing two Picard algorithms required for the FCεH implementation. The total time needed for the classification applying the FCεH is also higher when comparing to the LSVM. The main reason is the time needed for the clustering (the computing time of IRLS was studied in [13]).

Table 2. The comparison of the computing time

	BRE			DIA			HEA			SYN			THY			TIT		
L	4	20	40	4	20	40	4	20	40	4	20	40	4	20	40	4	20	40
t_1	0.3	3.1	5.6	5.5	2.9	5.9	2.7	2.9	2.1	26.1	110.5	116.8	186.8	195.0	36.0	5.5	5.1	70.5
t_2	7.6	41.6	161.7	5.2	26.5	148.6	3.9	22.9	39.6	9.1	475.0	948.7	82.4	363.0	741.2	3.0	11.1	29.2

5   Conclusions

The paper presented the new fuzzy clustering method with ε-hyperballs (FCεH) being the prototypes. It is based on the idea of regions of insensitivity described by the hyperballs of radius ε. The distances of objects located inside the insensitivity hyperball from its center are considered as equal to zero. The proposed clustering was applied to determine the Gaussian membership functions of fuzzy sets in antecedents of the fuzzy classifier. The efficiency of the classification based on the FCεH was evaluated using six benchmark datasets. The results were compared with three reference procedures: with two clustering methods combined with the same fuzzy classifier and with the Lagrangian Support Vector Machines (LSVM). All clustering methods outperformed the LSVM. Among them, the FCεH provided better classification accuracy for all or four datasets, depending on the reference method. Therefore we may conclude, that it is efficient method of determining the antecedents of the considered fuzzy classifier.

Acknowledgments. This work was partially supported by the Ministry of Science and Higher Education funding for: statutory activities of young researchers (BKM-508/RAu-3/2016) and statutory activities (BK-220/RAu-3/2016).

References

1. Aggarwal, C.C., Reddy, C.K.: Data Clustering. Algorithms and Applications. CRC Press, Boca Raton (2014)
2. Bezdek, J.C.: Pattern Recognition with Fuzzy Objective Function Algorithms. Plenum Press, New York (1981)
3. Cortes, C., Vapnik, V.: Support-vector networks. Mach. Learn. **20**(3), 273–297 (1995)
4. Doring, C., Lesot, M.-J., Kruse, R.: Data analysis with fuzzy clustering methods. Comput. Stat. Data Anal. **51**, 192–214 (2006)

5. Gorzalczany, M.B., Rudzinski, F.: Interpretable and accurate medical data classification - a multi-objective genetic-fuzzy optimization approach. Expert Syst. Appl. **71**, 26–39 (2017)

6. Ho, Y.-C., Kashyap, R.L.: An algorithm for linear inequalities and its applications. IEEE Trans. Electron. Comput. **14**(5), 683–688 (1965)

7. Jezewski, M., Czabanski, R., Horoba, K., Leski, J.M.: Clustering with pairs of prototypes to support automated assessment of the fetal state. Appl. Artif. Intell. **30**(6), 572–589 (2016)

8. Jezewski, M., Leski, J.M., Czabanski, R.: Classification based on incremental fuzzy $(1+p)$-means clustering. In: Gruca, A., Brachman, A., Kozielski, S., Czachórski, T. (eds.) Man–Machine Interactions 4. AISC, vol. 391, pp. 563–572. Springer, Cham (2016). doi:10.1007/978-3-319-23437-3_48

9. Kruse, R., Doring, C., Lesot, M.-J.: Fundamentals of fuzzy clustering. In: de Oliveira, J.V., Pedrycz, W. (eds.) Advances in Fuzzy Clustering and Its Applications, pp. 3–30. Wiley Ltd., Chichester (2007)

10. Leski, J.M.: An ε-insensitive approach to fuzzy clustering. Int. J. Appl. Math. Comput. Sci. **11**(4), 993–1007 (2001)

11. Leski, J.M.: Fuzzy $(c+p)$-means clustering and its application to a fuzzy rule-based classifier: toward good generalization and good interpretability. IEEE Trans. Fuzzy Syst. **23**(4), 802–812 (2015)

12. Leski, J.M.: Ho-Kashyap classifier with generalization control. Pattern Recogn. Lett. **24**(14), 2281–2290 (2003)

13. Leski, J.M.: Iteratively reweighted least squares classifier and its ℓ_2- and ℓ_1-regularized kernel versions. Bull. Polish Acad. Sci. Tech. Sci. **58**(1), 171–182 (2010)

14. Mangasarian, O.L., Musicant, D.R.: Lagrangian support vector machines. J. Mach. Learn. Res. **1**, 161–177 (2001)

15. Mika, S., Ratsch, G., Weston, J., Scholkopf, B., Muller, K.-R.: Fisher discriminant analysis with kernels. In: Proceedings of Neural Networks for Signal Processing IX, pp. 41–48 (1999)

16. Ratsch, G., Onoda, T., Muller, K.-R.: Soft margins for AdaBoost. Mach. Learn. **42**, 287–320 (2001)

17. Xu, R., Wunsch, II, D.C.: Clustering. Wiley Inc., Hoboken (2009)

18. Zadeh, L.A.: Fuzzy sets. Inf. Control **8**, 338–353 (1965)

Two Modifications of Yinyang
K-means Algorithm

Wojciech Kwedlo[✉]

Faculty of Computer Science, Bialystok University of Technology,
Wiejska 45a, 15-351 Bialystok, Poland
w.kwedlo@pb.edu.pl

Abstract. In the paper a very fast algorithm for K-means clustering problem, called Yinyang K-means, is considered. The algorithm
uses initial grouping of cluster centroids and the triangle inequality to
avoid unnecessary distance calculations. We propose two modifications of
Yinyang K-means: regrouping of cluster centroids during the run of the
algorithm and replacement of the grouping procedure with a method,
which generates the groups of equal sizes. The influence of these two
modifications on the efficiency of Yinyang K-means is experimentally
evaluated using seven datasets. The results indicate that new grouping
procedure reduces runtime of the algorithm. For one of tested datasets
it runs up to 2.8 times faster.

Keywords: Clustering · K-means algorithms · Yinyang K-means

1 Introduction

The problem of finding K clusters in data [8] is arguably the most important
unsupervised learning problem. It can be formulated as a search for a partition
of the learning set X consisting of N feature vectors, i.e. $X = x(1), \ldots, x(N)$,
$x(i) \in \mathbb{R}^M$, where M is the dimension of the feature space. The most popular
criterion used in clustering is the sum of square errors (SSE) defined as:

$$\text{SSE} = \sum_{i=1}^{N} \text{d}^2(x(i), c(a(i))), \tag{1}$$

where $\text{d}(x, y)$ is a distance (e.g., Euclidean), $a(i) \in \{1, \ldots, K\}$ denotes the index
of cluster, to which feature vector $x(i)$ is assigned, and $c(j) \in \mathbb{R}^M$, where $j \in
\{1, \ldots, K\}$, is the centroid (prototype) of the j-th cluster. The clustering problem
with SSE criterion and Euclidean distance is known to be NP-hard [1] and various
approximation heuristics have been proposed. The K-means method, the most
popular of these, is an iterative refinement algorithm, which, given an initial
solution (a set of centroids), generates a sequence of solutions with decreasing
values of SSE. Each iteration of the K-means method consists of two steps. In
the *assignment* step each feature vector is assigned to the cluster represented by

© Springer International Publishing AG 2017
L. Rutkowski et al. (Eds.): ICAISC 2017, Part II, LNAI 10246, pp. 94–103, 2017.
DOI: 10.1007/978-3-319-59060-8_10

the closest centroid (prototype). In the *update* step the centroid of each cluster is recalculated as the sample mean of the assigned feature vectors. The iterations of K-means proceed until local convergence, where improvements of the current solution cannot be found.

The simplest formulation of K-means method is Lloyd's algorithm [11]. In the assignment step it uses a brute-force approach and computes the distance between each feature vector and each cluster centroid. For applications of clustering to analysis of large datasets ('big data') the computational requirements of Lloyd's algorithm may be too high. Recently, a lot of research has been devoted to faster alternatives to Lloyd's algorithm. Efforts towards speeding up the Lloyd's algorithm include clever initialization techniques, e.g., K-Means++ [2], approximate algorithms [14,15], approaches using KD-trees [9,13] and triangle inequality [5–7]. Because they work well in high dimensional feature spaces (unlike KD-trees) and give the same results as the Lloyd's method (unlike approximate approaches), the algorithms using triangle inequality seem to be the most promising ones in clustering of big data.

The starting point of this work was a recent paper by Ding et al. [5], in which they proposed Yinyang K-means – a very fast K-means variant, which uses initial grouping of cluster centroids and the triangle inequality to skip unnecessary distance calculations. Their work used the standard K-means algorithm to obtain initial grouping of centroids, which does not change during the run of the algorithm. Our work extends the approach of Ding et al., by two modifications. The former consist in regrouping of cluster centroids during the run of the algorithm. The latter works by replacing initial grouping procedure with a method which generates the groups of equal sizes. We experimentally investigate the effect of these two changes on the efficiency of Yinyang K-means.

The rest of the paper is organized as follows. In the next section the Yinyang K-means algorithm is presented. Section 3 describes our modifications of the algorithm. Section 4 presents the results of computational experiments evaluating the influence of our modifications on the efficiency of the algorithm. The last section concludes the paper.

2   Yinyang K-means

The main idea of Yinyang K-means clustering [5] consists in partition of cluster centroids into t groups. In [5] it was proposed to set $t = K/10$ and to employ the standard K-means run (using cluster centroids as the data) to obtain this initial grouping. For each feature vector $x(i) \in X$ and j-th group the algorithm maintains a lower bound $l(i,j)$ on minimum distance between $x(i)$ and cluster centroids from the group, excluding the currently assigned centroid $c(a(i))$. The algorithm also maintains the upper bound $u(i)$ on the distance between $c(a(i))$ and $x(i)$. If $l(i,j) \geq u(i)$ then any centroid from the the j-th group cannot be closer to $x(i)$ than the currently assigned centroid $c(a(i))$. In this case, the calculations of distances to centroids from the group can be avoided.

The pseudocode of the algorithm, based on the formulation in [3], is shown in Algorithm 1. The algorithm starts with the partition of centroids (line 1).

Algorithm 1. Yinyang K-means algorithm.

Data: Initial cluster centroids $c(1), \ldots, c(K)$ and feature vectors $x(1), \ldots, x(N)$

1 Obtain t-partition of initial centroids; store cluster indexes in $G_1, \ldots, G_t$

2 **for** $i \leftarrow 1$ **to** N **do**

3   $\quad$ $a(i) \leftarrow \text{argmin}_{k=1,\ldots,K}\, d(x(i), c(k))$

4   $\quad$ $u(i) \leftarrow d(x(i), c(a(i))),\ \forall_{j=1,\ldots,t} l(i,j) \leftarrow \min_{\{k:k \in G_j \setminus a(i)\}} d(x(i), c(k))$

5 **repeat**

6   $\quad$ **for** $j \leftarrow 1$ **to** K **do**

7   $\quad\quad$ $c'(j) \leftarrow c(j),\ c(j) \leftarrow \text{mean}\{x(i) \in X : a(i) = j\},\ \delta(j) \leftarrow d(c'(j), c(j))$

8   $\quad$ $\forall_{j=1,\ldots,t} g(j) = \max_{k \in G_j} \delta(k)$

9   $\quad$ **for** $i \leftarrow 1$ **to** N **do**

10   $\quad\quad$ $u(i) \leftarrow u(i) + \delta(a(i))$

11   $\quad\quad$ $\forall_{j=1,\ldots,t} l(i,j) \leftarrow l(i,j) - g(j)$

12   $\quad\quad$ **if** $\min_{j=1,\ldots,t} l(i,j) \geq u(i)$ **then continue**

13   $\quad\quad$ $u(i) \leftarrow d(x(i), c(a(i)))$

14   $\quad\quad$ **if** $\min_{j=1,\ldots,t} l(i,j) \geq u(i)$ **then continue**

15   $\quad\quad$ $\hat{\mathcal{G}} \leftarrow \bigcup_{\{j \in 1,\ldots,t: l(i,j) < u(i)\}} G_j$

16   $\quad\quad$ Obtain $a(i)$ using centroids whose indexes are in $\hat{\mathcal{G}}$

17 **until** cluster assignment $a(1), \ldots, a(N)$ stabilizes

Result: Final cluster centroids $c(1), \ldots, c(K)$ and cluster assignment
$\quad\quad a(1), \ldots, a(N)$

The indexes of centroids are split into t groups. $G_1, \ldots, G_t$. Next, it performs one iteration of Lloyd's method (lines 2–4), which is necessary for tightly initializing the lower and upper bounds. The main loop of the algorithm is started (lines 6–7) with the computation of new centroids coordinates as sample means. In this process the drift (i.e., the distance moved) of each centroid $c(j)$ is calculated in $\delta(j)$. Next, (line 8) the drift of each group is calculated in $g(j)$ as a maximum drift of individual centroids. Afterwards the algorithm performs the assignment step, calculating $a(i)$ for all data items (lines 9–16).

The assignment steps begin with the update of lower and upper bounds (lines 10–11) using the drifts. Then, it tries to filter out as many distance calculations as possible using the bounds. It first checks if $l(i,j) \geq u(i)$ for all $j = 1, \ldots, t$ (line 12). If this condition, called the *global* filtering condition, holds, than no centroid can be closer to $x(i)$ then the currently assigned centroid $c(a(i))$ and calculation of all distances to $x(i)$ can be avoided. If this condition fails the algorithm tightens the upper bound $u(i)$ by the real distance (line 13) and re-checks the condition again (line 14). If this fails the algorithm enters the *group* filtering stage considering only the groups which did not pass the bounds comparison (line 15). The assignment $a(i)$ is obtained in line 16 with the help of the *local* filtering. The details of the local filtering [3,5] are omitted in this paper due to space limitation. However, it should be noted that our modifications of the algorithm do not influence the local filtering.

The pseudocode in Algorithm 1 is focused on the assignment step. Following [5,7] our implementation uses one important optimization of the update step. It retains vector sums and counts of objects assigned to each cluster. Using this approach the iteration over the whole dataset to compute the means (line 7) is not necessary. Instead it is sufficient to update the vector sums and the counts for feature vectors which changed the cluster membership. By doing so we incur the cost of one addition and one subtraction for each feature vector that changed the membership. Thus, this optimization is beneficial if less than half of feature vectors change cluster membership in an iteration.

3   Modifications of the Algorithm

3.1   Regrouping of Centroids

The aim of the first modification is to prevent a possible reduction of efficiency of the group filtering with the progress of Yinyang K-means. For maximal efficiency of the group filtering it may be beneficial to have similar centroids in each group. The centroid grouping (line 1 of Algorithm 1) step achieves this at initial iterations of the algorithm. However at later stages, because the centroids move, centroids in a group tend to be very distinct. Regrouping of centroids may alleviate this problem.

Because the cost of regrouping can be substantial, it is performed at iterations $I_s, 2*I_s, 4*I_s, 8*I_s, \ldots$, where I_s is an user-supplied parameter. It is performed before line 6 of Algorithm 1 and consists in running the standard K-means algorithm on cluster centroids $c(1), \ldots, c(K)$ (similarly to line 1).

After the regrouping, because some centroids have changed group membership, the lower bounds have to be updated. Denote by $\mathcal{G}' = G'_1, \ldots, G'_t$ the old (before regrouping) partition of centroids, and by $l'(i,j)$ (where $i = 1, \ldots, N$ and $j = 1, \ldots, t$) the old lower bounds. New lower bound is calculated as $l(i,j) = \min_{k \in \Delta(j)} l'(i,k)$, where $\Delta(j) \subset \mathcal{G}'$ denotes the indexes of source centroid groups (i.e., the old groups from which at least one centroid migrated to group j). More formally: $\Delta(j) = \{k : \exists_{l \in \{1, \ldots, K\}} l \in G'_k\}$.

3.2   Initial Grouping and Regrouping Using Same-Size K-means

The second modification replaces the clustering method used in the initial grouping of centroids (and in regrouping) by a variant of K-means algorithm, which, while performing local optimization of the SSE (1), generates clusters of equal sizes (same-size K-means). This change is motivated by a possible improvement of the efficiency of the group filter, when using groups with balanced cardinalities. The pseudocode of the same-size K-means is shown in Algorithm 2.

The same size K-means algorithm uses $c(1), \ldots, c(K)$ as the input data. The partition of the data into groups is represented by a variable b where $b(i) \in \{1, \ldots, t\}$ is the group index for $c(i)$. The algorithm requires such an initial partition of the data that cardinalities of groups differ by at most one. For an

---

Algorithm 2. Same-size K-means algorithm for centroid grouping

---

Data: Cluster centroids $c(1), \ldots, c(K)$ and group assignment $b(1), \ldots, b(K)$
1 **for** $i \leftarrow 1$ **to** t **do** $e(i) \leftarrow \text{mean}\{c(k) : b(k) = i\}$
2 **repeat**
3   **for** $i \leftarrow 1$ **to** K **do**
4   **for** $k \leftarrow 1$ **to** t **do** $D(i,k) \leftarrow \text{d}^2(c(i), e(k))$
5   **if** $min_{k \in \{1,\ldots,t\}} D(i,k)$ **then** $B(i) \leftarrow$ **true**
6   **else** $B(i) \leftarrow$ **false**
7   **for** $i \leftarrow 1$ **to** K **do**
8   **if** $B(i)$ **then continue**
9   **for** $j \leftarrow i+1$ **to** K **do**
10   **if** $D(i, b(i)) + D(j, b(j)) < D(j, b(i)) + D(i, b(j))$ **then**
11   swap $b(i)$ with $b(j)$
12   **for** $i \leftarrow 1$ **to** t **do** $e(i) \leftarrow \text{mean}\{c(k) : b(k) = i\}$
13 **until** group assignment $b(1), \ldots, b(K)$ stabilizes
Result: Final group assignment $b(1), \ldots, b(K)$

---

application of the algorithm to initial grouping the initial partition fulfilling this condition is obtained by greedy adding data items to the groups trying to minimize the SSE. For an application to regrouping the previous partition is used to initialize the algorithm.

The algorithm starts with the computation of centroids of groups in variable e (line 1). In a single iteration (lines 3–12) it first pre-computes in variable D (the loop in lines 3–6) the squared distance between each data item and each centroid. For each data item $c(i)$ it also sets the boolean flag $B(i)$ if the currently assigned centroid $e(b(i))$ is the closest to the data item $c(i)$. The flag $B(i)$ is later used to filter out such data items as the candidates for swap (line 8).

In the next loop (lines 7–11) the algorithm tries to lower the SSE as much as possible while preserving cardinalities of groups. It considers all pairs of data items, excluding the data items filtered out by the variable B. The two items exchange their groups (line 11) if and only if this operation lowers the SSE (line 10). The iteration is finished by recalculation of centroids (line 12).

Because both the group exchange operation and the recalculation of centroids lower the SSE of the solution, the convergence of Algorithm 2 is assured. The computational complexity of the algorithm is $O(K^2)$.

4 Experimental Results

The aim of the experimental study reported in this section was to evaluate whether our modifications of Yinyang K-means reduce its runtime. We performed clustering experiments using seven large datasets. The characteristics of the datasets are shown in Table 1.

Caltech101, notredame, tiny, and ukbench came from an content-based image retrieval application [16]. Uscensus is a well known large dataset from UCI

Table 1. The description of the datasets used in the experiments.

	caltech101	mix300	notredame	tiny	ukbench	urand300	uscensus
Dimension M	128	300	128	384	128	300	68
#vectors N	1000K	1000K	410K	1000K	1000K	1000K	2458K
Source	[16]	synthetic	[16]	[16]	[16]	synthetic	[10]

machine learning repository [10]. The last two datasets are artificially generated. Urand300 was obtained by random sampling from the uniform distribution on the interval $[0, 1]$. Mix300 was sampled, using a generator described in [12], from a Gaussian mixture consisting of 40 very well separated clusters.

All the experiments were performed on a system consisting of two 14-core Intel Xeon E5-2697 v3 processors with 64GB of RAM. The algorithms were implemented in C++ and compiled by Intel C++ compiler version 15.0.4 using the optimization options recommended for the processor. The algorithms were parallelized using the OpenMP shared memory programming standard [4] and employed all 28 cores in the system.

In the experiments we have tested four variants of Yinyang K-means obtained by using two initial clustering methods and by using/not using regrouping. These four variants are denoted by Y and Y_1–Y_3 and described in Table 2. While variant Y is the original Yinyang algorithm, the remaining three include at least one of two modifications proposed in this paper.

Table 2. The four variants of Yinyang K-means algorithm used in the paper.

Variant	Initial grouping	Regrouping
Y	K-means	No
Y_1	K-means	Yes
Y_2	same-size K-means	No
Y_3	same-size K-means	Yes

In the experiments we used $K \in \{64, 256, 1024, 4096\}$. In all the experiments we set $I_S = 5$. For each dataset and each value of K we generated ten random initial clustering solutions. Next, each of four variants of Yinyang K-means was run ten times using these initializations. The result of a single experiment was the average (over ten initializations) wall clock runtime of each algorithm and the average number of iterations. The results are reported in Table 3.

The results indicate that, although all four variants should give the same solution (and thus the number of iterations) given the same initialization, this is not always the case. For five real life datasets the average number of iterations for two methods (Y and Y_1) using the K-means centroid grouping is different from the average number of iterations for two methods (Y_2 and Y_3) using same-size

Table 3. Execution times in seconds (before '/' sign) and numbers of iterations (after '/' sign) of four variants (Y, Y_1, Y_2, Y_3) of Yinyang K-means. Each result is the average over 10 different random initializations. The fastest algorithm is indicated in bold.

Dataset	K	Y	Y_1	Y_2	Y_3
caltech101	64	2.9/355.3	2.85/355.3	**2.81/367.1**	2.82/367.1
	256	**7.46/468.3**	7.95/468.3	7.73/466.8	8.17/466.8
	1024	21.9/380.9	23.2/380.9	**20.9/397.4**	22.7/397.4
	4096	**49.3/197.9**	55.8/197.9	50.4/216.6	56.9/216.6
mix300	64	6.16/254.6	6.28/254.6	4.58/254.6	**4.57/254.6**
	256	36.9/405.9	29.2/405.9	13.7/405.9	**12.9/405.9**
	1024	94/352.9	63.4/352.9	43.9/352.9	**39.8/352.9**
	4096	204/277.1	172/277.1	138/277.1	**134/277.1**
notredame	64	1.44/282.3	**1.35/282.3**	1.38/317.0	1.36/317.0
	256	3.98/317.0	4.11/317.0	**3.69/303.5**	3.75/303.5
	1024	9.68/183.7	9.45/183.7	**8.09/176.7**	8.96/176.7
	4096	19.4/76.0	21.6/76.0	**19.2/76.8**	21.1/76.8
tiny	64	16.2/735.1	14.7/735.1	14.1/742.9	**13.9/742.9**
	256	33/654.2	28.4/654.2	**28.2/730.6**	28.4/730.6
	1024	76.1/458.0	69.1/458.0	**63.9/464.3**	64.4/464.3
	4096	174/232.4	171/232.4	**150/207.8**	157/207.8
ukbench	64	2.93/348.7	**2.88/348.7**	2.91/374.8	2.91/374.8
	256	**7.39/434.0**	7.8/434.0	8.07/497.1	8.51/497.1
	1024	21.1/340.4	22.4/340.4	**20.2/357.2**	22.3/357.2
	4096	55.6/257.4	62.5/257.4	**54.9/254.9**	61.1/254.9
urand300	64	43.2/1761.7	43.3/1761.7	40.2/1761.7	**40.1/1761.7**
	256	68.9/725.7	69.7/725.7	**61.3/725.7**	61.5/725.7
	1024	128/243.7	130/243.7	**112/243.7**	113/243.7
	4096	293/80.1	316/80.1	**251/80.1**	253/80.1
uscensus	64	1.28/82.4	1.29/82.4	**1.22/81.0**	1.27/81.0
	256	3.99/155.6	4.44/155.6	**3.98/144.9**	4.42/144.9
	1024	14.8/196.7	16.4/196.7	**14.7/207.0**	16.4/207.0
	4096	**42.3/156.3**	50.9/156.3	47.1/177.3	54/177.3

K-means centroid grouping. In this case we have also observed the differences in cluster assignments. The most likely cause of this behavior is different order of iteration over cluster centers. In case of equal distance between a feature vector and two centroids this order may influence the outcome of the algorithm. This behavior has not manifested itself for artificially generated datasets because the real life datasets were stored using limited precision (e.g., original features in image processing datasets have a single byte representation).

In order to make a correction for the above behavior of the algorithms, we also present results based on the average time of a single iteration. For each method $A \in \{Y, Y_1, Y_2, Y_3\}$ the average time of a single iteration is given by $t_i(A) = t(A)/I(A)$ where $t(A)$ and $I(A)$ denote the average execution time and average number of iterations over 10 runs, respectively. The speedup of the variant A over the unmodified Yinyang K-means, is then defined as $S(A) = t_i(Y)/t_i(A)$, where $t_i(Y)$ is the average iteration time of the unmodified method Y. These speedups are presented in Table 4.

Table 4. Speedups of three variants (Y_1, Y_2, Y_3) of Yinyang K-means over the unmodified method Y. The speedups are computed on the basis of the average iteration time. The results are average over 10 different random initialization. The fastest algorithm, if faster then unmodified Yinyang K-means variant, is indicated in bold.

Dataset	K	Y_1	Y_2	Y_3
caltech101	64	1.02	**1.06**	1.06
	256	0.938	0.962	0.91
	1024	0.945	**1.09**	1.01
	4096	0.884	**1.07**	0.948
mix300	64	0.981	1.35	**1.35**
	256	1.27	2.7	**2.87**
	1024	1.48	2.14	**2.36**
	4096	1.19	1.48	**1.52**
notredame	64	1.06	1.17	**1.18**
	256	0.968	**1.03**	1.02
	1024	1.02	**1.15**	1.04
	4096	0.899	**1.02**	0.927
tiny	64	1.1	1.16	**1.17**
	256	1.16	**1.31**	1.3
	1024	1.1	**1.21**	1.2
	4096	1.02	**1.04**	0.989
ukbench	64	1.02	**1.08**	1.08
	256	0.948	**1.05**	0.995
	1024	0.939	**1.09**	0.991
	4096	0.889	**1**	0.901
urand300	64	0.998	1.08	**1.08**
	256	0.989	**1.13**	1.12
	1024	0.982	**1.14**	1.14
	4096	0.926	**1.17**	1.16
uscensus	64	0.993	**1.03**	0.991
	256	0.898	0.935	0.841
	1024	0.902	**1.06**	0.95
	4096	0.831	**1.02**	0.889

The examination of the results from Table 4 reveals that:

- Addition of regrouping alone does not significantly reduce runtime of the algorithm. In most cases the runtime is actually increased. The only exception to this rule is mix300 dataset, for $K > 64$.
- Replacement of initial grouping method by same size K-means results in mostly fastest variant of the algorithm (column Y_2 in Table 4).
- Combination of two modifications proposed in the paper (Y_3), degrades the performance in comparison to Y_2. The only exception is again mix300 dataset.
- For the Y_2 variant speedup for all the datasets, except one, decreases when K is increased from 1024 to 4096. We conjecture that $O(K^2)$ computational complexity of our same size K-means implementation is limiting the speedup in this case.

5   Conclusions

In this paper we proposed two modifications of Yinyang K-means. We experimentally evaluated the impact of these modifications on the efficiency of the algorithm using five real life and two synthetic datasets. Although the results indicated that no method always dominated the others, they also pointed out, that the efficiency of Yinyang K-means can be further improved. In particular, for high dimensional datasets, replacing a K-means grouping method with its same-size counterpart significantly improved the efficiency.

An apparent direction of future work is the improvement of the same size K-means procedure. Its $O(K^2)$ computational complexity seems to degrade the efficiency for large number of clusters K. We are also going to develop a version of the algorithm for clusters consisting of dozens of multi-core compute nodes and evaluate its performance.

Acknowledgments. This work was supported by the Bialystok University of Technology grant S/WI/2/2013 funded by the Polish Ministry of Science and Higher Education. This research was carried out with the support of the Interdisciplinary Centre for Mathematical and Computational Modelling (ICM) University of Warsaw under grant no G65-12. The author is grateful to Dr. Jingdong Wang for providing access to caltech101, notredame, tiny and ukbench datasets.

References

1. Aloise, D., Deshpande, A., Hansen, P., Popat, P.: NP-hardness of euclidean sum-of-squares clustering. Mach. Learn. **75**(2), 245–248 (2009)
2. Arthur, D., Vassilvitskii, S.: K-means++: The advantages of careful seeding. In: Proceedings of the 18th Annual ACM-SIAM Symposium on Discrete Algorithms (SODA 2007), pp. 1027–1035 (2007)
3. Bottesch, T., Bühler, T., Käechele, M.: Speeding up k-means by approximating Euclidean distances via block vectors. In: Proceedings of The 33rd International Conference on Machine Learning, pp. 2578–2586 (2016)

4. Dagum, L., Menon, R.: OpenMP: an industry standard API for shared-memory programming. IEEE Comput. Sci. Eng. **5**(1), 46–55 (1998)
5. Ding, Y., Zhao, Y., Shen, X., Musuvathi, M., Mytkowicz, T.: Yinyang k-means: a drop-in replacement of the classic k-means with consistent speedup. In: Proceedings of the 32nd International Conference on Machine Learning (ICML 2015), pp. 579–587 (2015)
6. Elkan, C.: Using the triangle inequality to accelerate k-means. In: Proceedings of the 20th International Conference on Machine Learning (ICML 2003), vol. 3, pp. 147–153. AAAI Press (2003)
7. Hamerly, G., Drake, J.: Accelerating Lloyds algorithm for k-means clustering. In: Celebi, M.E. (ed.) Partitional Clustering Algorithms, pp. 41–78. Springer, Heidelberg (2015)
8. Jain, A.K.: Data clustering: 50 years beyond K-means. Pattern Recogn. Lett. **31**(8), 651–666 (2010)
9. Kanungo, T., Mount, D.M., Netanyahu, N.S., Piatko, C.D., Silverman, R., Wu, A.Y.: An efficient k-means clustering algorithm: analysis and implementation. IEEE Trans. Pattern Anal. Mach. Intell. **24**(7), 881–892 (2002)
10. Lichman, M.: UCI machine learning repository (2013). http://archive.ics.uci.edu/ml
11. Lloyd, S.: Least squares quantization in PCM. IEEE Trans. Inf. Theory **28**(2), 129–137 (1982)
12. Maitra, R., Melnykov, V.: Simulating data to study performance of finite mixture modeling and clustering algorithms. J. Comput. Graph. Stat. **19**(2), 354–376 (2010)
13. Pelleg, D., Moore, A.: Accelerating exact k-means algorithms with geometric reasoning. In: Proceedings of the Fifth ACM SIGKDD International Conference on Knowledge Discovery and Data Mining, pp. 277–281 (1999)
14. Philbin, J., Chum, O., Isard, M., Sivic, J., Zisserman, A.: Object retrieval with large vocabularies and fast spatial matching. In: 2007 IEEE Conference on Computer Vision and Pattern Recognition, pp. 1–8. IEEE (2007)
15. Wang, J., Wang, J., Ke, Q., Zeng, G., Li, S.: Fast approximate k-means via cluster closures. In: Baughman, A.K., Gao, J., Pan, J.Y., Petrushin, V.A. (eds.) Multimedia Data Mining and Analytics, pp. 373–395. Springer, Heidelberg (2015)
16. Wang, J., Wang, N., Jia, Y., Li, J., Zeng, G., Zha, H., Hua, X.S.: Trinary-projection trees for approximate nearest neighbor search. IEEE Trans. Pattern Anal. Mach. Intell. **36**(2), 388–403 (2014)

Detection of the Innovative Logotypes on the Web Pages

Marcin Mirończuk[✉], Michał Perełkiewicz, and Jarosław Protasiewicz

Laboratory of Intelligent Information Systems,
National Information Processing Institute,
al. Niepodległości 188b, 00-608 Warsaw, Poland
marcin.mironczuk@opi.org.pl
http://lis.opi.org.pl

Abstract. The aim of this study was to describe a found method for detection of logotypes that indicate innovativeness of companies, where the images originate from their Internet domains. For this purpose, we elaborated a system that covers a supervised and heuristic approach to construct a reference dataset for each logotype category that is utilized by the logistic regression classifiers to recognize a logotype category. We proposed the approach that uses one-versus-the-rest learning strategy to learn the logistic regression classification models to recognize the classes of the innovative logotypes. Thanks to this we can detect whether a given company's Internet domain contains a innovative logotype or not. Moreover, we find a way to construct a simple and small dimension of feature space that is utilized by the image recognition process. The proposed feature space of logotype classification models is based on the weights of images similarity and the textual data of the images that are received from HTMLs ALT tags.

Keywords: Logotypes classification · Logotypes recognition · Images classification · Images matching · Images feature construction · Feature construction

1 Introduction

Challenges. It is believed that the innovative businesses play a crucial role in the development of modern economies. One may want to locate them in order to establish a fruitful cooperation. It is a non-trivial task because such creative companies may exist in various fields of human activity. Fortunately, most of them have public web pages on the Internet. As it is impossible to search the whole Internet manually, we can recognize such companies by using the combination of web crawling techniques and machine learning methods. In our previous work [13], we developed a domain classification system that categorizes the Internet domains of companies into *innovative* and *non-innovative* classes.

Many features may indicate innovativeness of a company. Most importantly, it should launch/develop a product or service that is new or significantly

© Springer International Publishing AG 2017
L. Rutkowski et al. (Eds.): ICAISC 2017, Part II, LNAI 10246, pp. 104–115, 2017.
DOI: 10.1007/978-3-319-59060-8_11

improved - it may be a product, process, marketing campaign, or organizational innovation [1,11]. Such innovations may receive local, national, or international awards. They can also result from participation in research projects sponsored by various funds. To better promote these achievements, innovations and companies are usually advertised by different graphical signs. Thus, we can assume that the Internet domains of companies may include images, i.e. logotypes like trademarks or institution logotypes, which could contain valuable hints about these businesses' innovativeness.

Objectives. We assume that logotypes might improve the classification quality of the domain categorization system [13], which establishes whether a company is innovative or not based on the analysis of its web page. They may provide new, important features to the classification system closely related to the pure textual data. Therefore, we decided to work on the detection and recognition issue of logotypes on the Internet domains and propose an innovative logo detection system as a part and extension of the domain classification system. More specifically, the objectives of the study are as follows: (1) to design an innovative logo detection system that utilizes the supervised classification method to detect whether a given image indicates innovativeness or not, and (2) to check if we can use a new type of classification features, which values are created by the image similarity functions, to create the classification models that can recognize of the appropriate pre-selected logotypes indicating innovativeness.

We have to note that, in this study, we use short terms 'an innovative image' or 'an innovative logotype' for graphical signs that may are somehow connected with an innovative company, i.e. indicate its innovativeness.

Proposed approach. To *detect* if a given image is innovative or not, we *recognize* if this image matches at least one of the preselected groups (sets) of the innovative logotypes. For each group of preselected logotypes, we have built a classification model in a supervised manner. This model utilizes the similarity functions to create the values of the classification features. We compute the appropriate similarities between the given image and the preselected logotypes and texts from the reference datasets. After that, we use these values as the features of the classification model. Thanks to classification model, we can recognize whether the given image matches to the appropriate set of innovative logotypes. If the image is classified to at least one of the innovative logotype sets, then we assume that this image is innovative. Otherwise, the image is not innovative.

It has to be highlighted that we consider the logotypes as a particular subset of images, and we do not consider the images containing embedded logotypes, for example, photos from photo-cameras that may contain a logotype somewhere in the background. In this paper, we do not resolve the task of the logo detection or logo recognition (the logo spotting) in the mixed images (the natural images) mentioned above, where logotypes are somewhere in given images [3,8,20]. The above limitations imply that, in this study, we try to build a classification system that can recognize whether an image under examination contains or is similar to any logotype from a preselected set of logotypes and then classify the picture into one of two following groups: the innovative logotype and other logotypes.

The paper is structured as follows. Section 2 presents the selected previous works related to the image processing field. Sections 3 and 4 show an overview of the proposed innovative logotypes classification system and its implementation respectively. The next Sect. 5 describes the evaluation process of the proposed approach and results obtained during the experiments. Finally, Sect. 6 concludes the findings.

2   Related Works vs. Proposed Approach

There are many works concerning the issues of logo detection and recognition [2–4,6–8,14–16,19–21]. Article [10] presents a great review of these methods. It contains the comparative summary of (i) methods for logo recognition in document images, and (ii) image and video logo detection and retrieval methods. As we can notice, the problem of image recognition has been widely discussed in the literature. Conceptually, our approach to the logotypes recognition is nearly related to Rusinol et al. [17], Li et al. [12], and Baratis's [2,14,19] works.

Like the Rusinol et al. and Li et al., we use the reference images dataset for each logotype class to build classification model(s) that recognize(s) the logotype class of each unknown image. To determine an image class, Rusinol et al. used the SIFT local descriptors and a bag-of-visual-words model that based on the nearest neighbour ranking. They did not use the additional text data as a new feature and the machine learning classification techniques. Their work focused on the recognition of logotypes from the document images. Li et al., in the first research, used multiple types of image features to build a classification images model in a discriminative way. Their solution used multi-layer perceptrons. Unfortunately, their work focused on the development of the classification methodology for the automatic annotation of outdoor scene images. They did not resolve the logotype recognition problem. The last work (Baratis et al.) used the images from the Internet. In this work, the features were received from the different histograms, and the machine learning techniques were utilized to learn a classification model. Nonetheless, their work focused only on the logotype and trademark images detection. They did not propose the logotype recognition. In contrast to the works mentioned above, our approach is dedicated to innovative logotype images detection on the web pages by the utilization of the logotype recognition (an image is matched to the appropriate class of logotypes). We realize this matching by the proposed classification system. In this system, we use the features that are based on the values of similarity functions. Moreover, our approach uses one-versus-the-rest learning strategy to learn the logistic regression classification models to recognize the classes of innovative logotypes.

3   Problem Description

Suppose that we have an image im from the set of images I, i.e. $im \in I$ and the $im = (i_{Rel}, i_{Alt}, idDom)$. The i_{Rel} is the value of bytes of the image stream, i.e. the image representation such as a RGB matrix, a HSV matrix, a grayscale matrix, etc. The i_{Alt} is an image label that was extracted from an HTMLs' ALT tag, and $idDom$ is an identifier of an Internet domain, where we found the given image. We split the I set into two disjoint subsets, i.e. the logotype I_L and the other one I_O ($I = I_L \cup I_O$). Additionally, we divided the I_L set into two disjoint subsets, i.e. the innovative logotype I_{IL} and the non-innovative logotype $I_{\neg IL}$ ($I_L = I_{IL} \cup I_{\neg IL}$, $I_{IL} \not\subseteq I_{\neg IL}$, $|I_{IL}| \ll |I_{\neg IL}|$, and $I_{IL} \not\subseteq I_O$ but $I_{\neg IL} \subseteq I_O$). In our case, the I_{IL} set consists of the pre-selected sets of logotypes. We determined the fourteen classes of such logotypes, i.e. the set of innovative class labels $C_{IL} = \{forbes'\ diamond, business\ gazelle, business\ cheetah, innovative\ technology, innovative\ economy, laurels'consumer, ministry\ of\ science\ and\ higher\ education, polish\ agency\ for\ enterprise\ development, the\ national\ centre\ for\ research\ and\ development, poland\ now, european\ union, patent\ office, gold\ medal, integrated\ regional\ operational\ program\}$ shortly $C_{IL} = \{FD, BG, BC, IT, IE, LC, MOSAHE, PAFED, TNCFRAD, PN, EU, PO, GM, IROP\}$, and $I_{IL} = I_{BG} \cup I_{FD} \cup \cdots I_{c \in C_{IL}} = \bigcup_{c \in C_{IL}} I_c$. Figures 1 and 2 show the examples set of the innovative logotypes and other images respectively.

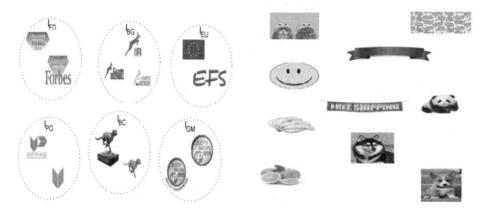

Fig. 1. The set of innovative logotypes I_{IL} with the example subsets of the innovative logotypes, such as $I_{FD}, I_{BG}, I_{EU}, I_{PO}, I_{BC}$, and I_{GM}.

Fig. 2. The set of the another image examples from Internet that may also includes the non-innovative logotypes, i.e. $I_O = I \setminus I_{IL}$.

As we can see (Fig. 1), the innovative logotypes from each subset have several variations. For example, the business gazelle logotype has different colours, text, etc. For this reason, in the learning phase of the proposed algorithm, we try to find and establish a small subset of reference logotypes set for each logotype class ($I_{Ref,c \in C_{IL}} \subset I_{c \in C_{IL}}$ and $|I_{Ref,c \in C_{IL}}| \ll |I_{c \in C_{IL}}|$) that well generalize a given class of logotypes (see Sect. 4 for more details). Figure 2 shows the set of the remaining images from Internet that also may include the other logotypes (I_O).

In our approach, we assume that we can create a process/method that recognizes if a given image im belongs to the appropriate subset of I_{IL} which implies that the image im is innovative. The innovative logotypes detection system obtains the image im in input and produces a decision $d_{Final} = \{innovative, non - innovative\}$ as output. The system labels the image im by the innovative label if image im matches the appropriate subset of the I_{IL} set. Otherwise, it assigns the non-innovative label to the image im.

4   System for Detection of Innovative Logotypes

4.1   System Overview

This subsection describes the proposed classification system of innovative logotypes. Figure 3 depicts a flow chart outlining the system structure. We use the *function interfaces* to describe how the system works. These interfaces are considered in the terms of computer science.

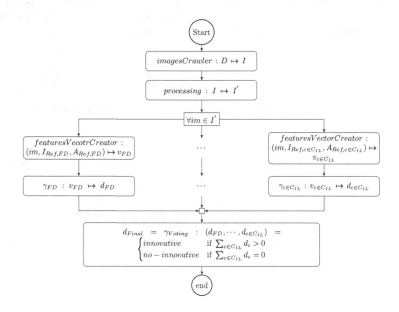

Fig. 3. The system for detection of innovative logotypes.

In the first step of the proposed process (Fig. 3), we use the *imagesCrawler* function to crawl a set of images I from the Internet. This function as an input receives the set of the hosts D and returns the set I that was described in Sect. 3. After that, the *processing* function processes the set of images I and returns the extended set of pre-processed images I'. This function is important because the authors recognized (based on empirical observations) that domains may contain images that include some sub-logotypes (one image/logotype can include many another appropriate logotypes, i.e. sub-logotypes). For this reason, we must recognize and decompose this main logotype to several logotypes.

For example, we may have the image (the main logotype) that includes the three sub-logotypes (The Innovative Economy, Polish Agency for Enterprise Development, and European Development Fund). We use the detection method based on boundary extension of feature rectangles [21] to recognize and extract these sub-logotypes.

After the realization of the *processing* function, we use the *features Vector-Creator* function to construct the vector of features and their values $v_{c \in C_{IL}}$ for each innovative logotypes class $c \in C_{IL}$ (Sect. 4.2 describes in detail this vector). This vector is constructed based on the functions' input parameters, i.e. the image im, the reference set of images $I_{Ref,c \in C_{IL}}$, and the reference set of alts' text $A_{Ref,c \in C_{IL}}$ for the given class $c \in C_{IL}$. After features construction, the classification function $\gamma_{c \in C_{IL}}$ utilizes this vector to compute the decision $d_{c \in C_{IL}} = \{0, 1\}$. The decision is binary, i.e. the classifier sets 1 if the given image im is matched (classified) to the given class $c \in C_{IL}$. Otherwise, the classifier sets 0. In our case, we use the logistic regression classification method to create a logistic regression classification function [18] (Subsect. 4.2 describes in details this research aspect).

In the last step, when we have decisions of all classifiers, we use the γ_{Voting} classification function that is based on the simple voting process. For each classifier, we take the decision $d_{c \in C_{IL}}$ and based on these decisions we create the final decision $d_{Finall} = \{innovative, non - innovative\}$. If the sum of decisions is greater than 0, then the image im is innovative. Otherwise, the image im is not innovative.

4.2   Learning Phase

The main contribution of our proposition is the way in which we construct the vector of features for the image classification system. For each logotype class $c \in C_{IL}$ (shortly c) we conduct the following steps:

1. The set of logotypes I_c is constructed. We utilized an image retrieval system [9] to finish this step, i.e. we sent the queries q_c to the retrieval system, and thanks to this, we collected the appropriate subset of logotypes, for example, we created the $q_{BG} = \{business\ gazelle, award\ gazelle\}$ to receive the I_{BG} logotypes.
2. The reference set of logotypes $I_{Ref,c}$ (the initial set of logotypes that is updated in the next steps of the algorithm) and the reference set of alts' texts $A_{Ref,c}$ are created. The $I_{Ref,c}$ set consists of two subsets $I_{RefHist,c}$ and $I_{RefPhash,c}$ ($I_{Ref,c} = I_{RefHist,c} \cup I_{RefPhash,c}$ and $|I_{Ref,c}| \ll |I_c|$). In the initial phase (the first phase of the algorithm), we manually selected logotypes that have good properties for the (i) similarity function (the images histogram correlation), and (ii) distance function that is based on pHash (the perceptual hash [5], for each image we compute its hash and after that we use the Hamming distance function to calculate the distance between images). The pre-selected logotypes are put to appropriate subsets that are mentioned above. Also, the $A_{Ref,c}$ set is created manually, i.e. we add to this set the n most frequent alts' texts for the given c.

3. The training dataset D_c is created. The D_c is utilized by the learning method Γ to create the classification function γ_c ($\Gamma : D_c \mapsto \gamma_c$) (the logistic regression method is used to create the classification function [18]). The D_c contains vectors $d = ((f_{simHist}, w_{simHist}), (f_{simPhash}, w_{simPhash}), (f_{isAlt}, w_{isAlt}), (classLabel, \{0,1\}))$ (where f is a feature name, $w \in R$ is a computed weight of the appropriate feature, and $classLabel$ is the label of a class), and it is created in the way explained below:

   3.1. The set of positive training logotype examples $I_{PL,c}$ for the given class c is constructed ($I_{PL,c} = I_c$).

   3.2. The set of negative training logotype examples $I_{NL,c}$ for the given class c is created ($I_{NL,c} = I_{IL} \setminus I_c$).

   3.3. Since sometimes the $|I_{PL,c}| \ll |I_{NL,c}|$, we replicate and add random noise to the images from $I_{PL,c}$ to balance the above set, i.e. to obtain $|I_{PL,c}| \approx |I_{NL,c}|$.

   3.4. The positive training dataset of positive training vectors is created. In our case, $D_P = \{(((f_{simHist}, \arg\max_{rp \in I_{RefHist,c}} simHist(\pi_0 : im \mapsto i_{Rel}, rp)), (f_{simPhash}, \arg\max_{rp \in I_{RefPhash,c}} simPhash(\pi_0 : im \mapsto i_{Rel}, rp)), (f_{isAlt}, isRefAlt(\pi_1 : im \mapsto i_{Alt}, A_{Ref,c})), (classLabel, 1)) : im \in I_{PL,c}\}$. The $simHist$ is a similarity function, i.e. histogram correlation of two images. The images are converted from the RGB color model to the HSV color model. Two channels H and S of the HSV are used to compute the histogram correlation function. The $simPhash$ is a function that computes the distance between two images based on their perceptual hashes. The $isRefAlt$ function returns the 1 if the $A_{Ref,c}$ contains the alt text of the given image im. Otherwise, it returns 0.

   3.5. The negative training dataset of negative training vectors is created. In our case, $D_N = \{(((f_{simHist}, \arg\max_{rp \in I_{RefHist,c}} simHist(\pi_0 : im \mapsto i_{Rel}, rp)), (f_{simPhash}, \arg\max_{rp \in I_{RefPhash,c}} simPhash(\pi_0 : im \mapsto i_{Rel}, rp)), (f_{isAlt}, isRefAlt(\pi_1 : im \mapsto i_{Alt}, A_{Ref,c})), (classLabel, 0)) : im \in I_{NL,c}\}$. The $simHist$, $simPhash$, and $isRefAlt$ function were mentioned above.

   3.6. The $D_c = D_P \cup D_N$ set is created, and this set is split on the 10 folds, i.e. we split D_c set into 10 separate datasets ($D_c = D_{c,1} \cup D_{c,2} \cup \cdots \cup D_{c,10}$ and $D_{c,1} \cap D_{c,2} \cap \cdots \cap D_{c,10} = \emptyset$). In each dataset, we learn single logistic regression classification function $\Gamma_1 : D_{N,1} \mapsto \gamma_1, \cdots, \Gamma_{10} : D_{N,10} \mapsto \gamma_{10}$. When the classification functions are determined, we create a final weighted classification function, i.e. $\gamma_c = \dfrac{\sum\limits_{i=1}^{10} \gamma_i}{10}$. The parameters of the logistic regression classification function are averaged. The averaged parameters create the new parameters of the logistic regression function γ_c.

   3.7. When we have the classification function γ_c, the classification process $\gamma_c : D_c \mapsto R$ is realized. The R is a set of classification results, i.e. the labelled set of images. Based on the R the confusion matrix is created. Thanks to this matrix, we compute the appropriate indicators such as precision, recall, F-measure, accuracy [18].

3.8. We analyse the indicators and the logotypes from the false positive and false negative groups. If the classification results are satisfied (the chosen indicator F-measure has achieved the required value), and we do not create all classifiers, then we return to the first step (the point 1). Otherwise, based on our observation of the false positive and false negative groups, we manually type the new logotypes that would better generalize the groups, i.e. the logotypes that have better properties for function similarities. Next, we put the new chosen logotypes into the reference set $I_{Ref,c}$. The algorithm returns to the 3.4 point and realizes the steps again for the given c.

4. The learned process is ended when all the classification functions are created.

At the end of this description, two things are worth mentioning. We require only 2–3 iteration of learning algorithm for each class (the steps from 3.4 to 3.8 we realize 2–3 times) to achieve well diversified and generalized reference logotypes sets. The applied phase, i.e. the function $featuresVecotrCreator$: $(im, I_{Ref,c}, A_{Ref,c}) \mapsto v_c$ (Fig. 3) uses the final reference sets of logotypes $I_{Ref,c}$ that were constructed in the learning phase described above. Furthermore, the feature vector to classify of an image $im \in I'$ is computed in the same manner as the positive or the negative training vector that was mentioned above (see point 3.4 or 3.5 of the learning procedure described above).

5 Evaluation

5.1 Experiments and Their Results

The evaluation process of the system is composed of three phases, namely (i) dataset construction and assessment, (ii) classification of images originating from the Internet, and (iii) classification of images coming from the domains of innovative companies. For the evaluation of the proposed algorithm the following indicators are used [18]: precision (positive predictive value, PPV), recall (true positive rate, TPR), accuracy (ACC) and F-measure.

In the learning phase, we had the input of a set of logotypes that contained 648 logotypes ($|I_{IL}| = 648$, on average 46 of logotype examples per class). From the learning phase, we received the final reference dataset of logotypes that contained 108 logotypes ($|\bigcup_{c \in C_{IL}} I_{RefPhash,c}| = 55$ logotypes for the pHash, $|\bigcup_{c \in C_{IL}} I_{RefHist,c}| = 53$ logotypes for the Histogram, and both reference sets for each logotype class include an average of 4 logotypes). After multiplication in the learning phase, each category contained on average 10,000 examples of logotypes; however, the number of these examples may vary among categories from about 9,856 to 33,660 except the TNCFRAD logotype, which contains only 31 examples.

Table 1 contains the results of dataset assessment carried out for each category respectively with the use of the system. The assessment depends on proper recognition of images as belonging to a specified category or not. We achieved almost perfect values of precision, recall, accuracy and F1-measure. They are mostly higher than 0.9, and only in two categories the indicators are a little lower, but they are still greater than 0.7. The results achieved on the training set indicate that the logotypes are properly selected and labelled. Thus, it was possible to use these logotype categories, logotype reference sets and the created classification models in the real classification process.

Table 1. The statistics of datasets containing reference logotypes and the indicators of the logotypes classification from the learning phase for each logotypes class ($c \in C_{IL}$).

Set name	Indicators								
-	TP	TN	FP	FN	SUM	PPV	TPR	ACC	F-measure
FD	8,658	6,770	525	481	16,434	0.9428	0.9474	0.9388	0.9451
BG	14,850	11,867	2	0	26,719	0.9999	1	0.9999	0.9999
BC	14,028	10,290	4,666	4,676	33,660	0.7504	0.75	0.7225	0.7502
IT	10,808	8,644	1	0	19,453	0.9999	1	0.9999	0.9999
IE	16,950	13,477	67	0	30,494	0.9961	1	0.9978	0.998
LC	11,514	8,550	654	0	20,718	0.9462	1	0.9684	0.9724
MOSAHE	6,872	5,459	35	0	12,366	0.9949	1	0.9972	0.9975
PAFED	11,008	8,751	34	0	19,793	0.9969	1	0.9983	0.9985
TNCFRAD	20	8	3	0	31	0.8696	1	0.9032	0.9302
PN	11,220	9,079	65	255	20,619	0.9942	0.9778	0.9845	0.9859
EU	10,176	7,947	514	424	19,061	0.9519	0.96	0.9508	0.9559
PO	5,278	4,208	167	203	9,856	0.9693	0.963	0.9624	0.9661
GM	8,170	6,442	89	0	14,701	0.9892	1	0.994	0.9946
IROP	9,516	7,606	6	0	17,128	0.9994	1	0.9996	0.9997

The second phase of experiments involved classification of 24,165 images originating from the Internet into two classes, i.e. an innovative or non-innovative logotype class. We have to mention that there are only 0.8% of images that were labelled as the innovative class. The results are presented in Table 2. There are two separate experiments regarding the use of a *processing* function in the system, which simply splits an image under examination into sub-logotypes. The value true in the column sub-logotypes recognition indicates that the function is used, whereas false means classification without this function. We can assume that the classification quality is satisfactory in case of using the *processing* function because recall and F1-score are higher than 0.7. They are almost equal; thus, we conclude that the classification model is neither under-fitted nor over-fitted. When the *processing* function is switched off, recall and F1-measure are low, which indicates that many innovative logotypes in the dataset were not properly classified. The conclusions are that the process of splitting an image to sub-logos sufficiently improves the system performance, and the system may be used for recognition of images that indicate innovativeness.

Table 2. The statistics of datasets containing classified images and the indicators of the classification from the apply phase, i.e. classification of images into the final logotypes class (c_{Final}) with and without *processing* function.

A	B	C	Indicators								
-	-	-	TP	TN	FP	FN	SUM	PPV	TPR	ACC	F-measure
True	201	23,964	148	23,924	40	53	24,165	0.7872	0.7363	0.9961	0.7609
False	201	23,964	9	23,958	6	192	24,165	0.6	0.04478	0.9918	0.0833

Abbreviations: A - Is sub logotypes recognition; B - the number of the innovative logotypes; C - the number of other images, i.e. not innovative logotypes

The last phase of the experiments tackles classification of images coming from the domains of companies selected on the Internet. According to conclusions from the previous experiment, we use the system with the *processing* function turned on. The results are depicted in Table 3. There are 1,385 domains in the dataset, and 14% of them belong to the categories of innovative domains, i.e. the domains that include at least one innovative logotype. The images are extracted from 1,385 domains of companies, considering documents up to the third level of each domain. Precision, recall and F1-score are equal to about 0.85, which indicates that 85% of the system decisions that the domains include innovative logotypes are correct, and simultaneously 85% of all innovative domains are selected as innovative because they include correctly detected innovative logotype. Yet we have to explain the differences in performance between the second and the third experiment. F1-score in the second test (Table 2) is lower by about 0.1 than in the third test (Table 3). Although in both cases the images are acquired from the Internet, in the second experiment there are images from various types of websites, whereas in the third experiment there were only images acquired and aggregated from the domains of the companies. Thus, the use of different datasets and way to aggregate results and their evaluation are the reasons for these differences.

Table 3. The statistics of the classified domains based on logotypes classification to the final logotypes class (c_{Final}) with the *processing* function.

A	B	Indicators								
-	-	TP	TN	FP	FN	SUM	PPV	TPR	ACC	F-measure
195	1,190	166	1,160	30	29	1,385	0.8469	0.8513	0.9574	0.8491

Abbreviations: A - the number of the domains containing innovative logotypes; B - the number of the domains without innovative logotypes.

6   Conclusion

We elaborated the system for detection of innovative logotypes on the web pages. It covers the supervised and heuristic method of constructing a reference dataset for each logotype category and the logistic regression classifiers that use values of similarity functions as the classifier features.

The experimental results proved that system could detect logotypes indicating innovativeness under varying conditions that occur on the Internet. Its performance is well balanced as it ensures sufficient level of correct decisions that the logotypes are innovative, and at the same time a similar level of selection of innovative images from their whole set.

Moreover, several interesting solutions are worth mentioning. Despite using a relatively small feature space, the system gives satisfactory results. The increase in the number of features would lead to higher computational complexity of the algorithm and consequently may decrease the overall performance. However, it is possible to add more features, as the system is flexible and can utilize new properties that may be distinguished in the future. The regression models are trained by a greedy algorithm, which compares a training example with examples from reference dataset, where the reference collection contains selected logotypes representative for each considered category. Thus, it is possible to calculate the features being inputs of the regression model easily.

On the contrary, the proposed approach has some weaknesses that have to be discussed. It requires some human commitment to select examples in reference sets for each logotype category separately. Some calculation errors may be caused by the low quality of images like small and blurry pictures, huge frames around a logotype, a small number of colors on a white background, appearance of the same logotype category with different colors in their background or with the various shades of a color.

We can use some automatic methods to resolve the problem related to human commitment. We can try to boost the logotypes sets randomly and check the results, or stop iteration if a classifications error is low. Presumably, it is possible to circumvent other problems by improving the system with an additional algorithm recognizing contours of images. Additionally, the classification quality may be enhanced by the application of a focused crawling to locate the areas of a domain, which usually contains logos, for instance, sub-pages containing awards, projects, partners, products, etc. It can be concluded that the proposed system is well suited for the recognition of images of a particular type originating from the Internet domains. The evaluation of this approach indicated well the balanced performance.

References

1. OECD/Eurostat Oslo Manual. http://dx.doi.org/10.1787/9789264013100-en
2. Baratis, E., Petrakis, E., Milios, E.: Automatic website summarization by image content: a case study with logo and trademark images. IEEE Trans. Knowl. Data Eng. **20**(9), 1195–1204 (2008)
3. Boia, R., Florea, C., Florea, L., Dogaru, R.: Logo localization and recognition in natural images using homographic class graphs. Mach. Vis. Appl. **27**(2), 287–301 (2016)
4. Cesarini, F., Francesconi, E., Gori, M., Marinai, S., Sheng, J., Soda, G.: A neural-based architecture for spot-noisy logo recognition. In: Proceedings of the Fourth International Conference 1997 on Document Analysis and Recognition, vol. 1, pp. 175–179 (1997)

5. Christoph, Z.: Implementation and Benchmarking of Perceptual Image Hash Functions, 1st edn. Standard, Cincinnati (2010)
6. Cyganek, B.: Hybrid ensemble of classifiers for logo and trademark symbols recognition. Soft Comput. **19**(12), 3413–3430 (2015)
7. Escalera, S., Fornes, A., Pujol, O., Escudero, A., Radeva, P.: Circular blurred shape model for symbol spotting in documents. In: ICIP (2009)
8. Farajzadeh, N.: Exemplar-based logo and trademark recognition. Mach. Vis. Appl. **26**(6), 791–805 (2015)
9. Fauzi, F., Belkhatir, M.: Image understanding and the web: a state-of-the-art review. J. Intell. Inf. Syst. **43**(2), 271–306 (2014)
10. Kesidis, A., Karatzas, D.: Logo and trademark recognition. In: Doermann, D., Tombre, K. (eds.) Handbook of Document Image Processing and Recognition, pp. 591–646. Springer, Heidelberg (2014)
11. Kotsemir, M.N. Abroskin, A., Meissner, D.: Innovation Concepts and Typology An Evolutionary Discussion. Higher School of Economics Research (2013)
12. Li, Y., Shapiro, L., Bilmes, J.: A generative/discriminative learning algorithm for image classification. In: Computer Vision 2005, ICCV 2005 (2005)
13. Mirończuk, M., Protasiewicz, J.: A diversified classification committee for recognition of innovative internet domains. In: Kozielski, S., Mrozek, D., Kasprowski, P., Małysiak-Mrozek, B., Kostrzewa, D. (eds.) Beyond Databases, Architectures and Structures. Advanced Technologies for Data Mining and Knowledge Discovery. CCIS, vol. 613, pp. 368–383. Springer, Cham (2016)
14. Petrakis, E.G.M., Voutsakis, E., Milios, E.E.: Searching for logo and trademark images on the web. In: Proceedings of the 6th ACM ICIVR, pp. 541–548 (2007)
15. Psyllos, A.P., Anagnostopoulos, C.N.E., Kayafas, E.: Vehicle logo recognition using a SIFT-based enhanced matching scheme. IEEE Trans. Intell. Transp. Syst. **11**(2), 322–328 (2010)
16. Romberg, S., Pueyo, L.G., Lienhart, R., van Zwol, R.: Scalable logo recognition in real-world images. In: Proceedings ICMR, pp. 25:1–25:8 (2011)
17. Rusinol, M., Llados, J.: Logo spotting by a bag-of-words approach for document categorization. In: ICDAR 2009, pp. 111–115, July 2009
18. Sammut, C., Webb, G.I.: Encyclopedia of Machine Learning, 1st edn. Springer Publishing Company, Inc., Heidelberg (2011)
19. Voutsakis, E., Petrakis, E., Milios, E.: Weighted link analysis for logo and trademark image retrieval on the web. In: Web Intelligence, pp. 581–585 (2005)
20. Wang, F., Qi, S., Gao, G., Zhao, S., Wang, X.: Logo information recognition in large-scale social media data. Multimedia Syst. **22**(1), 63–73 (2016)
21. Wang, H., Chen, Y.: Logo detection in document images based on boundary extension of feature rectangles. In: Document Analysis and Recognition 2009, pp. 1335–1339, July 2009

Extraction and Interpretation of Textual Data from Czech Insolvency Proceedings

Iveta Mrázová[✉] and Peter Zvirinský

Department of Theoretical Computer Science and Mathematical Logic,
Faculty of Mathematics and Physics, Charles University,
Prague, Czech Republic
iveta.mrazova@mff.cuni.cz, peter.zvirinsky@gmail.com

Abstract. Recently, the Czech Insolvency Register covers about 200000 insolvency proceedings. In order to better assess the real impact of indebtedness across the Czech society, the data about creditors or reasons for debt might be of great value. Unfortunately, the vast majority of such information is contained only in scanned document copies attached to the insolvency proceedings. Therefore, this study aims at finding efficient pre-processing, clustering and classification techniques capable of extracting the wanted information from these cca 1200000 pdf-files.

Keywords: Data pre-processing · Text processing · Classification · Clustering · Knowledge extraction · Semantics assignment

1 Introduction

On 1 January 2008, the Czech government launched a new information system called Insolvency Register of the Czech Republic (IR) that is publicly accessible at http://isir.justice.cz/. Recently, IR covers a detailed up-to-date information about 200000 insolvency proceedings (IPs). The data stored in IR comprises the information about the debtor such as his/her name, domicile, birth certificate number, etc. If the debtor is a legal entity, the company name, registered office and identification number are added. Further information includes the number of judicial senate handling the IP and a list of its administrators. The creditors involved in the IP are available only since October 2011, which means that creditors' names are missing for most IPs commenced between 2008 and 2011.

Our previous research [9] has been focused mainly on processing structured entries of the IPs that are easier to handle and comprehend. The vast majority of the data remains, however, available only in the form of several scanned document copies attached to each respective IP (i.e., approximately 1200000 of such documents in all). In particular this unstructured data represents often an

The first author was partially supported by the Czech Science Foundation under Grant No. 15-04960S. The second author was partially supported by the Charles University, project GA UK No. 120616.

L. Rutkowski et al. (Eds.): ICAISC 2017, Part II, LNAI 10246, pp. 116–125, 2017.
DOI: 10.1007/978-3-319-59060-8_12

invaluable source of important information [3]. The specific information we are looking for in this study are the names of the creditors involved in the insolvencies and even more importantly the reasons for the debt at hand. In this respect, efficient clustering and classification techniques could help us identify the circumstances influencing both the indebtedness and its possible solutions.

Our goal is thus first to extract accurately textual data from scanned pdf-files containing the so-called applications of receivables (AR). Afterwards, we will train a classifier that is able to assign actual creditors to pre-processed ARs commenced after October 2011. The trained classifier shall then be applied to find the missing creditors' names also for IPs commenced before October 2011.

With regard to huge amounts of data to be processed quickly, we will prefer computationally more efficient classifiers like, e.g., Naive Bayesian classifier, logistic regression or support vector machines. In order to better understand the overall indebtedness structure of the Czech society, we will finally aim at identifying the main reasons for debt by grouping ARs with similar debt origins together. In this case, the main problems refer to an appropriate assessment of the right number of data clusters necessary (cluster validity) and an adequate representation of the found characteristics (semantics assignment).

2 Related Works

In order to extract textual information from the scanned ARs, the documents shall be first properly pre-processed. Failure to, e.g., correct the document skew, can namely impact a serious performance degradation. For skew detection, various methods based on the Hough transform [4] have been developed. Their basic idea consists in finding reference lines in the image and using them to calculate the skew angle of the document. Afterwards, the image is rotated and aligned in the horizontal manner.

When detecting lines in images, the original Hough algorithm assumes that each of their points can lie on an infinite number of lines. A computationally more efficient variant with negligible performance degradation represents the so-called Probabilistic Hough transform [6] that involves only a small subset of the points selected at random. After pre-processing, optical character recognition (OCR) can be applied as usual.

In computational linguistics, the so-called n-grams are used to model natural language based on statistical properties of the text. An n-gram is a consecutive sequence of n items that might occur in a given type of text. The items can be letters, words but also white space and punctuation according to the application. Each of the n-grams corresponds to an attribute weighted according to the TF-IDF scheme in the so-called vector space model [8]. TF stands for the term frequency and IDF for the inverse document frequency.

2.1 Document Classification

Given an example vector $v = (v_1, ..., v_{|V|})$ with $|V|$ being the vocabulary size of the document collection, Naive Bayesian classifier (NBC, [8]) calculates conditional probabilities for all possible classes m_0 and then chooses that class c^* yielding maximum posterior probability $P(C_{c^*}|\boldsymbol{v})$. Assuming conditional independence of attribute values v_i given the class C_c, the formula for NBC sounds:

$$P(C_c|\boldsymbol{v}) \;=\; \frac{P(C_c)}{P(\boldsymbol{v})} \prod_{i=0}^{a} P(v_i|C_c). \tag{1}$$

The task of the NBC learning algorithm is thus to estimate prior class probabilities $P(C_c), c = 1, ..., m_0$, and conditional class probabilities $P(v_i|C_c)$ for the values v_i of the a attributes $A_i, i = 1, ..., a$.

Standard logistic regression (LR, [16]) is a classification method used for two-class problems. The classifier defined by the weight vector $\boldsymbol{w}$ predicts the likelihood of the classes based on the presented vector $\boldsymbol{v}$. To maximize the sample likelihood, gradient descent method can be utilized. For the sigmoid function $y = P(C = 1|\boldsymbol{v}) = 1/(1+exp(-\boldsymbol{w}^T\boldsymbol{v}))$, the weight vector update $\Delta\boldsymbol{w}$ corresponds to $\eta(d-y)\boldsymbol{v}$, where d stands for the correct class label and η for the learning rates. After the weights $\boldsymbol{w}$ are found, LR is used for classification so that the class label $C_c, c \in \{0, 1\}$ with the highest likelihood $P(C_c)/(1 - P(C_c))$ is predicted.

Support vector machines (SVM, [14]) aim at finding an optimal class separating hyperplane in the attribute space. An optimal hyperplane is equally distant from the nearest examples out of both classes called support vectors. Classes that cannot be linearly separated in the original space, may, however, become linearly separable in a higher-dimensional attribute space. Yet for the so-called linear kernel SVMs, even the original attribute space remains preserved. In the case of multiclass classification, the one-against-rest approach can be taken, where m_0 different binary classifiers are formed, each one of them for one particular task.

2.2   Clustering of Text Documents

To reveal the intrinsic structure of high-dimensional data, various clustering techniques can be applied [1]. In principle, the clusters are formed such that objects from the same cluster are very similar one to each other, and objects from different clusters are very distinct. The k-means clustering, for example, groups the data into k mutually exclusive clusters. Each of the k clusters in the partition is defined by its centroid, i.e., the point to which the sum of Euclidean distances from all objects in that cluster is minimized.

The so-called Self-organizing feature map (SOFM, [7]) is a neural network that learns to capture both the distribution and topology of the input data. Its output neurons are arranged on a topological grid (usually a rectangular one). Initial weights of a trained SOFM are spread across the input space. After presenting the network with the actual input pattern vector $\boldsymbol{x}(t)$ at time t, the SOFM identifies the winning neuron i^* that has the shortest distance to $\boldsymbol{x}(t)$. Next, the weights $\boldsymbol{w}_i$ of all the neurons i within a certain neighborhood $N(i^*)$ of the winning neuron are adjusted according to:

$$\boldsymbol{w}_i(t) \;=\; \boldsymbol{w}_i(t-1) \;+\; \alpha(t) \; (\; \boldsymbol{x}(t) \;-\; \boldsymbol{w}_i(t-1) \;). \tag{2}$$

Both the learning rates $\alpha(t)$ and the size of $N(i^*)$ decrease during training.

Hierarchical clustering groups the data by creating a multilevel hierarchy, where clusters at one level are merged together at the next level [15]. Agglomerative clustering first assesses the similarity between every pair of objects in the data set as their mutual distance. Afterwards, it groups together similar objects into a binary, hierarchical cluster tree called dendrogram. Finally, it is necessary to determine where to cut the formed tree into clusters and assign all the objects below each respective cut to a single cluster.

2.3   Cluster Validity and Semantics Assignment

The optimum number of clusters to partition the considered data can be determined by increasing step by step the number of clusters and evaluating cluster validity for each respective clustering. Most of the known validity measures are based on the ratio between the so-called separation of the clusters and their compactness. The Dunn index [5] can be, for example, calculated as:

$$D = \frac{\min_{1 \leq i < j \leq n} d(i,j)}{\max_{1 \leq k \leq n} d'(k)} \tag{3}$$

for each respective cluster partition, where $d(i,j)$ represents the distance between the clusters i and j, and $d'(k)$ measures the intra-cluster distance of cluster k. $d(i,j)$ can be determined, e.g., as the distance between the centroids of the clusters, and $d'(k)$ as the maximum distance between any pair of the elements from the cluster k. Higher values of the Dunn index indicate a better clustering.

The so-called Silhouette measure [12] also relates separation to compactness, but unlike Dunn by subtraction rather than division. Silhouette is based on the mean score $1/N \sum_{i=0}^{N} s_{x_i}$ over all N patterns x_i from the data set. Each pattern's individual score s_{x_i} evaluates the difference between the minimum $b_{q,i} = \min d_{q,i}$ average distance $d_{q,i}$ between the vector $x_i \in p$ and patterns from another cluster q and the average distance $a_{p,i}$ between vector x_i and every other pattern from cluster p. This difference is then divided by a normalizing term, which is the bigger one of the two averages $a_{p,i}$ and $b_{p,i}$: $s_{x_i} = (b_{q,i} - a_{p,i}) / \max\{a_{p,i}, b_{p,i}\}$. As clustering improves, the score will approach 1.

Semantics can be assigned to the found high-dimensional cluster representation of the data by identifying its most discriminative features. For categorical variables, e.g., the χ^2-statistics, can be used for this purpose [1]. Let $v_1, ... v_r$ be the r possible values of a particular attribute, and let a_{ij} be the number of data points containing the attribute value v_i that belong to cluster $j \in \{1, .., k\}$. Further, let n_i be the number of data points that take on the value v_i for the attribute. Then, for a data set containing n data points, the χ^2-statistics for the considered attribute is defined with:

$$\sum_{i=1}^{r} n_i = n \quad \text{and} \quad P_i = \frac{n_i}{n} \quad \text{as} \quad \chi^2 = \sum_{i=1}^{r} \sum_{j=1}^{k} \frac{(a_{ij} - P_i \sum_{i=1}^{r} a_{ij})^2}{P_i \sum_{i=1}^{r} a_{ij}} \tag{4}$$

Higher values of the χ^2-statistics indicate stronger cluster characteristics. Other applicable measures include, e.g., the Gini index or information gain.

3   Document Pre-processing

As of today, IR comprises about 1.2 million of scanned ARs attached to the stored IPs as pdf-documents. Due to a varying quality of the scanned documents, a lot of important information might get lost during the actual text mining. To retain as much information as possible, our aim will be to enhance the reliability of text extraction in particular. The process of automatic text extraction called optical character recognition (OCR) consists of three main steps: image preprocessing, image segmentation and character recognition.

The free Tesseract OCR software [10] we use in our study provides, unfortunately, only limited segmentation capabilities, especially when it concerns documents with lines, forms or tables. Therefore, we have used ImageMagick [13] to remove the background and deskew the image. Afterwards, OpenCV [2] was used to automatically detect and remove the lines by means of the so-called Probabilistic Hough Transform. The original document and its cleaned preprocessed version can be seen in Fig. 1(a) and (b) below. Finally, the Tesseract OCR can be used to extract the text from the preprocessed documents. Figure 1(c) clearly illustrates the extent to which preprocessing rises the quality of OCR.

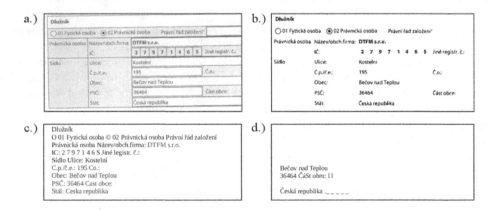

Fig. 1. OCR process: (a) original (slightly skewed) document, (b) pre-processed document, (c) text extracted from the pre-processed document (b), (d) text extracted from the original document (a).

To compare the quality, price and speed of the chosen free OCR software with commercial solutions, we resorted to Abby Fine Reader[1] for the tests. Abby Fine Reader belongs to the best on the market and does not require any advanced pre-processing. Its trial limit comprises 50 documents. For 50 randomly selected documents we thus checked if the involved software correctly extracted the names of the creditors and the value of ARs. Our approach built on free software was

---

[1] http://www.abbyy.com/finereader/.

Table 1. 10-fold CV results along with 95% confidence interval: Recall is a relative frequency of correctly classified positive examples. Precision estimates the proportion of correctly classified examples that were classified as positive. The F-measure is defined as $F = 2 \times Recall \times Precision/(Recall + Precision)$. For the tested creditor classifiers, the shown values correspond to the average over 20 classes and 10 folds.

Classifier	Recall	Precision	F-measure	Training(s)
NBC	0.974 ± 0.006	0.958 ± 0.002	0.955 ± 0.002	0.862
LR	$\mathbf{0.994 \pm 0.001}$	$\mathbf{0.994 \pm 0.001}$	$\mathbf{0.994 \pm 0.001}$	189.392
SVM (lin. kernel)	$\mathbf{0.994 \pm 0.001}$	$\mathbf{0.994 \pm 0.001}$	$\mathbf{0.994 \pm 0.001}$	**13.446**

able to yield both information correctly for 42 (84%) documents. Similarly, Abby Fine Reader managed to extract this data in 41 (82%) cases.

Altogether, more than 1200000 documents had to be processed in reasonable time, ideally in parallel. With the multi-core support, the Abby Fine Reader license costs more than $10000 (and about $2500 without it). When using our approach that involves free software tools, it takes on average 28 s to process a single document whereas 10 seconds are required by Abby Fine Reader. As a single document contains on average 6.5 pages, the overall processor run time necessary to process all documents is approximately $1200000 \times 6.5 \times 10\,s =$ 903 days, i.e., almost 2.5 years with Abby Fine Reader. With our solution, this time would be approximately $1200000 \times 6.5 \times 28\,s = 2528$ days.

To further speed up the processing, we re-implemented our solution so that it could process the documents in parallel. As our solution is based on free software only, it is possible to distribute it on multiple machines without any restrictions (this is, however, not possible with commercial tools). In this way, we were able to gain an even more significant time reduction. To benefit from these advantages, we harvested the cheapest computational power provided by Amazon Web Services, called Spot Instances. The biggest instances can be rented for cca $0.2 per hour and provide 32 cores for parallel execution. At the peak of our processing we were running our software on 4 such instances, i.e., processing documents on $4 \times 32 = 128$ cores. This kind of massive parallelization allowed us to reduce the time required for processing of all 1200000 documents from 2528 days to only 20 days (and $384).

4   Supporting Experiments

Within the framework of the further IP data analysis, we will be interested in information contained in the attached pdf-documents. In our opinion, especially this unstructured data could help us answer the following two questions:

1. Is it possible to deduce from the processed data even information not explicitly available, e.g., on creditors' names, and how reliable is such an inference?
2. What circumstances cause indebtedness and how much does its overall structure vary across the Czech society?

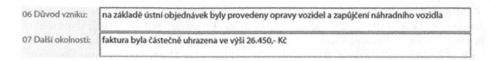

Fig. 2. An example of stated debt origin (line 06 Reason of origin: based on verbal orders, car repairs were done and a replacement vehicle was provided; line 07 Further circumstances: the invoice has been partially paid by the amount of 26.450,- Kč).

4.1   The First Set of Experiments - Inference of Creditors' Names

The information about the creditors is explicitly stated only in the ARs submitted after October 2011. For IPs commenced before ($\sim$ 270000 documents), we wanted to infer the names of the creditors by means of a classifier trained on later documents where the creditors are explicitly known and the documents were pre-processed in a similar way.

Further, we will consider only 19 most frequent creditors as class labels. For each of them, the training set contains 1000 randomly sampled AR-documents. The class of all other unknown creditors will be represented by additional 2000 randomly selected AR-documents. Altogether, the training set consists of 21000 sample documents out of 20 classes. All the letters were converted to lower case and all (Czech) stop-words and diacritical marks were removed. Afterwards, we transformed the text documents to a bag-of-words n-gram model ($2 \leq n \leq 3$) with each of the 41 937 resulting n-grams weighted by their TF-IDF score.

The above data set was used to train an NBC, an LR model and a linear kernel SVM. When first approaching this task with a significantly worse pre-processing [9], we obtained an overall accuracy of about 96.5% for LR. With improved pre-processing, the achieved precision was almost 3% higher, i.e., 99.4% for both the SVM and LR. SVM was, however, 14 times quicker than LR, see Table 1. For training, we used the scikit-learn and scipy python libraries [11] on Intel Core i7 920 (8 M Cache, 2.66 GHz) with 6x RAM 4096 MB DDR3.

4.2   The Second Set of Experiments - Debt Structure and Origin

For this experiment, we randomly selected 100 000 ($\sim$ 8.3% of all) ARs. Each scanned AR comprises a unique field that loosely describes the reason for the

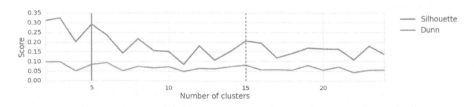

Fig. 3. Evaluation of the (normalized) Dunn and Silhouette indicators: an appropriate number of clusters has been set to 15, where both indicators reach local maximum.

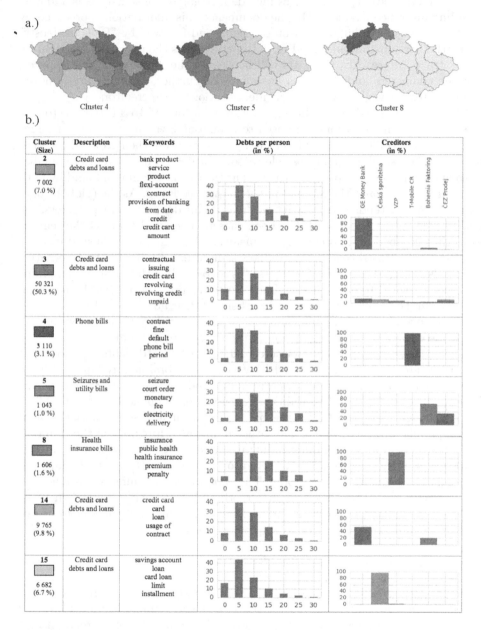

Fig. 4. An overview of the clustering results including the cluster identifiers and their brief (prototype) description by a set of keywords. The maps in Fig. (a) illustrate the distribution of cluster members over the country. Further, the percentage of debtors with the corresponding number of ARs who have at least one of their ARs in the considered cluster as well as the main creditors involved are shown in Fig. (b).

claimed receivables. Typical causes include mortgages, consumer loans and out-standing insurance payments. In the documents, this information is preceded by the text "Duvod vzniku" ($\sim$ reason of origin) and followed by "Další okolnosti" ($\sim$ further circumstances) – see Fig. 2. In between lies the text on debt origin we will extract. Afterwards, the texts will be cleaned, i.e., redundant white spaces, newline characters and diacritics are removed. Further, all digits are removed from the text as well as any explicit identification of the creditors.

The extracted strings of characters were transformed to TF-IDF vectors of the size 5000. These vectors were used to train a SOFM with 900 neurons organized into a 30 × 30 grid. The found weight vectors were then grouped together based on their similarities in possible reasons for debt. The appropriate number of such higher-level groups was estimated as the (local) maximum found by both the Dunn method and the Silhouette coefficient, see Fig. 3. As the (global) maximum equal to 2 tends to group distinct reasons for debt into larger clusters, the next viable option equal to 15 has been chosen. The weights of SOFM neurons were thus grouped into 15 clusters by means of Agglomerative Clustering.

To interpret the revealed indebtedness structure, a small set of representative keywords selected by means of the χ^2-statistics over all of the considered n-grams has been assigned to each cluster. Each cluster can be also labeled by a typical debt origin. For 7 selected clusters, Fig. 4 shows their absolute and relative size, assigned semantics and the first few representative creditors involved with both the keywords and a brief description of the cluster (prototype).

The biggest Cluster 3 is characterized by credit card debts and loans. It comprises around 50% of ARs filed by various creditors, mainly financial institutions. The remaining 14 clusters are smaller and correspond to rather specific reasons of debt. Cluster 4 contains, e.g., outstanding phone bills for T-Mobile ČR. Although T-Mobile ČR is the biggest operator and serves about 60% of Czech population, these ARs come rather from Eastern regions of the country.

Cluster 5 includes both outstanding electricity bills and seizures from western regions of the country. Typical creditors for these ARs are the companies Bohemia Faktoring and ČEZ Prodej. The debtors tend to suffer under a heavier load of ARs more than is common for other clusters. ARs grouped in Cluster 8 are quite specific for outstanding health insurance bills mainly from two northwestern regions with high unemployment rates.

5   Conclusions

Our long-term goal in the analysis of Czech IP data consists in understanding the structure and role of mutual relationships formed among subjects involved in IPs. Our past studies have already revealed several interesting relationships among the debtors, creditors, judicial senates and insolvency administrators. Unfortunately, a lot of information is attached to the IPs in an unstructured form as scanned AR-documents only. For this reason, both advanced OCR and reliable image processing techniques are beneficial. An improvement of 3% yields, namely, 8 thousand new creditor names in the context of 270 thousand ARs commenced before 2011. The main contribution of this paper thus consists in:

– proposing and implementing an OCR solution that matches commercial tools in terms of time and quality. A convenient combination of appropriate tools available for free and cheap computational means thus allowed for an extremely efficient pre-processing and analysis of millions of documents;
– extracting important information from the obtained (unstructured) texts such as missing creditor names with supervised machine learning techniques and segmentation of debts based on their origin with unsupervised training;
– semantics assignment based on the found (prototype) ARs. The found results confirmed that indebtedness structure varies across the country and may include region-specific reasons for debt.

Quite naturally, also the information on the amount of the claimed debt would add a lot to the analysis of the underlying social structure. Of a similar value might be data from various other publicly accessible resources like the value of the owned real estates, company shares, etc., too. We plan to deal with these issues within the framework of our further research.

References

1. Aggarwal, C.C.: Data Mining: The Textbook. Springer, Berlin (2015)
2. Bradski, G., Kaehler, A.: Learning OpenCV. O'Reilly, Sebastopol (2008)
3. Chen, C.L.P., Zhang, C.Y.: Data-intensive applications, challenges, techniques and technologies. Inf. Sci. **275**, 314–347 (2014)
4. Duda, R.O., Hart, P.E.: Use of the Hough transformation to detect lines and curves in pictures. Commun. ACM **15**, 11–15 (1972)
5. Dunn, J.C.: Well-separated clusters and optimal fuzzy partitions. J. Cybern. **4**(1), 95–104 (1974)
6. Kiryati, N., Eldar, Y., Bruckstein, A.M.: A probabilistic Hough transform. Pattern Recogn. **24**, 303–316 (1991)
7. Kohonen, T.: Self-Organizing Maps. Springer, Berlin (2001)
8. Liu, B.: Web Data Mining: Exploring Hyperlinks, Contents, and Usage Data. Springer, Berlin (2007)
9. Mrázová, I., Zvirinský, P.: Czech insolvency proceedings data: social network analysis. Procedia Comput. Sci. **61**, 52–59 (2015)
10. Patel, C., Patel, A.: Optical character recognition by open source OCR tool tesseract: a case study. Int. J. Comput. Appl. **55**, 50–56 (2012)
11. Pedregosa, F., et al.: Scikit-learn: machine learning in Python. JMLR **12**, 2825–2830 (2011)
12. Rousseeuw, P.J.: Silhouettes: a graphic aid to the interpretation and validation of cluster analysis. J. Comput. Appl. Math. **20**(1), 53–65 (1987)
13. Still, M.: The Definitive Guide to ImageMagick. Apress, Berkeley (2005)
14. Vapnik, V.: The Nature of Statistical Learning Theory. Springer, Berlin (1995)
15. Vesanto, J., Alhoniemi, E.: Clustering of the self-organizing map. IEEE Trans. Neural Netw. **11**, 586–600 (2000)
16. Zhang, T., Oles, F.: Text categorization based on regularized linear classifiers. Inf. Retrieval **4**, 5–31 (2001)

Spectral Clustering for Cell Formation with Minimum Dissimilarities Distance

Yessica Nataliani[1,2] and Miin-Shen Yang[1(✉)]

[1] Department of Applied Mathematics,
Chung Yuan Christian University, Chung-Li 32023, Taiwan
msyang@math.cycu.edu.tw
[2] Department of Information Systems,
Satya Wacana Christian University, Salatiga 50711, Indonesia

Abstract. Group Technology (GT) is a useful tool in manufacturing systems. Cell formation (CF) is a part of a cellular manufacturing system that is the implementation of GT. It is used in designing cellular manufacturing systems using the similarities between parts in relation to machines so that it can identify part families and machine groups. Spectral clustering had been applied in CF, but, there are still several drawbacks to these spectral clustering approaches. One of them is how to get an optimal number of clusters/cells. To address this concern, we propose a spectral clustering algorithm for machine-part CF using minimum dissimilarities distance. Some experimental examples are used to illustrate its efficiency. In summary, the proposed algorithm has better efficiency to be used in CF with a wide variety of machine/part matrices.

Keywords: Spectral clustering · Group technology · Cell formation · Minimum dissimilarity · Number of cells · Grouping efficacy

1 Introduction

Clustering is a method for finding the cluster structure of a data set such that objects within the same cluster demonstrate maximum similarity and objects within different clusters demonstrate maximum dissimilarity [5, 6]. There are many clustering methods proposed in the literature, especially for partitional methods that suppose the data set represented by finite cluster prototypes with their own objective functions [7, 13, 19, 21]. Recently, spectral clustering becomes more popular [17]. The spectral clustering method uses graph construction to represent data points with their similarities, where data points are represented by nodes, while similarities between points are represented as weights of edge between nodes. Spectral clustering had been widely applied in various areas (see [9, 10, 15, 17]).

Most manufacturing industries face more competition for them in the pursuit of profit. High manufacturing profit comes from lower cost and higher product quality where various kinds of machines with complicated work procedures are needed for finishing the manufacturing. Group technology (GT) has been a useful tool for creating a more flexible manufacturing process. The GT can generally divide the manufacturing facilities into small groups, called *cells*, such that the similar things can be done similarly. By dividing into cells, it can exploit similarities between components with lower cost and

© Springer International Publishing AG 2017
L. Rutkowski et al. (Eds.): ICAISC 2017, Part II, LNAI 10246, pp. 126–136, 2017.
DOI: 10.1007/978-3-319-59060-8_13

achieve more productivity with high quality. Cell formation (CF) is a key step in GT. It is used to design a good cellular manufacturing system by grouping parts and machines to have manufacturing units (cells). There are many CF methods proposed in the literature (see [2, 14, 16, 20]). Oliveira et al. [12] applied spectral clustering to solve CF problem based on the min-max cut for partitioning the parts and machines into groups to obtain manufacturing cells. In their algorithm, the partition is done recursively until a given number of cells obtained. The common problem in clustering is how to determine an optimal number of clusters, as well as in CF.

There are two ways to get an optimal number of cells. First, by comparing performance measurement (e.g., grouping efficiency, grouping efficacy, global efficiency, bond efficiency, etc.) for some number of cells and then choose the largest value to decide an optimal number of cells. The other one is to use a predefined (maximum) number of cells, then repeat a process with different number of cells (less than the predefined number), until the optimal solution is found. In this paper, we will focus on how to get an optimal number of cells by proposing an algorithm to solve the CF problem using minimum dissimilarity distance. Using part-machine matrix, we first get an optimal number of cells and part families (machine groups). We then apply spectral clustering using the cell number obtained from the first step to get the machine groups (part families). The remainder of this paper is organized as follows. In Sect. 2, we provide the background on spectral clustering, CF, and p-median problem. Section 3 describes our proposed spectral clustering for CF with minimum dissimilarity distance algorithm. Some experimental examples and the comparison of the proposed algorithm result to the original result are presented in Sect. 4. Finally, Sect. 5 concludes this paper.

2 Spectral Clustering and Cell Formation

2.1 Spectral Clustering

Let $X = \{x_1, x_2, \cdots, x_n\}$ be a set of n data points in a d-dimensional space. Clustering of X is to partition the data set X into k subsets that can well represent the data structure of X. Intuitively, points and similarities between points can be represented as graph. The n data points are represented as n nodes and an edge that connects two nodes is weighted by s_{ij}, where $s_{ij} = s(x_i, x_j)$ is the similarity measure between nodes x_i and x_j. Note that similarity matrix $S = (s_{ij})_{n \times n}$ is symmetric and non-negative. Thus, we can say that a clustering problem has the same meaning with partitioning graph, i.e., grouping nodes such that edges in the same cluster have high weights, but edges between different clusters have low weights. This clustering is known as spectral clustering (see [9, 11, 15, 17]). In this paper, we follow spectral clustering algorithm proposed by Ng et al. [11].

2.2 Cell Formation

Cellular manufacturing (CM) is an application of group technology (GT) that identifies the similarity of parts by grouping them into one cell [16]. The design of a CM system is called CF. In CF, part/machine matrix is represented by binary matrix [parts × machine]. The rows indicate p parts and the columns indicate m machines. Each element in the matrix represents a relationship between parts and machines, where '1' ('0') indicates part p_i is (not) carried out by machine m_j. The objective is to group parts and machines in cells based on their similarities. Figure 1(a) shows an example of [parts × machines] where the results obtained by CF are shown in Fig. 1(b).

MACHINE

		1	2	3	4	5	6
P	1	1	0	1	0	0	0
A	2	0	1	0	0	0	0
R	3	1	1	0	0	1	0
T	4	0	1	0	0	0	1
	5	0	0	0	1	0	1

(a) Original

MACHINE

		2	1	5	3	6	4
P	4	1	0	0	0	1	0
A	2	1	0	0	0	0	0
R	1	0	1	0	1	0	0
T	3	1	1	1	0	0	0
	5	0	0	0	0	1	1

(b) Clustering result

Fig. 1. Matrix of [parts × machines]

A '1' outside the diagonal block is called an exceptional element. A part which has an exceptional element is called an exceptional part, because it works on two or more machine groups. A machine is called a bottleneck machine if it process two or more part families. A '0' inside the diagonal block is called void. From Fig. 1(b), the part 3 is called an exceptional part, because it works on the machine 2 and machine group of {1, 3, 5}. The machine 2 is called a bottleneck machine because it process part 3 and part family of {2, 4}. In general, an optimal result for a machine/part matrix by a CF clustering method is desired to satisfy the following two conditions:

(1) To minimize the number of 0 s inside the diagonal blocks (i.e., voids);
(2) To minimize the number of 1 s outside the diagonal blocks (i.e., exceptional elements). Oliveira et al. [12] proposed a spectral clustering algorithm for CF. They used bipartite graph cut on two disjoint components, i.e. parts and machines, and recursively bipartition the graph until the given number of cell is reached.

2.3   A *p*-Median Problem

A *p*-median problem (PMP) is generally used for finding the location of *p* facilities (warehouses) on a network, such that the total distance of serving all demands can be minimized [3]. This problem can be applied to cell formation. Goldengorin and Krushinsky [1] proposed mixed Boolean pseudo-Boolean formulation (MBpBM) to apply PMP in cell formation. The goal is to locate machines that process the same parts to be closed to each other, such that the time spent by parts on moving from one machine to another will be reduced. They use commonality score proposed by Wei and Kern [18] to construct the distance of part and machine (dissimilarity measure) with

$$d(i,j) = r(r-1) \sum_{k=1}^{r} (a_{ik}, a_{jk}) \quad \text{where} \quad \Gamma(a_{ik}, a_{jk}) = \begin{cases} r-1, & \text{if } a_{ik} = a_{jk} = 1 \\ 1, & \text{if } a_{ik} = a_{jk} = 0 \,. \\ 0, & \text{if } a_{ik} \neq a_{jk} \end{cases}$$

3   The Proposed Algorithm

Based on the p-median problem and commonality score, we propose a new algorithm using dissimilarity measures to get the number of cells directly, and then solve the CF problem. The proposed algorithm can be described as follows:

Algorithm Spectral clustering for CF with minimum dissimilarity distance
Input: [parts x machines]
Step 1

1. Construct a pre-dissimilarity matrix

   $$P_i = (a_{i1}, a_{i2}, \hbar, a_{im})$$
   $$P_j = (a_{j1}, a_{j2}, \hbar, a_{jm})$$, where $a_{ij} = \begin{cases} 0, & \text{if part } p_i \text{ requires machine } m_j \\ 1, & \text{otherwise} \end{cases}$

2. Construct a dissimilarity matrix of parts, defined by $d(P_i, P_j) = \sum_{k=1}^{m} u(a_{ik}, a_{jk})$

   where $i, j = 1, \hbar, p$ and $u(a_{ik}, a_{jk}) = \begin{cases} 1.1, & \text{if } a_{ik} \neq a_{jk} \\ 0, & \text{if } a_{ik} = a_{jk} = 0 \\ -1, & \text{if } a_{ik} = a_{jk} = 1 \end{cases}$.

   Note that $u(a_{ik}, a_{jk})$ for $a_{ik} \neq a_{jk}$ must be a little higher than $|u(a_{ik}, a_{jk})|$ for $a_{ik} = a_{jk} = 1$, to avoid the elimination of calculation.

3. Construct $\text{min\_diss}(j) = \min_{j} d(P_i, P_j)$, $i = 1, \hbar, p$.

4. FOR $i = 1$ to $p - 1$ do

   FOR $j = i + 1$ to p do

   P_i will be put in the same group with P_j if $d(P_i, P_j) = \text{min\_diss}(j)$.

   ENDFOR
   ENDFOR

5. Group G_k will be merged with group G_l if G_k has the same elements at least half from the number of elements of G_l, $k = 1, \hbar, p-1, l = k+1, \hbar, p$.

6. If P_i is included in more than one group ($G_1, \hbar, G_n$), choose group with

   $\min(\text{min\_diss}(G_k))$, $k = 1, \hbar, n$ to put P_i.

7. We have that
   a. number of cells (i.e., number of different groups that have been constructed).
   b. part families (i.e., different groups that have been constructed).

Step 2
   1. Use a spectral clustering algorithm to cluster machines with the number of parts that have been obtained from Step 1.
   2. Assign each part families with each machine groups.

Step 3 Apply Step 1 and Step 2 for machine groups.
Step 4 If the part-machine cell results between two cases are different, compare the group efficacy of those cases. The result with larger group efficacy is the obtained result.

Output: number of cells and part-machine cells.

4 Experimental Results

In this section, we present some examples to validate the proposed algorithm, where all matrices presented in the numerical examples are [parts × machines]. We only present '1' for simplicity (i.e. the empty cells have value '0'). Here, we have two cases, (1) apply the minimum dissimilarity for parts, then spectral clustering for machines; (2) apply the minimum dissimilarity for machines, then spectral clustering for parts. If different numbers of cells are obtained from the two cases, then in order to choose the optimal result, we use grouping efficacy μ, proposed by Kumar and Chandrasekaran [8], as follows: $\mu = \frac{o-e}{o+v}$ here o is number of 1 s in the part/machine matrix; e is number of 1 s outside the diagonal block (i.e., exceptional elements); v is number of 0 s in the diagonal block (i.e., void).

Example 1. We consider a real application from Yang et al. [22] to the production of cast aluminum alloys and forging steels based on eight machines process. There are seven parts, i.e., ZL207, 50SiMn, 35SiMn, ZL205A, ZL402, 42SiMn, and ZL202, and eight machines, i.e., fusion, air impermeability, tensile test, casting, impact test, crack-arresting feature, heat treatment, and cooling. Figure 2(a) shows the part-machine matrix for that production where N: non-responsive; S: sand-casting; J: alloy-casting; T: tempering; OC: oil-cooled. We set the value 1 if the part is processed by the machine and 0 otherwise, as shown in Fig. 2(b).

		MACHINE							
		1	2	3	4	5	6	7	8
P	ZL207 (1)	603-637	N	N	S	N	N	N	N
A	50SiMn (2)	N	N	14	N	39	N	T	OC
R	35SiMn (3)	N	N	14	N	39	N	T	OC
T	ZL205A (4)	544-633	N	3	S	N	N	N	N
	ZL402 (5)	N	[40,70]	4	S	N	[20,50]	N	N
	42SiMn (6)	N	N	14	N	29	N	T	OC
	ZL202 (7)	N	[60,90]	N	S, J	N	[20,50]	N	N

(a)

		MACHINE							
		1	2	3	4	5	6	7	8
P	1	1			1				
A	2			1		1		1	1
R	3			1		1		1	1
T	4	1		1	1				
	5		1	1	1		1		
	6			1		1		1	1
	7		1		1		1		

(b)

Fig. 2. (a) Part-machine matrix input; (b) Binary part-machine matrix corresponding to (a)

After implementing the proposed algorithm as shown in Table 1, the part-machine outputs are shown in Fig. 3. By comparing the two cases, the grouping efficacy of the clustering result from Result 1 (three cells) is 83.33%, while from Result 2 (two cells) is 73.33%, and so we use three cells for the solution. Thus, the three group parts are {ZL402, ZL202}, {ZL207, ZL205}, {50SiMn, 35SiMn, 42SiMn} and the three group machines are {2, 6}, {1, 4}, {3, 5, 7, 8}. As shown in Fig. 4, these three cells are:

Cell 1: part {ZL402, ZL202} with machine {2, 6}; Cell 2: part {ZL207, ZL205} with machine {1, 4}; Cell 3: part {50SiMn, 35SiMn, 42SiMn} with machine {3, 5, 7, 8}.

Table 1. Clustering results for Fig. 2

Result 1 (step 1: part, step 2: machine)	Result 2 (step 1: machine, step 2: part)
Step 1: Minimum dissimilarity distance by part, obtains 3 part families: • Part family 1: 1, 4 • Part family 2: 2, 3, 6 • Part family 3: 5, 7	Step 1: Minimum dissimilarity distance by machine, obtains 2 machine groups: • Machine group 1: 1, 2, 4, 6 • Machine group 2: 3, 5, 7, 8
Step 2: Spectral clustering to get machine with 3 cells • Machine group 1: 2, 6 • Machine group 2: 1, 4 • Machine group 3: 3, 5, 7, 8	Step 2: Spectral clustering to get machine with 2 cells • Part family 1: 1, 4, 5, 7 • Part family 2: 2, 3, 6

MACHINE (a)

PART	2	6	1	4	3	5	7	8
5	1	1		1	1			
7	1	1		1				
1			1	1				
4			1	1	1			
2					1	1	1	1
3					1	1	1	1
6					1	1	1	1

MACHINE (b)

PART	2	6	1	4	3	5	7	8
1			1	1				
4			1	1	1			
5	1	1		1	1			
7	1	1		1				
2					1	1	1	1
3					1	1	1	1
6					1	1	1	1

Fig. 3. Part-machine matrix output as described in Table 1 for (a) Result 1; (b) Result 2

MACHINE

PART	2	6	1	4	3	5	7	8
ZL402 (5)	[40,70]	[20,50]	N	S	4	N	N	N
ZL202 (7)	[60,90]	[20,50]	N	S, J	N	N	N	N
ZL207 (1)	N	N	603-637	S	N	N	N	N
ZL205A (4)	N	N	544-633	S	3	N	N	N
50SiMn (2)	N	N	N	N	14	39	T	OC
35SiMn (3)	N	N	N	N	14	39	T	OC
42SiMn (6)	N	N	N	N	14	29	T	OC

Fig. 4. Part-machine matrix corresponding to Result 1

Example 2. This example is taken from Harhalakis et al. [4]. There are 20 machines and 20 parts. The part-machine matrix is shown in Fig. 5 where elements of the matrix indicate operation numbers. We compare the two cases in Table 2, where the grouping efficacies of these two clustering results are 69.89% and 63.33% for Result 1 and Result 2, respectively. Thus, Result 1 is the solution, as shown in Fig. 6.

MACHINE

PART \	1	2	3	4	5	6	7	8	9	10	11	12	13	14	15	16	17	18	19	20
1	2								3		1							4		5
2		3	2								1									
3							1												3	2
4		3	1									4	2							
5				1		3	4								2					
6						5					1		2			3	4			
7						1										2	3			
8					5			3		4			2	1						
9	4								2		3	5						1		
10								3										1	2	
11			3								1		2							
12	5				3					1		4				2				
13							1	2							3		4			
14	3	4						1		2										
15													1	2		3	4			
16							3	2							1			4		
17	2							1					3							
18								1		4									2	3
19			2	1		4							3							
20	3									2		4						1		

Fig. 5. Part-machine matrix input

Table 2. Clustering results for Fig. 5

Result 1 (step 1: part, step 2: machine)	Result 2 (step 1: machine, step 2: part)
Step 1: Minimum dissimilarity distance by part, obtains 5 part families: • Part family 1: 1, 9, 12, 17, 20 • Part family 2: 2, 4, 11, 19 • Part family 3: 3, 10, 14, 18 • Part family 4: 5, 8, 13, 16 • Part family 5: 6, 7, 15	Step 1: Minimum dissimilarity distance by machine, obtains 6 machine groups: • Machine group 1: 1, 9, 12, 18 • Machine group 2: 2, 3, 10, 11 • Machine group 3: 4, 13 • Machine group 5: 5, 14, 16, 17 • Machine group 6: 6, 7, 15 • Machine group 7: 8, 19, 20
Step 2: Spectral clustering to get machine with 5 cells • Machine group 1: 1, 9, 12, 18 • Machine group 2: 2, 3, 11 • Machine group 3: 8, 10, 19, 20 • Machine group 4: 4, 6, 7, 15 • Machine group 5: 5, 13, 14, 16, 17	Step 2: Spectral clustering to get machine with 6 cells • Part family 1: 2, 4, 11, 19 • Part family 2: 14, 20 • Part family 3: 1, 9, 12, 17 • Part family 4: 6, 7, 15 • Part family 5: 5, 8, 13, 16 • Part family 6: 3, 10, 18

Example 3. We use larger data with 35 parts and 28 machines, taken from Yang and Yang [23]. The original data of part-machine matrix is shown in Fig. 7. Table 3 shows the comparisons of two cases, which has the same results. Figure 8 shows these cell formation results.

Finally, we analyze its computational complexity. To get the cell number and part families (machine groups) using minimum dissimilarity distance where p is the number of parts and m is the number of machines, it takes $O(\max(p^2, m^2))$. For the spectral clustering algorithm, it takes $O(\max(p, m) k^2 I)$ where k is number of clusters and I is

MACHINE

PART	4	6	7	15	5	13	14	16	17	1	9	12	18	2	3	11	8	10	19	20
5	1	3	4	2																
8	5		3	1	2					4										
13		1	2	3				4												
16		3	2	1																4
6					5		2	3	4			1								
7					1			2	3											
15						1	2	3	4											
1										2	3	1	4							5
9										4	2	5	1		3					
12							3			5	1	4	2							
17										2	1	3								
20										3		4	1				2			
2														3	2	1				
4														3	1	2	4			
11					2										3	1				
19				4										2	1	3				
3																	1		3	2
10																	3		1	2
14					3								4				1	2		
18																	1	4	2	3

Fig. 6. Part-machine matrix output from Table 2 (Result 1)

MACHINE

PART	1	2	3	4	5	6	7	8	9	10	11	12	13	14	15	16	17	18	19	20	21	22	23	24	25	26	27	28
1		1														1								1		1		
2																								1		1		
3				1						1		1													1	1		
4			1								1						1	1							1	1		
5			1								1					1				1								1
6	1					1								1	1					1								
7		1													1									1		1		
8	1					1		1						1	1					1								1
9										1		1	1										1		1			
10				1							1		1										1		1			
11													1							1			1					1
12			1								1					1	1	1					1			1		
13			1		1											1		1					1			1		
14	1		1	1										1	1					1								
15	1					1								1	1	1					1							
16		1			1		1						1								1							
17		1		1					1	1		1				1					1							
18				1						1			1					1			1							
19									1					1					1				1				1	
20			1							1		1								1			1	1		1		1
21		1														1				1				1		1		
22		1														1								1		1		
23	1		1							1								1					1	1				1
24	1			1				1				1								1			1		1			
25		1														1								1		1		
26	1		1	1	1		1						1								1							
27			1	1		1	1													1		1						
28	1			1	1		1													1								
29			1									1		1			1	1			1			1				1
30	1				1	1							1	1						1								
31		1	1														1						1	1		1	1	
32			1	1							1						1	1			1							1
33												1							1				1		1			1
34	1					1							1	1						1								
35			1										1		1	1												1

Fig. 7. Part-machine matrix input

Table 3. Clustering results for Fig. 7

Result 1 (step 1: parts, step 2: machines)	Result 2 (step 1: machines, step 2: parts)
Step 1: Minimum dissimilarity distance by part, obtains 6 part families: • Part family 1: 1, 2, 7, 13, 21, 22, 25, 31 • Part family 2: 3, 9, 10, 17, 18, 24 • Part family 3: 4, 11, 19, 27, 33 • Part family 4: 5, 12, 20, 23, 29, 32, 35 • Part family 5: 6, 8, 14, 15, 30, 34 • Part family 6: 16, 26, 28	Step 1: Minimum dissimilarity distance by machine, obtains 6 machine groups: • Machine group 1: 1, 7, 14, 15, 21 • Machine group 2: 2, 6, 8, 13, 20 • Machine group 3: 3, 16, 24, 26 • Machine group 4: 4, 10, 17, 18, 22, 28 • Machine group 5: 5, 9, 12, 25 • Machine group 6: 11, 19, 23, 27
Step 2: Spectral clustering to get machine with 6 cells • Machine group 1: 3, 16, 24, 26 • Machine group 2: 5, 9, 12, 25 • Machine group 3: 11, 19, 23, 27 • Machine group 4: 4, 10, 17, 18, 22, 28 • Machine group 5: 1, 7, 14, 15, 21 • Machine group 6: 2, 6, 8, 13, 20	Step 2: Spectral clustering to get machine with 6 cells • Part family 1: 1, 2, 7, 13, 21, 22, 25, 31 • Part family 2: 3, 9, 10, 17, 18, 24 • Part family 3: 4, 11, 19, 27, 33 • Part family 4: 5, 12, 20, 23, 29, 32, 35 • Part family 5: 6, 8, 14, 15, 30, 34 • Part family 6: 16, 26, 28

MACHINE

PART	2	6	8	13	20	11	19	23	27	1	7	14	15	21	5	9	12	25	3	16	24	26	4	10	17	18	22	28
16	1	1	1	1	1																							
26	1	1	1	1	1											1					1							
28	1	1	1		1																							
4						1				1	1			1										1	1			
11						1	1	1	1																			
19						1	1	1	1						1													
27						1	1								1													
33						1	1	1	1																			
6										1	1	1	1	1														
8										1	1	1	1	1		1												
14										1		1	1	1	1		1											
15			1							1	1	1	1								1							
30		1								1	1	1	1	1														
34										1	1	1	1	1														
3															1		1	1	1		1							
9							1		1							1	1	1										
10										1					1	1	1	1										
17	1														1	1	1			1					1			
18		1													1	1										1		
24	1						1								1	1	1	1										1
1																			1	1	1	1						
2																					1	1						
7																			1	1	1	1						
13		1																	1	1		1					1	1
21																			1	1	1	1						
22																			1	1	1	1						
25																			1	1	1	1						
31	1						1	1											1	1	1	1						
5																							1	1	1		1	1
12																		1					1	1	1	1	1	
20							1								1	1							1	1		1	1	1
23	1														1				1				1	1	1		1	1
29									1						1				1				1		1		1	1
32																	1						1	1	1	1	1	
35														1									1		1	1		1

Fig. 8. Part-machine matrix output from Table 3 (Result 2)

number of iterations, but for matching each group of parts and machines, it needs $O(k^2)$. Thus, the computational complexity of the proposed algorithm is $O(\max(p^2, m^2) \, k^2 \, I)$.

5 Conclusions

From this research, we can conclude that minimum dissimilarity distance can be implemented to get an optimal number of cells for machine-part cell formation. Using spectral clustering with the obtained number of cells from the minimum dissimilarity computation, we can find good machine-part cells. Several experiment results actually show that the output can give good results. For a future work, we will consider to improve the proposed algorithm, especially on the construction of dissimilarity matrix.

References

1. Goldengorin, B., Krushinsky, D.: A computational study of the pseudo-boolean approach to the p-median problem applied to cell formation. In: Pahl, J., Reiners, T., Voß, S. (eds.) INOC 2011. LNCS, vol. 6701, pp. 503–516. Springer, Heidelberg (2011). doi:10.1007/978-3-642-21527-8_55
2. Goldengorin, B., Krushinsky, D., Pardalos, P.M.: Cell Formation in Industrial Engineering: Theory. Algorithms and Experiments. Springer, New York (2013)
3. Hakimi, S.L.: Optimum locations of switching centers and the absolute centers and medians of a graph. Oper. Res. 12, 450–459 (1964)
4. Harhalakis, G., Nagi, R., Proth, J.: An efficient heuristic in manufacturing cell formation for group technology applications. Int. J. Prod. Res. 28, 185–198 (1990)
5. Jain, A.K., Dubes, R.C.: Algorithms for Clustering Data. Prentice Hall, New Jersey (1988)
6. Kaufman, L., Rousseeuw, P.J.: Finding Groups in Data: An Introduction to Cluster Analysis. Wiley, New York (1990)
7. Krishnapuram, R., Keller, J.M.: A possibilistic approach to clustering. IEEE Trans. Fuzzy Syst. 1, 98–110 (1993)
8. Kumar, C.S., Chandrasekaran, M.P.: Grouping efficacy: a quantitative criterion for goodness of block diagonal forms of binary matrices in group technology. Int. J. Prod. Res. 28, 233–243 (1990)
9. Nascimento, M.C.V., de Carvalho, A.C.P.L.F.: Spectral methods for graph clustering-a survey. Eur. J. Oper. Res. 211, 221–231 (2011)
10. Newman, M.: Finding community structure in networks using the eigenvectors of matrices. Phys. Rev. E 74, 36–104 (2006)
11. Ng, A.Y., Jordan, M.I., Weiss, Y.: On spectral clustering: analysis and an algorithm. In: Advances in Neural Information Processing Systems (2001)
12. Oliveira, S., Ribeiro, J.F.F., Seok, S.C.: A spectral clustering algorithm for manufacturing cell formation. Comput. Ind. Eng. 57, 1008–1014 (2009)
13. Pollard, D.: Quantization and the method of k-means. IEEE Trans. Inf. Theor. 28, 199–205 (1982)
14. Sahin, Y.B., Alpay, S.: A metaheuristic approach for a cubic cell formation problem. Expert Syst. Appl. 65, 40–51 (2016)
15. Shi, J., Malik, J.: Normalized cuts and image segmentation. IEEE Trans. Pattern Anal. Mach. Intell. 22, 888–905 (2000)

16. Singh, N., Rajamani, D.: Cellular Manufacturing Systems. Chapman & Hall, New York (1996)
17. von Luxburg, U.: A tutorial on spectral clustering. Stat. Comput. **17**, 395–416 (2007)
18. Wei, J.C., Kern, G.M.: Commonality analysis: a linear cell clustering algorithm for group technology. Int. J. Prod. Res. **27**(12), 2053–2062 (1989)
19. Wu, K.L., Yang, M.S.: Mean shift-based clustering. Pattern Recogn. **40**, 3035–3052 (2007)
20. Xambre, A.R., Vilarinho, P.M.: A simulated annealing approach for manufacturing cell formation with multiple identical machines. Eur. J. Oper. Res. **151**, 434–446 (2003)
21. Yang, M.S.: A survey of fuzzy clustering. Math. Comput. Model. **18**, 1–16 (1993)
22. Yang, M.S., Hung, W.L., Cheng, F.C.: Mixed-variable fuzzy clustering approach to part family and machine cell formation for GT applications. Int. J. Prod. Econ. **103**, 185–198 (2006)
23. Yang, M.S., Yang, J.H.: Machine-part cell formation in group technology using a modified ART1 method. Eur. J. Oper. Res. **188**, 140–152 (2008)

Exercise Recognition Using Averaged Hidden Markov Models

Aleksandra Postawka$^{(\boxtimes)}$

Faculty of Electronics, Wroclaw University of Science and Technology,
Wroclaw, Poland
aleksandra.postawka@pwr.edu.pl

Abstract. This paper presents a novel learning algorithm for Hidden Markov Models (HMMs) based on multiple learning sequences. For each activity a few left-to-right HMMs are created and then averaged into singular model. Averaged models' structure is defined by a proposed Sequences Concatenation Algorithm which has been included in this paper. Also the modification of action recognition algorithm for such averaged models has been described.

The experiments have been conducted for the problem of modeling and recognition of chosen 13 warm-up exercises. The input data have been collected using the depth sensor Microsoft Kinect 2.0. The experiments results confirm that an averaged model combines the features of all component models and thus recognizes more sequences. The obtained models do not confuse modeled activities with others.

Keywords: Action recognition · Left-to-right Hidden Markov Models · Kinect · Averaged models

1 Introduction

Action recognition became a very important issue in many fields of life. The problem concerns, inter alia, gesture control, rehabilitation and monitoring systems. Controlling systems by gestures is useful for remote control applications, such as Kinect games, where remote control devices are unnecessary. Action recognition is also inappreciable in rehabilitation systems which often use exercise tracking for precision evaluation of movements [1]. In the case of monitoring systems the information about recognized action is needed for the purpose of e.g. life logging systems [2] or diagnosing serial diseases [3]. More detailed literature review on action recognition can be found in [4].

In recent years the technology of Microsoft Kinect depth sensor has been significantly improved and therefore increasingly used for problems associated with human movements tracking [2,3,5]. Moreover, using depth sensors with skeletal tracking feature takes an advantage of respecting observed persons privacy [3]. As this data acquisition method is quite precise it has been also applied in this work.

The Hidden Markov Models (HMMs) have been widely used for action recognition. In particular, HMMs have been successfully applied for recognition of:

© Springer International Publishing AG 2017
L. Rutkowski et al. (Eds.): ICAISC 2017, Part II, LNAI 10246, pp. 137–147, 2017.
DOI: 10.1007/978-3-319-59060-8_14

gestures in driving [6], tennis strokes [7], dance gestures [8], exercises [9] etc. Based on the results this approach seems to be very promising in the task of action recognition.

The left-to-right HMM model structure is usually applied for the activity modeling problem [7–10]. The states correspond to the activity stages and therefore such a model structure reflects the activity complexity. This procedure lessens the number of false-positives, however a low number of learning sequences may lead to overfitting.

Many different learning methods are used for HMMs. Sometimes it is the basic Baum-Welch algorithm with a single learning sequence [1] or Baum-Welch with Gaussian distribution for observation symbols distribution [9,11,12]. In other applications the Baum-Welch algorithm has been adopted for multiple learning sequences [5,13]. The different approach is based on averaging multiple models based on different learning sequences [14]. In [15] the profile HMMs are created from multiple learning sequences. The latter preserve the symbol order despite of different learning sequences.

In this paper the averaged HMM left-to-right models are investigated. In order to describe the activity in the best possible way one of the averaged models is used to define the structure of the rest of component models. The process of the component models creation based on one chosen model has been described in this paper as a Sequences Concatenation Algorithm. With this approach the association of the model state and activity stage is still preserved as in profile HMMs. The experiments have been conducted for the problem of modeling and recognition for chosen 13 warm-up exercises.

In Sect. 2 the general idea of averaging models with similar structure is presented. Section 3 presents the algorithm of component models creation based on one base observation sequence. Section 4 concerns the models averaging. The action recognition issue is described in Sect. 5. In Sect. 6 the conducted experiments and obtained results have been presented. Section 7 contains the overall conclusions and future work.

2   HMM - Based Activity Models

The following notation will be used for the Hidden Markov Models:

$\lambda = \{A, B, \pi\}$ - the complete parameter set for HMM,
N - the number of states,
M - the number of observation symbols,
T - the length of observation sequence,
$A = \{a_{ij} : i, j \in \{1 \ldots N\}\}$ - the state transition matrix,
$B = \{b_{ij} : i \in \{1 \ldots N\}, j \in \{1 \ldots M\}\}$ - the probability distribution matrix for observed symbols,
$\pi = \{\pi_i, i \in \{1 \ldots N\}\}$ - initial state distribution vector,
$O = O_1, O_2, \cdots, O_T$ - observation sequence.

The activities are usually modeled by left-to-right HMMs where subsequent states correspond to the activity's stages in time. In this model structure it is not possible to step back to previous states. An example of left-to-right HMM is presented in Fig. 1.

The examined here final averaged model for a given activity is created based on a few *similar* left-to-right HMM models, including a *base model*, which defines the model structure, and *child models*. The base model reflects its learning observation sequence - the number of states is equal to the number of *unique* symbols (repeating symbols are concatenated into a single one, i.e. 13514 is unique) in the observation sequence. The observation symbols probability distribution for each state s_i contains value $b_{ij} = 1.0$ for the respective symbol j and 0.0 for others. The other models for given activity (child models) are created based on the base model and a new learning observation symbol sequence (Sect. 3). In effect, the symbol probability distribution matrix is changed and some additional transitions are added in order to describe the new sequence in the best possible way.

A single model is very unlikely to be capable of generating a sequence of observation symbols for a new realization of a given activity (i.e. to recognize this activity). Therefore for each base model (each activity) many similar child models are created and merged into one resultant model (see Sect. 4).

In Fig. 1 an example of such process is presented. First the base model for the pattern observation sequence is created (Fig. 1(a)). The states are assigned to symbols (as can be seen in attached probability distributions of observed

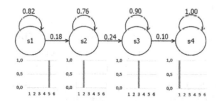

(a) Base model for 5..52..23..31..1 observation sequence

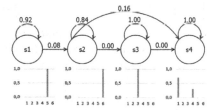

(b) Child model for 5..56..61..14..4 observation sequenceandbasemodel(a)

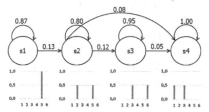

(c) An averaged HMM created from a base model(a)anditsonechildmodel(b)

Fig. 1. An example of averaged HMM generation process

symbols) thus the base model is also a Markov model. Next, the child models are created based on the base model and a new observation sequence like in Fig. 1(b). If a symbol from pattern sequence does not occur in the new sequence then additional transitions are added. If new symbols appear in the new sequence they are added to the observation symbols probability distribution for the given state. Finally, an averaged HMM is generated (Fig. 1(c)) based on the base model and all child models - in this case there is only one child model. The averaged model recognizes both pattern and new sequences. The further details are presented in next sections of this paper.

3   Child Models

Child models are created in four stages. The first task is to retrieve the sequence of leading HMM symbols in subsequent states from the base model. Each state is searched for the most probable observation symbol which is added to the *pattern* vector. In the second stage the resultant model structure for the base model and a new observation sequence is created (Sect. 3.1). Thirdly, based on the information obtained from the previous stage the child model is created by adding some noises or additional transitions to the base model (Sect. 3.2). Finally, such obtained matrices are used as the initial condition in the Baum-Welch learning algorithm [16] so that the probability value that the symbol sequence has been generated by this model is possibly highest (for given model structure).

3.1   Sequences Concatenation Algorithm

States corresponding to the same activity stages should coincide with each other in the base model and in the child model. However, usually the new sequence significantly differs from the pattern sequence. It can be noticed that the pattern sequence can always be obtained from the new sequence by a few operations of insertion or deletion of singular symbols. The information about the positions where symbols are inserted or deleted corresponds to adding some noises to the symbol distribution or adding some transitions in the base model, respectively. A child model obtained with minimized number of operations is most *similar* to the base model (the same stages of activity are represented by the same states in the base and child models) Therefore minimizing the number of insert- or delete-operations is equivalent to maximizing the number of overlapping activity stages.

The Sequences Concatenation Algorithm (SCA) proposed in this paper is a heuristic tree-based algorithm designed for the presented problem. The algorithm obtains one of the best solutions (if there are more than one). The input data are two *unique* symbol sequences (see Sect. 2) for pattern and new sequences. Each node contains an information about the gain, indices of current elements in the pattern and new sequences, pointer to the parent node (bidirectional tree) and pointers to the child nodes. The pseudocode for the SCA algorithm is presented in Table 1. After the algorithm is finished the best path (leading to the node with highest gain) has to be retrieved (Sect. 3.2).

In the situation when overlapping symbols are divided by many not overlapping symbols in both pattern and new sequences (Fig. 2, top image) it is

| Pattern sequence | 1 | 6 | 7 | 3 | 2 | 5 | 7 | 8 |
| New sequence | 1 | 3 | 4 | 9 | 6 | 8 | | |

| Case 1: | 1 | 6 | 7 | 3 | | 2 | | 5 | | 7 | 8 |
| | 1 | | | 3 | 4 | | 9 | | 6 | | 8 |

| Case 2: | 1 | 6 | 7 | 3 | 2 | 5 | 7 | | | | 8 |
| | 1 | | | 3 | | | | 4 | 9 | 6 | 8 |

Fig. 2. Case 1: an example of SCA algorithm. Case 2: Contrast unwanted situation.

Table 1. Sequences concatenation algorithm for left-to-right HMMs

```
SCA
create a new node
set pattern_index = −1, new_index = −1
set gain = 0
add the node to the active node list
while node list is not empty
    for all nodes in the active node list
        if pattern[pattern_index] == new[new_index]
            create a new child node
            set pattern_index = pattern_index_old + 1, new_index = new_index_old + 1
            set gain = gain_old + 10
            add child node to the new active node list
        else
            if pattern[pattern_index] == new[new_index + 1]
                create a new child node
                set pattern_index = pattern_index_old, new_index = new_index_old + 1
                set gain = gain_old + 9
                if in previous iteration only new_index has been increased
                    set gain− = 1
                endif
                add child node to the new active node list
            endif
            if pattern[pattern_index + 1] == new[new_index]
                create a new child node
                set pattern_index = pattern_index_old + 1, new_index = new_index_old
                set gain = gain_old + 9
                if in previous iteration only pattern_index has been increased
                    set gain− = 1
                endif
                add child node to the new active node list
            endif
        endif
    end
    sort new node list descending by gain
    set first element from the new node list as a winner node (best)
    for each element from the new node list
        if pattern_index ≤ pattern_index_best and new_index ≤ new_index_best
            delete node from the list
        endif
    end
    set new active node list as active node list
end
```

preferred to choose the symbols alternately (Fig. 2 Case 1). Otherwise pattern adjacent symbols would mean skipping many states (Fig. 2 Case 2, symbols 2–7) and new sequence's adjacent symbols would stand for adding many noises to the singular state (Fig. 2 Case 2, symbols 4–6). The information about symbols order would be lost and therefore the gain is decreased by 1. It is also preferred to increase both indices simultaneously and therefore in such a situation the gain is higher.

3.2   Model Creation

The node path obtained in the SCA algorithm has to be translated into the child model's structure. This process involves the definition of initial conditions for Baum-Welch algorithm. Firstly the transition matrix A is defined as upper triangular in order to describe each left-to-right model. Secondly the symbol probability distribution matrix B is generated based on SCA winning path. There are 3 cases for each iteration (state s_i):

1. only new sequence index has been increased
   The symbol $new[new\_index]$ do not coincide with such a symbol in the pattern sequence. As no new states are added to the base model structure thus the symbol is not assigned to any state - it will be added to the next state as noise.
2. only pattern sequence index has been increased
   The symbol $pattern[pattern\_index]$ do not coincide with such a symbol in the new sequence, so the value 1.0 is assigned to $B[s_i][pattern\_index]$. If there are some pending noises from previous iterations then some small values are assigned for $B[s_i][noises[j]]$ and the value of $B[s_i][pattern\_index]$ is decreased so as to sum with noises probabilities to 1.0. If no noises are added then this state will be omitted - the skip transition will be added in the Baum-Welch algorithm.
3. both pattern and new sequences indices have been increased
   The symbol coincides in both sequences ($new[new\_index] = pattern$ $[pattern\_index]$) so the value 1.0 is assigned to $B[s_i][pattern\_index]$. If there are some pending noises then the same procedure as in previous case is used.

Such an initial model is forwarded to the Baum-Welch algorithm which adjusts the parameters for the new sequence. The point is that this algorithm changes only the non-zero values. The non-zero values might be changed to zero, yet zero remains unchanged.

4   Averaged HMM Activity Models

An averaged activity model is created based on many similar HMM models of the same activity (including base model). Such a resultant model combines features of all investigated models. Therefore it recognizes all sequences that could be recognized by each model separately and even a few sequences that would not be recognized by singular models. Moreover, if the component models are properly created then the false positive rate should be smaller than in the case of activity model with Gaussian distribution often used in the literature, e.g. [11,12] - as in the averaged model only real cases have been included and in the Gaussian distribution all adjacent symbols are considered.

The lemma presented below is quite obvious, however it is very important for this research. The point is that non-negative weights that sum to 1, that is, the convex combination of weights, preserve the model's probability structure while averaging.

Lemma. If the D component models are HMMs and the vector of weights fulfills the condition (1) the resultant model given by the Eqs. (2) and (3) is a HMM too.

$$w = \{w_d : d \in \{1, \ldots, D\} \wedge w_d \geq 0 \wedge \sum_{d=1}^{D} w_d = 1\} \tag{1}$$

$$\bar{a}_{ij} = \sum_{d=1}^{D} w_d \cdot a_{ij}^{(d)} \tag{2}$$

$$\bar{b}_{ij} = \sum_{d=1}^{D} w_d \cdot b_{ij}^{(d)} \tag{3}$$

Proof. Based on the HMM definition the logical sentences $\sum_{j=1}^{N} a_{ij}^{(d)} = 1$ and $\sum_{j=1}^{M} b_{ij}^{(d)} = 1$ are true for each component model d. Then:

$$\sum_{j=1}^{N} \bar{a}_{ij} = \sum_{j=1}^{N} \sum_{d=1}^{D} w_d \cdot a_{ij}^{(d)} = \sum_{d=1}^{D} w_d \sum_{j=1}^{N} a_{ij}^{(d)} = \sum_{d=1}^{D} w_d = 1$$

$$\sum_{j=1}^{M} \bar{b}_{ij} = \sum_{j=1}^{M} \sum_{d=1}^{D} w_d \cdot b_{ij}^{(d)} = \sum_{d=1}^{D} w_d \sum_{j=1}^{M} b_{ij}^{(d)} = \sum_{d=1}^{D} w_d = 1$$

$\square$

The weights w_d may be adjusted depending on some model evaluation algorithm, but they can be equal as well. In the second case all models are of the same importance.

5   Action Recognition

The action recognition is based on forward algorithm [16] usually used for model evaluation in HMMs. In theory the posterior probability $Pr(O|\lambda)$ is calculated - the probability that the observation sequence O has been generated by model λ. In practice the logarithmic value $P_O = log[Pr(O|\lambda)]$ is calculated instead because the value of forward variable, used for the posterior probability calculation, tends to 0 exponentially along with the number of observations.

As the value of forward variable, used for the posterior probability calculation, tends to 0 exponentially along with the number of observations the logarithmic value $P_O = log[P(O|\lambda)]$ is calculated.

However, such an algorithm is insufficient in the real-time recognition, i.e. the situation where the test set contains also some unmodeled activities. Therefore the action recognition algorithm has been modified and the N_{real} variable has been added to each model. The variable describes the real number of states used in the model. Some observation sequences for the given activity are longer and some are shorter than the pattern sequence. It is important to skip the recognition of sequences which are only a part of an activity despite the fact that the posterior probability is greater than zero. While averaging models the N_{real} variable takes the lowest value from N_{real} in all component models.

If the P_O value is different than $-\infty$ then the most probable state sequence is decoded using the Viterbi algorithm [16]. The last decoded state number is compared with the N_{real} value and if it is equal or greater then the sequence is marked as recognized.

6   Experiments and Results

The data have been collected by Microsoft Kinect 2.0 depth sensor using the skeleton tracking feature. 60 repetitions for each of the 13 investigated warm-up exercises have been recorded. The exercises have been performed by three persons, 20 repetitions of each exercise per person. The actions have been listed and numbered in Table 2. According to the previous work [1] for each frame (singular file record) the HMM symbol for hands has been calculated. The exercise sequences have been manually cut out from the recording. The observation sequences for given activity are not of the same length.

The data set has been divided into the training set and the test set in proportion 3:2 within the series performed by different actors. In effect, the training set contains 36 sequences for each activity where each 12 sequences have been performed by a different person. The test set contains 24 sequences for each activity.

For each exercise one of the sequences has been chosen as a base sequence and a base model has been created. The other 35 training sequences have been used for child models creation. Finally, for each exercise the 36 component models have been averaged and the resultant model has been obtained. In both provided experiments $w_d = \frac{1}{36}$ for each d.

Experiment 1. In this experiment the efficiency of averaged HMMs for the learning set has been investigated. Each averaged model has been tested for each of the 468 learning sequences. The results have been shown in Table 3.

The Table 3 presents the number of training sequences that have been recognized using the given averaged HMM. All the diagonal elements of the matrix should be equal to 36 as it is the number of learning sequences used for each model. Any other sequence should not be recognized excluding one-hand activities which have their counterparts in both-hands activities. For the presented results such a situation occurs for raising and lowering the hand and for arm twisting forward or backward (which has the common part also with crawl).

The experiment confirms that the averaged left-to-right HMM combines the features of all component models because all component model's learning

Table 2. The list of modeled exercises

Number	Exercise
0	Right arm twisting forward
1	Right arm twisting backward
2	Raising and lowering right hand
3	Left arm twisting forward
4	Left arm twisting backward
5	Raising and lowering left hand
6	Both hands twisting forward
7	Both hands twisting backward
8	Raising and lowering both hands
9	Clapping hands
10	Clapping hands over the head
11	Crawl forward
12	Crawl backward

Table 3. Confusion matrix for the learning set

Action	0	1	2	3	4	5	6	7	8	9	10	11	12
Model 0	36	0	0	0	0	0	27	0	0	0	0	4	0
Model 1	0	36	0	0	0	0	0	19	0	0	0	0	4
Model 2	0	0	36	0	0	0	0	0	24	0	0	0	0
Model 3	0	0	0	36	0	0	31	0	0	0	0	0	0
Model 4	0	0	0	0	36	0	0	24	0	0	0	0	0
Model 5	0	0	0	0	0	36	0	0	27	0	0	0	0
Model 6	0	0	0	0	0	0	36	0	0	0	0	0	0
Model 7	0	0	0	0	0	0	0	36	0	0	0	0	0
Model 8	0	0	0	0	0	0	0	0	36	0	0	0	0
Model 9	0	0	0	0	0	0	0	0	0	36	0	0	0
Model 10	0	0	0	0	0	0	0	0	0	0	36	0	0
Model 11	0	0	0	0	0	0	0	0	0	0	0	36	0
Model 12	0	0	0	0	0	0	0	0	0	0	0	0	36

Table 4. Confusion matrix for the test set

Action	0	1	2	3	4	5	6	7	8	9	10	11	12
Model 0	21	0	0	0	0	0	18	0	0	0	0	3	0
Model 1	0	20	0	0	0	0	0	14	0	0	0	0	3
Model 2	0	0	21	0	0	0	0	0	19	0	0	0	0
Model 3	0	0	0	24	0	0	21	0	0	0	0	0	0
Model 4	0	0	0	0	24	0	0	20	0	0	0	0	0
Model 5	0	0	0	0	0	23	0	0	21	0	0	0	0
Model 6	0	0	0	0	0	0	20	0	0	0	0	0	0
Model 7	0	0	0	0	0	0	0	21	0	0	0	0	0
Model 8	0	0	0	0	0	0	0	0	15	0	0	0	0
Model 9	0	0	0	0	0	0	0	0	0	24	0	0	0
Model 10	0	0	0	0	0	0	0	0	0	0	15	0	0
Model 11	0	0	0	0	0	0	0	0	0	0	0	15	0
Model 12	0	0	0	0	0	0	0	0	0	0	0	0	6

sequences were recognized. Moreover, the noises included by multiple learning sequences did not increase the false positive recognition rate.

In the case of models 0–2 and 3–5 also the corresponding both-hand movements (6–8) were recognized with the average recognition rate of 64.8% and 75.9%, respectively. In the case of crawl only few sequences (not included in the learning set) were recognized. The probable reason for this is the fact that in both-hand activities the motion is a bit different than in one-hand activities.

Experiment 2. The second experiment was designed to check the averaged left-to-right HMMs efficiency for new sequences (test set). The results have been shown in Table 4.

In the case of one-hand movements the average recognition rate is higher than 92% (matrix diagonal). Similarly to Experiment 1 also a few both-hand sequences were recognized by one-hand models (78.5% without crawl). The crawl movement seems to be much different from twisting one's arm because only 6 motions were recognized.

The average recognition rate for both-hand activities is 69%. The most frequently recognized both-hand exercise is clapping hands with 100% recognition rate, while the least likely recognized is crawl backward with the result of 25%. In the case of crawl the Kinect recordings were very inaccurate and many points were inferred or temporarily untracked. This inconvenience results from the restrictions of Microsoft skeletal tracking.

Summarily, both-hand exercises are less recognizable than one-hand movements. The reason is that both-hand movements are more difficult to perform in repetitive way. Slight differences in right and left hands motions order cause numerous HMM symbol changes. In order to obtain recognition accuracy similar to one-hand movements, the learning set for both-hand movements needs to be much greater.

7   Conclusions and Future Work

In this paper the averaged Hidden Markov Models with left-to-right model structure have been investigated. The structure of component models used for averaging is defined by one of the models so as to describe the activity in the best possible way. The algorithm for component models generation has been presented in this paper.

In the real world it is very unlikely that the action is performed in the exact way so that the observation symbol sequence has the same pattern as in the left-to-right model obtained from a single learning sequence. Tests reveal that by disposing a sufficient number of different learning sequences it is possible to create a model that is much more resistant to noises and thus better than each model based on a singular sequence. The obtained models do not confuse modeled activities with others as only the real cases are taken into account. The average recognition rate is 92% and 69% for one-hand and both-hands movements, respectively.

In the near future the developed algorithms will be used to generate models for modeling observed persons behavior and further for emotions recognition. The methods will be also used for rehabilitation purposes - based on the most probable symbols in subsequent states next activity stage may be displayed. The movements precision can be evaluated using the posterior probability.

Acknowledgment. This work was supported by the statutory funds of the Faculty of Electronics 0402/0104/16, Wroclaw University of Science and Technology, Wroclaw, Poland.

References

1. Postawka, A., Śliwiński, P.: A kinect-based support system for children with autism spectrum disorder. In: Rutkowski, L., Korytkowski, M., Scherer, R., Tadeusiewicz, R., Zadeh, L.A., Zurada, J.M. (eds.) ICAISC 2016. LNCS, vol. 9693, pp. 189–199. Springer, Cham (2016). doi:10.1007/978-3-319-39384-1_17

2. Jalal, A., Kamal, S., Kim, D.: A depth video sensor-based life-logging human activity recognition system for elderly care in smart indoor environments. Sensors **14**, 11735–11759 (2014)

3. Pal, M., Saha, S., Konar, A.: Distance matching based gesture recognition for healthcare using microsofts kinect sensor. In: International Conference on Microelectronics, Computing and Communications, pp. 1–6 (2016)

4. Postawka, A., Śliwiński, P.: Recognition and modeling of atypical children behavior. In: Rutkowski, L., Korytkowski, M., Scherer, R., Tadeusiewicz, R., Zadeh, L.A., Zurada, J.M. (eds.) ICAISC 2015. LNCS, vol. 9119, pp. 757–767. Springer, Cham (2015). doi:10.1007/978-3-319-19324-3_68

5. Hai, P.T., Kha, H.H.: An efficient star skeleton extraction for human action recognition using hidden Markov models. In: IEEE Sixth International Conference on Communications and Electronics, pp. 351–356 (2016)

6. Siddiqui, S.A., Snober, Y., Raza, S., Khan, F.M., Syed, T.Q.: Arm gesture recognition on microsoft Kinect Using a Hidden Markov Model-based representations of poses. In: International Conference on Information and Communication Technologies, pp. 1–6 (2015)

7. Petkovic, M., Jonker, W., Zivkovic, Z.: Recognizing strokes in tennis videos using hidden Markov models. In: IASTED International Conference on Visualization, Imaging and Image Processing, pp. 512–516 (2001)

8. Anbarsanti, N., Prihatmanto, A.S.: Dance modelling, learning and recognition system of aceh traditional dance based on hidden Markov model. In: International Conference on Information Technology Systems and Innovation, pp. 86–92 (2014)

9. Feng-Shun Lin, J., Kulić, D.: Segmenting human motion for automated rehabilitation exercise analysis. In: Annual International Conference of the IEEE Engineering in Medicine and Biology Society, pp. 2881–2884 (2012)

10. Kamal, S., Jalal, A.: Hybrid feature extraction approach for human detection, tracking and activity recognition using depth sensors. Arab. J. Sci. Eng. **41**, 1043–1051 (2016)

11. Yin, J., Yang, Q., Junfen Pan, J.: Sensor-based abnormal human-activity detection. IEEE Trans. Knowl. Data Eng. **20**, 1082–1090 (2008)

12. Ángeles Mendoza, M., Pérez de la Blanca, N.: HMM-based action recognition using contour histograms. In: Martí, J., Benedí, J.M., Mendonça, A.M., Serrat, J. (eds.) IbPRIA 2007. LNCS, vol. 4477, pp. 394–401. Springer, Heidelberg (2007). doi:10.1007/978-3-540-72847-4_51

13. Jiang, M., Chen, Y., Zhao, Y., Cai, A.: A real-time fall detection system based on HMM and RVM. In: Proceedings of Visual Communications and Image Processing, pp. 1–6 (2013)

14. Davis, R.I.A., Walder, C.J., Lovell, B.C.: Improved classification using hidden Markov averaging from multiple observation sequences. In: Proceedings of the Fourth Australasian Workshop on Signal Processing and Applications, pp. 89–92 (2002)

15. Eddy, S.R.: Profile hidden Markov models. Bioinformatics **14**, 755–763 (1998)

16. Rabiner, L.R., Juang, B.H.: An introduction to hidden Markov models. IEEE ASSP Mag. **3**, 4–16 (1986)

A Study of Cluster Validity Indices
for Real-Life Data

Artur Starczewski[1($\boxtimes$)] and Adam Krzyżak[2]

[1] Institute of Computational Intelligence, Częstochowa University of Technology,
Al. Armii Krajowej 36, 42-200 Częstochowa, Poland
artur.starczewski@iisi.pcz.pl
[2] Department of Computer Science and Software Engineering,
Concordia University, Montreal, Canada
krzyzak@cs.concordia.ca

Abstract. In this paper a study of several cluster validity indices for real-life data sets is presented. Moreover, a new version of validity index is also proposed. All these indices can be considered as a measure of data partitioning accuracy and the performance of them is demonstrated for real-life data sets, where three popular algorithms have been applied as underlying clustering techniques, namely the *Complete–linkage, Expectation Maximization* and *K-means* algorithms. The indices have been compared taking into account the number of clusters in a data set. The results are useful to choose the best validity index for a given data set.

Keywords: Cluster validity index · Complete-linkage clustering · K-means clustering · Expectation maximization algorithm

1 Introduction

Clustering algorithms are very popular and found use in various fields such as, e.g., bioinformatics, exploration data, data mining etc. Among clustering algorithms many different approaches can be distinguished, e.g., partitioning, grid-based, density-based, hierarchical or fuzzy clustering. Partitioning algorithm K-means [6,31] is widely used algorithm and can be implemented to different practical problems. Other partitioning methods are also very popular, for example, K-Medoids [20], PAM [23] or BIRCH [45]. The next sort of algorithms, grid-based methods, includes such ones as, e.g. STING [42] or WaveCluster [38]. Among density-based clustering algorithms we can mention, e.g. DBScan [12] or DenClue [21]. It should be noted that the expectation maximization algorithm (EM) is a general-purpose maximum likelihood algorithm for missing-data problems [26]. The researchers often use hierarchical clustering, here are such methods as Single, Average, and Complete linkage [22,27,32], CURE [16] or ROCK [17]. In fuzzy clustering each data instance is associated with every cluster using some type of membership function. Thus, the membership of the instance in a cluster is partial. The representative examples of the fuzzy clustering methods include,

© Springer International Publishing AG 2017
L. Rutkowski et al. (Eds.): ICAISC 2017, Part II, LNAI 10246, pp. 148–158, 2017.
DOI: 10.1007/978-3-319-59060-8_15

among others, *Fuzzy C-Means* [2] or *Gustafson-Kessel FCM* algorithm [18]. It should be noted that nowadays a large number of new clustering algorithms appears, e.g. [14, 15].

Unfortunately, for the same data the results of a clustering algorithm can vary dramatically when the input parameters of the algorithm are different. A significant input parameter of many clustering algorithms is the number of clusters, which is often selected in advance. Thus, additional techniques should be used to properly evaluate results of data clustering. For this purpose cluster validity indices are often used. In many validity indices two properties of clusters are taken into account, i.e., compactness and separability [19]. There is a large number of various validity indices and some of them are often used by researchers for determining the right partition of data sets. Among them are well-known cluster validity indices such as, e.g., *Dunn* [11], *Davies-Bouldin (DB)* [9], *PBM* [28] or *Silhouette (SIL)* indices [33]. It needs to be noted that unlike most of the indices, the *Silhouette (SIL)* index is used for clusters of arbitrary shapes. Nowadays, in the literature new interesting solutions have been proposed for cluster evaluation [13, 24, 30, 37, 39–41, 44]. However, the existing validity indices have limitations and lack generalization in evaluation of clustering results [1]. It should be noted that clustering methods in conjunction with cluster validity indices can be used to designing various neural networks [3–5] and neuro-fuzzy systems [7, 8, 10, 34–36].

In this paper, a cluster validity index called the $STRv1$ is proposed. This new index contains a product of two components, which determine changes of cluster compactness during a partitioning process (see Eq. (1)). In order to present effectiveness of the new validity index several experiments were performed for various data sets. This paper is organized as follows: Sect. 2 presents a detailed description of the new $STRv1$ index. Section 3 illustrates experimental results of a study of cluster validity indices for real-life data. Finally, Sect. 4 presents conclusions.

2   A New Cluster Validity Index

Below the detailed description of the new index called $STRv1$ is presented. It is based on the approach described in paper [40], where the original STR index has been proposed. Let us look at the $STRv1$ index, which consists of the product of two components. As mentioned above, the index determines changes of cluster compactness in a partitioning scheme and is defined as follows:

$$STRv1 = |(E1(K) - E1(K + 1)) \cdot (E2(K + 1) - E2(K))| \qquad (1)$$

where $E1(K)$ is the ratio of the total scatter of all patterns to the total scatter of the within clusters and it is expressed as:

$$E1(K) = \frac{E_0}{E_K} \qquad (2)$$

where

$$E_0 = \sum_{\mathbf{x} \in X} \|\mathbf{x} - \mathbf{v}\| \tag{3}$$

$$E_K = \sum_{k=1}^{K} \sum_{\mathbf{x} \in C_k} \|\mathbf{x} - \mathbf{v}_k\| \tag{4}$$

The next factor $E2(K)$ is defined as the square of the product of the maximum distance between cluster centers and the ratio of the total scatter of the within clusters to the total scatter of all patterns. It can be written as:

$$E2(K) = \left(D_{K\,max} \cdot \frac{E_k}{E_0} \right)^2 \tag{5}$$

and

$$D_{Kmax} = \max_{i,k=1}^{K} \|\mathbf{v}_i - \mathbf{v}_k\| \tag{6}$$

where K is a number of clusters, C_k is the k_{th} cluster, $\mathbf{v}$ is the center of the data set X, $\mathbf{v}_i$ and $\mathbf{v}_k$ are the centers of the i_{th} and k_{th} clusters. Furthermore, $E1(K+1)$ and $E2(K+1)$ are calculated for the $K+1$ partition scheme. The maximum value of the $STRv1$ index is found to determine the right data partitioning. Note that we need to increase K by 1 to calculate the correct number of clusters. This is due to the use of the measure of cluster compactness (see Eqs. (2) and (5)).

3   Experiments

In this section experiments were conducted to show how effectively the several cluster validity indices work. All experiments have been carried out on real-life data using different clustering algorithms. It should be noted that a number of clusters of data sets is very important parameter for most of clustering methods. For these experiments three well-known algorithms were selected for the clustering of data sets, namely, the *Complete-linkage*, *K-means* and *EM* methods. For example, the *Complete-linkage* method combines the two clusters with the smallest maximum pairwise distance, whereas the *K-means* clustering approach consist in that elements of a data set are associated with a cluster by finding the nearest centre of this cluster. On the other hand, the *EM* algorithm creates clusters by using statistical distributions, such as multivariate normal distributions. In the performed tests, the number of clusters K was varied from $K_{max} = \sqrt{n}$ to $K_{min} = 2$. It is the best range of the number of clusters for data clustering analysis [29]. Note that for the hierarchical clustering the number varies from $K_{max} = n$ to $K_{min} = 2$. Moreover, it is assumed that the values of the validity indices are equal to 0 for $K = 1$. Furthermore, in all the experiments the Euclidean distance and the min-max data normalization have been used. This approach is often applied in different clustering tools, e.g., in the Weka machine learning toolkit [43], which was also used in these experiments. Below the several cluster validity indices are described. They are used in the conducted experiments.

Dunn index: It can be expressed as follows:

$$D = \min_{1 \leq i \leq K} \left(\min_{1 \leq j \leq K,\, i \neq j} \left(\frac{d\left(c_i, c_j\right)}{\max_{1 \leq k \leq K} \left(\delta\left(c_k\right)\right)} \right) \right) \tag{7}$$

where K is a number of clusters in a data set, $d(c_i, c_j)$ is the distance between two clusters c_i and c_j, $\delta(c_k)$ is the diameter of cluster c_k. For well-separable clusters, distances between clusters are large and their diameter is small and the maximum value of the index indicates the right partitioning of data.

Davies–Bouldin (DB) index: It is defined as the ratio of the sum of the within-cluster scatter to the inter-cluster separation and can be expressed as follows:

$$DB = \frac{1}{K} \sum_{i=1}^{K} R_i \tag{8}$$

where the factor R_i is written as:

$$R_i = \max_{j \neq i} \frac{S_i + S_j}{d_{ij}} \tag{9}$$

S_i and S_j denote the within-cluster scatter for i_{th} and j_{th} clusters, respectively, and the d_{ij} is the distance between the cluster centers, i.e., $d_{ij} = \|\mathbf{v}_i - \mathbf{v}_j\|$. The minimum of the DB index indicates the appropriate partitioning of a data set.

PBM index: It can be defined as follows:

$$PBM = \left(\frac{1}{K} \times \frac{E_o}{E} \times D \right)^2 \tag{10}$$

where E_0 represents total scatter of all patterns belonging to one cluster in the data set, E identifies the total within-cluster scatter and can be expressed:

$$E = \sum_{k=1}^{K} \sum_{j=1}^{n} \mu_{kj} \|\mathbf{x}_j - \mathbf{v}_k\| \tag{11}$$

n is a number of elements in the data set, $[\mu_{kj}]$ is a partition matrix of the data, $\mathbf{v}_k$ is the center of the k_{th} cluster. The last factor D is a measure of cluster separation and is defined as a maximum distance between cluster centers. The maximum value of the index corresponds to the best partitioning of a given data set.

Silhouette (SIL) index: It can be defined as:

$$SIL = \frac{1}{K} \sum_{k=1}^{K} SIL(C_k) \tag{12}$$

where $SIL(C_k)$ is the *Silhouette width* for the given cluster C_k and can be expressed as follows:

$$SIL(C_k) = \frac{1}{n_k} \sum_{\mathbf{x} \in C_k} \frac{b(\mathbf{x}) - a(\mathbf{x})}{max\left(a(\mathbf{x}), b(\mathbf{x})\right)} \tag{13}$$

where n_k is a number of patterns in C_k, $a(\mathbf{x})$ is the within-cluster mean distance and it is defined as the average distance between $\mathbf{x}$ and the rest of the patterns belonging to the same cluster, $b(\mathbf{x})$ is the smallest of the mean distances of $\mathbf{x}$ to the patterns belonging to the other clusters. The values of the index are from the range -1 to 1 and a maximum value (close to 1) provides the best partitioning of the data set.

SILA index [41]: It is the modification of the Silhouette and it involves using an additional component, which improves a performance of the new index. It can be defined as follows:

$$SILA = \frac{1}{n} \left(\sum_{\mathbf{x} \in X} \left(\frac{b(\mathbf{x}) - a(\mathbf{x})}{\max{(a(\mathbf{x}), b(\mathbf{x}))}} \cdot \frac{1}{(1 + a(x))} \right) \right) \tag{14}$$

A maximum value of the index indicates the right partition scheme.

3.1 Real-Life Data Sets

These data sets were downloaded from the UCI machine learning repository [25]. In Table 1, a brief overview of these 11 sets is given. It should be noted that they have a different number of clusters, attributes and elements. For example, the first set called *Breast cancer* is the Wisconsin Breast Cancer data. It consists of 683 patterns belonging to two classes: Benign (444 instances) and Malignant (239 instances). Each pattern is characterised by nine features. The next set called *Car* includes features extracted from vehicle silhouettes. This set includes 946 elements, which are located in 4 classes, and each sample is described by 18 features. The *Diabetes* data set includes results of studies relating to the signs of diabetes in patients. This set includes 768 instances belonging to 2 classes and each item is described by 8 features. The third set is *Ecoli* data set consisting of 336 instances, and the number of attributes equals 7. It has 8 classes, which represent the protein localization sites. Next comes the *Glass* data set, which contains information about 6 types of glass defined in terms of their oxide content. In more detail, the set has 214 instances and each of them is described by 9 attributes. The well-known *Iris* data are extensively used in many comparisons of classifiers. This set has three classes *Setosa*, *Virginica* and *Versicolor*, which contain 50 instances per class. Moreover, each pattern is represented by four features, and two classes *Virginia* and *Versicolor* overlap each other. On the other hand, the third class *Setosa* is well separated from the others. The next set is the *Parkinsons* data set and it consists of 195 cases, which are described by 22 attributes. These data are composed of biomedical voice measurements from a group of people and are used to discriminate healthy people from those with Parkinson's disease (2 classes). The *Spectf* data set describes diagnosing of cardiac Single Proton Emission Computed Tomography images. This set includes 267 instances, which are characterized by 44 features and the patients are classified into 2 classes. The next set is *Transfusion* data and it consists of 748 elements, which are described by 4 attributes. These data sets are taken from the Blood Transfusion Service Center and contain blood donors' data. They are

Table 1. Description of real-life data sets

Data set	No. of elements	Features	Classes
Brestcancer	683	9	2
Car	946	18	4
Diabetes	768	8	2
Ecoli	336	7	8
Glass	214	9	6
Iris	150	4	3
Parkinsons	195	22	2
Spectf	267	44	2
Transfusion	748	4	2
Vertebralcolumn	310	6	3
Wine	178	13	3

classified into 2 classes. The following set, *Vertebral column*, contains values of six biomechanical features, which are used to classify orthopaedic patients into 3 classes. In this set, the total number of cases equals 310. Finally, the *Wine* data set shows the results of a chemical analysis of wines. It comprises three classes of wines, which consist of 59, 71 and 48 samples per class, respectively. Altogether, the data set contains 178 patterns represented by 13 features.

3.2   Comparison of Validity Indices

A comparison of the *Dunn, Davies-Bouldin (DB), PBM, Silhouette (SIL), SILA* and *STRv1* indices has been presented. As mentioned above, some of these indices are frequently used to compare results of different clustering algorithms. The *Dunn* index is the ratio of the minimum distance between clusters to the maximum diameter of clusters, and the maximum value of the index provides the appropriate partitioning of data. The *Davies-Bouldin (DB)* index is the ratio of the sum of the within-cluster scatter to the inter-cluster separation, while the minimum of the *DB* index provides the right partitioning of data. On the other hand, the *PBM* index is a composition of three factors, namely, the number of clusters, the measure of cluster compactness and the measure of cluster separation. The index is proposed to be used to form a small number of compact clusters, whereas the maximum value of the index corresponds to the right partitioning of data. The *Silhouette (SIL)* and *SILA* use so-called the *silhouette width* to compute the right partitioning of data. Whereas, the new index *STRv1* consists of the product of two components, which determine changes of cluster compactness. Additionally, the *Accuracy rate* was defined to determine the accuracy of a validity index in detecting the number of clusters. In detail, this *rate* is expressed as follows:

$$\frac{1}{M} \sum_{m=1}^{M} \frac{|p_m - o_m|}{p_m} \tag{15}$$

where M is the number of the tested data sets, the factor p_m denotes the actual number of clusters present in a given data set, and the other factor o_m is the number of clusters provided by the validity index. Notice that when the index ensures perfect results the proposed *rate* is close to 0.

Table 2 presents a comparison of the six indices while taking into account the number of clusters. Note that the *Complete-linkage* method creates compact clusters with diameters roughly equal and it is sensitive to outliers. As it can be seen from the table, the *SILA* index provides the right number of clusters for most data sets, and the *Accuracy rate* is equal to 0.14. For the other indices, i.e. the *Dunn, DB, PBM, SIL* and *STRv1* it equals 2.42, 1.01, 0.20, 0.21 and 0.24, respectively. Thus, the results confirm very good effectiveness of the *SILA*, but the *STRv1* index also has good results when compared to the results provided by the other indices.

Table 3 provides a comparison of these indices for the *EM* method, which looks for Gaussian-shape clusters. Here, the *STRv1* index provides the best results for these data and the *Accuracy rate* value is equal to 0.17. So, the *STRv1* index was an excellent result and it means that this index determines the number of clusters very precisely. For the other indices, i.e. the *Dunn, DB, PBM, SIL* and *SILA*, the *Accuracy rate* equals 2.15, 0.24, 0.34, 0.30 and 0.30, respectively.

Table 2. Comparison of the number of clusters obtained when using the *Complete-linkage* algorithm in conjunction with the *Dunn, DB, PBM, SIL, SILA* and *STRv1* indices. N denotes the actual number of clusters in the data sets. The *Accuracy rate* determines the accuracy of the validity index in detecting the proper number of clusters.

Data set	N	Number of clusters obtained					
		Dunn	*DB*	*PBM*	*SIL*	*SILA*	*STRv1*
Cancer	2	15	2	2	2	2	2
Car	4	29	3	2	3	3	2
Diabetes	2	2	3	2	2	2	3
Ecoli	8	2	5	2	2	2	4
Glass	6	4	10	2	2	5	5
Iris	3	12	3	3	3	3	3
Parkinsons	2	2	2	2	2	2	2
Spectf	2	14	13	2	2	2	2
Transfusion	2	3	3	2	2	2	4
Vertebralcolumn	3	2	2	2	2	2	3
Wine	3	12	12	3	2	3	3
Accuracy rate		2.42	1.01	0.20	0.21	0.14	0.24

Table 3. Comparison of the number of clusters obtained when using the *EM* algorithm in conjunction with the *Dunn*, *DB*, *PBM*, *SIL*, *SILA* and *STRv1* indices. *N* denotes the actual number of clusters in the data sets. The *Accuracy rate* determines the accuracy of the validity index in detecting the proper number of clusters.

Data set	N	Number of clusters obtained					
		Dunn	*DB*	*PBM*	*SIL*	*SILA*	*STRv1*
Cancer	2	2	2	2	2	2	2
Car	4	12	2	2	2	2	2
Diabetes	2	18	4	4	4	4	2
Ecoli	8	4	4	10	4	4	10
Glass	6	7	6	2	2	2	6
Iris	3	2	2	3	2	2	2
Parkinsons	2	7	2	3	2	2	3
Spectf	2	2	2	2	2	2	2
Transfusion	2	13	2	3	2	2	2
Vertebralcolumn	3	14	2	4	2	2	2
Wine	3	6	3	3	3	3	3
Accuracy rate		2.15	0.24	0.34	0.30	0.30	0.17

Table 4. Comparison of the number of clusters obtained when using the *k-means* algorithm in conjunction with the *Dunn*, *DB*, *PBM*, *SIL*, *SILA* and *STRv1* indices. *N* denotes the actual number of clusters in the data sets. The *Accuracy rate* determines the accuracy of the validity index in detecting the proper number of clusters.

Data set	N	Number of clusters obtained					
		Dunn	*DB*	*PBM*	*SIL*	*SILA*	*STRv1*
Cancer	2	2	2	2	2	2	2
Car	4	15	2	2	2	2	2
Diabetes	2	2	4	2	2	2	2
Ecoli	8	3	3	5	3	5	3
Glass	6	4	3	3	5	5	3
Iris	3	2	2	3	2	2	2
Parkinsons	2	4	2	3	2	2	3
Spectf	2	2	2	2	2	2	2
Transfusion	2	15	2	6	2	2	2
Vertebralcolumn	3	2	2	3	2	2	2
Wine	3	4	3	3	3	3	3
Accuracy rate		1.11	0.30	0.35	0.18	0.16	0.25

In Table 4 the results for *K-means* method are shown. It is known that this algorithm often gets stuck at suboptimal configurations. In order to overcome this problem, several re-initializations are used for different initial cluster centers. This algorithm looks for compact clusters around a mean. From this Table it can be seen that the *SILA*, *SIL* and *STRv*1 indices provide the good results and the *Accuracy rate* equals 0.16, 0.18 and 0.25, respectively. For the other indices the *rate* equals 1.11, 0.30 and 0.35, respectively.

To summarize, regardless which one of the three underlying clustering algorithms was used, the *SIL*, *SILA* and *STRv*1 provide the right of clusters for most data sets in terms of the correct indication of the number of clusters.

4   Conclusions

Many cluster validity indices are used to assess data partitioning, but providing the right number of clusters is the most important issue in cluster analysis. However, due to a great variety of data sets many different clustering methods with various properties are applied. Moreover, there is no single validity index which works well in all cases. Hence, development of efficient validity indices is a constant need. To investigate the behaviour of the validity indices, three well-known methods characterizing different approaches to the partitioning of data sets were selected as the underlying clustering algorithms. They are the *Complete-linkage*, *K-means* and *EM* methods. In these experiments, several real-life data sets were used, where the real-life data was from three to forty four dimensional. Moreover, the number of clusters varied within a wide range. The conducted tests on real data sets have proven the efficiency of the *SIL*, *SILA* and *STRv*1 indices compared to the other indices.

References

1. Arbelaitz, O., Gurrutxaga, I., Muguerza, J., Pérez, J.M., Perona, I.: An extensive comparative study of cluster validity indices. Pattern Recogn. **46**, 243–256 (2013)
2. Bezdek, J.C.: Pattern Recognition with Fuzzy Objective Function Algorithms. Plenum Press, New York (1981)
3. Bilski, J., Smoląg, J.: Parallel architectures for learning the RTRN and Elman dynamic neural networks. IEEE Trans. Parallel Distrib. Syst. **26**(9), 2561–2570 (2015)
4. Bilski, J., Wilamowski, B.M.: Parallel learning of feedforward neural networks without error backpropagation. In: Rutkowski, L., Korytkowski, M., Scherer, R., Tadeusiewicz, R., Zadeh, L.A., Zurada, J.M. (eds.) ICAISC 2016. LNCS, vol. 9692, pp. 57–69. Springer, Cham (2016). doi:10.1007/978-3-319-39378-0_6
5. Bilski, J., Kowalczyk, B., Żurada, J.M.: Application of the givens rotations in the neural network learning algorithm. In: Rutkowski, L., Korytkowski, M., Scherer, R., Tadeusiewicz, R., Zadeh, L.A., Zurada, J.M. (eds.) ICAISC 2016. LNCS, vol. 9692, pp. 46–56. Springer, Cham (2016). doi:10.1007/978-3-319-39378-0_5
6. Bradley, P., Fayyad, U.: Refining initial points for k-means clustering. In: Proceedings of the Fifteenth International Conference on Knowledge Discovery and Data Mining, New York, pp. 9–15. AAAI Press (1998)

7. Cpałka, K., Rebrova, O., Nowicki, R., Rutkowski, L.: On design of flexible neuro-fuzzy systems for nonlinear modelling. Int. J. Gen. Syst. **42**(6), 706–720 (2013)
8. Cpałka, K., Rutkowski, L.: Flexible Takagi-Sugeno fuzzy systems. In: Proceedings of the 2005 IEEE International Joint Conference on IJCNN Neural Networks (2005)
9. Davies, D.L., Bouldin, D.W.: A cluster separation measure. IEEE Trans. Pattern Anal. Mach. Intell. **1**(2), 224–227 (1979)
10. Duch, W., Korbicz, J., Rutkowski, L., Tadeusiewicz, R. (eds.): Biocybernetics and Biomedical Engineering 2000. Neural Networks, vol. 6. Akademicka Oficyna Wydawnicza EXIT (2000)
11. Dunn, J.C.: Well separated clusters and optimal fuzzy partitions. J. Cybernetica **4**, 95–104 (1974)
12. Ester, M., Kriegel, H.-P., Sander, J., Xu, X.: A density-based algorithm for discovering clusters in large spatial data sets with noise. In: Proceedings of the Second International Conference on Knowledge Discovery and Data Mining, Portland, pp. 226–231 (1996)
13. Fränti, P., Rezaei, M., Zhao, Q.: Centroid index: cluster level similarity measure. Pattern Recogn. **47**(9), 3034–3045 (2014)
14. Gabryel, M.: A bag-of-features algorithm for applications using a NoSQL database. Inf. Softw. Technol. **639**, 332–343 (2016)
15. Gabryel, M., Grycuk, R., Korytkowski, M., Holotyak, T.: Image indexing and retrieval using GSOM algorithm. In: Rutkowski, L., Korytkowski, M., Scherer, R., Tadeusiewicz, R., Zadeh, L.A., Zurada, J.M. (eds.) ICAISC 2015. LNCS, vol. 9119, pp. 706–714. Springer, Cham (2015). doi:10.1007/978-3-319-19324-3_63
16. Guha, S., Rastogi, R., Shim, K.: CURE: an efficient clustering algorithm for large databases. In: Proceedings of the 1998 ACM-SIGMOD International Conference Management of Data (SIGMOD 1998), pp. 73–84 (1998)
17. Guha, S., Rastogi, R., Shim, K.: ROCK: a robust clustering algorithm for categorical attributes. In: The Proceedings of the IEEE Conference on Data Engineering (1999)
18. Gustafson, E., Kessel, W.: Fuzzy clustering with a fuzzy covariance matrix. In: Proceedings of IEEE CDC (1978). doi:10.1109/CDC.1978.268028
19. Halkidi, M., Batistakis, Y., Vazirgiannis, M.: Clustering validity checking methods: part II. ACM SIGMOD Record **31**(3), 19–27 (2002)
20. Hastie, T., Tibshirani, R., Friedman, J.: The Elements of Statistical Learning. Data Mining, Inference and Prediction. Springer, New York (2001)
21. Hinneburg, A., Keim, D.A.: An efficient approach to clustering in large multimedia databases with noise. In: Knowledge Discovery and Data Mining (1998)
22. Jain, A., Dubes, R.: Algorithms for Clustering Data. Prentice-Hall, Englewood Cliffs (1988)
23. Kaufman, L., Rousseeuw, P.J.: Finding Groups in Data: An Introduction to Cluster Analysis. Wiley, New York (1990)
24. Lago-Fernández, L.F., Corbacho, F.: Normality-based validation for crisp clustering. Pattern Recogn. **43**(3), 782–795 (2010)
25. Lichman, M.: UCI machine learning repository. University of California, School of Information and Computer Science, Irvine, CA (2013). http://archive.ics.uci.edu/ml
26. Meng, X., van Dyk, D.: The EM algorithm - An old folk-song sung to a fast new tune. J. R. Stat. Soc. Ser. B (Methodol.) **59**(3), 511–567 (1997)
27. Murtagh, F.: A survey of recent advances in hierarchical clustering algorithms. Comput. J. **26**(4), 354–359 (1983)

28. Pakhira, M.K., Bandyopadhyay, S., Maulik, U.: Validity index for crisp and fuzzy clusters. Pattern Recogn. **37**(3), 487–501 (2004)
29. Pal, N.R., Bezdek, J.C.: On cluster validity for the fuzzy c-means model. IEEE Trans. Fuzzy Syst. **3**(3), 370–379 (1995)
30. Pascual, D., Pla, F., Sánchez, J.S.: Cluster validation using information stability measures. Pattern Recogn. Lett. **31**(6), 454–461 (2010)
31. Pelleg, D., Moore, A.W.: X-means: extending k-means with efficient estimation of the number of clusters. In: Proceedings of the Seventeenth International Conference on Machine Learning, pp. 727–734 (2000)
32. Rohlf, F.: Single-link clustering algorithms. In: Krishnaiah, P.R., Kanal, L.N. (eds.) Handbook of Statistics, vol. 2, pp. 267–284 (1982)
33. Rousseeuw, P.J.: Silhouettes: a graphical aid to the interpretation and validation of cluster analysis. J. Comput. Appl. Math. **20**, 53–65 (1987)
34. Rutkowski, L., Cpałka, K.: Compromise approach to neuro-fuzzy systems. In: Sincak, P., Vascak, J., Kvasnicka, V., Pospichal, J. (eds.) Intelligent Technologies - Theory and Applications. New Trends in Intelligent Technologies. Frontiers in Artificial Intelligence and Applications, vol. 76, pp. 85–90 (2002)
35. Rutkowski, L., Przybył, A., Cpałka, K., Er, M.J.: Online speed profile generation for industrial machine tool based on neuro-fuzzy approach. In: Rutkowski, L., Scherer, R., Tadeusiewicz, R., Zadeh, L.A., Zurada, J.M. (eds.) ICAISC 2010. LNCS, vol. 6114, pp. 645–650. Springer, Heidelberg (2010). doi:10.1007/978-3-642-13232-2_79
36. Rutkowski, L., Cpałka, K.: A neuro-fuzzy controller with a compromise fuzzy reasoning. Control Cybern. **31**(2), 297–308 (2002)
37. Saha, S., Bandyopadhyay, S.: Some connectivity based cluster validity indices. Appl. Soft Comput. **12**(5), 1555–1565 (2012)
38. Sheikholeslami, G., Chatterjee, S., Zhang, A.: Wave cluster: a multiresolution clustering approach for very large spatial databases. In: Proceedings of the 1998 International Conference on Very Large Data Bases (VLDB 1998), pp. 428–439 (1998)
39. Shieh, H.-L.: Robust validity index for a modified subtractive clustering algorithm. Appl. Soft Comput. **22**, 47–59 (2014)
40. Starczewski, A.: A new validity index for crisp clusters. Pattern Anal. Appl. (2015). doi:10.1007/s10044-015-0525-8
41. Starczewski, A., Krzyżak, A.: A modification of the silhouette index for the improvement of cluster validity assessment. In: Rutkowski, L., Korytkowski, M., Scherer, R., Tadeusiewicz, R., Zadeh, L.A., Zurada, J.M. (eds.) ICAISC 2016. LNCS, vol. 9693, pp. 114–124. Springer, Cham (2016). doi:10.1007/978-3-319-39384-1_10
42. Wang, W., Yang, J., Muntz, M.: STING: a statistical information grid approach to spatial data mining. In: Proceedings of the 1997 International Conference on Very Large Data Bases (VLDB 1997), pp. 186–195 (1997)
43. Weka 3: Data Mining Software in Java. University of Waikato, New Zealand. http://www.cs.waikato.ac.nz/ml/weka/
44. Zhao, Q., Fränti, P.: WB-index: a sum-of-squares based index for cluster validity. Data Knowl. Eng. **92**, 77–89 (2014)
45. Zhang, T., Ramakrishnan, R., Linvy, M.: BIRCH: an efficient data clustering method for very large data sets. Data Min. Knowl. Discov. **1**(2), 141–182 (1997)

Improvement of the Validity Index for Determination of an Appropriate Data Partitioning

Artur Starczewski[1]([✉]) and Adam Krzyżak[2,3]

[1] Institute of Computational Intelligence, Częstochowa University of Technology,
Al. Armii Krajowej 36, 42-200 Częstochowa, Poland
artur.starczewski@iisi.pcz.pl
[2] Department of Computer Science and Software Engineering,
Concordia University, Montreal, Canada
krzyzak@cs.concordia.ca
[3] Department of Electrical Engineering, Westpomeranian University of Technology,
70-313 Szczecin, Poland

Abstract. In this paper a detail analysis of an improvement of the *Silhouette* validity index is presented. This proposed approach is based on using an additional component which improves clusters validity assessment and provides better results during a clustering process, especially when the naturally existing groups in a data set are located in very different distances. The performance of the modified index is demonstrated for several data sets, where the *Complete linkage* method has been applied as the underlying clustering technique. The results prove superiority of the new approach as compared to other methods.

Keywords: Clustering · Cluster validity index · Complete–linkage clustering technique

1 Introduction

Clustering aims at grouping data into homogeneous subsets (called clusters), inside which elements are similar to each other and dissimilar to elements of other clusters. The purpose of clustering is to discover natural existing structures in a data set. These techniques are widely used in various fields such as pattern recognition, image processing, data exploration, etc. It should be noted that due to a large variety of data sets different clustering algorithms and their configurations are formed. Generally, among clustering methods two major categories are distinguished: partitioning and hierarchical clustering. Partitioning clustering relocates elements of a data set between clusters iteratively until a given clustering criterium is obtained. For example, the well-known algorithms of

A. Krzyżak—Carried out this research at WUT during his sabbatical leave from Concordia University.

© Springer International Publishing AG 2017
L. Rutkowski et al. (Eds.): ICAISC 2017, Part II, LNAI 10246, pp. 159–170, 2017.
DOI: 10.1007/978-3-319-59060-8_16

this type include *K-means* and its variations [5,24] or *Expectation Maximization* (*EM*) [19]. On the other hand, hierarchical clustering is based on the agglomerative or divisive approach. The method known as the agglomerative hierarchical clustering starts from many clusters, which are then merged into larger ones until only one cluster has been formed. However, the divisive clustering methods start from a single cluster, which includes all elements of a data set, and then it is split into smaller clusters. For instance, well-known agglomerative hierarchical clustering methods include the *Single-linkage, Complete-linkage* or *Average-linkage* [16,20,25]. Nowadays, a large number of new clustering algorithms appears, e.g., [13,14]. But, for a wide variety of data sets a single clustering algorithm producing optimal data partitions does not exist. Moreover, the same clustering algorithm can also create different partition schemes of data depending on the choice of input parameters. Thus, the question asking how to find the best fit of a partition scheme to a data set is still very relevant.

The process of evaluation of partitioned data is a very difficult task and it is known as cluster validation. In this evaluation process, an estimation of the occurrence of the right clusters is very frequently realized by validity indices. In the literature on the subject, cluster validation techniques are often classified into three groups–external, internal and relative validation [16,31]. The external validation techniques are based on previous knowledge about data. On the other hand, the internal methods use only the intrinsic properties of the data set. The relative techniques compare partition schemes of a data set, which are created by changing values of input parameters of a clustering algorithm. The key parameter for many clustering methods is the number of clusters and this is most frequently changed. Next, the partitions are compared, i.e. depending on the approach used, the maximum or the minimum value of a validity index is used to determine the best fit of a partition scheme to the data set. So far, a number of authors have proposed different validity indices or modifications of existing indices, e.g., [1,11,12,15,17,18,32,36]. In the literature new interesting solutions for cluster evaluation are constantly suggested. For example, a stability index based on variation on some information measures over partitions generated by a clustering model is in [23], a new measure of distances between clusters is proposed in [30]. Papers [33,37] present indices which use the *knee-point* detection. It should be noted that cluster validity indices such as, e.g., the *Dunn* [10], *Davies-Bouldin (DB)* [8], *PBM* [21] or *Silhouette (SIL)* [26] indices are very frequently used to evaluate the efficacy of the new proposed validity approaches in detecting the right data partitioning. It is important to note that clustering algorithms in conjunction with cluster validity indices can be used during a process of designing various neural networks [2–4] and neuro-fuzzy structures [6,7,9,27–29].

In this paper, an improvement of the *Silhouette* index is described. For this purpose the new versions of this cluster validity index called the *SILA* and *SILAv1* have been presented. The first version of the index, i.e. *SILA* is described in paper [34]. The next version is called *SILAv1* and it uses an exponent defined by (9). A detailed explanation of the modifications involving the use of the component is presented in Sect. 2. In order to present effectiveness of the validity indices several experiments were performed for various data sets.

This paper is organized as follows: Sect. 2 presents a detailed description of the *Silhouette* index and an explanation of the proposed modifications of this index. Section 3 illustrates experimental results on data sets. Finally, Sect. 4 presents conclusions.

2   Improvement of the Silhouette Index

First, in this section the *Silhouette* index is described in more detail. Next, a modification of the index and an explanation of this change are presented.

2.1   The Detail Description of the Silhouette Index

Let us denote K-partition scheme of a data set X by $C = \{C_1, C, ..., C_K\}$, where C_k indicates k_{th} cluster, $k = 1, .., K$. Cluster compactness is measured based on a mean of within-cluster distances. The average distance $a(\mathbf{x})$ between element $\mathbf{x}$ and the other elements $\mathbf{x}_k$ belonging to the same cluster is defined as:

$$a(\mathbf{x}) = \frac{1}{n_k - 1} \sum_{\mathbf{x}_k \in C_k} d(\mathbf{x}, \mathbf{x}_k) \tag{1}$$

where n_k is the number of elements in C_k and $d(\mathbf{x}, \mathbf{x}_k)$ is a function of the distance between $\mathbf{x}$ and $\mathbf{x}_k$.

Furthermore, the mean of distances of $\mathbf{x}$ to the other elements $\mathbf{x}_l$ belonging to cluster C_l, where $l = 1, ..., K$ and $l \neq k$, can be written as:

$$\delta(\mathbf{x}, \mathbf{x}_l) = \frac{1}{n_l} \sum_{\mathbf{x}_l \in C_l} d(\mathbf{x}, \mathbf{x}_l) \tag{2}$$

where n_l is the number of elements in C_l. Thus, the smallest distance $\delta(\mathbf{x}, \mathbf{x}_l)$ can be defined as:

$$b(\mathbf{x}) = \min_{\substack{l,k=1 \\ l \neq k}}^{K} \delta(\mathbf{x}, \mathbf{x}_l) \tag{3}$$

The so-called *silhouette width* of element $\mathbf{x}$ can be expressed as follows:

$$S(\mathbf{x}) = \frac{b(\mathbf{x}) - a(\mathbf{x})}{max\,(a(\mathbf{x}), b(\mathbf{x}))} \tag{4}$$

Finally, the *Silhouette* (*SIL*) index is defined as:

$$SIL = \frac{1}{n} \sum_{\mathbf{x} \in X} \frac{b(\mathbf{x}) - a(\mathbf{x})}{max(a(\mathbf{x}), b(\mathbf{x}))} \tag{5}$$

where n is the number of elements in the data set X.

The value of the index is from the range -1 to 1 and a maximum value (close to 1) indicates the right partition scheme. Unfortunately, the index can detect incorrect data partition if differences between cluster distances are large [34].

2.2   Modification of the Silhouette Index

As mentioned above the modification of the *Silhouette* index was proposed in paper [34] and it involves using an additional component $A(\mathbf{x})$, which corrects values of the index. Thus, the new index, called the *SILA* index, in that paper was defined as follows:

$$SILA = \frac{1}{n}\left(\sum_{\mathbf{x}\in X}(S(\mathbf{x})\cdot A(\mathbf{x}))\right) \tag{6}$$

where the $S(x)$ is the *silhouette width* (Eq. (4)), whereas the additional component $A(\mathbf{x})$ was expressed as:

$$A(\mathbf{x}) = \frac{1}{(1+a(\mathbf{x}))} \tag{7}$$

or it can be written as follows:

$$A(\mathbf{x}) = \frac{1}{(1+a(\mathbf{x}))^q} \tag{8}$$

where the exponent $q = 1$.

Note that the value of exponent $q = 1$ can be insufficient for large difference of distances between clusters. Hence, the new modification of the index includes the component $A(\mathbf{x})$ in which the q exponent is defined as below:

$$q = 2 + \frac{K^2}{n} \tag{9}$$

where n is the number of elements in a data set. Thus, the new version of the index so-called $SILAv1$ can be presented in the following way:

$$SILAv1 = \frac{1}{n}\left(\sum_{\mathbf{x}\in X}\left(\frac{b(\mathbf{x})-a(\mathbf{x})}{\max(a(\mathbf{x}),b(\mathbf{x}))}\cdot\frac{1}{(1+a(x))^q}\right)\right) \tag{10}$$

where the q is expressed by (9).

This approach can ensure a better performance of this index than that previous version called the *SILA* and its efficiency was proved based on the experiments carried out on different data sets. In the next section a detailed explanation of the modifications involving the use of the additional component is presented.

2.3   Remarks

As mentioned above, the *Silhouette* index takes values between -1 and 1. Appropriate data partitioning is identified by a maximum value of the index, which can be close to 1. Notice that the definition of the *silhouette width* can be also expressed as follows [26]:

$$S(\mathbf{x}) = \begin{cases} 1 - \frac{a(\mathbf{x})}{b(\mathbf{x})} & if \quad b(\mathbf{x}) > a(\mathbf{x}) \\ 0 & if \quad b(\mathbf{x}) == a(\mathbf{x}) \\ \frac{b(\mathbf{x})}{a(\mathbf{x})} - 1 & if \quad b(\mathbf{x}) < a(\mathbf{x}) \end{cases} \tag{11}$$

Here, it is clear that when $b(\mathbf{x})$ is much greater than $a(\mathbf{x})$, the ratio of $a(\mathbf{x})$ to $b(\mathbf{x})$ is very small, and $S(\mathbf{x})$ is close to 1. But in the modified version of the index, the $SILAv1$ (or $SILA$), the additional component $A(\mathbf{x})$ makes it possible to correct the value of the *silhouette width*. In $A(\mathbf{x})$ a measure of cluster compactness $a(\mathbf{x})$ is used and plays a very important role. For instance, when a clustering algorithm greatly increases sizes of clusters, the factor $a(\mathbf{x})$ also increases and the ratio of $1/(1 + a(x))^q$ decreases significantly. It decreases the value of the index and thus, the large differences of distances between clusters do not affect the final result so much. This modified *silhouette width* can be expressed as follows:

$$S_m(\mathbf{x}) = \begin{cases} \left(1 - \frac{a(\mathbf{x})}{b(\mathbf{x})}\right) \cdot \frac{1}{(1+a(\mathbf{x}))^q} & if \quad b(\mathbf{x}) > a(\mathbf{x}) \\ 0 & if \quad b(\mathbf{x}) == a(\mathbf{x}) \\ \left(\frac{b(\mathbf{x})}{a(\mathbf{x})} - 1\right) \cdot \frac{1}{(1+a(\mathbf{x}))^q} & if \quad b(\mathbf{x}) < a(\mathbf{x}) \end{cases} \quad (12)$$

Let us look at the first situation. When $b(\mathbf{x})$ is greater than $a(\mathbf{x})$, the ratio of $a(\mathbf{x})$ to $b(\mathbf{x})$ is less than 1 and the value of $S_m(\mathbf{x})$ is positive (see Eq. (12)). Notice that when the number of clusters K decreases from K_{max} to a correct number of clusters c^*, then the clusters newly created by a clustering algorithm become larger and the value of $a(x)$ increases. However, the value of $a(\mathbf{x})$ is not very great and the factor $A(\mathbf{x})$ does not decrease so much. Thus, the value of $S_m(\mathbf{x})$ increases and it is only slightly reduced by $A(\mathbf{x})$. Generally, for compact clusters subdivided into smaller ones, when they are merged in larger clusters, the changes of their compactness and separability are not very significant. On the other hand, when the number of clusters K is equal to the right number c^*, the separability of clusters increases abruptly due to relatively large distances between clusters and now $b(\mathbf{x})$ is much larger than $a(\mathbf{x})$. Hence, when $K = c^*$, the component $S_m(\mathbf{x})$ increases considerably. Notice that $A(\mathbf{x})$ does not change significantly, since the changes of clusters compactness are still small and so $a(\mathbf{x})$ does not increase so much. Thus, the value of $S_m(\mathbf{x})$ is not considerably reduced by $A(\mathbf{x})$. In turn, when the number of clusters $K < c^*$, then cluster sizes can be really large and now the factor $a(\mathbf{x})$ strongly increases. Consequently, $A(\mathbf{x})$ decreases significantly and reduces the value of the index. It overcomes problems with too great differences of distances between clusters, and allows for indication of the appropriate data partitioning by the validity index.

The other situation takes place when $a(\mathbf{x})$ and $b(\mathbf{x})$ are equal. This means that it is not clear to which clusters the element should belong. In this case, the $SILAv1$ index (or $SILA$ and *Silhouette* indices) equals 0 (see Eqs. (11) and (12)). The last situation occurs when the factor $a(\mathbf{x})$ is larger than $b(\mathbf{x})$. In this case, the values of $S_m(\mathbf{x})$ (or $S(\mathbf{x})$) are negative. Thus, $\mathbf{x}$ should be assigned to another cluster. Notice that when $b(\mathbf{x})$ is equal to 0, then $S_m(\mathbf{x}) = -1/(1 + a(x))^q$.

As mentioned above, the $SILA$ index uses $q = 1$. However, such value q can cause that the $A(\mathbf{x})$ is too small to appropriately correct the *silhouette width*. However, when q is too large, the influence of $A(\mathbf{x})$ can be very strong and then the value of the index greatly decreases. Hence, the issue of the choice of the exponent q for $A(\mathbf{x})$ is a very significant problem. The new version of the index called $SILAv1$ contains a formula of the change of the exponent q depending on the number of clusters and it is expressed by (9). It should be noted that in this definition the important role is played by the ratio K^2/n, which makes that for large K the value of q is greater than 2 (close to 3) and for small K it is close to 2. This approach causes that the index does not obtain too large values for high K (q is close 3). It is very important because this index is strongly decreased by the component $A(\mathbf{x})$ with q close to 2 for small K, when values of $a(\mathbf{x})$ are large. Thus, component $A(\mathbf{x})$ has now a suitable influence on the index and it makes it possible to overcome the drawback of the $Silhouette$ index, where large differences of distances between clusters can provide incorrect results. It should be observed that the new index can take values between 1 and $-1/(1 + a(x))^q$.

In the next section the results of the experimental studies are presented to confirm effectiveness of this approach.

3   Experimental Results

In this section several experiments were carried out to verify effectiveness of the new index in detecting correct clusters. The experiments have been conducted on different data sets using hierarchical clustering. It should be noted that proper clustering of data is not possible without the knowledge of the right number of clusters occurring in the given data set. Thus, this parameter is a very important for most of the clustering algorithms but it is not usually known in advance. Cluster validity indices are often used to determine this parameter.

The experiments relate to determining the number of clusters in data sets when the $Complete\text{-}linkage$ hierarchical clustering is applied as the underlying clustering method. In each step this algorithm combines the two clusters with the smallest maximum pairwise distance. Furthermore, three validity indices, i.e. the $Silhouette$ (SIL), $SILA$ and $SILAv1$ are used to indicate the right number of clusters. Note that the best range of the number of clusters for data clustering analysis should be varied from $K_{max} = \sqrt{n}$ to $K_{min} = 2$ [22]. However, for the hierarchical clustering the number varies from $K_{max} = n$ to $K_{min} = 2$. To show the efficacy of the proposed approaches the values of validity indices are also presented on the plots, where the number of clusters was from the range $K_{max} = \sqrt{n}$ to $K_{min} = 2$. Moreover, it is assumed that the values of the validity indices are equal to 0 for $K = 1$.

In all the experiments the Weka machine learning toolkit [35] has been used, where the Euclidean distance and the min-max data normalization have been also applied.

3.1   Data Sets

Eight generated artificial data sets are used in the experiments. These data consist of various cluster structure, densities and dimensions. For instance, the first four of them called $Data\ 1$, $Data\ 2$, $Data\ 3$ and $Data\ 4$ are 2- dimensional with

3, 5, 8 and 15 clusters, respectively. The scatter plot of these data is presented in Figs. 1 and 2. Additionally, Table 1 shows a detailed description of all these artificial data. As it can be observed on the plots the distances between clusters are very different and some clusters are quite close. Generally, clusters are located in groups and some of clusters are very close and others quite far. Moreover, the sizes of the clusters are different and they contain various number of elements. Hence, many clusters validity indices can provide incorrect partitioning schemes.

Table 1. Detailed description of the artificial data sets

Data sets	No. of elements	Features	Classes	No. of elements per class
Data 1	300	2	3	50,100,150
Data 2	170	2	5	10,20,30,50,60
Data 3	495	2	8	25,30,50,50,60,80,100,100
Data 4	429	2	15	31,39,38,18,29,30,32,27,10,39,22,27,39,20,28
Data 5	550	3	4	100,100,150,200
Data 6	820	3	6	100,100,100,150, 170,200
Data 7	800	3	7	70,80,100,100,100, 150,200
Data 8	460	3	9	30,30,40,40,50,50, 50,70,100

Experiments. The hierarchical *Complete-linkage* method as the underlying clustering method was used for partitioning of these data. In Figs. 3 and 4 a comparison of the variations of the *Silhouette*, *SILA* and *SILAv1* indices with respect to the number of clusters are presented. As mentioned above, on the plots the maximal value of the number of clusters K_{max} is equal to $\sqrt{n}$ and values of the validity indices are equal 0 for $K = 1$. It can be seen that the *SILAv1* index provides the correct number of clusters for all the data sets. However, the previous index *SILA* indicates incorrect partition schemes for two sets, i.e., *Data* 3 and *Data* 6. On the contrary, the *Silhouette* index incorrectly selects all partitioning schemes and this index mainly provides high distinct peaks when the number of clusters $K = 2$. This means that when the clustering method combines clusters into larger ones and differences of distances between them are large, influence of the separability measure is significant and consequently, this index provides incorrect results. On the other hand, despite the fact that the differences of distances between clusters are large, the *SILAv1* index provides the correct partitioning for all these data. Notice that the component $A(\mathbf{x})$ (in the *SILAv1* or *SILA* indices) poorly reduces values of this index when the number of clusters $K > c^*$, because then they are not so large and have a compact structure.

4   Conclusions

As mentioned above, the *Silhouette* index can indicate an incorrect partitioning scheme when there are large differences of distances between clusters in a

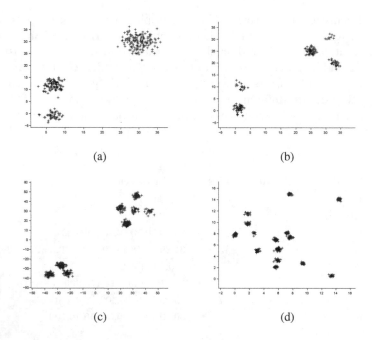

(a)   (b)

(c)   (d)

Fig. 1. 2-dimensional artificial data sets: (a) *Data* 1, (b) *Data* 2, (c) *Data* 3, and (d) *Data* 4

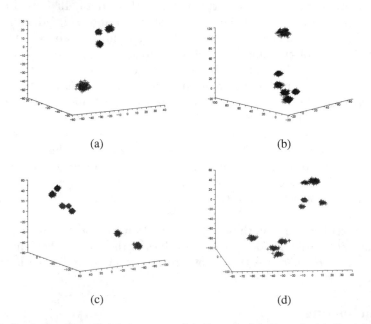

(a)   (b)

(c)   (d)

Fig. 2. 3-dimensional artificial data sets: (a) *Data* 5, (b) *Data* 6, (c) *Data* 7, and (d) *Data* 8

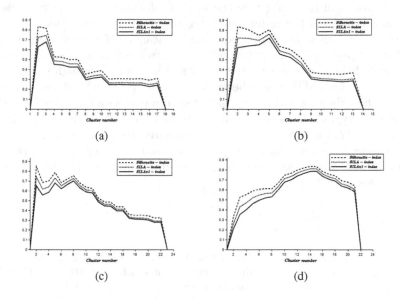

(a)   (b)

(c)   (d)

Fig. 3. Variations of the *Silhouette*, *SILA* and *SILAv*1 indices with respect to the number of clusters for 2-dimensional data sets: (a) *Data* 1, (b) *Data* 2, (c) *Data* 3, and (d) *Data* 4 partitioned by the *Complete-linkage* method.

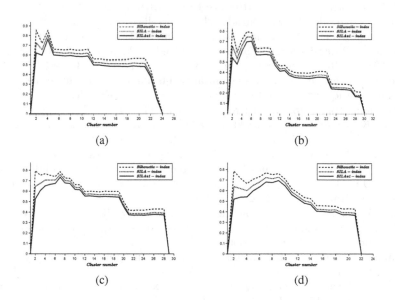

(a)   (b)

(c)   (d)

Fig. 4. Variations of the *Silhouette*, *SILA* and *SILAv*1 indices with respect to the number of clusters for 3-dimensional data sets: (a) *Data* 5, (b) *Data* 6, (c) *Data* 7, and (d) *Data* 8 partitioned by the *Complete-linkage* method.

data set. Consequently, to improve the index performance and to overcome the drawback, a change of the index has been proposed. It is based on the use of the additional component, which contains a measure of cluster compactness. The value of this measure increases when a cluster size increases considerably. Hence, the additional component decreases and it reduces the high values of the index caused by large differences between clusters. As the underlying clustering algorithms the *Complete-linkage* was selected to investigate the behaviour of the proposed validity indices. The conducted tests have proven the advantages of the proposed $SILA$ and $SILAv1$ indices compared to the *Silhouette* index. In these experiments, several data sets were used and the number of clusters varied within a wide range. All the presented results confirm high efficiency of the $SILAv1$ index.

References

1. Arbelaitz, O., Gurrutxaga, I., Muguerza, J., Prez, J.M., Perona, I.: An extensive comparative study of cluster validity indices. Pattern Recogn. **46**, 243–256 (2013)
2. Bilski, J., Smolag, J.: Parallel architectures for learning the RTRN and Elman dynamic neural networks. IEEE Trans. Parallel Distrib. Syst. **26**(9), 2561–2570 (2015)
3. Bilski, J., Wilamowski, B.M.: Parallel learning of feedforward neural networks without error backpropagation. In: Rutkowski, L., Korytkowski, M., Scherer, R., Tadeusiewicz, R., Zadeh, L.A., Zurada, J.M. (eds.) ICAISC 2016. LNCS, vol. 9692, pp. 57–69. Springer, Cham (2016). doi:10.1007/978-3-319-39378-0_6
4. Bilski, J., Kowalczyk, B., Żurada, J.M.: Application of the givens rotations in the neural network learning algorithm. In: Rutkowski, L., Korytkowski, M., Scherer, R., Tadeusiewicz, R., Zadeh, L.A., Zurada, J.M. (eds.) ICAISC 2016. LNCS, vol. 9692, pp. 46–56. Springer, Cham (2016). doi:10.1007/978-3-319-39378-0_5
5. Bradley, P., Fayyad, U.: Refining initial points for K-Means clustering. In: Proceedings of the Fifteenth International Conference on Knowledge Discovery and Data Mining, pp. 9–15. AAAI Press, New York (1998)
6. Cpałka, K., Rebrova, O., Nowicki, R., Rutkowski, L.: On design of flexible neuro-fuzzy systems for nonlinear modelling. Int. J. Gen. Syst. **42**(6), 706–720 (2013)
7. Cpałka, K., Rutkowski, L.: Flexible Takagi-Sugeno fuzzy systems. In: Proceedings of the 2005 IEEE International Joint Conference on Neural Networks IJCNN (2005)
8. Davies, D.L., Bouldin, D.W.: A cluster separation measure. IEEE Trans. Pattern Anal. Mach. Intell. **1**(2), 224–227 (1979)
9. Duch, W., Korbicz, J., Rutkowski, L., Tadeusiewicz, R. (eds.): Biocybernetics and Biomedical Engineering 2000. Neural Networks, vol. 6. Akademicka Oficyna Wydawnicza EXIT, Warsaw (2000)
10. Dunn, J.C.: Well separated clusters and optimal fuzzy partitions. J. Cybernetica **4**, 95–104 (1974)
11. Fränti, P., Rezaei, M., Zhao, Q.: Centroid index: cluster level similarity measure. Pattern Recogn. **47**(9), 3034–3045 (2014)
12. Fred, L.N., Leitao, M.N.: A new cluster isolation criterion based on dissimilarity increments. IEEE Trans. Pattern Anal. Mach. Intell. **25**(8), 944–958 (2003)
13. Gabryel, M.: A bag-of-features algorithm for applications using a NoSQL database. Inf. Softw. Technol. **639**, 332–343 (2016)

14. Gabryel, M., Grycuk, R., Korytkowski, M., Holotyak, T.: Image indexing and retrieval using GSOM algorithm. In: Rutkowski, L., Korytkowski, M., Scherer, R., Tadeusiewicz, R., Zadeh, L.A., Zurada, J.M. (eds.) ICAISC 2015. LNCS, vol. 9119, pp. 706–714. Springer, Cham (2015). doi:10.1007/978-3-319-19324-3_63

15. Halkidi, M., Batistakis, Y., Vazirgiannis, M.: Clustering validity checking methods: part II. ACM SIGMOD Rec. **31**(3), 19–27 (2002)

16. Jain, A., Dubes, R.: Algorithms for Clustering Data. Prentice-Hall, Englewood Cliffs (1988)

17. Kim, M., Ramakrishna, R.S.: New indices for cluster validity assessment. Pattern Recogn. Lett. **26**(15), 2353–2363 (2005)

18. Lago-Fernández, L.F., Corbacho, F.: Normality-based validation for crisp clustering. Pattern Recogn. **43**(3), 782–795 (2010)

19. Meng, X., van Dyk, D.: The EM algorithm - an old folk-song sung to a fast new tune. J. R. Stat. Soc. Ser. B (Methodol.) **59**(3), 511–567 (1997)

20. Murtagh, F.: A survey of recent advances in hierarchical clustering algorithms. Comput. J. **26**(4), 354–359 (1983)

21. Pakhira, M.K., Bandyopadhyay, S., Maulik, U.: Validity index for crisp and fuzzy clusters. Pattern Recogn. **37**(3), 487–501 (2004)

22. Pal, N.R., Bezdek, J.C.: On cluster validity for the fuzzy c-means model. IEEE Trans. Fuzzy Syst. **3**(3), 370–379 (1995)

23. Pascual, D., Pla, F., Sánchez, J.S.: Cluster validation using information stability measures. Pattern Recogn. Lett. **31**(6), 454–461 (2010)

24. Pelleg, D., Moore, A.W.: X-means: extending k-means with efficient estimation of the number of clusters. In: Proceedings of the Seventeenth International Conference on Machine Learning, pp. 727–734 (2000)

25. Rohlf, F.: Single-link clustering algorithms. In: Krishnaiah, P.R., Kanal, L.N. (eds.) Handbook of Statistics, vol. 2, pp. 267–284. Amsterdam, North-Holland (1982)

26. Rousseeuw, P.J.: Silhouettes: a graphical aid to the interpretation and validation of cluster analysis. J. Comput. Appl. Math. **20**, 53–65 (1987)

27. Rutkowski, L., Cpałka, K.: Compromise approach to neuro-fuzzy systems. In: Sincak, P., Vascak, J., Kvasnicka, V., Pospichal, J. (eds.) Intelligent Technologies - Theory and Applications. New Trends in Intelligent Technologies. Frontiers in Artificial Intelligence and Applications, pp. 85–90. IOS Press, Amsterdam (2002)

28. Rutkowski, L., Przybyl, A., Cpałka, K., Er, M.J.: Online speed profile generation for industrial machine tool based on neuro-fuzzy approach. In: Rutkowski, L., Scherer, R., Tadeusiewicz, R., Zadeh, L.A., Zurada, J.M. (eds.) ICAISC 2010. LNCS, vol. 6114, pp. 645–650. Springer, Heidelberg (2010). doi:10.1007/978-3-642-13232-2_79

29. Rutkowski, L., Cpałka, K.: A neuro-fuzzy controller with a compromise fuzzy reasoning. Control Cybern. **31**(2), 297–308 (2002)

30. Saha, S., Bandyopadhyay, S.: Some connectivity based cluster validity indices. Appl. Soft Comput. **12**(5), 1555–1565 (2012)

31. Sameh, A.S., Asoke, K.N.: Development of assessment criteria for clustering algorithms. Pattern Anal. Appl. **12**(1), 79–98 (2009)

32. Shieh, H.-L.: Robust validity index for a modified subtractive clustering algorithm. Appl. Soft Comput. **22**, 47–59 (2014)

33. Starczewski, A.: A new validity index for crisp clusters. Pattern Anal. Appl. 1–14 (2015). doi:10.1007/s10044-015-0525-8

34. Starczewski, A., Krzyżak, A.: A modification of the silhouette index for the improvement of cluster validity assessment. In: Rutkowski, L., Korytkowski, M., Scherer, R., Tadeusiewicz, R., Zadeh, L.A., Zurada, J.M. (eds.) ICAISC 2016. LNCS, vol. 9693, pp. 114–124. Springer, Cham (2016). doi:10.1007/978-3-319-39384-1_10
35. Weka 3: Data Mining Software in Java. University of Waikato, New Zealand. http://www.cs.waikato.ac.nz/ml/weka/
36. Wu, K.L., Yang, M.S., Hsieh, J.N.: Robust cluster validity indexes. Pattern Recogn. **42**, 2541–2550 (2009)
37. Zhao, Q., Fränti, P.: WB-index: a sum-of-squares based index for cluster validity. Data Knowl. Eng. **92**, 77–89 (2014)

Stylometric Features for Authorship Attribution of Polish Texts

Piotr Szwed[✉]

AGH University of Science and Technology, Kraków, Poland
pszwed@agh.edu.pl

Abstract. Authorship attribution aims at distinguishing texts written by different authors using text features representing their styles. In this paper we investigate stylometric features for the Polish language based on Part of Speech (POS) tagging (including POS bigrams) and function words. Due to high inflection level of Polish language the feature space tends to be very large. This in particular concerns POS n-grams. Focusing on POS bigrams, we propose their simplified representation allowing to keep the feature space compact. We report experiments, in which authorship attribution was conducted for varying in lengths documents, with use of classifiers from the Weka library. We evaluate classification results for combinations of the following features: POS tags, POS bigrams, function words and simple document statistics. Experiments indicate that the developed features provide good classification performance.

Keywords: Authorship attribution · Polish texts · Part of speech tagging

1 Introduction

The goal of authorship attribution is to identify the author of an unseen text document having text samples from different authors. In contrast to text categorization, which aims at assigning a topic or a list of topics to a text based on its content, authorship attribution uses domain independent text properties that may be characteristic for a particular author. The problem, apart from literary studies, has many practical application in various fields: intelligence, criminal law, computer forensic, plagiarism detection or author profiling [12].

A common approach to authorship attribution consists in extracting *stylometric features* from documents. Then documents represented as vectors in a feature space can be assessed according to their similarity or the original problem can be transformed into a typical classification task suitable for the machine learning techniques.

Various stylometric features has been proposed and applied in authorship attribution. They can be considered either language independent, e.g. word and punctuation statistics, character n-grams or bound to a particular language as: function words, n-grams of Part of Speech (POS) tags and syntactic rewrite rules [12].

© Springer International Publishing AG 2017
L. Rutkowski et al. (Eds.): ICAISC 2017, Part II, LNAI 10246, pp. 171–182, 2017.
DOI: 10.1007/978-3-319-59060-8_17

Although authorship attribution for text in various languages has been extensively researched, the number of studies addressing specifically Polish language is limited. Our goal is to find stylometric features that would provide good performance for Polish texts. We start with cases with controlled complexity, before turning to harder tasks, as analyzing streams of short messages written by multiple authors.

In our previous work [16] we investigated features based on POS tags and a limited set of their patterns. In this paper we extend them with POS bigrams and function words. Such features were successfully used in authorship attribution of English texts producing compact feature spaces (e.g. up to 1000 bigrams and 50 function words) [3,7]. In contrast to English, Polish is a highly inflected language and the number of POS tags that are used to describe words ranges at 1000. This potentially results in a high-dimensional feature space representing POS n-grams, which would be computationally inefficient in classification tasks. Focusing on POS bigrams, we propose their simplified representation allowing to keep the feature space compact. The level of inflection also influences the size of the dictionary of function words.

We investigate the proposed stylometric features in experiments performed on a documents extracted from Polish novels by four authors. Applying typical machine learning technique, we use selected classifiers from the Weka library and compare their performance. We also report experiments aiming at feature selection.

The paper is organized as follows: next Sect. 2 provides a short review of authorship attribution problems. It is followed by Sect. 3, which describes the datasets used. Section 4 presents sets of features used in classification. It is followed by Sect. 5 reporting results of conducted experiments. Section 6 provides concluding remarks.

2   Related Works

An important problem in authorship attribution is the choice of stylometric features that are "linguistic expressions" of particular authors [3]. Sets of proposed features may vary, depending on available data, the intended generality of their extraction method and applicability to specific languages.

The simplest features describe *statistical properties* of documents: word length, sentence length and vocabulary richness. *Function words* are features based on word frequencies. In contrast to text categorization problems, where the most frequent words are regarded useless or even harmful for classification, in authorship attribution problems they are often used as personal style markers. However, not all the most frequent words are good candidates to be included to that set of features: an important characteristic is *instability* [6], i.e. the possibility to be replaced by another word from the dictionary.

Another word-based features are *word sequences* (n-grams). An example of this approach can be found in [1], where classification using word sequences was tested on 350 poems in Spanish by 5 authors giving about 83% accuracy.

Features, which usually give very high accuracy measures are *character n-grams*, i.e. sequences of n characters extracted from words appearing in documents. They are considered language independent, i.e. they can be extracted from texts in various languages regardless of character sets used. See for example [5] for reports on authorship attribution of English, Greek and Chinese texts. In our opinion very good results of their application should be treated with caution: there is an obvious functional dependence between document content and character n-grams, so they may constitute and alternative representation of function words (what is probably good) or they may just render document content (what seems to be worse).

Feature that apparently abstract away from topic related vocabulary are *part of speech tags* and their n-grams. They were used in experiments, where POS n-grams were used in authorship attribution of English documents [3,7]. Another type of *syntactic features* are rewrite rules and chunks [12].

An issue related to authorship attribution is the difficulty of the particular case [7]. Generally, the problems, where the number of authors is small and large amounts of data samples are available are considered easy and high accuracy is expected. This regards artificially generated data, e.g. extracts from full novels [3,13]. On the other side, the difficulty increases with growing number of authors and smaller data size [9], what results in inferior accuracy measures.

Research related to stylometry in various languages, including Polish, was conducted by members of Computational Stylistcs Groups[1]. In their works they used frequent words [11], and character n-grams [2].

Typical Machine Learning approach for authorship attribution of Polish newspaper articles was repoted in [8]. Authors compared classifier performance for various feature sets including character n-grams, word n-grams, lemmas, POS tags and function words. This publication is is particularly interesting in the context of this work, as it is probably the single source of information on usage of POS tags and function words in classification of Polish texts. The experiments showed that the best accuracy was achieved with use of character n-grams and lemmas.

3   Datasets

Datasets used in the experiments were extracted from five classic Polish novels written at the turn of twentieth century: *Ziemia obiecana* (Eng. *Promised Land*) by Władysław Reymont, *W pustyni i w puszczy* (Eng. *In Desert and Wilderness*) and *Rodzina Połanieckich* (Eng. *Połaniecki family*) both by Henryk Sienkiewicz, *Syzyfowe prace* (Eng. *Sisyphus works*) by Stefan Żeromski and *Na srebrnym globie* (Eng. *On the Silver Globe*) by Jerzy Żuławski.

The texts were split into sentences and each sentence was marked as either narrative (N) or dialog (D). Then input documents were created by merging n consecutive sentences, where $n = 5, 10, 20, ..., 100$. The group of datasets

---

[1] https://sites.google.com/site/computationalstylistics/.

marked as A* (A5, A10,..., A100) comprise documents formed from all sentences. Datasests belonging to the group N* (N5, N10,..., N100) contain only documents build from narrative sentences, whereas D* only dialogs.

Such division was made to verify a hypothesis that author's style manifests itself primarily in narrative sentences, whereas dialogs rather mimic the way of speaking of portrayed persons, often they employ expressions observed in the real life and used in less formal communication. It was expected, that authorship attribution should give the best results for narrative documents, and the worse for dialog.

The statistics for the source data is given in Table 1. The number of instances in a particular dataset An can be derived by dividing total number of sentences by n.

Table 1. Source data summary: $\#s$ - number of sentences, $\#w$ - number of words, $\frac{\#w}{\#s}$ - average sentence length.

Author	A* (all)			N* (narrative)			D* (dialog)		
	$\#s$	$\#w$	$\frac{\#w}{\#s}$	$\#s$	$\#w$	$\frac{\#w}{\#s}$	$\#s$	$\#w$	$\frac{\#w}{\#s}$
Reymont	13 649	179 515	13.2	8 244	127 385	15.5	5 405	52 130	9.6
Sienkiewicz	22 522	320 515	14.2	17 085	268 737	15.7	5 437	51 778	9.5
Żeromski	4 323	63 359	14.7	3 600	56 901	15.8	723	6 458	8.9
Żuławski	3 932	57 100	14.5	3 558	53 967	15.2	374	3 133	8.4
Total	44 426	620 489	14.0	32 487	506 990	15.6	11 939	113 499	9.5

4   Features

Based on the documents' content four sets of features were computed: *POS* – part of speech tags, *2POS* – POS bigrams, *FW* - function words and *Stat* – 10 features capturing such document properties, as numbers of words and punctuation marks (including mean values, variance and standard deviation).

4.1   Part of Speech Tags

As POS tagger we used Morfologik [10], which is both a a comprehensive dictionary of Polish inflected forms based on PoliMorf [17], and a software library written in Java offering stemming and tagging functions. The dictionary distinguishes 25 basic part of speech elements, however each of them can be assigned with supplementary attributes indicating case, conjugation form, gender, etc. Hence, the total number of various tags appearing in the dictionary is equal to 1170. It should be noted, that Morfologik frequently returns multiple lemmas and sets of POS tags for a given inflected word form. An example is given in Table 2. The first tag component denotes the basic POS element (*adj* and *subst*

Table 2. Sample entry from the Morfologik dictionary.

Inflected form	Stem (lemma)	POS tags
Czarnym	Czarna	subst:pl:dat:f
	Czarny	adj:pl:dat:m1.m2.m3.f.n1.n2.p1.p2.p3:pos
		adj:sg:inst:m1.m2.m3.n1.n2:pos
		adj:sg:loc:m1.m2.m3.n1.n2:pos
		subst:pl:dat:m1
		subst:sg:inst:m1
		subst:sg:loc:m1

stands for adjective and noun), whereas other components represent attributes (*pl* – plural, *sg* – singular, *dat, inst, loc* - declination case, *m1, m2, m3, f, n1, n2* - gender, etc.) Attributes are specific with respect to elements, e.g. declination case can be attributed to nouns and adjectives, whereas reflexity to verbs.

During extraction of POS features from a document, a large string of all possible POS tags attributed to words. Then the string was converted to numeric attributes representing tag frequencies with Weka filter StringToWordVector configured to apply IDF transform. As the number of POS tags used in the documents was over 800, hence the corresponding number of POS features was extracted.

4.2 POS Bigrams

The specificity of Polish language makes the problem of finding adequate representation of POS *n*-grams hard. Firstly, the POS tagging is ambiguous, i.e. each word can be attributed with a set of POS tags and observed transitions occur rather between sets and not just individual tags. Secondly, the number of possible POS tags (ranging at 1000) may potentially result in very high dimensional feature space, which would be computationally inefficient.

While extracting n-gram features, we decided to consider bigrams only and introduced a simplification that resulted in a relatively compact representation counting about 1000 attributes.

Let us denote by by $\mathcal{P}$ a set of POS tags. For analysis purposes we assume a unified structure of tags as tuples of basic POS elements and *all* attributes:

$$\mathcal{P} = E \times A_1 \times \cdots, \times A_k,$$

where E is the basic set of 25 POS tags (see nodes in Fig. 1a) and A_i are sets of attributes. Each set A_i contains an additional element ϵ_i (equivalent to a missing attribute) that should be used, if an attribute is not applicable to a given element.

Let $POS : W \to \mathcal{P}$ be a function that returns a set of tags for a given word $w \in W$. We define also two functions $elm : \mathcal{P} \to E$ and $att_i : \mathcal{P} \to A_i$. They allow to extract POS elements or *i*-th attributes from tags.

For two consecutive words (w_1, w_2) the transitions in E and $A_i, i = 1, k$ are processed separately using the following procedure:

1. Calculate $POS(w_1)$ and $POS(w_2)$
2. Compute the set of transitions in E as:
   $T_E(w_1, w_2) = \bigcup_{p \in POS(w_1)} elm(p) \times \bigcup_{p \in POS(w_2)} elm(p)$
3. For $i = 1, k$: compute transitions in A_i as:
   $T_i(w_1, w_2) = \bigcup_{p \in POS(w_1)} att_i(p) \times \bigcup_{p \in POS(w_2)} att_i(p)$
4. Assign the frequency $\frac{1}{|T_E(w_1, w_2)|}$ for each transition $t_{ej} \in T_E(w_1, w_2)$
5. Analogously, assign frequencies $\frac{1}{|T_i(w_1, w_2)|}$ for transitions $t_{ij} \in T_i(w_1, w_2)$

The transition frequencies are summed for all pairs of consecutive words appearing in the document.

Figure 1 shows an example of POS bigrams representation for the whole novel by Reymont. Figure 1a presents frequencies for transitions in E (POS elements) and Fig. 1b for declination cases. A line (wedge) width corresponds to the observed frequency. It can be seen that certain transitions are more frequent, eg. between verbs, nouns, adjectives and prepositions. The term *UDecl* in Fig. 1b stands for missing declination attribute. The diagram shows that that after a word, for which a declination case does not apply, occurrences of nouns or adjective in nominative, accusative or genitive case are more likely.

The described representation yields about 1000 attributes, hence, it is much more compact than full POS bigrams model that would potentially comprise $800^2 = 640000$ features. Apparently, it is a simplification, as it does not capture dependencies between POS elements and applicable attributes. This also applies to grammar rules, which would potentially constrain their values, eg. a noun case governs the case of an adjective used as a complement. This approach can be compared to the Naïve Bayes model, where correlations between variables are ignored for the sake of computational efficiency.

4.3   Function Words

We performed a few experiments with authorship attribution based on documents' content using decision trees. Analysis of the obtained trees revealed that most relevant features selected by the classifier are proper names and words with no particular reference to the topic like prepositions or pronouns. Whereas the first group of words is apparently domain specific, the second seems to constitute authors' personal style markers.

A set of function words was built based on three criteria: part of speech element type, frequency and domain independence.

The basic dictionary of function words included: prepositions (*do*, *w*, *dookoła*), pronouns (*ja*, *ty*, *on,my*), conjunctions linking relative clauses (*że*, *aby*), numerals (*pierwszy*, *drugi*), predicates (*trzeba*) and interjections (words of phrases used as exclamations: *och*, *ach*).

We extracted the most frequent words from the input documents and cross-checked it with the set of function words. This allowed us to collect about 90

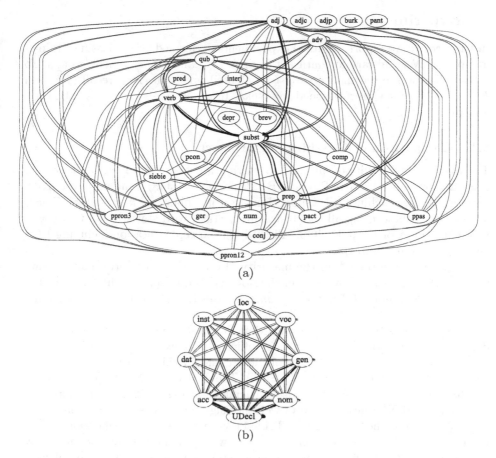

(a)

(b)

Fig. 1. Illustration of Part of speech bigrams extracted from documents by Reymont; the edge width corresponds to the observed frequency: (a) basic POS tags (b) declination case bigrams (*UDecl* stands for missing declination attribute)

additional words (lemmas) that can be considered domain independent. According to the Morfologik dictionary they are classified (sometimes surprisingly) as nouns or adjectives. Examples of such words are: *to, ten, tamto, tamten, tędy, tamtędy, pan, pani, jako, pewny,* etc.

The dictionary of function words is not stored explicitly, it is rather represented as a procedure. When the procedure was applied to the whole Morfologik dictionary it yielded 3321 words. The number of terms appearing in the processed documents classified as function words was equal to 1007. This number is large, if compared with typical settings for English texts classification, where about 50 most frequent words were used [4]. An examination of the determined set revealed that many of the words are inflected forms of a common lemmas, abbreviations, numerals and interjections.

5   Experiments

In this section we report results of experiments performed using classifiers implemented in Weka library for datasets from A*, N* and D* document groups. To assess classifiers performance we used weighted precision (Pr), recall (Rc) and $F1 = \frac{2 \cdot Pr \cdot Rc}{Pr + Rc}$ measures returned by Weka. In all cases 10-fold cross-validation was preformed.

5.1   Comparison of Classifier Performance

Table 3 presents results of authorship attribution tests, in which the following classifiers were used: Naïve Bayes (NB), decision tree (J48), random forest (RF), support vector machines (SMO) and k nearest neighbors for $k = 5$ (IBk). Calculations were made using all sets of features: Stat (word statistics), POS (tags), FW (function words) and 2POS. All (full bigram representation including POS elements and attributes).

As it can be observed, in the majority of cases classifier performance grows with n (the number of sentences in the documents), in spite of the fact that the number of documents (observations) m decreases (m it is inversely proportional to n, i.e. $m = \frac{\#sentences}{n}$).

In most cases SMO classifier with default parameters (polynomial kernel of degree 1) turned out to be the most efficient. As expected, for D* datasets the performace measures were visibly lower and in this case the J48 classifier turned out to be competitive to the SMO. The differences between classification results for documents in A* datasets (being mixtures of dialog and narrative sentences) and N* (narrative only) were not significant. It should be noted that the imbalance in the datasets (c.f. Table 1) was reflected in classification results for particular authors (they were generally worse in the case of documents by Żeromski and Żuławski). Another observation is related to the time complexity. Typically, SMO and J48 were by order of magnitude slower than Naïve Bayes, especially for large datastes (e.g. A5, which comprises 8887 documents).

5.2   Feature Selection: Sets of Attributes

We compared performance of two classifiers Naïve Bayes Multinomial (NBM) and Support Vector Machines (SMO) for various combinations of sets of features. The first was selected due to its learning speed, the second due to its proven accuracy. The analyzed sets of features included POS tags, 2POSA (full bigram representation), 2POSB (representation of bigrams covering only transitions between POS elements) and FW (function words). In all cases statistical features (sentence length, punctuation marks) were used.

Figure 2 shows F1 measure evaluated in 10-fold cross validation for various document lengths and combinations of features. It can be observed that in all cases the best results were obtained in experiments, where function words (FW) were used. The performance can be augmented if POS and POS bigrams are used. Basically, SMO outperforms NBM for shorter documents, however, for longer texts the performance of both classifiers is comparable.

Table 3. Comparison of classifier performance. NB - Naïve Bayes, J48-decision tree, RF - random forest, SMO - support vector machines, IBk - k nearest neighbors ($k = 5$)

Data	NB			J48			RF			SMO			IBk		
	Pr	Rc	F1	Pr	Rc	F1	Pr	Rc	F1	Pr	Rc	F1	Pr	Rc	F1
A5	0.595	0.529	0.545	0.684	0.690	0.686	0.735	0.661	0.596	0.825	0.825	**0.824**	0.517	0.501	0.484
A10	0.661	0.616	0.629	0.736	0.740	0.738	0.773	0.711	0.656	0.880	0.880	**0.879**	0.643	0.572	0.559
A20	0.720	0.693	0.702	0.787	0.788	0.787	0.809	0.768	0.722	0.946	0.946	**0.946**	0.764	0.673	0.668
A30	0.757	0.743	0.749	0.811	0.814	0.812	0.819	0.788	0.744	0.975	0.974	**0.974**	0.813	0.743	0.732
A50	0.802	0.796	0.798	0.826	0.829	0.827	0.858	0.834	0.805	0.988	0.988	**0.988**	0.885	0.843	0.833
A80	0.828	0.832	0.828	0.853	0.855	0.854	0.877	0.859	0.835	0.989	0.989	**0.989**	0.910	0.880	0.870
A100	0.878	0.877	0.874	0.871	0.873	0.871	0.880	0.862	0.838	0.993	0.993	**0.993**	0.935	0.920	0.913
N5	0.645	0.582	0.600	0.684	0.688	0.686	0.718	0.630	0.556	0.826	0.825	**0.825**	0.549	0.520	0.509
N10	0.708	0.677	0.687	0.726	0.728	0.727	0.765	0.682	0.625	0.892	0.892	**0.891**	0.671	0.600	0.590
N20	0.748	0.732	0.737	0.787	0.792	0.789	0.809	0.754	0.718	0.957	0.956	**0.956**	0.803	0.711	0.717
N30	0.796	0.793	0.793	0.795	0.798	0.796	0.832	0.794	0.765	0.977	0.977	**0.977**	0.844	0.789	0.792
N50	0.833	0.832	0.830	0.868	0.871	0.869	0.857	0.829	0.807	0.990	0.989	**0.989**	0.909	0.882	0.880
N80	0.890	0.890	0.887	0.859	0.850	0.853	0.888	0.873	0.859	0.998	0.998	**0.998**	0.947	0.934	0.932
N100	0.892	0.893	0.890	0.895	0.896	0.895	0.910	0.899	0.891	1.000	1.000	**1.000**	0.956	0.948	0.948
D5	0.587	0.396	0.454	0.795	0.808	**0.801**	0.680	0.749	0.713	0.731	0.738	0.733	0.495	0.523	0.503
D10	0.591	0.512	0.543	0.825	0.835	**0.830**	0.700	0.771	0.734	0.751	0.762	0.751	0.549	0.575	0.557
D20	0.643	0.643	0.643	0.840	0.845	0.842	0.707	0.778	0.741	0.867	0.868	**0.856**	0.651	0.655	0.645
D30	0.693	0.708	0.696	0.862	0.863	0.862	0.770	0.850	0.808	0.899	0.895	**0.882**	0.663	0.673	0.647
D50	0.698	0.731	0.707	0.868	0.864	0.865	0.782	0.864	0.821	0.899	0.901	**0.879**	0.719	0.748	0.718
D80	0.732	0.757	0.732	0.878	0.888	**0.883**	0.794	0.875	0.832	0.838	0.888	0.853	0.818	0.809	0.795
D100	0.672	0.730	0.694	0.850	0.885	0.864	0.799	0.877	0.836	0.853	0.902	**0.868**	0.872	0.885	0.866

Table 4. Feature selection: BF-CFS: BestFirst-CfsSubsetEval, Rank-Corr: Ranker-CorrelationAttributeEval, Rank-IG: Ranker-InfoGainAttributeEval

Data	Rank-Corr			Rank-IG			BF-CFS		
	Pr	Rc	F1	Pr	Rc	F1	Pr	Rc	F1
A5	0.820	0.820	0.819	0.820	0.820	**0.820**	0.779	0.784	0.773
A10	0.883	0.883	**0.883**	0.883	0.883	**0.883**	0.874	0.876	0.873
A20	0.952	0.952	**0.951**	0.952	0.952	**0.951**	0.934	0.934	0.933
A30	0.975	0.975	**0.975**	0.975	0.975	**0.975**	0.961	0.961	0.961
A50	0.988	0.988	**0.988**	0.988	0.988	**0.988**	0.979	0.979	0.979
A80	0.989	0.989	**0.989**	0.989	0.989	**0.989**	0.986	0.986	0.986
A100	0.993	0.993	0.993	0.993	0.993	0.993	0.996	0.996	**0.995**
N5	0.828	0.827	0.827	0.828	0.828	**0.828**	0.776	0.779	0.773
N10	0.900	0.899	**0.899**	0.900	0.899	**0.899**	0.884	0.886	0.885
N20	0.961	0.961	**0.960**	0.961	0.961	**0.960**	0.937	0.937	0.937
N30	0.979	0.979	**0.979**	0.979	0.979	**0.979**	0.956	0.957	0.956
N50	0.988	0.988	0.988	0.988	0.988	0.988	0.991	0.991	**0.991**
N80	0.998	0.998	0.998	0.998	0.998	0.998	0.998	0.998	**0.998**
N100	1.000	1.000	**1.000**	1.000	1.000	**1.000**	1.000	1.000	**1.000**
D5	0.730	0.740	0.734	0.730	0.740	0.734	0.841	0.843	**0.813**
D10	0.765	0.775	0.767	0.765	0.775	0.767	0.876	0.883	**0.868**
D20	0.860	0.862	0.850	0.860	0.862	0.850	0.870	0.877	**0.868**
D30	0.923	0.918	0.904	0.923	0.918	0.904	0.945	0.945	**0.942**
D50	0.893	0.897	0.872	0.893	0.897	0.872	0.952	0.950	**0.949**
D80	0.838	0.888	0.853	0.838	0.888	0.853	0.935	0.934	**0.933**
D100	0.852	0.902	0.868	0.852	0.902	0.868	0.948	0.943	**0.942**

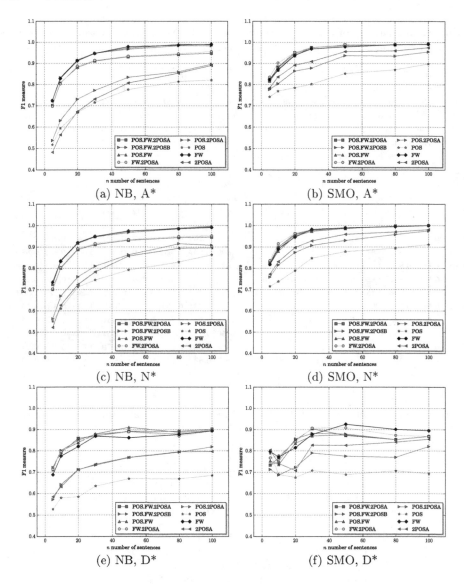

Fig. 2. F1 measure values for various various document lengths (n-number of sentences) and subsets of features. Classifiers: NB (Naïve Bayes Multinomial) and SMO (Support Vector Machines). Document gropus: A* (all sentences) N* (narrative sentences), D* (dialogs)

5.3   Feature Selection: Individual Attributes

We performed feature selection for individual attributes applying three combinations of search and evaluation methods available in Weka: BF-CFS (BestFirst, CfsSubsetEval), Rank-Corr (Ranker, CorrelationAttributeEval) and Rank-IG

(Ranker, InfoGainAttributeEval). The dataset with reduced set of features was further submitted to SMO (support vector machines) classifier. The results are gathered in Table 4. It can be observed that for datasets from A* and N* feature selection occurred ineffective (c.f. Table 3). However, for datasets from D* group, which are supposed to be noisy, application of BF-CFS improved the performance by almost 10%.

6 Conclusions

In this paper we report experiments of authorship attribution for Polish texts using features based on POS tags, their bigrams and function words. These features are language specific. In the case of the Polish language, which is characterized by high variability of inflected forms, the feature space is potentially very large. This in particular concerns POS bigrams, as more than 1000 POS tags are present in Polish dictionaries, whereas commonly used Brill tagger for English returns about 40 tags. Therefore, focusing on POS bigrams, we propose their simplified representation allowing to keep the feature space compact.

We performed evaluation of classifier performance and analyzed various combinations of sets of features. We also performed feature selection using three combinations of search and evaluation methods. Results for other experiments are not reported, because either they were partial due to unacceptable processing time or had inferior performance, as is the case of PCA.

SMO outperforms other classifiers, however, its training is very time consuming, especially if a great number of documents is to be processed. We also tried to use Multilayer Perceprton, which was reported to give promising results in authorship attribution of English texts [13]. Unfortunately, due to dimensionality of the feature space (about 2000 attributes) the method occurred too slow to be used.

Experiments with various sets of features showed that the presence of function words has the greatest impact on the performance. This is consistent with the results from our previous work [16], where we investigated features for the Polish language based on POS tags and their specific patterns used to extract concepts from documents [14,15]. Although the authorship attribution tasks reported in this paper were harder than in [16], the overall performance of classification was about 10% higher in all the cases, when function words were used.

References

1. Coyotl-Morales, R.M., Villaseñor-Pineda, L., Montes-y-Gómez, M., Rosso, P.: Authorship attribution using word sequences. In: Martínez-Trinidad, J.F., Carrasco Ochoa, J.A., Kittler, J. (eds.) CIARP 2006. LNCS, vol. 4225, pp. 844–853. Springer, Heidelberg (2006). doi:10.1007/11892755_87
2. Eder, M.: Style-markers in authorship attribution a cross-language study of the authorial fingerprint. Stud. Pol. Linguist. 6(1), 99–114 (2011)

3. Gamon, M.: Linguistic correlates of style: authorship classification with deep linguistic analysis features. In: Proceedings of the 20th International Conference on Computational Linguistics, COLING 2004. Association for Computational Linguistics, Stroudsburg (2004). http://dx.doi.org/10.3115/1220355.1220443

4. Juola, P.: Authorship attribution. Found. Trends Inf. Retriev. **1**(3), 233–334 (2006)

5. Kešelj, V., Peng, F., Cercone, N., Thomas, C.: N-gram-based author profiles for authorship attribution. In: Proceedings of the Conference Pacific Association for Computational Linguistics, PACLING, vol. 3, pp. 255–264 (2003)

6. Koppel, M., Akiva, N., Dagan, I.: Feature instability as a criterion for selecting potential style markers. J. Am. Soc. Inform. Sci. Technol. **57**(11), 1519–1525 (2006)

7. Koppel, M., Schler, J., Argamon, S.: Authorship attribution: what's easy and what's hard? J. Law Policy **21**, 317–331 (2013)

8. Kuta, M., Puto, B., Kitowski, J.: Authorship attribution of polish newspaper articles. In: Rutkowski, L., Korytkowski, M., Scherer, R., Tadeusiewicz, R., Zadeh, L.A., Zurada, J.M. (eds.) ICAISC 2016. LNCS (LNAI), vol. 9693, pp. 474–483. Springer, Cham (2016). doi:10.1007/978-3-319-39384-1_41

9. Luyckx, K., Daelemans, W.: The effect of author set size and data size in authorship attribution. Literary Linguist. Comput. **26**(1), 35–55 (2011)

10. Miłkowski, M.: Morfologik (2016). http://morfologik.blogspot.com/. Accessed December 2016

11. Rybicki, J.: Success rates in most-frequent-word-based authorship attribution. A case study of 1000 polish novels from ignacy krasicki to jerzy pilch. Stud. Pol. Linguist. **10**(2) (2015). http://www.ejournals.eu/SPL/2015/Issue-2/art/5409/

12. Stamatatos, E.: A survey of modern authorship attribution methods. J. Am. Soc. Inform. Sci. Technol. **60**(3), 538–556 (2009)

13. Stańczyk, U.: The class imbalance problem in construction of training datasets for authorship attribution. In: Gruca, A., Brachman, A., Kozielski, S., Czachórski, T. (eds.) Man–Machine Interactions 4. AISC, vol. 391, pp. 535–547. Springer, Cham (2016). doi:10.1007/978-3-319-23437-3_46

14. Szwed, P.: Concepts extraction from unstructured Polish texts: a rule based approach. In: 2015 Federated Conference on Computer Science and Information Systems (FedCSIS), pp. 355–364, September 2015

15. Szwed, P.: Enhancing concept extraction from polish texts with rule management. In: Kozielski, S., Mrozek, D., Kasprowski, P., Małysiak-Mrozek, B., Kostrzewa, D. (eds.) BDAS 2015-2016. CCIS, vol. 613, pp. 341–356. Springer, Cham (2016). doi:10.1007/978-3-319-34099-9_27

16. Szwed, P.: Authorship attribution for polish texts based on part of speech tagging. In: Mrozek, D., Kozielski, S., Malysiak-Mrozek, B., Kasprowski, P., Kostrzewa, D. (eds.) Proceedings of the 12th International Conference on Beyond Databases, Architectures and Structures. Advanced Technologies for Data Mining and Knowledge Discovery, BDAS 2017, Ustroń, Poland, 30 May–2 June 2017 (2017, to appear)

17. Wolinski, M., Milkowski, M., Ogrodniczuk, M., Przepiórkowski, A.: Polimorf: a (not so) new open morphological dictionary for polish. In: LREC, pp. 860–864 (2012)

Handwriting Recognition with Extraction of Letter Fragments

Michal Wróbel[1]([✉]), Janusz T. Starczewski[1,3], and Christian Napoli[2]

[1] Institute of Computational Intelligence, Częstochowa University of Technology,
Częstochowa, Poland
michal.wrobel@iisi.pcz.pl
[2] Department of Mathematics and Informatics, University of Catania, Catania, Italy
[3] Institute of Information Technology,
Radom Academy of Economics, Radom, Poland

Abstract. This paper is focused on intelligent character recognition of handwritten texts. We apply elements of the handwriting movement analysis in order to calculate possibilities of primitive character fragments called strokes. The key feature rely on the processing of uncertainty in the form of fuzzy quality values starting from the identification of strokes, through the construction of words and phrases, up to future application of language filters and possible contextual recognition.

Keywords: Handwriting recognition · Intelligent character recognition · Letter fragments · Strokes

1 Introduction

Although intelligent hand written character recognition (as a part of Optical Character Recognition *OCR*) has been under development for a quarter of a century, using i.a. convolutional or recurrent neural networks and deep learning techniques [5,6,12], there are still multiple challenges with respect to the quality of intelligent recognition algorithms, as for example announced by the International Conference on Document Analysis and Recognition.

Concerning off-line conversion of text from an image into letter characters, handwriting recognition is different than printed character recognition. In printed text, we can extract each letter, since spaces between letters, words and lines are regular. Handwriting is less precise as many times we are not able to separate a word into letters, especially if there is no language model in opposite to present trends in this area [15].

Our proposal is focused on recognition of handwritten casual cursive of the Latin derived alphabets, i.e. a combination of strokes joined with ligatures end separated with pen lifts. The method must be deprived of any lexicon, as it is primarily intended to discover new words or phrases as well as proper nouns in structured text like eg. archival birth records.

© Springer International Publishing AG 2017
L. Rutkowski et al. (Eds.): ICAISC 2017, Part II, LNAI 10246, pp. 183–192, 2017.
DOI: 10.1007/978-3-319-59060-8_18

The main idea of the method presented in this paper is to apply elements of the handwriting movement analysis, similarly to the on–line OCR techniques, in order to calculate possibilities of primitive character fragments. First of all, a handwritten text is separated into a set of such small fragments. It is intended that each fragment should be perceived as a single "move of a hand", for example an oval or a vertical line. We refer to this fragment of a letter as a "stroke". Then, we are able to search for possible letters strings fitting to a set of extracted strokes. If we find a vertical line next to an oval, it may be a small letter "a", but it may be for example "oi" or "d" as well. The key feature of this approach is the processing of uncertainty in the form of fuzzy quality values starting from the identification of strokes, through the construction of words and phrases, up to language filters and contextual recognition.

The possible secondary use of the proposed method is to improve recognition in standard intelligent character recognition systems based on lexicons or text corpora. Having a set of possible letters together with fuzzy quality and correctness values, we may try to guess a word. In this paper we consider all ambiguities with regard to possible letter strings, namely, our method has been developed as base stage of any dictionary recognition. Our primary simulations have indicated that the proposed method will increase recognition abilities of standard dictionary methods or even add an extra possibility of recognizing, in structured texts, proper nouns or words not found in the lexicon.

2   Preprocessing

In our approach, preprocessing of input image include two steps: binarization and skeletonization (see [16]). We are prohibited to reduce noise, since it may destroy important information desired in subsequent recognition steps. What is pivotal, single pixels will be removed in next steps only if they are out of the context.

3   Strokes Extraction

After binarization and skeletonization, we obtain a binary image, where each letter curve is of one-pixel width. Therefore, we can connect pixels into a lines. To connect each pixel with their nearest neighbour we may use algorithms building minimum spanning tree, for example Kruskall's algorithm. We must modify this algorithm by defining a maximal distance between pixels. So, we built a spanning forest, not a tree. Next, we should delete very small trees (including one or two pixels), since they considered to be a noise.

Afterwards, we can cut trees at pixels having more than two neighbors. Then we have achieve a set of strokes. At this moment, two problems are appeared: strokes which should be merged and strokes with "corners" which should be separated into two strokes.

To describe a merging procedure, we introduce a notion of an *end vector* defined as a vector connecting endpoint of a stroke with a point on the stroke

located at local neighborhood ($\frac{1}{5}$ of length away the endpoint), what is presented
in Fig. 1. Two strokes can be merged if two conditions are fulfilled. Firstly, the
distance between endpoints of these strokes must be small, not greater then maximal distance used in generating spanning forest. Secondly, the angle between
"end vectors" also must be small.

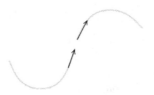

Fig. 1. Merging strokes.   **Fig. 2.** Splitting a stroke.

We have to verify whether the stroke should be split. We assume the stroke
is a one "move of a hand", hence it should not include sharp fragments. Therefore, we must detect splitting points. To detect these points we generate list of
vectors connecting next points in the stroke (first and second, second and third
ctc.). Next, the stroke should be smoothed. Each vector is smoothed by calculating weighted average of this vector (with weight $\frac{1}{2}$) and adjacent vectors (with
weights $\frac{1}{4}$).

$$v'_i = \begin{cases} \frac{3}{4}v_i + \frac{1}{4}v_{i+1} & \text{for } i = 1 \\ \frac{1}{4}v_{i-1} + \frac{1}{2}v_i + \frac{1}{4}v_{i+1} & \text{for } 1 < i < n \\ \frac{1}{4}v_{i-1} + \frac{3}{4}v_i & \text{for } i = n \end{cases} \qquad (1)$$

The sum of all vectors in the stroke will not change. The simple proof is as
follows:

$$\sum_{i=1}^{n} v'_i = \frac{3}{4}v_1 + \frac{1}{4}v_2 + \sum_{i=2}^{n-1}(\frac{1}{4}v_{i-1} + \frac{1}{2}v_i + \frac{1}{4}v_{i+1}) + \frac{1}{4}v_{n-1} + \frac{3}{4}v_n = \sum_{i=1}^{n} v_i \quad (2)$$

After smoothing we can check the angle between vectors, which are located
close, but not adjacent (for example with three vectors between them). Checking
angles between adjacent vectors brings possibility of detecting very local curva-
ture. If the angle is large (greater than right angle), the stroke should be cut. It
is presented in the Fig. 2.

4   Strokes Classification

Our purpose is to describe each stroke by the following characteristic values:

- length (l),
- distance between endpoints d_e,

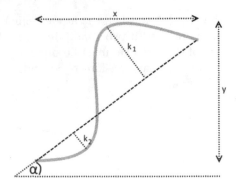

Fig. 3. Values used in stroke description.

- angle of line connecting endpoints (α),
- vertical and horizontal size (x_c, y_c),
- center of mass (average coordinates of all points, x_i, y_i for $i = 1, 2, \ldots$),
- point between endpoints (x_m, y_m).

These values are presented in Fig. 3. With the use of these values we are able to determine some additional values in range $[0, 1]$ describing the shape of the stroke, which are easy to interpret.

$\boldsymbol{d}$ – relative distance, i.e. distance between endpoints divided by the length ($d \approx 1$ for straight stroke, $d \approx 0$ for a circle),
$\boldsymbol{p}$ – ratio of vertical and horizontal size ($p \approx \frac{1}{2}$ for a square),
$\boldsymbol{s}$ – shape characteristic value $\frac{1}{2}sin(2\alpha) + \frac{1}{2}$,
$\boldsymbol{c}$ – shape characteristic value $\frac{1}{2}cos(2\alpha) + \frac{1}{2}$,
$\boldsymbol{h}$ – $\frac{1}{2}tanh((x_m - x_c) * \frac{2}{7}) + \frac{1}{2}$,
$\boldsymbol{v}$ – $\frac{1}{2}tanh((y_m - y_c) * \frac{2}{7}) + \frac{1}{2}$,

The way to calculate value p is following:

$$p = \begin{cases} \frac{x}{2y} & \text{for } x \leq y \\ 1 - \frac{y}{2x} & \text{for } x > y \end{cases} \tag{3}$$

The slope is expressed by sine and cosine 2α, since:

- it's important to get similar values for $\alpha \approx 0°$ and $\alpha \approx 360°$,
- for angles different by $180°$ values c and s should be the same.

A geometric relation between angle and values s and c is presented in Fig. 4.

Values d, p, s, c, h, v do not depend neither on a size nor a location of the stoke, but only depend on its shape. We may compute two strokes using these values. The result of computing r is fuzzy. If the value d is large (stroke is quite straight), the more important is a slope (shape values s and c). If the value d is small, more important is value p. Values h and v are more important if value d

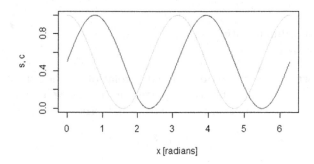

Fig. 4. Slope characteristic values c (bright) and s (dark)

is close to $\frac{1}{2}$. Now we can define a *shape similarity* as a value r calculated as a weighted arithmetic mean

$$r = 1 - \frac{d_d + d_a(s_d + c_d) + (1 - d_a)p_d + d_w \frac{h_d + v_d}{2}}{2 + d_w} \tag{4}$$

where distances are specified as:

$$d_d = |d_1 - d_2|,$$
$$p_d = |p_1 - p_2|,$$
$$s_d = |s_1 - s_2|,$$
$$c_d = |c_1 - c_2|,$$
$$h_d = |h_1 - h_2|,$$
$$v_d = |v_1 - v_2|,$$
$$d_a = \frac{d_1 + d_2}{2},$$
$$d_w = 1 - 2\left|\frac{1}{2} - d_w\right|.$$

Note that values $d_1, p_1, s_1, c_1, h_1, v_1$ describe first stroke while values $d_2, p_2, s_2, c_2, h_2, v_2$ describes the second stroke.

5 Pattern Search

Since each written letter is made up of a few pen strokes, a letter pattern consists of one or more stroke patterns. We may assume one of the strokes for a letter to be the most important, hence called a *dominant stroke*, which is usually the largest one. A dominant stroke pattern can be characterized only by four values d, p, k, s, c, while other stroke patterns are supposed to keep the size and location information:

r_l, length ratio,
r_x, horizontal coordinate ratio,
r_y, vertical coordinate ratio,

All these features should be determined in relation to the size of a dominant stroke. As we find the dominant stroke pattern in a text, we may calculate, how large and where should be placed other stroke patterns.

$$l_p = l_d * r_{lp} \tag{5}$$

$$x_p = x_d + l_d * r_{xp} \tag{6}$$

$$y_p = y_d + l_d * r_{xp} \tag{7}$$

Value l_d is the length of the given dominant stroke, values r_{xp}, r_{yp} are included in each non-dominant stroke pattern.

The shape similarity of a stroke to a non-dominant stroke pattern is defined by Formula 4, thus we can calculate the length and location similarities as following:

$$r_{length} = e^{-\pi(\frac{l_s - l_p}{l_d})^2} \tag{8}$$

$$r_{localization} = e^{-\pi(\frac{\sqrt{(x_s - x_p)^2 + (y_s - y_p)^2}}{l_d})^2} \tag{9}$$

A similarity between non-dominant strokes is the minimum value of the shape, length and localization similarities. Consequently, a letter similarity is the minimum of all stroke similarities.

At this point, a new problem appears. There is much more easier to find simple letter patterns, including only one stroke (as "l" or "o"), than more complex patterns (like "w"). To normalize this situation, we are able to increase similarity of complex patterns with the use of the following formula:

$$s' = s * i^{n-1} \tag{10}$$

where n is a number of strokes in the letter pattern, i is defined as an *multiplicity factor*, i.e. a constant slightly bigger than 1.

Summarizing, an algorithm for searching a letter pattern in a text is following:

- For each stroke in the text:
  - calculate a dominant stroke similarity,
  - for each non-dominant stroke pattern:
    * find the most similar stroke in the text,
  - calculate the minimum all stroke similarities and set it as a letter similarity,
  - if the letter similarity is greater than a given value (called a "minimal similarity"), store the found letter.

In all experiments presented in this paper, the multiplicity factor equal to 1.15 and the minimal similarity equal to 0.8 were selected by trial and error.

After searching for letters, we should erase extraordinary big and relatively small found patterns, since they likely do not represent correct letters and are formed by distortion and noise. To perform this, we may sort all found patterns by their size, e.g. a length of the diagonal, calculate the median and remove found letters, which diagonal is smaller than $\frac{m}{3}$ or greater than $3m$, where m is the length of the median of diagonals.

6   Construction of Text Variants

To avoid cases such that a single stroke belongs to more than one found characters, we explore sets of identified characters, which have no common strokes. Each explored set represent a possible variant of a text string. All these variants are constructed using a decision tree with a modified traversal of depth first search.

The root of the tree does not contain any letter L and has got all explored letters in a set denoted by S. Each node containing set S has got letters from an equivalent set of its parent excluding letter L and all letters having common strokes with L.

The tree is built deep-first. While we are in node N, we add letters from set S as children. To avoid generating identical variants, we should leave a letters from S, which have been already added to o sub-tree with root in the node N.

When the tree is complete, we may explore all variants. Hence, a number of variants equals to a number of leaves (nodes without children). Each variant contains all letters L along the path from the root to the considered leaf. An example of such tree is presented in Fig. 5.

After that, we have to sort letters in each variant by their horizontal coordinates to obtain a text string. If the space between adjacent letters is bigger than $2m$ (m is the length of the median of diagonals), we introduce the space character at this place. Each string has got its fuzzy quality value, which averages quality values of all found letters belonging to the text variant, i.e.,

$$q = \frac{1}{n} \sum_{i=1}^{n} s_i' \frac{1}{max_{s'}} \qquad (11)$$

where n is the number of found letters belonging to the text variant, $max_{s'}$ is the biggest possible value s', which depends on maximal number of strokes in the pattern and value i (see Eq. 10). We use this value to normalize q.

7   Experimental Results

Figure 6 presents initial glyphs of Latin letter patterns loaded to a program before experiments. The experiment was aimed at recognizing all possible variants of text strings collected in Fig. 7. Each variant was measured by the Levenshtein distance between the variant and the correct text string. Subsequently,

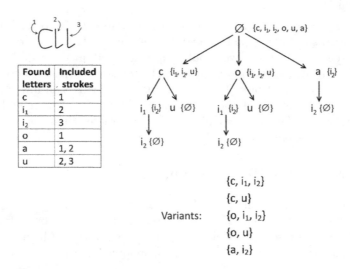

Fig. 5. Example of searching all phrase variants

the distance has been normalized by dividing it by a length of the longest string of a variant and the correct text and subtracting it from 1. As a result we have achieved percent values of correctness, i.e.,

$$c = 1 - \frac{L}{\max\{l_1, l_2\}} \tag{12}$$

where L is the Levenhstein distance and l_1, l_2 are lengths of text strings.

a	b	c	d	e
f	g	h	i	j
k	l	m	n	o
p	r	s	t	u
v	w	x	y	z

Fig. 6. Letter patterns used in experiments

Our results have been compared with the results obtained by FreeOCR software, see Table 1.

Fig. 7. Phrases, which have been recognizing during experiments

Table 1. Comparison of the proposed searching method with FreeOCR

Correct string	No. vars.	Best lexical variant	Qual. q	Corr. c	Most likely variant	Qual. q	Corr. c	FreeOCR result	Corr. c
ala ma kota	12600	aa ma kota	0.610	0.909	aa ma f oza	0.613	0.636	Cl/Lco 'mą |<oÅ3a«	0.333
napis	78	no is	0.585	0.600	mo is	0.590	0.400	'n«a«p'¿s	0.444
poczta	2720	pocztc	0.646	0.833	paztc	0.661	0.500	oczta.	0.667
problem	4500	erblu sm	0.567	0.375	tar lf yn	0.592	0.111	-P<>=e Í=›¿=e„¬<>=.-	0.045
telefon	224	teeom	0.578	0.571	e am	0.596	0.143		0.000
trzy napisy	64	t zj l ssj	0.554	0.364	i tj i sjj	0.568	0.091	W wawa	0.182
Average	33664	—	0.590	0.609	—	0.603	0.313	—	0.279

8 Conclusion

The first results of the proposed method indicate that it has no limitations in searching of non-dictionary expressions, like proper nouns in a structured text. Average correctness of the most likely variants is greater than the one of standard FreeOCR recognitions. Nevertheless, some additional effort should be made to develop and aggregate type-2 fuzzy quality measures to identify the best lexical variants. Although some inconveniences has been noted in recognizing letters p and l, the method has still a great potential with the use of an extensive glyph base. Our future work will be focused on a development of new methods for automatic learning of new letter glyphs (using computational intelligence methods as [1–4,7,8,11,13]) as well as on equipping the handwriting recognition system with a lexicon using for example bag-of-words models [9,10,14].

References

1. Bertini Junior, J.R., Nicoletti, M.D.C.: Enhancing constructive neural network performance using functionally expanded input data. J. Artif. Intell. Soft Comput. Res. **6**(2), 119–131 (2016)
2. Bilski, J., Smoląg, J.: Parallel architectures for learning the RTRN and Elman dynamic neural network. IEEE Trans. Parallel Distrib. Syst. **26**(9), 2561–2570 (2015)
3. Bilski, J., Smoląg, J., Galushkin, A.I.: The parallel approach to the conjugate gradient learning algorithm for the feedforward neural networks. In: Rutkowski, L., Korytkowski, M., Scherer, R., Tadeusiewicz, R., Zadeh, L.A., Zurada, J.M. (eds.) ICAISC 2014. LNCS, vol. 8467, pp. 12–21. Springer, Cham (2014). doi:10. 1007/978-3-319-07173-2_2

4. Bilski, J., Wilamowski, B.M.: Parallel learning of feedforward neural networks without error backpropagation. In: Rutkowski, L., Korytkowski, M., Scherer, R., Tadeusiewicz, R., Zadeh, L.A., Zurada, J.M. (eds.) ICAISC 2016. LNCS, vol. 9692, pp. 57–69. Springer, Cham (2016). doi:10.1007/978-3-319-39378-0_6

5. Burges, C., Ben, J., Denker, J., LeCun, Y.A.N.C.: Off line recognition of handwritten postal words using neural networks. Int. J. Pattern Recogn. Artif. Intell. **7**(4), 689–704 (1993)

6. Ciresan, D.C., Meier, U., Gambardella, L.M., Schmidhuber, J.: Deep big simple neural nets for handwritten digit recognition. Neural Comput. **22**(12), 3207–3220 (2010)

7. Damaševičius, R., Maskelinas, R., Venčkauskas, A., Woźniak, M.: Smartphone user identity verification using gait characteristics. Symmetry **8**(10), 100:1–100:20 (2016)

8. Damaševičius, R., Vasiljevas, M., Salkevicius, J., Woźniak, M.: Human activity recognition in AAL environments using random projections. Comput. Math. Methods Med. **2016**, 4073584:1–4073584:17 (2016)

9. Gabryel, M.: A bag-of-features algorithm for applications using a NoSQL database. In: Dregvaite, G., Damasevicius, R. (eds.) ICIST 2016. CCIS, vol. 639, pp. 332–343. Springer, Cham (2016). doi:10.1007/978-3-319-46254-7_26

10. Gabryel, M.: The bag-of-features algorithm for practical applications using the MySQL database. In: Rutkowski, L., Korytkowski, M., Scherer, R., Tadeusiewicz, R., Zadeh, L.A., Zurada, J.M. (eds.) ICAISC 2016. LNCS, vol. 9693, pp. 635–646. Springer, Cham (2016). doi:10.1007/978-3-319-39384-1_56

11. Harmati, I., Bukovics, D., Kóczy, L.T.: Minkowski's inequality based sensitivity analysis of fuzzy signatures. J. Artif. Intell. Soft Comput. Res. **6**(4), 219–229 (2016)

12. LeCun, Y., Bottou, L., Bengio, Y., Haffner, P.: Gradient-based learning applied to document recognition. Proc. IEEE **86**(11), 2278–2324 (1998)

13. Saitoh, D., Hara, K.: Mutual learning using nonlinear perceptron. J. Artif. Intell. Soft Comput. Res. **5**(1), 71–77 (2015)

14. Woźniak, M., Gabryel, M., Nowicki, R.K., Nowak, B.A.: An application of firefly algorithm to position traffic in NoSQL database systems. In: Kunifuji, S., Papadopoulos, G.A., Skulimowski, A.M.J., Kacprzyk, J. (eds.) Knowledge, Information and Creativity Support Systems. AISC, vol. 416, pp. 259–272. Springer, Cham (2016). doi:10.1007/978-3-319-27478-2_18

15. Zamora-Martínez, F., Frinken, V., España-Boquera, S., Castro-Bleda, M., Fischer, A., Bunke, H.: Neural network language models for off-line handwriting recognition. Pattern Recogn. **47**(4), 1642–1652 (2014)

16. Zhang, T.Y., Suen, C.Y.: A fast parallel algorithm for thinning digital patterns. Commun. ACM **27**(3), 236–239 (1984)

Multidimensional Signal Transformation Based on Distributed Classification Grid and Principal Component Analysis

Marcin Wyczechowski[1], Lukasz Was[1], Slawomir Wiak[1], Piotr Milczarski[2($\boxtimes$)],
Zofia Stawska[2], and Lukasz Pietrzak[1]

[1] Institute of Mechatronics and Information Systems, Technical University of Lodz,
Stefanowski str. 18/22, 90-924 Lodz, Poland
{marcin.wyczechowski,lukasz.was,slawomir.wiak,lukasz.pietrzak}@p.lodz.pl
[2] Faculty of Physics and Applied Informatics, University of Lodz,
Pomorska str. 149/153, 90-236 Lodz, Poland
{piotr.milczarski,zofia.stawska}@uni.lodz.pl
http://www.imsi.pl
http://www.wfis.uni.lodz.pl

Abstract. In the paper, the analysis of audio signal and spectral analysis based on sounds recorded by the authors are proposed. To perform the spectral analysis, the authors apply independent Principal Component Analysis. In this paper, we propose a novel approach to Distributed Classification Grid to improve performance and accelerate execution time.

Keywords: Principal Component Analysis · Spectral analysis · Distributed classification grid

1 Introduction

In this paper, the authors show an approach to audio signal classification process using image data representation. In this paper, we present a novel approach to classification problem using Distributed Classification Grid and algorithms to export from sound to image. From the resulting image, we extract the features that provide us with new interesting features.

We also describe Grid Computing Classification based on segmented signal of sounds recorded by the authors. Figure 1 shows a general form of algorithm that is presented and performed in the paper. Classification is one of computation tasks, nonetheless, the information about what kind of a signal we observe, possess or analyze is very important.

The paper is organized into six sections. In Sect. 2 a general description of the algorithm is presented. In Sect. 3 spectral analysis and estimation method are described. In Sect. 4 spectral analysis is given. In the next Section the research methodology and the results for the example engine sounds are shown and discussed. Conclusions are drawn in Sect. 6.

© Springer International Publishing AG 2017
L. Rutkowski et al. (Eds.): ICAISC 2017, Part II, LNAI 10246, pp. 193–205, 2017.
DOI: 10.1007/978-3-319-59060-8_19

2   Description of the Algorithm

Figure 2 shows a general form of algorithm that we perform. The algorithm used in the paper may be described shortly as follows.

In the first step, audio probes are divided into 1 second parts. Then, we perform frequency domain analysis performing FFT transformation on the segmented signals. Next, a spectrogram image of the sound and spectrogram images for each one second audio segment are built.

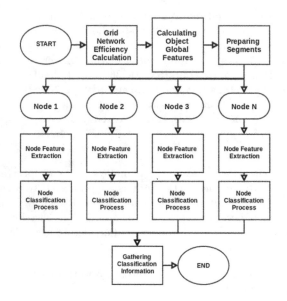

Fig. 1. General form of distributed classification grid.

On every spectrogram the authors use Principal Component Analysis to reduce covariance matrix but they are aware that this may produce some errors in classification afterwards. After that, histograms of each spectrogram are made. Then the heterogeneous distributed classification grids are compared using histogram similarity and correlation metrics. When classification grid using k-NN classification is finished, the weighted voting comes up to classify the signal probe.

Fast Fourier transform (FFT). Fast Fourier transform (FFT) is an algorithm to compute the discrete Fourier transform (DFT) and its inverse. Fourier analysis converts time (or space) to frequency (or wave number) and vice versa. FFT rapidly computes such transformations by factorizing the DFT matrix into a product of sparse (mostly zero) factors. Equation 1 shows how to calculate Discrete Fourier Transform:

$$\hat{x}_k = \frac{1}{N} \sum_{j=0}^{N-1} x_k e^{-2\pi i \frac{jk}{N}}, k = 0, ..., N-1 \qquad (1)$$

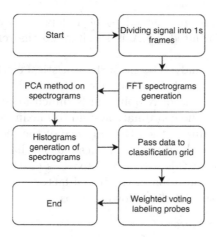

Fig. 2. Algorithm of calculation

and

$$\hat{x} = (\hat{x}_0, ..., \hat{x}_{N-1}) \tag{2}$$

A spectrogram is a visual representation of the spectrum of frequencies in a sound or other signal as they vary in time or some other variable. Spectrograms are sometimes called spectral waterfalls, voiceprints, or voicegrams. Spectrograms can be used to identify spoken words phonetically, and to analyze the various calls of animals. They are used extensively in the development of the fields of music, sonar, radar, and speech processing, seismology.

There are many methods for comparing histograms. We have selected correlation as a method of comparing histograms and it can be defined as in (3):

$$d(H_1, H_2) = \frac{\sum_I (H_1(I) - \bar{H}_1)(H_2(I) - \bar{H}_2)}{\sqrt{\sum_I (H_1(I) - \bar{H}_1)^2 \sum_I (H_2(I) - \bar{H}_2)^2}}, \tag{3}$$

where I means intensity and

$$\bar{H}_k = \frac{1}{N} \sum_J H_k(J), \tag{4}$$

where N is the total number of histogram bins.

Principal component analysis (PCA). Principal component analysis (PCA) is a mathematical procedure that uses an orthogonal transformation to convert a set of observations of possibly correlated variables into a set of values of linear variables called principal components. The number of principal components is less than or equal to the number of original variables. The transformation is defined as one in which the first principal component has the largest possible variance and each succeeding component in turn has the highest variance possible under the constraint that it be orthogonal to the preceding components.

Principal components are guaranteed to be independent only if the data set is jointly normally distributed. PCA is sensitive to the relative scaling of the original variables [11].

Principal component analysis is one of the statistical methods of factor analysis. Data set consisting of N observations, each of which includes the K variables can be interpreted as a cloud of N points in K-dimensional space. The purpose of PCA is the rotation of the coordinate system to maximize the variance in the first place the first coordinate, then the variance of the second coordinate, etc. [11]. The transformed coordinate values are called the charges generated factors (principal components). In this way a new space is constructed of observation, which explains the most variation of the initial factors.

PCA algorithm consists of the following steps:

- Determination of the average.
  This is the first step required to form a covariance matrix of the input matrix. Mathematically, this step can be define as in (5):

$$u[m] = \frac{1}{N} \sum_{n=1}^{N} X[m,n] \tag{5}$$

  The calculated mean values are the characteristics for all observations.
- Calculation of the deviation matrix.
  Since each element of the matrix we subtract the average for the line on which it is located [10]

$$a[i,j] = a[i,j] - u[i] \tag{6}$$

  The matrix obtained in this way will be further denoted as X'
- Determination of the covariance matrix.
  In general, the covariance matrix is calculated from the formula (7):

$$C = E[B \otimes B] = [B \cdot B^*] = \frac{1}{N} B \cdot B^*, \tag{7}$$

  where E is the expected value and B is the matrix of deviations. If the values are real, the matrix B used in the model Hermitian conjugation is identical with the normal transposition.
- The calculation of eigenvectors of the matrix V, which satisfies [10] the following condition (8):

$$V^{-1}CV = D, \tag{8}$$

  where D is the diagonal matrix of eigen values of C [10].

The use of PCA in our case will be described in Sect. 5.

3   Spectral Analysis and Estimation Method

In statistical signal processing, the goal of spectral density estimation (SDE) is introduced to estimate the spectral density of a random signal from a sequence of time samples of the signal. The spectral density is also known as the power spectral density (PSD). Intuitively speaking, the spectral density characterizes the frequency content of the signal. One purpose of estimating the spectral density is to detect any periodicities in the data, by observing peaks at the frequencies corresponding to these periodicities.

The various methods of spectrum estimation are categorized as follows:

- Nonparametric methods
- Parametric methods
- Subspace methods

Nonparametric methods are those in which the PSD is estimated directly from the signal itself. The simplest of a such method is a periodogram. Other nonparametric techniques such as Welch's method and the multitaper method (MTM) reduce the variance of the periodogram.

Parametric methods are those in which the PSD is estimated from a signal that is assumed to be output of a linear system driven by white noise [3,4,16,18]. Examples are the Yule-Walker autoregressive (AR) method and the Burg method. These methods estimate the PSD by estimating the parameters (coefficients) of the linear system that hypothetically generates the signal. They tend to produce better results than classical nonparametric methods when the data length of the available signal is relatively short. Parametric methods also provide smoother estimates of the PSD than nonparametric methods, but are also subject to error from model misspecification.

Subspace methods, also known as high-resolution methods or super-resolution methods, generate frequency component estimates for a signal based on an eigenanalysis or eigendecomposition of the autocorrelation matrix. Examples are the multiple signal classification (MUSIC) method or the eigenvector (EV) method [2,9,16].

These methods are best suited for line spectra that is, spectra of sinusoidal signals. They are effective in the detection of sinusoids buried in noise, especially when the signal to noise ratios are low. The subspace methods do not yield true PSD estimates: they do not preserve process power between the time and frequency domains, and the autocorrelation sequence cannot be recovered by taking the inverse Fourier transform of the frequency estimate.

Figures 3, 4a, b show a general form of performed FFT analysis [2,9,17].

4   Time-Frequency Analysis

In signal processing, time-frequency analysis comprises those techniques that study a signal in both the time and frequency domains simultaneously, using various time-frequency representations. Rather than viewing a 1-dimensional signal (a function, real or complex-valued, whose domain is the real line) and some transform (another function whose domain is the real line, obtained from the original via some transform), time-frequency analysis studies a two-dimensional

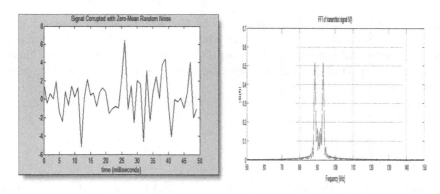

Fig. 3. The diagrams describe method mentioned in Sect. 3

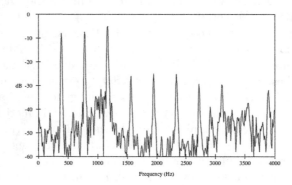

Fig. 4. Full FFT analysis of the signal

signal function whose domain is the two-dimensional real plane, obtained from the signal via a time-frequency transform [1,2,9,15].

The mathematical motivation for this study is that functions and their transform representation are often tightly connected, and they can be understood better by studying them jointly, as a two-dimensional object, rather than separately. A simple example is that the 4-fold periodicity of the Fourier transform and the fact that two-fold Fourier transform reverses direction can be interpreted by considering the Fourier transform as a 90° rotation in the associated time-frequency plane. For such rotations yield the identity, and two such rotations simply reverse direction (reflection through the origin) [6,12,14].

The practical motivation for time-frequency analysis is that classical Fourier analysis assumes that signals are infinite in time or periodic, while many signals in practice are of short duration, and change substantially over their duration. For example, traditional musical instruments do not produce infinite duration sinusoids, but instead they begin with a rapid sound that gradually vanishes. This is poorly represented by traditional methods, which motivates the use of time-frequency analysis [8,13,14].

One of the most basic forms of time frequency analysis is the short-time Fourier transform (STFT), but more sophisticated techniques have been developed, notably wavelets. In signal processing, time-frequency analysis is a body of techniques and methods used for characterizing and manipulating signals whose statistics vary in time, such as transient signals [6,7,14].

5 Results

5.1 Procedure Description

Described in previous sections general procedure was conducted on the sounds with the following steps.

1. First, the authors perform a parametric method in which the PSD is estimated from a signal that is assumed to be output of a linear system driven by white noise. In our case we have chosen as an input car engine sound. We perform time-frequency analysis; we choose the classical Fourier analysis. Then, we build spectrogram out of a sound.
2. We divide the spectrogram image into 1-second segments. We used only 5 seconds of the record from each sound, to have equal-dimension eigenvectors. Then for each segment, we create three normalized histograms in RGB color space, one for each color.
3. From each histogram we derive color values for two biggest local maxima. We obtain a vector of six features for each segment.
4. Then, the authors use the PCA method. We use the largest possible principal component in order to obtain the best classification results set which holds the best data analysis factors [3,10,18].
5. The final step, it is parameterizing the Nodes for weighted voting in Distributed Classification Grid.

5.2 Classification Results

The sounds recorded with two different engines were taken to the research. We labelled that classes as engine 1 and engine 2, correspondingly. We had 146 sound samples. The set of examples was insufficient to divide it to training and testing set, so the authors decided to use a cross-validation as a testing method. Cross-validation method was using 5 sets of 29–30 objects each.

Table 1 shows the results of classification based on Distributed Classification Grid using the signal analysis upon gathered and recorded audio signals of engine sounds. Classification grid was using 3-NN classifier. The classifiers in a grid were making decisions on the basis of different 1-second segments of a signal. Each grid classifier component had the same weight in the weighted voting. The error value of classification rate is due to inaccurate feature extraction and detection in audio signal processing and deficiencies of classification grid.

Table 1. Classification rate based on distributed classification grid

	Number	Percentage of
Correctly classified objects	107	73,19%
Incorrectly classified object	39	26,81%
Mean absolute error	0,237	

5.3 Testable Classification Grid - Network Efficiency

The authors' testable classification grid was built using different architecture machines. The decision not to use for example only x86 architecture was taken because in real world we can face with clients that operates on different types of machines. Testable Classification Grid includes three Intel Core family processors two I7 and two I5, two Raspberry Pi's, and two VPS platforms on OVH hosting. Core i7 computers and one i5 computer and Raspberry Pi were locally connected via 1 Gigabit network. However, Raspberry Pi's onboard Fast Ethernet speed is 100 Mbits.

Grid Elements operate also on Linux Platforms such as Ubuntu 14.04 LTS distribution for x86 machines and derivative of Debian Linux system for ARM machine (Raspberry PI) Raspbian distribution. Authors perform also standard web performance checks for how nodes act in communications with main machine of our structure which is a standard computer with core i7 processor. We conducted ping mean test for 100 times and mean values of measurements are presented in Table 2 as a baseline.

5.4 Graphs of Results

In this part we show the results of computation for the signals. All the results obtained are shown in the figures below for the left channel of the sounds recordings because the figures for the right channel are similar to the left ones.

Graphs of spectrograms. The diagrams presented in Figs. 5, 6, 7, 8, 9, 10, 11 and 12 summarize the approach that is described in the paper that is a part of

Table 2. Mean values of measurements

Machine	Mean ping
Laptop Core i5	0.8 ms
Laptop Core i7	0.7 ms
Raspberry Pi 256 MB RAM	0.9 ms
Raspberry Pi 512 MB RAM	0.9 ms
Core i7 Machine (Internet)	63.2 ms
VPS 1 (Internet)	46.8 ms
VPS 2 (Internet)	49.6 ms

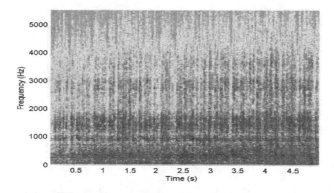

Fig. 5. Spectrogram of a sound of the engine 1 (Color figure online)

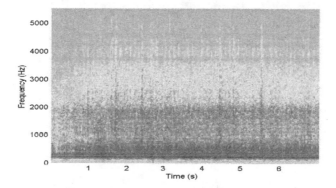

Fig. 6. Spectrogram of a sound of the engine 2 (Color figure online)

a general approach to the frequency analysis combined with parallel computing including factor analysis (PCA) [2,9,10].

The diagrams in Figs. 5 and 6 present the spectral analysis of different engine sounds.

Figures 7 and 8 show their corresponding histograms in the red channel for the time period between 0 a 1st second.

The distribution of red color describes the amount of specific frequencies in a period of time specific in a second time interval for periodicity analysis. The meaning of that is as follows. When we find red color we can establish a larger number of distribution per unit of time, where the distribution of green color stands for the amount of specific frequencies. Hence, when we find green color we can establish a smaller number of distribution per unit time in frequency time analysis [2,9].

Graphs of time-domain analysis. The diagrams in Figs. 9, 10, 11 and 12 describe the time domain and frequency analyses of two different engine sounds. The higher value peaks on the diagrams stand for a better

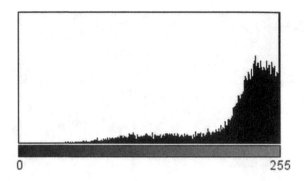

Fig. 7. Histogram of a sound of the engine 1 in the red channel for the time period between 0 a 1st second (Color figure online)

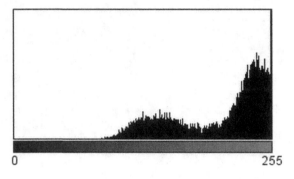

Fig. 8. Histogram of a sound of the engine 2 in the red channel for the time period between 0 a 1st second (Color figure online)

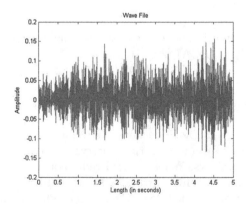

Fig. 9. Time-domain analysis of the sound of the engine 1

frequency distribution per unit of time. Specific loudness shown on the diagrams is a compressive nonlinearity that depends on a level and also on a frequency.

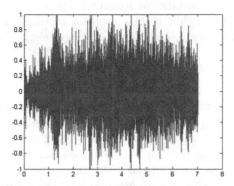

Fig. 10. Time-domain analysis of the sound of the engine2

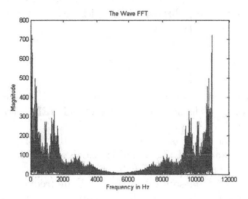

Fig. 11. Frequency analysis of the sound of the engine 1

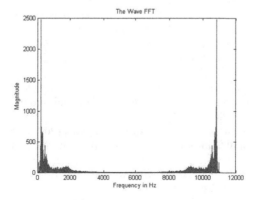

Fig. 12. Time-domain analysis of the sound of the engine 2

The time-corrected instantaneous frequency spectrogram can reasonably be viewed as an excellent solution to the problem of tracking the instantaneous frequencies of the components in a multicomponent signal [2,5,17].

6   Conclusions

The authors have combined two different fields of studies mainly parallel computing and signal analysis. We have shown the process of audio signal classification using image data representation. We performed frequency domain analysis using FFT transformation on the segmented signals. Then, we built the spectrogram images for audio signal and their initial histograms in R, G and B channels separately. We have used Principal Component Analysis to reduce covariance matrix in the spectrograms by analyzing their histograms. We are aware that this may produce some errors in classification afterwards. In the next step, the resulting after PCA enhanced histograms of each spectrogram have been made. Then the heterogeneous distributed classification grids are compared using histogram similarity and correlation metrics.

References

1. Auger, F., Flandrin, P.: Improving the readability of time-frequency and time-scale representations by the reassignment method. IEEE Trans. Signal Process. **43**, 1068–1089 (1995)
2. Bloomfield, P.: Fourier analysis of time series (2000). ISBN:978-0-471-88948-9
3. Briggs, W.L., Van Emden, H.: The DFT: An Owners' Manual for the Discrete Fourier Transform (2002). ISBN:978-0898713428
4. Flandrin, P., Auger, F., Chassande-Mottin, E.: Time-frequency reassignment: from principles to algorithms. In: Papandreou-Suppappola, A. (ed.) Applications in Time-Frequency Signal Processing, pp. 179–203. Boca Raton Press, FL (2003)
5. Friedman, D.H.: Instantaneous-frequency distribution vs. time: an interpretation of the phase structure of speech. In: Proceedings of the IEEEICASSP, pp. 1121–1124 (1985)
6. Fitz, K., Haken, L.: On the use of time-frequency reassignment in additive sound modeling. J. Audio Eng. Soc. **50**, 879–893 (2002)
7. Hainsworth, S.W., Macleod, M.D., Wolfe, P.J.: Analysis of reassigned spectrograms for musical transcription. In: IEEE Workshop on Applications of Signal Processing to Audio and Acoustics (2001)
8. Gardner, T.J., Magnasco, M.O.: Instantaneous frequency decomposition: an application to spectrally sparse sounds with fast frequency modulations. J. Acoustical Soc. Am. **117**, 2896–2903 (2005)
9. Grafakos, L.: Classical and modern Fourier analysis (2003). ISBN:978-0130353993
10. Jolliffe, I.T.: Principal Component Analysis. Springer Series in Statistics. Springer, NY (2002)
11. Krzanowski, W.J.: Principles of Multivariate Analysis: A User's Perspective. Oxford University Press, New York (2000)
12. Nakatani, T., Irino, T.: Robust and accurate fundamental frequency estimation based on dominant harmonic components. J. Acoustical Soc. Am. **116**, 3690–3700 (2004)

13. Nelson, D.J.: Instantaneous higher order phase derivatives. Digit. Sig. Process. **12**, 416–428 (2002)
14. Niethammer, M., Jacobs, L.J., Qu, J., Jarzynski, J.: Time-frequency representation of Lamb waves using the reassigned spectrogram. J. Acoustical Soc. Am. **107**, L19–L24 (2000)
15. Potter, R.K.: Visible patterns of sound. Science **102**, 463–470 (1945)
16. Roweis, S.: EM Algorithms for PCA and SPCA. In: Jordan, M.I., Kearns, M.J., Solla, S.A. (eds.) Advances in Neural Information Processing Systems. The MIT Press (1998)
17. Stein, E.M., Shakarchi, R.: Fourier Analysis: An Introduction (2003). ISBN:9780691113845
18. Wong, M.W.: Discrete Fourier Analysis (2011). ISBN:978-3-0348-0116-4

Artificial Intelligence in Modeling, Simulation and Control

The Concept on Nonlinear Modelling of Dynamic Objects Based on State Transition Algorithm and Genetic Programming

Łukasz Bartczuk[1(✉)], Piotr Dziwiński[1], and Vladimir G. Red'ko[2]

[1] Institute of Computational Intelligence,
Częstochowa University of Technology, Częstochowa, Poland
lukasz.bartczuk@iisi.pcz.pl
[2] Scientific Research Institute for System Analysis of the Russian
Academy of Sciences, Moscow, Russia
vgredko@gmail.com

Abstract. In this paper a new hybrid method to determine parameters of time-variant non-linear models of dynamic objects is proposed. This method first uses the State Transition Algorithm to create many local models and then applies genetic programming in order to join and simplify those models. This allows to obtain simply model which is not computationally demanding and has high accuracy.

Keywords: Nonlinear modelling · State transition algorithm · Genetic programming · Cooperative coevolution

1 Introduction

In this paper we describe a new method of nonlinear modelling of dynamic objects. Its purpose is to create a description of e.g. objects or physical phenomena in the form of a model with characteristic developed on the basis of observation of their responses to the input values.

If we assume that the structure of a model is known and can be presented in the form e.g. state-space representation [13,21], the identification process is restricted to finding parameters of selected structure. In literature we can find many different methods to solve various identification problems. Some of them are based on analytical methods (see e.g. [1,6,22]) and the others are based on computational methods like neural-networks (see e.g. [5]), fuzzy-systems (see e.g. [7–9,11,12,14,15,18,19,24–35]) or population based algorithms (see e.g. [4,10,20]). In our previous articles [2,3] we have proposed two approaches to determining the parameters of nonlinear models of dynamic systems. The first one [3] uses the neuro-fuzzy system to compute values of the state matrix for current operation point and allows to obtain models that have good accuracy, but for some systems can be computationally demanding. The second one [2] uses mathematical formulas discovered through genetic programming. This method

© Springer International Publishing AG 2017
L. Rutkowski et al. (Eds.): ICAISC 2017, Part II, LNAI 10246, pp. 209–220, 2017.
DOI: 10.1007/978-3-319-59060-8_20

allows to create models that have lower computational demanding and therefore is more suitable for real-time control systems, but it has also lower accuracy. The approach proposed in this paper is the two step method which allows to create a model in which values of parameters are also determined by mathematical formulas, however its accuracy is better than a model constructed using the method proposed in [2].

This paper is organized as follows. Section 2 describes the State Transition Algorithm [39], a simple but powerful method of non-constrained optimization. Section 3 presents how the State Transition Algorithm can be used to identification of time variant nonlinear systems and how we used genetic programming to simplify the created model. The simulation results are shown in Sect. 4. Finally, Sect. 5 presents our conclusions.

2   State Transition Algorithm

The State Transition Algorithm [36, 37, 39] is a simple but powerful method to solve unconstrained optimization problems in the form:

$$\min_{\mathbf{x} \in \mathbb{R}^n} \mathrm{ff}(\mathbf{x}) \tag{1}$$

where ff is an optimized objective function and $\mathbf{x}$ is a solution of the problem. In this algorithm a solution $\mathbf{x}$ is described as state which is transformed by several operators in order to find global optimum, which in general form can be given by the following equation:

$$\begin{cases} \mathbf{x}_{k+1} = \mathbf{A}_k \mathbf{x}_k \\ \mathbf{y}_{k+1} = f(\mathbf{x}_{k+1}) \end{cases} \tag{2}$$

where $\mathbf{x}_k$ is a current state of the solution, $\mathbf{A}_k$ and $\mathbf{B}_k$ are state transformation operators.

In paper [39] the following operators have been proposed to solve continuous function optimization problems:

1. Expansion

$$\mathbf{x}_{k+1} = \mathbf{x}_k + \gamma \mathbf{R}_e \mathbf{x}_k \tag{3}$$

   where $\gamma \in \mathbb{Z}^+$ is an expansion factor, $\mathbf{R}_e \in \mathbb{R}^{n \times n}$ is a random diagonal matrix in which elements are randomized according to the Gaussian distribution

2. Rotation

$$\mathbf{x}_{k+1} = \mathbf{x}_k + \alpha \frac{1}{n \|\mathbf{x}_k\|_2} \mathbf{R}_r \mathbf{x}_k \tag{4}$$

   where $\alpha \in \mathbb{Z}^+$ is a rotation factor, $\mathbf{R}_r \in \mathbb{R}^{n \times n}$ is a random matrix in which elements belong to $[-1, 1]$ range, and $\|\cdot\|_2$ is the Euclidean norm of a vector

3. Axesion

$$\mathbf{x}_{k+1} = \mathbf{x}_k + \delta \mathbf{R}_a \mathbf{x}_k \tag{5}$$

where $\delta \in \mathbb{Z}^+$ is an axesion factor $\mathbf{R}_a \in \mathbb{R}^{n \times n}$ is a random diagonal matrix in which one element is randomized according to the Gaussian distribution and others are equal to zero.

4. Translation

$$\mathbf{x}_{k+1} = \mathbf{x}_k + \beta \mathbf{R}_t \left\| \frac{\mathbf{x}_k - \mathbf{x}_{k+1}}{\mathbf{x}_k - \mathbf{x}_{k+1}} \right\|_2 \tag{6}$$

where $\beta \in \mathbb{Z}^+$ is a translation factor, $R_t \in \mathbb{R}$ is a random value from $[0,1]$ range.

During the application of the above operators SE new solutions are created from which only the best (in sense of optimization problem) is selected to further processing. The whole algorithm is presented in Fig. 1.

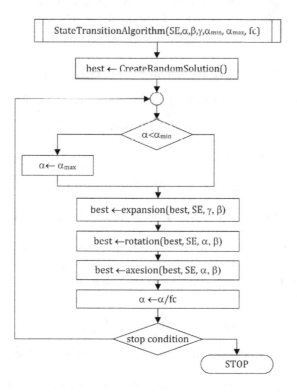

Fig. 1. Flowchart of the state transition algorithm

3   Nonlinear Modelling with State Transition Algorithm and Genetic Programming

The State Transition Algorithm presented in previous section was successfully used to solve many optimization problems [37,39] and to identification of time-invariant nonlinear systems [36] in the form:

$$\mathbf{x}(k+1) = \mathbf{A}\mathbf{x}(k) + \mathbf{B}\mathbf{u}(k) \qquad (7)$$

$$\mathbf{y}(k) = \mathbf{C}\mathbf{x}(k) + \mathbf{D}\mathbf{u}(k) \qquad (8)$$

where $\mathbf{x}(k)$ is the vector of state variables, $\mathbf{u}(k)$ is the vector of input values, $\mathbf{y}(k)$ is an output of model, $\mathbf{A}, \mathbf{B}, \mathbf{C}, \mathbf{D}$ are the state, input, output and feed forward matrices respectively and $k = 1, 2, \ldots$. In paper [36] State Transition Algorithm was used to determine the matrices $\mathbf{A}$ and $\mathbf{C}$ with the assumption that these matrices are constant. Such assumption can be too strong for some systems where values of elements of the state matrix can vary in each different operating point. In this paper we propose a simple extension of this method that allows to apply the State Transition Algorithm to such system and also we show how to reduce complexity of obtained models with the use of genetic programming.

3.1   Nonlinear Modelling with the State Transition Algorithm

Let us consider nonlinear systems that can be described by the following equation:

$$\mathbf{x}(k+1) = \mathbf{A}\mathbf{x}(k) \qquad (9)$$

and in particular by the formula:

$$\mathbf{x}(k+1) = (\mathbf{A} + \mathbf{P_A}(k))\,\mathbf{x}(k) \qquad (10)$$

where $\mathbf{A}$ is a known state matrix of approximated linear system, and $\mathbf{P_A}(k)$ is a correction matrix which should be estimated in such a way that error of linear approximation is as small as possible. Because we assume that values of $\mathbf{P_A}(k)$ may vary in each step k the method proposed in paper [36] is not suitable. However we can treat a nonlinear system (11) as a composition of multiple local approximated linear models:

$$\mathbf{x}(k+1) = (\mathbf{A} + \mathbf{P_{A\,s}}(k))\,\mathbf{x}(k) \qquad (11)$$

where $\mathbf{P_{A\,s}}(k)$ is a correction matrix determined for the s-th local approximated linear model $s = 1, \ldots, S$.

In order to create local models we split the reference data set $\mathbf{Y}$ into S subsets of size M ($\mathbf{Y} = \mathbf{Y}_1 \cup \ldots \cup \mathbf{Y}_S$) and apply the State Transition Algorithm to each of these subsets. In order to preserve continuity between those models, only for the first one the initial solution $best_1^1$ of the STA algorithm is created in random way. For other models the STA algorithm is initialized with the best solution

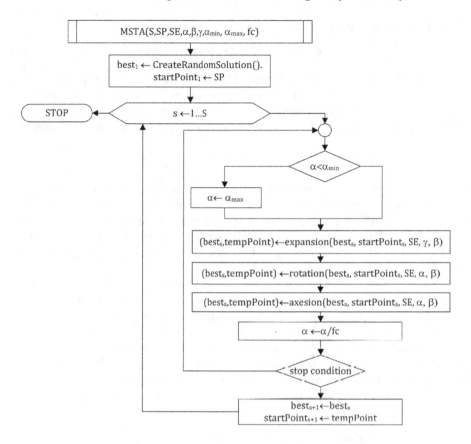

Fig. 2. The block diagram of extended State Transition Algorithm, allowing the modelling time-variant nonlinear model

found in the previous subset ($best_s = best_{s-1}$). The schema of this method is presented in Fig. 2.

As a results of this algorithm we obtain the triple $\langle \overline{\mathbf{Y}}, \overline{\mathbf{X}}, \overline{\mathbf{P_A}} \rangle$ where $\overline{\mathbf{Y}} = \overline{\mathbf{Y}}_1 \cup \ldots \cup \overline{\mathbf{Y}}_S$ is the output of created model ($\overline{\mathbf{Y}}_s = [\overline{\mathbf{y}}_{s,1}, \ldots, \overline{\mathbf{y}}_{s,M}]$), $\overline{\mathbf{X}} = \overline{\mathbf{X}}_1 \cup \ldots \cup \overline{\mathbf{X}}_S$ is the ordered set of vectors of state variables ($\overline{\mathbf{X}}_s = [\overline{\mathbf{x}}_{s,1}, \ldots, \overline{\mathbf{x}}_{s,M}]$) and $\overline{\mathbf{P_A}} = [\overline{\mathbf{P_A}}_1, \ldots, \overline{\mathbf{P_A}}_S]$ is a set of correction matrices.

3.2 Simplifying Model with the Genetic Programming Paradigm

The approach presented in previous section allows to obtain very good modelling results, however large number of models can be hard to understand and analyze as well as to implement the controller. However we can use this model to create its functional description. In this case the elements of matrix $\mathbf{P_A}$ have to be functions that take into account the current state $\mathbf{x}$ of the modeled system, so

Eq. (11) can be rewritten in the form:

$$\mathbf{x}(k+1) = (\mathbf{A} + \mathbf{P_A}(\mathbf{x}(k)))\,\mathbf{x}(k) \tag{12}$$

In order to find elements of matrix $\mathbf{P_A}(\mathbf{x}(k))$ we use genetic programming (see e.g. [16,17,38]) treating sets $\overline{\mathbf{X}}$ and $\overline{\mathbf{P_A}}$ as inputs data. However, because the values of correction matrices are the same for all elements of $\overline{\mathbf{X}}^s$, we can reduce each set $\overline{\mathbf{X}}^s$ to one point which can be computed by the following equation:

$$\widetilde{\mathbf{x}}_s = [\;\operatorname*{avg}_{m=1,\dots,M}\,(\overline{x}^1_{sm}),\dots,\;\operatorname*{avg}_{m=1,\dots,M}\,(\overline{x}^n_{sm})] \tag{13}$$

This operation allows to reduce the number of calculations that must be executed in order to compute fitness value of each individual. The use of sets $\widetilde{\mathbf{X}} = [\widetilde{\mathbf{x}}_1,\dots,\widetilde{\mathbf{x}}_s]$ and $\overline{\mathbf{P_A}}$ allows to search for each element of the matrix $\mathbf{P_A}(\mathbf{x}(k))$ independently. Unfortunately, despite the satisfactory results of approximation of individual functions, their independent determination not yielded good results of modelling.

From this reason we decided to use method similar to the cooperative coevolutionary approach presented in paper [23]. This method uses several different subpopulations (called species) that cooperate together to solve an optimization problem. In our case each species is responsible to find the functional dependency which approximate the appropriate elements from matrices $\overline{\mathbf{P_A}}$ in the best way (we call this the local optimization). Moreover, all species must cooperate to obtain high modelling accuracy (we call this the global optimization). To ensure cooperation between species we used the following function to evaluate each individual in population:

$$\mathrm{ff}(\mathbf{ch}_i) = w \cdot \mathrm{ff}_{AccI}(\mathbf{ch}_i) + (1 - w) \cdot \mathrm{ff}_{AccG}(\mathbf{ch}_i) \tag{14}$$

where: $\mathrm{ff}_{AccI}(\mathbf{ch}_i)$ is the accuracy of the approximation of the points $P_\mathbf{A}$, $\mathrm{ff}_{AccG}(\mathbf{ch}_i)$ is the accuracy of nonlinear modelling and w is a weight that determines the influence of particular components. Such form of fitness function allows to obtain a good compromise between local and global accuracy.

4 Simulation Results

To examine the effectiveness of the proposed method, we considered the problem of harmonic oscillator. Such oscillator can be defined using the following formula:

$$\frac{d^2x}{dt^2} + 2\zeta\frac{dx}{dt} + \omega^2 x = 0, \tag{15}$$

where ζ and ω are oscillator parameters and $x(t)$ is a reference value of the modelled process as function of time. We used the following state variables: $x_1(t) = dx(t)/dt$ and $x_2(t) = x(t)$. In such a case the system matrix $\mathbf{A}$ and the matrix of corrections coefficients $\mathbf{P_A}$ are described as follows:

$$\mathbf{A} = \begin{bmatrix} 0 & \omega \\ -\omega & 0 \end{bmatrix} \qquad \mathbf{P_A} = \begin{bmatrix} 0 & p_{12}(\mathbf{x}) \\ p_{21}(\mathbf{x}) & 0 \end{bmatrix}.$$

In our experiments the parameter ω was modified according to the formula:

$$\omega(x_1) = 2\pi - \frac{\pi}{(1 + |2 \cdot x_1|^6)}. \tag{16}$$

The parameters of the extended version of the State Transition Algorithm are presented in Table 1.

Table 1. Parameters of the extended version of the State Transition Algorithm used in simulations

Name	Value
S	50
M	40
SE	30
A_{min}	0.0001
A_{max}, α, β, γ, δ	1
Fc	2

According with parameters of the STA, the algorithm generating model that is composed of 50 local submodels. Its accuracy was RMSE = 0,0008. Signals and errors obtained from this model are presented in Fig. 3. On the basis of this model the data sets describing dependencies between the values of state variables and the parameters of the model were generated. Graphical representation of these data sets are shown in Fig. 4.

As explained in Sect. 3 these data sets were used to determine, by using the genetic programming, the equations that describe functional dependency between current state and parameters of correction matrix. The parameters of genetic programming algorithm are presented in Table 2.

Table 2. Parameters of genetic programming used in simulations

Functions set F	$\{+, -, \cdot, /^a, \text{neg}, \text{pow}, \text{inv}^a$
The number of species	2
Number of constants	61
Constants range	$[1, 7]$
Number of epochs	1000
Population size μ	20
Probability of crossover p_c	0.5
Probability of mutation p_m	0.5
Weight w	0.5

[a] In case of division and multiplicative inverse operator we used their safe versions

a)   b)

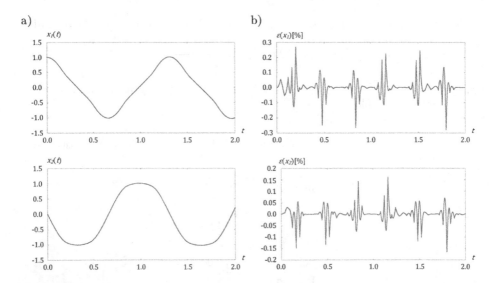

Fig. 3. Graphical illustration of modelling the harmonic oscillator by the extended version of State Transition Algorithm (a) the values of signals and (b) the errors obtained for signals x_1 and x_2, respectively.

a)   b)

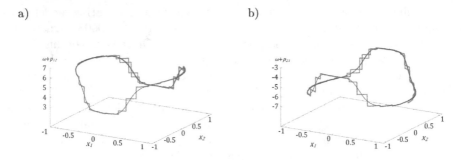

Fig. 4. The graphical representation of the dependency between state variables and values of elements of state matrices for reference data and data obtained by the State Transition Algorithm.

The best model discovered by genetic programming can be written as follows:

$$\begin{cases} p_{12}(\mathbf{x}) = \dfrac{-6.6}{4.7^{5.7} \cdot x_1^4 \cdot 2.1} \\ p_{21}(\mathbf{x}) = \dfrac{1}{(1.4 - x_2^{2 \cdot (1.4 - x_2^2)}) \cdot (0.7142/x_2^4)} \end{cases} \tag{17}$$

It should be noted that in the second equation of system (17) is placed a different variable than in Eq. (16). This is because the variables x_1 and x_2 are strongly correlated and the algorithm automatically chose the shorter version of this formula. The accuracy of this model is RMSE = 0.004. Signals and errors

a)

b)

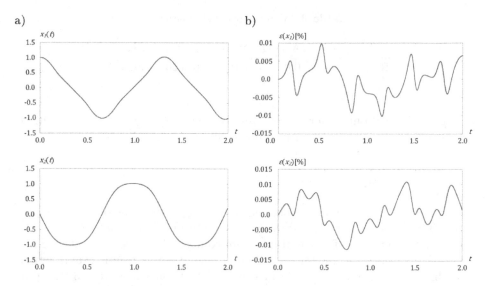

Fig. 5. Graphical illustration of modelling the harmonic oscillator by the method proposed in this paper (a) the values of signals and (b) the errors obtained for signals x_1 and x_2, respectively.

obtained from this model are presented in Fig. 5 and the graphical representation of dependencies between state variables and values of elements of the state matrix in Fig. 6.

The summary of simulations and its comparison with results obtained by methods described in [2,3] are presented in Table 3.

This table shows that proposed method allows to obtain much better results than the method based on semantic version of gene expression programming algorithm [2]. The results obtained by the method that used fuzzy systems [3]

a)

b)

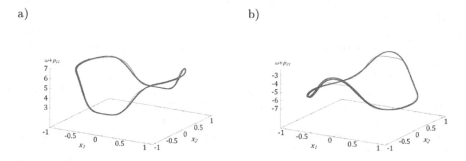

Fig. 6. The graphical representation of dependency between state variables and values of elements of state matrices for reference data and data obtained by the method proposed in this paper.

Table 3. A summary of simulations

	Best	Average	Worst	Best [3]	Best [2]
RMSE	0.0045	0.01695	0.04686	0.0026	0.017

are slightly better, but under the assumption that they are intended to achieve high accuracy at the expense of interpretability.

5   Conclusions

In this paper a new method of nonlinear modelling of dynamic objects was presented. This method first applied the State Transition Algorithm to generate the initial model composed from multiple approximated linear local models. Then it used data obtained from that model and implement the genetic programming algorithm to simplify and transform this model. As shown in Sect. 3, the proposed method gives very good results for selected problems. This proves the efficiency of our approach.

Acknowledgment. The project was financed by the National Science Center on the basis of the decision number DEC-2012/05/B/ST7/02138.

References

1. Jordan, A.J.: Linearization of non-linear state equation. Bull. Pol. Acad. Sci. Tech. Sci. **54**(1), 63–73 (2006)
2. Bartczuk, Ł.: Gene expression programming in correction modelling of nonlinear dynamic objects. In: Borzemski, L., Grzech, A., Świątek, J., Wilimowska, Z. (eds.) Information Systems Architecture and Technology: Proceedings of 36th International Conference on Information Systems Architecture and Technology – ISAT 2015 – Part I. AISC, vol. 429, pp. 125–134. Springer, Cham (2016). doi:10.1007/978-3-319-28555-9_11
3. Bartczuk, Ł., Przybył, A., Cpałka, K.: A new approach to nonlinear modelling of dynamic systems based on fuzzy rules. Int. J. Appl. Math. Comput. Sci. **26**(3), 603–621 (2016)
4. Bello, O., Holzmann, J., Yaqoob, T., Teodoriu, C.: Application of artificial intelligence methods in drilling system design and operations: a review of the state of the art. J. Artif. Intell. Soft Comput. Res. **5**(2), 121–139 (2015)
5. Bertini, J.R., Nicoletti, M.D.C.: Enhancing constructive neural network performance using functionally expanded input data. J. Artif. Intell. Soft Comput. Res. **6**(2), 119–131 (2016)
6. Caughey, T.K.: Equivalent linearization techniques. J. Acoust. Soc. Am. **35**(11), 1706–1711 (1963)
7. Cpałka, K.: On evolutionary designing and learning of flexible neuro-fuzzy structures for nonlinear classification. Nonlinear Anal. Ser. A: Theor. Methods Appl. **71**, 1659–1672 (2009)

8. Cpałka, K., Łapa, K., Przybył, A., Zalasiński, M.: A new method for designing neuro-fuzzy systems for nonlinear modelling with interpretability aspects. Neuro-computing **135**, 203–217 (2014)

9. Cpałka, K.: Design of Interpretable Fuzzy Systems. Springer (2017)

10. Cpałka, K., Łapa, K., Przybył, A.: A new approach to design of control systems using genetic programming. Inf. Technol. Control **44**(4), 433–442 (2015)

11. Cpałka, K., Rutkowski, L.: Flexible Takagi-Sugeno. Fuzzy systems, neural networks. In: Proceedings of the 2005 IEEE International Joint Conference on IJCNN 2005, vol. 3, pp. 1764–1769 (2005)

12. Cpałka, K., Rebrova, O., Nowicki, R., Rutkowski, L.: On design of flexible neuro-fuzzy systems for nonlinear modelling. Int. J. Gen. Syst. **42**(6), 706–720 (2013)

13. Freeman, R., Kokotovic, P.V.: State-space and Lyapunov techniques. Springer Science & Business Media, New York (2008)

14. Korytkowski, M.: Novel visual information indexing in relational databases. In: Integrated Computer-aided Engineering, pp. 1–10 (2016). doi:10.3233/ICA-160534

15. Korytkowski, M., Rutkowski, L., Scherer, R.: Fast image classification by boosting fuzzy classifiers. Inf. Sci. **327**, 175–182 (2016)

16. Koza, J.R.: On the Programming of Computers by Means of Natural Selection, vol. 1. MIT Press, Cambridge (1992)

17. Krawiec, K.: Behavioral Program Synthesis with Genetic Programming, vol. 618. Springer, Switzerland (2016)

18. Łapa, K., Przybył, A., Cpałka, K.: A new approach to designing interpretable models of dynamic systems. In: Rutkowski, L., Korytkowski, M., Scherer, R., Tadeusiewicz, R., Zadeh, L.A., Zurada, J.M. (eds.) ICAISC 2013. LNCS, vol. 7895, pp. 523–534. Springer, Heidelberg (2013). doi:10.1007/978-3-642-38610-7_48

19. Łapa, K., Cpałka, K., Wang, L.: New method for design of fuzzy systems for nonlinear modelling using different criteria of interpretability. In: Rutkowski, L., Korytkowski, M., Scherer, R., Tadeusiewicz, R., Zadeh, L.A., Zurada, J.M. (eds.) ICAISC 2014. LNCS, vol. 8467, pp. 217–232. Springer, Cham (2014). doi:10.1007/978-3-319-07173-2_20

20. Łapa, K., Szczypta, J., Venkatesan, R.: Aspects of structure and parameters selection of control systems using selected multi-population algorithms. In: Rutkowski, L., Korytkowski, M., Scherer, R., Tadeusiewicz, R., Zadeh, L.A., Zurada, J.M. (eds.) ICAISC 2015. LNCS, vol. 9120, pp. 247–260. Springer, Cham (2015). doi:10.1007/978-3-319-19369-4_23

21. Nelles, O.: Nonlinear System Identication: From Classical Approaches to Neural Networks and Fuzzy Models. Springer Science & Business Media (2013)

22. Nonaka, S., Tsujimura, T., Izumi, K.: Gain design of quasi-continuous exponential stabilizing controller for a nonholonomic mobile robot. J. Artif. Intell. Soft Comput. Res. **6**(3), 189–201 (2016)

23. Potter, M.A., Jong, K.A.: A cooperative coevolutionary approach to function optimization. In: Davidor, Y., Schwefel, H.-P., Männer, R. (eds.) PPSN 1994. LNCS, vol. 866, pp. 249–257. Springer, Heidelberg (1994). doi:10.1007/3-540-58484-6_269

24. Prasad, M., Liu, Y.-T., Li, D.-L., Lin, C.-T., Shah, R.R., Kaiwartya, O.M.: A new mechanism for data visualization with Tsk-Type preprocessed collaborative fuzzy rule based system. J. Artif. Intell. Soft Comput. Res. **7**(1), 33–46 (2017)

25. Przybył, A., Er, M.J.: The idea for the integration of neuro-fuzzy hardware emulators with real-time network. In: Rutkowski, L., Korytkowski, M., Scherer, R., Tadeusiewicz, R., Zadeh, L.A., Zurada, J.M. (eds.) ICAISC 2014. LNCS, vol. 8467, pp. 279–294. Springer, Cham (2014). doi:10.1007/978-3-319-07173-2_25

26. Przybył, A., Er, M.J.: The method of hardware implementation of fuzzy systems on FPGA. In: Rutkowski, L., Korytkowski, M., Scherer, R., Tadeusiewicz, R., Zadeh, L.A., Zurada, J.M. (eds.) ICAISC 2016. LNCS, vol. 9692, pp. 284–298. Springer, Cham (2016). doi:10.1007/978-3-319-39378-0_25

27. Rutkowski, L., Cpałka, K.: Compromise approach to neuro-fuzzy systems. In: Proceedings of the 2nd Euro-International Symposium on Computation Intelligence. Frontiers in Artificial Intelligence and Applications, vol. 76, pp. 85–90 (2002)

28. Rutkowski, L., Cpałka, K.: A neuro-fuzzy controller with a compromise fuzzy reasoning. Control Cybern. **31**(2), 297–308 (2002)

29. Rutkowski, L., Przybył, A., Cpałka, K.: Novel online speed profile generation for industrial machine tool based on flexible neuro-fuzzy approximation. IEEE Trans. Ind. Electr. **59**(2), 1238–1247 (2012)

30. Rutkowski, L., Przybył, A., Cpałka, K., Er, M.J.: Online speed profile generation for industrial machine tool based on neuro-fuzzy approach. In: Rutkowski, L., Scherer, R., Tadeusiewicz, R., Zadeh, L.A., Zurada, J.M. (eds.) ICAISC 2010. LNCS, vol. 6114, pp. 645–650. Springer, Heidelberg (2010). doi:10.1007/978-3-642-13232-2_79

31. Starczewski, J., Rutkowski, L.: Connectionist structures of Type 2 fuzzy inference systems. In: Wyrzykowski, R., Dongarra, J., Paprzycki, M., Waśniewski, J. (eds.) PPAM 2001. LNCS, vol. 2328, pp. 634–642. Springer, Heidelberg (2002). doi:10.1007/3-540-48086-2_70

32. Zalasiński, M.: New algorithm for on-line signature verification using characteristic global features. Adv. Intell. Syst. Comput. **432**, 137–146 (2016)

33. Zalasiński, M., Cpałka, K.: New algorithm for on-line signature verification using characteristic hybrid partitions. Adv. Intell. Syst. Comput. **432**, 147–157 (2016)

34. Zalasiński, M., Cpałka, K., Rakus-Andersson, E.: An idea of the dynamic signature verification based on a hybrid approach. In: Rutkowski, L., Korytkowski, M., Scherer, R., Tadeusiewicz, R., Zadeh, L.A., Zurada, J.M. (eds.) ICAISC 2016. LNCS, vol. 9693, pp. 232–246. Springer, Cham (2016). doi:10.1007/978-3-319-39384-1_21

35. Zalasiński, M., Cpałka, K., Rutkowski, L.: A new algorithm for identity verification based on the analysis of a handwritten dynamic signature. Appl. Soft Comput. **43**, 47–56 (2016)

36. Zhou, X., Yang, C., Gui, W.: Nonlinear system identification and control using state transition algorithm. Appl. Math. Comput. **226**, 169–179 (2014)

37. Zhou, X., Gao, D.Y., Yang, C., Gui, W.: Discrete state transition algorithm for unconstrained integer optimization problems. Neurocomputing **173**, 864–874 (2016)

38. Yin, Z., O'Sullivan, C., Brabazon, A.: An analysis of the performance of genetic programming for realised volatility forecasting. J. Artif. Intell. Soft Comput. Res. **6**(3), 155–172 (2016)

39. Zhou, X., Yang, C., Gui, W.: Initial version of state transition algorithm. In: 2011 Second International Conference on Digital Manufacturing and Automation (ICDMA), pp. 644–647. IEEE (2011)

A Method for Non-linear Modelling Based on the Capabilities of PSO and GA Algorithms

Piotr Dziwiński[1(✉)], Łukasz Bartczuk[1], and Huang Tingwen[2]

[1] Institute of Computational Intelligence,
Częstochowa University of Technology, Częstochowa, Poland
piotr.dziwinski@iisi.pcz.pl
[2] Texas A&M University at Qatar, Doha, Qatar
tingwen.huang@qatar.tamu.edu

Abstract. The most nonlinear dynamic objects have their Approximate Nonlinear Model (ANM). Their parameters are known or can be determined by one of the typical identification procedures. The model obtained in this way describes well the main features of the identified dynamic object only in some Operating Point (OP). In this approach we use hybrid model increasing accuracy of the modeling. The hybrid model is composed of two parts: base ANM and Takagi-Sugeno (TS) fuzzy system. A Particle Swarm Optimization with Genetic Algorithm (PSO-GA) was used for identification of the parameters of the ANM and TS fuzzy system. An important advantage of the proposed approach is the obtained characteristics of the unknown parameters of the ANM described by the Fuzzy Rules (FR) of the TS fuzzy system. They provide the valuable knowledge for the experts about the nature of the unknown phenomena.

Keywords: Nonlinear modeling · Non-invasive identification · Significant operating point · Particle swarm optimization · Genetic algorithm · Permanent magnet synchronous motor · Takagi-Sugeno fuzzy system

1 Introduction

The most nonlinear dynamic objects have their ANM. Their parameters are known or can be determined by one of the typical identification procedure. The model obtained in this way describes well the main features of the identified dynamic object only in some OP [14]. Between them there are many secondary phenomena that are not described precisely enough by the mathematical model. The observed phenomena must be reproduced in order to obtain the model precise enough for the practical application.

A large number of mathematical models which can describe the linear or nonlinear systems in universal way were proposed in the literature, among others, neural networks [23,36] treated as black box models, fuzzy systems [9,20], flexible fuzzy systems [24,27], neuro-fuzzy systems [28,29,43], flexible neuro-fuzzy systems [6,7,10,26,39–42], interval type 2 neuro-fuzzy systems [33,34],

© Springer International Publishing AG 2017
L. Rutkowski et al. (Eds.): ICAISC 2017, Part II, LNAI 10246, pp. 221–232, 2017.
DOI: 10.1007/978-3-319-59060-8_21

Takagi-Sugeno systems [12], flexible Takagi-Sugeno systems [11]. The methods mentioned earlier enable modeling in an universal way but do not provide enough precision of the reproduction of the reference values.

Much better result can be obtained by using a hybrid approach [3,15]. The approximate linear or nonlinear model can be used in the hybrid approach. It enables reproduction of the reference values with a sufficient precision only in the OP, whereas the universal model can determine the values of the parameters of the approximate model in different OP and between them. This approach ensures to obtain a sufficient precision of the identification in all states of the nonlinear dynamic object. In this paper we propose a new representation of the approximate state and input matrices by including the sparse corrections $\Delta\hat{\mathbf{g}}(\mathbf{x}(t))$ and $\Delta\hat{\mathbf{q}}(\mathbf{x}(t))$ of the known or estimated parameters $\mathbf{g}$ and $\mathbf{q}$. It allows to obtain characteristics of the unknown parameters of the ANM described by the FR of the TS fuzzy system. This approach provides the valuable knowledge for experts in order to identify better mathematical model of the ANM.

The remainder of this paper is organized as follows. Section 2 describes approximate modeling of nonlinear dynamic objects by the algebraic equations and on the basis of the state variable technique using sparse corrections of the known or estimated parameters in the operating points. Section 3 deals with fuzzy modeling of the corrections of the parameters in the operating points using the TS fuzzy system. Section 4 presents the algorithm for online identification of the OP described by FR of the TS fuzzy system. Section 5 describes the Permanent Magnet Synchronous Motor (PMSM) working in the two operating points. Finally, Sect. 6 shows simulation results which proves the effectiveness of the proposed method.

2   Approximate Modeling of the Nonlinear Dynamic Object

Let us consider the nonlinear dynamic stationary object described by the algebraic equations and based on the state variable technique [24]

$$\frac{d\mathbf{x}}{dt} = \mathbf{A}(\mathbf{x}(t))\mathbf{x}(t) + \mathbf{B}(\mathbf{x}(t))\mathbf{u}(t), \tag{1}$$

$$\mathbf{y}(t) = \mathbf{C}\mathbf{x}(t), \tag{2}$$

where $\mathbf{A}(\mathbf{x}(t))$, $\mathbf{B}(\mathbf{x}(t))$ are the system and input matrices respectively, $\mathbf{u}(t)$, $\mathbf{y}(t)$ are the input and output signals respectively, $\mathbf{x}(t)$ is the vector of the state variables. The algebraic equations based on the state variable technique, delivered by the experts, describe the dynamic nonlinear object with a sufficient precision only in some characteristic work state called operating point. Beyond the OP there are phenomena that are not included in the mathematical model. Overall accuracy of such a model may be too low for many practical applications.

In this work we propose the hybrid method which increases effectiveness of the modeling of the nonlinear dynamic object. It is done by the modeling of the system and input matrices parameters which are not described precisely enough by the mathematical model. The entire approximate model can be described by algebraic equations and on the basis of the state variable technique, where unknown linear or nonlinear part can be modeled by the $\hat{\mathbf{A}}(\mathbf{x}(t), \mathbf{g}+\Delta\hat{\mathbf{g}}(\mathbf{x}(t)))$ and $\hat{\mathbf{B}}(\mathbf{x}(t), \mathbf{q}+\Delta\hat{\mathbf{q}}(\mathbf{x}(t)))$ approximate matrices. The unknown parameters change and can be described by the correction values $\Delta\hat{\mathbf{g}}(\mathbf{x}(t))$ and $\Delta\hat{\mathbf{q}}(\mathbf{x}(t))$. For example, consider the specific nonlinear dynamic deterministic system with the element of the system matrix $a_{23} = -i_d - \lambda_m/L$ with the unknown or estimated value of the λ_m. The element a_{23} can be written as $a_{23} \approx -i_d - (\lambda_m + \Delta\hat{\lambda}_m)/L$. The parameter λ_m has constant value in the OP but changes in unknown way between them and can be modeled by the $\Delta\hat{\lambda}_m$ correction values. So by using the approximate matrices, we obtain the following form of the Eq. (1)

$$f(\mathbf{x}(t), \mathbf{u}(t)) = \hat{\mathbf{A}}\left(\mathbf{x}(t), \mathbf{g} + \Delta\hat{\mathbf{g}}(\mathbf{x}(t))\right)\mathbf{x}(t) \tag{3}$$
$$+ \hat{\mathbf{B}}\left(\mathbf{x}(t), \mathbf{q} + \Delta\hat{\mathbf{q}}(\mathbf{x}(t))\right)\mathbf{u}(t),$$

where $\hat{\mathbf{A}}$, $\hat{\mathbf{B}}$ are approximate state and input matrices respectively, $\mathbf{g}$, $\mathbf{q}$ are known parameters, $\Delta\hat{\mathbf{g}}(\mathbf{x}(t))$, $\Delta\hat{\mathbf{q}}(\mathbf{x}(t))$ are the sparse corrections of parameters $\mathbf{g}$ and $\mathbf{q}$, respectively.

3   Fuzzy Modeling of the Identified Parameters

The changes of the correction of the parameter values $\Delta\dot{\mathbf{g}}(\mathbf{x}(t))$ and $\Delta\dot{\mathbf{q}}(\mathbf{x}(t))$ take place between OP does not occur rapidly usually, but in a smooth unknown manner which is difficult to describe by using the mathematical model. The values of the parameters in the operating points pass fluently among themselves and overlap. Fuzzy systems are frequently used by many researches to fuzzy modeling and classification [9,18–20,35]. So, for modeling of the sparse corrections $\Delta\hat{\mathbf{g}}(\mathbf{x}(t))$, $\Delta\hat{\mathbf{q}}(\mathbf{x}(t))$ of the parameters $\mathbf{g}$ and $\mathbf{q}$, respectively, the TS fuzzy system is perfectly suitable as the universal approximator.

The construction of the most neuro-fuzzy structures [11] is based on the Mamdani reasoning type described by using t-norm, for example product or minimum. They require defuzzification of the output values, thus they cannot be applied easily for modeling of the corrections of the parameters opposed to TS fuzzy system [17]. This system includes dependence between a premise **IF** and a consequent **THEN** of the rule in the form

$$R^{(l)} : \text{ IF } \bar{\mathbf{x}} \text{ is } \mathbf{D}^l \text{ THEN } \mathbf{y}^l = \mathbf{f}^{(l)}(\mathbf{x}), \tag{4}$$

where: $\bar{\mathbf{x}} = [\bar{x}_1, \bar{x}_2, \ldots, \bar{x}_N] \in \bar{\mathbf{X}}$, $\mathbf{y}^l \in \mathbf{Y}^l$, $\mathbf{D}^l = D_1^l \times D_2^l \times \ldots \times D_N^l$, $D_1^l, D_2^l, \ldots, D_N^l$, are the fuzzy sets described by the membership functions $\mu_{D_i^l}(\bar{x}_i)$, $i = 1, \ldots, N$, $l = 1, \ldots, n$, L is the number of the rules and N is the number of the inputs of the TS fuzzy system, $\mathbf{f}^{(l)}$ are the functions describing values of the system matrix or input matrix for the l-th fuzzy rule. In case of the a_{23} element of the state matrix, the $f^{(l)}$ function from the consequent takes the form: $f^{(l)}(t) = -i_d(t) - (\hat{\lambda}_m + \Delta\hat{\lambda}_m(t))/L$.

Assuming the aggregation method as weighted average, using the Eq. (3) and Euler integration method with time step T_s, we obtain the discrete approximate hybrid model described by Eq. (5)

$$f(\mathbf{x}(k), \bar{\mathbf{x}}(k), \mathbf{u}(k+1)) =$$

$$\left(\mathbf{I} + \left(\hat{\mathbf{A}} \left(\mathbf{x}(k), \mathbf{g} + \frac{\sum_{l=1}^{L} \hat{\mathbf{g}}^l \cdot \mu_{\mathbf{D}}{}^l(\bar{\mathbf{x}}(k))}{\sum_{l=1}^{L} \mu_{\mathbf{D}}{}^l(\bar{\mathbf{x}}(k))} \right) \right) T_s \right) \mathbf{x}(k)$$

$$+ \left(\hat{\mathbf{B}} \left(\mathbf{x}(k), \mathbf{q} + \frac{\sum_{m=L+1}^{L+M} \hat{\mathbf{q}}^m \cdot \mu_{\mathbf{D}}{}^m(\bar{\mathbf{x}}(k))}{\sum_{m=L+1}^{L+M} \mu_{\mathbf{D}}{}^m(\bar{\mathbf{x}}(k))} \right) \right) \mathbf{u}(k+1), \qquad (5)$$

where $\bar{\mathbf{x}}(k)$ is the vector of the fuzzy values obtained from the vector $\mathbf{x}(k)$ using singleton fuzzification, $\mathbf{g}^l$, $\mathbf{q}^m$ are the sparse vectors containing the correction values for the changing parameters in the l and m OP, $l = 1, \ldots, L$, $m = L+1, \ldots, L+M$, L, M - number of the rules describing the OP for the state and input matrices respectively, $\mu_{\mathbf{D}}{}^m(\bar{\mathbf{x}}(k))$ and $\mu_{\mathbf{D}}{}^l(\bar{\mathbf{x}}(k))$ are the membership functions describing activation levels of the operating point and $\mathbf{I}$ is the identity matrix.

The Eq. (5) represents the discrete hybrid model describing the dynamic nonlinear deterministic system. The Euler integration method was selected for simplicity but there should be chosen better one in the practical application.

So, the local linear or nonlinear model in the operating point l and m is defined through the set of the parameters $\theta_l = \{\hat{\mathbf{A}}(\mathbf{x}(k)), \mathbf{g}, \hat{\mathbf{g}}^l, \mathbf{D}^l\}$ and $\theta_m = \{\hat{\mathbf{B}}(\mathbf{x}(k)), \mathbf{q}, \hat{\mathbf{q}}^m, \mathbf{D}^m\}$. The parameters are determined by hybrid operating point identification method using PSO and GA algorithms.

4 Online Identification of the Operating Point

The automatic detection of the OP in nonlinear modeling is a very hard and time-consuming task. In the most researches, authors focus on solutions using grouping and classification algorithms to discover potential areas that can be good candidates for operating points. The mentioned methods require a complete data set for estimating good candidate areas for OP.

In many researches there were used hybrid and evolutionary approach for find solution of the very hard and time-consuming tasks. Among others, Eftekhari [16] has used subtractive clustering algorithm [5,13] to discover potential areas of applying local linear models which were identified subsequently by Ant Colony Algorithm (ACO). Brasileiro [4] applied ACO to the problem of choosing the best combination path in transparent optical networks. Aghdam [1] used PSO for feature selection in text categorization. Szczypta [32] has used evolutionary approach for design optimal controllers.

Algorithm 1. Pseudocode of the algorithm for identification of the operating points.

1	**Algorithm IdentifyOP** $(\mathbf{u}(t), \mathbf{x}(t), \mathbf{y}(t), T_{max}, T_s, E_{min}, e_{max}, \Delta E)$

Data: $\mathbf{u}(t), \mathbf{x}(t), \mathbf{y}(t)$ $t = 0, T_s, ..., t_{max}^{(e)}$, T_s - integration steep, t_{max} - total time of the measurements, $E_{m}in$ - minimal RMSE error, e_{max} - maximum epoch number, ΔE - the minimum value of the improve of the solution in the z epoch.

Result: $\theta_l = \{\hat{\mathbf{A}}, \mathbf{g}, \hat{\mathbf{g}}^l, \mathbf{D}^l\}$, $\theta_l \in \Theta$, $l = 1, ..., L$, L - number of the detected OP.

2   Set initially: $\Theta = \emptyset$, $L = 0$, $e = 0$, e - the current epoch in the PSO-GA algorithm;

3   **repeat**

4   Add new OP: ⎫

5      $L \leftarrow L + 1, \Theta \leftarrow \Theta \cup \theta_L$;

6   Determine the initial time interval: ⎬ The initialization

7   **repeat**   of the OP (stage 1)

8        $t_{max}^{(e)} \leftarrow t_{max}^{(e)} + T_s$

9   **until** $d(x(0), x(t_{max}^{(e)})) < d_{start}$;   ⎭

10   **repeat**

11   Run the hybrids algorithm: ;   ⎫

12        $e \leftarrow e + 1$;

13        $\mathbf{S}^{(e)} = \text{PSO-GA}(\Theta)$;   ⎬ The parameter identifica-

14        $\mathbf{E}^{(e)} = \text{Evaluate}(\mathbf{S}^{(e)}, \mathbf{u}(t), \mathbf{x}(t), t_{max}^{(e)})$;   tion (stage 2)

15        $\Theta_{best}^{(e)} = \text{Get-Best}(\mathbf{S}^{(e)}, \mathbf{E}^{(e)})$;   ⎭

16   **if** $E_{best}^{(e-1)} > E_{best}^{(e)}$ **then**

17   Extend the time interval $t_{max}^{(e)}$:   ⎫ The acquisition

18   **while** $\epsilon(\Theta_{best}^{(e)}, t_{max}^{(e)}) < \epsilon_{min}$ **do**   ⎬ the new data

19            $t_{max}^{(e)} \leftarrow t_{max}^{(e)} + T_s$;   samples (stage 3)

20   **if** $t_{max}^{(e)} > t_{max}^{(e-1)}$ **then**   ⎫

21   Update the fuzzy set $\mathbf{D}^u$:   ⎬ The update of

22          $u = \arg \max_{i=1,...,L} \mu_{\mathbf{D}^i}(\bar{\mathbf{x}}(t_{max}^{(e)}))$;   fuzzy sets (stage 4)

23   Update$(\mathbf{D}^u, \mathbf{x}(t), t_{max}^{(e)})$;   ⎭

24   Evaluate the obtained solution $\Theta_{best}^{(e)}$:   ⎫ The desicion for

         $E_{best}^{(e)} = \text{Evaluate}(\Theta_{best}^{(e)}, \mathbf{x}(t), \mathbf{u}(t), t_{max}^{(e)})$;   ⎬ adding the new

25   **until** $\left(t_{max}^{(e)} > t_{max}^{(e-z)}\right) \mid \left(E_{best}^{(e)} - E_{best}^{(e-z)}\right) < \Delta E$;   ⎭ OP (stage 5)

26   **until** $\left(t_{max}^{(e)} < T_{max}\right)$ & $\left(E_{best}^{(e)} > E_{min}\right)$ & $(e < e_{max})$;

27

Lapa [22] and Szczypta [31] selected structure and parameters of the control system using Multi-Population Algorithms. Stanovov and Semenkin [30] proposed self-configuring hybrid evolutionary algorithm for fuzzy Imbalance classification. Bartczuk [3] proposed a new method for nonlinear fuzzy correction modeling of dynamic objects. He applied gene expression programming [2]. Przybył [25] has used genetic algorithm for observer parameter tuning. Yang [37] used genetic algorithm combined with local search method for identifying susceptibility Genes.

In a new proposed hybrid evolutionary method based on [15] we identify the unknown parameters of the dynamic nonlinear stationary system and parameters of the FR step by step on the basis of incoming data samples. It is done by extending the measurement area as long as identified local linear or nonlinear model can reproduce of a reference values with sufficient precision. This a is similar approach as in the case Evolving Fuzzy Systems (EFS).

A new algorithm contains four main stages: the initialization of the OP (stage 1), the parameter identification by the PSO-GA (stage 2), the acquisition of the new data samples (stage 3), the update of fuzzy sets (stage 4) and key decision stage for adding the new OP (stage 5).

A new OP is added in the first initialization stage. The initial time $t_{max}^{(e)}$ for used data samples is determined according to distance criterion $d(\mathbf{x}(0), \mathbf{x}(t_{max}^{(e)})) < d_{start}$, where d_{start} is the maximum distance between measurements determined by the expert or from the experiments. Then, the initial parameters for the fuzzy set $\mathbf{D}^1$ are estimated. For the trapezoidal membership function described by the Eq. (6), the initial parameters are determined using Eq. (7):

$$\mu_{\mathbf{D}}(x; a, b, c, d) = \begin{cases} 1 & if \ (b \leq x \leq c) \\ \frac{x-a}{b-a} & if \ (a \leq x < b) \\ \frac{d-x}{d-c} & if \ (c < x \leq d) \\ 0 & otherwise \end{cases}, \tag{6}$$

where a, b, c, d are parameters of the trapezoidal membership function,

$$b_i^1 = \min_{t<t_{max}^{(e)}} x_i(t), \quad c_i^1 = \max_{t<t_{max}^{(e)}} x_i(t),$$
$$a_i^1 = b_i - (c_i - b_i)\rho_{init}, \quad d_i^1 = c_i + (c_i - b_i)\rho_{init}, \tag{7}$$

where ρ_{init} is the initial fuzzy factor for the fuzzy set.

The one epoch of the hybrid swarm algorithm (PSO-GA) is performed in the parameters identification stage (2). The PSO-GA algorithm determines the unknown parameters of the nonlinear deterministic system and parameters of the fuzzy sets. If algorithm gives better results $E_{best}^{(e)} > E_{best}^{(e+1)}$ then the algorithm goes to the acquisition of the new data samples (stage 3) used for the parameters identification. The new data samples are included, if they meet the error criterion presented in Eq. (8)

$$\epsilon(\Theta_{best}^{(e)}, t_{max}^{(e)}) = (\mathbf{y}'(\Theta_{best}^{(e)}, t_{max}^{(e)}) - \mathbf{y}(t_{max}^{(e)}))^2, \tag{8}$$

where: $\mathbf{y}'(\Theta_{best}^{(e)}, t_{max}^{(e)})$ is the output obtained for the best created model so far $\Theta_{best}^{(e)}$ in the time $t_{max}^{(e)}$ of the simulation, $\mathbf{y}(t_{max}^{(e)})$ is the measured reference value.

In the result of added the new data samples, there are small changes of the domain for the fuzzy sets. So in the stage (4), the update of the most activated fuzzy set $u = \arg \max\limits_{i=1,...,L} \mu_{\mathbf{D}^i}(\bar{\mathbf{x}}(t^{(e)}_{max}))$ is needed and is performed by equations:

$$\left.\begin{array}{l} b^u_i = x_i(t^{(e)}_{max}) \\ a^u_i = b^u_i - (c^u_i - b^u_i)\rho_{up} \end{array}\right\} \quad \text{if } (b^u_i > x_i(t^{(e)}_{max})),$$

$$\left.\begin{array}{l} c^u_i = x_i(t^{(e)}_{max}) \\ d^u_i = c^u_i + (c^u_i - b^u_i)\rho_{up} \end{array}\right\} \quad \text{if } (c^u_i < x^u_i(t^{(e)}_{max})). \tag{9}$$

The reassessment of the obtained solutions is needed and is performed in the stage (5) in the consequence of updating of the fuzzy set and using the new data samples. In the stage (5), we check if there getting a new measurement data $t^{(e)}_{max} > t^{(e-z)}_{max}$ for predefined number of epochs or the obtained error $E^{(e)}_{best}$ for the best solution $\Theta^{(e)}_{best}$ decreases $(E^{(e)}_{best} - E^{(e-z)}_{best}) \leq \Delta E$. If not, the algorithm proceeds to add a new OP.

The algorithm finishes the work, when all measurement data $t^{(e)}_{max} = T_{max}$ were used and the error criterion $E^e_{best} \leq E_{min}$ has been meet. As a criterion of the error we use Root Mean Square Error measure (RMSE).

5   The Permanent Magnet Synchronous Motor

The simulations were performed for the nonlinear model of the Permanent Magnet Synchronous Motor (PMSM). The PMSM can be described by algebraic equations based on the state variable technique using Eqs. (10) and (11)

$$\begin{bmatrix} i_d(k+1) \\ i_q(k+1) \\ \omega_r(k+1) \\ \theta_r(k+1) \end{bmatrix} = \begin{bmatrix} a_{11} & 0 & a_{13} & 0 \\ 0 & a_{22} & a_{23} & 0 \\ 0 & a_{32} & a_{33} & 0 \\ 0 & 0 & a_{43} & 0 \end{bmatrix} \begin{bmatrix} i_d(k) \\ i_q(k) \\ \omega_r(k) \\ \theta_r(k) \end{bmatrix} + \begin{bmatrix} b_{11} & 0 & 0 \\ 0 & b_{22} & 0 \\ 0 & 0 & b_{33} \\ 0 & 0 & 0 \end{bmatrix} \begin{bmatrix} V_d(k) \\ V_q(k) \\ T_L(k) \end{bmatrix}, \tag{10}$$

$$\begin{bmatrix} i_d(k) \\ i_q(k) \end{bmatrix} = \begin{bmatrix} 1 & 0 & 0 & 0 \\ 0 & 1 & 0 & 0 \end{bmatrix} \begin{bmatrix} i_d(k) \\ i_q(k) \\ \omega_r(k) \\ \theta_r(k) \end{bmatrix}, \tag{11}$$

where: $i_d(k)$, $i_q(k)$ - d-axis and q-axis current component, $\omega_r(k)$ - rotor speed, Θ_r - rotor position, $V_d(k)$, $V_q(k)$ - d-axis and q-axis voltage, $T_L(k)$ - load torque, k - integration step.

The parameters of the system matrix $\mathbf{A}$ and input matrix $\mathbf{B}$ are described by Eqs. (12–14)

$$a_{11} = a_{22} = 1 - T_s\frac{R}{L}, \quad a_{13} = i_q(k)T_s, \quad a_{23} = -T_s(i_d(k) + \frac{\lambda_m}{L}), \tag{12}$$

where: R - stator resistance (1.456 Ω), L - stator inductance (0.008 H),

$$a_{32} = 1.5T_sP^2\frac{\lambda_m}{J}, \quad a_{33} = 1 - T_s\frac{F}{J}, \tag{13}$$

where: λ_m - rotor flux linkage $(0.175Vs)$, F - friction coefficient,

$$a_{43} = T_s, \quad b_{11} = b_{22} = \frac{T_s}{L}, \quad b_{33} = -T_s\frac{P}{J}, \tag{14}$$

where: P - number of pole pairs (3), J - moment of inertia $(0.06 \ JKgm^2)$.

6   Experimental Results

The experiments are performed for the PMSM with unknown values of the friction coefficient F and moment of inertia J. The approximate initial values of the

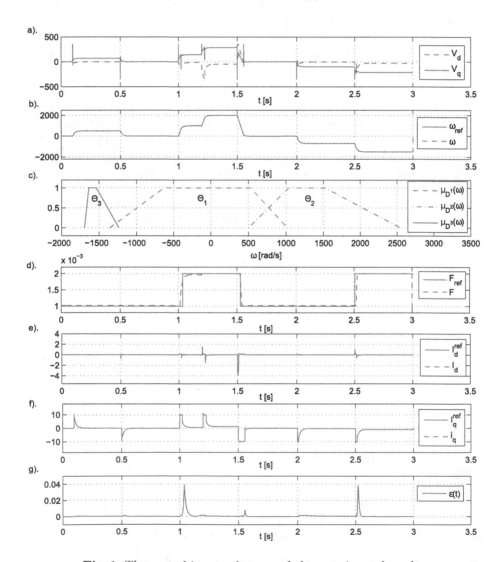

Fig. 1. The control input voltages and the experimental results.

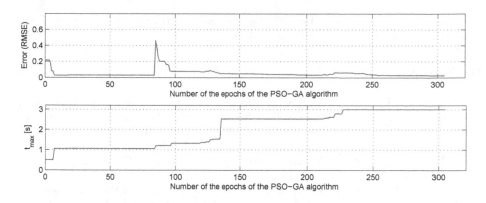

Fig. 2. The progress of the new method for non-linear modeling.

parameters obtained from the expert are: $F \approx 0.0015$ and $J \approx 0.0015$. Other parameters are known and are not identified in the experiments. The learning data set was prepared using mathematical model of the PMSM with known values of the J_{ref} (const) and F_{ref}. The values of the parameter F_{ref} change in the three operating points according to the Fig. 1d. The control voltages V_q and V_d are presented in the Fig. 1a. The goal of the experiment is reproduction of the reference values ω_{ref}, I_d^{ref} and I_q^{ref} with the smallest error measure (RMSE). It is done by discovering the moment of inertia J (constant) and friction coefficient F values in the entire work area by the identification of the nonlinear stationary object in all operating points.

Figure 1d presents the discovered characteristic of the parameter F described by the membership functions $\mu_{D_1}(\omega)$, $\mu_{D_2}(\omega)$, $\mu_{D_3}(\omega)$ of the TS fuzzy system. Each membership function $\mu_{D_i}(\omega)$ describes the identified operating point Θ_i. We select the important inputs from the measured values on the basis of expert's knowledge. Figure 1e and f contain the obtained d-axis current I_d and q-axis current I_q. Figure 1b and g, prove the effectiveness of the proposed method. The identified value of the moment of inertia $J = 0.001$ does not change in the entire work area as we expected.

Finally Fig. 2a and b present the progress of the method in the function of the epochs number. The observed local growth of the error in Fig. 2a is consequence of the acquisition of the new measurements presented in the Fig. 2b.

7 Conclusions

The proposed method for non-linear modeling using PSO and GA algorithms obtains very good results. Moreover, modeling of the OP in the form of the fuzzy rules of TS fuzzy system, provides valuable knowledge for the experts about the nature of the omitted phenomena. In future works the techniques developed in this paper will be extended to cope with interpretability aspects [8, 21].

Acknowledgment. The project was financed by the National Science Centre (Poland) on the basis of the decision number DEC-2012/05/B/ST7/02138. Also, this publication was made possible by NPRP grant #8-274-2-107 from the Qatar National Research Fund (a member of Qatar Foundation).

References

1. Aghdam, M.H., Heidari, S.: Feature selection using particle swarm optimization in text categorization. J. Artif. Intell. Soft Comput. Res. **5**(4), 231–238 (2015)
2. Bartczuk, Ł.: Gene expression programming in correction modelling of nonlinear dynamic objects. In: Borzemski, L., Grzech, A., Świątek, J., Wilimowska, Z. (eds.) ISAT 2015–Part I. Advances in Intelligent Systems and Computing. Springer, Cham (2016)
3. Bartczuk, Ł., Przybył, A., Koprinkova-Hristova, P.: New method for nonlinear fuzzy correction modelling of dynamic objects. In: Rutkowski, L., Korytkowski, M., Scherer, R., Tadeusiewicz, R., Zadeh, L.A., Zurada, J.M. (eds.) ICAISC 2014. LNCS, vol. 8467, pp. 169–180. Springer, Cham (2014). doi:10.1007/978-3-319-07173-2_16
4. Brasileiro, Í., Santos, A., Rablo, R., Mazullo, F.: Ant colony optimization applied to the problem of choosing the best combination among m combinations of shortest paths in transparent optical networks. J. Artif. Intell. Soft Comput. Res. **6**(4), 231–242 (2016)
5. Chiu, S.: Fuzzy model identification based on cluster estimation. J. Intell. Fuzzy Syst. **2**(3), 267–278 (1994)
6. Cpałka, K.: A method for designing flexible neuro-fuzzy systems. In: Rutkowski, L., Tadeusiewicz, R., Zadeh, L.A., Żurada, J.M. (eds.) ICAISC 2006. LNCS, vol. 4029, pp. 212–219. Springer, Heidelberg (2006). doi:10.1007/11785231_23
7. Cpałka, K.: On evolutionary designing and learning of flexible neuro-fuzzy structures for nonlinear classification. Nonlinear Anal. Ser. A: Theor. Methods Appl. **71**, 1659–1672 (2009). Elsevier
8. Cpałka, K.: Design of Interpretable Fuzzy Systems. Springer, Heidelberg (2017)
9. Cpałka, K., Łapa, K., Przybył, A.: A new approach to design of control systems using genetic programming. Inf. Technol. Control **44**(4), 433–442 (2015)
10. Cpałka, K., Rebrova, O., Nowicki, R., Rutkowski, L.: On design of flexible neuro-fuzzy systems for nonlinear modelling. Int. J. Gen. Syst. **42**(6), 706–720 (2013)
11. Cpałka, K., Rutkowski, L.: Flexible Takagi-Sugeno fuzzy systems. In: Proceedings of the 2005 IEEE International Joint Conference on IJCNN 2005, vol. 3, pp. 1764–1769 (2005)
12. Cpałka, K., Rutkowski, L.: Flexible Takagi Sugeno neuro-fuzzy structures for nonlinear approximation. WSEAS Trans. Syst. **4**(9), 1450–1458 (2005)
13. Dziwiński, P., Rutkowska, D.: Algorithm for generating fuzzy rules for WWW document classification. In: Rutkowski, L., Tadeusiewicz, R., Zadeh, L.A., Żurada, J.M. (eds.) ICAISC 2006. LNCS, vol. 4029, pp. 1111–1119. Springer, Heidelberg (2006). doi:10.1007/11785231_116
14. Dziwiński, P., Bartczuk, Ł., Przybył, A., Avedyan, E.D.: A new algorithm for identification of significant operating points using swarm intelligence. In: Rutkowski, L., Korytkowski, M., Scherer, R., Tadeusiewicz, R., Zadeh, L.A., Zurada, J.M. (eds.) ICAISC 2014. LNCS, vol. 8468, pp. 349–362. Springer, Cham (2014). doi:10.1007/978-3-319-07176-3_31
15. Dziwiński, P., Avedyan, E.D.: A new method of the intelligent modeling of the nonlinear dynamic objects with fuzzy detection of the operating points. In: Rutkowski, L., Korytkowski, M., Scherer, R., Tadeusiewicz, R., Zadeh, L.A., Zurada, J.M. (eds.) ICAISC 2016. LNCS, vol. 9693, pp. 293–305. Springer, Cham (2016). doi:10.1007/978-3-319-39384-1_25

16. Eftekhari, M., Zeinalkhani, M.: Extracting interpretable fuzzy models for nonlinear systems using gradient-based continuous ant colony optimization. Fuzzy Inf. Eng. **5**, 255–277 (2013). Springer
17. Gabryel, M., Cpałka, K., Rutkowski, L.: Evolutionary strategies for learning of neuro-fuzzy systems. In: Proceedings of the I Workshop on Genetic Fuzzy Systems, Granada, pp. 119–123 (2005)
18. Korytkowski, M.: Novel visual information indexing in relational databases. preprint, Integrated Computer-Aided Engineering, pp. 1–10 (2016)
19. Korytkowski, M., Rutkowski, L., Scherer, R.: Fast image classification by boosting fuzzy classifiers. Inf. Sci. **327**, 175–182 (2016)
20. Łapa, K., Cpałka, K., Wang, L.: New method for design of fuzzy systems for nonlinear modelling using different criteria of interpretability. Artif. Intell. Soft Comput. **8467**, 217–232 (2014)
21. Łapa, K., Przybył, A., Cpałka, K.: A new approach to designing interpretable models of dynamic systems. In: Rutkowski, L., Korytkowski, M., Scherer, R., Tadeusiewicz, R., Zadeh, L.A., Zurada, J.M. (eds.) ICAISC 2013. LNCS, vol. 7895, pp. 523–534. Springer, Heidelberg (2013). doi:10.1007/978-3-642-38610-7_48
22. Łapa, K., Szczypta, J., Venkatesan, R.: Aspects of structure and parameters selection of control systems using selected multi-population algorithms. In: Rutkowski, L., Korytkowski, M., Scherer, R., Tadeusiewicz, R., Zadeh, L.A., Zurada, J.M. (eds.) ICAISC 2015. LNCS, vol. 9120, pp. 247–260. Springer, Cham (2015). doi:10.1007/978-3-319-19369-4_23
23. Muhammad, A., Helon, V., Muhammad, A.: Nonlinear system identification using neural network. In: Chowdhry, B.S., Shaikh, F.K., Hussain, D.M.A., Uqaili, M.A. (eds.) Emerging Trends and Applications in Information Communication Technologies. Communications in Computer and Information Science, vol. 281, pp. 122–131. Springer, Heidelberg (2012)
24. Przybył, A., Cpałka, K.: A new method to construct of interpretable models of dynamic systems. In: Rutkowski, L., Korytkowski, M., Scherer, R., Tadeusiewicz, R., Zadeh, L.A., Zurada, J.M. (eds.) ICAISC 2012. LNCS, vol. 7268, pp. 697–705. Springer, Heidelberg (2012). doi:10.1007/978-3-642-29350-4_82
25. Przybył, A., Jelonkiewicz, J.: Genetic algorithm for observer parameters tuning in sensorless induction motor drive. In: Rutkowski, L., Kacprzyk, J. (eds.) Neural Networks and Soft Computing. Advances in Soft Computing, vol. 19, pp. 376–381. Physica, Heidelberg (2003)
26. Rutkowski, L., Cpałka, K.: Flexible structures of neuro-fuzzy systems. Quo Vadis Computational Intelligence, Studies in Fuzziness and Soft Computing, vol. 54, pp. 479–484. Springer, Heidelberg (2000)
27. Rutkowski, L., Cpałka, K.: Flexible weighted neuro-fuzzy systems. In: Proceedings of the 9th International Conference on Neural Information Processing (ICONIP 2002), Orchid Country Club, Singapore, 18–22 November 2002
28. Rutkowski, L., Cpałka, K.: Compromise approach to neuro-fuzzy systems. In: Proceedings of the 2nd Euro-International Symposium on Computation Intelligence, Frontiers in Artificial Intelligence and Applications, vol. 76, pp. 85–90 (2002)
29. Rutkowski, L., Cpałka, K.: Neuro-fuzzy systems derived from quasi-triangular norms. In: Proceedings of the IEEE International Conference on Fuzzy Systems, Budapest, vol. 2, pp. 1031–1036, July 26–29 2004
30. Stanovov, V., Semenkin, E., Semenkina, O.: Self-configuring hybrid evolutionary algorithm for fuzzy imbalanced classification with adaptive instance selection. J. Artif. Intell. Soft Comput. Res. **6**(3), 173–188 (2016)

31. Szczypta, J., Łapa, K., Shao, Z.: Aspects of the selection of the structure and parameters of controllers using selected population based algorithms. In: Rutkowski, L., Korytkowski, M., Scherer, R., Tadeusiewicz, R., Zadeh, L.A., Zurada, J.M. (eds.) ICAISC 2014. LNCS, vol. 8467, pp. 440–454. Springer, Cham (2014). doi:10.1007/978-3-319-07173-2_38

32. Szczypta, J., Przybył, A., Cpałka, K.: Some aspects of evolutionary designing optimal controllers. In: Rutkowski, L., Korytkowski, M., Scherer, R., Tadeusiewicz, R., Zadeh, L.A., Zurada, J.M. (eds.) ICAISC 2013. LNCS, vol. 7895, pp. 91–100. Springer, Heidelberg (2013). doi:10.1007/978-3-642-38610-7_9

33. Starczewski, J., Rutkowski, L.: Interval type 2 neuro-fuzzy systems based on interval consequents. In: Rutkowski, L., Kacprzyk, J. (eds.) Neural Networks and Soft Computing. Advances in Soft Computing, vol. 19, pp. 570–577. Springer, Heidelberg (2003)

34. Starczewski, J., Rutkowski, L.: Connectionist structures of type 2 fuzzy inference systems. In: Wyrzykowski, R., Dongarra, J., Paprzycki, M., Waśniewski, J. (eds.) PPAM 2001. LNCS, vol. 2328, pp. 634–642. Springer, Heidelberg (2002). doi:10.1007/3-540-48086-2_70

35. Scherer, R.: Multiple Fuzzy Classification Systems. Springer, Heidelberg (2012)

36. Xinghua, L., Jiang, M., Jike, G.: A method research on nonlinear system identification based on neural network. In: Zhu, R., Ma, Y. (eds.) Information Engineering and Applications. LNEE, vol. 154, pp. 234–240. Springer, Heidelberg (2012). doi:10.1007/978-1-4471-2386-6_193

37. Yang, C.H., Moi, S.H., Lin, Y.D., Chuang, L.Y.: Genetic algorithm combined with a local search method for identifying susceptibility genes. J. Artif. Intell. Soft Comput. Res. **6**(3), 203–212 (2016)

38. Zalasiński, M., Cpałka, K.: A new method of on-line signature verification using a flexible fuzzy one-class classifier. pp. 38–53. Academic Publishing House EXIT (2011)

39. Zalasiński, M.: New algorithm for on-line signature verification using characteristic global features. Adv. Intell. Syst. Comput. **432**, 137–146 (2016). doi:10.1007/978-3-319-28567-2_12

40. Zalasiński, M., Cpałka, K.: New algorithm for on-line signature verification using characteristic hybrid partitions. Adv. Intell. Syst. Comput. **432**, 147–157 (2016). doi:10.1007/978-3-319-28567-2_13

41. Zalasiński, M., Cpałka, K., Hayashi, Y.: A New Approach to the Dynamic Signature Verification Aimed at Minimizing the Number of Global Features. In: Rutkowski, L., Korytkowski, M., Scherer, R., Tadeusiewicz, R., Zadeh, L.A., Zurada, J.M. (eds.) ICAISC 2016. LNCS, vol. 9693, pp. 218–231. Springer, Cham (2016). doi:10.1007/978-3-319-39384-1_20

42. Zalasiński, M., Cpałka, K., Rakus-Andersson, E.: An Idea of the Dynamic Signature Verification Based on a Hybrid Approach. In: Rutkowski, L., Korytkowski, M., Scherer, R., Tadeusiewicz, R., Zadeh, L.A., Zurada, J.M. (eds.) ICAISC 2016. LNCS, vol. 9693, pp. 232–246. Springer, Cham (2016). doi:10.1007/978-3-319-39384-1_21

43. Zalasiński, M., Cpałka, K., Rutkowski, L.: A new algorithm for identity verification based on the analysis of a handwritten dynamic signature. Appl. Soft Comput. **43**, 47–56 (2016)

Linguistic Habit Graphs Used for Text Representation and Correction

Marcin Gadamer[(✉)]

Department of Automatics and Biomedical Engineering,
AGH University of Science and Technology,
Mickiewicza Av. 30, 30–059 Krakow, Poland
gadamer@agh.edu.pl

Abstract. This paper introduces a novel associative way of storing, compressing, and processing sentences. The Linguistic Habit Graphs (LHG) are introduced as graph models that could be used for spell checking, text correction, proof–reading, and compression of sentences. All the above mentioned functionalities are always available in the constant computational complexity as a result of the associative way of text processing, special kinds of connections and graph nodes that enable to activate various important relations between letters and words simultaneously for any given contexts. Furthermore, using the proposed graph structure, new algorithms have been developed to provide effective text analyzes and contextual text correction. These new algorithms can properly locate and often automatically correct typical mistakes in texts written in a given language for which the graph was build.

Keywords: NLP · Text correction · Proof–reading · Linguistic Habit Graph

1 Introduction

The understanding of a natural language is a very difficult task from the computer science perspective. The algorithms need to understand the language structure, the words, and how these two things influence each other in order to proof–read texts. On the other hand when user uses Google for searching phrase "imeges", it is prompting "Did you mean: images". It shows that there are algorithms, which help us with spell checking and text correction [22]. This kind of algorithms can be classified into a few groups. Spelling correction algorithms in search engines are similar to those of spell checkers in word processors. The first one checks every query term in a dictionary. If a term is not found in the dictionary, then those words from the dictionary, which are most similar to the query term in a question are shown as spelling suggestions. The next group of algorithms uses similarity computed as an edit distance. The similarity measure between two words is usually represented by the Damerau–Levenshtein distance [6]. Another type of this group is a "weighted edit distance". Using it, we can give a higher priority to the pairs which sound

© Springer International Publishing AG 2017
L. Rutkowski et al. (Eds.): ICAISC 2017, Part II, LNAI 10246, pp. 233–242, 2017.
DOI: 10.1007/978-3-319-59060-8_22

similar or which are close to each other on the keyboard layout. Soundex is an example of an algorithm, which uses phonetics for indexing names by sound [21].

There are different approaches for calculating the edit distance between dictionary terms and query terms:

1. Naive Approach – calculating the edit distance between a query term and every dictionary term. This method is very expensive and slow.
2. Peter Norvig's Approach – deriving all possible terms with an edit distance ≤ 2 from the query term and looking them up in the dictionary. This is an improvement to the first one, although the algorithm remains still expensive. For 9 characters word with edit distance equals 2 it returns 114 324 terms [12].
3. Faroo's Approach – deriving deletion only with an edit distance ≤ 2 both from a query term and each dictionary term. It is three orders of magnitudes faster comparing to previous one [1].

These kinds of algorithms may use a few other aspects as:

– Sorting – suggestions are sorted first by the (weighted) edit distance then by static word frequency or a number of results which suggested query would return for the index.
– Language detection – the selected or detected language of the query might be taken into account, e.g. by using a language specific spell checking dictionary.
– Dictionary – the spell checking dictionary might be static or dynamic. The second one might be generated or supplemented from the search engine index or the queries entered by the users. In this case, if the word frequency for a specific term is above a given threshold then that word is added to the spell checking dictionary. It is also possible to use the search index itself as a spell checking dictionary.
– Markov Model – An alternative to the dictionary based on spelling are statistical methods, e.g. the Hidden Markov Model [3].

This paper is focused on the specialized graph constructions that are able to reflect, memorize and weigh natural relations between letters and words in many human languages. The natural letter and word order are used to construct a Linguistic Habit Graph that is able to associate and store the natural human linguistic behaviors. A few spell checking and text correction methods are presented based on the proposed graph model, which resembles functioning of a brain. Each vertex in this graph could be active as well as neurons are in a human brain. The active graph vertices are called neurons, and edges, are called connections. They are used to represent and then to recall the order and contextual dependencies between words in corrected texts.

2   Linguistic Habit Graphs

The main goal of this paper is to demonstrate the use of the LHG graphs for text correction. The intelligent semi–automatic text correction is one of the common, important and practical computational tasks that can be solved by the mentioned graph. The graph structure is able to gather linguistic habits of many individuals and use them to define language components, interconnect them in a way people usually do, trigger off and propose the most probable following words in common contexts, and help to correct mistakes in texts. The presented neural graph represents many active connections between letters and words, that can be automatically triggered off as a result of activation of any combination of other neurons representing a given phrase of the internal context. There is also a possibility to build up a graph for an individual person and use it to recognize if a given text has been written by that person [11]. The LHG graphs can suggest various correction options in the context of other words in the sentences which were previously read. Moreover, various suggested options are weighted by the frequency of their usage by other people in the past and by a given context of the other words in the analyzed sentences. As a result of the associative connections in this graph, the checking process and determination of the subgraph of the most probable corrections are always available in constant time.

The LHG is a graph. It consists of vertices that represent letters, apostrophes and some special characters like the beginning of each word. To build such a structure the algorithm needs to look through a text corpora only a few times. Each word is represented by an interconnected sequence of letter neurons (Fig. 1). The letter neuron representing the last letter of each word is also called the word neuron. In the next step, the word neurons are interconnected in many ways to reflect word orders and previous word contexts in processed sentences. Each word is represented exactly by a single word neuron in this graph.

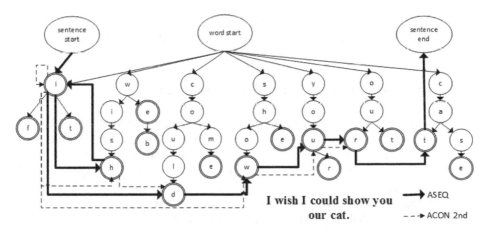

Fig. 1. The small piece of the LHG graph structure with ASEQ and ACON 2nd interconnections between the following words in the presented sentence.

The word neurons are interconnected by ASEQ and ACON connections, where ASEQ means an associative sequence and ACON stands for an associative context. First of all, the word neurons are naturally ASEQ connected after the sequences words from read sentences of the text corpora. All connections (edges of the graph) are directed and weighted by the frequency of these word sequences. To unambiguously read sentences using this graph, it is necessary to add some extra context connections (ACON) interconnecting more previous words with the next word to explicitly and contextually point to the next word in the sequence and the given word order. The associative context connections are added only if the context of the previous word neuron is ambiguous, in order to determine which neuron should be activated next (in the context of the previously activated word neurons after the already read text corpora). The LHG graph uses ACON edges from 2nd to 5th level, where ACON 2nd level edge means that connection is between $(n-2)$–word and (n)–word in the sentence.

Figure 2 shows the ASEQ and ACON connections in the LHG after reading sentences: "Alice has a cat.", "Bob has a dog.", "Mary's dog is called Star.".

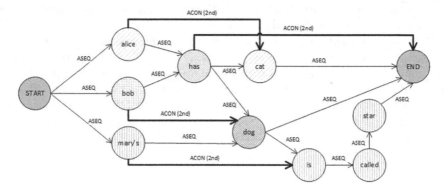

Fig. 2. The part of the LHG graph with ASEQ and ACON 2nd connections.

3   Associative Text Correction Methods

There are four main types of mistakes in written language: spelling, punctuation, grammar, and usage. The Linguistic Habit Graph could be widely used for developing spell checking and text correction methods for all of these main types of common mistakes. Each word neuron contains information about a word and its properties. Each word neuron property contains its own counter which is updated while constructing this graph. As an example after reading sentence "Alice has a dog called Star!" The LHG's word neurons will contain information described in Table 1.

3.1   Checking Punctuation in Sentences

Based on the properties described in the previous section a new algorithm was constructed for checking punctuation in sentences. The main task for this algorithm is to check if a word should start with a small or a capital letter, if the comma is mandatory and also if the sentence starts with a capital letter and

Table 1. Word neuron's properties after reading sentence "Alice has a dog called Star!"

Word neuron	Word neuron's properties
alice	startASentence, startWithSmallLetter
has	startWithSmallLetter
dog	followedByA, startWithSmallLetter
called	startWithSmallLetter
star	startWithBigLetter, endWithExclamationMark

ends with a full-stop, an exclamation mark or a question mark. It relies on the frequency of occurrence properties of the word. For example the "alice" word is represented by following counter properties:

- $word_C = 131$ – how many times the word was analyzed,
- $startSentence_C = 19$ – how many times the word starts a sentence,
- $smallLetter_C = 19$ – how many times the word starts with small letter (when the word is at the beginning of a sentence it is always marked as startWith-SmallLetter),
- $bigLetter_C = 112$ – how many times the word starts with a capital letter.

Based on these properties the punctuation correction algorithm can assume that the "alice" word always starts with a capital letter. It is done by checking inequality (1). After series of experiments, the $bigLetterFactor$ was set to 0.87 the $endWithExclamationMark$ was set to 0.84 and the $endWithQuestionMark$ was set to 0.79 the to give the best correction results. Moreover, there are many similar inequalities that check other possible punctuation errors.

$$\frac{startWithBigLetter_{COUNTER}}{(word_{COUNTER} - startSentence_{COUNTER})} \geq bigLetterFactor \qquad (1)$$

3.2 Semi–automatic Text Correction Methods

The text correction methods could be split into two groups. The first group contains methods that could check the text while entering it. If the context of a sentence is incorrect the correction algorithms should immediately mark all incorrect words and propose the correct replacements. The second group of methods looks for errors after entering the whole text. After entering sentences the algorithms should check if all words are correctly placed. If the words are in an incorrect order, the semi–automatic text correction methods should find and mark them. Then they try to automatically correct a sentence by replacing the incorrect words to the correct ones. If the automatic text correction is impossible or many the same way similar changes are available then propose a few possible replacements.

These two groups of methods are strictly connected not only with the spell correction of words but even more with the context of words. The biggest problem in correction tools is that they do not address the problem with word contexts. The main purpose of constructing the LHG graph is to store the ASEQ and ACON connections and rebuild a correct sentence context from them. The algorithms of semi–automatic correction using the LHG graph work well under the assumption that this graph has been constructed after a possibly huge training text corpora. Generally, the context of the previously activated word neurons displays the next word neurons that usually appear in a given context.

There are two commonly used groups of methods for the text correction task. The first one measures the edit distances [15]. The second one uses the language modeling by n–gram models [2]. These two kinds of methods were developed based on the LHG graph.

The naive approach to the "edit distance" metric is done by calculating the distance between a query term and every dictionary term. These methods are very expensive and very slow. Thanks to the ASEQ connections checking the edit distance is much faster. First, the ASEQ connections are analyzed for each word. If the ASEQ connection is not found between two neighboring word neurons then the algorithm checks the Damerau–Levenshtein distance only between a potentially wrong word and all successive words from the previous word neuron connected with the ASEQ connection (Fig. 3). This approach returns only candidates, which could exist in a given context.

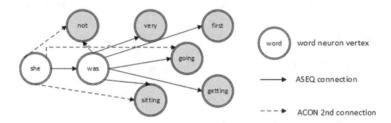

Fig. 3. The next possible words for context "she was ...".

The second group of methods uses the n–gram models. The goal of these methods is to assign a probability of correctness to a given sentence. This probability might be used in machine translation, spell correction and speech recognition. An example of the sentence probability is shown in a inequality (2) [17].

$$P(\text{I saw a van}) > P(\text{eyes awe of an}) \tag{2}$$

The simplest case uses the unigram model for measuring whole sentence correctness. In that model of the sentence, probability is counted as shown in formula (3) where w_i means a word on the i position.

$$P(w_1 w_2 ... w_n) \approx \prod_i P(w_i) \tag{3}$$

The more advanced model uses the bigram model instead of the unigram model. The probability of a sentence using the bigram model is defined as it is shown in formulas from (4) to (6) where w_i means a word on the i position and $c(w_{i-1}, w_i)$ is a counter how many times the "w_{i-1}, w_i" sequence occurred in all analyzed texts.

$$P(w_i|w_1 w_2...w_{i-1}) \approx P(w_i|w_{i-1}) \tag{4}$$

$$P(w_i|w_{i-1}) = \frac{c(w_{i-1}, w_i)}{c(w_{i-1})} \tag{5}$$

$$P(s) = \prod_{i=1}^{l+1} P(w_i|w_{i-1}) \tag{6}$$

The LHG could easily check the existence of bigrams by checking frequency of ASEQ connections between word neurons. On the other hand, the n–gram models (unigrams, bigrams etc.) are many times insufficient to model a language because of long–distance dependencies. For example in sentence *"The computer, which I had just put into the machine room on the fifth floor, crashed"* the verb *"crashed"* is strictly connected with the subject "computer" [13]. The next problem with using n–gram models is a case when $c(w_{i-1}, w_i) = 0$ or $c(w_{i-1}) = 0$.

The conducted research showed, that using the n–gram, the models sometimes need the whole sentence context to perform corrections. It is very difficult to store sentences with a whole unambiguous context in the LHG graph. Based on that, new methods that give a better result of text correction than using the n–gram models methods have been developed.

First method uses all possible ASEQ paths between words. The first step is to measure the path distance between words in a few other words surroundings. By the default, the algorithm is checking the path distance between two nearest neighbors for each node. Paths with the maximum length of 3 are only taken into consideration.

The next step is to count the occurrence of path's length for each word. The algorithm needs to check:

- the cardinality of direct connections between words (a number of direct paths $node_1 \rightarrow node_2$)
- the cardinality of connections passing through the neighboring vertices (paths $node_1 \rightarrow neighbor_1 \rightarrow node_2$ and $node_1 \rightarrow neighbor_1 \rightarrow neighbor_2 \rightarrow node_2$)

After determining the cardinality for each word the next algorithm calculates for them the "priority", by which they can occur in a given context. This algorithm allows determining, which words are in an order and which ones are in an incorrect order and need to be corrected by another algorithm. The word priority in a given position is calculated as a sum of its number of occurrences (o_{nb}) in that position multiplied by the difference in the distance between neurons that stimulated each occurrence (7).

$$w_p(i) = \sum_{n=i-2}^{i+2} o_{nb}(i) * \left(\frac{1}{2}\right)^{|(i-n)|} \tag{7}$$

Error type	Sentence to correct (with errors)	Sentence after correction
big letter word	'What a curious feeling!' said alice.	'What a curious feeling!' said **Alice**.
wrong article	Oh, Joe, you're a angel.	Oh, Joe, you're **an** angel.
missing ending mark	what does this mean	What does this mean**?**
missing apostrohe	Well, try to recollect - **cant** you?	Well, try to recollect - **can't** you?
missing comma mark	Look here what does this mean?	Look here**,** what does this mean?
putting two words together	Well, **iknow** that.	Well, **I know** that.
putting two words together	Did you? Why, **its** funny I didn't see you.	Did you? Why, **it is** funny I didn't see you.
spelling error	They caught fish, **coked** supper and ate it.	They caught fish, **cooked** supper and ate it.
wrong word	After several turns, he sat **on** again.	After several turns, he sat **down** again.
ommiting one word	Alice had no idea what do.	Alice had no idea what **to** do

Fig. 4. The correction of the typical errors by the developed algorithms.

The second method tries to set a correct order of words in a given sentence. The conducted research has shown that this algorithm performs very well especially for a few kinds of errors:

- when a word has been omitted in a sentence,
- when a word has been excessively introduced in a sentence,
- when two words were combined into one.

3.3 Comparison of the Text Correction Methods

After constructing the LHG graph model and semi–automatic text correction methods several experiments were conducted in order to check how the new

Correction tools	Sentence to correct: She was always **redy** to talk about her pet. Correct sentence: She was always **ready** to talk about her pet.
Word 2010	reedy, red, **ready**, reds, rely
Libre Office Writer 5	red, redye, reedy, **ready**, reds, redo, rely
OnlineCorrection.com	red, **ready**, rely, Rudy, reds, redo, reedy
Google search	**ready**
LHG graph	**ready**

	Sentence to correct: Do you **thinki** can listen? Correct sentence: Do you **think I** can listen?
Word 2010	think, thinks, thinking, thank
Libre Office Writer 5	think, thinks, **think I**, thin, thank
OnlineCorrection.com	think, thinks, **think i** (small letter)
Google search	**think i** (small letter)
LHG graph	**think I**

	Sentence to correct: Alice could think nothing else to say. Correct sentence: Alice could think **of** nothing else to say.
Word 2010	no correction
Libre Office Writer 5	no correction
OnlineCorrection.com	no correction
Google search	no correction
LHG graph	**think of**

Fig. 5. The comparison of the correction results obtained from different applications.

algorithms will be able to process and correct the sentences with mistakes. For the purpose of the experiments the LHG graph was constructed, which contained 3 millions of word neurons, and more than 16,5 millions of ASEQ and ACON connections.

Figure 4 shows the typical kind of errors and how developed spell checking algorithms have corrected them.

These experiments have shown that the presented idea is valuable. The next step was to compare the correction results with other contemporary applications. The result of the working sentence recreation algorithm are presented in Fig. 5.

The conducted research and experiments have shown that the proposed LHG graphs combine the abilities to automatically check and correct various texts with high efficiency and new algorithms can properly locate and often automatically correct typical mistakes in texts.

4 Summary and Conclusion

In this paper the novel associative way of storing, compressing and processing sentences by the Linguistic Habit Graphs and their use for semi–automatic correction have been described. Thanks to the use of the associative mechanisms implemented in the presented graph structure the new correction methods based on a word context have been achieved. This research has shown, that a context of a sentence is crucial for the text correction methods and tools. The introduced LHG graph is a new model that stores words in the contexts of other words. The experiments have shown that the correction can be more accurate if the context of the previous words is used. The obtained results are usually more adequate than corrections provided by commonly used applications.

References

1. 1000x Faster Spelling Correction algorithm, June 2012. http://blog.faroo.com/2012/06/07/improved-edit-distance-based-spelling-correction/. Online; Visited 03 Oct 2016
2. Brown, P.F., et al.: Class-based n-gram models of natural language. Comput. Linguist. **18**(4), 467–479 (1992)
3. Blunsom, P.: Hidden Markov Models. Lecture notes, vol. 15, pp. 18–19, August 2004
4. Chen, S.F., Goodman, J.: An empirical study of smoothing techniques for language modeling. In: Proceedings of the 34th Annual Meeting on Association for Computational Linguistics. Association for Computational Linguistics (1996)
5. Ide, N., Pustejovsky, J. (eds.): Handbook of Linguistic Annotation. Springer, Netherlands (2017). doi:10.1007/978-94-024-0881-2
6. Deepak, P., Deshpande, P.M.: Operators for Similarity Search. Semantics, Techniques and Usage Scenarios. Springer International Publishing, Cham (2015)
7. Gadamer, M., Horzyk, A.: Text analysis and correction using specialized linguistic habit graphs LHG. Image Process. Commun. **17**(4), 245–250 (2012). Bydgoszcz

8. Horzyk, A.: Artificial Associative Systems and Associative Artificial Intelligence, pp. 1–280. Academic Publishing House EXIT, Warsaw (2013)
9. Horzyk, A.: How does generalization and creativity come into being in neural associative systems and how does it form human-like knowledge? Neurocomputing **144**, 238–257 (2014). Elsevier
10. Horzyk, A.: Innovative types and abilities of neural networks based on associative mechanisms and a new associative model of neurons. In: Rutkowski, L., Korytkowski, M., Scherer, R., Tadeusiewicz, R., Zadeh, L.A., Zurada, J.M. (eds.) ICAISC 2015. LNCS, vol. 9119, pp. 26–38. Springer, Cham (2015). doi:10.1007/978-3-319-19324-3_3
11. Horzyk, A., Gadamer, M.: Associative text representation and correction. In: Rutkowski, L., Korytkowski, M., Scherer, R., Tadeusiewicz, R., Zadeh, L.A., Zurada, J.M. (eds.) ICAISC 2013. LNCS (LNAI), vol. 7894, pp. 76–87. Springer, Heidelberg (2013). doi:10.1007/978-3-642-38658-9_7
12. How to write a spelling corrector. http://norvig.com/spell-correct.html. Online; Visited: 05 Oct 2016
13. Jurafsky, D.: Speech & language processing. Pearson Education India, Noida (2000)
14. Kohonen, T., Somervuo, P.: Self-organizing maps of symbol strings. Neurocomputing **21**(1), 19–30 (1998)
15. Manning, C.D., Schütze, H.: Foundations of Statistical Natural Language Processing, vol. 999. MIT Press, Cambridge (1999)
16. Manning, C.D., Surdeanu, M., Bauer, J., Finkel, J.R., Bethard, S., McClosky, D.: The stanford CoreNLP natural language processing toolkit. In: ACL (System Demonstrations), pp. 55–60 (2014)
17. Martin, D., Jurafsky, J.H., Jurafsky, D.: Speech and Language Processing. An Introduction to Natural Language Processing, Computational Linguistics, and Speech Recognition. Prentice Hall, Upper Saddle River (2000)
18. Savary, A., Piskorski, J.: Lexicons and Grammars For Named Entity Annotation in the National Corpus of Polish, pp. 141–154. Intelligent Information Systems, Siedlce (2010)
19. Stolcke, A.: SRILM-an extensible language modeling toolkit. In: Proceedings International Conference on Spoken Language Processing, pp. 257–286 (2002)
20. Täckström, O., Das, D., Petrov, S., McDonald, R., Nivre, J.: Token and type constraints for cross-lingual part-of-speech tagging. Trans. Assoc. Comput. Linguist. **1**, 1–12 (2013)
21. The soundex indexing system. https://goo.gl/5BUcR5 (May 2007). Online; Visited 04 Oct 2016
22. What are some algorithms of spelling correction that were used by search engine? https://goo.gl/8xhvpQ. Online; Visited 03 Oct 2016
23. Whitelaw, C., Hutchinson, B., Chung, G.Y., Ellis, G.: Using the web for language independent spellchecking and autocorrection. In: Proceedings of the 2009 Conference on Empirical Methods in Natural Language Processing, vol. 2, pp. 890–899. Association for Computational Linguistics (2009)

Dynamic Epistemic Preferential Logic of Action

Krystian Jobczyk[1,2]([✉]) and Antoni Ligeza[2]

[1] University of Basse-Normandie of Caen, Caen, France
krystian.jobczyk@unicaen.fr
[2] AGH University of Science and Technology of Kraków, Kraków, Poland

Abstract. H.P. van Ditmarsch, W. van der Hoek and B.P. Kooi proposed in 2003 some complete formalism for representation of actions for Multi-Agent Systems. This paper is aimed at proposing a new preferential extension of this formalism in terms of dynamic-epistemic logic supported by a unique multi-valued logic. This new system is interpreted in the interval fibred semantics on a base of earlier ideas of D. Gabbay.

1 Introduction

A notion of action constitutes a key concept of classical and temporal planning and reasoning. It has already been discussed in many contexts and works – such as: [5,18,19]. Different types of logic – suitable to render actions – is currently known. Some of them are very expressive such as: temporal logic of action of L. Lamport from [16] or description logic of Baader from [2]. Finally, action dynamism may be represented in terms of a variant of dynamic-epistemic logic such as in [1] – in particular – in the so-called concurrent logic of actions for Multi-Agent Systems of Ditmarsch from [3]. Some possibility to deal with actions – as associated to time intervals – might also rendered in terms of Halpern-Shoham logic from [10] and its preferential extensions – discussed in [11,12]. Finally, a capability of expressing actions consitutes a one of criterions considered in some evaluation attempts of temporal and fuzzy logic systems in [13,14].

1.1 Paper Motivation

Unfortunately, both actions and preferences – due to [20,21] – are represented and semantically interpreted in a *point-wise manner* in the framework of these approaches. Meanwhile – as earlier mentioned – both entities require rather to be referred to time periods than time points. Unfortunately, no attempt to deal with them in this way is known. The Halpern-Shoham logic only partially respect this expectation. In fact, it allows us to consider actions, but only *implicitly* – by intervals associated to them. Even the so-called *fibred semantics* – invented by D. Gabbay in [6] as a comfortable and flexible tool to interpret mixed formal systems – is a type of a point-wise semantics.

All these shortcomings and lacks form the main motivation factor of this work.

© Springer International Publishing AG 2017
L. Rutkowski et al. (Eds.): ICAISC 2017, Part II, LNAI 10246, pp. 243–254, 2017.
DOI: 10.1007/978-3-319-59060-8_23

1.2   The Paper Objectives and Paper Organization

According to these facts, this paper is aimed at:

- construction of a hybrid multi-modal logic system for actions a preferences in- terpreted in an interval-based semantics.
- proposing an outline a fibred semantics for such a mixed system,
- explanation how this system might be exploited in modeling of some temporal version of Traveling Salesman Problem.

On the one hand, this approach make use of results based on modal logic – such as: [1,3], but it is also supported by ideas of fuzzy logic systems from [7–9]. This solution is dictated by an intermediate nature of the proposed multi-valued logic system for actions and preferences. This approach – from Sects. 3 and 5 in particular – develops ideas from [11,15].

The rest of the paper is organized as follows. In Sect. 2 a terminological background of analyses is given. Section 3 presents the formalism of Multi-Valued Preferential Logic. Dynamic Epistemic Preferential Logic of Action ($\mathcal{DEPLA}$) – interpreted in fibred semantics is introduced in Sect. 4. Section 5 shows how this semantics type works in a practice. Section 6 contains concluding remarks.

2   Preliminaries

We preface the proper paper analysis by an introducing a terminological and technical background concerning a point-wise fibring semantics.

Fibring (pointwise) Semantics. The fibring semantics method – invented in [6] – allows us to form a new uniform semantics for combined systems – interpreted in separate semantics. To illustrate this methods consider a mixed formula $\phi = \Diamond_1 \Box_2 \chi$ and assume also that ϕ is considered as some language, say $\mathcal{L}_1$ -formula. From the point of view of $\mathcal{L}_1$ has a form $\Diamond_1 p$, where $p = \Box \chi$ is atomic formula in $\mathcal{L}_1$ since it does not recognize $\Box_2$-modality.

We define now a model $\mathcal{M}_1$ for $\Diamond_1 p$ as $\mathcal{M}_1 = \langle S_1, R_1, a, h_1 \rangle$, where S_1 is a set of possible worlds, $a \in S$ is an actual world such that $a \models \Diamond p$; $R_1 \subseteq S \times S$ is an accessibility relation and h is such a binary assignment function that $h(t, p) \in \{0, 1\}$ for any $t \in S$ and atomic p. The satisfaction condition in $\mathcal{M}$ is as follows:

$$a \models \Diamond p \iff \exists \in S_1 (aR_1 t \text{ and } t \models p, \text{i.e.} t \models \Box_2 q). \tag{1}$$

Since $\Box_2 q$ does not belong to $\mathcal{L}_1$ and, therefore, it (generally) cannot be evaluated in $\mathcal{M}_1$ – one needs a new model $\mathcal{M}_2$ of a new language, say $\mathcal{L}_2$, to evaluate $\Box_2 q$. Therefore, we associate with a state t a new model $\mathcal{M}_2^t = \langle S_2^t, R_2^t, a_2^t, h_2^t \rangle$ – in such a way that

$$t \models_1 \Box_2 q \iff a_2^t \models_2 \Box q. \tag{2}$$

We can also establish that some mapping **F** associates a model $\mathcal{M}_2^t$ with t and than the fibred semantics for the language $\mathcal{L}_{1,2}$ has a model

$\mathcal{M} = \langle S_1, R_1, a, h_1, \mathbf{F} \rangle$. This reasoning introduces the main conceptual body of Gabbay's *fibring(fibred) semantics*.

We also adopt the following definitions of the interval-based interpreted system and an interval.

Definition 1. *An interval-based interpreted system IBIS is a tuple* (S, s_0, t, L) *such that: S is a finite set of (ordered) global states (that can be points or intervals) accessible from an initial state s_0; t is a standard transition relation between states and $L : S^2 \mapsto 2^{Var}$. An interval is a finite path in IBIS, or as a sequence $I = s_1 s_2 s \ldots s_k$ such that $s_i t s_{i+1}$ for $1 \leq i \leq k$ and a transition t. In other words, IBIS forms an unravelled Kripke frame. Each discrete (finite) path of (global) states of IBIS will be called an interval. Anyhow, outside of IBIS-system we also admit continuous intervals.*

2.1 Traveling Salesman Problem as a Motivating Problem

At the end of this section, we introduce some useful version of Traveling Salesman Problem in a role of a motivation problem for this paper. This problem will be modeled in terms of fibred semantics, later introduced.

Thus, let us consider a salesman K, which intends to deliver a pocket A from a city C_1 to C_2 (his strong preference). Because of a temporal distance between C_1 and C_2 his preference could be satisfied not earlier than 3 h and in some interval in the city C_2. It has been mentioned that the fact that ist strongly preferable (by i) to (possibly) deliver the pocket A in C_1 can be rendered by a modal formula of (4):

$$[\text{K (strongly) prefers}] \langle \text{Deliver} \rangle A_{C_2}. \qquad (3)$$

(read: "K strongly prefers, (possibly deliver) the pocket A in a city C_2"). Recall that the outer operator [K(strongly)prefers]ϕ plays a role of a box-type operator for representation of preference of K and $\phi = \langle \text{Deliver} \rangle \psi$ plays a role of (an additionally specified) $\mathcal{L}(\mathcal{DEPLA})$. Finally, $\psi = A_{C_2}$ is a unique atomic formula.

3 Multi-Valued Preferential Logic (MVPL)

We intend to adopt a common convention to interpret preferences as transitive relations, hence we will semantically interpret them by partial orders. Thus, S4-system (containing axioms K, 4 and T) is suitable for their syntactic representation. In addition, we intend to interpret them in the interval-based semantics and in a fuzzy manner.

These postulates will be reflected in two types of operators: of a $\square$-type $[\text{Pref}]_i^\alpha \phi$ read: "an agent i (strongly) prefers ϕ with a degree α belonging to a finite set $G \subset [0, 1]$, and $\langle \text{Pref} \rangle_i^\alpha \phi$ "an agent i weakly prefers ϕ with a degree α" as a $\Diamond$-operator.

Language of MVPL. The language of MVPL, $\mathcal{L}(MVPL)$, is given by a grammar:

$$\phi := p \,|\, \neg \phi \,|\, \phi \wedge \psi \,|\, [\text{Pref}]_i^\alpha \phi \,|\, \langle \text{Pref} \rangle_i^\alpha \phi,$$

where $i \in \mathcal{A}$, $\alpha \in G$ and G is a finite subset of $[0,1]$. The following definitions and axioms are adopted in the MVPL-syntax.

Def.: $[\mathrm{Pref}]_i^\alpha \phi \iff \neg \langle \mathrm{Pref} \rangle_i^\alpha \neg \phi$ for each $\alpha \in G$.
Axioms: The axioms of MVPL are:

1. axioms of Boolean propositional calculus
2. $[\mathrm{Pref}]_i^\alpha (\phi \to \chi) \to ([\mathrm{Pref}]_i^\alpha \phi \to [\mathrm{Pref}]_i^\alpha \chi)$ (axiom K)
3. $[\mathrm{Pref}]_i^\alpha \phi \to [\mathrm{Pref}]_i^\alpha [\mathrm{Pref}]_i^\alpha \phi$ (axiom 4)
4. $[\mathrm{Pref}]_i^\alpha \phi \to \phi$ (axiom T)

As *inference rules* we adopt *Modus Ponens*, substitution and a necessitation rule for the $[\mathrm{Pref}]_i^\alpha$-operator: $\frac{\phi \to \psi}{[\mathrm{Pref}]_i^\alpha \phi \to [\mathrm{Pref}]_i^\alpha \psi}$ for each $\alpha \in G \subset [0,1]$, G is finite.

For a use of further analysis we will consider α as restricted to a finite $G \subset [0,1]$ (even if it is not mentioned) – due to [7,8] and to our arrangements – in order to ensure a context of a multi-valued logic. All these arrangements leads to the following definition of MVPL.

Definition 2. *MVPL is defined as the smallest theory in $\mathcal{L}(MVPL)$, which contains axioms 1–4 and closed on the above inference rules.*

3.1 Interval-Based Semantics for MVPL

The introduced MVPL with preferential operators will be interpreted now in an interval-based semantics. A "core" of this semantics construction is to introduce the appropriate accessibility relation between intervals, denoted later by $\precsim_i$ and to specify it by α in the next construction stage.

Accessibility relation $\precsim_i$. For this reason, assume that intervals $I = s_1 s_2 \ldots s_k$ and $I' = s_1' s_2' \ldots s_l'$ are given for some k, l and establish also an agent $i \in \mathcal{A}$. Let us define a new accessibility relation $\precsim_i \subseteq \mathcal{P}(S \times S)$ between I and I' as follows: $I \precsim_i I' \iff k \leq l$ and $l_i(s_j) = l_i(s_j')$ for all $j < k$, i.e. an agent i cannot distinguish between the corresponding states of I and I' up to j, between j-prefixes of both intervals. In other words, $I \precsim_i I' \iff I|_j \sim_i I'|_j$, or if and only if the behavioral equivalence condition holds for j-prefixes of both intervals.

Accessibility relation $\precsim_i^\alpha$. In order to grasp the similarity degree between I and I' introduce a new relation between these intervals, denoted later by $\precsim_i^\alpha$. Thus, establish a finite $G \subset [0,1]$ and introduce some new function $\| \bullet \| : \mathcal{P}(S \times S) \times \mathcal{A} \mapsto \mathcal{G}$ defined as: $\| I \precsim_i I' \| = \alpha$, for $\alpha \in G$. Intuitively, this new function associates the earlier relation $I \precsim_i I'$ to some α from G. The appropriate methods of achieving of α will be discussed later. Independently of this, one can define a desired relation $\precsim_i^\alpha$ in the product $\mathcal{P}(S \times S) \times G \times \mathcal{A}$ in the following way: $I \precsim_i^\alpha I' \iff \| I \precsim_i I' \| = \alpha$. It remains to describe how α can be found. A possible method is presented in the example below (Figs. 1 and 2).

Example 1. Consider a pair of discrete intervals $(I_1, I_2) : I_1 \precsim_i I_2$ having $j = 10$ common points and define $\alpha = \| I_1 \precsim_i I_2 \| = \frac{j(I_1, I_2)}{K}$ for some $K = 200$ ¿ length

$$\| I_1 \preccurlyeq_K I_2 \| = 10/K, \quad \text{for some } K$$

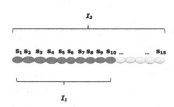

Fig. 1. Two discrete intervals I_1 and I_2 having 10 common points with an example of computing a value $\| I_1 \precsim_K I_2 \|$ for some K.

of each of intervals and each $i \in \mathcal{A}$. Thus, their similarity $\alpha = \| I_1 \precsim_i I_2) \| = \frac{10}{200} = \frac{1}{20}$. Moreover, $\| sI_1 \precsim_i I_2 \| = \frac{1}{20} = \| I_1 \precsim_k I_2 \|$ for another agent $k \in \mathcal{A}$, since the common j-prefix of both intervals is equally recognized by both agents because of the behavioral equivalence assumption for all agents in $\mathcal{A}$.

Definition 3 *(Satisfaction). Given a formula $\phi \in \mathcal{L}(MVPL)$ with a set of propositions Prop, an IBIS, an interval I, a labeling function Lab, and a finite set $G \subset [0,1]$ we define inductively the fact that ϕ is satisfied in IBIS and in an interval I (symb.$I \models \phi$) as follows:*

- *for all $p \in$ Prop, we have IBIS, $I \models p$ iff $p \in Lab(I)$.*
  IBIS, $I \models \neg\phi$ iff it is not such that IBIS, $I \models \phi$.
- *IBIS, $I \models \phi \wedge \psi$ iff IBIS, $I \models \phi$ and IBIS, $I \models \psi$.*
- *IBIS, $I \models [\mathrm{Pref}]_i^\alpha \phi$, where $i \subset \Lambda$, iff for all $I \precsim_i^\alpha I'$ we have IBIS, $I' \models \phi$ for each $\alpha \in G$.*
- *IBIS, $I \models \langle \mathrm{Pref} \rangle_i^\alpha \phi$, where $i \in A$, iff there is I' such that $I \precsim_i^\alpha I'$ and IBIS, $I' \models \phi$ for each $\alpha \in G$.*

The key clause in the above definition is this one referring to the modal operators $[\mathrm{Pref}]_i^\alpha \phi$ and $\langle \mathrm{Pref} \rangle_i^\alpha \phi$. These conditions assert that such modal formulas are satisfied in an interval I and model IBIS iff the same formula ϕ holds in all intervals accessible from this I *via* $\precsim_i^\alpha$-relation.

Relation $\precsim_i^\alpha$ for representation of preferences. Nevertheless, not all relations $\precsim_i^\alpha$ are suitable to represent preferences – intentionally represented by partial orders (MVPL forms a unique S4-system). It is not difficult to observe that belonging to the appropriate relation class depends on a method of α-achieving. For example, $\alpha = \frac{j}{K}$ ensures reflexivity and transitivity of $\precsim_i^\alpha$, but unfortunately – symmetry. In order to ensure a desired anti-symmetry for $\precsim_i^\alpha$ one need distinguish the relation $I_1 \precsim_i^\alpha I_2$ from its inverse relation by means of α-degree like in this example.

Example 2. $\alpha = \begin{cases} \frac{1}{jK} & \text{if } length(I_1) \leq length(I_2), \ 2 \leq j \\ \frac{j}{K} & \text{if } length(I_1) > length(I_2), \ 2 \leq j \end{cases}$

Corollary 1. *The accessibility relation $I_1 \precsim_i^\alpha I_2$ defined as above forms a partial order.*

Proof: Reflexivity is obvious. Anti-symmetry follows from the fact that if $I \precsim_i^\alpha I'$ holds, then does not hold $I' \precsim_i^\alpha I$ as $\alpha = \frac{1}{jK}$ holds in mutually disjoint cases.

In order to show transitivity assume that $I_1 \precsim_i^\alpha I_2$ and $I_2 \precsim_i^\beta I_3$ and $\alpha = \frac{1}{jK}$ and $\beta = \frac{1}{JK}$ for some established K and j – denoting a number of common points of I_1 and I_2 and J – denoting a number of common points of I_2 and I_3 (*resp.*). Thus, obviously, $I_1 \precsim_i^A I_3$, where $A = \frac{1}{jK}$ as $I_1 (\subseteq I_2 \subseteq I_3)$ has just j common points with I_3. Hence, $I_1 \precsim_i^\alpha I_3$. The same reasoning may be repeated for α defined alternatively.   □

Remark 1. *Note that the assumption $2 \leq j$ is essential in the above example. To illustrate this fact let us consider $I_1 = s_1$ and $I_2 = s_1 s_2$, so $j = 1$ and – in result – $\frac{1}{jK} = \frac{1}{2}$ and $\frac{j}{K} = \frac{1}{2}$, what implies that $\precsim_i^\alpha$ is symmetric.*

In this way we have just obtained a semantic interpretation of the sentence "an agent i (weakly) prefers ϕ with a degree α". It exactly means that there exists (at least one) such a pair of α-similar intervals having identical prefixes, which are observational recognized by an agent i as such ones.

4   Dynamic Epistemic Logic for Actions

In this section we introduce the dynamic-epistemic logic for actions, shortly denoted $\mathcal{DELA}$ in order to interpret it in the interval-based semantics. The proposed system forms a syntactic restriction of Ditmarsch's system from [3]. This restriction consists in a fact that we omit an assumption of a *common knowledge* of agents (with the appropriate modal operator and its axiomatic properties) and we restrict some conditions imposed on actions. Finally, we omit a detailed presentation of the proof system for $\mathcal{DELA}$.

4.1   Syntax and Semantics of $\mathcal{DELA}$

Assume that a set A of agents is given. We define a language $\mathcal{L}(\mathcal{DELA})$ inductively – as a fusion of two sub-languages: $\mathcal{L}(\mathcal{DELA})^{Actions}$ and $\mathcal{L}(\mathcal{DELA})^{Epistemic}$. The first sub-language contains action symbols, the second one – the epistemic symbols. Both sub-languages are based on the same set of propositions *Prop*.

Actions. Let us begin with a definition of $\mathcal{L}(\mathcal{DELA})^{Actions}$.

Definition 4. *Language $\mathcal{L}(\mathcal{DELA})^{Actions}$ is given by the grammar:*

$$\phi := ?\phi \,|\, R_B \phi \,|\, (a\,!\,a) \,|\, (a; b) \,|\, (a \cup a) \,|\, (a \cap a)\,, \tag{4}$$

where $\phi \in Prop$ and $B \subseteq A$ (Sense of all expressions will be explained below).

Definition 5. *The language of* $\mathcal{L}(\mathcal{DELA})^{Epistemic}$ *is given by the grammar:*

$$\phi := p \,|\, \neg\phi \,|\, (\phi \wedge \phi) \,|\, K_n\phi, \bar{K}_n\phi, \tag{5}$$

where $\bar{K}_n$ *is a 'diamond'-type dual operator for* $K_n\phi$, $n \in A$.

Definition 6. *The language of* $\mathcal{L}(\mathcal{DELA})$ *is defined by the grammar:*

$$\phi := p \,|\, \neg\phi \,|\, (\phi \wedge \phi) \,|\, K_n\phi \,|\, [a]\phi, \tag{6}$$

where $n \in A$ *and* $[a]\phi \in \mathcal{L}(\mathcal{DELA})^{Actions}$ *and may be one of the action symbols listed in Definition 4.*

The program operator $R_B\phi$ is called a *realization* and can be read as: 'A group B of agents performs ϕ.[1] Action $?\phi$ is a *test-action*; $(a\,; a')$ is classically understood as a *sequential execution*, $(\alpha \cap \alpha')$ – as a *concurrent execution* and $(\alpha \cup \alpha')$ is a syntactic representation of a *nondeterministic choice*. Finally, $(a\,!a)$ is a so-called local choice that can be pronounced: 'from a to a', choose the first one'. See [3].

Example 3. The fact that a sequential choice (performing) of actions a and b by agents 1,2 should always lead to ϕ may be rendered in $\mathcal{L}(\mathcal{DELA})$ by $[a; b]_{1,2}\phi$. By contrast, the fact that a non-deterministic choice of actions a and b by agents 1,2,3 may sometimes lead to ϕ may be rendered in $\mathcal{L}(\mathcal{DELA})$ by $\langle a \sup b \rangle_{1,2,3}\phi$.

Proof system: Since actions classically are defined by their preconditions (denoted: *pre*), the proof system for $\mathcal{DELA}$ should contains them and some general conditions of action execution.

- **Preconditions.** $pre(?\phi) := \phi; pre(a\,; b) = pre(a)^p re(b); pre(a \cup b) = pre(a) \cup pre(b); pre(a \cap b) = pre(a) \cap pre(b); pre(a\,!b) = pre(a).$
- **Conditions for actions:**

- *all propositional tautologies*
- $K_n(\phi \rightarrow \chi) \rightarrow (K_n\phi \rightarrow K_n\chi)$,
- $K_n\phi \rightarrow \phi, K_n\phi \rightarrow K_n\phi K_n\phi, \neg K_n\phi \rightarrow K_n K_n\phi$ (**S5** for K_n),
- $[?\phi]\phi \iff (\phi \rightarrow \chi)$.

As *inference rules* we adopt:

- if ϕ and $\phi \rightarrow \chi$, then χ,
- if ϕ, then $K_n\phi$,
- if $\phi \rightarrow \chi$, then $[\alpha]\phi \rightarrow [\alpha]\phi$.

Semantics. We intend to interpret $\mathcal{DELA}$ in the interval-based semantics. Assume, thus, that some interval-based Kripke model $\mathcal{M}$ is given with a domain W and $Nor \subseteq W \neq \emptyset$ is a set of normal worlds and let R be an accessibility relation on $W \times W$.

---

[1] Set of agents associated with actions is defined as follows $set(\phi) = \emptyset, set(R_B) = B$ and for above admitted combinations of actions α and α' by $set(a) \cap set(a')$.

Definition 7 *(Satisfaction of formulas).* *Let* $\mathcal{M} = \langle W, Nor, \subseteq, R, Val \rangle$ *be a modal with Nor and R defined as above. We define a satisfaction relation inductively as follows:*

$$\mathcal{M}, \mathcal{I} \models p \iff V(p, I) = 1$$
$$\mathcal{M}, \mathcal{I} \models \neg\phi \text{ for every } I' \text{ such that } I' \subseteq I \text{ it holds not } \mathcal{M}, \mathcal{I}' \models \phi$$
$$\mathcal{M}, \mathcal{I} \models \phi \wedge \chi \iff \mathcal{M}, \mathcal{I} \models \phi \text{ and } \mathcal{M}, \mathcal{I} \models \chi$$
$$\mathcal{M}, \mathcal{I} \models \phi \vee \chi \iff \mathcal{M}, \mathcal{I} \models \phi \text{ or } \mathcal{M}, \mathcal{I} \models \phi$$
$$\mathcal{M}, \mathcal{I} \models \phi \rightarrow \chi \iff \text{ for every } I' \text{ such that } I' \subseteq I \text{ if } \mathcal{M}, \mathcal{I} \models \phi \text{ then } \mathcal{M}, \mathcal{I}' \models \phi$$
$$\mathcal{M}, \mathcal{I} \models \Box\phi \iff I' \in Nor \text{ and for every } I' \text{ such that } I' \subseteq I \text{ and } I'' \text{ such that } I' RI'', \text{ it holds } \mathcal{M}, \mathcal{I}'' \models \phi \text{ for a box-type operator } \Box\phi \in \{K_n\phi, [\alpha]\phi\}.$$

The key clause is the last one that admits a satisfaction relation for modal operators only in normal interval-worlds of the set Nor. Indeed, it has been said that *no* formula of a box-type is satisfied in queer worlds set $W - Nor$.

The interpretation of actions is classical and may be easily found in [3]. The only difference is that we exchange arbitrary point-states for interval-states in this interpretation. Because this interpretation will not be used in further analysis, we omit its presentation.

5 $\mathcal{DEPLA}$ for Actions and Preferences

In last sections $\mathcal{DELA}$ for actions was introduced and it completeness with respect to the interval-based semantics has been proved. In this section a required dynamic-epistemic logic for actions and preferences – denoted later as $\mathcal{DEPLA}$ – will be defined and interpreted in the interval version of the fibred semantics.

Syntax of $\mathcal{DEPLA}$. The language $\mathcal{L}(\mathcal{DEPLA})$ includes the following components:

$$Prop \,|\, \phi \rightarrow \chi \,|\, \neg\phi \,|\, \langle \mathrm{Pref} \rangle_i^\alpha(\phi) \,|\, [\mathrm{Pref}]_i^\alpha \phi \,|\, [a]\phi \,|\, [K_n]\phi \,,$$

where *Prop* is a countable set of propositional variables of $\mathcal{L}(\mathcal{DEPLA})$, $\alpha \in G \subset [0,1]$, where G is finite, and $\langle \mathrm{Pref} \rangle_i^\alpha(\phi) \in \mathcal{L}(MVPL)$ and $[\alpha]\phi \in \mathcal{L}(\mathcal{DELA})^{Actions}$, $[K_n]\phi \in \mathcal{L}(\mathcal{DELA})$ and their duals are specified as earlier in Definition 4.

The interval-based semantics for each component of $\mathcal{DEPLA}$ has already proposed in Sects. 2 and 3. The main focus of this section is to proposed a *fibred interval-based semantics* for mixed components.

5.1   Fibring Interval-Based Semantics for $\mathcal{DEPLA}$

In this section we demonstrate how the mechanism of fibred semantics works with respect to combined modalities that cannot be modeled by models $\mathcal{M}_1$ and $\mathcal{M}_2$, which – in general case – cannot recognize modal operators interpreted in the second model. Consider therefore:

– a formula $\psi = \langle \text{Pref} \rangle_i^\alpha \langle a \rangle \phi$ and
– a model $\mathcal{M}_1, I^1 \models \langle \text{Pref} \rangle_i^\alpha p$, where p atomic.

It can arise a natural question what about satisfiability of an atomic formula p in $\mathcal{M}_1, I^1$, if $p = \langle L \rangle \phi$. More precisely:

– what about $\mathcal{M}_1, I^1 \models p$, if $p = \langle a \rangle \phi$?

As it has been mentioned – in a general case it may hold the following case:

– $\mathcal{M}_1, I^1$ can not to 'recognize' $p = \langle a \rangle \phi$!

Fibring mapping. This difficulty generates a natural question how to deal with this fact? In order to evaluate $[L]\psi$ at $\mathcal{M}_1$ one need some mapping between $\mathcal{M}_1$ and $\mathcal{M}_2$ in order to transfer the validity checking from this first model to the validity checking within the second one. Thus, we introduce such a new function – called a *fibring mapping* $\mathbf{F}$ – that $\mathbf{F}(\mathcal{M}_1, I_1) = (\mathcal{M}_2^{I_1}, I_2)$ and the following equality holds:

$$\mathcal{M}_1, I_1 \models [a]\psi \iff \mathcal{M}_2^{I_1}, I_2 \models [a]\psi$$

for some interval I_2 of $\mathcal{M}_2$.

Since a model $\mathcal{M}_2$ is characterized by the interval I_2, we can identify $\mathbf{F}(I_1)$ with this new interval I_2 of the associated model $\mathcal{M}_2^{I_1}$. It allows us to formulate a new satisfaction condition in the form:

$$\mathcal{M}_1, I_1 \models [a]\psi \iff \mathcal{M}_2^{I_1}, \mathbf{F}(I_1) \models [a]\psi$$

As in a point-wise fibring semantics, one can impose on the fibred mapping $\mathbf{F}$ a condition of "switching semantics", i.e. for each $I \in \mathcal{M}_1$, it holds $\mathbf{F}(I) \in \mathcal{M}_2$ and for each $I \in \mathcal{M}_2$: $\mathbf{F}(I) \in \mathcal{M}_1$. We also assume that if $I_1 \neq I_2$, than also $\mathbf{F}(I_1) \neq \mathbf{F}(I_2)$. ($\mathbf{F}$-images of two different intervals are different, too). The same reasoning may be carried out for $[K_n]\phi$-formulas of $\mathcal{L}(\mathcal{DEPLA})$.

Fibred models. Assume now that the following interval-based Kripke models are given:

- $\mathcal{M}_1 = \langle S_1, R_1, h_1, \mathbf{F} \rangle$, where: $S_1 = \{J_1, J_2 \ldots, J_{2k}\}$ for some fixed $2k$, $J_j R_i J_l \iff J_j \precsim_1^\alpha J_l$ in IBIS (for $j, l \in \{1, \ldots, 2k\}, i \leq m$), h_1 is an assignment function and $\mathbf{F}$ is the fibring mapping defined as above;

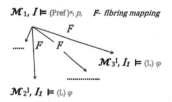

$\mathcal{M}_1, I \models \langle \text{Pref} \rangle_i^{\alpha} p,$ $F\text{- fibring mapping}$

$\mathcal{M}_3^I, I_1 \models \langle L \rangle \varphi$

$\mathcal{M}_2^I, I_1 \models \langle L \rangle \varphi$

Fig. 2. A fibring mapping which should not necessary be a function

- $\mathcal{M}_2 = \langle S_2, a, h_2, \mathbf{F} \rangle$, where: $S_2 = \{I_1 \ldots I_{2k}\}$ for some fixed $2k$, a is an action admissible in $\mathcal{DEPLA}$-formalism. h_2 and $\mathbf{F}$ are defined as earlier. Then a *fibred models* $\mathcal{M}$ will be defined as a tuple:

$$\mathcal{M} = \langle S_1 \otimes S_2, R_L \otimes R_i, h_1 \otimes h_2, \mathbf{F} \rangle, \tag{7}$$

where: $\mathbf{F}$ is the same fibring mapping and $\otimes$ denotes a simple sum or a fusion of the appropriate components. It means that elements of $\mathcal{M}$ cannot be 'mixed' – as in a usual set-theoretic sum of components – without considering the fibring mapping $\mathbf{F}$.

6   Fibring Semantics in Use

The fibred interval semantics will be exploited now in modeling of the initial *Traveling Salesman Problem*. Thus, return now to the TSP with our salesman K, which intends to deliver a pocket A from a city C_1 to C_2 (his strong preference). Because of a temporal distance between C_1 and C_2 his preference could be satisfied not earlier than $3\,\mathrm{h}$ and in some interval in the city C_2. It has been mentioned that the fact that ist strongly preferable (by i) to (possibly) deliver the pocket A in C_1 can be rendered by a modal formula of (4):

$$[\mathrm{K}\ (\mathrm{strongly})\ \mathrm{prefers}]\langle \mathrm{Deliver}\rangle \mathrm{A}_{C_2}. \tag{8}$$

(read: "K strongly prefers, (possibly deliver) the pocket A in a city C_2"). Recall that the outer operator $[\mathrm{K}(\mathrm{strongly})\mathrm{prefers}]\phi$ plays a role of a box-type operator for representation of preference of K and $\phi = \langle \mathrm{Deliver}\rangle \psi$ plays a role of (an additionally specified) $\mathcal{L}(\mathcal{DEPLA})$. Finally, $\psi = \mathrm{A}_{C_2}$ is already a unique atomic formula.

Model for the preferential component. In order to find the appropriate model for a preferential component of the formula (4) assume that P and $P^{Deliver}$ are some discrete intervals interpreting the Salesman's preference such that:

- P is an interval where the preference is "expressed" and
- $P^{Deliver}$ is the interval, which the subject of the salesman's preference is materialized in. Formally: $P^{Deliver} \models \langle \mathrm{Deliver}\rangle \mathrm{A}_{C_2}$, or a fact of (possible) delivering of a packet A to C_2 holds in this interval.

Assume also that $\precsim_K^{strongly}$ is an accessibility (preference) relation between them, i.e. it holds $P \precsim_K^{strongly} P^{Deliver}$. Thus, a model for the preferential component is given as follows:

$$\mathcal{M}_1 = \langle \{P, P^{Deliver}\}, \precsim_K^{strongly}, h_1 \rangle \tag{9}$$

for some valuation h_1.

Model for the temporal component. Similarly, we find a model for a temporal component. For that reason consider two *temporal discrete* intervals: I_1 for a representation of "now" and I_2 for a representation of "sometimes in a future".

Thus, the appropriate model in this case can be given as follows:

$$\mathcal{M}_2 = \langle \{\text{now}, \text{sometimes in a future}\}, \text{Deliver}, h_2 \rangle \tag{10}$$

for some valuation h_2. The *fibred model* for the whole formula (2) is determined by the tuple:

$$\mathcal{M} = \langle S, R^*, h, \mathbf{F} \rangle, \tag{11}$$

where:

- $S = \{P, P^{Deliver}\} \otimes \{I_1 = \text{now}, I_2 = \text{sometimes in a future}\}$,
- $R^* = \{\precsim_K^{strongly}\} \otimes \{\text{Deliver}\}$,
- $h = h_1 \otimes h_2$, $\mathbf{F}(P^{Deliver}) = I_1(= now)$.[2]

7 Concluding Remarks and Possible Research Areas

It has just been shown how Ditmarsch's dynamic epistemic logic system may be combined with some multi-valued system for preferences. This combinations was proposed in terms of the new interval-based fibred semantics. This new and flexible semantics was applied to modeling of The Traveling Salesman Problem. Probably, an area of further extensions of this semantics type cannot be restricted to this problem. In fact, some utility of Ditmarsch's system to describing a dynamism of multi-agent systems (Muddy's children problem representation, playing cards etc.) seems to be a promising bridgehead for research on application area for fibred semantics.

Anyhow, fibred semantics required some further specification, even for the purely theoretic point of view. For example, it is still unclear – at least for the authors of this paper – what about such meta-logical properties of $\mathcal{DEPLA}$-system as completeness w. r. t the proposed semantics. It seems to be an excellent subject of further research.

References

1. Aucher, G., Schwarzentruber, F.: On the complexity of dynamic epistemic logic. In: TARK 2013, pp. 19–28 (2013)
2. Baader, F., Horrock, J., Sattler, U.: Description logics. In: Handbook of Knowledge Representation. Elsevier (2007)
3. Ditmarsch, H.P., wan der Hoek, D., Kooi, B.P.: Concurrent dynamic epistemic logic for MAS. In: AAMAS, pp. 289–304 (2003)
4. Fagin, R., Halpern, J., Moses, Y., Vardi, M.: Reasoning about Knowledge. The MIT Press, Cambridge (1995)
5. Fikes, P., Nilsson, N.: Strips: a new appraoch to the application of theorem proving to problem solving. Artif. Intell. **2**(3–4), 189–208 (1971)

---

[2] (**F** joins these two intervals such that the last preferential interval is connected with the first temporal one).

6. Gabbay, D., Shehtman, V.: Product of modal logic, part 1. Logic J. IGPL **6**(1), 73–146 (1998)
7. Godo, L., Esteva, H., Rodriquez, R.: A modal account of similarity-based reasoning. Int. J. Approxim. Reason. **4** (1997)
8. Hajek, P.: Metamathematics of Fuzzy Logic. Kluwer Academic Publishers, Dordrecht (1998)
9. Hajek, P., Hramancova, D.: A qualitative fuzzy possibilistic logic. Int. J. Approxim. Reason. **12**, 1–19 (1995)
10. Halpern, J., Shoham, Y.: A propositional modal logic of time intervals. J. ACM **38**, 935–962 (1991)
11. Jobczyk, K., Ligeza, A.: Fuzzy-temporal approach to the handling of temporal interval relations and preferences. In: Proceeding of INISTA, pp. 1–8 (2015)
12. Jobczyk, K., Ligeza, A.: Multi-valued halpern-shoham logic for temporal allen's relations and preferences. In: Proceedings of the Annual International Conference of Fuzzy Systems (FuzzIEEE) (2016). Page to appear
13. Jobczyk, K., Ligeza, A.: Systems of temporal logic for a use of engineering. Toward a more practical approach. In: Stýskala, V., Kolosov, D., Snášel, V., Karakeyev, T., Abraham, A. (eds.) Intelligent Systems for Computer Modelling. AISC, vol. 423, pp. 147–157. Springer, Cham (2016). doi:10.1007/978-3-319-27644-1_14
14. Jobczyk, K., Ligeza, A., Kluza, K.: Selected temporal logic systems: an attempt at engineering evaluation. In: Rutkowski, L., Korytkowski, M., Scherer, R., Tadeusiewicz, R., Zadeh, L.A., Zurada, J.M. (eds.) ICAISC 2016. LNCS (LNAI), vol. 9692, pp. 219–229. Springer, Cham (2016). doi:10.1007/978-3-319-39378-0_20
15. Jobczyk, K., Ligęza, A., Bouzid, M., Karczmarczuk, J.: Comparative approach to the multi-valued logic construction for preferences. In: Rutkowski, L., Korytkowski, M., Scherer, R., Tadeusiewicz, R., Zadeh, L.A., Zurada, J.M. (eds.) ICAISC 2015. LNCS (LNAI), vol. 9119, pp. 172–183. Springer, Cham (2015). doi:10.1007/978-3-319-19324-3_16
16. Lamport, L.: The temporal logic of actions. ACM Trans. Program. Lang. Syst. **16**, 872–923 (1991)
17. Lomuscio, A., Michaliszyn, J.: An epistemic halpern-shoham logic. In: Proceedings of IJCAI 2013, pp. 1010–1016 (2013)
18. Nilsson, N.J.: Principles of Artificial Intelligence. Tioga Publishing, Palo Alto (1980)
19. Traverso, P., Ghallab, M., Nau, D.: Automated Planning: Theory and Practice. Elsevier, Amsterdam (1997, 2004)
20. van Benthem, J.: Dynamic logic for belief revision. J. Appl. Non-Class. Logics **17**(2), 119–155 (2007)
21. van Benthem, J., Gheerbrant, A.: Game: Solution, epistemic dynamics, and fixed-point logics. Fundamenta Informaticae **100**, 19–41 (2010)
22. von Wright, J.: The Logic of Preference. Edinburgh University Press, Edinburgh (1963)

Proposal of a Multi-agent System for a Smart Outdoor Lighting Environment

Radosław Klimek$^{(\boxtimes)}$

AGH University of Science and Technology,
Al. Mickiewicza 30, 30-059 Kraków, Poland
`rklimek@agh.edu.pl`

Abstract. Outdoor smart lighting is more and more popular since it is
regarded as a significant example of an ideal and friendly environment.
Systems controlling outdoor lighting, considered as context-aware software, are challenging. A multi-agent system for outdoor street lighting,
dealing with intelligent software applications in pervasive computing,
has been proposed. Smart scenarios, typical for such an intelligent environment, are presented. An agent-based architecture for a multi-agent
system, dealing with the well-known framework JADE, is proposed. It
allows further testing of these smart scenarios. This is the first proposal
and the beginning of a greater work for implementation and testing of
smart lighting scenarios carried out in the agent systems. This work
describes the rationale of efforts for achieving ecosystems also working
in the IoT paradigm by focusing on a rural environment, featuring data
collection, as well as event detections and coordinated reactions.

Keywords: Smart outdoor lighting · Smart scenario · Multi-agent system · JADE · IoT · Pervasive computing

1 Introduction

The outdoor lighting systems have changed greatly over the past time. To the
smart outdoor lighting we include guidelines, solutions and lighting techniques
which create the most energy efficient and inhabitants friendly lighting solution.
Smart outdoor lighting solutions allow cities, rural environments or facility owners to reduce their power consumption, as well as to add intelligence to their
newly installed lighting systems in a very economical way. Smart lighting control has never been easier. The smart outdoor lighting system uses information
technology to monitor the environment, control the electric appliance and to
communicate with the outer world.

Soft computing agents are applied in many areas including process control,
engineering, data mining, web-computing and others. Multi-agent systems are
also one of the most quickly developing branches of the artificial intelligence
research. At the same time, they are applied in the particular industrial and
business fields such as [8]:

© Springer International Publishing AG 2017
L. Rutkowski et al. (Eds.): ICAISC 2017, Part II, LNAI 10246, pp. 255–266, 2017.
DOI: 10.1007/978-3-319-59060-8_24

- solving problems of the dispersed nature, the agent approach enables modelling the systems working in the dispersed environment and deal with problems which complexity goes far beyond the capacity of an individual entity;
- simulation of reality, especially it applies to the groups composed of the autonomous entities, it is used in the social and biological sciences
- management of knowledge, finding, collecting and analyzing the information stored in the Internet network;
- mobile agents, moving through the network and doing the tasks for users, the agent can be an interface between the user and the device;
- a subject of control in robotics, the reactive agents can influence the surrounding on the basis of signals which they receive from it;
- the agent technologies influence software modeling concepts providing the high-level abstraction such as agent.

The main purpose of this work is to propose the complex architecture of the multi-agent system designed to implement the intelligent street lightning system as well as to build the agent society and carry its initial tests, working simulation for a few real situations and scenarios. It is also necessary to create tools for rules of interaction making it possible to control the outdoor lighting system. Its implementation will be possible thanks to the use of a well-known framework JADE. Among the designed functions of the system there are:

- controlling the particular street lights;
- turning on and off as well as controlling the power of the particular lights;
- the automatic light intensity modification dependent from the time of a day or the current driving conditions;
- possibility of creating the independent rules for each street light located in the places where the special caution is advisable, for example near pedestrian crossings, schools or other institutions where paying special attention is necessary;
- keeping the statistics about work of the particular elements of the system.

The advantages of such system are not only related to the higher level of comfort of everyday life but also to the safety of the system for its users, saving the energy and many others. Creation and implementation of the JADE environment enables continuing the research over much more advanced scenarios.

This work follows paper [6] where context-aware and pro-active, as well as agent-ready approach for smart lighting was postulated but not yet introduced. In other words, this work extends the previous one introducing an agent architecture and a prototype and conceptual software system. The term context-awareness in ubiquitous computing was introduced by Schilit [9]. Context aware devices may also provide user's current situation. Dey and Abowd [1] define context as "any information that can be used to characterize the situation of an entity". A great deal of research on intelligent home environment and context-aware architectures has been done. In [10] they have proposed a design for automatic room light detection and control, the light controls the light using a fixed threshold value resulting in inefficiency through this system controls. In [7] the problem of information exchange among agents maintaining graph-based systems is discussed. This paper also deals with works [4,5] which concern observing behaviors of users/inhabitants, as well as modeling agent architecture for an appopriate systems.

2 Smart Lighting Scenarios

The proposed system combines heterogeneous appliances which can adjust themselves correspondingly to the various scenarios, see Fig. 1. The examples of the situations are presented on Fig. 2, and scenarios are verbalized in Tables 1, 2 and 3. They include: road intersections, dangerous turn, a social event, or the place of a higher risk (for example school).

Fig. 1. Smart outdoor lighting environment

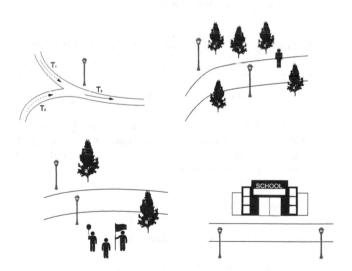

Fig. 2. Sample lighting situations: **top left**: flows $T1$ and $T2$ aggregate to $T3$; **top right**: a sharp curve; **bottom left**: a social event; **bottom right**: a school

In Table 1, there is presented the illustration of a scenario related to operation on the sharp turn. The vehicle enters the area covered by the system and moves in the direction of a dangerous part of the road. If only a pedestrian appears on the turn, it does not matter if he/she moves or not, the current sensor values are

Table 1. The use case scenario for a guest recognition

UC name: "Handling a sharp turn"
Precondition: Evening hours, lightning is switched on
Scenario: 1. A pedestrian appears near a sharp wooded turn; 2. Checking the current sensor values related to the presence of the vehicles in the area 3. Checking the fixed bus timetable for expected vehicles 4. Checking the private vehicles and their routes for the possibility of getting closer to the analyzed turn 5. Increasing the amount of light near the sharp turn, if a pedestrian is there
Postcondition: Possibility of increasing the light intensity

checked for the presence of vehicles, as well as the historical data, which enables detection of the public transport vehicles expected in the nearest future.

Table 2 shows the case of the important object which requires stronger light intensity in the moment when it is passed by vehicles. Table 3 presents the case of detecting the social event, understood as the amount of people bigger than normal, which was detected near the road. In such case, the light intensity is increased regardless of the presence of the vehicles. The higher light intensity is kept at the same level until all people leave the place.

Table 2. The use case scenario for an important place (school)

UC name: "Handling an important place"
Precondition: Evening hours, lightning is switched on
Scenario: 1. Constant monitoring of the current movement and prediction of the moment when vehicles will get closer to the important and monitored object 2. Constant checking of the fixed bus timetable for the predicted vehicles together with the moment when the place is passed by them 3. Increasing the amount of light near the monitored area
Postcondition: Possibility of increasing the light intensity

Table 3. The use case scenario for a social event

UC name: "Handling an important place"
Precondition: Evening hours, lightning is switched on
Scenario: 1. Detecting the bigger group of people on the roadside, probably a social event 2. Increasing the amount of light in the surrounding area as well as near the main road and side streets in the neighboring area
Postcondition: Possibility of increasing the light intensity

3 Multi-agent System for Outdoor Lighting

Informally, an *agent* is a piece of software located in some environment and able to act in order to meet its design objectives. A *Multi-agent system* MAS is a system which consists of many agents communicating and cooperating together. Its main base is the single agent which is understood as an autonomous identity, set in a particular surrounding which the agent is able to detect and influence, and which has an ability to communicate with all other agents. The multi-agent system consists of many agents which communicate together staying autonomous, both in working and in decision taking by the particular agents, at the same time. Such an understanding of MAS system enables better and more natural intelligent system modelling taking into consideration the complex context conditions.

In the system considered, there exist two types of objects/actors which operate in the environment; objects for which some actions, related to light controlling, are planned:

Pedestrian – a person who interacts, usually involuntarily, with the system by entering within its reach. The object is identified and the actions undertaken by the system are dependent from the present and past behavior of the pedestrian within the reach of the system. The data related to the pedestrian will be stored after every interaction with the system.

Vehicle – a mean of transport possessing the markings which enable the unambiguous identification, for example car or motorbike. The actions taken by the system are based on the present and past vehicle behavior within the range of the system. The data related to the vehicle will be stored after every interaction with the system.

The proposed multi-agent system for an intelligent street lightning creates a hierarchical system with the specific roles, and is shown on Fig. 3. Three levels constitute the hierarchy. The *physical layer*, or *low level* agents, consists of agents

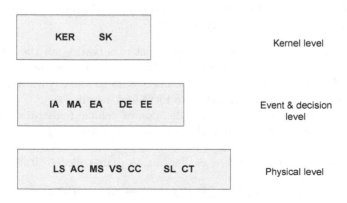

Fig. 3. Agent hierarchy

located in the particular devices, responsible for recording the elementary events which later are interpreter on the higher level of the system:

- *Light sensor* (LS) – natural light sensor, a photocell located at the top of the electric lightning fitting – it provides the information about the current level of the natural light.
- *Astronomical clock* (AC) – the agent which, on the basis of the information about the current date and geographical coordinates, is able to determine the daily on-and-off scheme for the clock in accordance with the sunrise and sunset.
- *Movement sensor* (MS) – radar or passive infrared sensors which detect the emergence of the user on the lit area and recognize his/her individual features. The sensor detects only the important objects, ignoring the remaining ones.
- *Velocity sensor* (VS) – the sensor of velocity and movement direction, provides the information about presence of the objects as well as about the velocity and direction of movement of all objects within its range. It enables creating different lightning scenarios related to the objects behavior prediction.
- *Camera* (CC) – Closed-circuit television camera, or shortly CCTV camera, provides the visual data useful for events and actors identification in the system. The collected data are sent to the higher level agents.

Two other agents belong to the low layer, that is the *executive agents*, or *light managers*:

- *Street lamp* (SL) – a physical implementation which covers transfer controlling from the driver, responsible for switching on and off, brightening and darkening of the physical lamp produced by a particular manufacturer.
- *Controller* (CT) – the agent responsible for a direct control over the "street light" agent. The controlling signals are transmitted to the controlling module in a group of agents which identify the particular events in the system. The received signals are verified together with the historical data stored in the database.

Agents from *event and decision level* are grouped into two sub-groups. The first one, called *event level*, identify events:

- *Image analyzer* (IA) – reacts on the current observation on the basis of the environment condition (condition reactive agent). The agent processes the data in order to achieve the information necessary to identify the actors within a range of the system. The actors are identified as follows:
  - a vehicle: the analysis and identification of vehicle registration plate;
  - a pedestrian: identification of the biometric features (face) of the actor (shape, pupils spacing etc.).
  The data which enable this type of identification are collected from the "Communication Camera" agent.
- *Movement analyzer* (MA) – analyses the movement within the area of the system, reacts on the current situation on the basis of the environment condition (condition reactive agent). Moreover, it processes the data in order to

achieve the information which helps to identify the movement in the system. It determines the movement parameters such as: direction, velocity or acceleration. The module is supported by "Image analyzer" helping to identify the event dependent from the actor which has already been registered in the system. The observation is taken from "Movement sensor", "Velocity sensor" and "Communication camera" agents.

- *Environment analyzer* (EA) – the analyzer of the physical parameters, reacts on the current observation on the basis of the environment condition (condition reactive agent). The agent processes observations in order to gain the data which identifies the event of activating or deactivating the street light within the range of the system. The observation is collected form "Light sensor" and "Astronomical clock" agents.

The second sub-group, called *decision level*, consists of

- *Decision Engine* (DE) – develops decisions based on the simple facts collected in the system as well as the compared events, handling rules.
- *Execution Engine* (EE) – carries out events of the decision agent.

The *kernel level*, or *top level* aggregates the entire system. Agents are not isolated but related to a central resource, or database.

- *Storekeeper* (SK) – a data storekeeper and database integration module. The agent is an intermediary between agents which identify the events and the database. It is also responsible for identification of data migration to the database as well as for storage of the data gained form the system
- *Kernel* (KER) – the core of the system, it allocates resources, supervises its efficiency and controls everything which happens in the system.

Figure 4 shows the system architecture. Basic sensors are mapped to individual componnets: LS, AC, MS, VS, CC. Some basic lifecycle operations are encapsulated into other components. "Events" component controls all different physical layer agents gathering data (IA, MA, EA). "Events" components build complex events based on facts. "Decisions" component manages physical devices from physical layer (DD, EE). "Decisions" component handles rules and facts in the system (DE, EE). Components generate an object/event list which are transmitted to the system. "Light managers" executes (SL CT) the agent actions.

4   Simulation Environment

The following three multi-agent environments were considered:

- JADE (Java Agent Development Framework) – is a framework fully created in Java. The other systems based on it can work on different systems (dispersion) at the same time and share their configuration thanks to the remote GUI. Moreover, the configuration can be changed during the whole working

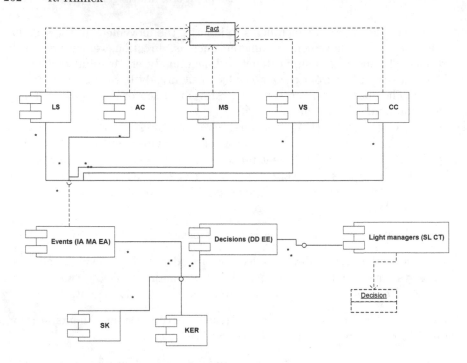

Fig. 4. Agent architecture

time of the system. It provides the simple methods of running the tasks as well as the communication between them. It has a big, active society which informs about mistakes on an ongoing basis and helps to solve problems for beginners. The exchange of information between agents is compatible with ACL standard (Agent Communication Language). Messages are synchronized and arranged for each agent separately. Sending the interface implementing objects is automatic. They are serialized at the sender and deserialized in the target object by framework.

- Cougaar (Cognitive Agent Architecture) – is the open source framework based on Java and designed for the big scalable agent-oriented applications. Its architecture uses the newest agent-oriented components and possesses numerous functionalities.
- MASON (Multi-Agent Simulator Of Neighborhoods) – fast multi-agent library designed for simulation. It has many devices which simplify the event simulation including the visualization libraries both 2D and 3D. The library is written in Java. Its models are independent from the components responsible for visualization. The models can be dynamically added, deleted or changed. Simulation results are independent from the platform on which they are run. Additionally, the simulation results can be saved as the print screens, short films or diagrams. On the ground that the device is so advanced and has a support of the users society, JADE has been chosen.

Every physical street lamp in the system has the following set of sensors providing basic information to the system:

- Velocity sensor – a sensor directed towards the road, detects the movable objects;
- Infrared sensor – a sensor which detects pedestrians, differently form velocity sensor it can also detect the person who does not move;
- Light sensor – a sensor which measures the light intensity in a surrounding of the particular street lamp.

The data flow structure consists of modules. Their task is to transform the data provided by agents collecting data. Among them are:

- movement analysis module: collects and analyses data received from sensors which detect the objects important from the simulation point of view (vehicles, pedestrians);
- parameter analysis module: collects and analyses data related to the simulation surrounding;
- controlling module: receives the pre-processed data from the rest of modules and, on their basis, determines the lightning power of the particular lamps.

All agents are created in a hierarchical order, starting from the lowest level agents including "simulation clock" agent which is created at the beginning. The next agents are created via the already existing agents. It can be achieved thanks to this code:

```
AgentContainer kontener = getContainerController();
AgentController kontroler = kontener.createNewAgent
                (NAZWA_AGENTA,PELNA_NAZWA_KLASY_AGENTA,null);
kontroler.start();
```

The function `getContainerController()` is used to download the controller to the container where the current agent exists. Creating all agents in one container simplify work because it does not require the configuration of network interface from all agents in order to make the communication between them possible and efficient. The next step is creating the new agent, within the container, using the `createNewAgent()` method.

The agents look for another agents to which they send the already processed data. Sending data to the agent is based on the automatic object serialization. Additionally, every message has a special field which informs about the type of message and its important parameters. Receiving data is based on picking up those messages from the queue which fit to the certain pattern. If such message does not exist, there can be evoked the blocking method which results in agent sleep until receiving the next message.

The graphic interface of the agent system enables tracing the simulation progress and the possible change of parameters. The simulation graph shows the analyzed system where the extremities match to the particular street lights and the edges symbolize the ways of moving of the physical environment objects. Every street light has two parameters:

1. First one, marked by "N" letter, provides the information about the light intensity in the surrounding of the particular street light.
2. Second one, marked by "M" letter shows the current light intensity of the street lamp.

The given below two graphs present the movement of the vehicle in the environment as well as the presence of the pedestrian near the road within the range of sensors/detectors.

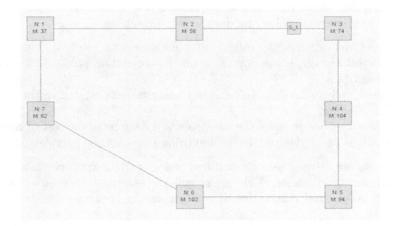

Fig. 5. Movement of a car in the simulation environment

The changes in the presence of the objects (cars, pedestrians), see Figs. 5 and 6, in the environment are detected and, in accordance to them, the street lamp parameters are modified – lamps are brightened or darkened. It results in increasing the safety level, users comfort and saving the energy. Not all lamps need to lit all the time with a maximum power.

The changes in the presence of the objects (cars, pedestrians) in the environment are detected and, in accordance to them, the street lamp parameters are modified – lamps are brightened or darkened. It results in increasing the safety level, users comfort and saving the energy. Not all lamps need to lit all the time with a maximum power. The total image of the street lamps work is presented on Fig. 7. Different consumption rate of the particular lamps results in huge savings and is the cause of lamp configurations on the vehicle routes or the pedestrian crossings. The graph presents data from all lamps which enable to visualize the relative change of light intensity for each of them. Every lamp works in a much more effective way than when it would lit with the same intensity power equal 50% of its utmost value when there is a lack of events on the street or the roadside. In spite of that, the driver who is passing the street lamp can notice much better lightening of the street as well as the roadside.

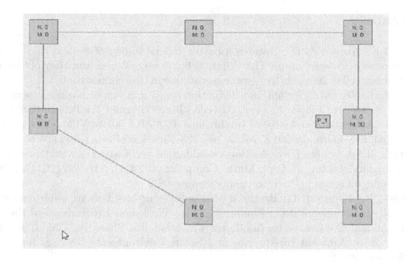

Fig. 6. The pedestrian near the road in the simulation environment

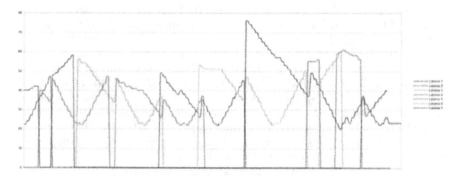

Fig. 7. The work chart overlapping for street lamps in an intelligent environment

5 Conclusions

In this paper we propose a multi-agent system for smart outdoor lighting, and implemented using JADE environment. The hierarchy of agents as well its architecture is presented. This work opens a research area which is of crucial importance for the idea of smart lighting.

This is the first attempt to provide an agent system for modern outdoor lighting. Future works may include more complex smart scenarios including workflows [2,3], as well as and simulations.

Acknowledgments. I would like to thank my students Maciej Ciurej, Anna Włudyka, Konrad Pietras, and Mateusz Zadka (AGH UST, Kraków, Poland) for their cooperation when preparing this work.

References

1. Dey, A.K., Abowd, G.D.: Towards a better understanding of context and context-awareness. In: Workshop on The What, Who, Where, When, and How of Context-Awareness (CHI 2000). http://www.cc.gatech.edu/fce/contexttoolkit/
2. Klimek, R.: Towards formal and deduction-based analysis of business models for soa processes. In: Filipe, J., Fred, A. (eds.) Proceedings of 4th International Conference on Agents and Artificial Intelligence (ICAART 2012), Vilamoura, Algarve, Portugal, 6–8 February 2012, vol. 2, pp. 325–330. SciTePress (2012)
3. Klimek, R.: A system for deduction-based formal verification of workflow-oriented software models. Int. J. Appl. Math. Comput. Sci. **24**(4), 941–956 (2014). http://www.amcs.uz.zgora.pl/?action=paper&paper=802
4. Klimek, R., Kotulski, L.: Proposal of a multiagent-based smart environment for the IoT. In: Augusto, J.C., Zhang, T. (eds.) Workshop Proceedings of the 10th International Conference on Intelligent Environments, Shanghai, China, 30 June–1 July 2014. Ambient Intelligence and Smart Environments, vol. 18, pp. 37–44. IOS Press (2014)
5. Klimek, R., Kotulski, L.: Towards a better understanding and behavior recognition of inhabitants in smart cities. A public transport case. In: Rutkowski, L., Korytkowski, M., Scherer, R., Tadeusiewicz, R., Zadeh, L.A., Zurada, J.M. (eds.) ICAISC 2015. LNCS, vol. 9120, pp. 237–246. Springer, Cham (2015). doi:10.1007/978-3-319-19369-4_22
6. Klimek, R., Rogus, G.: Modeling context-aware and agent-ready systems for the outdoor smart lighting. In: Rutkowski, L., Korytkowski, M., Scherer, R., Tadeusiewicz, R., Zadeh, L.A., Zurada, J.M. (eds.) ICAISC 2014. LNCS (LNAI), vol. 8468, pp. 257–268. Springer, Cham (2014). doi:10.1007/978-3-319-07176-3_23
7. Kotulski, L., Sędziwy, A., Strug, B.: Conditional synchronization in multi-agent graph-based knowledge system. Procedia Comput. Sci. **51**, 1043–1051 (2015)
8. Nawarecki, E., Kozlak, J.: Building multi-agent models applied to supply chain management. Control Cybern. **39**(1), 149–176 (2010)
9. Schilit, B., Adams, N., Want, R.: Context-aware computing applications. In: Proceedings of the 1994 First Workshop on Mobile Computing Systems and Applications (WMCSA 1994), pp. 85–90. IEEE Computer Society (1994). http://dx.doi.org/10.1109/WMCSA.1994.16
10. Ying-Wen, B., Yi-Te, K.: Automatic room light intensity detection and control using a microprocessor and light sensors. IEEE Trans. Consum. Electron. **54**, 1173–1176 (2008)

Understanding Human Behavior in Intelligent Environments: A Context-Aware System Supporting Mountain Rescuers

Radosław Klimek[✉]

AGH University of Science and Technology,
Al. Mickiewicza 30, 30-059 Kraków, Poland
rklimek@agh.edu.pl

Abstract. Intelligent environments provide people-centered computing to support people in their daily lifes. Understanding human behavior and context information is crucial to provide context-aware and pro-active services for all actors of smart spaces. On the other hand, mobile phone network data, collected by suppliers, provide valuable information about human locations and behaviors. This paper presents a unified approach comprising both informal (use cases) and more formal (algorithms) elements which enable obtaining a common framework that use information encoded into pervasive datasets to generate, through context-based reasoning, decisions which support actors operating in a smart space. The system is designed to support mountain rescuers. It provides pro-active decision taking or warning about dangerous situations on the mountain trails. In this way, the system supports rescuers and makes tourist staying in the mountains more safe.

Keywords: Intelligent environment · Mobile phone network · Base transiver station · Pro-activity · Context reasoning · Mountain rescuer

1 Introduction

Smart spaces around us play more and more important role. Not only they make our environment more friendly for its inhabitants and users, but also much safer and often more energetically effective. The context system is the one which takes decision on the basis of the context in which it appears. The system possesses the descriptions of different situations (contexts) which can appear and the possible operation scenarios for every specific context. It is essential that the decisions are taken autonomously and pro-actively and the whole system is transparent for the residents making their life safer and more comfortable.

The aim of this paper is to show how to use information about mountain tourist activity for context-aware smart decisions in an intelligent environment. This paper contributes to obtaining the unified approach consisting of use cases and algorithms for software agents which are an evidence and an argument validating the proposed system. Another contribution is an innovative method

© Springer International Publishing AG 2017
L. Rutkowski et al. (Eds.): ICAISC 2017, Part II, LNAI 10246, pp. 267–279, 2017.
DOI: 10.1007/978-3-319-59060-8_25

of mapping and filtering the pervasive streams of datasets into a collection of individual and anonymized tourist activities located in a particular tourist destination. To the best of our knowledge, this research paper presents the first study for the mentioned area as well as for the tourist movement case. This study opens some new directions especially related to system implementation and experiments in particular.

There are many works considering behavior analysis in intelligent environments. A survey for human activity is provided in [1], which is comprehensive and discusses many important aspects for the domain. In [3] a hierarchical framework for human activity recognition is presented, the framework focuses on video based activity recognition. Sensing and monitoring activities basing on mobile phone datasets seems hot and relatively new [2,11]. Calabresse et al. [2] describe a real time monitoring system, where vehicles and pedestrians movements, are positioned providing urban mobility. In work by Gonzalez et al. [7] trajectories of anonymized mobile phone owners are discussed, which are characterized by a high degree of both temporal and spatial regularity. The context issues are crucial for contex-aware and pro-active systems, and work [5] proposes patterns for a property specification. Zimmermann et al. [14] identified the context categories, user and role, process and task, location, time and others to cover a broad variety of scenarios. This paper follows works [10,12] which concern observing behaviors of users/inhabitants and modeling logical specifications understood as user preferences.

2   Preliminaries

2.1   Babia Góra

Babia Góra[1] is a mountain peak, or rather a mountain range, located on the Polish-Slovak border which is a perfect illustration for the foregoing considerations and, honestly speaking, was also an inspiration to plan and design the information system.

Babia Góra is a very popular destination of numerous mountain expeditions as well as family and school trips. On the one hand, the mountain seems to be quite easy to climb what, apart from its numerous attractions and stunning landscape, would justify its popularity. On the other hand, every year there is noted a huge number of emergency interventions. Getting lost and suffering from serious injuries caused by accidents are not rarity. Apart from getting lost or having frostbite there are also a few fatal accidents which cannot be avoided because of the ineffective help, too late report of the problem or even because of the unawareness that something tragic is happening.

Babia Góra is situated in the National Park. It is one of the highest summits, except of the alpine type mountains, in Poland. At its top there is located the spacious and popular mountain shelter. Most of the incidents take place during climbing or going down the mountain. The weather conditions are changeable,

---

[1] See http://en.wikipedia.org/wiki/Babia_Gora.

whimsical and dependent on the current situation, time of a day, season of a year and many other factors. Winters are marked by the snow hurricanes, summers are characterized by dangerous and violent storms. There is also a danger of avalanches as well as strong winds, especially in the uncovered parts of the mountains.

Enormous variability of the weather conditions, from the point of view of the planned information system, is the main part of the carefully observed and analyzed context. The present-day technological progress provides the opportunity of saturation the natural environment with non-invasive equipment which allows us to trace behaviors of the people on the mountain trails. An already existing technological infrastructure such as the mobile telecommunication network together with the widespread possession of mobile phones helps us to trace tourists' behaviors, their location, individual and collective behaviors etc. Summing up, this special variabilty of environment conditions is a justification for the creation a rescuer supporting system.

2.2 Basic Nomenclature

Here we present the basic nomenclature and devices used in the system as well as the assumptions necessary to create the proactive context-aware software which enables the context assumptions as a consequence.

Basic transivcr station (BTS) base transmitter receiver station enables the bilateral communication with the mobile device in the GSM frequency. Reading the position report of the mobile devices by the system allows counting the distance between the device and the station on the basis of the response time and the signal strength parameters. As a result, the fairly accurate positioning of mobile phones and their users becomes possible.

Call detail record (CDR) contains data recorded by telecommunications equipment. It contains data that is specific in any single case of a phone call or other communication transaction. The structure of CDR is relatively complex. CDRs, as collections of information, have a special format [6]. Bclow thcrc is a sample fragment of a CDR text decoded from the binary format. The first row must contain a header row which includes the field names:

```
''Call Type'',''Call Cause'',''Customer Identifier'',''Telephone Number Dialled'',
''Call Date'',''Call Time'',''Duration'',''Bytes Transmitted'',''Bytes Received'',
''Country of Origin'',''Network'', ''Retail tariff code'',''Remote Network'',
''APN'',''Diverted Number'',''Ring time'',''RecordID'',''Currency''
```

The meaning of the columns is not analyzed here since they are intuitive and the detailed discussion is too broad for the scope of this paper.

Mountain trails are the designated routes where tourists are liable to move in the area considered. Trails have different levels of difficulty: from the easy to very challenging ones meant only for experienced climbers. The difficulty level of a trail in the context of different weather factors has a significant influence on the rescue operations or their lack. In the area covered by the system there were designed eleven trail types classified according to their difficulty level, see Table 5.

Types of risks are the conditions defined by the mountain rescuers as those which influence the safety of wandering through the mountain trails and require specific behaviors, tracing or monitoring. Among them are

- weather conditions, see Table 2, dependent from: wind, rain, fog, temperature,
- risk of avalanches, see Table 1, as well as
- presence of wild animals.

The detailed description of the risks enumerated above, as well as the factors which are their key drivers is described and explained below.

Table 1. Avalanche risk table

Risk level	Avalanche risk
5	Even on gentle slopes, many large spontaneous avalanches are likely to occur
4	Avalanches are likely to be triggered on many slopes even if only light loads are applied. In some places, many medium or sometimes large spontaneous avalanches are likely
3	Avalanches may be triggered on many slopes even if only light loads are applied. On some slopes, medium or even fairly large spontaneous avalanches may occur
2	Avalanches may be triggered when heavy loads are applied, especially on a few generally identified steep slopes. Large spontaneous avalanches are not expected
1	Avalanches are unlikely except when heavy loads are applied on a very few extreme steep slopes. Any spontaneous avalanches will be minor sloughs. In general, safe conditions

Wind. It is very difficult to maintain the standing position if average wind speed increases to 80–100 km/h. In such conditions wandering through very high parts of the mountains becomes impossible and increases the risk of accidents involving falls in the mountain gap. When traversing the wooden areas tourists are also exposed on crushing due to broken branches, uprooted trees or falling rocks.

Fog. Its presence not only prevents from admiring the view but first of all, it may significantly hamper our orientation in the field. The most dangerous is the winter fog when the snow whiteness blends with the fog whiteness. The reference points are invisible and we cannot define our location, direction of the march or even evaluate the steepness of the slope.

Temperature. At peaks temperature is usually lower than at the bottom of the mountains. With every 100 m the temperature falls down about 1°. Feeling of cold is additionally exacerbated by wind. A weak wind, when the thermometer shows 0°, causes the feeling of cold equal to −3°, airflow to 5 m/s almost −9°, wind up to 15 m/s even −18°. During the moderate frost (−10°) the same wind can cause the feeling of 33° below zero.

Rain, storm. In the mountains the storm is especially dangerous because of its frequency, the height and the lack of shelter. The additional danger is caused by railings, chains and ladders and the lack of conductive layers. The storm is dangerous when the time between lightning and thunder is shorter than 515 s. It means that lightning hit closer than 25 km from us. Additionally, the heavy rainfall increases the difficulty of walking along the trails.

Avalanche. The natural phenomenon which causes the loss of stability and the rapid movement of ice and snow masses. The avalanche risk is presented in Table 1.

2.3   Towards Context Reasoning

Context is understood as a set of circumstances or facts surrounding a particular event, situation, etc. in which something happens. In other words, it is impossible to understand what happened without looking at the context. It is also the type of place, time, manner, etc., that accompany or influence an event or condition.

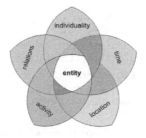

Fig. 1. Categories of information [14] for context reasoning.

The description of the context information fall into some categories [4]. The fundamental categories for context information are shown in Fig. 1. Those basic categories can be referred to the current system:

1. Identity – everyone possesses an unique identifier, in case of this system it is the private phone number.
2. Time – recent event time, duration of the episode, in case of the system, also season of year and time of day.
3. Location – position or spatial relations calculated on the basis of GSM network data, possibly also from the additional measurements undertaken by drones which serve as the mobile BTS stations, that is normal data are not available.
4. Activity – what exactly takes place in a particular situation: following the prescribed route, having rest for a longer time in the mountain shelter etc.

5. Relations – relations to other members of the group like moving away, coming closer (also to the potentially dangerous objects such as wild animals).

The weather conditions which have been mentioned above will be projected on the scale of the values from Table 2. Analyzing the context of the particular situation will allow us to take into consideration the specific values. In other words: the real life conditions, such as wind, will be projected on the scale values.

Table 2. Weather conditions

	Scale	Description
Wind	1–3	1: up to 30 km/h; 2: 30–80 km/h; 3: above 80 km/h
Fog	1–3	1: observable fog without an influence; 2: weak visibility; 3: poor visibility, significant difficulties in field orientation
Temperature	1–3	1: above 15 or below 5; 2: above 24 or below −3; 3: above 30 or below −15
Rain, strom	1–3	1: rain – from 15 to 40 mm, wind – from 0 to 30 km/h; 2: rain – from 41 to 70 mm, wind – from 31 to 80 km/h; 3: rain – above 70 mm, wind – above 80 km/h

Every tourist is understood as a user of the unique mobile phone which was identified and is being tracked by the BTS station. The application operates in many emergency situations, as for example:

- a tourist remains in one place for a long time – possibility of losing the consciousness.
- a tourist appeared in a place where is no admission because of the weather conditions such as the avalanche risk.
- a tourist moved away from the group – possibility of losing. The group can be registered before going on a mountain trip, additionally the phone number of the group leader is identified.

·3   System Description

The model of the system presented below shows both the diagram of the use cases as well as the activity diagrams which illustrate the basic algorithms connected to data processing inside the system.

3.1   Uses Cases for the System

The main use case diagram for the supporting system is shown in Fig. 2. The following use cases are considered:

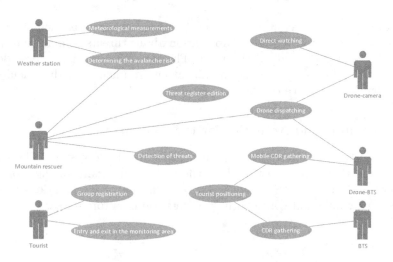

Fig. 2. The main use case diagram for the system

- **Meteorological measurements** – the set of all measurements which represent weather conditions related to wind, rain, fog etc.
- **Determining the avalanche risk** – it enables to predict the risk level as well as its area. This type of analysis takes into consideration not only the meteorological measurements but also the rescuers' tips and their on-site verification.
- **Entry and exit in the monitoring area** – the exact place where tourists go in and go out from the National Park which remains under monitoring of the system. **Group registration** – keeping the exact record of the groups is a form of additional and optional registration. It includes both the group leader and all its members and helps to prevent the unexpected scenarios such as moving away of one member of the group.
- **CDR gathering** – collecting CDR data about the mobile devices or phones which are within the BTS station range.
- **Mobile CDR gathering** – gathering the CDR data by the BTS station installed on a drone.
- **Drone dispatching** – decision about sending a drone in order to gather more accurate data – made by the rescue teams.
- **Tourist positioning** – precise positioning of the tourists based on the BTS station data, all positions are applied on the map.
- **Direct watching** – enables direct observation of the event and preparing a photographic record of the place which is an issue of a special surveillance, for example a place where is a tourist who is far away from the group and does not move.
- **Threat register edition** – edition of the database which includes the information about the type of activities undertaken in certain categories of danger. This type of record can be edited only by the mountain rescuer and once set, does not have to be modified.

– **Detection of threats** – the automatic detection of the particular types of threats on the monitored area, information about threats is immediately sent to rescuers and plotted on the map. The proactive service is provided as a result of the existence of the intelligent environment and the calculations based on the pervasive computing paradigm.

3.2 Algorithms and Activities for the System

In this subsection we present the main algorithms of the system. The way in which they work will be illustrated by the activity diagram. In the monitored area there work a few BTS stations. These is the set of BTS stations which is analyzed as a whole and which entirely covers the monitored area.

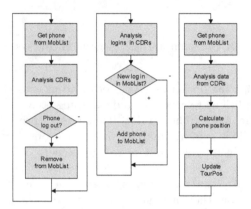

Fig. 3. Activity diagrams for use cases "Entry and exit in the monitoring area" (left, middle) and "Tourist positioning" (right)

Activity diagrams for use cases "Entry and exit in the monitoring area" and "Tourist positioning" are shown in Fig. 3. MobList stores data about all mobile phones existing on the monitored area. The analysis of information accumulated in CDR, particularly noticing that there was logging out from BTS without logging in to another station belonging to BTS set, means leaving the area and as a consequence, removing the phone from MobList. The analysis of records in CDR data belonging to BTS set enables to discover the totally new logins to the monitored area and as a result, adding the phone to MobList.

Activity diagrams for use case "Detection of threats" are shown in Fig. 4. The use case is crucial for the whole system and is carried out by many processes, each of them detects different threats. If any threats are detected, they are registered and displayed on the list of the potential dangers:

– lack of movement of a tourist for a period of time (it takes into account such possibilities as staying in a shelter, the time of day and other aspects);

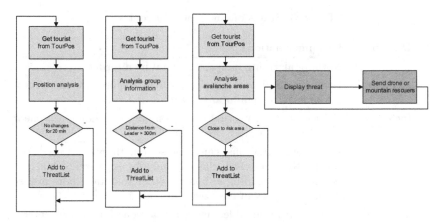

Fig. 4. Activity diagram for use case "Detection of threats" (from left to right): the no-movement process; the too-big-gap process; the risk-area process; and threat continuous handling

- too long distance between a tourist and a group leader which can signify his/her loss (it applies only to the groups which registered their presence such as school trips or any other registered groups);
- moving too close to the dangerous areas, especially to the avalanche areas and all other types of areas previously included in the system by the rescuers.

Taking all further measures depends on the rescuers' decision. They can send the drone in order to examine the situation profoundly or decide about starting the rescue operation.

3.3 Reasoning

Decision about the type of rescue operation is based on the current context in which the tourist is the actor of the system. In this case the context includes both the place and the external factors (weather). Drawing conclusions about the threats on the monitored areas is based on three main factors: (1) weather conditions, (2) avalanche risk, (3) difficulty level of the route. While the difficulty level of the route is centrally-defined, the information about the weather conditions and their components is collected from the sensors located in different parts of the monitored area. Among them are: wind, rain, fog, temperature. Each of them has its own scale. For the weather conditions we use 1 to 3 scale, for avalanche risk 1 to 5 scale and for the difficulty level of the route there is 1 to 4 scale, see Tables 2, 1, and 5, respectively.

The result is an indicator of the threat level which is presented on the scale from 1 to 5 in Table 3. The system suggests the particular solution on the basis of the gained result. As presented in Table 4, there are separate types of actions according to the threat level. The content of the table is fixed/unchangeable and has been created by the mountain rescuers. The data obtained from the

Table 3. States of emergency for routes

Level	Description	Recommendations
1	Low	Nothing alarming happens, normal monitoring
2	Medium	Partially disadvantageous conditions, requires the assessment of the local or temporary threat
3	Increased	Existence of risky situations, possibility of sending a drone in order to obtain more precise data
4	High	Dangerous situations, can cause serious threat and need to be constantly monitored, climbing only for experienced hikers, possibility of sending a drone in order to obtain more precise data, possibility of emergency operation
5	Very high	Walking is impossible, emergency action is necessary, entering the routes by other tourists is strictly forbidden

monitored area are analyzed according to the conditions described in the table which are based on the certain factors. The separate factors, which are mentioned here in order to avoid the overcomplicated model, are analyzed for example for the wild animals. The examples of the system reactions are presented below:

- on the route there is the 5th level of threat — tourists from the whole route should be evacuated, the presence of rescuers is almost always necessary;
- a tourist entered the restricted zone (avalanche or other threats), we send an SMS message asking him/her to exit the zone. In case of lack of reaction we can send the rescue team;
- a tourist goes in the direction of a wild animal (the animals such as bears have the positioning transmitters), we send SMS information to the tourist asking him to come back or follow another route. The rescuers also get the information about the incident.

It is planned to make decisions expressed in Table 4 basing on SAT solvers.

Table 4. Defined reaction levels

Level	Logical functions
5	$diffLevel = 4 \wedge ((wind \geq 2) \vee (rain \geq 2) \vee (fog \geq 1) \vee (avalanche \geq 2))$
	$diffLevel = 3 \wedge ((wind \geq 3) \vee (rain \geq 3) \vee (fog \geq 2) \vee (avalanche \geq 3))$
	$diffLevel = 2 \wedge ((wind \geq 3) \vee (rain \geq 3) \vee (fog \geq 2) \vee (avalanche \geq 2))$
	....
4	....

Table 5. Mountain hiking trails (for Babia Góra)

Num	Color	Hiking trail	Transition time	Trail length	Difficulty level
1	Yellow	Markowe Szczawiny – Sucha Kotlinka – Diablak	1 h 30 min	3 km	1
2	Yellow	Zawoja Czatoża – Fickowe Rozstaje – Górny Płaj – Markowe Szczawiny	3 h 20 min	5 km	1
3	Blue	Zawoja Czatoża – Markowe Rowienki – Zawoja Markowa – Ryzowana – Zawoja Policzne	3 h 30 min	7,50 km	2
4	Blue	Zawoja Policzne – Polana Krowiarki – Markowe Szczawiny	3 h	11 km	2
5	Green	Zawoja Markowa – Pośredni Bór – Markowe Szczawiny	1 h 20 min	3,6 km	3
6	Green	Górny Płaj – Sokolica	45 min	1,5 km	3
7	Green	Przełęcz Jałowiecka – Mała Babia Góra – Przełęcz Brona	2 h	4 km	3
8	Green	Przywarówka – Głodna Woda – Diablak	2 h 30 min	2,3 km	3
9	Green	Polana Krowiarki – Hala Śmietanowa – Zubrzyca Górna	2 h	3 km	3
10	Red	Polana Krowiarki – Sokolica – Kępa – Gówniak – Diablak – Przełęcz Brona – Markowe Szczawiny – Fickowe Rozstaje – Przełęcz Jałowiecka	6 h	14,5 km	4
11	Black	Podryzowana – Ryzowana – Markowe Szczawiny	2 h 30 min	3,5 km	4

4   Conclusion

In this paper, the problem of sensing tourist activities on mountain trails are considered. Behaviors are mined from BTS networks. A system supporting mountain rescuers is proposed. This work opens a research area which is of crucial importance for the idea of intelligence environments.

Future works may include more detailed algorithms and software architecture, formally verified due to business and activity diagram workflows, architectural aspects [8,9,13], as well as implementation basing on information brokers to gather necessary data and logical solvers to provide decisions supporting rescuer teams.

Acknowledgments. I would like to thank my students Nikodem Kirsz, Anna Połeć, and Marlena Zdybel (AGH UST, Kraków, Poland) for their valuable cooperation when preparing this work.

References

1. Aggarwal, J., Ryoo, M.: Human activity analysis: a review. ACM Comput. Surv. **43**(3), 16:1–16:43 (2011)
2. Calabrese, F., Colonna, M., Lovisolo, P., Parata, D., Ratti, C.: Real-time urban monitoring using cell phones: A case study in rome. IEEE Trans. Intell. Transp. Syst. **12**(1), 141–151 (2011)
3. Chen, S., Liu, J., Wang, H., Augusto, J.C.: A hierarchical human activity recognition framework based on automated reasoning. In: IEEE International Conference on Systems, Man, and Cybernetics, Manchester, SMC 2013, United Kingdom, 13–16 October 2013, pp. 3495–3499 (2013)
4. Dey, A.K., Abowd, G.D.: Towards a better understanding of context and context-awareness. In: Workshop on The What, Who, Where, When, and How of Context-Awareness (CHI 2000). http://www.cc.gatech.edu/fce/contexttoolkit/
5. Dwyer, M.B., Avrunin, G.S., Corbett, J.C.: Patterns in property specifications for finite-state verification. In: Proceedings of the 21st International Conference on Software Engineering (ICSE 1999), Los Angeles, CA, USA, 16–22 May 1999, pp. 411–420 (1999)
6. Federation of Communication Services: UK Standard for CDRs. Standard CDR Format, January 2014
7. Gonzalez, M.C., Hidalgo, C.A., Barabasi, A.L.: Understanding individual human mobility patterns. Nature **453**(7196), 779–782 (2008)
8. Grobelna, I., Grobelny, M., Adamski, M.: Model checking of UML activity diagrams in logic controllers design. In: Zamojski, W., Mazurkiewicz, J., Sugier, J., Walkowiak, T., Kacprzyk, J. (eds.) Proceedings of the Ninth International Conference DepCoS-RELCOMEX. AISC, vol. 286, pp. 233–242. Springer, Cham (2014). doi:10.1007/978-3-319-07013-1_22
9. Klimek, R.: Towards formal and deduction-based analysis of business models for SOA processes. In: Filipe, J., Fred, A. (eds.) Proceedings of 4th International Conference on Agents and Artificial Intelligence (ICAART 2012), Vilamoura, Algarve, Portugal, 6–8 February 2012, vol. 2, pp. 325–330. SciTePress (2012)

10. Klimek, R.: Behaviour recognition and analysis in smart environments for context-aware applications. In: Proceedings of the IEEE International Conference on Systems, Man, and Cybernetics (SMC 2015), City University of Hong Kong, Hong Kong, 9–12 October 2015, pp. 1949–1955. IEEE Computer Society (2015)

11. Klimek, R.: Mapping population and mobile pervasive datasets into individual behaviours for urban ecosystems. In: Rutkowski, L., Korytkowski, M., Scherer, R., Tadeusiewicz, R., Zadeh, L.A., Zurada, J.M. (eds.) ICAISC 2016. LNCS, vol. 9692, pp. 683–694. Springer, Cham (2016). doi:10.1007/978-3-319-39378-0_58

12. Klimek, R., Kotulski, L.: Proposal of a multiagent-based smart environment for the IoT. In: Augusto, J.C., Zhang, T. (eds.) Workshop Proceedings of the 10th International Conference on Intelligent Environments, Shanghai, China, 30 June–1 July 2014. Ambient Intelligence and Smart Environments, vol. 18, pp. 37–44. IOS Press (2014)

13. Klimek, R., Szwed, P.: Verification of archimate process specifications based on deductive temporal reasoning. In: Proceedings of Federated Conference on Computer Science and Information Systems (FedCSIS 2013), Kraków, Poland, 8–11 September 2013, pp. 1131–1138. IEEE Xplore Digital Library (2013)

14. Zimmermann, A., Lorenz, A., Oppermann, R.: An operational definition of context. In: Kokinov, B., Richardson, D.C., Roth-Berghofer, T.R., Vieu, L. (eds.) CONTEXT 2007. LNCS, vol. 4635, pp. 558–571. Springer, Heidelberg (2007). doi:10.1007/978-3-540-74255-5_42

TLGProb: Two-Layer Gaussian Process Regression Model for Winning Probability Calculation in Two-Team Sports

Max W.Y. Lam[✉]

Department of Computer Science and Engineering,
The Chinese University of Hong Kong, Shatin, Hong Kong
maxwylam@cuhk.edu.hk

Abstract. Sports analytics is gaining much attention in the research community nowadays. This paper deals with a prominent problem in sports analytics, namely, *winning probability calculation*. In particular, we focus on the *two-team sports*. A novel model called *TLGProb* is proposed by stacking a non-linear regression model – Gaussian process regression (GPR) to address complex association between match outcomes and players' performances. For evaluation, we selected a popular sports event around the world – National Basketball Association (NBA) as the domain for experiments. Finally, using *TLGProb*, we correctly predicted 85.28% of outcomes among 1,230 matches in NBA 2014/2015 season.

Keywords: Gaussian process · Sports analytics · Winning probability calculation

1 Introduction

Sports analytics has received much attention in recent years. Its success greatly contributes to the business of professional sports, which nowadays is a multi-billion dollar industry. Well-developed analytical techniques can, to a large extent, benefit many parties including the athletes, the sports team owners, the coaches and even the fans. For instance, it is advantageous for a sports team to have well-explicated coaching or tailor-made player acquisition [6].

In this work, we particularly focus on the analytics of *two-team sports*, which is a kind of sports that practices between opposing teams, and covers most popular sports nowadays, including football, basketball, volleyball, and handball. Specifically, we tackle a prominent problem in sports analytics – winning probability calculation. Without a doubt, careful calculation of winning probability is vital in the business of sports wagering, which is another multi-billion dollar industry on its own. In fact, to determine the winning odds, most of the wagering market nowadays adopt parimutuel schemes [5], in which the winning odds are mainly driven by public preferences to the sports teams, and will occasionally fluctuate on real-time matches. Although it appears that public preferences are

© Springer International Publishing AG 2017
L. Rutkowski et al. (Eds.): ICAISC 2017, Part II, LNAI 10246, pp. 280–291, 2017.
DOI: 10.1007/978-3-319-59060-8_26

efficient estimates, the general public is likely to have heavy favorites, which resort to over-estimation of winning probabilities [12].

An accurate predictive model for winning probability calculation is demanding. This is particularly conspicuous for some popular sports such as football and basketball, where many sports lotteries and all kinds of analytic reports spread around the world trying to provide the best possible prediction of the winners of sports events. To facilitate human in making rational forecasts, we naturally make use of historical data and carry out machine learning. There are some researches dealing with sports results forecasting, where techniques ranging from traditional statistical models to machine learning models are used. One can see [4] for a review of applying data mining techniques on sports prediction. It is also exciting that we can build a real-time decision-making system [2] based on these predictive models to facilitate coaches in making tactical decisions before the match or during the match.

In this paper, a novel model called *TLGProb* is proposed. To address potentially high dimensionality and non-linearity in our problem, we propose to employ a powerful probabilistic regression method – Gaussian process regression (GPR) [13]. It is worthwhile to note that *TLGProb* is designed as a generic model so that it can be applied to any type of two-team sports.

National Basketball Association (NBA), being one of the most popular sports events around the world, has comprehensive matches data that is open to public for analysis. To evaluate our system for real matches, we select a regular season of NBA as the testing dataset for *TLGProb*. Finally, using *TLGProb*, we correctly predicted 85.28% of outcomes among 1,230 matches in NBA 2014/2015 season.

The remainder of this paper is organized as follows. In Sect. 2, we define the problem that is concerned in this paper. In Sect. 3, we present the design of *TLGProb* with reasoning. In Sect. 4, we show an experimental evaluation of our system. in Sect. 5, we conclude our work and identify some potential research directions.

2   Problem Definition

In this paper, our goal is to analytically derive the winning probability of two teams competing in the coming match while only the historical data before the match are used. For the ease of reference, we denote the two competing teams by *Home Team* and *Visiting Team*. In other words, we wish to answer how likely is the Home Team or the Visiting Team to win the next game. We coin this problem as *one-match-ahead forecasting*. In our perspective, past performances of the two teams are sufficient to infer the result of next match, since the ability of each team player should not differed a lot.

Formally, match result is regarded as an ordered pair, $\left(G_k^H, G_k^V\right)$, which denotes the game points attained by the Home Team and the Visiting Team respectively on the kth match. For the coming match where the match result is unknown, it is natural four us to use random variables, yielding $\left(\mathbf{g}_{k+1}^H, \mathbf{g}_{k+1}^V\right)$.

To calculate the winning probabilities of the coming match, we only need to concern the signed deviation of the game points attained by two teams, i.e.,

$\mathbf{g}_{k+1} = \mathbf{g}_{k+1}^H - \mathbf{g}_{k+1}^V$, since the winning probabilities of the Home Team and the Visiting Team can be directly determined by $Pr(\mathbf{g}_{k+1} > 0)$ and $Pr(\mathbf{g}_{k+1} < 0)$.

If we know the result and players' performances of the first k matches, then essentially the random variable $\mathbf{g}_{k+1}$ can be estimated. We insist that the distribution of this random variable is very much dependent on the past performances of both the Home Team and the Visiting Team. In fact, from the empirical evidence, we show that this dependency is strong enough for us to construct an accurate predictive model.

With the dependency assumption, we formally define an abstract regression problem:

$$\mathbf{g}_{k+1} = f(\mathbf{p}_{1:k}) + \varepsilon_{k+1}, \quad \varepsilon_{k+1} \sim \mathcal{N}(0, \gamma_{k+1}), \tag{1}$$

where $\mathbf{p}_{1:k}$ is the performances of team's players in first k matches. Now, our goal is to find a "model" f so that it best explain the dependency between winning probabilities of the coming match and past performances of the two competing teams.

3   Model Description

For the ease of explanation, we use NBA as an example of two-team sports throughout this section. In NBA, historical data is divided into multiple seasons. Here, we define a season as a sequence of matches:

$$\mathcal{M}_y = (\mathbf{M}_1^y, \mathbf{M}_2^y, ..., \mathbf{M}_k^y, ...),$$

where $\mathcal{M}_y$ denotes the season that starts in y and $\mathbf{M}_k^y$ denotes the kth scheduled match of the season $\mathcal{M}_y$. In each match, we have two teams, labelled as the *Home Team* and the *Visiting Team*, competing each other. To refer to a specific team on $\mathbf{M}_k^y$, we define $\mathbf{T}_k^H$ and $\mathbf{T}_k^V$ for the Home Team and the Visiting Team respectively. The set of all teams in the season $\mathcal{M}_y$ is defined as $\mathcal{T}_y$. For simplicity, in the remainder of this section we use $\mathbf{T}_k$ for the derivation that can be applied to both $\mathbf{T}_k^H$ and $\mathbf{T}_k^V$.

In our notion, a team $\mathbf{T}_k$ on the kth match is modeled as a set of n players $\{\mathbf{P}_1, \mathbf{P}_2, ..., \mathbf{P}_n\}$, in which the order of players does not matter. Usually, for team sports, players are assigned to different positions. For example, in basketball, there are 3 main positions – *Center*, *Forward* and *Guard*, and each position can be interpreted a set of players. It is notable that some players may take 2 positions at the same time, so these sets are not necessarily disjoint. To encode the information of player positions, we define $\mathcal{G}_k^{(p)}$, the set of players on the kth match that are assigned to the pth position, such that

$$\mathbf{T}_k = \bigcup_{p=1}^P \mathcal{G}_k^{(p)} \tag{2}$$

where P is the number of positions in the team according to the type of sports. To enable a mapping from a player to his position, we define a function that returns the position index:

$$p = I_k(\mathbf{P}_i). \tag{3}$$

Table 1. The set of attributes that describes a player's performance

Attribute	Description
FG	Field goals per minute
FGA	Field goals attempts per minute
FG%	Percentage of successful field goals
3P	3-point field goals per minute
3PA	3-point field goals attempts per minute
3P%	Percentage of successful 3-point field goals
FT	Free throws per minute
FTA	Free throw attempts per minute
FT%	Percentage of successful free throws
ORB	Offensive rebounds per minute
DRB	Defensive rebounds per minute
TRB	Total rebounds per minute
AST	Assists per minute
STL	Steals per minute
BLK	Blocks per minute
TOV	Turnovers per minute
PF	Personal fouls per minute

Suppose we know the performances of the players participated on the kth match, then for any player $\mathbf{P}_i \subset \mathbf{T}_k$, we describe his performance by a D-dimensional vector $\mathbf{p}_k^i$. We call this the *player's performance*. For example, in NBA, we use 17 attributes, as shown in Table 1, to describe players' performances. In general, we define a time-dependent function ρ:

$$\mathbf{p}_k^i = \rho(\mathbf{P}_i, \mathbf{T}_k), \tag{4}$$

which gives the performance vector of the player $\mathbf{P}_i$ in the team $\mathbf{T}_k$.

Now, we can express all players' past performances in vector form. This seems fascinating because we may apply techniques in time series forecasting to predict players' performances on the coming match. It is essential that the values of players' performances usually fluctuate a lot from matches to matches. One example is illustrated on Fig. 1, where LeBron James' performances on 3-point field goals and field goals are investigated. It is suspected that directly using players' performance for prediction is not likely to meaningful. Therefore, we introduce a concept of *player's ability*. We argue that a player's performance should not equated to his ability since players sometimes make good use of their talents but sometimes not. Essentially, player's performance is not only affected by players' abilities but also affected by other factors, such as opponents' abilities and team strategies. For practical concern, it is claimed that a player's performance is an estimate of his ability, while the other factors are treated as

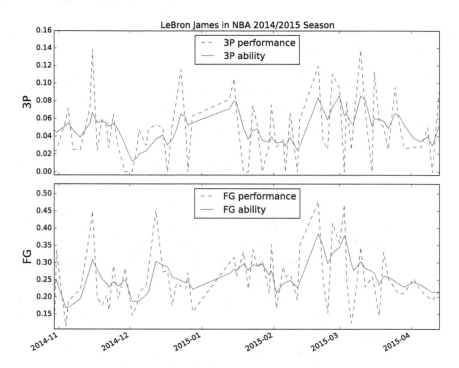

Fig. 1. Time series plot of 3-point field goals (3P) and field goals (FG) of LeBron James in NBA 2014/2015 season. (Color figure online)

noise on estimation. In fact, it is reasonable because player's performance on a match is expected to be consistent with his ability at that time.

Again, player's ability $\mathbf{a}_k^i$ is represented by a D-dimensional vector to preserve the consistency with player's performance. Since players are trained from time to time and gaining experiences across the season of matches, obviously players' abilities should be changing with time. In this regard, we define another time-dependent function α:

$$\mathbf{a}_k^i = \alpha(\mathbf{P}_i, \mathbf{T}_k), \tag{5}$$

which returns the ability of the player $\mathbf{P}_i$ in the team $\mathbf{T}_k$ after the kth match is finished. While a player's performance is treated as an estimate of his ability, we get

$$\mathbf{a}_k^i = \mathbb{E}\left[\mathbf{p}_k^i\right]. \tag{6}$$

Under this assumption, it is now possible to infer player's ability by applying time-series techniques like smoothing [15]. In *TLGProb*, we employ a simple but effective method – exponential smoothing [3]. We made a small change to the standard exponential smoothing procedure to enhance the adaptiveness of the model for different kinds of sports. To formulate our smoothing method, we define a count-up index set of participated matches before the kth match for player $\mathbf{P}_i$:

$$\mathcal{Q}_k^i = \{q \in \mathbb{Z}_+ : \mathbf{P}_i \in (\mathbf{T}_k \cap \mathbf{T}_{k-q})\}. \tag{7}$$

Then, for any player $\mathbf{P}_i$, the inferred ability is determined by

$$\alpha(\mathbf{P}_i, \mathbf{T}_k) = \begin{cases} \alpha_0(\mathbf{P}_i, \mathbf{T}_k) & \text{if } \mathcal{Q}_k^i = \emptyset; \\ (1 - c^\lambda)\rho(\mathbf{P}_i, \mathbf{T}_k) + c^\lambda \alpha(\mathbf{P}_i, \mathbf{T}_{k-r}) & \text{else,} \end{cases} \qquad (8)$$

where $c \in [0, 1)$ is a constant that control the time dependency on the past performances, $r = \min \mathcal{Q}_k^i$ is the number of matches that have been passed since last match participated by $\mathbf{P}_i$, is the initialization function of player's ability while no past performances is recorded in our data: $\alpha_0(\mathbf{P}_i, \mathbf{T}_k)$

$$\alpha_0(\mathbf{P}_i, \mathbf{T}_k) = \frac{1}{\left| \mathcal{G}_k^{(p)} \right|} \sum_{\mathbf{P}_j \in \mathcal{G}_k^{(p)}} \rho(\mathbf{P}_j, \mathbf{T}_k), \quad p = I_k(\mathbf{P}_i), \qquad (9)$$

and λ is a scaling factor that is proportional to the time difference between the $(k - r)$th match and the kth match. Note that, the only difference between our method and the standard exponential smoothing is the presence of λ. When λ is large, $\alpha(\mathbf{P}_i, \mathbf{T}_k)$ will have less dependence on previous ability $\alpha(\mathbf{P}_i, \mathbf{T}_{k-r})$, but having more dependence on current performance $\rho(\mathbf{P}_i, \mathbf{T}_k)$. This is reasonable because more recent performances are more relevant estimators of current ability.

For NBA games, the time difference of the same player joining any two consecutive matches is usually within 1–6 days. Therefore we simply set $\lambda = d/3$, where d is the time difference in days. After setting λ, we can obtain c by optimization:

$$c^* = \operatorname{argmin}_{c \in [0,1)} \| \alpha(\mathbf{P}_i, \mathbf{T}_{k-r}) - \rho(\mathbf{P}_i, \mathbf{T}_k) \|_2^2. \qquad (10)$$

Two examples of inferring player's ability from player's performance are shown on Fig. 1. In these two plots, it is observed that the red solid line has significant clearer trend than the blue dotted line. For instance, we can witness an improvement of 3-point field goals from early 2014-12 to late 2015-01 from the red solid line, whereas, it is difficult to be witnessed from the dotted blue line due to several sudden drops in that period.

Having above qualities defined for players, we now come up with an estimation of teams' abilities so that we can compare the competing teams quantitatively. To avoid the abuse of terminology, we refer *team's strength* to the actual ability of the team. For any team $\mathbf{T}_k$, the team's strength is denoted by $s(\mathbf{T}_k)$. Recall that, in our model a team modeled as a set of players. A sensible approach to measure team's strength is to look at its players' abilities. A strictforward practice is to concatenate the ability vectors of team players:

$$s(\mathbf{T}_k) = \bigoplus_{\mathbf{P}_i \in \mathbf{T}_k} \alpha(\mathbf{P}_i, \mathbf{T}_k), \qquad (11)$$

where $\oplus$ is the operator of vector concatenation. Though, in most cases, vectors' concatenation is problematic for the later training of regression model.

As shown on Table 1, in NBA games, player's ability and performance are represented by 17-dimensional vectors. If we carry out vectors' concatenation, the number of attributes to represent team's strength will be huge. In NBA games, we have 13 players participating in each match including the reserves, so a simple concatenation makes $s(\mathbf{T}_k)$ a 221-dimensional vector. In this case, our regression model will be likely to suffer from the curse of dimensionality. A possible solution is to employ some well-known dimensionality reduction methods, such as principal component analysis (PCA) and linear discriminant analysis (LDA). Nonetheless, these methods do not encode the domain knowledge from the sport itself.

Alternatively, we designed a dimensionality reduction method that is domain-specific for sports games. The main idea is to train an embedding function for each player position that maps player's ability to a general point estimator of player's contribution to his team. In fact, there are a number of such estimators proposed in the research community of sports analytics [11]. Examples include the plus-minus score, the player efficiency rating, and the individual offensive and defensive ratings. The estimator we used in *TLGProb* is called *adjusted plus-minus* (APM) score [16]. One main advantage of using APM score is that each player's APM score is independent of the abilities of that player's opponents and teammates, by contrast, the traditional plus-minus score is highly opponent-dependent. In this sense, APM score is a good measure of a player's individual contribution. Formally, we denote the APM score of player $\mathbf{P}_i$ by $z(\mathbf{P}_i, \mathbf{T}_k)$. Using this point estimate, we can define the embedding function g_p for the pth position:

$$z(\mathbf{P}_i, \mathbf{T}_k) = g_p(\alpha(\mathbf{P}_i, \mathbf{T}_k)). \tag{12}$$

Here, it is vital that we have different embedding functions for different player positions because players in different positions usually have distinctly distributed ability vectors. For example, in basketball, Guard is likely to get higher values on 3P, while Forward is likely to get higher values on FG.

To find a suitable embedding function, we treat it as a regression problem: for all $\mathbf{T}_k \in \mathcal{T}_y$, $\mathbf{P}_i \in \mathcal{G}_k^{(p)}$, our inferred function $\hat{g}_p(\cdot)$ is defined by

$$z(\mathbf{P}_i, \mathbf{T}_k) = \hat{g}_p(\alpha(\mathbf{P}_i, \mathbf{T}_k)) + \varepsilon_p, \ \ \varepsilon_p \sim \mathcal{N}(0, \sigma_p^2). \tag{13}$$

Here, we apply Gaussian process regression (GPR),

$$\hat{g}_p(\mathbf{a}_k^i) \sim \mathcal{GP}\left(m(\mathbf{a}_k^i), \ k(\mathbf{a}_k^i, \mathbf{a}_k^j)\right) \ \ s.t. \ I_k(\mathbf{P}_i) = I_k(\mathbf{P}_j), \tag{14}$$

where the mean function $m(\cdot)$ and the covariance kernel function $k(\cdot, \cdot)$ can be flexibly determined with respect the error measure. The use of GPR in this part is regarded as the first layer of *TLGProb*.

Using the convention of $z(\mathbf{P}_i, \mathbf{T}_k)$, we now determine team's strength by

$$s(\mathbf{T}_k) = \bigoplus_{\mathbf{P}_i \in \mathbf{T}_k} z(\mathbf{P}_i, \mathbf{T}_k), \tag{15}$$

which yields an n-dimensional vector, while n is the number of players in the team. This is acceptable since n is usually small in team sports, and the curse of dimensionality can be prevented. In addition, similar to the definition of player's performance, we define a variable addressing the actual performance of the team.

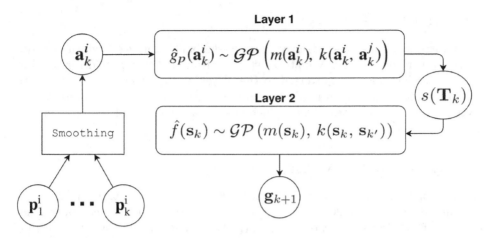

Fig. 2. Block diagram of TLGProb

Naturally, it would be the *game points* scored by the team. Similar to the assumptions that are made for the players, it is claimed that game point attained by the team is dependent on team's strength. Yet, as mentioned in Sect. 3, we only want to predict the game points differential $\mathbf{g}_k$. In this case, we solve a regression problem with the computed team's strength:

$$\mathbf{g}_{k+1} = \hat{f}\left(\mathbf{s}_k\right) + \varepsilon_{k+1}, \quad \varepsilon_{k+1} \sim \mathcal{N}\left(0, \sigma^2\right), \tag{16}$$

where

$$\mathbf{s}_k = s(\mathbf{T}_k^H) \oplus s(\mathbf{T}_k^V). \tag{17}$$

Again, we adopt GPR to infer $\hat{f}$, i.e.,

$$\hat{f}(\mathbf{s}_k) \sim \mathcal{GP}\left(m(\mathbf{s}_k), k(\mathbf{s}_k, \mathbf{s}_{k'})\right), \tag{18}$$

while $m(\cdot)$ and $k(\cdot, \cdot)$ can be determined by empirical analysis. This regression task is viewed as the second layer of *TLGProb*. It is trained subsequently after the GPRs in the first layer are well-trained and selected with mean squared error measure. A block diagram of *TLGProb* is shown on Fig. 2.

4   Experiments

In this section, we analyze the results of applying *TLGProb* on NBA 2014/2015 season, and compare our model with other similar works. The implementation that we present in this section is available on Github.[1]

---

[1] https://github.com/MaxInGaussian/TLGProb.

4.1 NBA Dataset

NBA historical data was retrieved from Basketball-Reference,[2] which is a site that stores records of NBA matches starting from 1948. It provides box score statistics, which can be converted to the player performance statistics that we use in this paper as shown in Table 1. For the ease of data collection, we used a Python library called *Selenium*,[3] which enables automation of the browser and captures HTML tables.

4.2 Training of Our Model

Before applying *TLGProb* on NBA 2014/2015 season, it is trained first with matches data in NBA 2013/2014 regular season. To avoid the computational burden of training GPR, we used a variant of GPR model – sparse spectrum Gaussian process regression (SSGPR) [8] to speed up the training and also to enhance the regression performance. Hyperparameters were repeatedly tuned with 5-fold cross validation. After the training of 50 randomly initialised SSGPR using Adam [7], the one that gives the least mean squared error was selected for our evaluation.

4.3 Evaluation Metrics

Our model *TLGProb* was evaluated over 1,230 matches in NBA 2014/2015 season. Since we only have two teams competing on each match, winning team prediction is nothing more than a binary classification problem. Therefore, to measure the performance of *TLGProb*, we use accuracy measure, which is calculated given the acceptance threshold of winning probability τ:

$$\text{accuracy} = \frac{\sum_{k=1}^{|\mathcal{M}_{2014}|} \mathbb{1}\left(\left\{\left(G_k^H > G_k^V\right) \wedge \left(Pr\left(\mathbf{g}_k^H > \mathbf{g}_k^V\right) > \tau\right)\right\}\right)}{|\mathcal{M}_{2014}|}, \qquad (19)$$

where $\mathbb{1}(A)$ is an indicator function that gives 1 only if $A \neq \emptyset$ otherwise it gives 0.

4.4 Comparing to Other Predictive Models

Without considering much detail of the game, randomly guessing naturally would achieve a baseline of 50% accuracy in the long run. Though, we conceived that rational people normally would not guess randomly. Therefore, a new set of baseline models was examined. To simulate how people predict the game winner, we can simply look back previous win-or-lose records and predict that the team with more winning records is the winner. Predictors based on the number of victories apparently worked better than random guessing, while the best achieved a modest accuracy up to 67%.

---

[2] http://www.basketball-reference.com.
[3] http://www.seleniumhq.org.

Table 2. Accuracies attained by various predictive models

NBA results prediction model	Accuracy
Baseline model based on random guessing	50%
Baseline model based on win-or-lose records of last 5 games	64.64%
Baseline model based on win-or-lose records of last 10 games	65.67%
Baseline model based on win-or-lose records of last 20 games	64.21%
Baseline model based on win-or-lose records of last 50 games	67.09%
One-layer model based on naive bayes [10]	67%
One-layer model based on support vector machine	67.91%
One-layer model based on multilayer perceptron	65.36%
Human experts [9]	71%
NBA oracle [9]	73%
TLGProb	**85.28%**

Several works applying machine learning models on NBA results prediction were also investigated. It is notable that [10] uses naive Bayes to captures teams' performances for the prediction of coming match's result, whereas individual performances of team players are neglected. We refer it as a *one-layer model*, as opposed to two-layer structure in *TLGProb*. In addition to naive Bayes, We also tried support vector machine and artificial neural network for the one-layer model.

As shown on Table 2, *TLGProb* surpassed the others in terms of accuracy. When acceptance threshold on the winning probability τ is naturally set to be 0.5, *TLGProb* correctly predicted the winning teams in 1,049 games out of 1,230 games, giving 85.28% accuracy. Noted that it is hard to improve the accuracy even just a small amount since the result of each game is controlled by many sources of non-deterministic factors, including player injuries, player attitudes, team rivalries and subjective officiating. We claim that the success of our model is very much due to the flexibility of GPR as well as the modeling of two layers – associating players' performances with team's strength and associating two teams' performances with the match outcome.

4.5 Benefits of Getting Winning Probability

It is vital that we obtain winning probability from *TLGProb* instead of a binary prediction. In the realistic scenario, this is a highly appealing feature, especially for determination of winning odds [5]. Simply concerning the task of one-match-ahead prediction, we can see that the winning probability returned from *TLGProb* is beneficial, as it provides us an option to reject the prediction. Here, the acceptance threshold τ acts as our confidence on the prediction result so that matches with winning probability smaller than τ are rejected.

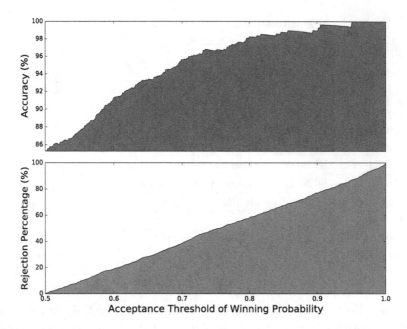

Fig. 3. Plot of accuracies and rejection percentages when different acceptance thresholds of winning probability are used

It is expected that, if the calculated winning probability is a reasonable measure of uncertainty, then the accuracy should increase with the acceptance threshold. In our experiment, it is exciting to see that the outputs obtained from *TLGProb* truly satisfy our expectation, as plotted on Fig. 3. For example, if we set $\tau = 0.6$, then the accuracy will be improved to 91%, while only 20% of the matches are rejected.

Taking advantage of the probabilistic framework, *TLGProb* is suited to work with Bayesian decision theory. That is, practitioners can optionally pick the match results where they have a strong belief in, and then ignore the matches with less confidence. For a realistic example, in wagering market, to carefully determine the winning odds, we should give high risks on matches that high winning probability co-occurs high winning odds.

5   Conclusion and Future Work

In this paper, we describe a model for winning probability calculation by which we can accurately predict the winning team of the next match. We conceive that the success of our proposed model is mainly due to the flexibility of GPR as well as the modeling of two layers – associating players' performances with team's strength and associating two teams' performances with the match outcome. Our system is shown to achieve 85% accuracy in the domain of basketball. In the future, it is potential to examine the performance of *TLGProb* in other domains of two-team sports such as football and baseball. Also, to further investigate the effectiveness of *TLGProb*, it is rational to compare GPR with the state-of-the-art regression methods such as random forests [1] and kernel ridge regression [14].

References

1. Breiman, L.: Random Forests. Mach. Learn. **45**(1), 5–32 (2001)
2. Polese, G., Troiano, M., Tortora, G.: A data mining based system supporting tactical decisions. In: 14th International Conference on Software Engineering and Knowledge Engineering, pp. 681–684. ACM (2002)
3. Gardner, E.S.: Exponential smoothing: the state of the art. J. Forecast. **4**(1), 1–28 (1985)
4. Haghighat, M., Rastegari, H., Nourafza, N.: A review of data mining techniques for result prediction in sports. Adv. Comput. Sci. Int. J. **2**(5), 7–12 (2013)
5. Hausch, D.B., Ziemba, W.T.: Handbook of Sports and Lottery Markets. Elsevier, Amsterdam (2011)
6. Bhandari, I., Colet, E., Parker, J., Pines, Z., Pratap, R., Ramanujam, K.: Advanced scout: data mining and knowledge discovery in NBA data. Data Min. Knowl. Discov. **1**(1), 121–125 (1997)
7. Kingma, D., Ba, J.: Adam: A method for stochastic optimization (2014)
8. Lázaro-Gredilla, M., Quiñonero-Candela, J., Rasmussen, C.E., Figueiras-Vidal, A.R.: Sparse spectrum Gaussian process regression. J. Mach. Learn. Res. **11**, 1865–1881 (2010)
9. Beckler, M., Wang, H., Papamichael, M.: NBA Oracle (2013). https://www.mbeckler.org/coursework/2008-2009/10701_report.pdf
10. Miljković, D., Gajić, L., Kovačević, A., Konjović, Z.: The use of data mining for basketball matches outcomes prediction. In: 2010 IEEE 8th International Symposium on Intelligent Systems and Informatics (SISY), pp. 309–312. IEEE (2010)
11. Oliver, D.: Basketball on Paper: Rules and Tools for Performance Analysis. Potomac Books Inc., Washington (2004)
12. Ottaviani, M., Sørensen, P.N.: Surprised by the parimutuel odds? Am. Econ. Rev. **99**(5), 2129–2134 (2009)
13. Rasmussen, C.E., Williams, C.K.: Gaussian Processes for Machine Learning. MIT Press, Cambridge (2006)
14. Saunders, C., Gammerman, A., Vovk, V.: Ridge regression learning algorithm in dual variables. In: 15th International Conference on Machine Learning, pp. 515–521 (1998)
15. Simonoff, J.S.: Smoothing Methods in Statistics. Springer Science & Business Media, Heidelberg (2012)
16. Winston, W.L.: Mathletics: How Gamblers, Managers, and Sports Enthusiasts Use Mathematics in Baseball, Basketball, and Football. Princeton University Press, Princeton (2012)

Fuzzy PID Controllers with FIR Filtering and a Method for Their Construction

Krystian Łapa[1][✉], Krzysztof Cpałka[1], Andrzej Przybył[1], and Takamichi Saito[2]

[1] Institute of Computational Intelligence,
Częstochowa University of Technology, Częstochowa, Poland
{krystian.lapa,krzysztof.cpalka,andrzej.przybyl}@iisi.pcz.pl
[2] Department of Computer Science, Meiji University, Tokyo, Japan
saito@cs.meiji.ac.jp

Abstract. In this paper a new structure of fuzzy PID controllers with FIR filters and a method for selecting its parameters is presented. The proposed solution can be particularly important in solving problems with noise of the object's feedback signals. To confirm the effectiveness of the proposed method a typical control problem was tested.

Keywords: Controller · PID · FIR · Parameter selection · Genetic algorithm

1 Introduction

The control is a process that affects an object in a purpose to achieve expected behavior or state of the object (in contrast to the identification of the object [58–62]). For this aim controllers are used and their purpose is to generate control signal (on the basis of the input signals) that affects (interacts with) the object (see e.g. [14]). The controllers input signals might include different types of signals. Usually it is the control error, i.e. the difference between the set and the current state of the object. Each signal might be additionally filtered. The properly designed controller allows us to: (a) ensure the proper rise time, (b) mitigate overshooting, (c) reduce settling time, (d) reduce steady-state error and (e) provide stability [38]. All this makes the design of the controllers for special applications (for which there is no clear methodology in the literature and which is usually solved by trial and error method) a complex issue [39].

The controllers most often used in practice are PID controllers (see e.g. [2,72]). These controllers are based on three elements (three-term controllers): proportional (P), integral (I) and differential (D). Each element has a number parameter assigned, which is interpreted as gain or time constant. Another often used controllers are the ones based on a computational intelligence (see e.g. [1,13,22,24–26,28,29,33,35,40–45,57,69,70,73–75]) that includes neural networks (see e.g. [7–11,20,23,31,34]), fuzzy systems (see e.g. [19,21,27,37,47,64–68,76–82]), genetic programming (see e.g. [3–6,18]), etc. The use of such controllers provides good opportunities for: (a) processing of multiple input signals,

© Springer International Publishing AG 2017
L. Rutkowski et al. (Eds.): ICAISC 2017, Part II, LNAI 10246, pp. 292–307, 2017.
DOI: 10.1007/978-3-319-59060-8_27

(b) creation of complex mechanisms for generating control signals (see e.g. [63]),
(c) extraction of interpretable knowledge about the controller (e.g. in the form
of fuzzy rules [17,32,48,49]).

In this paper a new structure of fuzzy-PID controllers with FIR filters and
a method for selecting its parameters is presented. Proposed structure is an
ensemble of PID controller and fuzzy systems controller with FIR filters. The
method for parameters selection requires knowledge about the object and allows
us to automate the process of selecting the controller parameters. The purpose
of using filters is to improve controller efficiency in cases of noise occurrence
in controller input signals. The noise is usually caused by imperfection of used
sensors to monitor the state of the object or low resolution of the signals.

This paper structure is as follows: Sect. 2 contains description of PID controllers, fuzzy systems and FIR filters, Sect. 3 includes description of proposed
structure of the controller and in the Sect. 4 a method for selection of its parameters is presented. In the Sect. 5 obtained results are shown and in the Sect. 6
conclusions are drawn.

2 PID Controllers, Fuzzy Systems and Filters

In this section three components that are used in proposed method are described:
PID controllers, fuzzy systems and filters.

2.1 PID Controllers

As mentioned, the PID controllers are built from elements P, I and D and each
element has a number parameter assigned (K^P, K^I and K^D). The influence of ·
changes of parameters K^P, K^I and K^D on PID controller behavior is presented
in Table 1. The output signal $u(t)$ from the controller is determined as follows:

$$u(t) = K^P \cdot e(t) + K^I \cdot \int_0^t e(t)dt + K^D \cdot \frac{de(t)}{dt}. \tag{1}$$

Formula (1) in discrete form can be written as follows:

$$u(k) = K^P \cdot e(k) + K^I \cdot \sum_{k2=0}^{k2=k} e(k2) + K^D \cdot (e(k) - e(k-1)), \tag{2}$$

where $K^I = \frac{T_s}{T^I}$, T^I stands for integral time constant, $K^D = \frac{T^D}{T_s}$, T^D stands for
differential time constant and T_s stands for control time step.

In the literature many controllers that use combination of P, I and D elements can be found: PI with feed-forward [46], PID with additional low-pass [52],
PID with anti-windup and compensation mechanism [56], pseudo-derivative feedback with feed-forward gain (PDFF) [15] etc. To increase controller efficiency,
controllers that allow us to process multiple inputs (e.g. feedback signals) are
proposed (see e.g. Cascade PID [55], MIMO PID [12]). However, such solutions
usually require structures designed by experts. Furthermore, in these solutions
precise knowledge concerning the problem under consideration is essential.

Table 1. Effects of increasing parameters K^P, K^I and K^D on PID controller [38].

Controller parameter	Controller properties				
	Rise time	Overshoot	Settling time	Steady-state error	Stability
K^P	Decrease	Increase	Small change	Decrease	Degradation
K^I	Decrease	Increase	Increase	Decrease	Degradation
K^D	Minor change	Decrease	decrease	No effect	Improve*

*-if value of the variable is small enough

2.2 Fuzzy Systems

Typical fuzzy system is a multi-input, multi-output system that maps $\mathbf{X} \to \mathbf{Y}$, where $\mathbf{X} \subset \mathbf{R}^n$ and $\mathbf{Y} \subset \mathbf{R}^m$. It works on the basis of fuzzy rules, definition of which depends on the type of system (see e.g. [63]). In case of considered in this paper Mamdani system with singleton fuzzification (see e.g. [63]), the fuzzy rules notation can be defined as follows:

$$R^b : \left(\text{IF } e_1(k) \text{ is } A_1^b \text{ AND ... AND } e_n(k) \text{ is } A_n^b \text{ THEN } u(k) \text{ is } B^b \right), \qquad (3)$$

where b stands for fuzzy rule index ($b = 1, ..., N$), $e_i(k)$ stands for controller input signals ($i = 1, ..., n$), $u(k)$ stands for controller output signal, A_i^b stands for input fuzzy sets determined by the membership functions $\mu_{A_i^b}(\bar{x}_i)$ ($i = 1, ..., n$), B^b stands for output fuzzy sets determined by the membership functions $\mu_{B^b}(\bar{y})$. Output signal $\bar{u}(k)$ calculated with center of area method takes the following form (details can be found in our previous works, see e.g. [51]):

$$\bar{u}(k) = \frac{\sum\limits_{r=1}^{N} \bar{y}_r \cdot \overset{N}{\underset{b=1}{S}} \left\{ T \left\{ T \left\{ \mu_{A_1^b}(e_1(k)), ..., \mu_{A_n^b}(e_n(k)) \right\}, \mu_{B^b}(\bar{y}_r) \right\} \right\}}{\sum\limits_{r=1}^{N} \overset{N}{\underset{b=1}{S}} \left\{ T \left\{ T \left\{ \mu_{A_1^b}(e_1(k)), ..., \mu_{A_n^b}(e_n(k)) \right\}, \mu_{B^b}(\bar{y}_r) \right\} \right\}}, \qquad (4)$$

where $\bar{y}_r$ stands for centers of output fuzzy sets ($r = 1, ..., N$), $S\{\cdot\}$ and $T\{\cdot\}$ stand for triangular norms (t-norms and t-conorms [36]).

The fuzzy systems can be used as controllers (see e.g. [51]) with the following advantages: (a) the possibility of tuning their parameters in evolutionary learning, (b) the ability to adapt to changing work conditions, (c) capabilities of processing multiple input signals, (d) possibilities of extracting interpretable knowledge of the object, etc.

2.3 Filters

In the control process it is important to deal with noise of the signals coming from the object (see e.g. [71]). Such signals usually come from sensors, which have their own noise (for example at level 0.1%–1%) and specified digital resolution. In order to minimize the impact of this noise, filters can be used [71].

One of the most commonly used filters are finite impulse response filters (FIR, see e.g. [2]). The feature of such a filter is that when its input is given as a signal with finite length (in a time domain) the output value has also finite length. This is due to the fact that the FIR filters (as opposed to infinite impulse response-IIR) have no feedbacks.

The purpose of FIR filters, in the practice, is smoothing the signal values. It is achieved by weighted average of signal values from consecutive time steps. The output value of the filter depends on the filter length and its parameters. Although FIR filters are used to improve the quality of control, an inappropriate selection of their parameters can cause significant decrease in the controller efficiency. This is due to the fact (among others) that the filtration adds a side effect-phase delay of the filtered signal, which can result in the excitation of undesirable oscillations in the system.

3   Description of Proposed Controller Structure

The proposed controller structure (FIR+PID+FS) uses FIR filters, PID elements and fuzzy system (in a form (4)) to generate the control signal. The proposed structure is shown in Fig. 1 and its properties can be summed up as follows:

- It contains four layers: filtration layer, PID layer, normalization layer, inference layer (description can be found in further part of this section).
- It can process any number of input signals.
- It uses the possibilities of P, I and D elements.
- It uses FIR filters.
- It can extract interpretable knowledge in a form of fuzzy rules (3).
- Its parameters might be tuned by any gradient or population-based algorithm. In this paper a genetic algorithm was used (see Sect. 4).

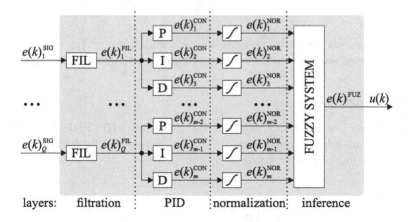

Fig. 1. Proposed controller structure (FIR+PID+FS).

3.1   Filtration Layer

In the filtration layer of the proposed FIR+PID+FS controller (see Fig. 1) FIR filters are used (in the practice any type of filter can be used). The output values from filtration layer are calculated as follows:

$$e_q^{\text{FIL}}(k) = \sum_{l=0}^{s_q^{\text{FIL}}-1} b_{q,l}^{\text{FIL}} \cdot e_q^{\text{SIG}}(k-l),\tag{5}$$

where s_q^{FIL} stands for odd length of the filter (filter has to have a middle element), $q = 1, ..., Q$ stands for input signal index, Q stands for number of input signals, $e_q^{\text{SIG}}(k-l)$ stands for input signal from $k-l$ time step, $b_{q,l}^{\text{FIL}}$ stands for weight of the signal for $t-l$ time step calculated as follows:

$$b_{q,l}^{\text{FIL}} = \begin{cases} \frac{\sin\left(2\cdot\pi\cdot ft_q^{\text{FIL}}\cdot\left(l-0.5\cdot\left(s_q^{\text{FIL}}-1\right)\right)\right)}{\pi\cdot\left(l-0.5\cdot\left(s_q^{\text{FIL}}-1\right)\right)} & \text{for } l \neq 0.5 \cdot \left(s_q^{\text{FIL}}-1\right) \\ 2 \cdot ft_q^{\text{FIL}} & \text{for } l = 0.5 \cdot \left(s_q^{\text{FIL}}-1\right) \end{cases},\tag{6}$$

where ft_q^{FIL} stands for frequency of the filters.

3.2   PID Layer

In the PID layer of the proposed FIR+PID+FS controller (see Fig. 1) for each input signal (of this layer) a three outputs (equivalent to P, I and D elements) are calculated as follows:

$$\begin{cases} e_{3q-2}^{\text{CON}}(k) = K_{3q-2}^{\text{CON}} \cdot e_q^{\text{FIL}}(k) \\ e_{3q-1}^{\text{CON}}(k) = K_{3q-1}^{\text{CON}} \cdot \sum_{k2=0}^{k2=k} e_q^{\text{FIL}}(k2) \\ e_{3q}^{\text{CON}}(k) = K_{3q}^{\text{CON}} \cdot \left(e_q^{\text{FIL}}(k) - e_q^{\text{FIL}}(k-1)\right), \end{cases}\tag{7}$$

where K_i^{CON} stands for parameters of K^P, K^I and K^D ($i = 1, ..., m$), m stands for the number of parameters multiplied by number of input signals ($m = 3 \cdot Q$).

3.3   Normalization Layer

The purpose of normalization layer of the proposed FIR+PID+FS controller (see Fig. 1) is to adjust (normalize) the signal values to the fuzzy layer. This process can be calculated as follows:

$$e_i^{\text{NOR}}(k) = \frac{p3_{d_2(i)}^{\text{NOR}}}{1 + \exp\left(-\left(p1_{d_2(i)}^{\text{NOR}} \cdot e_i^{\text{CON}}(k) - p2_{d_2(i)}^{\text{NOR}}\right)\right)} + p4_{d_2(i)}^{\text{NOR}},\tag{8}$$

where $p1_q^{\mathrm{NOR}}$, $p2_q^{\mathrm{NOR}}$, $p3_q^{\mathrm{NOR}}$, $p4_q^{\mathrm{NOR}}$ stand for parameters of normalization function, calculated as follows:

$$
\begin{cases}
p1_q^{\mathrm{NOR}} = \dfrac{-\log\left(-\frac{y^{\mathrm{CST}}}{y^{\mathrm{CST}}-e_q^{\mathrm{MAX}}+e_q^{\mathrm{MIN}}}\right)+\log\left(-\frac{y^{\mathrm{CST}}-e_q^{\mathrm{MAX}}+e_q^{\mathrm{MIN}}}{y^{\mathrm{CST}}}\right)}{e_q^{\mathrm{MAX}}-e_q^{\mathrm{MIN}}} \\[3mm]
p2_q^{\mathrm{NOR}} = \dfrac{-e_q^{\mathrm{MIN}}\cdot\log\left(-\frac{y^{\mathrm{CST}}}{y^{\mathrm{CST}}-e_q^{\mathrm{MAX}}+e_q^{\mathrm{MIN}}}\right)+e_q^{\mathrm{MAX}}\cdot\log\left(-\frac{y^{\mathrm{CST}}-e_q^{\mathrm{MAX}}+e_q^{\mathrm{MIN}}}{y^{\mathrm{CST}}}\right)}{e_q^{\mathrm{MAX}}-e_q^{\mathrm{MIN}}} \\[3mm]
p3_q^{\mathrm{NOR}} = y^{\mathrm{MAX}} - y^{\mathrm{MIN}} \\[1mm]
p4_q^{\mathrm{NOR}} = y^{\mathrm{MIN}},
\end{cases}
\qquad (9)
$$

where e_q^{MIN} and e_q^{MAX} stand for minimum and maximum of acceptable values of the signals $e_q^{\mathrm{SIG}}(k)$, y^{CST} stands for tiny value resulting from an infinite medium of sigmoid function, y^{MIN} and y^{MAX} stand for lower and upper values of normalization function.

3.4   Inference Layer

In the inference layer of the proposed FIR+PID+FS controller (see Fig. 1) a Mamdani fuzzy system described in Sect. 2.2 is used. Output of this system is calculated according to the Eq. (4). In the proposed system a Gaussian membership functions (see e.g. [63]) were used with the following parameters: centers and sizes of input fuzzy sets $\bar{()}xA_{i,k}^{\mathrm{SYS}}$ and $\sigma A_{i,k}^{\mathrm{SYS}}$), centers (and simultaneously discretization points) and sizes of output fuzzy sets $\bar{()}yB_k^{\mathrm{SYS}}$ and $\sigma B_k^{\mathrm{SYS}}$).

4   Description of Designing Method of Proposed Controller

Method of deigning the proposed controller (FIR+PID+FS) is based on a genetic algorithm. This algorithm belongs to computational intelligence methods that are used mostly to solve optimization problems. Their main characteristic is that they are capable of finding approximate solutions in acceptable time. Genetic algorithms are included in the population-based algorithms (see e.g. [63]), where process of finding solution is based on modification (processing) of the group of individuals (population). Each individual encodes a complete solution to the problem under consideration.

4.1   Encoding of the Individuals

The proposed encoding of the individuals is based on Pittsburgh approach, where single individual $\mathbf{X}_{ch}$ encodes all information about controller parameters. Thus, each individual $\mathbf{X}_{ch}$ takes the following form:

$$
\mathbf{X}_{ch} = \left\{
\begin{array}{c}
s_1^{\mathrm{FIL}}, ..., s_Q^{\mathrm{FIL}}, ft_1^{\mathrm{FIL}}, ..., ft_Q^{\mathrm{FIL}}, \\
K_1^{\mathrm{CON}}, ..., K_n^{\mathrm{CON}}, \\
\bar{x}A_{1,1}^{\mathrm{SYS}}, ..., \bar{x}A_{n,1}^{\mathrm{SYS}}, ..., \bar{x}A_{1,N}^{\mathrm{SYS}}, ..., \bar{x}A_{n,N}^{\mathrm{SYS}}, \\
\sigma A_{1,1}^{\mathrm{SYS}}, ..., \sigma A_{n,1}^{\mathrm{SYS}}, ..., \sigma A_{1,N}^{\mathrm{SYS}}, ..., \sigma A_{n,N}^{\mathrm{SYS}}, \\
\bar{y}B_1^{\mathrm{SYS}}, ..., \bar{y}B_N^{\mathrm{SYS}}, \sigma B_1^{\mathrm{SYS}}, ..., \sigma B_N^{\mathrm{SYS}}
\end{array}
\right\},
\qquad (10)
$$

where $K_1^{\mathrm{CON}}, ..., K_n^{\mathrm{CON}}$ represents all parameters from Eq. (1)-K^{P}, K^{I} and K^{D}. The minimum and maximum values of parameters from Eq. (10) were experimentally set as follows: $s_q^{\mathrm{FIL}} \in [5, 25]$, $ft_q^{\mathrm{FIL}} \in [0.1, 0.5]$, $K_i^{\mathrm{CON}} \in [-1, 1]$, $\bar{x}A_{i,b}^{\mathrm{SYS}} \in [-1, 1]$, $\bar{y}B_b^{\mathrm{SYS}} \in [-1, 1]$, $\sigma A_{i,b}^{\mathrm{SYS}} \in [0.1, 0.3]$, and $\sigma B_b^{\mathrm{SYS}} \in [0.1, 0.3]$.

4.2   Processing of the Individuals

Encoding of the individuals (10) allows us to use any population-based algorithm to find satisfactory set of parameters. In this paper a genetic algorithm is used. In the first step of the algorithm a population of N^{init} individuals is generated randomly. Next, these individuals are evaluated by fitness function that assigns to them the value that defines their adaptation (quality) in the population. In the second step offspring population is generated. For this purpose parent individuals are selected from the base population (by any type of selection mechanism, see e.g. [63]) and on their basis the offspring individuals are formed. To create them a the crossover and mutation genetic operators are used (see e.g. [63]). The offspring individuals are evaluated by fitness function as well. Next, the individuals for the next generation (iteration) of the algorithm are selected. The common approach is the selection of best N^{pop} individuals (according to fitness function values) from both parent and offspring population. In the last step of the algorithm a stop condition is checked. If this condition is met (for example specified number of iterations was achieved) the algorithm stops and the best individual is presented. In the other case the algorithm goes back to the second step. More details can be found in e.g. [53].

5   Simulations

In the simulations a set of different cases was considered (see Table 2). The goal of the simulations was to show: (a) problems that arise from noise in the controller input signals (1A vs 1B, 2A vs 2B, 3A vs 3B), (b) differences in use of proposed controller FIR+PID+FS and fuzzy system controller (FS) (1A vs 2A, 1B vs 2B, 1C vs 2C) and (c) differences between proposed controller FIR+PID+FS and PID+FS controller (2A vs 3A, 2B vs 3B, 2C vs 3C).

5.1   Simulation Problem

In the simulations the Mass-Spring-Dump problem was used [51] with the following criteria used for evaluation: (a) RMSE between expected state of the object and actual state, (b) oscillations of the controller output (OSC), (c) overshooting of the actual object state (OVH). These criteria were integrated in the fitness function used to evaluate the individuals (encoded according to Eq. (10)):

$$\mathrm{ff}\,(\mathbf{X}_{ch}) = w_1 \cdot \mathrm{RMSE} + w_2 \cdot \mathrm{OSC} + w_3 \cdot \mathrm{OVH}, \tag{11}$$

where $w_1 = 1.00$, $w_2 = 0.05$, $w_3 = 0.20$ stands for weights of fitness function components. The smaller values of the fitness function mean better performance of the individual (that encodes the controller). More details about simulation problem can be found in [51].

Table 2. Simulation cases. The noise level was set to 0.2% (in cases with noise).

Case	System	Noise in the learning phase	Noise in the testing phase
1A	FS	-	-
1B		-	yes
1C		yes	yes
2A	PID+FS	-	-
2B		-	yes
2C		yes	yes
3A	FIR+PID+FS	-	-
3B		-	yes
3C		yes	yes

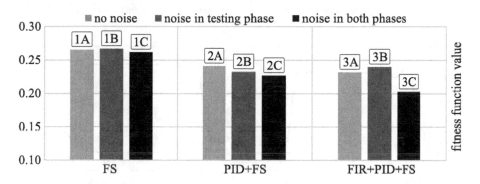

Fig. 2. Simulation results (averaged) for all cases 1A–3C presented in Table 2 (lower values mean better results).

5.2   Simulation Parameters

For the simulations the parameters of controller FIR+PID+FS (see Fig. 1) were set as follows: $N = 3$, $Q = 3$ ($e_1^{SIG}(k) = s^1 - s^*$, $e_2^{SIG}(k) = s^1$, $e_3^{SIG}(k) = s^*$), $T_s = 0.0001s$, type of triangular norms: algebraic.

For the genetic algorithm (described in Sect. 4), the following parameters were set: number of individuals $N^{init} = N^{pop} = 100$, number of iterations: 1000, crossover probability: 0.90, mutation probability: 0.30, mutation range: 0.15, selection method: roulette wheel, number of simulations for each case: 50 (the results were averaged).

5.3   Simulation Results

The obtained results are presented in Table 3 and in Fig. 2. The comparison of the results with other methods is presented in Table 4. Moreover, the best obtained controllers are presented on Fig. 3.

Table 3. Simulation results (the best results for each case were marked in bold).

System	Case	Average				The best (by ff)			
		ff	RMSE	OSC	OVH	ff	RMSE	OSC	OVH
FS	1A	0.2655	0.1812	**1.0213**	0.1661	**0.2440**	0.1608	**0.9891**	**0.1689**
	1B	0.2668	0.1852	1.0292	**0.1508**	0.2452	0.1612	1.0000	0.1700
	1C	**0.2618**	**0.1709**	1.0370	0.1951	0.2571	**0.1568**	1.0380	0.2420
	Avg.	0.2647	0.1791	1.0292	0.1707	0.2488	0.1596	1.0090	0.1936
PID+FS	2A	0.2413	0.0913	2.2138	0.1964	0.3201	**0.0566**	4.1825	0.2717
	2B	0.2326	0.1001	1.9022	**0.1871**	0.3489	0.0778	4.5653	0.2144
	2C	**0.2270**	**0.0907**	**1.9628**	0.1907	**0.1557**	0.0784	**1.0198**	**0.1314**
	Avg.	0.2336	0.0940	2.0263	0.1914	0.2749	0.0709	3.2559	0.2058
FIR+PID+FS	3A	0.2320	**0.0997**	1.6702	0.2438	0.3612	**0.0647**	5.0096	**0.2302**
	3B	0.2399	0.1121	1.5815	0.2437	0.3642	0.0674	5.0102	0.2313
	3C	**0.2026**	0.1067	**1.3509**	**0.1422**	**0.1707**	0.0723	**1.0040**	0.2407
	avg	**0.2248**	0.1061	1.5342	0.2099	0.2987	0.0681	3.6746	0.2341

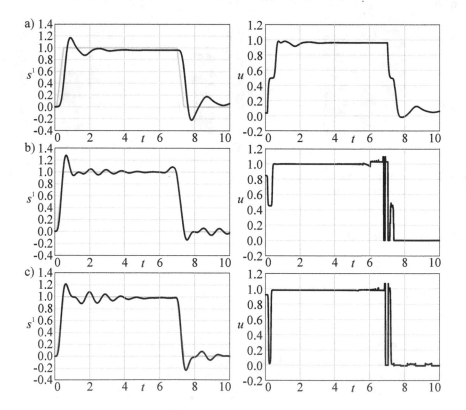

Fig. 3. The best obtained simulation results for case: (a) 1A–1C, (b) 2A–2C, (c) 3A–3C. The gray line stands for expected state of the object (signal s^*).

Table 4. Comparison of simulation results with other methods.

system	Average				The best (by RMSE)			
	ff	RMSE	OSC	OVH	ff	RMSE	OSC	OVH
PID cascade [50]	-	-	-	-	0.6902	0.0510	12.3150	0.1170
PID+FS [51]	0.2895	0.1280	2.4500	0.1950	0.2741	0.0900	2.6570	0.2560
FS (case 1A)	0.2655	0.1812	1.0213	0.1661	0.2440	0.1608	0.9891	0.1689
PID+FS (case 2A)	0.2413	0.0913	2.2138	0.1964	0.3201	0.0566	4.1825	0.2717
FIR+PID+FS (case 3A)	**0.2320**	0.0997	1.6702	0.2438	0.3612	0.0647	5.0096	0.2302

5.4 Simulation Conclusions

Simulation conclusions can be summed up as follows:

- Considering noise of controller input signals only in testing phase (case 1B, 2B and 3B) caused decrease of the controller accuracy in comparison with cases 1A, 2A i 3A (RMSE was degraded by 2%, 10% and 12%)-see Table 3.
- Considering noise of controller input signals in testing and learning phases (case 1C, 2C and 3C) caused increase of the controller accuracy in comparison of cases 1B, 2B and 3B (RMSE was improved by 8%, 10% i 5%)-see Table 3.
- The use of PID+FS controller caused increase of the controller overall efficiency (fitness value improved by 13%), RMSE improved twice with acceptable increase of oscillations-see Table 3.
- The use of FIR+PID+FS controller caused increase of the controller overall efficiency (fitness function value improved by another 4%) with small improvement of RMSE (1%) and decrease of oscillations (by 25%)-see Table 3.
- The achieved results for FIR+PID+FS controller are better than results obtained by other methods (see Table 4).

6 Conclusions

In this paper a new structure of fuzzy-PID controllers with FIR filters was proposed. Moreover, a method for tuning the controller parameters based on a genetic algorithm using proposed encoding of the controller, was presented. The controllers obtained in simulations are characterized by high precision, acceptable level of oscillations and low overshoot. The proposed solution can be particularly important in solving problems with noise of the object feedback signals.

The further planned studies in the field of designing controllers includes, among the others: (a) the development of a method for automatic design of the structure and parameters of the controller and its filters, (b) the development of dedicated evaluation criteria for readability of knowledge accumulated in the structure of the controller.

Acknowledgment. The project was financed by the National Science Centre (Poland) on the basis of the decision number DEC-2012/05/B/ST7/02138.

References

1. Abbas, J.: The bipolar choquet integrals based on ternary-element sets. J. Artif. Intell. Soft Comput. Res. **6**(1), 13–21 (2016)
2. Alia, M.A.K., Younes, T.M., Alsabbah, S.A.: A design of a PID self-tuning controller using LabVIEW. J. Softw. Eng. Appl. **4**, 161–171 (2011)
3. Bartczuk, Ł., Przybył, A., Koprinkova-Hristova, P.: New method for non-linear correction modelling of dynamic objects with genetic programming. In: Rutkowski, L., Korytkowski, M., Scherer, R., Tadeusiewicz, R., Zadeh, L.A., Zurada, J.M. (eds.) ICAISC 2015. LNCS, vol. 9120, pp. 318–329. Springer, Cham (2015). doi:10. 1007/978-3-319-19369-4_29
4. Bartczuk, Ł.: Gene expression programming in correction modelling of nonlinear dynamic objects. In: Borzemski, L., Grzech, A., Świątek, J., Wilimowska, Z. (eds.) Information Systems Architecture and Technology: Proceedings of 36th International Conference on Information Systems Architecture and Technology – ISAT 2015 – Part I. AISC, vol. 429, pp. 125–134. Springer, Cham (2016). doi:10.1007/ 978-3-319-28555-9_11
5. Bartczuk, Ł., Łapa, K., Koprinkova-Hristova, P.: A new method for generating of fuzzy rules for the nonlinear modelling based on semantic genetic programming. In: Rutkowski, L., Korytkowski, M., Scherer, R., Tadeusiewicz, R., Zadeh, L.A., Zurada, J.M. (eds.) ICAISC 2016. LNCS, vol. 9693, pp. 262–278. Springer, Cham (2016). doi:10.1007/978-3-319-39384-1_23
6. Bartczuk, Ł., Galushkin, A.I.: A new method for generating nonlinear correction models of dynamic objects based on semantic genetic programming. In: Rutkowski, L., Korytkowski, M., Scherer, R., Tadeusiewicz, R., Zadeh, L.A., Zurada, J.M. (eds.) ICAISC 2016. LNCS, vol. 9693, pp. 249–261. Springer, Cham (2016). doi:10. 1007/978-3-319-39384-1_22
7. Bilski, J., Rutkowski, L.: Numerically robust learning algorithms for feed forward neural networks. In: Advances in Soft Computing-Neural Networks and Soft Computing, pp. 149–154. Physica-Verlag, A Springer-Verlag Company (2003)
8. Bilski, J., Smoląg, J.: Parallel realisation of the recurrent RTRN neural network learning. In: Rutkowski, L., Tadeusiewicz, R., Zadeh, L.A., Zurada, J.M. (eds.) ICAISC 2008. LNCS, vol. 5097, pp. 11–16. Springer, Heidelberg (2008). doi:10. 1007/978-3-540-69731-2_2
9. Bilski, J., Smoląg, J.: Parallel realisation of the recurrent elman neural network learning. In: Rutkowski, L., Scherer, R., Tadeusiewicz, R., Zadeh, L.A., Zurada, J.M. (eds.) ICAISC 2010. LNCS, vol. 6114, pp. 19–25. Springer, Heidelberg (2010). doi:10.1007/978-3-642-13232-2_3
10. Bilski, J., Smoląg, J.: Parallel realisation of the recurrent multi layer perceptron learning. In: Rutkowski, L., Korytkowski, M., Scherer, R., Tadeusiewicz, R., Zadeh, L.A., Zurada, J.M. (eds.) ICAISC 2012. LNCS, vol. 7267, pp. 12–20. Springer, Heidelberg (2012). doi:10.1007/978-3-642-29347-4_2
11. Bilski, J., Smoląg, J., Galushkin, A.I.: The parallel approach to the conjugate gradient learning algorithm for the feedforward neural networks. In: Rutkowski, L., Korytkowski, M., Scherer, R., Tadeusiewicz, R., Zadeh, L.A., Zurada, J.M. (eds.) ICAISC 2014. LNCS, vol. 8467, pp. 12–21. Springer, Cham (2014). doi:10. 1007/978-3-319-07173-2_2
12. Boyd, S., Hast, M., Åström, K.J.: MIMO PID tuning via iterated LMI restriction. Int. J. Robust Nonlinear Control **26**, 1718–1731 (2016)

13. Brester, C., Semenkin, E., Sidorov, M.: Multi-objective heuristic feature selection for speech-based multilingual emotion recognition. J. Artif. Intell. Soft Comput. Res. **6**(4), 243–253 (2016)
14. Chen, Q., Abercrombie, R.K., Sheldon, F.T.: Risk assessment for industrial control systems quantifying availability using mean failure cost (MFC). J. Artif. Intell. Soft Comput. Res. **5**(3), 205–220 (2015)
15. Cheng, S., Li, C.W.: Fuzzy PDFF-IIR controller for PMSM drive systems. Control Eng. Pract. **19**, 828–835 (2011)
16. Cierniak, R., Rutkowski, L.: On image compression by competitive neural networks and optimal linear predictors. Sig. Process. Image Commun. **156**, 559–565 (2000)
17. Cpałka, K.: Design of Interpretable Fuzzy Systems. Springer, Heidelberg (2017)
18. Cpałka, K., Łapa, K., Przybył, A.: A new approach to design of control systems using genetic programming. Inf. Technol. Control **44**(4), 433–442 (2015)
19. Cpałka, K., Rebrova, O., Nowicki, R., Rutkowski, L.: On design of flexible neuro-fuzzy systems for nonlinear modelling. Int. J. Gen Syst **42**(6), 706–720 (2013)
20. Cpałka, K., Rutkowski, L.: Flexible takagi-sugeno, fuzzy systems, neural networks. In: Proceedings of the 2005 IEEE International Joint Conference on IJCNN 2005, vol. 3, pp. 1764–1769 (2005)
21. Cpałka, K., Zalasiński, M., Rutkowski, L.: A new algorithm for identity verification based on the analysis of a handwritten dynamic signature. Appl. Soft Comput. **43**, 47–56 (2016)
22. Duda, P., Jaworski, M., Pietruczuk, L.: On pre-processing algorithms for data stream. In: Rutkowski, L., Korytkowski, M., Scherer, R., Tadeusiewicz, R., Zadeh, L.A., Zurada, J.M. (eds.) ICAISC 2012. LNCS, vol. 7268, pp. 56–63. Springer, Heidelberg (2012). doi:10.1007/978-3-642-29350-4_7
23. Er, M.J., Duda, P.: On the weak convergence of the orthogonal series-type kernel regresion neural networks in a non-stationary environment. In: Wyrzykowski, R., Dongarra, J., Karczewski, K., Waśniewski, J. (eds.) PPAM 2011. LNCS, vol. 7203, pp. 443–450. Springer, Heidelberg (2012). doi:10.1007/978-3-642-31464-3_45
24. Dziwiński, P., Avedyan, E.D.: A new approach to nonlinear modeling based on significant operating points detection. In: Rutkowski, L., Korytkowski, M., Scherer, R., Tadeusiewicz, R., Zadeh, L.A., Zurada, J.M. (eds.) ICAISC 2015. LNCS, vol. 9120, pp. 364–378. Springer, Cham (2015). doi:10.1007/978-3-319-19369-4_33
25. Dziwiński, P., Avedyan, E.D.: A new approach for using the fuzzy decision trees for the detection of the significant operating points in the nonlinear modeling. In: Rutkowski, L., Korytkowski, M., Scherer, R., Tadeusiewicz, R., Zadeh, L.A., Zurada, J.M. (eds.) ICAISC 2016. LNCS, vol. 9693, pp. 279–292. Springer, Cham (2016). doi:10.1007/978-3-319-39384-1_24
26. Gabryel, M.: A bag-of-features algorithm for applications using a NoSQL database. In: Dregvaite, G., Damasevicius, R. (eds.) ICIST 2016. CCIS, vol. 639, pp. 332–343. Springer, Cham (2016). doi:10.1007/978-3-319-46254-7_26
27. Gabryel, M., Cpałka, K., Rutkowski, L.: Evolutionary strategies for learning of neuro-fuzzy systems. In: Proceedings of the I Workshop on Genetic Fuzzy Systems, Granada, pp. 119–123 (2005)
28. Gabryel, M., Grycuk, R., Korytkowski, M., Holotyak, T.: Image indexing and retrieval using GSOM algorithm. In: Rutkowski, L., Korytkowski, M., Scherer, R., Tadeusiewicz, R., Zadeh, L.A., Zurada, J.M. (eds.) ICAISC 2015. LNCS, vol. 9119, pp. 706–714. Springer, Cham (2015). doi:10.1007/978-3-319-19324-3_63

29. Gabryel, M.: The bag-of-features algorithm for practical applications using the MySQL database. In: Rutkowski, L., Korytkowski, M., Scherer, R., Tadeusiewicz, R., Zadeh, L.A., Zurada, J.M. (eds.) ICAISC 2016. LNCS, vol. 9693, pp. 635–646. Springer, Cham (2016). doi:10.1007/978-3-319-39384-1_56

30. Gałkowski, T., Rutkowski, L.: Nonparametric fitting of multivariate functions. IEEE Trans. Autom. Control **31**(8), 785–787 (1986)

31. Hagan, M.T., Demuth, H.B., Jesús, O.D.: An introduction to the use of neural networks in control systems. Int. J. Robust Nonlinear Control **12**(11), 959–985 (2002)

32. Hayashi, Y., Tanaka, Y., Takagi, T., Saito, T., Iiduka, H., Kikuchi, H., Bologna, G.: Recursive-rule extraction algorithm with J48graft and applications to generating credit scores. J. Artif. Intell. Soft Comput. Res. **6**(1), 35–44 (2016)

33. Held, P., Dockhorn, A., Kruse, R.: On merging and dividing social graphs. J. Artif. Intell. Soft Comput. Res. **5**(1), 23–49 (2015)

34. Jaworski, M., Er, M.J., Pietruczuk, L.: On the application of the parzen-type kernel regression neural network and order statistics for learning in a non-stationary environment. In: Rutkowski, L., Korytkowski, M., Scherer, R., Tadeusiewicz, R., Zadeh, L.A., Zurada, J.M. (eds.) ICAISC 2012. LNCS, vol. 7267, pp. 90–98. Springer, Heidelberg (2012). doi:10.1007/978-3-642-29347-4_11

35. Kapustianyk, V., Shchur, Y., Kityk, I., Rudyk, V., Lach, G., Laskowski, Ł., Tkaczyk, S., Swiatek, J., Davydov, V.: Resonance dielectric dispersion of TEA-CoCl2Br 2 nanocrystals incorporated into the PMMA matrix. J. Phys. Condens. Matter **20**(36), 365215–365223 (2008). IOP Publishing

36. Klement, E.P., Mesiar, R., Pap, E.: Triangular Norms. Springer, Heidelberg (2000)

37. Korytkowski, M., Rutkowski, L., Scherer, R.: Fast image classification, by boosting fuzzy classifiers. Inf. Sci. **327**, 175–182 (2016)

38. Kurien, M.: Overview of different approach of PID controller tuning. Int. J. Res. Advent Technol. **2**(1), 167–175 (2014)

39. Lan, K., Sekiyama, K.: Autonomous viewpoint selection of robot based on aesthetic evaluation of a scene. J. Artif. Intelli. Soft Comput. Res. **6**(4), 255–265 (2016)

40. Laskowska, M., Laskowski, Ł., Jelonkiewicz, J.: SBA-15 mesoporous silica activated by metal ions-verification of molecular structure on the basis of Raman spectroscopy supported by numerical simulations. J. Mol. Struct. **1100**, 21–26 (2015). Elsevier

41. Laskowski, Ł.: A novel hybrid-maximum neural network in stereo-matching process. Neural Comput. Appl. **23**(7–8), 2435–2450 (2013). Springer

42. Laskowski, Ł., Laskowska, M., Jelonkiewicz, J., Boullanger, A.: Spin-glass implementation of a hopfield neural structure. In: Rutkowski, L., Korytkowski, M., Scherer, R., Tadeusiewicz, R., Zadeh, L.A., Zurada, J.M. (eds.) ICAISC 2014. LNCS, vol. 8467, pp. 89–96. Springer, Cham (2014). doi:10.1007/978-3-319-07173-2_9

43. Laskowski, Ł., Laskowska, M., Jelonkiewicz, J., Boullanger, A.: Molecular approach to hopfield neural network. In: Rutkowski, L., Korytkowski, M., Scherer, R., Tadeusiewicz, R., Zadeh, L.A., Zurada, J.M. (eds.) ICAISC 2015. LNCS, vol. 9119, pp. 72–78. Springer, Cham (2015). doi:10.1007/978-3-319-19324-3_7

44. Laskowski, Ł., Laskowska, M., Jelonkiewicz, J., Dulski, M., Wojtyniak, M., Fitta, M., Balanda, M.: SBA-15 mesoporous silica free-standing thin films containing copper ions bounded via propyl phosphonate units-preparation and characterization. J. Solid State Chem. **241**, 143–151 (2016). Elsevier

45. Laskowski, Ł., Laskowska, M., Jelonkiewicz, J., Gałkowski, T., Pawlik, P., Piech, H., Doskocz, M.: Iron doped SBA-15 mesoporous silica studied by Mössbauer spectroscopy. J. Nanomaterials **2016**, 1–6 (2016). Hindawi Publishing Corp
46. Leva, A., Papadopoulos, A.V.: Tuning of event-based industrial controllers with simple stability guarantees. J. Process Control **23**, 1251–1260 (2013)
47. Li, X., Er, M.J., Lim, B.S., Zhou, J.H., Gan, O.P., Rutkowski, L.: Fuzzy regression modeling for tool performance prediction and degradation detection. Int. J. Neural Syst. **2005**, 405–419 (2010)
48. Łapa, K., Cpałka, K., Wang, L.: New method for design of fuzzy systems for nonlinear modelling using different criteria of interpretability. In: Rutkowski, L., Korytkowski, M., Scherer, R., Tadeusiewicz, R., Zadeh, L.A., Zurada, J.M. (eds.) ICAISC 2014. LNCS, vol. 8467, pp. 217–232. Springer, Cham (2014). doi:10.1007/978-3-319-07173-2_20
49. Łapa, K., Przybył, A., Cpałka, K.: A new approach to designing interpretable models of dynamic systems. Artif. Intell. Soft Comput. **7895**, 523–534 (2013)
50. Łapa, K., Szczypta, J., Venkatesan, R.: Aspects of structure and parameters selection of control systems using selected multi-population algorithms. Artif. Intell. Soft Comput. **9120**, 247–260 (2015)
51. Łapa, K., Szczypta, J., Saito, T.: Aspects of evolutionary construction of new flexible PID-fuzzy controller. Artif. Intell. Soft Comput. **9692**, 450–464 (2016)
52. Maggio, M., Bonvini, M., Leva, A.: The PID+p controller structure and its contextual autotuning. J. Process Control **22**, 1237–1245 (2012)
53. Melanie, M.: An Introduction to Genetic Algorithms. MIT Press, Massachusetts (1999)
54. Nobukawa, S., Nishimura, H., Yamanishi, T., Liu, J.: Chaotic states induced by resetting process in izhikevich neuron model. J. Artif. Intell. Soft Comput. Res. **5**(2), 109–119 (2015)
55. Pamar, K., Arvapalli, R., Sadhu, Y., Viswaraju, S.: Cascaded PID controller design for heating furnace temperature control. IOSR J. Electr. Commun. Eng. **5**(3), 76–83 (2013)
56. Ribića, A.I., Mataušek, M.R.: A dead-time compensating PID controller structure and robust tuning. J. Process Control **22**, 1340–1349 (2012)
57. Rivero, C.R., Pucheta, J., Laboret, S., Sauchelli, V., Patio, D.: Energy associated tuning method for short-term series forecasting by complete and incomplete datasets. J. Artif. Intell. Soft Comput. Res. **7**(1), 5–16 (2017)
58. Rutkowski, L.: On-line identification of time-varying systems by nonparametric techniques. IEEE Trans. Autom. Control **27**(1), 228–230 (1982)
59. Rutkowski, L.: On nonparametric identification with prediction of time-varying systems. IEEE Trans. Autom. Control **29**(1), 58–60 (1984)
60. Rutkowski, L.: Nonparametric identification of quasi-stationary systems. Syst. Control Lett. **6**(1), 33–35 (1985)
61. Rutkowski, L.: A general approach for nonparametric fitting of functions and their derivatives with applications to linear circuits identification. IEEE Trans. Circ. Syst. **33**(8), 812–818 (1986)
62. Rutkowski, L.: Adaptive probabilistic neural networks for pattern classification in time-varying environment. IEEE Trans. Neural Netw. **15**(4), 811–827 (2004)
63. Rutkowski, L.: Computational Intelligence. Springer, Heidelberg (2008)
64. Rutkowski, L., Cpałka, K.: Compromise approach to neuro-fuzzy systems. In: Proceedings of the 2nd Euro-International Symposium on Computation Intelligence, Frontiers in Artificial Intelligence and Applications, vol. 76, pp. 85–90 (2002)

65. Rutkowski, L., Cpałka, K.:, Flexible weighted neuro-fuzzy systems. In: Proceedings of the 9th International Conference on Neural Information Processing (ICONIP 2002), Orchid Country Club, Singapore, November 18–22, 2002, CD (2002)

66. Rutkowski, L., Cpałka, K.: Neuro-fuzzy systems derived from quasi-triangular norms. In: Proceedings of the IEEE International Conference on Fuzzy Systems, Budapest, July 26–29, vol. 2, pp. 1031–1036 (2004)

67. Rutkowski, L., Przybył, A., Cpałka, K.: Novel Online Speed Profile Generation for Industrial Machine Tool Based on Flexible Neuro-Fuzzy Approximation. IEEE Trans. Ind. Electron. **59**(2), 1238–1247 (2012)

68. Rutkowski, L., Przybył, A., Cpałka, K., Er, M.J.: Online speed profile generation for industrial machine tool based on neuro-fuzzy approach. In: Rutkowski, L., Scherer, R., Tadeusiewicz, R., Zadeh, L.A., Zurada, J.M. (eds.) ICAISC 2010. LNCS, vol. 6114, pp. 645–650. Springer, Heidelberg (2010). doi:10.1007/978-3-642-13232-2_79

69. Saitoh, D., Hara, K.: Mutual learning using nonlinear perceptron. J. Artif. Intell. Soft Comput. Res. **5**(1), 71–77 (2015)

70. Sakurai, S., Nishizawa, M., Soft, C.R.: A new approach for discovering top-k sequential patterns based on the variety of items. J. Artif. Intell. Soft Comput. Res. **5**(2), 141–153 (2015)

71. Segovia, R.V., Hägglund, T., Aström, K.J.: Noise filtering in PI and PID control. In: American Control Conference, pp. 1763–1770 (2013)

72. Szczypta, J., Łapa, K., Shao, Z.: Aspects of the selection of the structure and parameters of controllers using selected population based algorithms. In: Rutkowski, L., Korytkowski, M., Scherer, R., Tadeusiewicz, R., Zadeh, L.A., Zurada, J.M. (eds.) ICAISC 2014. LNCS, vol. 8467, pp. 440–454. Springer, Cham (2014). doi:10.1007/978-3-319-07173-2_38

73. Tabellout, M., Kassiba, A., Tkaczyk, S., Laskowski, Ł., Świątek, J.: Dielectric and EPR investigations of stoichiometry and interface effects in silicon carbide nanoparticles. J. Phys. Condens. Matter **18**(4), 11–43 (2006). IOP Publishing

74. Tezuka, T., Claramunt, C.: Kernel analysis for estimating the connectivity of a network with event sequences. J. Artif. Intell. Soft Comput. Res. **7**(1), 17–31 (2017)

75. Yamamoto, Y., Yoshikawa, T., Furuhashi, T.: Improvement of performance of Japanese P300 speller by using second display. J. Artif. Intell. Soft Comput. Res. **5**(3), 221–226 (2015)

76. Zalasiński, M., Cpałka, K.: A new method of on-line signature verification using a flexible fuzzy one-class classifier, pp. 38–53. Academic Publishing House EXIT (2011)

77. Zalasiński, M., Cpałka, K.: Novel algorithm for the on-line signature verification using selected discretization points groups. In: Rutkowski, L., Korytkowski, M., Scherer, R., Tadeusiewicz, R., Zadeh, L.A., Zurada, J.M. (eds.) ICAISC 2013. LNCS, vol. 7894, pp. 493–502. Springer, Heidelberg (2013). doi:10.1007/978-3-642-38658-9_44

78. Zalasiński, M., Cpałka, K., Er, M.J.: New method for dynamic signature verification using hybrid partitioning. In: Rutkowski, L., Korytkowski, M., Scherer, R., Tadeusiewicz, R., Zadeh, L.A., Zurada, J.M. (eds.) ICAISC 2014. LNCS, vol. 8468, pp. 216–230. Springer, Cham (2014). doi:10.1007/978-3-319-07176-3_20

79. Zalasiński, M., Cpałka, K., Hayashi, Y.: New method for dynamic signature verification based on global features. In: Rutkowski, L., Korytkowski, M., Scherer, R., Tadeusiewicz, R., Zadeh, L.A., Zurada, J.M. (eds.) ICAISC 2014. LNCS, vol. 8468, pp. 231–245. Springer, Cham (2014). doi:10.1007/978-3-319-07176-3_21

80. Zalasiński, M., Cpałka, K., Hayashi, Y.: A new approach to the dynamic signature verification aimed at minimizing the number of global features. In: Rutkowski, L., Korytkowski, M., Scherer, R., Tadeusiewicz, R., Zadeh, L.A., Zurada, J.M. (eds.) ICAISC 2016. LNCS, vol. 9693, pp. 218–231. Springer, Cham (2016). doi:10.1007/978-3-319-39384-1_20

81. Zalasiński, M., Cpałka, K., Rakus-Andersson, E.: An idea of the dynamic signature verification based on a hybrid approach. In: Rutkowski, L., Korytkowski, M., Scherer, R., Tadeusiewicz, R., Zadeh, L.A., Zurada, J.M. (eds.) ICAISC 2016. LNCS, vol. 9693, pp. 232–246. Springer, Cham (2016). doi:10.1007/978-3-319-39384-1_21

82. Zalasiński, M., Łapa, K., Cpałka, K.: New algorithm for evolutionary selection of the dynamic signature global features. In: Rutkowski, L., Korytkowski, M., Scherer, R., Tadeusiewicz, R., Zadeh, L.A., Zurada, J.M. (eds.) ICAISC 2013. LNCS, vol. 7895, pp. 113–121. Springer, Heidelberg (2013). doi:10.1007/978-3-642-38610-7_11

The Use of Heterogeneous Cellular Automata to Study the Capacity of the Roundabout

Krzysztof Małecki[✉]

West Pomeranian University of Technology,
Zolnierska Str. 52, 71-210 Szczecin, Poland
kmalecki@wi.zut.edu.pl

Abstract. This article presents a research study analysing the impact of changing the roundabout island diameter on the roundabout capacity. The study was based on the developed Cellular Automata Model and the implemented simulation system. The developed CA Model takes into account various types of vehicles (cars, trucks and motorcycles) and various sizes of roundabouts; also, it reflects the actual technical conditions of those vehicles (acceleration and braking depending on the vehicle dimensions and function, as well as driving on the roundabout with different speeds that are adequate to the vehicle size). The study was based on the example of a two-lane roundabout with four two-lane feeder roads.

Keywords: Capacity of roundabout · Cellular Automata (CA) · CA roundabout model · Roundabout traffic simulation

1 Introduction

A roundabout is an intersection where traffic moves in a circle around a central island. Any vehicles approaching a roundabout must give way to the vehicles moving around it. It was found that roundabouts decrease the number of collision points at an intersection [1], and studies carried out in the USA have shown that following introduction of roundabouts the number of accidents fell by 29% and the number of the injured by 81% [2]. As determined in [3] on the basis of a simulation, roundabouts are particularly recommended in places where the traffic is equally distributed among all the feeder roads. Factors affecting the capacity of roundabouts are specified in [4], also, it was found that a three-lane roundabout does not contribute to increasing the capacity compared to a two-lane roundabout.

The aim of this article is to investigate a roundabout capacity in relation to its diameter. To achieve this aim, a CA model was developed, and a computer simulation was run using software developed for the purposes of road traffic simulation.

2 Related Work

Due to the complexity of the issue in question and its stochastic nature, road traffic analysing and modelling is a challenge currently considered by many research centres. Not wanting to expose drivers to any inconveniences connected with attempts to adjust

© Springer International Publishing AG 2017
L. Rutkowski et al. (Eds.): ICAISC 2017, Part II, LNAI 10246, pp. 308–317, 2017.
DOI: 10.1007/978-3-319-59060-8_28

a specific area (e.g. an intersection or a few intersections located near each other), road administrators and researchers resort to various kinds of simulation solutions. The available literature describes mathematical models [5, 6], methods and algorithms [7], large simulation software systems: VISSIM, AIMSUN, SUMO, applications [8, 9], hardware simulators [10–12] and hardware-software simulators [13], as well as various analyses [14–16].

Due to their stochastic nature, cellular automata are perfect for traffic flow modelling. The Nagel–Schreckenberg (N-Sch) model is one of the basic models of cellular automata to simulate the cars' movement. It was developed in 1992 by Nagel and Schreckenberg [5]. This model describes the one-lane car's movement and is the basis for testing various traffic scenarios [17]. Another proven model is the model developed by Biham [18]. This is a simple cellular automaton model showing the traffic into two intersecting directions. Each array cell can be occupied with vehicle traveling in one of two directions (north or east). The vehicle moves by one cell to the chosen direction, when it is empty. In another case, it remains in its position. The most important model showing traffic within the intersection is the model by Chowdhury and Schadschneider [19]. The authors model the traffic on one-way single-lane roads. The development of this model to the version for two-lane and two-way roads is presented in [9]. The new model expands the original idea of the intersections with the mechanism of induction loops activating traffic light to eliminate congestion (keeping smooth movement at the intersection). An interesting modification of the N-Sch model is presented in [20]. The author has underlined that the study was intended for urban traffic flow simulation, involving vehicles of various sizes.

The traffic rules for roundabouts are addressed in [21], single-lane roundabout modelling in [22–24], and multi-lane roundabout modelling in [25–28]. For the purposes of this publication, an original model for multi-lane roundabouts was used, which applies the current road traffic regulations and makes it possible to study roundabout capacity in various aspects, including the impact of a roundabout dimensions on its capacity.

3 CA Model of Roundabout

In simple terms, a cellular automaton is a multi-dimensional (e.g. 2-dimensional) grid consisting of cells. Each cell may switch into one of many states, depending on the states of the adjacent cells. Although behaviour of each cell is deterministic and depends only on its neighbours, the behaviour of the automaton as a whole (depending on the initial state of the cells and their modification rules) may be very complicated and hard to foresee. For example, it is possible to observe movement of some structures (groups) of cells, their replication, division, destruction, increase in complexity, reciprocal interactions of organised systems of cells, etc. Thus, there is emergence of qualitatively new, unexpected behaviours, which arise as a result of known interactions between simple elements.

The model applied in this study was based on the Nagel-Schreckenberg model (N-Sch) [5] and its modification, i.e. "leading head algorithm" [20]. In the N-Sch model, the length of the automaton single cell was assumed to correspond to 7.5 m of the road, which represents an average length of a car together with its surrounding space. As the model was developed for the purposes of motorway traffic modelling, the velocity unit in the N-Sch model corresponds to the actual speed of 27 km/h. The Hartman's modification, in turn, pertained to urban speed traffic, which is relevant for the model presented in this paper.

Basic models discussed above relate to traffic on a straight one-way road. Based on these models, author developed a model of traffic on the multi-lane roundabout with multiple entrance and exit roads. In order to discuss the details, the following structure was assumed: a two-lane roundabout with four entrance and exit roads, where each of the feeder roads has two lanes. The model describes right-hand traffic. The lanes on the roundabout are numbered from 0 (inner lane) to $n - 1$, where n corresponds to the total number of lanes on the roundabout. The feeder road consists of the entrance road and the exit road. The right-hand lane of the entrance road and the right-hand lane of the exit road are to first lanes to be numbered starting from zero. The right-hand lane is the lane on the right-hand side from the point of view of a driver driving on the road in the given direction. The lane numbering is shown in Fig. 1.

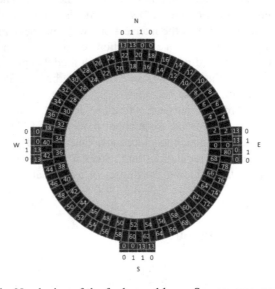

Fig. 1. Numbering of the feeder road lanes. Source: own research.

Due to different lengths of the roundabout lanes, it is impossible to divide them into the same number of cells. Assuming that the cell length is constant and amounts to 2.5 m, as in [20], the outer lanes will have more cells that the inner ones, reflecting the actual differences in the lanes lengths. The lane length, taking its inner (shorter) edge, may be determined using the circumference formula:

$$l_{lane} = 2 * \pi * (r_{island} + n * w_{lane}) \tag{1}$$

where: l_{lane} – lane length, r_{island} – island radius, n – subsequent lane number, starting from the inner lane numbered 0, w_{lane} – single lane width.

The developed model posits that the width of each lane w_{lane} is the same. By parameterising the model in relation to the actual size of the roundabout, the number of cells in each lane is determined in relation to its length:

$$n_{cells} = \left\lfloor \frac{l_{lane}}{2.5} \right\rfloor \tag{2}$$

The resulting value is rounded down, as the number of cells is an integral number, and there may not be more cells physically squeezed into the lane. Assuming that the radius of a sample roundabout amounts to 28 m, and the lane width is 4.5 m, the number of cells on subsequent lanes will then be, respectively:

$$n_{cells} = \left\lfloor \frac{2 * \pi * (28 \text{ m} + 0 * 4.5)}{2.5 \text{ m}} \right\rfloor = \lfloor 70.37 \rfloor = 70 \tag{3}$$

$$n_{cells} = \left\lfloor \frac{2 * \pi * (28 \text{ m} + 1 * 4.5)}{2.5 \text{ m}} \right\rfloor = \lfloor 81.68 \rfloor = 81 \tag{4}$$

The outer lane will be divided into 81 cells, and the inner lane – into 70 cells. Dividing the roundabout into cells starts on the diameter line, from the feeder road on the eastern side. To simplify the model presentation it was assumed that all the feeder roads have the same length equalling 35 m, which makes it possible to calculate, using the previous formula, that they consist of $n_{cells} = \left\lfloor \frac{35 \text{ m}}{2.5 \text{ m}} \right\rfloor = 14$ cells. The feeder roads cells are numbered in compliance with the traffic direction in the given lane. Thus, the entrance roads are numbered from the outer side, and the exit roads from the inner side, as shown in Fig. 2.

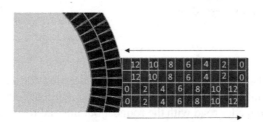

Fig. 2. Dividing the feeder road lanes into cells. Source: own research.

The structure of a roundabout is aimed at forcing vehicles to decelerate, which increases traffic safety. However, different types of vehicles are able to enter or exit a roundabout at different speeds. A car may take a turn with a greater velocity than a truck, which is mainly due to their dimensions, weight and dynamics. Two kinds of maximum speeds were assumed for vehicles in urban traffic:

- the velocity at which a vehicle may move on the road in the absence of other vehicles in front of it – v_{max},
- the velocity at which a vehicle may safely enter or exit an intersection v_{turn_max} (Table 1).

Table 1. Maximum velocities established for various vehicles. Source: own research.

Vehicle type	v_{max} [cells/s]	v_{turn_max} [cells/s]
Motorcycle, car	5	2
Van	5	2
Minibus	3	2
Bus, truck	2	1

This means that (taking into account the values shown in [20]) a truck will decelerate to enter or exit the intersection at a speed of no more than 9 km/h, whereas a car will slow down to 18 km/h. The assumed values coincided with the velocity values obtained in the studies described in [29]. In order to make this model more realistic, a function of gradual braking was introduced, unless sudden braking was necessary to avoid an accident. In the tested model, vehicles decelerate gradually as they approach the roundabout. It is important to make sure that vehicles decelerate approaching the right exit:

1. On the roundabout: If distance d to be covered by a vehicle that decreases its velocity v in each iteration by 1, until reaching safe turning speed v_{turn_max}, is greater or equal the distance from the exit d_{exit}, then the velocity is decreased by 1.
2. On an entrance road: If distance d to be covered by a vehicle that decreases its velocity v in each iteration by 1, until reaching safe turning speed v_{turn_max}, is greater or equal the distance from the entrance $d_{entrance}$, then the velocity is decreased by 1.
3. Points 1 and 2 are expressed by:

$$v \to v - 1 \tag{5}$$

Calculation of the cell value in the subsequent iteration for vehicles entering and exiting the roundabout is dependent on information on adjacent cells. Adjacent cells are those, which link the feeder road with the roundabout internal road. Adjacency does not need to be of the first degree. For example, cells in the inner lane of the roundabout are not directly adjacent to any cell in the feeder road. However, there are cells located between the aforementioned cells, which link the two lanes and enable movement of vehicles. When a vehicle exits the roundabout, an adjacent cell is the last one, which the vehicle covers on its lane before moving into a cell located in the feeder road lane. Depending on the lane on which the vehicle is currently located and the lane onto which it is going to move, these will be various cells corresponding to the natural turns

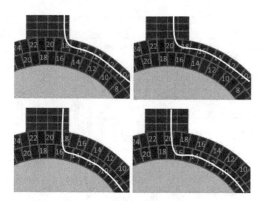

Fig. 3. Trajectories of vehicles exiting the roundabout. Source: own research.

taken by vehicles. Figure 3 shows the possible trajectories of vehicles that exit the roundabout, taking into account all the possible combinations:

- exiting from the outer lane into the right-hand lane – last cell: 18,
- exiting from the outer lane into the left-hand lane – last cell: 19,
- exiting from the inner lane into the right-hand lane – last cell: 16,
- exiting from the inner lane into the left-hand lane – last cell: 17.

The values of cells from which vehicles may exit the roundabout are determined by the formula:

$$c_{exit} = \left\lceil \frac{n_{cells}}{4} * r - 2 + 1 * l_{exit} \right\rceil, \tag{6}$$

where:

c_{exit} – the cell from which a vehicle may exit the roundabout, n_{cells} – the quantity of cells in the roundabout lane in which the vehicle is travelling, r – multiplier depending on the road, N-1, E-2, S-3, W-4, l_{exit} – the exit lane.

In summary, developed model allows the modelling of traffic on the multi-lane roundabout with multiple entrance and exit roads, modelling the movement of different vehicles classified according to their (possible) speed, studying the effects of changes in the diameter on the capacity of the roundabout, and enables the study of the behaviour of participants on roundabout according to the rules of the road (not used in this work).

4   The System Developed for the Simulation

The application was developed in JavaScript and it can be operated in web browsers as well as by means of a console. Running the application by means of a console is possible via NodeJS runtime environment based on V8 engine in the Chrome browser. Running the application this way is considerably faster, as the program may operate without the graphics layer. The application makes use of many free tools enabling the programming works. Figure 4 shows the structure of the application together with the tools applied.

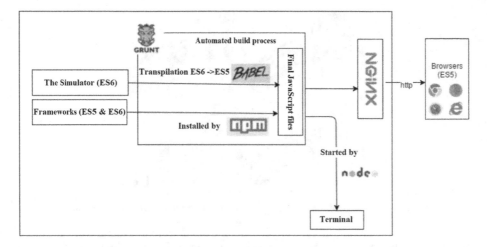

Fig. 4. The application structure. Source: own research.

Frequent problems with running the application are due to different runtime environments. The problems are solved by the combination of Vagrant and Ansible tools. Vagrant enables management of virtual machines, offering identical runtime conditions for applications. Ansible is used to ensure that the virtual machine is always equipped with any indispensable libraries. Upon starting, the program compares the current state of the machine with the expected one and carries out any necessary set-ups. Before that, it is necessary to prepare the configuration files that define the dependencies.

The application is written in accordance with the latest standards of ECMAScript 6. To enable correct operation of the application in web browsers, the Babel transpiler was used in order to change the code into the one compliant with ECMAScript 5. The process of transpilation and providing the application to the www server was automated by means of the Grunt program.

5 Experimental Results

The main goal of this paper was to develop a model based on cellular automata, aimed at investigating the impact of a roundabout island diameter on the roundabout capacity. Naturally, the bigger the size of the island, the bigger the roundabout as well as the number of cars that may travel on the roundabout concurrently. In the case of a roundabout with four feeder roads arranged perpendicularly, the distance between them will increase as well. The author of this paper has investigated whether increasing the dimensions of a roundabout translates into an increase in its capacity and whether there is a threshold value above which any further increase in the roundabout diameter does not have an impact on its capacity. The investigated scope of diameters ranged from 35 to 115 m, where each subsequent experiment involved an increment of 1 m. The lane width on the roundabout was constant and amounted to 4.5 m. The outer lane was

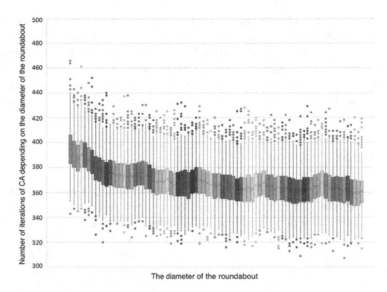

Fig. 5. Distribution of the number of the automaton iterations depending on the roundabout island diameter. Source: own research.

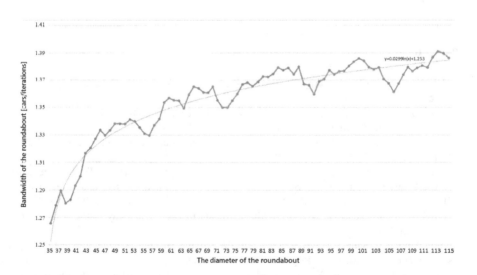

Fig. 6. Correlation between the island diameter and the roundabout capacity. Source: own research.

therefore from 167 to nearly 418 m long. The distribution of the received results is presented in Fig. 5. The received differences between the biggest and the smallest island are not considerable. Increasing the diameter has only an insignificant effect on the number of the automaton iterations, and consequently on the roundabout capacity.

The graph (Fig. 6) presents the correlation between the island diameter and the roundabout capacity. Increasing the diameter from 35 to 115 m raised the capacity by 10%. This is a logarithmic function, therefore the initial increases in the roundabout size result in greater increments. The bigger the roundabout, the smaller the increment resulting from the increased diameter. Nevertheless, even the initial increments are not very significant. For example, increasing the diameter from 35 to 45 m raised the capacity by 4.7%.

6   Conclusion

This paper focused on applying cellular automata in modelling a roundabout capacity in relation to different diameters of the roundabout island. A model was developed and implemented in the form of a simulation system which served to carry out the research study described herein. The received results showed a small (10%) increase in the roundabout capacity as a result of changing the island diameter from 35 to 115 m. It must be noted that it is not feasible to carry out such research studies in reality, mainly due to technical and location-related aspects, traffic impediments caused by a round-about modification, or a possibility of traffic collisions. Therefore, a computer simu-lation was used, and the results of the simulation system operation are shown in respective graphs.

References

1. https://nextstl.com/2013/10/mythbusters-tackles-four-way-stop-v-roundabout-traffic-throughput/
2. Transportation Research Board of the National Acad: National Cooperative Highway Research Program Report 572 - Roundabouts in the Unites States (2007)
3. Sisiopiku, V.P., Oh, H.-U.: Evaluation of roundabout performance using SIDRA. J. Transp. Eng. **127**(2), 143–150 (2001)
4. Wang, R., Liu, M.: A realistic cellular automata model to simulate traffic flow at urban roundabouts. In: Sunderam, V.S., Albada, G.D., Sloot, P.M.A., Dongarra, J.J. (eds.) ICCS 2005. LNCS, vol. 3515, pp. 420–427. Springer, Heidelberg (2005). doi:10.1007/11428848_56
5. Nagel, K., Schreckenberg, M.: A cellular automata model for freeway traffic. J. Phys. I **2**, 2221–2229 (1992)
6. Chowdhury, D., Santen, L., Schadschneider, A.: Statistical physics of vehicular traffic and some related systems. Phys. Rep. **329**, 199–329 (2000)
7. Andrzejewski, G., Zając, W., Kołopieńczyk, M.: Time dependencies modelling in traffic control algorithms. In: Mikulski, J. (ed.) TST 2013. CCIS, vol. 395, pp. 1–6. Springer, Heidelberg (2013). doi:10.1007/978-3-642-41647-7_1
8. Esser, J., Schreckenburg, M.: Microscopic simulation of urban traffic based on cellular automata. Int. J. Mod. Phys. **8**(5), 1025–1036 (1997)
9. Małecki, K., Iwan, S.: Development of cellular automata for simulation of the crossroads model with a traffic detection system. In: Mikulski, J. (ed.) TST 2012. CCIS, vol. 329, pp. 276–283. Springer, Heidelberg (2012). doi:10.1007/978-3-642-34050-5_31
10. Popescu, M.C., Ranea, C., Grigoriu, M.: Solutions for traffic lights intersections control. In: Proceedings of the 10th WSEAS (2010)

11. Han, X., Sun, H.: The implementation of traffic signal light controlled by PLC. J. Changchun Inst. Opt. Fine Mech. **4**, 029 (2003)
12. Kołopieńczyk, M., Andrzejewski, G., Zając, W.: Block programming technique in traffic control. In: Mikulski, J. (ed.) TST 2013. CCIS, vol. 395, pp. 75–80. Springer, Heidelberg (2013). doi:10.1007/978-3-642-41647-7_10
13. Jaszczak, S., Małecki, K.: Hardware and software synthesis of exemplary crossroads in a modular programmable controller. Przeglad Elektrotechniczny **89**(11), 121–124 (2013)
14. Macioszek, E.: Relationship between vehicle stream in the circular roadway of a one-lane roundabout and traffic volume on the roundabout at peak hour. In: Mikulski, J. (ed.) TST 2014. CCIS, vol. 471, pp. 110–119. Springer, Heidelberg (2014). doi:10.1007/978-3-662-45317-9_12
15. Macioszek, E., Sierpiński, G., Czapkowski, L.: Problems and issues with running the cycle traffic through the roundabouts. In: Mikulski, J. (ed.) TST 2010. CCIS, vol. 104, pp. 107–114. Springer, Heidelberg (2010). doi:10.1007/978-3-642-16472-9_11
16. Macioszek, E.: Analysis of significance of differences between psychotechnical parameters for drivers at the entries to one-lane and turbo roundabouts in Poland. In: Sierpiński, G. (ed.) Intelligent Transport Systems and Travel Behaviour. AISC, vol. 505, pp. 149–161. Springer, Cham (2017). doi:10.1007/978-3-319-43991-4_13
17. Nagel, K., Wolf, D.E., Wagner, P., Simon, P.M.: Two-lane traffic rules for cellular automata: a systematic approach. Phys. Rev. E **58**(2), 1425–1437 (1998)
18. Biham, O., Middleton, A.A., Levine, D.: Self-organization and a dynamical transition in traffic-flow models. Phys. Rev. A **4**(6), 6124 (1992)
19. Chowdhury, D., Schadschneider, A.: Self-organization of traffic jams in cities: effects of stochastic dynamics and signal periods. Phys. Rev. E **59**, 1311–1314 (1999)
20. Hartman, D.: Head leading algorithm for urban traffic modeling. Positions **2**, 1 (2004)
21. Belz, N.P., Aultman-Hall, L., Montague, J.: Influence of priority taking and abstaining at single-lane roundabouts using cellular automata. Transp. Res. Part C Emerg. Technol. **69**, 134–149 (2016)
22. Wang, R., Ruskin, H.: Modeling traffic flow at a single-lane urban roundabout. Comput. Phys. Commun. **147**, 570–576 (2002)
23. Lakouari, N., Ez-Zahraouy, H., Benyoussef, A.: Traffic flow behavior at a single lane roundabout as compared to traffic circle. Phys. Lett. Sect. A Gen. At. Solid State Phys. **378** (43), 3169–3176 (2014)
24. Belz, N.P., Aultman-Hall, L., Lee, B.H.Y., Gårder, P.E.: An event-based framework for non-compliant driver behavior at single-lane roundabouts. Transp. Res. Rec. J. Transp. Res. Board Nat. Academies **2402**, 38–46 (2014). Washington, D.C.
25. Wagner, P., Nagel, K., Wolf, D.: Realistic multilane traffic rule for cellular automata. Phys. A **234**, 687–698 (1997)
26. Wang, R., Ruskin, Heather J.: Modelling traffic flow at a multilane intersection. In: Kumar, V., Gavrilova, M.L., Tan, C.J.K., L'Ecuyer, P. (eds.) ICCSA 2003. LNCS, vol. 2667, pp. 577–586. Springer, Heidelberg (2003). doi:10.1007/3-540-44839-X_62
27. Wang, R., Ruskin, H.J.: Modelling traffic flow at multi-lane urban roundabouts. Int. J. Mod. Phys. C **17**(5), 693–710 (2006)
28. Schroeder, B., Rouphail, N., Salamati, K., Bugg, Z.: Effect of pedestrian impedance on vehicular capacity at multilane roundabouts with consideration of crossing treatments. Transp. Res. Rec. J. Transp. Res. Board Nat. Acad. **2312**(10), 14–24 (2012)
29. Macioszek, E.: Geometrical determinants of car equivalents for heavy vehicles crossing circular intersections. In: Mikulski, J. (ed.) TST 2012. CCIS, vol. 329, pp. 221–228. Springer, Heidelberg (2012). doi:10.1007/978-3-642-34050-5_25

A Method for Design of Hardware Emulators for a Distributed Network Environment

Andrzej Przybył[1]($^{(\boxtimes)}$) and Meng Joo Er[2]

[1] Institute of Computational Intelligence, Częstochowa University of Technology,
Częstochowa, Poland
andrzej.przybyl@iisi.pcz.pl
[2] School of Electrical and Electronic Engineering,
Nanyang Technological University, Singapore, Singapore
emjer@ntu.edu.sg

Abstract. This paper describes the method for hardware implementation of the emulator of nonlinear dynamic objects in FPGA technology. In order to ensure high-fidelity of emulation it has been proposed a new architecture of the arithmetic unit used to operations on real numbers in digital systems. The method allows us to obtain high processing performance similar to that obtained in fixed-point systems, while offering a wide range of numbers as in a floating-point notation. Based on this idea it has been proposed a super-scalar architecture of the digital processing unit. The described approach provides powerful processing of a matrix state equation with variable coefficients, which are calculated in real-time by fuzzy systems. Obtained and presented results confirm the high performance of the developed solution.

Keywords: Hardware-in-the-loop emulation · Hardware implementations · Fixed-point arithmetic · FPGA

1 Introduction

In recent years a method named hardware-in-the-loop (HIL) is used more and more often for testing and development of control systems [24,31,36,38,40]. Testing procedure in this method is carried out on the system working with closed control loop, in which functionally equivalent emulator was inserted in place of the actual object to carry out the necessary tests. At the same time, the main controller of the system (i.e. a device which in this way has been tested) works in its normal operating mode.

Simulators used in HIL systems must meet a number of conditions to be able to reliably replace the actual object. Above all, they must be equipped with the same communication interface. In addition, they must simulate the operation of the actuators, the object and the measuring equipment. A very important feature is also the fact that they must be able to work in real-time. A device that meets these requirements is called emulator, while HIL testing method is referred to as closing a control-loop by the emulator.

© Springer International Publishing AG 2017
L. Rutkowski et al. (Eds.): ICAISC 2017, Part II, LNAI 10246, pp. 318–336, 2017.
DOI: 10.1007/978-3-319-59060-8_29

HIL testing method is the most commonly used to validate the operation of the main functions of a complex control system (both hardware and software) in the range of normal operation and response to unusual situations [40]. To make this possible, emulators must be properly designed to simulate the work of the replaced equipment with high fidelity.

A particularly important issue related to the design of emulators is the selection of proper modeling algorithm. Practical objects subjected to modeling are usually strongly non-linear. However, it should be said that nonlinear models are difficult to analyse and implement in comparison to the linear models. Therefore, in many situations it is convenient to have an approximate linear model of a non-linear system. It is especially true if the linear model, in terms of the modeling accuracy, is not significantly differ from the nonlinear model.

One method of linearization of non-linear dynamics is called an equivalent linearization technique. Using this method [4,5,34] an equation describing the nonlinear dynamics of the object can be presented in the following form:

$$\mathbf{x}(k+1) = \mathbf{A_d}(k) \cdot \mathbf{x}(k) + \mathbf{B_d}(k) \cdot \mathbf{u}(k). \tag{1}$$

In Eq. (1) subscript $_\mathbf{d}$ means that the matrices $(\mathbf{A_d})$ and $(\mathbf{B_d})$ are presented in the discrete form, tailored to work with a fixed time step T_S.

The model presented above can be regarded as an accurate only under the condition that used linearization is done locally, i.e. in a short distance from the current point. This is due to the fact that the current form of the matrix can vary quite significantly as you progress through the action of the modeled phenomenon. However, according to what was mentioned earlier, in the described method, the matrix coefficients are constantly updated for each new operating point [19–21] according to the formula $\mathbf{A_d}(k) = \mathbf{A_d}(\mathbf{x}(k))$ oraz $\mathbf{B_d}(k) = \mathbf{B_d}(\mathbf{x}(k))$. The method is very universal and can be applied to simulate many different kind of dynamic object. Such simulators can be very useful, for example, in automatic process of controllers tuning with the use of evolutionary algorithms [29,17].

The test using the method of the HIL is very convenient but it is not always cost-effective from an economic point of view. One of the reasons is that emulators are often used for simulation of power electronic devices, which often have a switched, i.e. strongly non-linear dynamic characteristics. In this case, in order to ensure sufficient precision, the emulation time step must be very low, for example less than 500 ns that is achievable in commercial available emulators [9]. Such high requirements can be fulfilled only by hardware implementation of a specially designed, dedicated to the application, and usually a rather complex modeling algorithm. For this reason, commercially available emulator devices for HIL systems are very complex and expensive.

This paper proposes an efficient method of implementation in FPGA technology the equation describing the nonlinear dynamics (1) of a wide class of objects. In contrast to the solutions used in commercially available emulators for the system HIL, presented solution is designed for use in the distant emulators used in the remote-hardware-in-the-loop (RHIL) systems [35,36], which are

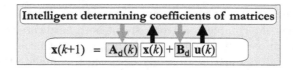

Fig. 1. The general idea of the used intelligent modeling method.

characterized by several times higher reaction time. A typical area of application of RHIL-type systems is therefore slightly different. In such a systems emulators are used to simulate the work of distant objects in a complex distributed control systems [10]. Thus, the emulator can simulate the operation of an integrated set of electromechanical device (consisting of a controller power electronics, electric motor and the driven mechanical components) which is subordinated to a master controller device [45,46] as part of a distributed control system, for example, of a computer numerically controlled (CNC) machine tool.

In the above-mentioned reasons, the emulation time step in RHIL systems may be even several times higher than in HIL systems. While, in HIL systems it is often required the emulation step less than $1\,\mu s$, in RHIL systems it is usually sufficient value less than $10\,\mu s$. Such less restrictive requirements on the one hand allows us for the use of digital controllers which are characterized by a lower performance and cost which in turn reduces the total cost of the emulator. On the other hand, it becomes feasible to realize the modeling, in which the matrix coefficients describing the dynamics of the system can be accurately modeled as a nonlinear function of certain variables (Fig. 1). This allows us to build an emulator reproducing the modeled phenomenon with higher fidelity than would be possible with the use of a matrix with fixed or switchable coefficients. Approximation of such nonlinear relationships is possible, for example, by one of the many known methods of artificial intelligence, such as neural networks [16,17,25], fuzzy systems [11], neuro-fuzzy systems [12,13,22,41–44], or by using other methods [23]. One advantage of using fuzzy systems is the potential ability to interpret the knowledge accumulated in them, which may be useful in some applications [15,27,28,34]. In addition, the high usability of fuzzy structures is confirmed by successful applications in various areas, including, for example, the recognition and classification of images [26]. In the method described in this paper, as suggested in the work [35,36], a hyper radial basis function (HRBF) system, which is functionally equivalent to some class of fuzzy systems, is used to update the coefficients of the matrices.

The proposed method of implementation is well suited to the characteristics of FPGA technology [52]. This enables realization of a simulator that works in real time and with a small time step on programmable systems from the lower price range. As a result, the use of solutions such HIL may also be justified in applications for which the economic aspect is important.

This paper is organized into 4 sections. Section 2 contains an idea of new methods of digital processing of real numbers as applied to modeling the nonlinear dynamics. Implementation results are presented in Sect. 3. Conclusions are drawn in Sect. 4.

2   A New Method of Digital Processing of Real Numbers in Modeling the Nonlinear Dynamics

In this section a method of efficient and accurate processing of real numbers within the matrix state equation will be presented. This method combines the advantages of two different standards used to binary representation of real numbers, i.e. fixed-point and floating-point notation. It allows us to achieve the high processing performance and high precision, at the same time offering a wide range of numbers.

2.1   Analysis of the Issue of the Digital Processing of Real Numbers in the Matrix State Equation

In order to analyze the problem, let us see equations of the dynamics of an exemplary electromechanical system with PMSM motor written in a matrix form. Such a system is commonly used in industrial applications. Having a reliable emulation of such a system brings great benefit, because (according to what was said in the introduction) it allows us to perform any tests based on the HIL methodology in a safe manner. Model of such an electromechanical system model is widely known and used in many works, including work [37].

However, for the purpose of analysis in this paper we consider the slightly upgraded model that is able to represent the phenomena in electromechanical system more accurately. The new description takes into account a number of nonlinear phenomena in the magnetic circuit of the motor. In particular, it is possible to recreate non-linear influence of the electric current and the angular position of the rotor shaft of the motor windings and the inductance of the magnetic induction, i.e. $L_d = L_d(i_d, \theta)$, $L_q = L_q(i_q, 0)$ and $\lambda_m = \lambda_m(i_q, \theta)$ in a manner compatible with non-linear dependencies outlined in the papers [18, 24, 38]. Forms of state variables vector and input vector are represented by the Formula (2).

$$\mathbf{x} = \begin{bmatrix} i_d & i_q & \omega & \theta \end{bmatrix}^T; \mathbf{u} = \begin{bmatrix} u_d & u_q & T_L \end{bmatrix}^T \tag{2}$$

While the system matrix $\mathbf{A_d}(\mathbf{x}(k))$ and input matrix $\mathbf{B_d}(\mathbf{x}(k))$ have the following form:

$$\mathbf{A_d}(k) = \begin{bmatrix} a_{11}^v & 0 & a_{13}^v & 0 \\ 0 & a_{22}^v & a_{23}^v & 0 \\ 0 & a_{32}^v & a_{33} & 0 \\ 0 & 0 & a_{43} & a_{44} \end{bmatrix}, \mathbf{B_d}(k) = \begin{bmatrix} b_{11}^v & 0 & 0 \\ 0 & b_{22}^v & 0 \\ 0 & 0 & b_{33} \\ 0 & 0 & 0 \end{bmatrix}. \tag{3}$$

The coefficients of these matrices that have been marked with a superscript "v" are dependent on the values of several state variables, which form a vector of state variables $\mathbf{x}$. In accordance with the used model these coefficients have the following form: $a_{11}^v = 1 - T_S \frac{R}{L_d(i_d,\theta)}$, $a_{13}^v = T_S \frac{L_q(i_q,\theta)}{L_d(i_d,\theta)} i_q$, $a_{22}^v = 1 - T_S \frac{R}{L_q(i_q,\theta)}$,

$a_{23}^v = -T_S \left(\frac{L_d(i_d,\theta)}{L_q(i_q,\theta)} i_d - \frac{\lambda_m}{L_q(i_q,\theta)} \right)$, $a_{32}^v = 1.5 T_S P^2 \frac{\lambda_m + (L_d(i_d,\theta) - L_q(i_q,\theta)) i_d}{J}$, $b_{11}^v = \frac{T_s}{L_d(i_d,\theta)}$, $b_{22}^v = \frac{T_s}{L_q(i_q,\theta)}$. The remainder of the non zero coefficients has a constant value: $a_{33} = 1 - T_S \frac{F}{J}$, $a_{43} = T_S$; $a_{44} = 1$, $b_{33} = -T_S \frac{P}{J}$, where R, L_d, L_q, λ_m, F represent fixed or variable parameters of the modeled electromechanical system.

It should be noted that in the general case an analytical formula describing the dependence of the coefficients of the matrix $\mathbf{A_d}$ and $\mathbf{B_d}$ of the vector of state variables has a complex form, unsuitable for processing in real-time, or such a formula is not known [18,24,38]. In many cases, these data can only be provided in the form of tables of data obtained from simulations or measurements. As a result, the use of non-linear approximation of the matrix coefficients is advantageous in terms of computational complexity and memory usage of digital controllers. As mentioned previously, this can be done, for example, according to certain methods of artificial intelligence. Such an approach was described, for example in papers [4–8], where it was also demonstrated that it offers the good accuracy of nonlinear dynamics modeling. In the next section we describe the method allows us for the efficient hardware implementation of described solutions.

Using the solution described in the above mentioned works, it was proposed that the matrix coefficients were generated by the intelligent systems. For their representation a widely known feature of artificial neural networks and fuzzy systems has been used. This is the ability to work as a universal approximator of any continuous, nonlinear relationships. Wherein, from the standpoint of the described implementation method it is not important to know the exact form of these approximated relationships. Some methods that can be used to determine these dependencies can be found, for example in papers [2,3,19–21]. Methods described in cited references are based on so called operating points, that affect the dynamics of the modeled object. Generally speaking, the determination of the main features of dynamics is used in various practical applications, for example in a verification of a handwritten signature [14,48–51]. As shown in [36] to identify the main features of the modeled object, i.e. its operating points, the radial basis-function type of HRBF are well suited. The issue of appropriate determine the operating points of the modeled object goes beyond this study and will not be analyzed in the current work. However, in order to determine the degree of complexity of the problem, it is important to specify the number of signals on which the given value depends. Therefore, it will be presented some guidance on which it will be possible to estimate the efficiency of the proposed solution used for the modeling of nonlinear dynamical systems.

It can be readily estimated as described in the work [37] that a non-zero coefficients of system and the input matrices are in the range of $\langle T_S \ldots 1 \rangle$. In line with the methodology of this study it is necessary to use time step in the model at level of $T_S \sim 1 - 10\,\mu s$. Therefore, it must be assumed that the smallest absolute value of matrix coefficients are at the level of $1 \cdot 10^{-6}$. Furthermore, given some rough estimation we can assume that we expect from the binary representation of such elements mapping accuracy at a level not worse than 1% of their true value. These requirements can be achieved by choosing an appropriate scale and the number of bits of binary word used for fixed-point representation of the real numbers. In our case it was used a 32-bit binary word in the fixed-point format S:3.28, i.e. one bit for the sign, 3-bits for the value of the integer part and 28 bits for the fractional part.

It should be noted that the format of the fixed-point binary representation of real numbers in comparison to the corresponding 32-bit floating-point format has a significant advantages. First, it is much easier to implement in FPGA technology, and therefore have a much greater performance and a lower consumption of hardware resources. Secondly, it offers higher accuracy. While the first of these advantages seems obvious, the second requires some explanation. It is generally known that floating-point numbers are commonly used in modern computer systems because they offer a relatively good accuracy both for coding the numbers of very small and very large absolute value. This is due to the possibilities offered by automatically adjusting the scale to the current value of the absolute numbers. This is realized based on the separation of 8 bits (in the case of 32-bit standard called single precision) to represent the value of the exponent. The remaining bits are used to store the sign (1-bit) and 23-bit mantissa, which represents appropriately scaled absolute value of the real number.

Automatic scaling used in the floating-point notation allows us for a fairly broad and dynamically matched range of variation of coded real numbers. Unfortunately, the adverse effect of this method is greatly reduced the number of bits allocated to the storage of significant digits representing the real number. Compared to the fixed-point notation, in which all the 32-bit binary word is used to store significant digits, the 32-bit floating-point standard is characterized by poorer precision of real numbers representation. In some situations, particularly in the modeling algorithm with a very small time step T_S, resolution offered by floating-point single precision representation may be insufficient. On the other hand, the use of precise enough 64-bit (i.e. double precision) floating-point standard would require a much more efficient digital system and could prove to be economically unreasonable.

2.2 The New Architecture of the Arithmetic Unit

In this paper we propose a new approach based on the use of fixed-comma encoding of real numbers and specially organized method of scaling. As mentioned earlier, this approach combines the advantages of fixed-and floating-comma encoding of real numbers.

In the classical approach, scaling real numbers, stored in binary floating-point occurs based on their current absolute value. Scaling is done automatically during the processing of the numbers by the arithmetic [31]. This approach brings the advantage that the scale is automatically adjusted to the current value of the absolute number of processed without involving a programmer creating code. However, the main drawback of this method is the large complexity and the relatively low performance of floating-point arithmetic unit as compared to fixed-point units.

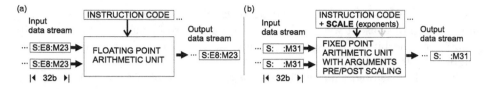

Fig. 2. Classical (a) and the proposed new (b) methodology for processing real numbers in digital systems.

In this paper we propose a new architecture of the arithmetic unit, developed to achieve high processing performance of real numbers. This unit, as opposed to either floating-point unit, is not provided with any mechanism for detecting the current absolute value of processed numbers and the automatic scaling. The role of determining the appropriate exponent takes compiler, which equips each arithmetic instruction with additional information on the arbitrarily allocated scale for the numbers being processed at a given stage of the implemented algorithm (Fig. 2).

In this method, information about the exponent values of the processed real numbers is not stored in the binary code. In return, this information is transmitted in a binary code of executed arithmetic instructions. As a result, more bits (31 compared to 23) is used to represent the significant digits of processed real number, thereby increasing its accuracy. However, it should be noted that the binary word representing the arithmetic instructions (i.e. operation code) of the executed computer program has to be extended. It should be clear that in the case of realization of repetitive arithmetic operations on large amounts of the same type of data, such solution brings significant benefits. These conditions are satisfied, inter alia, for the implementation of operations' matrix used to model the dynamic objects with the use of state variables technique.

As explained above, scaling procedure based on the method shown in Fig. 2 is not dependent on the current absolute value of the processed numbers. Scaling is based only on arbitrarily assigned range (to each processed arguments) at a given stage of the algorithm. This is due to the fact that, as shown in the previous section, during the implementation of operations' matrix, it is possible to plan the appropriate ranges for the processed number at a given stage of processing algorithm. This is done by the so-called worst case analysis. The ranges for each stage of the algorithm should be selected to be narrow to ensure the highest accuracy. On the other hand, ranges should be wide enough to not become overrun due to overflow of binary word at a given stage of the algorithm. At each stage of processing, all the numbers have statically assigned scale (i.e. value of the exponent). On this basis the compiler equips a binary code of each arithmetic instruction in the relevant information about the scales of the arguments and the required scale of the result of the executed operation.

Three most important features of the proposed processing algorithm of the real numbers are: significant simplification of the unit used to performing arithmetic operations, processing performance increasing and raising the effective resolution of processed numbers.

An additional and important feature of this system is that the information about the scale of processed numbers need not be stored in the form of 8-bit exponents but it can be successfully replaced by a much reduced information. In the sample application analyzed in this study (modeling electromechanical system) found that use of only two arbitrarily established exponents of real numbers is sufficient. Of course, in the case of modeling other objects it may be required a larger number ranges of used fixed-point arithmetic. In this case, the structure of a scalable arithmetic unit should be slightly expanded. In the analyzed example the format S:3.28 (low range, LR) was used to encode the coefficients of the matrix. This fixed scale corresponds to the floating point notation with the relative value of the exponent equal to +3. On the other hand different binary format, i.e. S:12.19 (high range, HR), was used to storage the vectors of state variables and input values. In turn, this fixed scale corresponds to the floating point notation with the relative value of the exponent equal to +12. These arbitrarily assigned scales are consistent with the calculations presented in the work [37].

The algorithm of the hardware implementation of the state equations is based primarily on the realization of the matrix by vector multiplication according to Eq. (1). In addition, it is necessary to perform a regular update of the matrices coefficients. For efficient and accurate implementation of these operations, as well as for immunization for unusual situations (for example, characterized by an increase in the processed numbers value outside the expected ranges) two solutions were applied. First, in order to achieve the precise execution of the operation type multiply and accumulate (MAC) a dedicated accumulation register was used. While processed arguments are represented by 32-bit words the accumulation register use 48-bits to store the intermediate results. In this way both the range and the resolution of processed numbers have increased. Secondly, the above mentioned MAC unit was equipped with saturation mechanism, in order to prevent dangerous effects of a possible overflow of numbers in binary notation. These two techniques are derived from solutions commonly employed in fixed-point digital signal processors which is an additional confirmation of their high relevance.

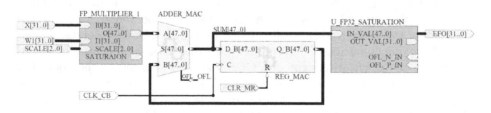

Fig. 3. The hardware implementation of the multiply and accumulate arithmetic unit.

Figure 3 shows how the FIXP_MAC block was realized which is the basic unit used to processing real numbers in the proposed solution. On the left side is visible fixed-point unit carrying out the operation of multiplication of numbers witch scaling the result. Scaling arithmetic operation applies in the present case only to the scaling operation has comma-multiplying two numbers implemented in the MAC. According to what has been said before, scaling is made based on 3-bit vector SCALE[2..0] taken from a fragment of binary code that defines performed arithmetic operations. Two bits define the scale of the two arguments and the third bit defines the scale of the result. According to this methodology, the value of the three bits of scale vector SCALE [2..0] defines how the 48-bit result is selected from a wider, i.e. 64-bit fixed-point output in multiplying unit (FP_MULTIPLIER_1) shown on the left side of Fig. 3.

The middle part of the Fig. 3 contains a 48-bit adder unit and a corresponding accumulation register. On the right side there is a block of saturation, which detailed construction for clarity of presentation is not shown in the figure. Applied saturation mechanism detects any over-range of the binary word. In this case, the result is set to the proper limit and it is generated information corresponding to the control unit.

It should be noted that the description of the scaling process presented above relates only to the MAC unit. Although, a similar procedure could be used to implement other basic arithmetic operations such as addition, subtraction, or division, however, in the case of the described implementation of a system for modeling of non-linear dynamics such operations appear to be unnecessary. All necessary mathematical operations (including addition) are performed as previously described by the MAC and by other arithmetic units, which will be described later in this work.

2.3   Super-Scalar Architecture for Efficient Digital Processing of Real Numbers in the Matrix State Equation

Based on the arithmetic unit presented in the previous section, this chapter proposes a super-scalar architecture of the digital processing unit. This architecture is well suited to the efficient processing of data stored in the form of a matrix equation and to update the matrix coefficients in real-time.

Fuzzy system was used to model the matrix coefficients used in Eq. (1). As shown in the work [36] for some kind of fuzzy systems, i.e. if we are dealing with a Gaussian input fuzzy sets with the product as the T-norms, the degree of activity of the fuzzy rule is possible by performing a series of operations such as MAC and one operation of determining the value of the exponential function. In turn, the defuzzyfication type of COGS needs, in addition to the operation of the type as mentioned above, an arithmetic division of real numbers, which can be done by multiplying the nominator by the inverse of the denominator. These actions are relatively easy to implement on FPGAs, they are carried at high speed and consume relatively small hardware resources.

The corresponding arithmetic unit (AU) suitable for performing such operations is shown in Fig. 4. It consists of three blocks: described earlier (Fig. 3) block

FIXP_MAC, units of FIX_03_12_GAUSS defining the value of the Gaussian function and block FIXP_03_12_RECIPR calculating the arithmetic inverse of the real number.

Block FIX_03_12_GAUSS is based on a simple lookup table method, in which values of Gaussian function are stored in the ROM memory. The memory is organized as 1024 words each with a width of 18-bits, which seems sufficient to store the value of the activation function of fuzzy sets. It should be noted that the resulting 18-bit resolution of stored values and 10-bit resolution of input domain is larger than the corresponding values (6–8 bits) declared in many other publications, eg. [1,39] what should be considered an advantage. The last processing block, i.e. the FIXP_03_12_RECIPR is constructed as described in the work [36] and offers a resolution similar to the resolution of the block described earlier.

According to the above-described requirements and based on the previously presented in Fig. 4 basic block of the arithmetic unit, in the following part of the paper it is proposed a hardware method for parallel data processing of matrix state Eq. (1). The general idea of this method is shown in Fig. 5. This method is based on the simultaneous operation of P specialized fixed-point, scalar arithmetic units (AU#1... AU#P). In accordance with generally accepted nomenclature architecture such architecture is called a super-scalar.

This system executes the program code which is defined by sequential commands. Each of the commands is defined by a 256-bits binary word using technology called very long instruction word (VLIW) [33]. Each long word consists of P control word CW#1... CW#P, one for each arithmetic unit. Each of the arithmetic units has unrestricted and concurrent with the other units access to

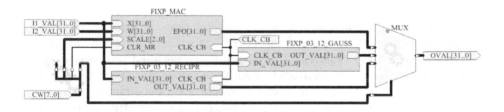

Fig. 4. The internal structure of the proposed arithmetic unit (AU).

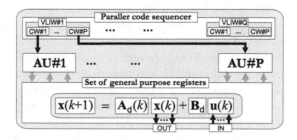

Fig. 5. The general idea of the proposed super-scalar processing system.

read any two registers from the group consisting of S general purpose registers. At the same time, each of these units has unlimited and concurrent with other units access to write any single register from this set. The individual registers perform the function assigned to them by the user. In particular, some of them are statically assigned to store coefficients of the vector $\mathbf{x}(k)$ or $\mathbf{u}(k)$. Other registers are dynamically used to store temporary values, including the coefficients of the matrices. In addition, the user code using the selected unit AU can substitute any constant numbers to any register, if this is necessary for the implementation of the algorithm.

The order of execution of the various arithmetic is not imposed by the architecture shown in Fig. 5. However, it is important, what is characteristic of the VLIW architecture, the order must be properly designed (by the programmer or automatically) in such a way that the required algorithm was implemented not only properly but also as fast as possible. In the case of implementation of the model with the same level of complexity as analyzed in this paper, the task can be successfully accomplished by man. However, for the modeling of more complex systems it is advisable to use an automatic procedure for designing an optimal code. This problem is not described in this paper and will be the subject of further study of the authors.

According to the calculations contained in the work [36] concerning described there serial method of data processing, to determine the output value of HRBF structure with N inputs and M radial sets, it is necessary the following number of clock cycles

$$c_s^{max} = 2M \cdot c_r + 2,\tag{4}$$

where c_r is the number of cycles needed to determine the value of the activation degree a single radial set.

The above calculations are based on the fact that the operation of determining the single output value of the HRBF structure, with using the COGS defuzzification method requires: M cycles for the weighted sum necessary to determine the value of the nominator, the same number of cycles the denominator, one cycle to determine the arithmetic inverse of the denominator and finally of one cycle to perform multiplication of the nominator by the inverse of the denominator. While the process of determining the activation degree value of a single radial function requires: one cycle for initialization of registers, N cycles for multiplications, $2N$ cycles for MAC operations and one cycle for determining the value of the exponent. The total number of clock cycles required for this purpose can be calculated as follows $t_r = 1 + 3N + 1$. In each single cycle in each AU block is performed one MAC operation, one exponential function or one arithmetic inverse of a real number in accordance with the functionality of the AU shown in Fig. 4. Because of the aforementioned calculations are executed in a serial manner, the value determined by Formula (4) is the upper limit of clock cycles needed to determine output value of HRBF system.

In the literature it is also presented in parallel different approaches that can be used to implement this type of operation. For example, in the work [32] parallel architecture is used for calculation of convolution. In the cited work it was assumed that in the used structure of the FPGA is available a sufficient number of resources, and all required by the implemented algorithm multiplication operations (i.e. N) can be performed in parallel in one cycle. While the accumulation of the particular multiplication results occurs in pairs, in subsequent $log_2\,(2N)$ cycles. Implementation of the calculation in such a way, however, has its major drawback. Namely, requires a very large amount of hardware resources, in particular fixed-point adders and multipliers.

By analogy to the above described method for determining the convolution function to determine the value of the activation degree of a single radial-basis function it is required $3N$ fixed-point multipliers. This is due to the fact that this action requires three operations of multiplication [36] for each input affecting the value of the function. On the other hand, it is easy to check the required number of adders is $2^{(N-1)} - 1$. Such a significant utilization of hardware resources disqualify, in many practical applications, the described method of parallel calculations.

In the remainder part of the work a mixed approach will be presented. It is based on the serial realization of the process of determining the activation degree of a radial set. By contrast, the different individual radial sets are processed in parallel by a fully independent arithmetic unit. In case if there is available a sufficient number of independent AUs, by analogy with the calculations presented in the paper [32], it is possible to determine the values of a single output HRBF system with the following number of clock cycles:

$$c_s^{min} = log_2\,(M) + c_r + 2. \tag{5}$$

It seems that the approach described by the Formula (5) is a reasonable compromise between the obtainable accuracy and speed. However, the actual number of clock cycles required will be limited by the number (P) available AU blocks, the number and complexity of the processed fuzzy structures. It will also depend on the "quality" of the VLIW code, which will be periodically executed by the described super-scalar system. Therefore, the actual number of cycles will be housed somewhere in the range of $c_s \in \langle c_s^{min} \ldots c_s^{max} \rangle$.

The rest of the work will present the obtained implementation results of an exemplary system, i.e. simulator of the electromechanical system described in the previous section. For clarity of presentation te details of implementation will not be presented, but only primary and most important features of this method.

3   Implementation Results

In our investigation it was considered a problem of hardware implementation of main functional blocks of the emulator. According to the description presented in the previous sections this task consists of determining the values of matrices coefficients (Fig. 1) and the matrix multiplication (1).

As it has been previously shown, selected coefficients of matrices $\mathbf{A_D}$ and $\mathbf{B_D}$ are non-linearly dependent on the value of the vector of state variables $\mathbf{x}$. It is assumed that they can be approximated by a properly designed fuzzy structure. In the present work, due to the limitation of its volume, the parameters of the fuzzy structure and the method used for the learning will not be analyzed in detail. It will only be roughly estimated the complexity of appropriate structures, which are based on the of HRBF method. On this basis it will be determined the approximate computing performance of the proposed method. Thus it can be evaluated the usefulness of the proposed method to realization of simulators working in real-time and which is suitable for use in systems such as RHIL.

After analyzing the form of the modeled coefficients it was noted that two of them (a_{11}^v, b_{11}^v) are simply dependent on nonlinear function $\frac{T_S}{L_d(i_d,\theta)}$. Two others (a_{22}^v, b_{22}^v) depend in a simple way from the $\frac{T_S}{L_q(i_q,\theta)}$. In contrast, the dependence of the other three coefficients $(a_{13}^v, a_{23}^v, a_{32}^v)$ does not depend simply on these values and must be expressed as unknown function of the following signals (i_d, i_q, θ).

Accordingly, it is necessary to apply three fuzzy type structures HRBF [36] employed as a nonlinear function approximators. The first two structures (FS1 and FS2) are designed to generate a single output value based on the two inputs. The third structure (FS3) generates a single output value based on the three inputs.

Further calculations were based on some estimate of assumptions that do not limit the universality of this method. It was assumed that to ensure the accuracy of approximation, structures named FS1 and FS2 should be built with four hyper-radial-basis-functions each. While, the structure named FS3 should be built with eight hyper-radial-basis-functions. Of course, these values can be adjusted at the learning process so as to ensure the required accuracy with a minimum of complexity.

The experiment concerning the implementation of the described structures has been completed on the Spartan FPGA from Xilinx XC6SLX45-3C by Means of Altium Designer and Xilinx ISE software. The device used for the experiment comes from the low-end family of a wide range of programmable devices manufactured by Xilinx. Positive results of the implementation obtained with this type of FPGA devices confirm the high usefulness of the used method. As mentioned in the introduction, by using the aforementioned solution it is possible to realize emulation system HIL in application areas in which the economic aspect is important.

Experimentally built super-scalar computing system consists of a $P = 8$ AU blocks and $S = 32$ general purpose registers. The timing analysis shows that the exemplary system in the FPGA device is able to work with maximal clock frequency of 32.4 MHz. The actual clock frequency was set to $f_F = 25.0$ MHz which is equivalent to a clock cycle length equal to $T_F = 40$ ns.

The structure of AUs used FS1#1 and AU#2 to determine the values of the activation function of their four radial sets. As stated earlier, this structure is built on the basis of four radial sets. The output value of each radial set is determined by one unit AU. To determine the output value of the whole FS1 structure are thus required two successive iterations. Each iteration requires a

c_r cycles of the system clock. The second structure, i.e. the FS2 uses the blocks AU#3 and AU#4 in an analogous manner.

The structure FS3 uses the remaining four blocks, i.e. AU#5..AU#8. Also in this case, two iterations are needed to determine the value of the activation function of eight fuzzy four AUs. Therefore, using all eight AUs, the number of cycles required to determine the value of the activation function of radial sets used by all three systems HRBF is $c_r^* = 2c_r$. With the largest number of inputs $N = 3$ occurring in the FS3 system, it gives a value of $c_r^* = 16$ clock cycles.

In the following steps, according to the description given in the work [36] the operations necessary in defuzzification process are executed. It proposed for this purpose a fairly simple solution. It do not guarantee top performance, however, is relatively easy to design. As mentioned in the previous section, the issue of the creation of the optimal code for super-scalar processing unit is a separate issue and will not be analyzed here.

In the experiment three pairs of AUs were used, one pair for each fuzzy system. The first three units, working in the mode MAC, calculate the weighted sum used to determine the nominator used in then COGS defuzzification method. The three remaining AU blocks accumulate of denominators for this operation. With so designed algorithm it is possible the parallel execution of this part of the calculation. The number of cycles needed to determine the output of all three fuzzy systems (FS1, FS2 and FS3) by analogy to the Formula (4) is equal to $c_s = M + c_r^* + 2 = 8 + 16 + 2 = 26$ clock cycles.

As described in the initial part of this chapter, the output fuzzy systems are used to determine the seven coefficients $(a_{11}^v, a_{22}^v, a_{23}^v, a_{32}^v, b_{11}^v, b_{22}^v)$ of the matrices in accordance with the general idea shown in Fig. 1. This simple task is performed by only a few activities such multiply-and-accumulate performed as described coefficients shown in the beginning of the chapter. Based on eight AUs, all of these operations are performed in parallel. It takes only four ticks of the clock, i.e. $c_2 = 4$.

After determining the coefficients of the matrices, the next operation is carried out. It is a matrix multiplication by a vector according to Eq. (1). Four AUs are used for this purpose, each for determining a weighted sum of the four rows of a matrix equation. $\mathbf{A_d}$ and $\mathbf{B_d}$ (3) are sparse matrices, so appropriate algorithm is used, wherein the zero coefficients of a matrix are omitted. The number of cycles required to perform this action in the described case takes only $c_3 = 4$ clock ticks, including the one to initiate the registers.

In summary, the total number of clock cycles required to carry out single-step of modeling method is equal to

$$c^{total} = c_s + c_2 + c_3 = 26 + 4 + 4 = 34. \tag{6}$$

Thus, the code executed by a super-scalar unit (Fig. 5) consists of $Q = c^{total}$ of VLIW instructions executed sequentially. Thus, this allows us to obtain the minimum value for the step of real-time simulation at level $T_S \geq c^{total} * T_f = 1.36\,\mu s$. Thus, it is possible implementation of fairly reliable emulator even in the case of simulations of strongly nonlinear dynamics.

Table 1. The FPGA resource usage of the super-scalar processing system.

Hardware resource name	DSP48A1	Registers	RAMB16B	LUTs
Used/available (percentage usage)	40/58 (68%)	3770/54576 (6%)	72/116 (62%)	16760/27288 (61%)

Obtaining such a low time step for matrix equation with variable coefficients confirms the high usefulness of the method described for the implementation of real-time simulators for the entire class of similar problems. In addition, it should be noted that the described method is highly scalable and can be applied even to perform much more complex models. Obviously, in this case, it may be necessary extension of the time step of the simulation. When the extension of the time step is not acceptable, it is necessary to use the FPGA with a larger number of resources and the expansion of the emulator with further arithmetic unit AU according to the idea shown in Fig. 5.

As a result of the carried out experiment it was determined also use the hardware resources of the FPGA shown in Table 1. As it can be seen, it was possible to realize this system at a reasonable utilization of hardware resources of the programmable unit. In the FPGA it was implemented the main block of the emulator (Fig. 5) and an universal 32-bit RISC soft-core processor named TSK3000A, supported by the Altium Designer environment. The purpose of this processor was to manage the process of real-time simulation and an operator interface. Moreover, as described in the work [35] a real-time Ethernet was intended to act as the I/O interface of the emulator.

4 Summary

This paper describes the modeling method of nonlinear dynamical systems, using the matrix equation with variable coefficients. To determine the changing values of the coefficients it was used a HRBF type structure working as a nonlinear approximator. The described approach enables the simulator to reproducing the behavior of non-linear objects with greater fidelity than in the case of using model with constant or switchable coefficients.

It should also be noted that the described method of determining the matrix coefficients is not only suitable for modeling the nonlinear dynamics. It can also be used to implement complex control systems, for example, working on the basis of direct feedback from the state with gain scheduling (see e.g. [30]).

The most important element proposed in this paper is a new method of processing real numbers in digital systems. As shown in the analyzed example, it allows the implementation on FPGA with modest parameters (i.e. low end) running in real-time simulation of nonlinear dynamics.

Based on the presented modeling technique and method of processing real numbers, it was also proposed a super-scalar architecture of a digital processing unit operated based on the VLIW methodology.

For such a comprehensive approach in the paper the results of implementation was presented, which confirm its high performance. The proposed approach makes the practical implementation of the emulator for hardware-in-the-loop systems not only easier but also cheaper.

Acknowledgment. The project was financed by the National Science Centre (Poland) on the basis of the decision number DEC-2012/05/B/ST7/02138.

References

1. Antonio-Mendez, R., de la Cruz-Alejo, J., Peñaloza-Mejia, O.: Fuzzy logic control on FPGA for solar tracking system. multibody mechatronic systems. In: Proceedings of the MUSME Conference, Huatulco, Mexico, vol. 25, pp. 11–21 (2014)
2. Bartczuk, Ł., Dziwiński, P., Starczewski, J.T.: A new method for dealing with unbalanced linguistic term set. In: Rutkowski, L., Korytkowski, M., Scherer, R., Tadeusiewicz, R., Zadeh, L.A., Zurada, J.M. (eds.) ICAISC 2012. LNCS, vol. 7267, pp. 207–212. Springer, Heidelberg (2012). doi:10.1007/978-3-642-29347-4_24
3. Bartczuk, Ł., Dziwiński, P., Starczewski, J.T.: New method for generation Type-2 fuzzy partition for FDT. In: Rutkowski, L., Scherer, R., Tadeusiewicz, R., Zadeh, L.A., Zurada, J.M. (eds.) ICAISC 2010. LNCS, vol. 6113, pp. 275–280. Springer, Heidelberg (2010). doi:10.1007/978-3-642-13208-7_35
4. Bartczuk, Ł., Przybył, A., Koprinkova-Hristova, P.: New method for non-linear correction modelling of dynamic objects with genetic programming. In: Rutkowski, L., Korytkowski, M., Scherer, R., Tadeusiewicz, R., Zadeh, L.A., Zurada, J.M. (eds.) ICAISC 2015. LNCS, vol. 9120, pp. 318–329. Springer, Cham (2015). doi:10.1007/978-3-319-19369-4_29
5. Bartczuk, Ł., Cpałka, K., Przybył, A.: A new approach to nonlinear modelling of dynamic systems based on fuzzy rules. Int. J. Appl. Math. Comput. Sci. **26**(3), 603–621 (2016)
6. Bartczuk, Ł., Łapa, K., Koprinkova-Hristova, P.: A new method for generating of fuzzy rules for the nonlinear modelling based on semantic genetic programming. In: Rutkowski, L., Korytkowski, M., Scherer, R., Tadeusiewicz, R., Zadeh, L.A., Zurada, J.M. (eds.) ICAISC 2016. LNCS, vol. 9693, pp. 262–278. Springer, Cham (2016). doi:10.1007/978-3-319-39384-1_23
7. Bartczuk, Ł.: Gene expression programming in correction modelling of nonlinear dynamic objects. Adv. Intell. Syst. Comput. **429**, 125–134 (2016)
8. Bartczuk, Ł., Galushkin, A.I.: A new method for generating nonlinear correction models of dynamic objects based on semantic genetic programming. In: Rutkowski, L., Korytkowski, M., Scherer, R., Tadeusiewicz, R., Zadeh, L.A., Zurada, J.M. (eds.) ICAISC 2016. LNCS, vol. 9693, pp. 249–261. Springer, Cham (2016). doi:10.1007/978-3-319-39384-1_22
9. Bélanger, J.: Real-time FPGA-based solutions for power electronics and power systems. FPGA-based HIL platform combining performance and flexibility, OPAL-RT Technologies (2014). 2014Q2, http://www.opal-rt.com/sites/default/files/Brochure_eFPGAsim_OPAL-RT(2).pdf
10. Chen, Q., Abercrombie, R.K., Sheldon, F.T.: Risk assessment for industrial control systems quantifying availability using mean failure Cost (MFC). J. Artif. Intell. Soft Comput. Res. **5**(3), 205–220 (2015)

11. Cpałka, K., Rutkowski, L.: Flexible Takagi-Sugeno fuzzy systems, neural networks. In: Proceedings of the 2005 IEEE International Joint Conference on IJCNN 2005, vol. 3, pp. 1764–1769 (2005)

12. Cpalka, K.: A method for designing flexible neuro-fuzzy systems. In: Rutkowski, L., Tadeusiewicz, R., Zadeh, L.A., Żurada, J.M. (eds.) ICAISC 2006. LNCS, vol. 4029, pp. 212–219. Springer, Heidelberg (2006). doi:10.1007/11785231_23

13. Cpałka, K., Rebrova, O., Nowicki, R., Rutkowski, L.: On design of flexible neuro-fuzzy systems for nonlinear modelling. Int. J. Gen. Syst. **42**(6), 706–720 (2013)

14. Cpałka, K., Zalasiński, M., Rutkowski, L.: A new algorithm for identity verification based on the analysis of a handwritten dynamic signature. Appl. Soft Comput. **43**, 47–56 (2016)

15. Cpałka, K.: Design of Interpretable Fuzzy Systems. Springer (2017)

16. Duda, P., Hayashi, Y., Jaworski, M.: On the strong convergence of the orthogonal series-type kernel regression neural networks in a non-stationary environment. In: Rutkowski, L., Korytkowski, M., Scherer, R., Tadeusiewicz, R., Zadeh, L.A., Zurada, J.M. (eds.) ICAISC 2012. LNCS, vol. 7267, pp. 47–54. Springer, Heidelberg (2012). doi:10.1007/978-3-642-29347-4_6

17. Er, M.J., Duda, P.: On the weak convergence of the orthogonal series-type kernel regresion neural networks in a non-stationary environment. In: Wyrzykowski, R., Dongarra, J., Karczewski, K., Waśniewski, J. (eds.) PPAM 2011. LNCS, vol. 7203, pp. 443–450. Springer, Heidelberg (2012). doi:10.1007/978-3-642-31464-3_45

18. Dufour, C., Yamada, T., Imamura, R., Bélanger, J.: FPGA permanent magnet synchronous motor floating-point models with variable-DQ and spatial harmonic finite-element analysis solvers. In: Proceedings of the 2011 14th European Conference on Power Electronics and Applications (EPE-2011), pp. 1–10 (2012)

19. Dziwiński, P., Avedyan, E.D.: A new approach to nonlinear modeling based on significant operating points detection. In: Rutkowski, L., Korytkowski, M., Scherer, R., Tadeusiewicz, R., Zadeh, L.A., Zurada, J.M. (eds.) ICAISC 2015. LNCS, vol. 9120, pp. 364–378. Springer, Cham (2015). doi:10.1007/978-3-319-19369-4_33

20. Dziwiński, P., Avedyan, E.D.: A new approach for using the fuzzy decision trees for the detection of the significant operating points in the nonlinear modeling. In: Rutkowski, L., Korytkowski, M., Scherer, R., Tadeusiewicz, R., Zadeh, L.A., Zurada, J.M. (eds.) ICAISC 2016. LNCS, vol. 9693, pp. 279–292. Springer, Cham (2016). doi:10.1007/978-3-319-39384-1_24

21. Dziwiński, P., Avedyan, E.D.: A new method of the intelligent modeling of the nonlinear dynamic objects with fuzzy detection of the operating points. In: Rutkowski, L., Korytkowski, M., Scherer, R., Tadeusiewicz, R., Zadeh, L.A., Zurada, J.M. (eds.) ICAISC 2016. LNCS, vol. 9693, pp. 293–305. Springer, Cham (2016). doi:10.1007/978-3-319-39384-1_25

22. Gabryel, M., Cpałka, K., Rutkowski, L.: Evolutionary strategies for learning of neuro-fuzzy systems. In: Proceedings of the I Workshop on Genetic Fuzzy Systems, Granada, pp. 119–123 (2005)

23. Gałkowski, T., Rutkowski, L.: Nonparametric fitting of multivariate functions. IEEE Trans. Autom. Control **31**(8), 785–787 (1986)

24. Inaba, Y., Cense, S., Ould Bachir, T., Yamashita, H., Dufour, C.: A dual high-speed PMSM motor drive emulator with finite element analysis on FPGA chip with full fault testing capability. In: Proceedings of the 2011 14th European Conference on Power Electronics and Applications (EPE-2011), pp. 1–10 (2011)

25. Jaworski, M., Er, M.J., Pietruczuk, L.: On the application of the parzen-type kernel regression neural network and order statistics for learning in a non-stationary environment. In: Rutkowski, L., Korytkowski, M., Scherer, R., Tadeusiewicz, R., Zadeh, L.A., Zurada, J.M. (eds.) ICAISC 2012. LNCS, vol. 7267, pp. 90–98. Springer, Heidelberg (2012). doi:10.1007/978-3-642-29347-4_11

26. Korytkowski, M.: Novel visual information indexing in relational databases. Integr. Comput.-Aided Eng. 24(2), 119–128 (2016)

27. Łapa, K., Przybył, A., Cpałka, K.: A new approach to designing interpretable models of dynamic systems. In: Rutkowski, L., Korytkowski, M., Scherer, R., Tadeusiewicz, R., Zadeh, L.A., Zurada, J.M. (eds.) ICAISC 2013. LNCS, vol. 7895, pp. 523–534. Springer, Heidelberg (2013). doi:10.1007/978-3-642-38610-7_48

28. Łapa, K., Cpałka, K., Wang, L.: New method for design of fuzzy systems for nonlinear modelling using different criteria of interpretability. In: Rutkowski, L., Korytkowski, M., Scherer, R., Tadeusiewicz, R., Zadeh, L.A., Zurada, J.M. (eds.) ICAISC 2014. LNCS, vol. 8467, pp. 217–232. Springer, Cham (2014). doi:10.1007/978-3-319-07173-2_20

29. Łapa, K., Szczypta, J., Venkatesan, R.: Aspects of structure and parameters selection of control systems using selected multi-population algorithms. In: Rutkowski, L., Korytkowski, M., Scherer, R., Tadeusiewicz, R., Zadeh, L.A., Zurada, J.M. (eds.) ICAISC 2015. LNCS, vol. 9120, pp. 247–260. Springer, Cham (2015). doi:10.1007/978-3-319-19369-4_23

30. Nonaka, S., Tsujimura, T., Izumi, K.: Gain design of quasi-continuous exponential stabilizing controller for a nonholonomic mobile robot. J. Artif. Intell. Soft Comput. Res. 6(3), 189–201 (2016)

31. Ould, B.T., Dufour, C., David, J.P., Bélanger, J., Mahseredjian, J.: A high-speed PMSM drive case study. In: Proceedings of Electrimacs (ELECTRIMACS-2011), Cergy-Pontoise, France, pp. 1–6 (2011)

32. Pamuła, W.: Method of decomposing image processing algorithms for implementation in FPGA (in Polish). Pomiary Automatyka Kontrola 57(6), 648–651 (2011)

33. Philips Semiconductors: An introduction to very-long instruction word (VLIW) computer architecture, vliw-wp.pdf, pp. 1–11 (2011). 2016Q4

34. Przybył, A., Cpałka, K.: A new method to construct of interpretable models of dynamic systems. In: Rutkowski, L., Korytkowski, M., Scherer, R., Tadeusiewicz, R., Zadeh, L.A., Zurada, J.M. (eds.) ICAISC 2012. LNCS, vol. 7268, pp. 697–705. Springer, Heidelberg (2012). doi:10.1007/978-3-642-29350-4_82

35. Przybył, A., Er, M.J.: A new approach to designing of intelligent emulators working in a distributed environment. In: Rutkowski, L., Korytkowski, M., Scherer, R., Tadeusiewicz, R., Zadeh, L.A., Zurada, J.M. (eds.) ICAISC 2016. LNCS, vol. 9693, pp. 546–558. Springer, Cham (2016). doi:10.1007/978-3-319-39384-1_48

36. Przybył, A., Er, M.J.: The method of hardware implementation of fuzzy systems on FPGA. In: Rutkowski, L., Korytkowski, M., Scherer, R., Tadeusiewicz, R., Zadeh, L.A., Zurada, J.M. (eds.) ICAISC 2016. LNCS, vol. 9692, pp. 284–298. Springer, Cham (2016). doi:10.1007/978-3-319-39378-0_25

37. Przybył, A., Szczypta, J.: Method of evolutionary designing of FPGA-based controllers. Przegląd Elektrotechniczny 7, 174–179 (2016)

38. Poon, J., Chai, E., Čelanović, I., Genić, A., Adzic, E.: High-fidelity real-time hardware-in-the-loop emulation of PMSM inverter drives. In: Conference: ECCE - Energy Conversion Congress and Exposition, At Denver, CO, USA, pp. 1–6 (2013)

39. Poplawski, M., Bialko, M.: Implementation of fuzzy logic controller in FPGA circuit for guiding electric wheelchair. In: Rutkowski, L., Korytkowski, M., Scherer, R., Tadeusiewicz, R., Zadeh, L.A., Zurada, J.M. (eds.) ICAISC 2012. LNCS, vol. 7268, pp. 216–222. Springer, Heidelberg (2012). doi:10.1007/978-3-642-29350-4_26
40. Schulte, T., Kiffe, A., Puschmann, F.: HIL simulation of power electronics and electric drives for automotive applications. Electronics **16**(2), 130–135 (2012)
41. Rutkowski, L., Cpałka, K.: Compromise approach to neuro-fuzzy systems. In: Proceedings of the 2nd Euro-International Symposium on Computation Intelligence. Frontiers in Artificial Intelligence and Applications, vol. 76, pp. 85–90 (2002)
42. Rutkowski, L., Cpałka, K.: A neuro-fuzzy controller with a compromise fuzzy reasoning. Control Cybern. **31**(2), 297–308 (2002)
43. Rutkowski, L., Cpałka, K.: Flexible weighted neuro-fuzzy systems. In: Proceedings of the 9th International Conference on Neural Information Processing (ICONIP 2002), Orchid Country Club, Singapore, 18–22 November 2002. CD
44. Rutkowski, L., Cpałka, K.: Neuro-fuzzy systems derived from quasi-triangular norms. In: Proceedings of the IEEE International Conference on Fuzzy Systems, Budapest, 26–29 July 2004, vol. 2, pp. 1031–1036 (2004)
45. Rutkowski, L., Przybył, A., Cpałka, K., Er, M.J.: Online speed profile generation for industrial machine tool based on neuro-fuzzy approach. In: Rutkowski, L., Scherer, R., Tadeusiewicz, R., Zadeh, L.A., Zurada, J.M. (eds.) ICAISC 2010. LNCS, vol. 6114, pp. 645–650. Springer, Heidelberg (2010). doi:10.1007/978-3-642-13232-2_79
46. Rutkowski, L., Przybył, A., Cpałka, K.: Novel on-line speed profile generation for industrial machine tool based on flexible neuro-fuzzy approximation. IEEE Trans. Ind. Electr. **59**, 1238–1247 (2012)
47. Szczypta, J., Łapa, K., Shao, Z.: Aspects of the selection of the structure and parameters of controllers using selected population based algorithms. In: Rutkowski, L., Korytkowski, M., Scherer, R., Tadeusiewicz, R., Zadeh, L.A., Zurada, J.M. (eds.) ICAISC 2014. LNCS, vol. 8467, pp. 440–454. Springer, Cham (2014). doi:10.1007/978-3-319-07173-2_38
48. Zalasiński, M., Łapa, K., Cpałka, K.: New algorithm for evolutionary selection of the dynamic signature global features. In: Rutkowski, L., Korytkowski, M., Scherer, R., Tadeusiewicz, R., Zadeh, L.A., Zurada, J.M. (eds.) ICAISC 2013. LNCS, vol. 7895, pp. 113–121. Springer, Heidelberg (2013). doi:10.1007/978-3-642-38610-7_11
49. Zalasiński, M., Cpałka, K., Hayashi, Y.: New method for dynamic signature verification based on global features. In: Rutkowski, L., Korytkowski, M., Scherer, R., Tadeusiewicz, R., Zadeh, L.A., Zurada, J.M. (eds.) ICAISC 2014. LNCS, vol. 8468, pp. 231–245. Springer, Cham (2014). doi:10.1007/978-3-319-07176-3_21
50. Zalasiński, M.: New algorithm for on-line signature verification using characteristic global features. Adv. Intell. Syst. Comput. **432**, 137–146 (2016)
51. Zalasiński, M., Cpałka, K., Hayashi, Y.: A new approach to the dynamic signature verification aimed at minimizing the number of global features. In: Rutkowski, L., Korytkowski, M., Scherer, R., Tadeusiewicz, R., Zadeh, L.A., Zurada, J.M. (eds.) ICAISC 2016. LNCS, vol. 9693, pp. 218–231. Springer, Cham (2016). doi:10.1007/978-3-319-39384-1_20
52. Xilinx Spartan-6 FPGA User Guides: UG389 v1.5, UG389 v1.2 (2014). 2016Q4, http://www.xilinx.com/support/documentation/user_guides/

Iterative Learning of Optimal Control – Case Study of the Gantry Robot

Ewaryst Rafajłowicz[(✉)] and Wojciech Rafajłowicz

Faculty of Electronics, Wrocław University of Technology, Wrocław, Poland
`ewaryst.rafajlowicz@pwr.wroc.pl`

Abstract. In [15] the authors proposed an iterative learning algorithm for searching for optimal control of linear dynamic systems. This algorithm has been preliminary tested on the laser power control for the cladding process. The aim of this paper is to present a case study of a similar algorithm when applied to control Z-axis of a gantry robot. The original algorithm from [15] has to be modified in order to cover the case when the tracking signal is the output of the system instead of its whole state, as in [15]. The obtained results indicate a fast rate of convergence of the learning algorithm. One can also observe how learning of the shapes of the optimal input and output signals are convergent.

Keywords: Iterative learning control · Gantry robot · Testing · Camera in the loop

1 Introduction

This paper deals with iterative learning control, which has been intensive area of research over the last two decades. We refer the reader to [3,7], [1,10,11,18,19] for survey papers and to [2,4,6,8,9,13,14,20] for recent papers on the iterative learning control (ILC) that are related to our paper.

The aim of this paper is to test the iterative learning of an optimal control (ILOC) algorithm that was proposed in [15], using the well known example of gantry robot. This example has been used in so many papers on iterative learning control (ILC) that it can be considered as a benchmark example (see, e.g., [5,17]).

For the same reason its very accurate mathematical model is known and we can use it safely for comparisons without having access to its hardware realisation. We shall consider learning in a functional space, but we shall also compare our results – after sampling – with those obtained for discrete time models.

The paper is organized as follows:

- we summarize the iterative learning control algorithm from [15], together with a modification that allows to track a reference signal by the output signal instead of formulating the tracking problem in the state space,
- then, a short description of the gantry model for Z-axis is provided and we discuss the methodology of simulations,

© Springer International Publishing AG 2017
L. Rutkowski et al. (Eds.): ICAISC 2017, Part II, LNAI 10246, pp. 337–346, 2017.
DOI: 10.1007/978-3-319-59060-8_30

– finally, we provide the results of simualtions and discuss possible improvements by using a camera in the control loop.

2   Problem Statement and ILOC Algorithm

In this section we recall the problem statement and ILOC algorithm from [15] – with some modifications. These modifications are necessary, because in the gantry robot case it is customary to set requirements on the system output, instead of on the system state.

2.1   Problem Statement

The same dynamic system is run many times at a finite time interval of the length $T > 0$. Each run is called a pass and it starts from the same initial conditions. It is convenient to consider the time variable t as changing from 0 to T and to add a pass number n to the vector of state variables. In other words, we model the system states as $x_n(t) \in R^d$, $t \in (0, T)$, $n = 1, 2, \ldots$. The initial pass profile $x_0(t)$, $t \in (0, T)$ is assumed to be given, as well as the initial condition $x_n(0) = x_0$. Notice that we must also have: $x_0(0) = x_0$. For n-th pass the system states evolve as follows: by

$$\dot{x}_n(t) = A\, x_n(t) + b\, u_n(t), \; t \in (0, T), \; x_n(0) = x_0, \tag{1}$$

where a control signal along n-th pass is denoted by $u_n(t)$, $t \in [0, T]$ and – for simplicity of formulas – it is assumed to be a real valued function from $L_2(0, T)$ space that is multiplied by vector $d \times 1$ vector b of amplifications. A denotes $d \times d$ matrix, which is also given.

In this paper the system description is extended by the following state-output relationship:

$$y_n(t) = c^{tr}\, x_n(t), \quad t \in (0, T), \quad n = 0, 1\, 2, \ldots, \tag{2}$$

where $c \in R^d$ is the given column vector, while $y_n(t)$ is n-th profile of the system output. We consider only a scalar output signal, since it suffices for the purposes of our case study.

The problem is to find a sequence of control signals $\breve{u}_n(.)$ and the corresponding sequences of the system states $\breve{x}_n(.)$ as well as outputs $y_n(.)$ such that they are convergent – in $L_2(0, T)$ norms – to control signal $\breve{u}(.) \in L_2(0, T)$ and the corresponding state $\breve{x}(.)$, $t \in (0, T)$ and output $\breve{y}((.)$ for which the following cost functional $\breve{J}(u)$ attains its minimum:

$$\breve{J}(u) = \int_0^T \left[\frac{1}{2} \left(y_{ref}(t) - c^{tr}\, x(t) \right)^2 + \delta\, u^2(t) \right] dt, \tag{3}$$

where y_{ref} is a known reference signal, while $\delta > 0$ is a weighting factor, while $x(t) \in R^d$ is linked with $u(t)$ by the following equations:

$$\dot{x}(t) = A\, x(t) + b\, u(t), \; t \in (0, T), \; x(0) = x_0. \tag{4}$$

Notice that A and b in (4) are the same as in (1).

The difference between the above problem statement and the one considered in [15] is in the control quality criterion, which in [15] had the following form:

$$J(u) = \int_0^T \left[\frac{1}{2} ||x_{ref}(t) - x(t||_d^2 + \delta u^2(t) \right] dt, \qquad (5)$$

where $||.||$ is the Euclidean norm in R^d and also the reference trajectory $x_{ref}(t)$ is defined in R^d. Clearly, (3) is less restrictive than (5), because it forces to track the reference signal in the space that has a smaller dimension.

2.2   Iterative Learning Algorithm

The proposed algorithm for iterative learning of optimal control has the following form:

Step 0: Select a starting control signal $\breve{u}_0(t)$, $t \in (0, T)$ and $\eta > 0$ – the parameter in the stopping condition. Set the pass number $n = 0$.

Step 1: Calculate the Frechet derivative $\breve{F}_n(t)$ of the control quality criterion as follows:

$$\breve{F}_n(t) = \left(2\delta \breve{u}_n(t) + b^{tr} \breve{\psi}_n(t) \right), \qquad (6)$$

where $\breve{\psi}_n(t)$ is obtained by solving the following set of differential equations:

$$\dot{\breve{\psi}}_n(t) = -A^{tr} \breve{\psi}_n(t) + c\,(y_{ref}(t) - c^{tr} \breve{x}_n(t)), \ t \in (0, T), \quad \psi_n(T) = 0.$$

$$\dot{\breve{x}}_n(t) = A\breve{x}_n(t) + b, \breve{u}_n(t)), \ t \subset (0, T), \quad x_n(T) = x_0. \qquad (7)$$

Step 2: Update the control signal according to the following rule:

$$\breve{u}_{n+1}(t) = \breve{u}_n(t) - \gamma \breve{F}_n(t), \quad t \in [0, T]. \qquad (8)$$

Step 3: If $\max_{t \in [0, T]} |\breve{F}_n(t)| < \eta$, then STOP ($\breve{u}_n(t)$ is a sufficiently accurate approximation to $\breve{u}(t)$). Otherwise, set $n := n + 1$ and go to Step 1.

Several remarks are in order concerning the above learning procedure.

Remark 1. *For $r > 0$ and for sufficiently small $\gamma > 0$ the convergence of the above algorithm – with a constant step size – can be proved, since we have a linear model and quadratic criterion, but this is outside the scope of this paper (see [16] for the proof and [12], Sect. 3.1, for the discussion of the Wolfe conditions in a finite dimensional case).*

Remark 2. *The formula for calculating the Frechet derivative (6) formally looks like $F_n(t)$ in [15], but this time the adjoint variables $\breve{\psi}_n(.)$ are calculated using the system output error instead of the state tracking error.*

Remark 3. *In our case study we take advantage of the fact that we have $\breve{x}_n(t)$ available during simulations. Notice that in practice the state variables $\breve{x}_n(t)$ are not always accessible and one has to use a state observer.*

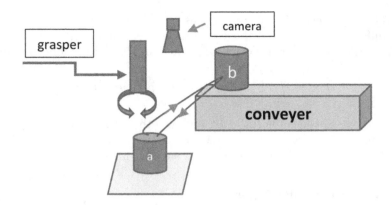

Fig. 1. The outline of the gantry robot performance.

3   Gantry Robot Models

The well known gantry robot (see Fig. 1) takes a can from position **a** and moves it along a specified trajectory in 3D space, puts it at a prescribed place (position **b**) and moves back for the next can – along the trajectory marked as backward arrow (see, e.g., [4,17] for a more detailed description). The possible role of a camera will be explained at the end of the paper, because most of ILC procedures work in the open loop and – for a fair comparison – we also consider the proposed method mainly as operating in the open loop way.

3.1   Continuous Time Model

Mathematical models of the gantry robot are examined for many years. At present the most accurate model, presented, e.g., in [17], has the form of three transfer functions that are expressed in terms of the Laplace transform. These transfer functions describe decoupled responses for each X, Y and Z axis of the gantry for an input signal. The decoupled models means that we can consider behavior of each axis separately. In our case study we concentrate on Z-axis only.

The following model has been identified for the gantry Z-axis (see [17])

$$G_z(s) = \frac{15.8869(s + 850.3)}{s(s^2 + 707.6s + 3.377 \times 10^5)},\tag{9}$$

where $G_z(s)$ is the transfer function for Z-axis. The desired trajectory for this axis is shown in Fig. 2 (left panel). It corresponds to moving a can from zero level (position **a** in Fig. 1) up to the conveyer (position **b** in Fig. 1).

3.2   Discrete Time Model of the Gantry Z-Axis

As a discrete time model of the gantry Z-axis we shall use the same model as in [17] in order to be able to compare the results of simulations. This model has

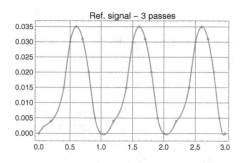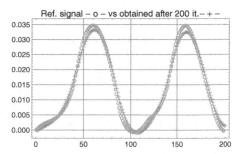

Fig. 2. Left panel – plot of $y_{ref}t)$ for Z-axis of the gantry robot – 3 passes. Right panel – plots of two passes of $y_{ref}t)$ and $y_{200}(t)$ for Z-axis of the gantry robot.

been obtained by constructing the minimal state space realization of $G_z(s)$ and then its discretization with the sampling rate 0.01 s. The state equations for n-th pass at discrete time instant k are the following:

$$x_{k+1}(n) = A_d\, x_k(n) + B_d\, u_k(n), \quad k = 1, 2, \ldots, \quad x_0(n) = x_0 \qquad (10)$$

$$y_k(n) = C_d\, x_k(n), \qquad (11)$$

where $x_k(n)$ is the state vector of the gantry at Z-direction, while the matrices A_d, B_d, C_d are defined as follows:

$$A_d = \begin{pmatrix} 0. & 1. & 0. & 0. \\ 0. & 1 & 1.0478 & 0. \\ 0. & 0. & -0.003 & 1. \\ 0. & 0. & -0.0008 & -0.003 \end{pmatrix} \qquad (12)$$

$$B_d = [0., 0., 0., 0.0313]^{tr} \qquad (13)$$

$$C_d = [0.0001, 0.0122, 0.0117, 0.]^{tr} \qquad (14)$$

In the simulations reported below, 200 samples and $x_0(n) = \bar{0}$, $n = 1, 2, \ldots$ have been used.

4 Results of Simulations and Comparisons

The following methodology of simulations has been applied.

- The ILOC algorithm has been applied to the gantry robot model with continuous time (c.t.). In fact, the simulations have been run using differential equations solvers but on a very fine grid.
- Then, the resulting (approximately) optimal input signal has been discretized with the step size .01 s. and fed as the input of the discrete time model. The reason for using such an approach is dictated by the fact that nowadays controllers work in discrete time and it is useful to know how the system would behave when the discretized version of the input signal is applied.

During the simulations the following parameter values have been used:

- the weighting factor in the control quality criterion – $\delta = 10^{-7}$,
- the step size of ILOC algorithm – $\gamma = 200$,
- time horizon – $T = 2$
- starting point – $u_0(t) \equiv 0$.

We have selected rather small $\delta = 10^{-7}$ in order to demonstrate tracking abilities of the proposed approach. Selecting larger δ would results in a control signal that is a compromise between the tracking accuracy and the energy of the input signal. The fact that we have a linear system, the quadratic criterion and a starting point relatively close to the optimal input signal, allow us to use a relatively large constant step size $\gamma = 200$ (see also Remark 1). In other cases it may be useful to use gradually decreasing step size $\gamma_k > 0$ such that the following conditions hold: $\lim_{k \to \infty} \gamma_k = 0$ and $\sum_{k=1}^{\infty} \gamma_k = \infty$, which are in the spirit of stochastic approximation algorithms.

The ILOC algorithm has been stopped after 200 passes and the following results have been obtained:

Tracking performance. In Fig. 2 (right panel) the reference signal is compared with the Z-axis output. The difference is visible near the turn point of the gantry. Notice however, that it is not too large. Indeed, the examination of the plot in Fig. 3 shows that the largest tracking error is equal to 2 mm. It can be reduced by allowing more energy usage for control, i.e., by reducing δ. We emphasise that our goal is to minimize J, which consists of the tracking error and the penalty for input signal energy. At the end of the paper we shall comment on how one can obtain even better tracking performance,

Optimal input signal – c.t. case. The approximation of the optimal input signal $u_{200}(t)$ is shown in Fig. 4 (left panel). The plot of the Frechet derivative in Fig. 3 (right panel) indicates that it is indeed near optimal signal $(\max |F_{200}(t)| = 1.5 \cdot 10^{-5})$. In Fig. 4 (left panel) this signal is compared to the signal that was used in [5] for control Z-axis of the gantry robot. As one can notice, a general shape is similar, although our signal has a higher amplitude, but it is smoother.

Sampled optimal input signal. Then, signal u_{200} was sampled with the rate 0.01 s. and fed as an input to the discrete time model (10). The resulting tracking error between (11) and the reference signal is plotted in Fig. 4 (right panel). The largest absolute difference is of the order 0.003.

Learning rate. The plot of $J(u_n)$ in the log-scale versus the number of passes (epochs of learning) is shown in Fig. 5 (left panel). It indicates that ILOC attains the linear rate of convergence except for the several first passes at which a large improvement of $J(u_n)$ can be observed. In general, this learning rate is faster than that of many ILC algorithms proposed so far.

Figure 5 (right panel) presents the mean squared error (MSE) of tracking in subsequent passes. The rate of decreasing of MSE is slightly slower than that in Fig. 5 (left panel) which means that the learning algorithm also puts emphasis on decreasing the energy of the input signal.

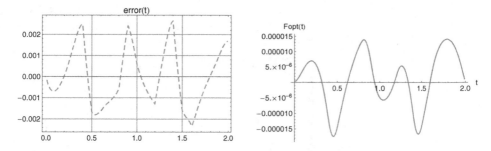

Fig. 3. Left panel – plot of the tracking error $e(t) = (y_{ref})(t) - y_{200}(t)$. Right panel – plot of the Frechet derivative after 200 passes. Simulations in continuous time in both cases.

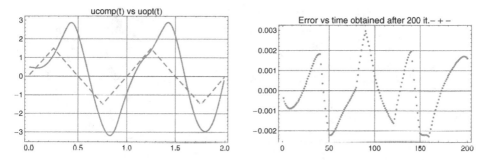

Fig. 4. Left panel – plot of $u_{200}(t)$ – approximately optimal input signal vs control signal from [5] (continuous time simulations). Right panel – plot of the tracking error obtained by sampling $u_{200}(t)$ and feeding it as the input of the discrete time model.

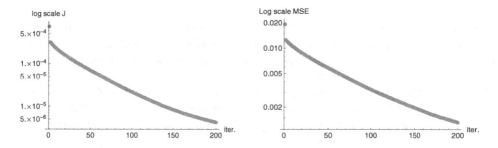

Fig. 5. ILOC learning rate. Left panel – control criterion $J(u_n)$ (in log scale) in subsequent passes. Right panel – MSE of tracking.

Learning the shape of the optimal input signal. In Fig. 7 (left panel) one can follow how the ILOC algorithm learns the shape of the optimal input signal. These shapes in subsequent passes are stuck and then plotted as a 3D plot. It is visible that the essential changes of shapes take place at the beginning of the learning process (up to about the 50-th iteration). A more detailed

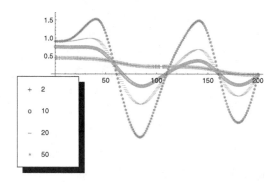

Fig. 6. Learning the shape of the input signal – initial passes.

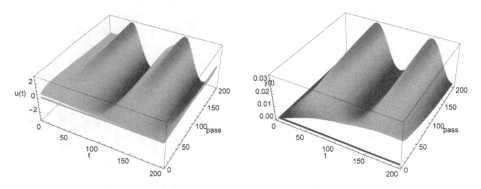

Fig. 7. Evolution of the shape of the input signal (left panel) and of the output signal (right panel).

description of the input signal shape at the beginning phase of learning can be observed in Fig. 6, which shows that the shape changes of u_n are relatively rapid from the initial passes up to pass 20 and then a fine tuning takes place (passes between 20 and 200).

Learning the shape of the optimal output signal. By smoothing properties of the system itself, even faster convergence can be observed for the system response – see the 3D plot in Fig. 7 (right panel). In general, the shape learning by ILOC is satisfactory.

5   Camera in the Loop and Conclusions

Our simulations show that the ILOC algorithm has a relatively fast rate of learning. Furthermore, the resulting (approximately) optimal input signal provides a sufficiently accurate output signal of the gantry robot grasper that tracks the reference signal with the accuracy not worse than 2 mm. Notice that the simulations have been run in the open loop, i.e., the output signals have been calculated using the mathematical model of the gantry. On the other hand, it is well known that the optimal input signal in the linear-quadratic problem can also be obtained in the closed loop realization: $\breve{u}(t) = K(t)\,(y_{ref})(t) - c^{tr}\,\breve{x}(t)$, where $K(t)$ is a solution of the corresponding Riccati equation. Thus, also in the case of iterative learning the following closed loop version can be applied $\breve{u}_n(t) = K_n(t)\,(y_{ref})(t) - y_n(t)$, where $y_n(t)$ is the system output. The fundamental difference between the open loop and closed loop realization is in that for the closed loop realization we have to supply $y_n(t)$ from a real system. In our case it is possible, when the gantry grasper trajectory is observed by a camera (see Fig. 1) , which operates in the closed loop. Such a configuration of the control loop has a potential to provide more accurate tracking, since it receives the true system output that is free from model errors. It may happen that measurement errors can reduce the benefits of the closed loop realization, but the detailed research in this direction is outside the scope of this paper.

Acknowledgements. This research has been supported by the National Science Center under grant: 2012/07/B/ST7/01216.

The authors would like to express their thanks to the anonymous referees.

References

1. Ahn, H.-S., Chen, Y.Q., Moore, K.L.: Iterative learning control: brief survey and categorization. IEEE Trans. Syst. Man Cybern. Part C Appl. Rev. **37**(6), 1099 (2007)
2. Bien, Z., Xu, J.-X.: Iterative Learning Control: Analysis, Design, Integration and Applications. Springer Science & Business Media, New York (2012)
3. Bristow, D., Tharayil, M., Alleyne, A.G., et al.: A survey of iterative learning control. IEEE Control Syst. **26**(3), 96–114 (2006)
4. Hladowski, L., Galkowski, K., Cai, Z., Rogers, E., Freeman, C.T., Lewin, P.L.: Experimentally supported 2D systems based iterative learning control law design for error convergence and performance. Control Eng. Pract. **18**(4), 339–348 (2010)
5. Hladowski, L., Galkowski, K., Cai, Z., Rogers, E., Freeman, E., Lewin Paul, L.: A 2D systems approach to iterative learning control for discrete linear processes with zero markov parameters. Int. J. Control **84**(7), 1246–1262 (2011)
6. Lee Jay, H., Lee Kwang, S., Kim, W.C.: Model-based iterative learning control with a quadratic criterion for time-varying linear systems. Automatica **36**(5), 641–657 (2000)
7. Moore, K.L.: Iterative learning control: an expository overview. In: Datta, B.N. (ed.) Applied and Computational Control, Signals, and Circuits, vol. 1, pp. 151–214. Springer, New York (1999)
8. Moore, K.L.: Iterative Learning Control for Deterministic Systems. Springer Science & Business Media, London (2012)
9. Moore, K.L., Chen, Y.Q., Bahl, V.: Monotonically convergent iterative learning control for linear discrete-time systems. Automatica **41**(9), 1529–1537 (2005)
10. Moore, K.L., Dahleh, M., Bhattacharyya, S.P.: Iterative learning control: a survey and new results. J. Robot. Syst. **9**(5), 563–594 (1992)
11. Moore, K.L., Xu, J.-X.: Editorial: special issue on iterative learning control (2000)
12. Nocedal, J., Wright, S.: Numerical Optimization, 2nd edn. Springer, New York (2006)

13. Owens David, H., Hätönen, J.: Iterative learning controlan optimization paradigm. Ann. Rev. Control **29**(1), 57–70 (2005)
14. Paszke, W., Rogers, E., Galkowski, K.: New KYP lemma based stability tests and control law design algorithms for differential linear repetitive processes. In: 52nd IEEE Conference on Decision and Control, pp. 2109–2114. IEEE (2013)
15. Rafajłowicz, E., Rafajłowicz, W.: Iterative learning in repetitive optimal control of linear dynamic processes. In: Rutkowski, L., Korytkowski, M., Scherer, R., Tadeusiewicz, R., Zadeh, L.A., Zurada, J.M. (eds.) ICAISC 2016. LNCS, vol. 9692, pp. 705–717. Springer, Cham (2016). doi:10.1007/978-3-319-39378-0_60
16. Rafajłowicz, E., Rafajłowicz, W.: Iterative learning in optimal control of linear dynamic processes. Int. J. Control Revised Vers. Under Rev. (2017)
17. Ratcliffe, J., van Duinkerken, L., Lewin, P., Rogers, E., Hatonen, J., Harte, T., Owens, D.: Fast norm-optimal iterative learning control for industrial applications. In: Proceedings of the 2005, American Control Conference, pp. 1951–1956. IEEE (2005)
18. Wang, Y., Gao, F., Doyle, F.J.: Survey on iterative learning control, repetitive control, and run-to-run control. J. Process Control **19**(10), 1589–1600 (2009)
19. Xu, J.-X.: A survey on iterative learning control for nonlinear systems. Int. J. Control **84**(7), 1275–1294 (2011)
20. Xu, J.-X., Tan, Y.: Linear and Nonlinear Iterative Learning Control, vol. 291. Springer, Berlin (2003)

An Approach to Robust Urban Transport Management. Mixed Graph-Based Model for Decision Support

Piotr Wiśniewski[(✉)] and Antoni Ligęza

AGH University of Science and Technology,
al. A. Mickiewicza 30, 30-059 Krakow, Poland
{wpiotr,ligeza}@agh.edu.pl

Abstract. In this paper, we present a mathematical model of public transport network, which can be used for generation of alternative routes during crisis situations. It is based on a mixed graph, where decision points are represented by vertices and track sections by edges. Route and vehicle definitions are also provided. We determine the objective function to select the most suitable route as well as the forbidden path set which contains paths that cannot be executed in real networks. The model definition is preceded by examples and analyses of different types of crisis situations.

Keywords: Graph theory · Decision support · Robust traffic management · Route planing · Public transport

1 Introduction

Traffic congestion in cities is constantly growing. At the same time, the level of environmental awareness is becoming higher. Because of these two major factors, we can observe that the significance of public transport in our everyday life is growing. This situation applies in particular to big agglomerations where traffic systems are highly developed and based on different means of transport. Crisis situations, which are usually caused by infrastructure conditions as well as random factors, are a significant challenge for entities responsible for urban transport management. In the most severe cases, the traffic on a section is interrupted, and an intervention of a traffic controller, which consists of an emergency disposition of vehicles, is necessary. Incorrect solution or lack of decisions in this case may destabilize the entire system and as a consequence cause significant financial losses and affect the image of a transport company. The problem of public transport continuity assurance was also taken into consideration by European Commission [1] which remarked dangers related to disruption of transport systems, giving as an example the eruption of Eyjafjallajökull volcano in Iceland in April 2010.

The subject of this work is related to several independent studies, such as road traffic analysis, graph theory and route planning. A similar approach is

© Springer International Publishing AG 2017
L. Rutkowski et al. (Eds.): ICAISC 2017, Part II, LNAI 10246, pp. 347–356, 2017.
DOI: 10.1007/978-3-319-59060-8_31

a robust route planning for passenger vehicles in city traffic [2] and its use in real-time systems [3]. The vehicle routing problem based on tree search can be used to generate optimal routes for vehicle drivers who need to reach their customers [4]. One of the proposed solutions with respect to public transport is a bus dispatching system based on stochastic processes [5]. Decision support systems for transportation purposes were also classified in a manner which includes market planning, vehicle fleet management and schedule creation [6]. Another related work presents a Petri Net based decision support system which can be used in case of temporal railway track closures [7]. Problems of optimal tram dispatching and scheduling in a depot were also discussed [8,9].

2   Overview of Crisis Situations in Urban Transport

According to the encyclopedic definition, the notion of public transport covers a set of activities related to the movement of people which is publicly accessible and performed using appropriate means. The basic means of transport used in urban areas are buses, trolleybuses as well as rail vehicles which include tramways, metro and commuter trains. In some cities, one can encounter traffic lines based on cable, sea or inland water transport. An example of an alternative means of transport is the Gdynia funicular railway, built in 2015. Figure 1a shows the track of the Gdynia funicular.

There are many factors that can negatively impact the functioning of urban transport. Because of its specifics this type of transit is exposed to different kinds of disruptions. The authors of [10] divide these crisis situations into four groups, regarding the factor which triggers the event: the whole transport system, rolling stock, infrastructure and human factor.

Regarding the scope of this work situations related to the whole transport system were not considered. The focus was put on events which can be allocated to groups 2–4 and which effect can be a temporal shutdown of a network section. Events described above can cause significant difficulties for urban rail transport, especially tramways which usually co-exist with road traffic. This situation is caused by many factors which include: lack of possibility to bypass a car that was broken down, small number of auxiliary tracks, dependency from road traffic as well as from external power supply. An example of such situation was shown in Fig. 1b, which presents the actions taken after the derailment of line 6 tramcar in Katowice. In the right part of the picture, a technical car is visible. The derailed section is lifted by the rail service using pneumatic cushions.

Rail vehicles and infrastructure producers are using various means which aim to minimize potential effects of a crisis situation. They include bi-directional cars as well as auxiliary batteries that enable to pass short distances during power shutdowns. Another solution is a monorail car on rubber tyres, equipped with a steering wheel and an auxiliary Diesel engine. These kind of vehicles are in service in Caen (France) and can move independently in case of a crisis situation [11]. Robust solutions used in infrastructure include mainly tracks separated from road traffic. The continuity of power delivery is assured by the possibility

(a) The Gdynia funicular.   (b) Tramcar no. 804 after derailment.

Fig. 1. An alternative means of transport and an example of a crisis situation.

to supply energy of one section from different power stations. The number of overhead line disruptions can be minimized by using compensated power line, where wire tension is automatically controlled. In order to describe tramway traffic, a notion of train is used. It can be explained as a car or set of cars that is moving according to a defined timetable or according to the orders of a traffic controller, on a determined route performing courses within a specific line or off-schedule.

Tramway networks usually contain single track routes where alternation occurs. The traffic on these tracks is controlled by the timetable, according to which cars meet at determined passing loops or by so called inter-loop signalling. This kind of network structure limits maximal route frequencies and may cause significant difficulties in case of any disruptions. Tramway traffic is managed by dispatchers whose task is to maintain correct and punctual operation of urban transport within the determined area. In particular places, e.g. on line junctions, traffic control posts may be located. In the traffic operator's head office usually central traffic and infrastructure control centres are situated as well as traffic control service whose employees dispose of dedicated cars.

3   Mathematical Model

In many cases, road and rail transport systems are represented by directed [12,13] or undirected graphs [14]. In this paper, both approaches were combined and a mixed graph was used. This kind of graph is one of the possibilities to represent a real traffic network and is given by Formula 1.

$$G = (V, E_1, E_2, \alpha, \omega, \lambda),\tag{1}$$

where:

- V is a finite set of vertices,
- E_1 is a set of directed edges,
- E_2 is a set of undirected edges,

- $\alpha : E_1 \longrightarrow V$ determines the beginning of a directed edge,
- $\omega : E_1 \longrightarrow V$ determines the end of a directed edge,
- $\lambda : E_2 \longrightarrow V \times V$ determines the endpoints of an undirected edge.

In order to simplify the notation, it was assumed that $E = E_1 \cup E_2$. This model can be applied to tramway networks. Undirected edges correspond to single tracks, which are operated in both directions, while directed edges correspond to tracks on double track sections. However, most of road networks contain more bi-directional streets, in tramway networks most of the tracks are operated in one direction.

While modelling a transport network, it is necessary to determine a weight function γ. It is a relation, where each edge is related to a non-negative number [15]. In real transport systems, each track section can be described by several parameters. Values of the weight function must be therefore multidimensional. Its general form is given by expression 2:

$$\gamma : E \longrightarrow \mathbb{R}^n_{\geq 0} \tag{2}$$

Unlike the approach where a mathematical model of a traffic network includes particular elements of the infrastructure and is based on its technical documentation [16], a decision was made to represent only selected decision points and connections between them. It is unnecessary to represent all the stops between decision points as their inclusion in the network model would not affect the generated path but only negatively impact the complexity of the analyzed problem. Stops and parking tracks can be occupied by a limited number of cars and thus function C, which assigns non-negative capacity to every vertex, was defined. The capacity can be expressed by an integer greater than or equal to zero, either in measurement units or in number of cars. The capacity function is given by Formula 3:

$$C : V \longrightarrow \mathbb{Z}_{\geq 0} \tag{3}$$

Edges of the model graph correspond to track sections which connect the decision points. According to the previous assumption, the model consists mainly of directed edges, while undirected edges represent sections utilized in both directions. In the analysed case, it was assumed that weight function γ defined by Formula 2 has values in form of a bi-dimensional vector which consists of the planed time of ride and length of a section, expressed in kilometers. In the network model it is specified by Formula 4.

$$\forall e \in E, \quad \gamma(e) = \{t, l\} \tag{4}$$

Similarly to the description of vertices, capacity function c, described by Formula 5 was defined. It determines the maximal number of trains which can occupy a section.

$$c : E \longrightarrow \mathbb{Z}_{\geq 0} \tag{5}$$

4   Model Creation

A graph model of a tramway network can be created basing on a network scheme. Taking into account the fact that this paper is related to path generating problem, modeling track systems in depots was omitted. Network model creation consists of the following steps:

1. Placing a graph vertex in every place where there is a: track junction, waiting point before a single track, terminus, parking track, initial stop, entrance or exit of a depot.
2. Connecting bi-directional track endpoints with undirected edges.
3. Linking the other vertices with directed edges regarding left-hand driving.

Distances in kilometers and times in minutes were assigned to each edge basing on current timetables. Edges connecting two vertices within the area of one stop were given a zero-weight. The same rule was applied to edges leading to single track sections.

A significant problem that appeared during model creation was the representation of single track endpoints. Model of a single track, based on the example of the section Bytom City Hall – Bytom Olimpijska Street, was shown in Fig. 2a. The presented fragment implies that there exists a path designated by a sequence of vertices $(2, 3, 1)$. In real situations following this path with an undirectional car is not admissible, because it would require a change of direction, which in this case is possible only on a loop or on a turning triangle. In order to eliminate the error of changing direction in a place where it is not physically possible, it was necessary to add to the model a set of forbidden paths. In this set, all paths consisting of two directed edges which are both incident to a single track endpoint were placed. If we define a path as a limited sequence of edges where every two edges are adjoining or the same [15] and $P(G)$ as a set of all admissible paths, the set of forbidden paths can be defined as all paths in form of:

$$P_z(G) = \{p \in P(G) : p = \{e_1, e_2\}\} \tag{6}$$

which run through sequence of vertices (v_1, v_2, v_3) and fulfill condition 7.

$$e_{1,2} \in E_1 \land \exists e_3 \in E_2 : v_2 \in \lambda(e_3) \tag{7}$$

Another case where it was necessary to define additional constraints to the model were junctions where double track sections cross with single tracks. Such a place can be illustrated with an example of Łagiewniki Targowisko stop in Bytom whose layout was presented in Fig. 2b. A car which turns left coming from the south or coming from the west towards the north needs to use a short part of a track operated in both directions. However, it is not a single track section so according to previous assumptions its weight vector is equal to zero. As in the case of the single track section shown in Fig. 2a there is a possibility to generate such paths in the graph that in real situations cannot be used because of track system constraints.

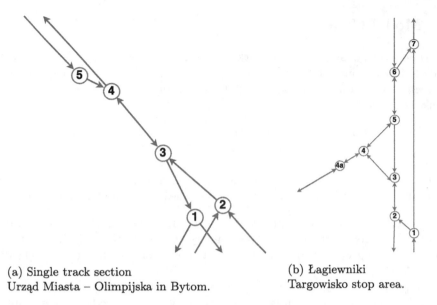

(a) Single track section
Urząd Miasta – Olimpijska in Bytom.

(b) Łagiewniki
Targowisko stop area.

Fig. 2. Model parts which caused difficulties.

Condition 7 excludes the forbidden change of direction when coming from the north through points 6 and 7 and coming from the south through points 1 and 2. In the analysed case, the track system does not allow also to generate paths running through the following sequences of vertices: $(3, 4, 5)$, $(4, 5, 3)$, $(5, 3, 4)$ and $(5, 4, 3)$. Solution to this problem was the definition of additional elements to the set of forbidden paths. The following conditions were formulated:

1. Entrance to an undirected edge with a non-zero weight vector from an edge whose weight is also different than zero.

$$e_2 \in E_2 \wedge \gamma(e_2) \neq 0 \wedge \gamma(e_1) \neq 0 \tag{8}$$

2. Entrance to a directed edge or an edge with a non-zero weight vector while the common vertex is incident to an edge with a weight vector equal to zero.

$$(\gamma(e_2) \neq 0 \vee e_2 \in E_1) \wedge \exists e_k \in E_2 : e_k \neq e_1 \wedge v_2 \in \lambda(e_k) \wedge \gamma(e_k) = 0 \tag{9}$$

Summarizing, a forbidden path is a path, whose any subsequence 6 fulfills at least one of conditions 7–9. In other words, during path generation the following rules should be obeyed:

- passing from one double-track section to another is not possible if there is an endpoint of a bi-directional track in between (condition 7),
- entry to a single track section may occur only directly from an edge with a weight vector equal to 0 (condition 8),
- if a vertex is an endpoint of an undirected edge any other edge cannot be chosen when generating path running through this vertex (condition 9).

According to condition 8 passing directly from one single track section to another is not possible. For this purpose every two neighbouring single tracks, between which there is a possibility to pass without changing the direction, were connected with an undirected edge with zero weights. Such a model modification required adding a new vertex for every such case. In Fig. 2b such vertex is named $4a$.

A tramway line can by defined as a route connecting terminuses, which is marked by numbers or letters [17]. In the graph model it corresponds to a set of paths which connect two vertices marked as terminuses. In order to identify terminuses a logical function $\tau : V \longrightarrow \{0, 1\}$ was defined on the set of vertices:

$$\forall v \in V, \quad \tau(v) = \begin{cases} 1, \text{if } v \text{ is a terminus,} \\ 0, \text{otherwise.} \end{cases} \tag{10}$$

Regarding the fact that the created model contains directed edges, a path generated between vertices v_1 and v_2 may be different from the path running in the opposite direction. In this purpose, a notion of variant which specifies the run of a selected route was introduced. In the graph model, a variant is therefore a path connecting two terminuses. Assuming that $p(v_s, v_t)$ is a path from v_s to v_t, a definition of a variant can be described by Formula 11.

$$w(v_s, v_t) = p(v_s, v_t) : \tau(v_s) = 1 \wedge \tau(v_t) = 1 \tag{11}$$

On the other hand, a tramway line can be determined as a set of variants marked by a specific number:

$$L(n) = \{w(v_s, v_t)\}, \quad n \in \mathbb{N}. \tag{12}$$

A train corresponds to a vehicle moving within the tramway network. This notion was explained in details in Sect. 2. This work covers planed trains, so the trains which operate on specific lines according to a predefined timetable. In the proposed model, a planed train is determined by the following parameters:

1. Train number, which can consist of a line number with a cardinal number.
2. Timetable, which contains departure times from initial stops. It can be presented as a set of pairs (t_d, w_i) where t_d is a departure time and w_i is an appropriate variant of a line determined by the train number.

5   Solving Method

In case of a crisis situation, it may be necessary to exclude a certain section from use. In the mathematical model, it is represented by a temporal removal of one or more graph edges. In this case, the trains whose routes contain the blocked section must be redirected to alternative routes, with respect to the closing time of the section. The route generation problem consists in finding the appropriate destination vertex and the shortest path to this point. Then, a return path to the original route is proposed.

Assuming that the network model was developed, lines and trains were defined along with the timetables, an alternative route can be generated. First, it is necessary to determine the blocked section e_x and its closing time t_x. The analysis is conducted for a selected day, therefore it is sufficient for the time to be expressed with one minute precision, starting from 12 AM that day. Secondly, the train for which an alternative route is generated must be localized. It may be done by a tracking system or basing on the timetable and edge weights. In the next step, it is necessary to determine subpath p_0 of variant w_{t_x}, that was found before, which begins in v_p. Generation of an alternative path is necessary if section e_x belongs to path p_0. The generated route should be as similar to the original path as possible. It is therefore crucial to define a profit function that include this assumption. The following form of the profit function is proposed:

$$Q(p_0, p_a, p_c, v_a) = \frac{d(p_c) + \phi(v_a)}{|d(p_0) - d(p_a)|},\tag{13}$$

where:

- p_a is the generated alternative path,
- p_c is the common part of paths p_0 i p_a,
- v_a is the endpoint of path p_a,
- $d(p)$ determines the length of path p,
- $\phi(v)$ is equal to 1 if v is the endpoint of path p_0 and 0 otherwise.

Embedding the length difference in the denominator enables to prevent situations where, in order to cover a maximal number of vertices, the alternative path would be excessively long in comparison with the original route. Next phase consists in removing edge e_x from the network model. Then, for each vertex v_{ti}, where $\tau(v_{ti}) = 1$, the following algorithm should be used:

1. Set $Q_{max} = 0$.
2. Generate the shortest path p_{ti} from v_p to v_{ti}.
3. Calculate time of ride t_{pt} from v_p to v_{ti}.
4. If in point v_{ti}, at time $t_x + t_{pt}$, there are other trains and their overall length added to the length of train n_p exceeds the capacity of vertex $C(v_{ti})$, choose next vertex and go to step 1.
5. Determine the value of profit function Q for path p_{ti}.
   (a) If $Q > Q_{max}$, mark p_{ti} as the selected path p_a and set $Q_{max} = Q$.
   (b) If $Q = Q_{max}$ and the length difference between paths p_{ti} and p_0 is lower than between paths p_a and p_0 or in point v_{ti} there are less trains than at the endpoint of path p_a, set $p_a = p_{ti}$,
6. The recommended route is p_a.

Because of the existence of the set of forbidden paths, defined by Formula 6 and conditions 7–9, it is not possible to directly use one of the available methods suitable for generating paths. One of the possible solutions is utilizing a modified Dijkstra algorithms which excludes forbidden paths by the time of searching.

Basing on the work [18] and assumptions made in Sect. 4, the algorithm determining the shortest route between two decision points, was formulated. The elimination of forbidden paths is done by assigning infinite distances to vertices which are reachable via such a path. In this purpose, membership conditions for the set of forbidden path were rewritten in such a manner that there is a possibility to verify it in reference to a path running through the selected vertices. It was admitted that z is the currently chosen vertex, $pred(z)$ is its determined predecessor and $d(v_1, v_2)$ expresses the length of the edge between two vertices.

A distance $l(u) = \infty$ is assigned to successor u of vertex z if at least one of the following conditions is satisfied:

1. Vertex z is incident to at least one undirected edge. Vertices $pred(z)$ and u are incident only to directed edges.
2. Edge weights $\{pred(z), z\}$ and $\{z, u\}$ are non-zero and $\{z, u\}$ is undirected.
3. Vertex z is incident to an undirected edge with a zero weight vector and edge $\{z, u\}$ is directed or its weight vector is different than zero.

Summarizing, the modified Dijkstra algorithm, which can be used to calculate route length from point s to t is as follows:

1. Assign to vertex s distance value $l(s) = 0$. For all other vertex, set temporary infinite distances. For each vertex $v \in V$, determine predecessor $pred(v) = 0$.
2. Determine the set of vertices which do not have a constant distance assigned $V_T = V \backslash \{s\}$. Set currently chosen vertex $z = s$.
3. For every successor u of vertex z, satisfying condition $l(u) > l(z) + d(z, u)$:
   (a) if at least one of forbidden path conditions is satisfied, then set $l(u) = \infty$,
   (b) otherwise assign new distance $l(u) = l(z) + d(z, u)$ to vertex u and determine its predecessor $pred(u) = z$.
4. From set V_T choose vertex x with the smallest distance l. Set $z = x$ and $V_T = V_T \backslash \{x\}$.
5. If $z = t$ then finish searching. Otherwise go back to step 3.

Length of such a path is expressed by distance $l(z)$, while vertices through which it is running are determined by checking predecessor until vertex s is reached, for which $pred(s) = 0$.

6 Conclusions

In the paper, we presented a new method of modelling tramway networks, which may be part of an intelligent decision support system for robust traffic management. In contrast to the existing methods, our approach is not limited to one specific transport system, but may be applied for different types of tramway networks. Thanks to the use of a mixed graph, it can represent systems containing both single and double track sections. As future works, we plan to extend the method by optimizing the route generation algorithm. In order to do so, the mathematical model may be extended by another structure. One of possible extensions could be an algebraic-logical meta model [19], which is used for solving discrete optimization problems.

References

1. European Commission: Continuity of Passenger Mobility Following Disruption of the Transport System, Brussels (2014)
2. Ernst, S., Ligęza, A.: A rule-based approach to robust granular planning. In: International Multiconference on Computer Science and Information Technology, Wisła, pp. 105–111 (2008)
3. Ernst, S.: Artificial intelligence techniques in real-time robust route plannings. Ph.D. thesis, AGH University of Science and Technology, Kraków (2009)
4. Mańdziuk, J., Nejman, C.: UCT-based approach to capacitated vehicle routing problem. In: Rutkowski, L., Korytkowski, M., Scherer, R., Tadeusiewicz, R., Zadeh, L.A., Zurada, J.M. (eds.) ICAISC 2015. LNCS (LNAI), vol. 9120, pp. 679–690. Springer, Cham (2015). doi:10.1007/978-3-319-19369-4_60
5. Adamski, A.: DISCON: public transport dispatching robust control. In: EWGT201 16th Meeting of the EURO Working Group on Transportation, Porto (2014)
6. Żak, J.: Decision support systems in transportation. In: Jain, L.C., Lim, C.P. (eds.) Handbook on Decision Making, pp. 249–294. Springer, Heidelberg (2010)
7. Fay, A.: A fuzzy petri net approach to decision-making in case of railway track closures. In: IFSA World Congress and 20th NAFIPS International Conference (2001)
8. Blasum, U., et al.: Scheduling trams in the morning. Mathe. Methods Oper. Res. **49**(1), 137–148 (1999)
9. Winter, U., Zimmerman, U.T.: Real-time dispatch of trams in storage yards. Ann. Oper. Res. **96**(1), 287–315 (2000)
10. Drdla, P., Buliček, J.: Crisis situations in urban public transport. Perner's Contacts **VII**(4), 35–40 (2012)
11. University College London: Innovative Technologies for Light Rail and Tram: A European Reference Resource. http://www.polisnetwork.eu/eu-projects/sintropher
12. Knaup, J., Homeier, K.: RoadGraph - graph based environmental modelling and function independent situation analysis for driver assistance systems. In: 13th International IEEE Conference on Intelligent Transportation Systems (ITSC), pp. 428–432 (2010)
13. Lückerath, D., Ullrich, O., Speckenmeyer, E.: Modeling time table based tram traffic. Simul. Notes Europe **22**(2), 61–38 (2012)
14. Schlechte, T.: Railway track allocation: models and algorithms. Ph.D. thesis, Technische Universität Berlin (2012)
15. Wilson, R.J.: Introduction to Graph Theory. Pearson, New Delhi (2007)
16. Lüttich, K., Brückner, K., Mossakowski, T.: Tramway networks as route graph. In: FORMS/FORMAT 2004 - Formal Methods for Automation and Safety in Railway and Automotive Systems, Braunschweig, pp. 109–119 (2004)
17. Chamber of Urban Transport: Urban Transport in Numbers. Explanation of Selected Definitions (In Polish). http://ssl.igkm.com.pl/ankieta/definicje.pdf
18. Jaworski, J., Palka, Z., Szymański, J.: Discrete Matematics for Computer Scientists. Part I: Elements of Combinatorics. Adam Mickiewicz University Press, Poznań (2007) (In Polish)
19. Grobler-Dębska, K., Kucharska, E., Dudek-Dyduch, E.: Idea of switching algebraic-logical models in flow-shop scheduling problem with defects. In: 18th International Conference on Methods and Models in Automation and Robotics (MMAR), pp. 532–537 (2013), pp. 35–40 (2012)

Street Lighting Control, Energy Consumption Optimization

Igor Wojnicki[✉] and Leszek Kotulski

Department of Applied Computer Science, AGH University of Science and
Technology, Al. Mickiewicza 30, 30-059 Krakow, Poland
{wojnicki,kotulski}@agh.edu.pl

Abstract. Using a graph formalism to model outdoor lighting
infrastructure has proven to be an efficient method for both design and
control. It has been tested not only on laboratory scale, but also in a city-
scale deployment. The paper proposes further energy usage optimization
if the streetlights are dynamically controlled. It is to alter the design
process taking into account influence of the control schemas. As a result
substantial energy consumption savings can be achieved. The introduced
optimization is also modeled with graphs and graph transformations.

1 Introduction

The most important in designing a lighting system is a proper selection of the
lighting class applied to a given area being illuminated. A *lighting class* is a set of
lighting requirements such as illumination, uniformity etc. The classes are enumerated e.g. *me2*, *me3*, *me4* etc. The lower the number the higher the class. In
general the higher the class the more light has to be provided by a light source.
According to CEN/TR 13201-1:2014 and CEN/TR 13201-1:2004 standard [1,2],
a lighting class can be dynamically chosen when the traffic intensity changes.
Thus as it is pointed out in [3,4] a control system that reacts on the information from sensors (eg. induction loops or ambient light sensors) and chooses a
proper lighting profile resulting in dimming of particular lamps, is feasible and
delivers energy savings [5,6]. A lighting profile ensures that at given circumstances a selected lighting class requirements are met on a given area. It defines
all luminaire parameters there. As a result the lighting level is adjusted to meet
particular needs. In case of an urbanized environment an area is some length of
a street – a segment.

Such a control system has been successfully tested in Krakow, Poland. It
covers 3,768 light points, and it gives more that 70% reduction of the energy
consumption, providing 1463,12 MWh energy savings annually. The main simplification of this project is an assumption that the lighting profiles for a given
segment are designed in isolation [7]. One segment is not influenced by light from
other segments.

© Springer International Publishing AG 2017
L. Rutkowski et al. (Eds.): ICAISC 2017, Part II, LNAI 10246, pp. 357–364, 2017.
DOI: 10.1007/978-3-319-59060-8_32

In this paper we show that this assumption results in some decrease of energy efficiency of the lighting system and we propose a solution. It occurs if neighboring segments have lighting profiles which meet different lighting classes. Without a dynamic control these loses are not significant, since there is not too many neighboring segments with different lighting profiles. However if a dynamic control is applied such a situation becomes more common.

The paper is organized as follows. Section 2 gives motivation and introduces lighting norms and lighting control opportunities. Section 3 presents the design process. Later, a graph-based control model is introduced in Sect. 4. The energy usage optimization is presented in Sects. 5 and 6.

2 Motivation

Outdoor lighting systems are usually designed and operated with several objectives in mind. These include conformance to lighting standards (i.e. provision of minimum required illuminance in given regions), minimization of power consumption, improvement of user comfort and aesthetic. In this paper we consider latter two.

The intensity of street lighting is defined by a lighting class applied to a given area being illuminated. There are different lighting class families for various types of areas: roads for motor vehicles, pavements, cycle lanes, areas, etc. According to the CEN/TR 13201-1:2004 standard [1,2] application of a given lighting class within a family depends on a number of parameters. Some of them, such as traffic volume or ambient light, usually change in time, and the standards allow for temporary change of the lighting class.

Each area, or more precisely a street segment in this case, is assigned one base lighting class. An alternative class can be applied temporarily when the values of parameters allow it, which results in less intense lighting. To assign a class is to perform appropriate photometric calculations to identify each of the luminaires parameters such as: tilt, rotation, overhang, spacing and dimming. All these parameters for a given area form a photometric design.

It needs to be noted that the quality of photometric design itself has direct influence on the efficiency of lighting, and the effort needed to prepare the design is one of the factors which determine the feasibility of such projects. Usually, design is performed *by hand* and verified using industry-standard software such as DIALux. To make such design process feasible, certain simplifications are applied. For instance, lamps in each series are assumed to be precisely in line and differences in their distances are ignored. This may cause the design not to fulfill the standards, or to be suboptimal with regard to energy efficiency. Fortunately, a new generation of software tools, which automatically calculate the optimum configuration, may remedy this problem. As shown in [7], they can save as much as 15% of energy consumed by lamps.

This issue becomes even more important when dynamic lighting is to be applied. It is because the designer must prepare not only the design for the base lighting class (e.g. *me2*), but also designs which assure fulfillment of values for each alternative lighting class (e.g. *me3* and *me4*). If other factors, such as the level of ambient light (significant during dusk or dawn), are to be taken into account, it may be necessary to prepare not one, but a few dozen of photometric designs for each street. The aforementioned automatic photometric design tools may prove even more useful in such scenario. Moreover, the performance of such tools can be improved e.g. using parallel computation [8].

3 Lighting System Design Process

The reports presented by consulting companies show that annual energy cost generated by 340 millions of streetlights worldwide is estimated at \$23,9 billion and grows. In [9] the following schemas of deployment for modern and efficient outdoor lighting systems are defined.

1. **Retrofit:** replacing HID (High-Intensity Discharge) (e.g., mercury-vapor, sodium-vapor, metal-halide) lamps with LEDs.
2. **Design:** optimization based on a uniform street layout, "customized" optimization relying on accurate coordinates (e.g., GIS-based) of the road and luminaries.
3. **Control:** dimming and scheduled dimming, scheduled lighting class reduction, adapting dimming level to actual environment state.
4. Other, unclassified methods, eg. improving reflective properties of road surface.

Such an approch has been adopted in a large pilot of the smart lighing system in one of the districts of Krakow, Poland. The sodium-vapor lamps have been exchanged by LEDs. Furthermore the lamps are remotely managed by *Owlet*, the Schreder mesh communication system which allows dynamically dim each luminaire.

The pilot area has been divided into 619 segments. A segment is a part of a street that has the same characteristics (eg. number of lanes, width, speed limit etc.). During the design phase for each segment a proper lighting profile is calculated for the base class and all the lower classes. No influence of any two segments on their design profiles is assumed.

In the real world some lamps might influence more than one segment. Formally according to the norm, the lamp influences a rectangle $12H \times 5H$ where H is height of the lamp and the longer side is parallel to the street. Practically at least a border lamp in the segment influences the next segment.

4 Control Graph

Let us discuss an example in Fig. 1, that is a Control Availability Graph (CAG), an excerpt from the one presented in [4] and practically used in the pilot deployment in Kraków, Poland. It represents luminaires, their configurations, lighting profiles and segments. To increase readability vertex indices represent particular vertex label (lower cased) followed by a number. For example $s1$ and $s2$ indicate two different vertices, both labeled S. There are two adjacent segments $s1$ and $s2$ and three luminaries: $l1$, $l2$, $l3$. The luminaries are adjacent to each other which is indicated by an edge labelled with adj. Due to the segment's adjacency the luminaire $l2$ illuminates partially both segments. For each of the segments there are two separate lighting configurations calculated: one supporting $me2$ and the other $me3$ lighting classes for each segment.

Because of the adjacency there might be multiple, even contradicting parameters to be sent to $l2$ which results in non determinism during control. In case if there is $me2$ required on both segments, the configuration $c1$ is engaged on $s1$ and $c3$ on $s2$. Because of $c1$ the luminaire $l2$ should be set at 100%, and because of $c3$ at 100%. There is no contradiction, the result is deterministic. However, in case of $c2$ and $c4$ engaged, or $c1$ and $c4$, or $c2$ and $c3$, the actual parameters for $l2$ are non deterministic being: $\{80\%, 70\%\}$, $\{100\%, 70\%\}$, $\{80\%, 100\%\}$ respectively.

In case of non deterministic parameters the control system selects the maximum from the set of multiple values. Regarding the above example cases it would be 80%, 100%, 100% respectively. Such an approach results in compliance with the lighting norms, with respect to the lighting levels, however some overlighting is present. In case of $c2$ and $c4$ engaged it is because $l2$ is dimmed less then defined by the configuration $c4$. As a result $l2$ is at 80%, instead of 70% which results in $s2$ being overlit.

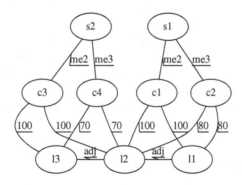

Fig. 1. General environment model, control availability graph.

5   Reduction of Overlighting

Up to now, we assumed that during the design phase there is no influence among luminaires in different segments. As a consequence the dimming values for particular luminaires providing a given profile are subject to optimization. They are selected minimizing a given goal function such as energy consumption. Usually these are the same or very similar values for the whole segment. However from the computing point of view the dimming can vary from luminaire to luminaire.

In the above example it is intuitive that if lamp $l2$ is dimmed at 80% instead of 70%, so the lamp $l3$ can be dimmed at a value less than 70% and the lighting class $me3$ will also be met in $s2$ then. We have to recalculate the profile for the $me3$ class in the segment $s2$ on a condition forced by a neighbouring segment $s1$ that power of the lamp $l2$ is set at 80%. After such a recalculation, and due to the luminaire adjacency and increased power of $l2$, $l3$ it is reduced from 70% to 61% in the case if the segments $s1$ and $s2$ meet the $me3$ class. It is reduced from 70% to 48% in the case if the segment $s2$ needs class $me2$. Analogically due to the luminaire adjacency and increased power of $l2$, $l1$ is reduced from 80% to 64% in the case when the segment $s1$ needs $me3$ and $s1$ needs $me2$. The above reduction has been defined by manual analysis of the of the graph presented in Fig. 1.

To reduce this overlighting and in consequence reduce energy usage by recalculating luminaire settings the following algorithm is proposed.

1. if there is a set of parameters P to be sent to a luminarire l for which $max(P) \neq min(P)$ do:
   (a) find a set of luminaries adjacent to l defined as

$$P_{adj} = \{l_{adj} : c, c_{adj}, s, s_{adj}, l, l_{adj} \in V,$$
$$(c, l), (s, c), (c_{adj}, l_{adj}), (s_{adj}, c_{adj}) \in E,$$
$$lab_V(c) - C, lab_V(c_{adj}) - C,$$
$$lab_V(s) = S, lab_V(s_{adj}) = S,$$
$$lab_V(l) = L, lab_V(l_{adj}) = L,$$
$$lab_E(l, l_{adj}) = adj,$$
$$s_{adj} \neq s,$$
$$att_V(c_{adj}, engaged) = true, att_V(c, engaged) = true,$$
$$lab_E(c, l) = max(P)\} \quad (1)$$

   where V is a set of vertices, E is a set of edges, $lab_E()$ and $lab_V()$ are edge and vertex labelling functions, and $att_V()$ is an edge attributing function,
   (b) for the luminaries in P_{adj} run photometric calculations decreasing power settings assuming that the power setting for l remains constant.

This approach, while simple and effective, poses one issue. The photometric calculations have to be run and enforced by the control system to ensure meeting the lighting norms. Such a recalculation might not be feasible, since it requires substantial computing power while controlling the lights. That is one of the main reasons to update the Control Availability Graph instead.

6   Recalculation

In order to prevent the above photometric calculations taking place during runtime the CAG has to be updated as it is shown in Fig. 2. However, it needs to be pointed out that such an action is a design process paradigm change. So far the control was driven by the design. Now, it is reversed and interaction between control and design is introduced. The design becomes driven by the control. In other words, all luminaire influences due to possible different control settings for neighboring segments have to be identified before the design process even starts. Each of the configurations *c1'*, *c2'*, *c3'*, *c4'*, in the Fig. 2, take into account lighting requirements of both *s1* and *s2* segments. *c1'* comes from merging *c1* and *c3*, *c2'* from *c2* and *c4*, *c3'* from *c1* and *c4*, *c4'* from *c2* and *c3* (see Fig. 1 for comparison). It enables to perform all necessary photometric calculations in advance.

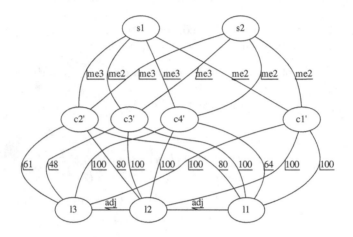

Fig. 2. General environment model, control availability graph, recalculated.

There is also a by-product of such an approach. The control system behaviour expressed with graph transformations becomes more complex. The transformations, which provide actual control, have to be altered in order to take into account multiple edges between a given configuration and a segment. Thus the definition of a lighting profile changes as well. It is no longer a set of parameters determining lighting conditions on a single segment. It becomes a set of parameters over multiple segments (see Fig. 2).

As a result there are multiple edges between a configuration and segments. This handles different lighting requirements for neighboring segments and influence of the luminaires on them. For example *c3'* is engaged if *me2* is required for *s1* and *me3* for *s2*. It would result in setting *l1*, *l2*, and *l3* to 100%, 100% and 48% respectively. Previously *l3* has been set at 70% so one can expect that the result of the complication of the control system generates energy savings. Unfortunately this in not a simple subtraction of the previous and current power of *l3*, because decreasing of the power of *l3* causes necessity of decreasing of the power of its successor luminaires in the segment *s1*, while increasing of the power of the next successor and so on. More formally let:

- $R(s1, me3)$ be an optimal power of all luminaries in the segment $s1$ for the lighting class $me3$ and
- $R(s1, me3)_{l2=80\%}$ be an optimal power of all luminaries in the segment $s1$ for the lighting class $me3$ on a condition that the luminaire $l3$ is set at 80%

then the energy savings is evaluated as $\Delta = R(s1, me3) - R(s1, me3)_{l2=80\%}$

Real savings taking into account the above under an assumption of 100 W luminaires are given in Table 1 for the example situation.

Table 1. Power savings

Segment	Lighting class			
s1	me2	me3	me2	me3
s2	me2	me3	me3	me2
Power savings for 100 W luminaires	0 W	8 W	20 W	14 W

For a pilot installation in Krakow, there is 3,768 luminaires and 619 segments. Let us assume that for a pair of neighboring segments there is only one common luminaire which illuminates both of them. A situation in which both segments meet the maximal norm statistically occurs in 27% of the time when outdoor lights are on. It is based on historical traffic intensity data. Having at least 8 W of savings at each lamp in such a case results in 1.7% of energy savings globally. Taking into account the annual energy consumption of the installation which is rated at 927 MWh, it leads to 16 MWh of savings. This is a lower bound estimation since often more than one luminaire would be influenced.

7   Conclusions

As it has been verified applying a dynamic control of street lighting results in substantial energy savings [3,4]. Use of the graph formalism enables it to be deployable over significant number of light points. However currently used design process leaves certain energy efficiency gaps which are subject to further optimization. These gaps regard a situation if there are different lighting requirements for neighboring segments. Such differences, while rare for a static lighting system, become common if a dynamic control is in place.

It has been observed that in order to provide even more energy efficient outdoor lighting, the design paradigm needs to be changed. Instead of providing all necessary photometric calculations first and applying a dynamic control later, the influence of the luminaires on the neighboring controlled areas has to be defined first. Having such knowledge the photometric design, taking into account this influence, can be established. It needs to be pointed out that without computer aided support during the design stage performing such calculations is not feasible.

Potential benefits in terms of energy savings coming from applying the proposed process are estimated at least 1.7% for a deployment of 3,768 luminaires. The actual number depends on the deployment scale and geographical features. It is expected to be much higher which is about to be confirmed by tracking of the performance in Krakow.

References

1. CEN: CEN/TR 13201–1: 2014, road lighting - Part 1: guidelines on selection of lighting classes. Technical report, European Commitee for Standardization, Brussels (2014)
2. CEN: CEN/TR 13201–1: 2004, road lighting. Selection of lighting classes. Technical report, European Commitee for Standardization, Brussels (2004)
3. Wojnicki, I., Ernst, S., Kotulski, L.: Economic impact of intelligent dynamic control in urban outdoor lighting. Energies **9**(5), 314 (2016)
4. Wojnicki, I., Ernst, S., Kotulski, L., Sdziwy, A.: Advanced street lighting control. Expert Syst. Appl. **41**(4), 999–1005 (2013)
5. Fan, S., Yang, C., Wang, Z.: Automatic control system for highway tunnel lighting. In: Li, D., Liu, Y., Chen, Y. (eds.) CCTA 2010. IAICT, vol. 347, pp. 116–123. Springer, Heidelberg (2011). doi:10.1007/978-3-642-18369-0_14
6. Guo, L., Eloholma, M., Halonen, L.: Intelligent road lighting control systems. Technical report, Helsinki University of Technology, Department of Electronics, Lighting Unit (2008)
7. Sędziwy, A.: A New Approach To Street Lighting Design. LEUKOS (2015)
8. Sedziwy, A.: On acceleration of multi-agent system performance in large scale photometric computations. In: Barbucha, D., Le, M.T., Howlett, R.J., Jain, L.C. (eds.) Advanced Methods and Technologies for Agent and Multi-Agent Systems, Proceedings of the 7th KES Conference on Agent and Multi-Agent Systems - Technologies and Applications (KES-AMSTA 2013), Hue City, Vietnam, 27–29 May 2013. Frontiers in Artificial Intelligence and Applications, vol. 252, pp. 58–67. IOS Press (2013)
9. Sędziwy, A., Kotulski, L.: Towards highly energy-efficient roadway lighting. Energies **9**(4), 263 (2016)

Various Problems of Artificial Intelligence

Patterns in Serious Game Design and Evaluation Application of Eye-Tracker and Biosensors

Jan K. Argasiński[(✉)] and Iwona Grabska-Gradzińska

Department of Games Technology, Jagiellonian University, ul. Łojasiewicza 11,
30-348 Kraków, Poland
{Jan.Argasinski,Iwona.Grabska}@uj.edu.pl

Abstract. In this paper, a general process of design and evaluation of serious games is presented. There are, obviously, qualitative and quantitative ways to analyze game dynamics (understood as mechanics activated by the user). Their general principles have been taken into consideration and a proposal of a modified pattern-based framework that allows for inclusion of data acquired from eye-tracker and biosensors used in affective computing. The paper concludes with a case study of design patterns in serious game that is currently in use as part of OHD training at the authors' Alma Mater.

1 Motivation

Serious video games are a relatively new way of training and learning. There is, of course, a rich and vivid history of teaching through immersion and play – from a multifaceted upbringing of ancient Greek paideia, through Prussian tabletop wargames (*Kriegsspiel*) in 19-th century, up to modern VR based training simulations. Varied influence of video games, especially for younger audience, was perceived very early. A good example is a game titled *Death Race* by Exidy from 1976, which was one of the first being criticized for brutality, even having regard, that it was just few monochromatic pixels crushing other pixels. It tells a lot about the amount of imagination invested in emerging new medium by its audience. This investment, a sense of presence, immersion and engagement is a cornerstone of serious games effectiveness. Through such mechanisms as procedural rhetoric, games constitute a new dimension of educational influence and are a very efficient cognitive catalyst of acquiring knowledge.

When it comes to evaluating video games, one of the most important factors is "playability" interpreted as "quality of gameplay". It is very an elusive factor yet crucial when it comes to overall rating of gaming experience. In the process of designing, game producers usually rely on designer's experience in the matter. There are a few practical methods for assuring "fun". Instead, there are some quantitative methods of post mortem evaluation of delivered product (i.e. questionnaire based on psychometry or collecting and interpreting "big" data about player's activity via game engine – especially accurate for MMO games).

© Springer International Publishing AG 2017
L. Rutkowski et al. (Eds.): ICAISC 2017, Part II, LNAI 10246, pp. 367–377, 2017.
DOI: 10.1007/978-3-319-59060-8_33

The main goal of work is to enhance area of research and evaluation of serious game design with intellectual framework supported by data from eye-tracker and affective computing oriented biosensors. The proposed sensor is based on conceptions of Björk and Holopainen's ("Patterns in game design" [1]) and *Mechanics Dynamics Aesthetics (Design Play Experience)* framework. *Knowledge, Skills, Abilities* attributes to organize conclusions on the effectiveness of analyzed game were also used.

2   Methods for Study of Players and Gameplay

There are, obviously, many methods of studying players and gameplay – some of them are qualitative and some quantitative. The main goal of the methods of first kind is to recognize and interpret the anatomy of games – focus on elements that construct ludic experience. A formal approach to qualitative study of playful phenomena includes study of actions that are (and that could have been) performed by the player in the light of apparent and hidden mechanics (called dynamics) and goals as conditions and outcomes that are important and valued by players (which makes them worth effort). The latter methods are usually concentrating around studying human behavior (both intentional and subconscious). Quantitative researchers believe that obtaining data about actual course of gameplay and its effects on the user (player) is the key to understand general principles of ludic attitude and building adequate models. The major problem here is the numerical framing of complex ideas. In later discussed case of serious game, it was easy to obtain quite well structured data but the problem is to correlate it with highly subjective and compound mental insights (such as affects). In the preliminary investigations, the research was based on measurements of blood volume pulse (by means of photo plethysmograph, PPG) and skin conductance (galvanic skin response, GSR). The interpretation of obtained data, relied on known and well documented procedures described in [19–21]. The authors of the study are aware of the imperfections of these methods and at the same time work on own solutions in this field – including those that use aid of eye-tracking devices.

3   Application of Patterns for Evaluation and Understanding of Players and Gameplay

3.1   Patterns in Game Design

Patterns are recurring themes that occur through the game's mechanics. Authors of book "Patterns in Game Design" [1] see them as "a language for talking about gameplay", where "each pattern describes a part of the interaction possible in games, and together with other patterns they describe the possible gameplay in the game". Every pattern contains a title, an example, a semi-formal description and (what is the most interesting) interrelated descriptions in terms of which other mechanics described pattern instantiates, modulates, is instantiated by, is modulated by, is potentially conflicting with. In this framework, the provided patterns are presented through their consequences, relations and references to the gameplay. Combining different mechanics is expected to lead to the formation of emergent entireness.

3.2   Serious Game Design

A popular way to represent the educational objectives is *Bloom's Taxonomy*. It locates the results of teaching in three areas: cognitive, psychomotoric and affective. A popular way to analyze games is an MDA framework that takes into account *Mechanics, Design and Aesthetic* features of the entertainment systems. Its expansion – DPE *(Design, Play, Experience)* point of view [see [11] for more information] is applied in research. The basic element in the process of designing serious game is the already discussed category of "fun" as a factor characteristic for the games while critical to the educational or therapeutic processes. In the paper, an approach to designing and evaluating affective video games that allows to capture (create metrics for) category of "fun" to design improvement purposes is presented.

3.3   Pattern Framework and DPE

The main elements one must take into consideration while designing a (serious) game are mechanics. They have already been considered (after Sicart [see: [9]) as elements that the player can interact with in order to change state of the system. As such, mechanics are main game-shaping elements. The first thing to do when analyzing serious games (in this case *Janek w opałach/the John in Distress* OSH training game) would be to extract main design patterns that govern the play. It can be done by application of Björk and Holopainen's before mentioned framework [1]. The most obvious selection includes patterns such as: Awatars, Buttons, Clues[e], Collecting, Controllers, Delayed Effects, Early Elimination, Exploration[e], Direct Information[e], Indirect Information[e], Extra-Game Information, Game State, Overview, Guard, Helpers, Irreversible Actions, Rescue and Survive. The conducted inquiry also showed presence of some misused patterns, such as: Tension[b], Identification, Illusion of Influence, Emotional Immersion[b], Consistent Reality Logic. What is interesting – the outcomes of some of the patterns can be evaluated by using biosensors (patterns marked[b]), and eye-tracking (marked[e]). For detailed description of patterns, their relations and consequences see [1].

The reconstructed gameplay overview can be interpreted in relation to DPE *(Design, Play, Experience)* model. It focuses on user experience seen as individual player's game story with accompanying affects and resulting engagement. These phenomena are associated with designer's practice (prepared story, mechanics, user interface) and particular course of play (resulting in storytelling, dynamics – "mechanics in motion" and interactivity). When the components are duly assembled, the result can be overall experience of "fun" (see Fig. 1).

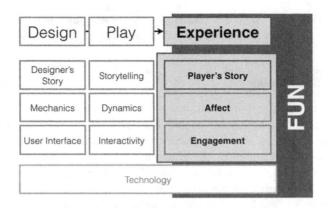

Fig. 1. DPE framework and category of "fun"

3.4   Affective Patterns

The ultimate goal of presented work is to propose the inclusion of the concept of patterns to DPE framework and a tool for describing mechanics, responsible for the "fun" factor in the games. Affective patterns would be "building blocks" from which emotional interactions with the user can be built in order to increase his/her involvement and ultimately control motivation and achieve educational and therapeutic effects.

Emotions are phenomena that are notoriously difficult to grasp. Traditionally, they are placed in areas outside the rationality, in the field of things inherently non-scientific [7]. Regardless that fact, systematic research on emotions is being conducted at least since the pragmatic turn in XIX-th century philosophy that gave psychological studies an equitable and scientific direction [22]. Theoretical concepts, developed by thinkers such as William James, established the foundation stones for the serious study of the phenomena of human emotions. What is important, from the very beginning they were interpreted in correlation with the physiological, bodily states [23]. Today, with the development of intelligent, the context aware technologies - interest in emotions returns to the sciences. Since the relatively recent time the tools that allow not only to look at the nature of emotive states but also for their modeling and simulation, have been developed.

In her book, considered to be foundational to the field of AfC, Rosalind Picard states that "affective computing is the study and development of systems and devices that can recognize, interpret, process and simulate human affects" [7]. The basic - not very difficult to accept nowadays - assumption of newly founded discipline was that emotions are both cognitive and physical, which might be the key to unlock quantitative empirical access to the significant range of mental processes. The main problem here is that "the range of means and modalities of emotion expression is so broad (...) [and] people's expression of emotion is so idiosyncratic and variable that there is little hope of accurately recognizing an individual's emotional state from the available data" [4]. Given that, there is a rather limited access to various bodily signals (in view of their overall quantity and frequency of changes) and it is highly possible that in the near future it will be possible to effectively distinguish only some types of affective phenomena. Context seems to be a great facilitator of interpretation here. Thus, context aware systems imply to be natural support of the work on emotional machines. Hence, video games - artificial, carefully controlled environments - prove to be the excellent testing ground for affective computing.

The project of building affective system requires some kind of definition of affective states and model of emotions to help distinguish from various psychophysiological events. Some kind of scheme for annexing different signals is necessary. In conducted investigations, wearable biosensors and typical gaming controllers (game pads, motion trackers, high dpi mouse) are used. Data obtained in this way has to be classified in a way that allows further interpretation. Available literature provides a multitude of proposals (such as well-known theories by James and Lange, Cannon and Bard, Schachter and Singer and many others). After a thorough analysis, the James-Lange based theory developed by Jesse Prinz was applied. The bottom-up concept of appraisal theory that comes from bodily phenomena seems more accurate for our purposes than highly cognitive-involved, intentional-oriented ones.

In the provided case study the Microsoft Band 2 and Cooking Hacks' Arduino-based e-Health sensor platform were used.

3.5   KSA: Knowledge, Skills, Abilities

The main purpose of the serious game analyzed in later part of this paper is to learn and test knowledge about conduct in case of fire. When evaluating educational media it is good to have some guidelines. The popular KSA statements, sometimes addressed as "quality rating factors", were taken into account. Definitions of individual elements are provided by (among others) U.S. Office of Personnel Management:

> Knowledge – is ready to use information content contained in the user direct access
> Skill – is applicable competence of performing learned psychomotor acts
> Ability – is available possibility to perform a behavior that has observable result.

The simplest way of evaluating serious games is to observe how the selected patterns lead to verify or learn knowledge, skills or abilities in the desired area.

4   Patterns Useful in the Serious Games

Immersion

Immersion is not relevant in the serious games as a value *per se*, but helps to gain all the educational aims, increase the motivation and memorization of achieved knowledge. Prior to investigation on the elements of Knowledge attribution, Skills learning, Ability training schema are taken into consideration.

There are four aspects of immersion distinguished in Björk and Holopainen's work: Cognitive immersion, Emotional immersion, Sensory-motoric immersion and Spatial immersion. In the serious games, cognitive or sensory-motoric abilities can be learned, depending of the aim of the game. For example, therapeutic games for people with neurologic disorders emphasizes sensory-motoric immersion. The OSH games usually try to reinforce the ability of making use of the knowledge and decision making under pressure of time. The emotional and spatial immersions are the means to make the learning process more effective and enjoyable.

Cognitive Immersion

Pattern: Clues. Clues, in the sense of Björk and Holopainen's pattern framework mean all game elements that give players information about how the goals of the game can be reached. Clues in *the John in Distress* are mainly the elements of HUD. Even the diegetic objects suggesting the action are strengthened by nearby non-diegetic elements. An additional tip is short distance between the problem and the solving element of the game.

The problem is that this convention is not implemented consequently. This interferes with cognitive priorities of surrounding objects and that is why the educational aim is not reached.

Fig. 2. Confusing diegetic and non-diegetic clues in OSH training game

The diegetic elements essential from the OSH point of view (in example: fire exit sign) are ignored by the player, because they are cognitively weaker than HUD elements. What is worse, this strategy gives the player reward (Fig. 2).

Pattern: Consistent Reality Logic. Interaction with clues is connected with getting close to it. Usually, the dialog box appears with further instructions and waits for player's decision. If the decision is wrong, the gravestone (R.I.P. sign) emerges followed by the end of the game. Another possible interaction after stepping the question mark sign is taking nearby object into inventory or, if inventory is not empty, replacing the content of the inventory with the current object (Fig. 3).

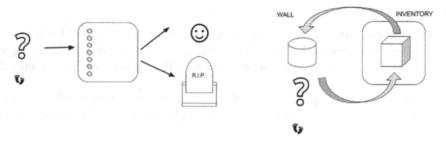

Fig. 3. Inconsistency in game reality logic (different things happen when player interacts with non-diegetic question marks)

The problem is that sometimes it does not work like that and getting close to the question mark causes death. It breaks the another Björk and Holopainen's pattern: Consistent Reality Logic. Taking something into inventory sometimes blocks the possibility of picking up anything else.

As can it be figured out, the consistency of the world logic is the main problem raised by responders in the questionnaires (Fig. 4).

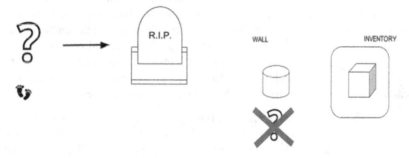

Fig. 4. Further inconsistencies.

Sensory-Motoric Immersion, Spatial Immersion
The John in Distress is a third-person perspective 3D game. The specific, unnatural perspective of isometric projection plus the way of controlling the character gives no chance to get the spatial immersion at all. Because of that, no motion camera based experiment was held, the study has been limited only to the observation of players during the gameplay. No significant motoric reaction was observed and no typical elements of identification were given. As participants noticed in the questionnaire, fact that all information about elements player interacted with are taken from the dialog boxes, not the zoom to the object itself, did not help with sensory-motoric immersion.

Emotional Immersion

Pattern: Identification. There are lot of game elements that fulfill the Identification pattern. The first of them is the status of the character. It is a student, similarly as target audience. Then, the player in the conversation with NPCs is forced to present himself by using the character's name.

Nevertheless, the changing fortunes of the character does not map in the affective sensors evaluation. Even the information of the death causes no measurable emotional reaction.

Pattern: Rescue. The same surprising lack of emotional engagement is noticed during getting into the rescue pattern. The player can save the life of five people and a cat, but - possibly because of no time pressure - success in this quest results in no signs of satisfaction and engagement.

Pattern: Tension. There are no examples of implementation of tension pattern in the game. This lack is affirmed in the affective sensors study. High initial level of excitation and engagement fall constantly during the gameplay, no matter what the final result is.

Knowledge Attribution

The right decisions in the game are rewarded with positive finale of the story. The problem is which elements of the gameplay cause the right reaction: whether they connected to the real word experience, or they are the element of the HUD or other object existing only in the game world. The player's decisions in the games, especially serious games, are usually based on the knowledge, both general or acquired through the game. Gaining knowledge during the game is important, as well as confirming and solidifying the knowledge achieved before. The important question is whether the game causal connections are translatable to the real word experiences.

Patterns: Direct and Indirect Information. The useful tool for investigating that problem is comparing the usage of patterns: Direct and Indirect Information. Pattern: Direct Information includes knowledge gained from elements of HUD – those objects which do not exist in the real word. With Indirect Information, the players obtain information about the game situation in a mediated way. They have to connect different information or enter into interactions with objects.

From this point of view, Indirect Informations are much more essential for the educational purposes than Direct ones, as they are relevant to the real word situations. Direct Information, especially this connected with HUD elements, lead the player to the success, but does not teach him the right reasoning.

- Example: HUD arrows vs. fire exit signs
  One of the most important decisions in the OSH game is the choice of the way to escape. In the real life, when the person does not know the building, he has to follow the escape sings rather than go in the headlong rush. The eye-tracker study showed that none of participants noticed the signs on the walls in the game world.
- Example: powder extinguisher vs. foam extinguisher.
  There are two types of extinguishers available in the game word, but they are distinguished on HUD level. Recognizing the type of extinguisher does not require

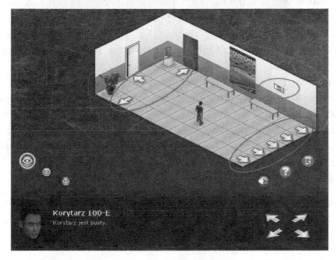

Fig. 5. No player noticed exit signs on the walls.

knowledge about standardized signs or is not available to gather during game interactions. Type of extinguishing agent is labeled in player's inventory (Fig. 5).

Patterns: Tools and Collecting. The only tools in the game available for collection are those mentioned before: powder extinguisher and foam extinguisher.

The usage of collected items is instantly connected with the place, where they were noticed. The effect is that the preoccupation of the inventory is rather low, because the need of usage is connected with the action of taking the element, not with the overview of the equipment. The small distance from the tool to the problem which tool is the solving one may have caused there is no eye-tracking mark of looking for the tool. It was hard to notice if any participant was looking around, while seeking for the needed objects.

Learning Skills
Björk and Holopeinen distinguished some patterns useful in the player's learning during the gameplay, e.g. Smooth Learning Curves, Experimenting, Gain Information, Gain Competence, Handicaps. Those kinds of activities are relevant in OSH games, as the situation of danger is hardly ever developed in typical way and very often people have to learn instantly and make use of objects around in the creative way.

In *the John in Distress*, none of the above-mentioned patterns are used. Only the already possessed knowledge can be used by the player. No elements of discovery, associating facts or getting knowledge from diegetic elements are implemented.

Training Ability

Pattern: The Show Must Go On. This pattern is present when the game state can change without any player's actions and especially when the player is forced to hurry up because of exhausting resources.

One of the crucial resources in OSH problems is the time itself, what is typified by the clock or by the exhaustion of the resources.

The John in Distress does not give the time to pass the elements. The fire is consuming the same area of the building all the time and NPCs are waiting for player's help for the whole eternity.

Using the affective computing methods, the curve of emotional engagement during the gameplay can be measured. After conducted study it was clear that when players realized there will be no consequences due to slow activity – the tension and engagement disappeared almost entirely.

Pattern: Deadly Trap. Some elements of rapid test of the player's knowledge using pattern of deadly trap were implemented. Unfortunately, there were no signs of emotional engagement during this action on the player's part.

5   Summary and Further Development

The main goal of the presented paper was to show a simple way of evaluating design of serious games. Some conceptual models for game design (MDA, DPE) and evaluation of educational effectiveness were used along with Björk and Holopainen's pattern framework to expose formal fabrication of selected serious game. It is proposed that this formal system should be expanded by usage of biosensors (in paradigm of affective computing) and eye-trackers. It is believed that along with further research, appropriate patterns could be developed.

Further work will be dedicated to disambiguate proposed method of inquiry and build set of unequivocal affective and perceptive patterns for evaluation of serious games.

Acknowledgement. The project has been (partially) supported by the grant of the Polish Ministry of Science and Higher Education number 7150/E-338/M/2015.

References

1. Björk, S., Holopainen, J.: Patterns in Game Design. Charles Rivers Media, Rockland (2004)
2. Cannon-Bowers, J., Bowers, C.: Serious Game Design and Development: Technologies for Training and Learning. Information Science Reference (2010)

3. Hudlicka, E.: Affective computing for game design. In: Proceedings of the 4th International North American Conference on Intelligent Games and Simulation (GAMEONNA), McGill University, Montreal, Canada (2008)
4. Hudlicka, E.: To feel or not to feel: the role of affect in human computer interaction. Int. J. Hum. Comput. Stud. **59**, 1–32 (2003)
5. Michael, D., Chen, S.: Serious Games: Games that Educate, Train and Inform. Thomson Course Technology, Boston (2006)
6. Pelachaud, C. (ed.): Emotion-Oriented Systems. Wiley, UK (2012)
7. Picard, R.: Affective Computing. MIT Press, Cambridge (1997)
8. Prinz, J.: Gut Reactions. A Perceptual Theory of Emotion. Oxford University Press, Oxford (2003)
9. Sicart, M.: Defining game mechanics. Game Stud. **8** (2008)
10. Yannakakis, G., Paiva, A.: Emotions in games. In: Calvo, R., D'Mello, S., Gratch, J., Kappas, A. (eds.) The Oxford Handbook of Affective Computing. Oxford University Press, Oxford (2014)
11. Ferdig, E.: Handbook of Research on Electronic Gaming in Education. Information Science Reference (2009)
12. Almeida, S.: Augmenting video game development with eye movement analysis. Universidade de Aveiro (2009)
13. Kiili, K., Koivisto, A., Finn, E., Ketamo, H.: Measuring user experience in tablet based educational game. In: Felicia, P. (ed.), Proceedings of the 6th European Conference on Games-Based Learning, ECGBL 2012, Cork, Ireland, 4–5 October 2012
14. Duchowski, A.T.: A breadth-first survey of eye-tracking applications. Behav. Res. Methods Instrum. Comput. **34**(4), 455–470 (2002)
15. Knoepfle, D.T., Wang, J.T., Camerer, C.F.: Studying learning in games using eye-tracking. J. Eur. Econ. Assoc. **7**, 388–398 (2009)
16. Richardson, D., Spivey, M.: Eye-tracking: characteristics and methods. In: Wnek, G., Bowlin, G. (eds.) Encyclopedia of Biomaterials and Biomedical Engineering. Marcel Dekker Inc., New York (2004)
17. Lindner, M.A., Eite, A., Thoma, G.-B., Dalehefte, I.-M., Ihme, J.M., Köller, O.: Tracking the decision-making process in multiple-choice assessment: evidence from eye movements. Appl. Cogn. Psychol. **28**(5), 738–752 (2016)
18. https://theeyetribe.com/. Accessed 01 Feb 2017
19. Calvo, R., D'Mello, S., Gratch, J., Kappas, A.: The Oxford Handbook of Affective Computing. Oxford University Press, New York (2015)
20. Scherer, K., Bazinger, T., Roesch, E.: Blueprint for Affective Computing. Oxford University Press, Oxford (2010)
21. Ortony, A., Clore, G., Collins, A.: The Cognitive Structure of Emotions. Cambridge University Press, Cambridge (1988)

Photo-Electro Characterization and Modeling of Organic Light-Emitting Diodes by Using a Radial Basis Neural Network

Shiran Nabha Barnea[1], Grazia Lo Sciuto[2], Nathaniel Hai[1], Rafi Shikler[1], Giacomo Capizzi[2], Marcin Woźniak[3(✉)], and Dawid Połap[3]

[1] Department of Electrical and Computer Engineering, Ben-Gurion University of the Negev, Beersheba, Israel
{nabha,mirilasn}@post.bgu.ac.il, rshikler@bgu.ac.il
[2] Department of Electric, Electronic and Informatics Engineering, University of Catania, Catania, Italy
glosciuto@dii.unict.it, gcapizzi@diees.unict.it
[3] Institute of Mathematics, Silesian University of Technology, Kaszubska 23, 44-100 Gliwice, Poland
{Marcin.Wozniak,Dawid.Polap}@polsl.pl

Abstract. In this paper we present a new RBFNNs neural networks based model to relate the overall OLEDs electroluminescent density as a function of the voltage and current at different wavelengths. The polymer-based OLEDs considered in this paper are realized in the Opto-electronic Organic Semiconductor Devices Laboratory at Ben Gurion University of the Negev. The simulation results show a good agreement between the experimental data and those obtained with the proposed model. This results prove that the model is capable of repeating and interpreting the experimental data.

Keywords: OLED · RBFNNs neural networks · Electroluminescent spectrum

1 Introduction

Organic light emitting diodes (OLEDs) are opto-electronic thin film devices that composed of an organic emitting polymer layer which sandwiched between two charged electrodes, one as a cathode and one as a transparent anode. When voltage is applied between the anode and cathode, the active polymer layer emits light termed electroluminescent (EL) [1].

Recent progress in OLEDs has attracted considerable attention because they have wide viewing angle, fast response, and the potential of being mechanically thin and flexible [2–4].

A typical OLED is composed of two electrodes and one or more active polymer layers enclosed between the electrodes. By modeling OLEDs, analysis of quantum and power efficiencies is made possible from the current-voltage and

© Springer International Publishing AG 2017
L. Rutkowski et al. (Eds.): ICAISC 2017, Part II, LNAI 10246, pp. 378–389, 2017.
DOI: 10.1007/978-3-319-59060-8_34

luminance behaviors. Several approaches have been followed to generate device level behavioral models. The numerical solutions of semiconductor equations of the devices are applied to accurately model the physics of devices [5]. Similarly by starting from basic device physics, microscopic or particle level simulation approaches provided the device electrical characteristics.

Recent examples of devoted computation techniques show that intelligent solutions can efficiently improve decision support systems [7]. Programming by the use of concern aware source code methodology makes the program running in more efficient way [6], it is also very important to use this coding for data extraction from images [8,9]. Similarly to these, device modeling is useful in many ways like optimization of design, integration with existing tools, prediction of problems in process control and better understanding of degradation mechanism.

Different types of theoretical models for OLEDs may be found in the literature [10–13]. The modeling of OLEDs were performed using complex models that consider each layer that composes the device, or by means of simplified models, which model only the OLED overall behavior [11].

In this paper we present an RBFNNs neural networks based model to relate the overall OLEDs electroluminescent density as a function of the voltage and current at different wavelengths.

The simulation results show a good agreement between the experimental data and those obtained with the proposed model. This results prove that the model is capable of repeating and interpreting the experimental data.

2   Experimental Set up

Device Architecture. The fabrication of the polymer-based OLED was realized in the Optoelectronic Organic Semiconductor Devices Laboratory (OOSDL) at Ben Gurion University of the Negev. The architecture of the OLED is shown in Fig. 1 and it is composed as follows:

- A Substrate (glass) having dimensions of 12 mm × 12 mm.
- An anode - a transparent conductor layer of Indium Tin Oxide (ITO) with dimensions of 6 mm × 12 mm and thickness of 100 ± 10 nm, and with resistance of $20\,\Omega^{-2}$.
- HIL/HTL - a conductive polymer layer (PEDOT:PSS).
- Active Layer (emissive) - a PFO layer with a thickness of 80–100 nm.
- Cathode - Calcium layer (30 nm) and a metal aluminum electrode (85 nm).

Fabrication of OLED Devices. The polymer used in this study was a blue-emitting polyfluorene, which is often employed in OLED devices due to its high luminescence quantum yield. The Energy levels of the OLED's structure are shown in Fig. 2.

From a chemical perspective, polyfluorenes are hydrocarbons consisting of fluorene units attached end-to-end, where a single fluorene composed of 6-5-6 carbon atom rings that are joined together.

We have realized 9 OLED devices, each of which were deposited on a glass substrate coated by 100 nm of ITO with resistance of $20\,\Omega^{-2}$. The device architecture is listed above and depicted in Fig. 1.

The substrates were precleaned in acetone, methanol and isopropanol using an ultrasonic bath (SK221OHP KUDOS) for 15 min. The cleaning process was followed by treatment with oxygen plasma (Plasma Preen II-862 Plasmatic Systems Inc.) for 5 min. On top of the ITO, a hole transporting and injecting material, (poly(3,4-ethylenedioxythiophene) polystyrene sulfonate) (PEDOT:PSS) (Ossila) was filtered in a $0.45\,\mu$m filter, and layered by spin-casting. The active layer was deposited by spin-casting from 1.5% xylene solution (15 mg PFO/1ml xylene). The electrode layer following the active layer composed of calcium (30 nm) aluminum (85 nm), was deposited using thermal evaporation through a shadow mask under high-vacuum ambient (10^{-7} Torr) and at rate of 0.1 kA/s.

Spin coating is a deposition fabrication technique for thin films. In this method:

1. the sample is coated with solution
2. the sample is rotated at controlled speed, acceleration and duration for spread the coating solution and removal of the majority of the solution
3. the solvent of the solution is evaporated and leave the molecules on the solved material on the sample surface.

The working principle of this PFO-based OLED is as follow: Electrons are injected from a low work function electrode (Aluminum) into the LUMO of the PFO, and Holes are injected from a high work function electrode (ITO) into the hole transport layer (PEDOT:PSS) that have a comparably high hole mobility. Then, holes are injected from the PEDOT:PSS layer into the HOMO of the PFO. The electron and the hole form an exciton in the PFO layer which can radiatively recombine under emission of a photons.

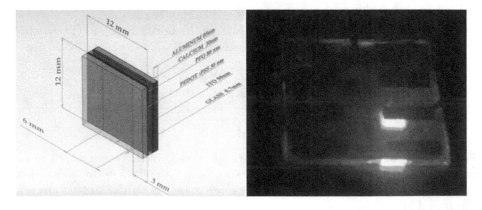

Fig. 1. The architecture of the OLED (left), PFO-based blue emitting OLED (right). (Color figure online)

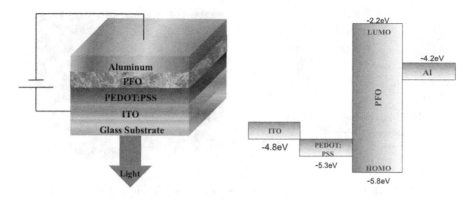

Fig. 2. Simplified layer structure of the OLED (left) with the related energy levels (right).

Electro-Optical Characterization. The electro-optical characterizations of the OLED devices were conducted using the following equipment: Agilent B1500A Semiconductor Device Analyzer (Keysight Technologies) for current-voltage (I-V) measurements, BLACK-Comet SR Spectrometer (Stellar Net Inc.) for EL spectra, and PD300 laser photodiode sensor (Ophir Photonics) for emission intensity.

The Agilent B1500A Semiconductor Device Analyzer is a tool for electrical characterization and evaluation of devices and materials. It provides a wide range of measurements, and supports major aspects of parametric test, from fundamental current-voltage (I-V) sweep to capacitance-voltage (C-V) characterization, and more. The measurements are done using probe-station with conducting needle-tip probes to provide contacts to the anode and cathode of the device. Each probe is connected to different source-measurement unit in the instrument that are controlled by the EasyEXPERT software. The Agilent B1500A Semiconductor Device Analyzer and illustration of the measurement using the needle-tip probes are depicted in Fig. 3.

The EL spectrum was recorded on a BLACK-Comet SR Spectrometer (Stellar Net Inc.). Stellar Net's BLACK-Comet SR are miniature spectrometers which equipped with concave gratings that deliver high performance for spectroscopy applications in the UV-VIS wavelength ranges, covering 190–850 nm. This spectrometer significantly improves spectral shapes by reducing comma and astigmatism found in plane grating spectrograph designs. The spectrograph architecture does not utilize mirrors, and therefore provides the lowest possible stray light in the UV with additional assistance from the holographic line grating. The concave grating produces a flat field on the CCD detector creating uniform resolution over the entire range. One of the main advantages of the Stellar Net is its ability to drastically reduce stray-light levels, 0.02% at 435 nm and 0.2% at 200 nm, the lowest values of any field spectrometer. By design, the holographic grating has a smoother surface than the normal ruled gratings used in competing spectrometer models. The results and stored data are displayed graphically on a PC.

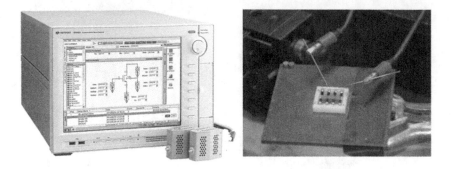

Fig. 3. Agilent B1500A semiconductor device analyzer (Keysight Technologies) (left), measurement probe station (right).

The Realized Devices. We have investigated the OLED's emitting area over 9 identical devices that were fabricated under the same conditions with the same structure. Each OLED contains 8 contacts while we measured only 6 of them in order to avoid shorts. Schematic illustration and practical realization of our devices are depicted in Fig. 4. As mention before, the current-voltage measurements were done using two conducting needle-tip probes in the range of 0 V to 18 V. In order to measure the EL spectra, a constant voltage was applied, and then the EL spectra were collected using the BLACK-Comet SR Spectrometer. We have checked 6 voltages in the emission range (10 V, 12 V, 14 V, 16 V and 18 V) and measured them 6 times each for different contacts. Because the spectrometer output is unitless, we used the PD300 laser photodiode sensor for the absolute emission intensity at these voltages.

Fig. 4. Geometrical model of the oled structure (left), Practical realization of our device (right)

Figure 5 shows the EL spectrum of PFO-based OLED. The emission spectrum for blue-emitting material shows different peaks, in particular at 438 nm, 466.5 nm and at 496 nm.

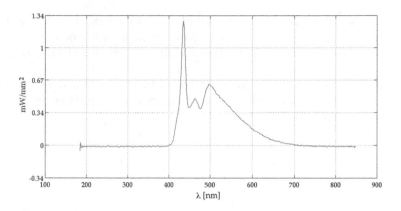

Fig. 5. The blue EL spectrum of PFO-based OLED (Color figure online)

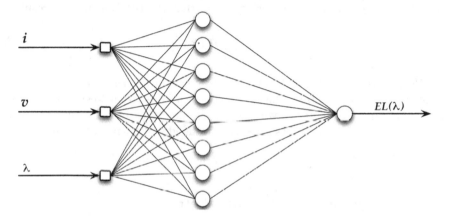

Fig. 6. The implemented radial basis neural network.

3   The Proposed RBFNN Based Model

The RBFNN proposed in this paper (Fig. 6) is developed in order to represent the electroluminescence density at different wavelengths as function of voltage and current for the realized OLEDs. The RBFNN is composed by three layers: an input layer, an hidden layer and an output layer. The input layer has three input (voltage, current, wavelength), the hidden layer is composed of 8 neurons, with Gaussian radial basis transfer function while the output layer is composed of one neuron with linear transfer function. The network has been trained by using the measured tension V and current I at different wavelengths as input and the measured electroluminescence density as target. The training of an RBFNNs network involves the determination of the number of neurons, the optimal values of the centers, the weights and biases. The criterion used to train the network has been that of minimize the sum of squared errors.

Note that a radial basis function network always has exactly three layers and that the input layer and the hidden layer are always fully connected because of the distance computation (that is, all coordinates are used to determine the distance).

In a RBFNN each pattern unit is activated by means of a RBF function f so that, given a centroids vector $\mathbf{c}$ and starting from the inputs vector $\mathbf{x} = \mathbf{x}^{(0)}$, the j-th pattern output $\mathbf{x}_j^{(1)}$ is

$$\mathbf{x}_j^{(1)} = f\left(\frac{\sqrt{\sum_k |\mathbf{x}_k^{(0)} - \mathbf{c}_j|^2}}{\beta}\right) \tag{1}$$

where β is a parameter that is intended to control the cluster distribution shape. Such an output is the given as input to another hidden layer (called summation layer) where a weighted sum is performed so that the i-th output $\mathbf{x}_i^{(2)}$ results

$$\mathbf{x}_i^{(2)} = \sum_j \mathbf{W}_{ij}\mathbf{x}_j^{(1)} \tag{2}$$

where $\mathbf{W}_{ij}$ represents the weight matrix. Such weight matrix consists of a weight value for each connection from the j-esime pattern units to the i-esime summation unit. These summation units work as in the neurones of a linear perceptron network. Incidentally such units give us also the global output of the network $\mathbf{y_i}$ therefore

$$\mathbf{y}_i = \mathbf{x}_i^{(2)} \tag{3}$$

3.1   Calculating and Adjusting the Centroids

The training process for a RBFNN is composed of two stages: find the number and the centers c_i and then find the weights $w_{i,j}$.

The number of basis functions (neurons of the hidden layer) is typically much less than the number of the training patterns. The centers of the basic functions don't necessarily have to belong to the set of training patterns, then the determination of the centers becomes part of the learning process.

To find the weights, the RBF Network must be trained with supervised techniques. Instead, regarding the centers, there are two strategies:

1. Use an unsupervised learning to determine a set of bump locations c_i and use LMS algorithm to train output weights $w_{i,j}$. This is a hybrid training scheme.
2. This strategy requires the training of all parameters: spread parameters σ_i (for our problem we use the same unit constant value for all the spread parameters), output weights $w_{i,j}$ and bump locations c_i through an supervised learning algorithm (often the backpropagation algorithm).

Training methods that separate the tasks of prototype determination and weight optimization (the first strategy) often do not use the input-output data from the training set for the selection of the prototypes. For instance, the random selection method and the k-means algorithm result in prototypes that are completely independent of the input-output data from the training set. Although this results in fast training, it clearly does not take full advantage of the information contained in the training set [14–16]. For this reason, in this paper, we have chosen the second strategy with the backpropagation algorithm. In fact the gradient descent training of RBF networks has proven to be much more effective than more conventional methods [17].

The Gaussian radial basis transfer function used for the neurons of the hidden layer is reported by the following equation:

$$\phi(r) = e^{-(r/\sigma)^2} \tag{4}$$

The radii σ_k are initialized to equal values according to the heuristics

$$\sigma_k = \frac{d_{max}}{\sqrt{2m}} \tag{5}$$

where d_{max} is the maximal distance between the input vectors of two training patterns.

Then we have calculated the weights, the radii and center vectors by iteratively computing the partials and performing the following updates:

$$E = \tfrac{1}{2} \sum_{j=1}^{m} (d_j - y_j)^2 = \tfrac{1}{2} \sum_{j=1}^{m} e_j^2$$

$$w_{i,j}^{new} = w_{i,j}^{old} - \eta_1 \frac{\partial E}{\partial w_{i,j}}$$

$$c_i^{new} = c_i^{old} - \eta_2 \frac{\partial E}{\partial c_i} \tag{6}$$

$$\sigma_i^{new} = \sigma_i^{old} - \eta_3 \frac{\partial E}{\partial \sigma_i}$$

where η_1, η_2 and η_3 are learning rate coefficients. While the optimal number of neurons in the hidden layer has been established by the trial and error method.

4   Experimental Data and Results

To investigate the operational limits of our basic OLED structure, current-voltage (I-V) and electroluminescence measurements have been performed at increasing driving voltages (10 V, 12 V, 14 V, 16 V and 18 V) for all the working devices (six).

Figure 7 shows the EL spectra of the realized OLEDs at different driving voltages. Notice that all the all the working devices has three peaks corresponding to 438 nm, 466.5 nm and 496 nm wavelengths as previously described.

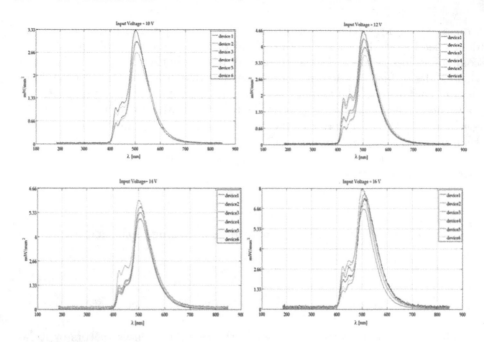

Fig. 7. The luminescence power of OLED emitting layer of the realized OLEDs at different driving voltages

More than ten parameters and characteristics describe the properties of OLEDs. Basic device characteristics are energy efficiency, OLED light intensity, maximum brightness, turn-on voltage, operating voltage, luminescence power etc. In this paper we are interested in modeling the dependence of the emission spectrum from the current and the wavelength at different voltages.

The measured EL spectra of the realized OLEDs shown a maximum spread between the different devices in all the operating conditions of about 10%.

The proposed model is able to accurately reproduce the spectra of the realized OLEDs (see Fig. 8). To verify our approach, the experimental data collected in laboratory have then been used to train the RBFNN and for testing. The extensive simulations show a good agreement between simulated and experimental spectra and also a maximum spread reduction between the different devices and the neural model in all the operating conditions of about 40%.

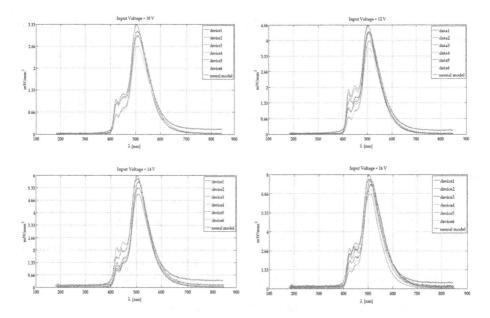

Fig. 8. The luminescence power of OLED emitting layer of the realized OLEDs and the simulated spectra by the implemented RBFNN at different driving voltages

5   Conclusion

The paper presents the structure, materials and operation principle of OLED with two organic layers. To investigate the device and predict its optical characteristics a new RBFNNs neural networks based model to describe the behavior of the OLEDs was proposed in this paper. The proposed model is able to accurately reproduce the spectra of the realized OLEDs and also to reduce the maximum spread reduction (use to the manufacturing process) between the different devices and the neural model in all the operating conditions.

The proposed RBFNNs neural model can be used as a tool to analyze and develop OLED drivers before manufacturing, saving time and cost in the design process. In future work, this model will be used for optimization and simulation of more complicated devices in comparison with experimental results and also emission angle are some issues that are likely to be addressed.

When a OLEDs is employed at high intensity, as required in many applications, the temperature of its substrate is a critical parameter and affects the luminous efficacy, the maximum light output, and the reliability. The thermal resistance of this layer should be as low as possible in order to facilitate heat transfer. For a given thermal conductivity of a material, the thermal resistance can be reduced by increasing the contact area or reducing the thickness. In both approaches to reduce the resistance, the mechanical stress of the layer will be proportionally larger, which can cause delamination [18,19]. However, it is impossible to directly measure the substrate temperature due to the encapsulation. For this reason the authors in future work will extend the model to include the thermal phenomena. So the extended model can be used for the temperature management of organic electronic devices and then as a design tool for OLED devices.

Acknowledgments. Authors acknowledge contribution to this project from the "Diamond Grant 2016" No. 0080/DIA/2016/45 funded by the Polish Ministry of Science and Higher Education.

References

1. Gómez-Bombarelli, R., et al.: Design of efficient molecular organic light-emitting diodes by a high-throughput virtual screening and experimental approach. Nat. Mater. **15**(10), 1120–1127 (2016)
2. Smith, J.T., Katchman, B.A., Kullman, D.E., Obahiagbon, U., Lee, Y.K., O'Brien, B.P., Raupp, G.B., Anderson, K.S., Christen, J.B.: Application of flexible oled display technology to point-of-care medical diagnostic testing. J. Disp. Technol. **12**(3), 273–280 (2016)
3. Kathirgamanathan, P., Bushby, L.M., Kumaraverl, M., Ravichandran, S., Surendrakumar, S.: Electroluminescent organic and quantum dot leds: the state of the art. J. Disp. Technol. **11**(5), 480–493 (2015)
4. Yang, X., Xu, X., Zhou, G.: Recent advances of the emitters for high performance deep-blue organic light-emitting diodes. J. Mater. Chem. C **3**(5), 913–944 (2015)
5. Shin, H.-J., Takasugi, S., Park, K.-M., Choi, S.-H., Jeong, Y.-S., Kim, H.-S., Oh, C.-H., Ahn, B.-C.: Technological progress of panel design and compensation methods for large-size UHD OLED TVs. SID Symp. Dig. Tech. Pap. **45**(1), 720–723 (2014)
6. Sulír, M., Nosál', M., Porubän, J.: Recording concerns in source code using annotations. Comput. Lang. Syst. Struct. **46**, 44–65 (2016)
7. Cpalka, K., Zalasinski, M., Rutkowski, L.: A new algorithm for identity verification based on the analysis of a handwritten dynamic signature. Appl. Soft Comput. **43**, 47–56 (2016). http://dx.doi.org/10.1016/j.asoc.2016.02.017
8. Grycuk, R., Gabryel, M., Nowicki, R., Scherer, R.: Content-based image retrieval optimization by differential evolution. In: IEEE Congress on Evolutionary Computation, CEC 2016, Vancouver, BC, Canada, 24–29 July 2016, pp. 86–93. IEEE (2016). http://dx.doi.org/10.1109/CEC.2016.7743782
9. Gabryel, M.: A bag-of-features algorithm for applications using a NoSQL database. In: Dregvaite, G., Damasevicius, R. (eds.) ICIST 2016. CCIS, vol. 639, pp. 332–343. Springer, Cham (2016). doi:10.1007/978-3-319-46254-7_26
10. Chen, H.-T., Choy, W.C., Hui, S.R.: Characterization, modeling, and analysis of organic light-emitting diodes with different structures. IEEE Trans. Power Electron. **31**(1), 581–592 (2016)
11. Bender, V.C., Barth, N.D., Mendes, F.B., Pinto, R.A., Alonso, J.M., Marchesan, T.B.: Modeling and characterization of organic light-emitting diodes including capacitance effect. IEEE Trans. Electron Devices **62**(10), 3314–3321 (2015)
12. Lin, R.L., Tsai, J.Y., Buso, D., Zissis, G.: OLED equivalent circuit model with temperature coefficient and intrinsic capacitor. In: IEEE Industry Application Society Annual Meeting 2014, pp. 1–8 (2014)

13. Savaidis, S.P., Stathopoulos, N.A.: Simulation of light emission from planar multi-layer OLEDs, using a transmission-line model. IEEE J. Quantum Electron. **45**(9), 1089–1099 (2009)
14. Capizzi, G., Sciuto, G., Napoli, C., Tramontana, E.: A multithread nested neural network architecture to model surface plasmon polaritons propagation. Micromachines **7**(7) (2016)
15. Lo Sciuto, G., Susi, G., Cammarata, G., Capizzi, G.: A spiking neural network-based model for anaerobic digestion process, pp. 996–1003 (2016)
16. Grazia, L.S., Capizzi, G., Salvatore, C., Shikler, R.: Geometric shape optimization of organic solar cells for efficiency enhancement by neural networks. In: Eynard, B., Nigrelli, V., Oliveri, S.M., Peris-Fajarnes, G., Rizzuti, R. (eds.) Advances on Mechanics, Design Engineering and Manufacturing. Lecture Notes in Mechanical Engineering, pp. 789–796. Springer International Publishing, Heidelberg (2017)
17. Karayiannis, N.B., Mi, G.W.: Growing radial basis neural networks: merging supervised and unsupervised learning with network growth techniques. IEEE Trans. Neural Netw. **8**(6), 1492–1506 (1997)
18. Chen, H.T., Tao, X.H., Hui, S.Y.R.: Estimation of optical power and heat-dissipation coefficient for the photo-electro-thermal theory for led systems. IEEE Trans. Power Electron. **27**(4), 2176–2183 (2012)
19. Qi, X., Forrest, S.R.: Thermal analysis of high intensity organic light-emitting diodes based on a transmission matrix approach. J. Appl. Phys. **110**(12), 124516-1–124516-11 (2011)

Conditioned Anxiety Mechanism as a Basis for a Procedure of Control Module of an Autonomous Robot

Andrzej Bielecki[1]([✉]), Marzena Bielecka[2], and Przemysław Bielecki[3]

[1] Chair of Applied Computer Science, Faculty of Automation,
Electrical Engineering, Computer Science and Biomedical Engineering,
AGH University of Science and Technology,
Al. Mickiewicza 30, 30-059 Kraków, Poland
azbielecki@gmail.com

[2] Chair of Geoinformatics and Applied Computer Science,
Faculty of Geology, Geophysics and Environmental Protection,
AGH University of Science and Technology,
Al. Mickiewicza 30, 30-059 Kraków, Poland
bielecka@agh.edu.pl

[3] Student Scientific Association AI LAB, Faculty of Automation,
Electrical Engineering, Computer Science and Biomedical Engineering,
AGH University of Science and Technology,
Al. Mickiewicza 30, 30-059 Kraków, Poland
bielecki01@yahoo.com

Abstract. This paper is devoted to the problem of self-control of autonomous robot in a complex, unknown environment. In such an environment it is impossible to predict all situations the robot could be faced with. Because of this it is necessary to equip the robot with control procedures that allow it to avoid dangerous scenarios. Mechanisms that serve to avoid threatening events have been worked out during evolution and living organisms are equipped with them. Conditioned anxiety is one of such mechanisms. In this paper the way in which this mechanism can be adapted to control of behaviour of autonomous robot, is presented. The effectiveness of the proposed approach has been verified by using V-REP simulator.

Keywords: Emotional robot · Autonomous agent · Conditioned anxiety · Artificial intelligence · Learning · Control

1 Introduction

Artificial intelligence (AI for abbreviation) systems are rooted in various originals. Most of these originals are connected with human cognitive abilities. AI systems based on various types of logic, fuzzy inference, syntactic algorithms, semantic nets and mathematical linguistics can be put as examples [12]. Furthermore, artificial neural networks (ANNs for abbreviation) are based on structures

© Springer International Publishing AG 2017
L. Rutkowski et al. (Eds.): ICAISC 2017, Part II, LNAI 10246, pp. 390–398, 2017.
DOI: 10.1007/978-3-319-59060-8_35

and functions of biological neural systems [26, 31, 35]. The aforementioned AI systems are universal ones in the context of solving a wide spectrum of problems. They are, however, time consuming and therefore possibilities of applying them to on-line problems are limited.

The idea of autonomous agent is the one in which functional possibilities of AI systems are used widely. On the one hand an agent is, usually, equipped with cognitive modules which enable the agent to perceive and understand the environment in which it acts [5–9, 11, 24, 25] and [12], Chap. 14. On the other hand, however, as it has been aforementioned, such types of systems are insufficient for solving on-line problems connected with the agent self-control. Therefore, a crucial stream of studies in cybernetics and robotics consists in searching effective mechanisms of on-line control. Biocybernetics, in particular neurocybernetics, is one of the areas of such studies [27–30] the more so because biological models are one of the crucial basis of cybernetics as a whole [34].

Autonomous robots as well as biological individuals are examples of embodied autonomous agents and can be studied in the frame of autonomous systems theory [4, 21]. They often operate in complex, unknown environments in which they can be faced with dangerous situations. Space missions, including exploration of Mars by using remote-controlled or autonomous vehicles, are examples of such risky ventures [2, 14, 15, 18, 33]. Therefore, there is a great demand for control procedures that enable to avoid by a robot situations that are potentially dangerous. Conditioned anxiety which has been created by evolution process is one of such mechanism [13, 16, 17, 19, 23]. It can be modelled and implemented in a robot. Such approach is situated in the studies concerning so called emotional agents in which models of emotions are important component of a decision making process [10, 22, 32].

In this paper a proposal of adaptation of the conditioned anxiety mechanism to control of autonomous robot is described. This mechanism enable the robot to learn quickly the circumstances of the threatening situation it has been faced with. As a consequence the robot will avoid in future similar circumstances because there is significant probability that they are correlated with the experienced danger. It should be mentioned that this paper is a continuation of the studies that concern scene analysis and self-control of autonomous agents operating in complex environment [5–9], in particular, the investigations concerning adaptation of conditioned mechanisms to AI systems [3].

This paper is organized in the following way. In the next section justification significance for carrying out this studies is presented. The way of adaptation of the mechanism of conditioned anxiety to control of autonomous robot is described in Sect. 3, whereas the tests of the proposed approach efficiency are described in Sect. 4.

2   Motivations

As it has been aforementioned, evolution worked out conditioned reactions that, among others, enables the living individuals to learn a proper reaction quickly in critical situations [17]. Such types of reactions are usually connected with strong emotions or, frequently, are triggered by them [17]. Autonomous robots lack of such mechanisms. Although emotions are tried to be modelled and implemented in robots as an element of a decision making algorithms - see [1,10,22,32] - the scheme which consists in modelling conditioned reactions and implementing them as computer and robotic algorithms seems to be a pioneering approach [3]. It seems, furthermore, that conditioned reactions have not been neither modelled cybernetically nor implemented as algorithms in connection with emotions.

Most of contemporary robots are remote-controlled. Such solution has, however, crucial drawbacks: communication can be broken especially in critical conditions, for instance where meteorological conditions are hard, during military operations or in space exploration. Furthermore, communication channels can suffer from low-bandwidth or high-latency, especially during interplanetary communication. Therefore, there is a great demand of high degree of autonomy for spacecrafts. This problem was articulated explicitly: "*The vehicles used to explore the Martian surface require a high degree of autonomy to navigate challenging and unknown terrain, investigate targets, and detect scientific events. Increased autonomy will be critical to the success of future missions.*" [2]. The Mars is explored intensively by using wheeled unmanned vehicles - until now three generates of vehicles can be constructed: Sojourner vehicle used during Pathfinder mission in 1997, Spirit and Opportunity - Mars Exploration Rover vehicles used during the mission in 2004 and Mars Science Laboratory vehicle used during the mission in 2009 [2]. Since it is impossible to predict all threatening events during a mission, the demand for control procedures that enables effective learning of proper reactions in critical situations is significant. On the other hand, studies concerning this topic are at the very initial stage. This paper is intended to fill partially this gap.

3   The Proposed Algorithm

In this section the proposed algorithm is described. In the first subsection the very idea rooted in neuropsychology is presented. In the next subsection the algorithm is proposed.

3.1   Conditioned Anxiety as a Mechanism of Control

As it has been aforementioned agents which act in a complex and unknown environment are not able to predict all situations they can be faced with. Therefore they should be equipped with mechanisms protecting them from dangers. Such mechanisms as conditioned aversion, conditioned anxiety and conditioned fear plays great role in evolution as a protecting mechanisms [13,16,17,19,23]. Let us consider their structure.

A biological individual confronted by a threatening event should avoid it in future. Such events are experienced together with some accompanying circumstances. Since there is probable that these circumstances are not accidental but, at least some of them coincide with the experienced danger, the strategy consists in avoiding them in future seems to be a good solution which should enable the individual to avoid the experienced danger again. Avoiding circumstances that proceeded the dangerous event is evidently even more efficient solution. The learning process which enables avoiding the dangerous events should be very quick because each such event can cause the death of the individual. Therefore, single experience of this type should be sufficient to learn effectively the proper schema of reaction. Conditioned anxiety is such a biological mechanism. If an animal was attacked by a predator in a strongly limited terrain, for instance in a cramped ravine, it should avoid cramped places in future provided that it managed to survive itself in confrontation with the attacker. Similarly, the animal which survived an attack of predatory bird on the field should avoid in future open spaces. Since, as it has been aforementioned, the learning process should be quick and effective, the victim associates the proceeding circumstances with the attack by generate the strong anxiety conditioned by these circumstances. In such a way claustrophobia or fear of open spaces is generated. On the one hand such mechanism is experienced by biological individuals as a great discomfort, especially if they have neurotic character. On the other hand, however, it is very effective as a source of a protective strategy that consists in steering clear of potentially dangerous places and situations.

3.2   The Proposed Algorithm

The conditioned anxiety generated in dangerous situation can be implemented in artificial systems in the following way. First of all let us remark that, from the cybernetic point of view, a mobile self-controlling robot has to be organized as a cybernetic autonomous system. The basis of the theory of such systems has been founded - see [20,21] - and is developed [4]. Thus, let us assume that the correlator, i.e. the module in which the knowledge of the autonomous system is stored and processed, is equipped with an array in which data registered by the robot receptors during a declared period of time, for instance last thirty seconds, are written and updated, let us say, every second. This array corresponds to short-term memory of biological individuals. If the robot got into dangerous situation, which can be recognized by it as threatening because of partial damage of the vehicle or serious problems with homeostasis or navigation after coming onto extremely difficult terrain, then the content of the short-term memory is written to the list of the arrays that encode the circumstances associated with the threatening events, one array for one event. This list corresponds to triggers that cause conditioned anxiety in biological systems. During the robot mission the list should be continuously compared to the contain of short-term memory in order to detect circumstances which are encoded as associated with a danger. If the current circumstances encoded in short-term memory are recognized as identical with one of the patterns written in the list, then the robot reversed itself as quickly as possible in order to change the circumstances. Since the procedure should act extremely quickly, the list should be implemented at the level of programmable electronics. Thus, the outline of the algorithm of the creating the mechanism which is analogous to conditioned anxiety can be written in the following way:

- analyze continuously the readings of the receptors;
- analyze continuously the readings of the detectors which track the inner state of the robot (let us call them: inner detectors);
- **if** readings from the inner detectors are alarming **then** *create* the new array in the list of arrays which encode the circumstances associated with the threatening events;
- *write* the readings from the receptors for the last time interval created array;

In the above notation of the algorithm, the crucial actions are written in italics. The *time interval* denotes the aforementioned declared period of time which defines uploading of the short-term memory.

The implemented conditioned anxiety-like mechanism acts in the following way (the outline):

- browse continuously the list of arrays which encode the circumstances associated with the threatening events;
- **if** the content of the short-term memory is similar to the content of one of the browsed arrays **then** immediately start to reverse;
- continue the reversion **until** the content of the short-term memory is similar sufficiently to the content of the array;

4   Experiments and Results

The algorithm proposed in the previous section has been tested by using V-REP (Virtual Robot Experimentation Platform) simulator.[1] The algorithm was implemented by using Robot Operating System (ROS).[2] The scene on which the robot operated had been created - see Fig. 1. Two types of obstacles were created on the scene. Big obstacles, which were destroying for the robot if it collides with them and the small ones, denoted on the scene as thin cubicoids with squares bases, which damaged the robot if it run onto it. It was assumed that the robot could be damaged two times by the cubicoids but the third time was destroying. It was also assumed that a damaged robot could continue its mission. Exploration of the scene in order to create its map was the task of the robot. A mission lasted 150 s. If the robot was not destroyed during performing the task, then the mission was regarded as successful. An example of the map created by the robot during a successful mission is shown in Fig. 2.

The robot, simulated as a three-wheeled vehicle, was equipped with two senses. The first one, the LIDAR (Light Detection and Ranging) can detect only big obstacles. The robot was equipped with the algorithm of navigation which allowed it to avoid collisions with big obstacles. Ultrasonic detector, fastened to

---

[1] http://www.coppeliarobotics.com.
[2] http://wiki.ros.org/.

Fig. 1. The scene.

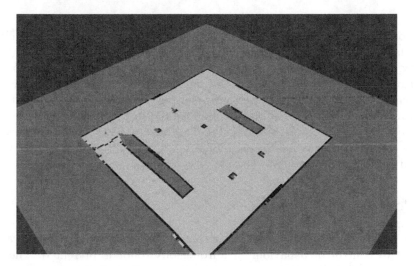

Fig. 2. The map of the environment - an obstacle floor map - created by the robot as a result of its mission - an example of a result of a successful mission.

the extension arm on the front of the robot was directed down and it can detect only cubicoids - see Fig. 3. The robot, however, was not a priori equipped with the knowledge that running onto cubicoids caused by its damage. The cognitive module of the robot had to associate detection of the cubicoid with the fact that the vehicle would be damaged a few seconds after a cubicoid detection. This association was created by using the algorithm described in the previous section. As it has been aforementioned, creating of the map of the scene was the task of the robot. The maps were created by using Simultaneous Localization and Mapping (SLAM) technique, realized by gmapping node available in ROS.

Fig. 3. The robot operating on the scene.

This node enables creation a 2-D occupancy grid map, for instance a building floor plan. The tests were performed both for the robot without conditioned anxiety, let us call it as the simple robot, and for the robot with the model of conditioned anxiety implemented in the way described in the previous section - let us call this type of robot as the emotional one. In each test the robot started at a randomly chosen point of the scene presented in Fig. 1. For twelve tests for the simple robot seven were unsuccessful. This means that 58% of tests ended as failures. For eleven tests for the emotional robot only three were failures which is equal to 27% of all tests.

5   Concluding Remarks

In this paper the way in which the mechanism of conditioned anxiety can be adapted to control of behaviour of autonomous robot is presented. The performed tests shown that the mechanism is effective. Probability of failure for the simple robot was over two times greater than for the emotional one. It should be stressed that the algorithm has been implemented in the extremely simple version. Therefore, there were no problems with associate the circumstances with the threatening event. Furthermore, the created scene was very simple as well. Thus, the obtained results should be regarded as preliminary.

References

1. Arbib, M.A., Fellous, J.M.: Emotions: from brain to robot. Trends Cogn. Sci. **8**, 554–561 (2004)
2. Bajracharya, M., Maimone, M.W., Helmick, D.: Autonomy for Mars rovers: past, present, and future. Computer **41**(12), 44–50 (2008)

3. Bielecki, A.: A model of human activity automatization as a basis of artificial intelligence systems. IEEE Trans. Auton. Mental Dev. **6**, 169–182 (2014)
4. Bielecki, A.: A general entity of life - a cybernetic approach. Biol. Cybern. **109**, 401–419 (2015)
5. Bielecki, A., Buratowski, T., Ciszewski, M., Śmigielski, P.: Vision based techniques of 3D obstacle reconfiguration for the outdoor drilling mobile robot. In: Rutkowski, L., Korytkowski, M., Scherer, R., Tadeusiewicz, R., Zadeh, L.A., Zurada, J.M. (eds.) ICAISC 2016. LNCS, vol. 9693, pp. 602–612. Springer, Cham (2016). doi:10.1007/978-3-319-39384-1_53
6. Bielecki, A., Buratowski, T., Śmigielski, P.: Syntactic algorithm of two-dimensional scene analysis for unmanned flying vehicles. In: Bolc, L., Tadeusiewicz, R., Chmielewski, L.J., Wojciechowski, K. (eds.) ICCVG 2012. LNCS, vol. 7594, pp. 304–312. Springer, Heidelberg (2012). doi:10.1007/978-3-642-33564-8_37
7. Bielecki, A., Buratowski, T., Śmigielski, P.: Recognition of two-dimensional representation of urban environment for autonomous flying agents. Expert Syst. Appl. **40**, 3623–3633 (2013)
8. Bielecki, A., Buratowski, T., Śmigielski, P.: Three-dimensional urban-type scene representation in vision system of unmanned flying vehicles. In: Rutkowski, L., Korytkowski, M., Scherer, R., Tadeusiewicz, R., Zadeh, L.A., Zurada, J.M. (eds.) ICAISC 2014. LNCS, vol. 8467, pp. 662–671. Springer, Cham (2014). doi:10.1007/978-3-319-07173-2_56
9. Bielecki, A., Śmigielski, P.: Graph representation for two-dimensional scene understanding by the cognitive vision module. Int. J. Adv. Robot. Syst. **14**(1), 1–14 (2017)
10. Camurri, A., Coglio, A.: An architecture for emotional agents. IEEE Multimedia **5**(4), 24–33 (1998)
11. Ferber, J.: Multi-Agent System: An Introduction to Distributed Artificial Intelligence. Addison Wesley Longman, Harlow (1999)
12. Flasiński, M.: Introduction to Artificial Intelligence. Springer International Publishing, Switzerland (2016)
13. Ganella, D.E., Kim, J.H.: Developmental rodent models of fear and anxiety: from neurobiology to pharmacology. Br. J. Pharmacol. **171**, 4556–4574 (2014)
14. Gat, E., Desai, R., Ivlev, R., Loch, J., Miller, D.P.: Behavior control for robotic exploration of planetary surfaces. IEEE Trans. Robot. Autom. **10**, 490–503 (1994)
15. Grotzinger, J.P., et al.: Mars science laboratory mission and science investigation. Space Sci. Rev. **170**, 5–56 (2012)
16. Kępiński, A.: Anxiety. PZWL (1972) (in Polish)
17. LeDoux, J.: The Emotional Brain. Simon and Schuster, New York (1996)
18. Lindemann, R.A., Bickler, D.B., Harrington, B.D.: Mars exploration rover mobility development. IEEE Robot. Autom. Mag. **13**(2), 19–26 (2006)
19. Maren, S.: Neurobiology of Pavlovian fear conditioning. Ann. Rev. Neurosci. **24**, 897–931 (2001)
20. Mazur, M.: Cybernetic Theory of Autonomous Systems. PWN, Warszawa (1966) (in Polish)
21. Mazur, M.: Cybernetics and Character. PIW, Warszawa (1976) (in Polish)
22. Maria, K.A., Zitar, M.A.: Emotional agents: a modeling and an application. Inf. Softw. Technol. **49**, 695–716 (2007)
23. Nesse, R.M.: Proximate and evolutionary studies of anxiety, stress and depression: synergy at the interface. Neurosci. Behav. Rev. **23**, 895–903 (1999)

24. Scheier C., Pfeifer R.: The embodied cognitive science approach. In: Tschacher, W., Dauwalde, J.P. (eds.) Dynamics Synergies Autonomous Agents. Studies in Nonlinear Phenomena in Life Science, vol. 8, pp. 159–179 (1999)
25. Tadeusiewicz, R.: Vision Systems of Industrial Robots. WNT, Warszawa (1992) (in Polish)
26. Tadeusiewicz, R.: Neural Networks. Akademicka Oficyna Wydawnicza, Warszawa (1993) (in Polish)
27. Tadeusiewicz, R.: Problems of Biocybernetics. PWN, Warszawa (1994) (in Polish)
28. Tadeusiewicz, R.: New trends in neurocybernetics. Comput. Methods Mater. Sci. **10**, 1–7 (2010)
29. Tadeusiewicz, R.: Place and role of intelligent systems in computer science. Comput. Methods Mater. Sci. **10**, 193–206 (2010)
30. Tadeusiewicz, R. (ed.) Theoretical Neurocybernetics. WUW, Warszawa (2009) (in Polish)
31. Tadeusiewicz, R., Chaki, R., Chaki, N.: Exploring Neural Networks with C. CRC Press, Boca Raton (2015)
32. Velásquez, J.D.: When robots weep: emotional memories and decision-making. In: Proceedings of the 15th National Conference on Artificial Intelligence, AAAI-98, pp. 70–75 (1998)
33. Volpe, R., Balaram, J., Ohm, T., Ivlev, R.: Rocky 7: a next generation Mars rover prototype. Adv. Robot. **11**, 341–358 (1997)
34. Wiener, N.: Cybernetics or Control and Communication in the Animal and the Machine. MIT Press, Cambridge (1947)
35. Żurada, J.M.: Introduction to Artificial Neural Systems. West Publishing Company, St. Paul (1992)

Framework for Benchmarking Rule-Based Inference Engines

Szymon Bobek[(✉)] and Piotr Misiak

AGH University of Science and Technology,
al. Mickiewicza 30, 30-059 Krakow, Poland
szymon.bobek@agh.edu.pl

Abstract. Rule-based systems constitute the state of the art solutions in the area of artificial intelligence. They provide fast, human readable and self explanatory mechanism for encoding knowledge. Due to large popularity of rules, dozens of inference engines were developed over last few decades. They differ in the reasoning efficiency depending on many factors such as model characteristics or deployment platform. Therefore, picking a reasoning engine that best fits the requirement of the system becomes a non-trivial task. The primary objective of the work presented in this paper was to provide a fully automated framework for benchmarking rule-based reasoning engines.

Keywords: Rule-based systems · Reasoning engine · Benchmark framework · Performance analysis

1 Introduction

Knowledge engineering aims at encoding expert knowledge in a form that can be processed by the machine to allow for automated inference. Among many formalisms that allow for such mapping from an expert domain to the machine domain, rules proven to be one of the most efficient and commonly used approach. Although nowadays artificial intelligence methods are dominated by machine learning-based solutions, rule-based knowledge representation is still used in areas such as mobile context-aware systems [15,20,26], recommender systems [2], or business process management [27].

The reason for constant popularity of this knowledge representation method can be explained by the fact that rules provide strong formal foundation in logics. Therefore, rule-based models can be formally evaluated in terms of completeness or consistency of knowledge [18]. Furthermore, rule-based inference is always traceable, which allows to provide explanations of systems decisions. Such feature is called intelligibility [19] and is one of the most important requirements for building personal mobile advisor systems [4]. The importance of the intelligibility of the artificial intelligence systems was recently highlighted by

© Springer International Publishing AG 2017
L. Rutkowski et al. (Eds.): ICAISC 2017, Part II, LNAI 10246, pp. 399–410, 2017.
DOI: 10.1007/978-3-319-59060-8_36

the European Union in General Data Protection Regulation (GDPR) proposal[1]. The GDPR states that every user of artificial intelligence system should have the *right to ask for explanation* of an automated algorithmic decision that was made about them [13]. This regulation aims at providing accountability of artificial intelligence, which can be achieved by rule-based knowledge representation.

Due to the long history of knowledge-based systems, dozens of different rule-based inference engines and rule languages were developed over last few decades. They differ in the type of formalism they are based on, ranging from propositional logic, to first-order logic. They use different pattern matching algorithms that have direct impact on the efficiency of reasoning in terms of CPU and memory usage. While there are several comparisons of the ontological reasoners, to the best of our knowledge there is no publicly available benchmark of production rule-based inference systems.

To address this issue, we proposed a toolkit for performing such a benchmark. We focused on developing a tool for measuring execution time and memory consumption of inference engines depending on the model characteristics, like number of rules, average length of the conditions within a single rule, type of attributes used in conditions, etc. The process of benchmarking is fully automated as presented in Fig. 1.

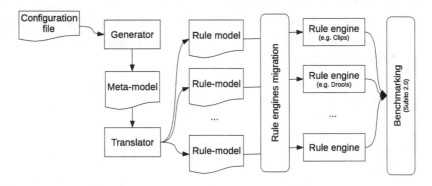

Fig. 1. Workflow of a benchmarking process of rule-based inference engines.

First, a meta-model is generated according to given configuration, that specifies the characteristics of the meta-model, such as number of rules, number of conditions, etc. The meta-model is an abstract rule model definition, that in the consecutive step is translated into specific rule notation, used by one of the inference engines that is benchmarked. If necessarily, benchmarking process may be preceded by the migration of the inference engine to other platform (e.g. Android or iOS operating systems).

---

[1] The full regulation text is available online. See: http://eur-lex.europa.eu/legal-content/EN/TXT/?uri=uriserv:OJ.L_.2016.119.01.0001.01.ENG\&toc=OJ:L:2016:119:TOC.

The rest of this paper is organized as follows. Section 2 presents the state of the art in the area of rule-based inference engines and most popular approaches for benchmarking them. Motivation for our work was given in Sect. 3. Section 4 presents the notation used in our approach for encoding rule-based meta-models. The process of automatic generation of the meta representation and its translation to specific rule languages is given in Sect. 5. The benchmarking process and sample results from the benchmarking tool are given in Sect. 6. Finally, Sect. 7 summarizes research presented in this paper and presents potential future works.

2 Related Works

Production rules are one of the most popular type of rules. They are defined in a form of IF <conditions> THEN <decisions> statements, where <conditions> represents a conjunction of logical expressions, and <deicsions> usually represents the assertion of a new fact to the knowledge base. The rule is selected for execution if all of its conditions are true. The reasoning engine is responsible for testing these conditions against the information that is located in a fact base and selecting rules that should be executed, if conditions of more than one rule were fulfilled.

The process of testing rules conditions is called *matching*, and the process of selecting rules for execution is called *conflict resolution*. These two steps of the inference process are critical in terms of resource consumption and vary in different reasoning engines implementations. Most of the implementations of rule inference engines use Rete-based algorithms for the matching phase. It is an efficient pattern matching algorithm for implementing production rule systems [11]. However, there are also other algorithms such as Treat [22] or Gator [14] that may outperform Rete in special cases. Different reasoning engines use different matching algorithms and reasoning strategies that may impact their performance.

One of the most popular rule-based engine based on Rete is Clips[2]. It provides its own programming language that supports rule-based, procedural and object-oriented programming [12]. The syntax of Clips rule language is based on Lisp. Clips supports knowledge modularization, which allows for creating separate Rete networks per module, and thus improves the memory efficiency of matching phase. Clips is written in C language, which makes it efficient and platform independent. Jess is another rule-based inference engine, and can be considered a Java implementation of Clips[3]. Jess introduces several enhancements with comparison to its ancestor in terms of modeling language. With respect to inference mechanism, Jess is also based on Rete. However, there is only one Rete network per knowledge base, not per module as it is in Clips.

One of the recent approaches for modeling and processing business rules is offered by Drools toolkit. It introduces the business logic integration platform which provides a unified and integrated platform for rules, workflow and event processing. The rule-based inference engine in Drools is called Drools Expert and

---

[2] See: http://clipsrules.sourceforge.net.
[3] See: http://herzberg.ca.sandia.gov.

is based on Rete-based algorithm – ReteOO. The main advantage of Drools is that it is written in Java and thus is highly portable between different operating systems, including mobile platforms.

In recent years, gain in popularity was also observed among engines that are dedicated to context-aware systems and those implemented in Java programming language, due to the high portability. HEARTDROID is a rule-based inference engine for Android, that is distributed under the GNU General Public License. It is not based on Rete, but instead optimizes the performance of matching by spiting the knowledge base into several modules that gathers similar rules to minimize the subsets for matching.

TUHEART shares the syntax with HEARTDROID, but the reasoning mechanism is based on the unification algorithm implemented in Prolog interpreter. It was not intentionally designed to work on the mobile devices, however due to recent advancements in Prolog and Java integration [28], it is possible now to execute TUHEART on Android-based platforms as well.

Among the vast number of other rule engines, there are also solutions that exploits Java programming language annotations, or external XML configuration files to transform regular Java objects into rules, or decision units. These are JRuleEngine[4], EasyRules[5] and ContextToolkit [10]. These inference engines do not use Rete or any other matching algorithm but depend solely on the mechanisms build in Java.

2.1   Frameworks for Benchmarking Performance

To the best of our knowledge, there is no comprehensive framework for benchmarking performance of the production rules systems. The existing comparative studies concern mostly the expressiveness power of languages supported by the reasoner, their portability capabilities, or other qualitative measures. However, there is lack of research that provides a full solution for comparison of production-based inference engines.

There are available public datasets of rules, that can be used for benchmarking rule-based engines, such as WaltzDB, or Manners[6]. These databases were used in several works to evaluate execution times of different matching algorithms and inference engines [8,21,30]. However, these works focus either on the evaluation of a pure matching algorithms, or on testing fixed subset of available reasoning engines and do not provide methodology for extending the benchmark to other systems. There exists also qualitative performance analysis of rule engines performed on different operating systems, like in [5], where authors discuss possible scenarios of migration of rule-based engines from desktop to mobile platform, and analyze potential negative impacts of this migration.

Although historically production rules systems were developed long before the ontology-based languages, there are a lot of benchmarks that tests the OWL

---

[4] See: http://jruleengine.sourceforge.net.
[5] See: http://www.easyrules.org.
[6] See: ftp://ftp.cs.utexas.edu/pub/ops5-benchmark-suite.

reasoners performance [6,29]. These reasoners use ontologies as their primary knowledge base instead of the production rules. Therefore, the discussion concerning them is out of the scope of this paper.

3   Motivation

The primary goal of the rule-based systems discussed in Sect. 2 is to support human or another artificial intelligence system in a decision process. However, the inference mechanism is rarely a standalone system, that is crafted to best fit the hardware requirements of the device it operates on. Most often, it is immersed in the larger artificial intelligence software as a submodule and works on top of an existing operating system.

In many cases, it is crucial to assure high performance of the reasoning engine in order not to slow other systems components, or not to waste valuable resources such as CPU time or memory. This is especially important on mobile platforms, where the resource efficiency is of the great value, and the reasoning needs to be performed usually under the soft real-time constraints. The available test sets for benchmarking rule-based systems, like aforementioned WaltzDB or Manners, contain fixed size of rules which are not modifiable. This allows only for measuring performance of engines against fixed criteria. On the other hand, the performance of the reasoners may be affected by many different factors such as the type of the operating system they operate on, the characteristics of the model such as size of the model, average number of conditions within a single rule, type of operators and type of attributes used in rules conditions.

Therefore, the primary motivation for the work presented in this paper was to provide a comprehensive framework for performing benchmarks of production rules inference engines. We aim at providing a comprehensive approach to measure the impact of the size of rules in the model, type of attributes (i.e. symbolic, numeric or mixed) and structure of the model (i.e. number of separate sets of rules that use the same attributes in conditional part). Such an analysis will allow for selecting inference engine that works best for the requirements of the specific systems. These requirements may vary on the countless factors, hence one of the objectives of our framework was to allow for automated generation of artificial models according to given specification and later provide a translation of these models into different rule-language used by different inference engines used in the benchmarking process.

4   Meta-model Representation

One of the first steps of the benchmarking process presented in Fig. 1 is a generation of a meta representation of rule-based model. The heterogeneity of different inference engines and rule notations makes it difficult to provide separate models for evaluation manually. Instead, we propose a meta-model notation, that can be generated with an automated tool and later translated into specific rule notation. This assures that every tested model shares the same characteristics.

There exists several approaches that allows for knowledge transfer among different rule-based notations. One of the most popular is Rule Markup Language (RuleML) [7] defined by the RuleML Initiative[7]. The objective of the initiative is to develop an open, vendor neutral XML/RDF-based rule language that will allow for exchange of rules between various systems. The more general solution was proposed in [16], where the author proposed a universal framework for knowledge interoperability and a knowledge base translation mechanism. However, both of these approaches use complex and not human readable formats. On the other hand, our goal was to provide a meta-representation that will be:

- Human readable – to allow reviews and changes in a meta-model by a person performing a benchmark and to allow for rapid prototyping of models for benchmarks.
- Modularized – to improve the maintenance of meta-models and to allow for their visualization.
- Formalized – to allow for automated generation and verification of meta-models (e.g. assuring that there exists a reasoning chain in the generated set of rules).

The notation we used is a subset of HMR+ language, which is the native rule language of HEARTDROID inference engine [3]. It is based on the Prolog syntax [9], and thus is easily readable by a human, but also ready for automatic processing by the machine. Furthermore, the language has also support of two visual editors: HQEd and HWEd[8]. The underlaying formalism used in HMR+ notation is the XTT2 rule representation [24]. Every HMR+ model contain rules definitions, attributes definitions used in rules and definitions of types of the attributes. Every type definition contains precise information on the domain of admissible values of the attribute. Additionally, every model contains *schema* definitions, which can be considered modules that gathers rules that use similar attributes in their conditional and decision parts. All of this allows for automatic generation and formal verification of models [23].

The simple example of a meta-model definition with a usage of HMR+ was given in Listing 1.1. It defines two types for two attributes respectively and one rule that depending on a day of a week decides weather it is a workday or a weekend. It is worth mentioning that the definition of type includes explicit enumeration of all domain elements. It also distinguishes the primitive type of the domain elements, which can be *numeric* for floating point and integer numbers and *symbolic* for strings and nominal values. This information is crucial for verification of completeness an consistency of a generated model. The HMR+ meta-model is generated with a tool, discussed in more details in the following section.

---

[7] See: http://wiki.ruleml.org/index.php/RuleML_Home.
[8] See: http://glados.kis.agh.edu.pl/doku.php?id=pub:software:hwed:start.

Listing 1.1. Sample meta-model definition

```
xtype [name: day_type, base: symbolic, ordered: yes,
      domain: [mon/1,tue/2,wed/3,thu/4,fri/5,sat/6,sun/7]].
xtype [name: today_type, base: symbolic, domain: [weekend,workday] ].

xattr [name: day, class: simple,  type: day_type].
xattr [name: today, class: simple,type: today_type].

xschm 'Today': [day] ==> [today].

xrule 'Today'/1:  [day in [mon to fri]] ==>  [today set workday]
```

5   Model Generation and Translation

The meta representation of rule-based knowledge presented in previous section is automatically generated and later translated to different rule-based notations used by the inference engines that are in consideration for benchmarking. This section discusses these two steps in more details.

5.1   Meta-model Configuration and Generation

The process of automatic generation of HMR+ meta-models is supported by a dedicated tool[9]. It uses a configuration file, which allows to specify the characteristics of a model to be generated. The description of selected fields of a configuration file that can be used to customize model was given in Table 1. The full specification is available on-line, on a project website[10].

Value of each field can be set with a usage of #define directive. For instance to force the generator to create 100 different types in the meta-model, with a ratio of 2 numeric types per each 3 symbolic types, one should include the following line in a configuration file:

Listing 1.2. Fragment of an configuration file for meta-model gerneator

#define TYPES_AMOUNT 100
#define BASES_RATIO (2/3)

Given the configuration file, the generator creates a rule-based model encoded with HMR+ notation. The model generation is a stochastic process, and the subsequent calls can produce different models. However, these differences are always within the bounds defined in the configuration file.

The generator tool also guarantees, that there always exists a reasoning path that covers all the rule schemas that were generated. This assures that the size of the model corresponds to the size of the reasoning chain. The output of the generator in the process of benchmarking is then passed to the translation module, discussed next.

---

[9] The complete documentation of the generation process, with sample models used for benchmarks in this paper, can be found on the project website: https://bitbucket.org/sbobek/xtt-generator.

[10] See: https://bitbucket.org/pm1234/xtt-generator.

Table 1. Fields of configuration file of meta-model generator tool.

Filed name	Description
TYPES_AMOUNT	The amount of different types to generate
BASES_RATIO	The ratio of a number of numeric types to number of symbolic types
DOMAIN_SIZE	Indicates how many elements should types' domains contain
ATTRIBUTES_AMOUNT	Number of attributes to be generates
SCHEMAS_AMOUNT	Demanded number of schemas to be create
MIN_ATTRS_AMOUNT	The minimal number of attributes that form a conditional and decision parts of rules
MAX_ATTRS_AMOUNT	The maximal number of attributes that form a conditional and decision parts of rules
MIN_RULES_PER_SCHEMA	Indicates the minimal number of rules within a single schema

5.2 Translating Models

The inference engines presented briefly in Sect. 2 are based on production rules. Although the specific rule languages used by them are different, they share the same concepts of the underlying rule definition [16]. This allows for relatively easy translation from one language to another. It is worth noting that the translation is performed on the level of single rule. Therefore, other components of the modeling language such as modules definition or inference control sequences are not translated.

Listings 1.3 and 1.4 represent the same rule, encoded with Jess and JRule notations respectively translated with a usage of the translation module[11].

Listing 1.3. Sample rule in Jess notation

```
( defrule today/1  ""
( day  ( value  ?v  $  : ( member$  ?v  ( create$  saturday  sunday  ))))
=>
  ( assert  ( today  ( value  [ weekend ] )))
)
```

Listing 1.4. Sample rule in JRuleEngine notation

```
< rule  name="today/1"  description="none"  >
    < if  leftTerm ="[ saturday  , sunday ]"  op=" containsatleastone "
        rightTerm =" engine . today . getDay"  />
    < then  method=" engine . today . setToday"  arg1=" weekend"  />
</rule>
```

Translated models are then passed to the benchmarking module discussed next.

---

[11] The translation module source code can be found on the project website: https://bitbucket.org/sbobek/hmr-converter.

6 Benchmarking

The benchmarking process is fully automated by the Subito 2.0 software[12] [17].
It reads the models delivered by the translation module and executes them in a
forward chaining mode. For each engine and each model, the execution time and
memory consumption is calculated using the build-in Unix `time` command. The
results from the execution are aggregated in CSV files. Additionally Octave[13]
scripts that plot the results are generated. The sample output from the benchmarking tool was presented in Figs. 2 and 3. The former presents the dependency
between the type of attributes used in the rules conditions and the execution
time of the reasoning process. The latter shows the differences in execution time
with respect to the number of rules in the model. Crosses on the plots represent
last successful reasoning before the time of inference exceeded a 30 s threshold,
or model failed to be loaded. It can be observed that the type of attributes
does not affect the reasoning performance for most of the engines, except for
TUHEART, HEARTDROID and Drools. This information may be valuable when
assuring stability in reasoning times is crucial. On the other hand, analyzing
performance of the engines with respect to the number of rules present in the
model, it is easier to select the most scalable solution.

As it was mentioned in Sect. 5, the generator module assures there exists
exactly one reasoning chain in the model, starting at a point defined by generated
initial state. The Subito 2.0 allows to artificially duplicate models, to simulate
disjoint reasoning chains. Additionally, it allows to specify number of runs that
should be performed before the results are averaged and aggregated. Currently,
Subito supports following reasoning engines: Clips, Jess, Drools, EasyRules,

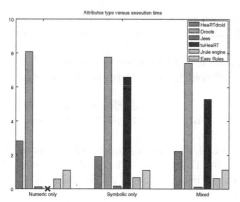

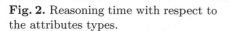

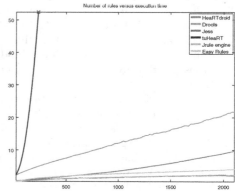

Fig. 2. Reasoning time with respect to
the attributes types.

Fig. 3. Reasoning times with respect to
the number of rules in the model.

---

[12] See: https://bitbucket.org/sbobek/subito.
[13] GNU Octave is a high-level interpreted language, primarily intended for numerical
computations. See: https://www.gnu.org/software/octave.

JRuleEngine, ContextToolkit, DTRules, HEARTDROID, and TUHEART. Including additional engines is straightforward, however, due to space limitation will not be discussed here.

7   Summary and Future Works

In this paper, we presented a complete framework for performing benchmarks of the rule-based inference engines. It allows for automatic generation of models with predefined characteristics, translation of these models into specific rule languages and executing them in a test environment that measures execution time and memory consumption.

In a future work, we plan to extend the translation mechanisms to include more complex language features like modules, or inference control mechanisms. For this purpose, the knowledge interoperability framework proposed in [16] is planned be used. We also plan to extend the generator of meta-models to allow for manipulation on the length, and numbers of reasoning chains available in the model. The testing tool could also be embedded in a web-based semantic wiki to allow for exchanging results between knowledge engineers [1,25].

References

1. Adrian, W.T., Bobek, S., Nalepa, G.J., Kaczor, K., Kluza, K.: How to reason by HeaRT in a semantic knowledge-based wiki. In: Proceedings of the 23rd IEEE International Conference on Tools with Artificial Intelligence, ICTAI 2011, Boca Raton, Florida, USA, pp. 438–441, November 2011. http://ieeexplore.ieee.org/xpls/abs_all.jsp?arnumber=6103361&tag=1
2. Bobadilla, J., Ortega, F., Hernando, A., Gutiérrez, A.: Recommender systems survey. Knowl. Based Syst. **46**, 109–132 (2013). http://www.sciencedirect.com/science/article/pii/S0950705113001044
3. Bobek, S.: Methods for modeling self-adaptive mobile context-aware sytems. Ph.D. thesis, AGH University of Science and Technology, April 2016. Supervisor: G.J. Nalepa
4. Bobek, S., Nalepa, G.J.: Uncertain context data management in dynamic mobile environments. Future Gener. Comput. Syst. **66**, 110–124 (2017). http://www.sciencedirect.com/science/article/pii/S0167739X1630187X
5. Bobek, S., Nalepa, G.J., Ślażyński, M.: Challenges for migration of rule-based reasoning engine to a mobile platform. In: Dziech, A., Czyżewski, A. (eds.) MCSS 2014. CCIS, vol. 429, pp. 43–57. Springer, Cham (2014). doi:10.1007/978-3-319-07569-3_4
6. Bock, J., Haase, P., Ji, Q., Volz, R.: Benchmarking OWL reasoners. In: van Harmelen, F., Herzig, A., Hitzler, P., Lin, Z., Piskac, R., Qi, G. (eds.) Proceedings of the ARea 2008 Workshop, vol. 350. CEUR Workshop Proceedings, June 2008. http://ceur-ws.org
7. Boley, H., Paschke, A., Shafiq, O.: RuleML 1.0: the overarching specification of web rules. In: Dean, M., Hall, J., Rotolo, A., Tabet, S. (eds.) RuleML 2010. LNCS, vol. 6403, pp. 162–178. Springer, Heidelberg (2010). doi:10.1007/978-3-642-16289-3_15. http://dx.doi.org/10.1007/978-3-642-16289-3_15

8. Brant, D., Grose, T., Lofaso, B., Miranker, D.: Effects of database size on rule system performance: five case studies. In: Proceedings of the 17th International Conference on Very Large Data Bases (VLDB) (1991)
9. Bratko, I.: Prolog Programming for Artificial Intelligence, 3rd edn. Addison Wesley, Redwood City (2000)
10. Dey, A.K.: Providing architectural support for building context-aware applications. Ph.D. thesis, Atlanta, GA, USA (2000). aAI9994400
11. Forgy, C.: Rete: a fast algorithm for the many patterns/many objects match problem. Artif. Intell. **19**(1), 17–37 (1982)
12. Giarratano, J.C., Riley, G.D.: Expert Systems. Thomson, Toronto (2005)
13. Goodman, B., Flaxman, S.: EU regulations on algorithmic decision-making and a "right to explanation" (2016). arXiv:1606.08813. Comment: Presented at 2016 ICML Workshop on Human Interpretability in Machine Learning (WHI 2016), New York, NY
14. Hanson, E.N., Hasan, M.S.: Gator: an optimized discrimination network for active database rule condition testing. Technical report 93–036, CIS Department University of Florida, December 1993
15. Jang, J.H., Yang, S.H.: Development of the rule-based inference engine for the advanced context-awareness. Int. J. Smart Home **9**(4), 195–202 (2015)
16. Kaczor, K.: Knowledge formalization methods for semantic interoperability in rule bases. Ph.D. thesis, AGH University of Science and Technology, February 2015. Supervisor: G.J. Nalepa
17. Kaczor, K.: Practical approach to interoperability in production rule bases with SUBITO. In: Rutkowski, L., Korytkowski, M., Scherer, R., Tadeusiewicz, R., Zadeh, L.A., Zurada, J.M. (eds.) ICAISC 2015. LNCS, vol. 9120, pp. 637–648. Springer, Cham (2015). doi:10.1007/978-3-319-19369-4_56
18. Ligęza, A.: Logical foundations for knowledge-based control systems - knowledge representation, reasoning and theoretical properties. Sci. Bull. AGH Autom. **63**(1529), 144 (1993). Kraków
19. Lim, B.Y., Dey, A.K.: Investigating intelligibility for uncertain context-aware applications. In: Proceedings of the 13th International Conference on Ubiquitous Computing, UbiComp 2011, New York, NY, USA, pp. 415–424 (2011). http://doi.acm.org/10.1145/2030112.2030168
20. Mahmud, U., Javed, M.Y.: Context inference engine (CIE): inferring context. Int. J. Adv. Pervasive Ubiquitous Comput. **4**(3), 13–41, July 2012. http://dx.doi.org/10.4018/japuc.2012070102
21. Miranker, D.P., Lofaso, B.J.: The organization and performance of a treat-based production system compiler. IEEE Trans. Knowl. Data Eng. **3**(1), 3–10 (1991)
22. Miranker, D.P.: TREAT: a better match algorithm for AI production systems, long version. Technical report 87–58, University of Texas, July 1987
23. Nalepa, G.J., Bobek, S., Ligęza, A., Kaczor, K.: HalVA - rule analysis framework for XTT2 Rules. In: Bassiliades, N., Governatori, G., Paschke, A. (eds.) RuleML 2011. LNCS, vol. 6826, pp. 337–344. Springer, Heidelberg (2011). doi:10.1007/978-3-642-22546-8_27. http://www.springerlink.com/content/c276374nh9682jm6/
24. Nalepa, G.J., Ligęza, A., Kaczor, K.: Overview of knowledge formalization with XTT2 Rules. In: Bassiliades, N., Governatori, G., Paschke, A. (eds.) RuleML 2011. LNCS, vol. 6826, pp. 329–336. Springer, Heidelberg (2011). doi:10.1007/978-3-642-22546-8_26

25. Nalepa, G.J.: Loki–semantic wiki with logical knowledge representation. In: Nguyen, N.T. (ed.) Transactions on Computational Collective Intelligence III. LNCS, vol. 6560, pp. 96–114. Springer, Heidelberg (2011). doi:10.1007/978-3-642-19968-4_5. http://www.springerlink.com/content/y91w134g03344376/
26. Nalepa, G.J., Bobek, S.: Rule-based solution for context-aware reasoning on mobile devices. Comput. Sci. Inf. Syst. **11**(1), 171–193 (2014)
27. Nalepa, G.J., Kluza, K., Kaczor, K.: Proposal of an inference engine architecture for business rules and processes. In: Rutkowski, L., Korytkowski, M., Scherer, R., Tadeusiewicz, R., Zadeh, L.A., Zurada, J.M. (eds.) ICAISC 2013. LNCS, vol. 7895, pp. 453–464. Springer, Heidelberg (2013). doi:10.1007/978-3-642-38610-7_42. http://www.springer.com/computer/ai/book/978-3-642-38609-1
28. Ostermayer, L.: Seamless cooperation of java and prolog for rule-based software development. In: Proceedings of the RuleML 2015 Challenge, the Special Track on Rule-based Recommender Systems for the Web of Data, the Special Industry Track and the RuleML 2015 Doctoral Consortium hosted by the 9th International Web Rule Symposium (RuleML 2015), Berlin, Germany, 2–5 August 2015 (2015). http://ceur-ws.org/Vol-1417/paper2.pdf
29. Pan, Z.: Benchmarking dl reasoners using realistic ontologies. In: Grau, B.C., Horrocks, I., Parsia, B., Patel-Schneider, P.F. (eds.) OWLED. CEUR Workshop Proceedings, vol. 188. CEUR-WS.org (2005). http://dblp.uni-trier.de/db/conf/owled/owled2005.html#Pan05
30. Weert, P.V.: Efficient lazy evaluation of rule-based programs. IEEE Trans. Knowl. Data Eng. **22**(11), 1521–1534 (2010)

Web-Based Editor for Structured Rule Bases

Szymon Bobek$^{(\boxtimes)}$, Grzegorz J. Nalepa, and Przemysław Babiarz

AGH University of Science and Technology,
al. Mickiewicza 30, 30-059 Krakow, Poland
{sbobek,gjn}@agh.edu.pl

Abstract. Knowledge engineering aims at providing methods for efficient knowledge encoding to allow for automatic reasoning. Most of the research in this field is devoted to the design of expressive modeling languages or effective reasoning mechanisms. We argue that powerful knowledge representation and inference mechanism is not enough to assure high quality knowledge bases. It is crucial to provide methods for creation and visualization of knowledge. This allows an engineer to focus on the task of building the knowledge without the distraction caused by the complexity of the representation, syntax, etc. The original contribution of this paper is a definition of three categories of requirements for visualization and editing software for structured rule bases. We propose the prototype implementation of such a tool and provide the evaluation that involves comparison with existing approaches and user test to measure the usability of the solution.

Keywords: Rule-based systems · Knowledge base design · Web editors

1 Introduction

Research in the area of knowledge engineering focuses mostly on the methods that allow for knowledge encoding and processing, leaving behind aspects of knowledge visualization and design. However, these two latter factors play invaluable role in the process of design, development and later in the maintenance of artificial intelligence software. The most popular knowledge engineering approaches are supported by the modeling tools. These include solutions such as Protege for ontology [4], Drools [3] or HQED [6] for rules. However, these tools are usually very complex systems, that require expert knowledge to install and setup them. Furthermore, their usability and intelligibility features are also limited, as it is implicitly assumed that their users are experts in the modeling approach these tools were made for.

In this paper we focus on the rule bases and tools that support building and visualizing them. When it comes to practical knowledge engineering with rule bases, several fundamental cases can be identified. In the most basic case, a rule-based shell, like CLIPS or Jess, is used. It does not offer any support for rule base building. Therefore, rules have to be written manually in text editor.

© Springer International Publishing AG 2017
L. Rutkowski et al. (Eds.): ICAISC 2017, Part II, LNAI 10246, pp. 411–421, 2017.
DOI: 10.1007/978-3-319-59060-8_37

More advanced tools include the Drools platform, as well as OpenRules[1] or DTRules[2]. They offer an Excel-based editing mode where prototypes of rules can be provided in a form of decision table. In fact visual modeling of rule base in the form of tables or trees seems to be the most scalable method. More recently a new notation for supporting the design of decision table based logic was provided by OMG called Decision Model and Notation (DMN) [7]. It is based on previous works of community cooperating by Jan Vanthiennen. In this work we focus mostly on the XTT2 knowledge representation that was designed as a part of the Semantic Knowledge engineering methodology for building Intelligent Systems [6].

We compared existing editors for structured knowledge bases. Using the analysis of their limitations we defined general requirements for visual editors for rule-based knowledge. Then, we proposed a prototype of HWED – a web-based tool for editing and visualization of structured rule bases. Finally, we compared it with other available editors with respect to previously defined requirements and also provided a usability study, based on Software Usability Measurement Inventory framework[3].

The rest of the paper is organized in the following way. In Sect. 2 our previous work in the area of knowledge base visualization is presented. Based on limitation of existing tools, in Sect. 3 we provide a set of requirements for the new editor. Then we describe its implementation in Sect. 4. The evaluation of the tool and directions for future works are given in Sect. 5.

2   HaDES Tools

Our previous work includes several tools supporting the visual design of rule bases. They were made available as HaDEs toolset as part of the Semantic Knowledge Engineering for Intelligent Systems (SKE) design approach [6]. The SKE introduces a systematic design process that includes three main phases. Their simplified description is provided below:

1. **Conceptual modeling and rule prototyping** where the system attributes are identified and prototypes of decision tables are build.
2. **Logical specification and analysis** where the structured rule base composed of number of decision tables connected into a decision network is designed.
3. **Physical design with prototype integration** where the design of rule base is automatically translated into an executable format and run by an inference engine.

HQEd (HeKatE Qt Editor) [6] provides support for the logical design with XTT2. It can import an HML file with a conceptual model. and generate an

---

[1] See: http://openrules.com.

[2] See: http://www.dtrules.com/newsite.

[3] See: http://sumi.uxp.ie.

XTT2 prototype. HQEᴅ allows for editing the XTT2 structure with support for syntax checking at the table level. Attribute values are checked with domain specification, and the use of proper relational operators is assured, so that potential anomalies are detected and eliminated.

HQEᴅ can automatically generate the textual executable representation of the XTT2 knowledge base in the HMR format (HᴇKᴀᴛE Meta Representation). HMR files can be directly executed by a custom inference engine called HᴇᴀRT (HᴇKᴀᴛE RunTime) [6]. The role of the engine is twofold: run the XTT2 rule logic designed with the use of the editor, and provide on-line formal analysis of the rule base using the HᴀLVA (HᴇKᴀᴛE Verification and Analysis) framework. The engine runs as a stand-alone application. Moreover, it can be embedded into a larger application. It also has a communication module which allows for an integration with HQEᴅ as well as provides network-based logic service. The HQEᴅ editor [6] allows for visual design of the XTT2 knowledge bases. Moreover, it provides a mechanism for checking the rule model against the syntax and logical anomalies. The tool is able to detect important syntax errors, such as: inaccessible rules, attribute values out of domain, etc. The logical anomalies are detected using HᴇᴀRT.

HQEᴅ was implemented in a platform independent way using the Qt programming library[4] a cross-platform application development framework. It also provides an intuitive and user friendly graphical interface. Each dialog window has appropriate controls preventing a user from entering incorrect data.

The HQEᴅ architecture includes three main components. The *controller* is the most important element, as it enables the data flow between the layers. The *model* consists of two layers that are responsible for the internal models representation. *ARD Model* stores an conceptual model and the *XTT Model* stores an XTT2 model. The remaining layers are included in the *view* that maps the model to other formats:

- User Interface – the visual model representation that is appropriate for the user. This layer renders the visual representation of the XTT2 diagram.
- XML mapping – translates the *model* data to HML. This layer allows for storing the state of the model using files.
- Plugins API – provide the communication to other services.

The editor can be integrated with HᴇᴀRT, which provides on-line execution and verification. Below, the main aspects of knowledge modeling with HQEᴅ are discussed.

The complete XTT2 diagram consists of tables which contain rules and links. The design starts with a definition of the attribute types. The sequence in defining the XTT2 model with HQEᴅ includes definitions of types, attributes, tables, rules, and links. The tool partially enforces this design sequence. For instance: a table can be defined only when there exists at least one attribute.

The logical design process starts with the specification of domains of attributes. The *Type Editor* window dialog assists in this process. The next

---

[4] See http://www.trolltech.com.

step involves a definition of the XTT2 tables and their rows representing rules. The table schema is defined first. For a given table, it groups the attributes in the condition or decision parts of the rule. The *Table editor* dialog allows for defining the table schemas. The last stage of the table definition includes rule specification. Every table row corresponds to a single rule. The last XTT2 design stage consists in linking the tables. There are two ways to define a link: using the *Connection editor* dialog, or *drag&drop*. The example of a complete XTT2 diagram is shown in Fig. 1.

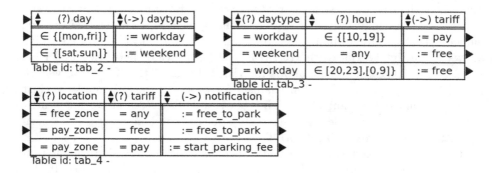

Fig. 1. Model exported from HQED. It aims at detecting if the driver should pay for a parking based on his or her location, time and a day.

The description given above concerns the case when the XTT2 model is designed from scratch. However, HQED is able to import the preliminary conceptual model. Such a model contains the preliminary definitions of attributes. It also contains rule schemas and inference links. In such a model, HQED allows for filling the XTT2 tables with rules. It also allows for refining the inference network.

The tool checks the model quality during the whole design process. It immediately notifies the user of the discovered anomalies. HQED supports XTT2 syntax checking, that includes detection of all anomalies except for the logical ones. It checks the entered attribute values for compatibility with the defined domain. Thus, it limits the possibility of entering incorrect values. When the tool detects an incorrect value, it shows appropriate message to the user, who can correct it. The logical rule analysis is performed with the use of HEART.

As an editing, tool HQED has number of limitations. It is a desktop tool that is run locally, on the host operating system, making it more difficult to assure portability that covers most of the available operating systems. It has a complex architecture that makes it difficult to deploy on the target operating system. The complexity of the architecture is also reflected by the complexity of the user interface. HQED provides a variety of functionalities, that goes beyond the set of features offered by the typical editor. This makes the software difficult to use by a non-expert users. It relies on an XML-based file format that is not human-readable, therefore any changes to the model created with HQED is possible only in the editor. It does not support new features of HMR+ notation such as time-based operators [2] or uncertainty modeling [1] which were introduced recently to XTT2 representation.

Based on the analysis of these limitations, a prototype of a new editor, called HWED was created. The next section discusses this in more details.

3 Requirements for Online Modeling Tool

To overcome the limitations of the HQED editor, we propose a set of requirements that every visualization and editing toolkit for structure rule-bases should fulfill. These requirements can be divided into three categories, depending on which aspect of the editing process they focus. These categories are:

- Visualization – that focuses on the aspects of presenting complex knowledge bases and to the user improving usability of the editor.
- Notation – that focuses on the aspects of quality of editing knowledge base in terms of intelligibility and user experience.
- System – that focuses on technical aspects such as simplicity of installation process and portability.

For *visualization*, it is crucial to provide an ability of the system to appropriately present *structure* and *hierarchy* of the knowledge. This will allow the user to faster discover dependencies between rules, and groups of rules.

The *notation* category, on the other hand can be understood in terms of assuring high user experience by providing methods for syntax verification and autocompletion. This makes the creation of knowledge bases faster, and increases the chance of discovery of mistakes at the level of design, rather than runtime. The last requirement is to assure that the representation of the model is human-readable. This improves the in-line editing of the model, even outside the editor, which may be useful for knowledge engineers.

Finally, the *system* category covers aspects of portability and architectural complexity of the editing software. The editor should be platform independent, at least in terms of accessibility, i.e. it should be available on every available operating system through Internet browser. It also should be lightweight so that the deployment is possible on any operating system, without the need to install any platform-dependent libraries.

These requirements were the guidelines for the creation of the HWED editor – a prototype of a new editor for XTT2. It is web-based, and has much simplified interface over HQED. It is implemented only in Javascript and runs in a web browser, which makes it lightweight, portable and much more usable on mobile devices. It does not support XML for the model representation, instead it generates only HMR files suitable for both editing and execution. It was developed mainly with HEARTDROID in mind but is a general purpose editor. It follows the specification of knowledge visualization principles given in SKE. The following section gives more details on the implementation and architecture of the HWED.

4   HWED

In this section the details about the HWED editor will be presented. First, we will describe its main functionalities that cover *visualisation* and *notation* related requirements given in Sect. 3. Finally, the insight into architecture will be provided to motivate the fulfillment of the *system-based* requirement.

4.1   Functionality

In order to make the usage of the editor as simple as possible, we limited its functionality to the most important features, designated by the requirements presented in Sect. 3. The editor consists of only five windows that are used for: (1) visualizing and editing rules and XTT2 tables, (2) defining types (3) defining attributes, (4) changing editors setting and (5) Instant HMR model preview. The screenshot of the main window of the editor was presented in Fig. 2.

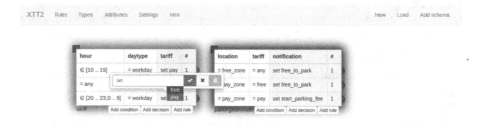

Fig. 2. Screenshot of HWED editor window

The visualization and edition window provide the most important functionalities with respect to the creation of the knowledge base. It visualizes the knowledge base, where rules are organized into XTT2 tables. It shows the hierarchy of the knowledge by highlighting tables that depend on each other (i.e. one of the table produces attributes that the other table uses). It allows for edition of the models providing verification and autocompletion mechanism, as presented in Fig. 2. The windows for instant HMR model preview shows the textual representation of the model presented in the visualization window. This allows experienced knowledge engineer to browse through source code and select parts of the model that can be used in HAQuNA [5] commandline interface for rapid prototyping and verification. Additionally, the user interface allows also to save and load HMR models from the hard drive of the user computer.

---

[5] HAQuNA is an interactive commandline shell to load, modify, run HMR models. See: http://glados.kis.agh.edu.pl/doku.php?id=pub:software:heartdroid:tutorials:haquna.

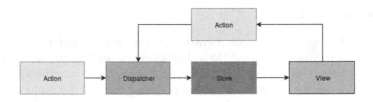

Fig. 3. Flux architecture diagram

The user interface – front-end of the application – lays on top of the Flux[6] architecture and is supported by the Redux[7] framework. Architecture helps to keep code more structured and maintainable. Framework on the other hand greatly enhances development speed, readability and lowers error capacity. The following sections discuss these two components in details.

4.2   Architecture

Flux is the application architecture developed by Facebook. It was first designed specifically to complement React framework[8] and its components by utilizing a unidirectional data flow. Flux consists of three major parts: dispatcher, stores, and views. Implementation of Flux used in this project is called Redux. It additionally enforces that application uses a single store, and adds one more major part to Flux update cycle - actions. As shown in Fig. 3, in a Flux application data flows in a single direction.

Dispatcher is a container for a registry of callbacks into the stores for the entire Flux application. The input of the dispatcher is an action. This action is then propagated to proper store. In case of Redux implementation it would be the main store.

Store holds the application data and business logic for data mutation and view updates. It has no direct setter methods, instead the only way of getting new data into its self-contained world is by sending action to dispatcher which initiates store update.

Views are composable and re-renderable React components. They get their state from parent component or in case they are on the top of the hierarchy they get it from store they are subscribed to.

Actions are wrappers composed of action type and payload data if needed. They are usually created in some view, but they can also be constructed in event handlers or any other utility or library as long as they do not violate unidirectional data flow. Once created actions are passed to dispatcher.

---

[6] See: https://facebook.github.io/flux.
[7] See: https://reduxframework.com.
[8] See: https://facebook.github.io/react.

4.3   Implementation

Figure 4 clearly shows that view hierarchy is the largest and most important module. It incorporates most of the other modules and provides interface for users to use them. On top of view hierarchy resides the controller-views called containers. Containers subscribe for store updates, retrieve data and action definitions proper for specific container. They pass received data and functions to another component. Further components incorporate some rendering logic and pass properties to their children. Usually, the lower component is in the hierarchy the lesser it becomes, all the way to components consisting of pure HTML like code.

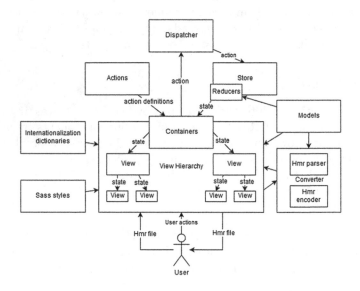

Fig. 4. HWED implementation diagram

Actions are most commonly created inside view hierarchy. Then action is passed to the dispatcher which forwards it to the store. Each action passed to store goes through hierarchy of reducers which contain logic for producing new state based on previous one, action type and its payload. Every state is immutable. After computing the new state, a change event is emitted to adjacent container-views. Converter is an interface for handling HMR files. HMR encoder is made using template string feature. HMR parser is generated using grammar and Peg.js library.

Following current state of Javascript a module bundler was necessary to build project. Application code and styles had to be transpiled because they were written using extended CSS language, ECMAScript 6 standard and JSX syntax. Additionally during transpilation, assets are copied and parser is generated using HMR grammar. All the files are then concatenated into several chunks of minified code and map files, and exported under a single folder ready to be copied to a server.

5   Evaluation and Future Works

In this section two types of evaluation will be given. First, we compare the existing editors, not limiting them to these supporting only XTT2 knowledge representation. We will perform the comparison with respect to the requirements presented in Sect. 3 Secondly we will show the results from the usability tests performed by the professional knowledge engineers. Finally, the direction of future works will be discussed.

5.1   Comparison of Tools for Knowledge Visualization and Editing

We compared the HWED editor with the popular systems for building rule knowledge bases with respect to the aspects of *visualization, notation* used and *system* portability and complexity. The summary of the comparison was given in Table 1. Although there exist solutions that partially fulfill the requirements defined by us in Sect. 3, only HWED provide the full support for all of them.

Table 1. Comparison of tools for knowledge visualization and editing

		HWED	HQED	DTRules	OpenRules	Drools
Visualisation	**Structure**	**Yes**	**Yes**	**Yes**	**Yes**	**Yes**
	Hierarchy	**Yes**	No	No	No	No
Notation	**Verification**	**Yes**	**Yes**	No	No	**Yes**
	Autocompletion	**Yes**	Partial	No	No	Partial
	Human-readable	**Yes**	Partial	**Yes**	**Yes**	No
System	**Portability**	**Yes**	Partial	Partial	Partial	Partial
	Lighweight	**Yes**	No	Partial	Partial	No

All of the compared systems partially support portability. This means that they are written in C++ or other programming language that is portable between operating systems, but still requires separate build for each of them. Partial support for autocompletion and verification features means that the systems either use external tools for editing knowledge base (like Excel in case of DTRules or OpenRules) or use their own build-in mechanism that forbids a user entering invalid values, like in case of HQED.

5.2   Usability Studies

In the usability studies, we aimed at comparing existing editor for creating XTT2 knowledge bases HQED with HWED prototype. We prepared a set of tasks that each of the participants had to perform[9]. This tasks included getting familiar with the HMR notation and try to build a very simple model for determining a parking fee presented in Fig. 5 in both editors. After performing aforementioned tasks, each participant was asked to fill out the form consisted of 50 questions according to Software Usability Measurement Inventory (SUMI) framework [5].

Fig. 5. Model generated with HWED, used for the user tests (equivalent to the model from Fig. 1)

The participants were experts in knowledge engineering, having at least 6 years of experience in the field. However, not all of them were familiar with HMR+ language, nor any of the investigated editors. The result shows that the overall impression of the usability of the editor was positive, although helpfulness and controllability of the software was not rated high[10]. The output summary of the SUMI report is presented in Fig. 6. However, it is worth noting that the HWED is still a prototype implementation.

5.3   Future Works

One of the main direction for future works is an integration of HWED with a library of callbacks in order to provide a comprehensive tool for building rule-driven mobile applications. Callbacks system is HMR+ feature that allows for

---

[9] The set of tasks can be found here: http://glados.kis.agh.edu.pl/doku.php?id=pub:software:hwed:usability_studies.

[10] The complete report from the SUMI framework can be found here: http://glados.kis.agh.edu.pl/doku.php?id=pub:software:hwed:usability_studies:results.

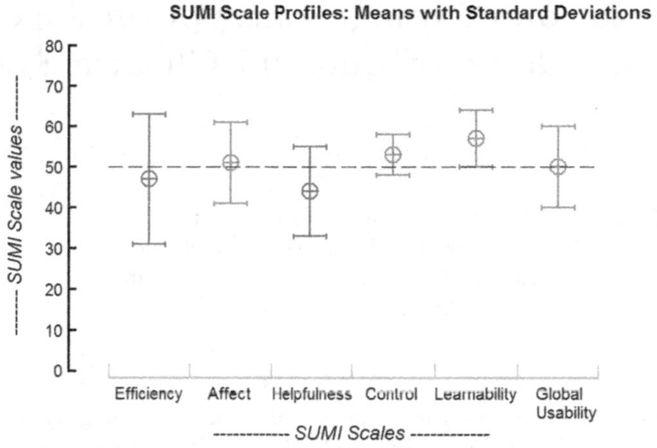

SUMI Scale Profiles: Means with Standard Deviations

Fig. 6. Graphical summaries of SUMI scales

integration of HMR+ models and external systems, in this case the Android system. These callbacks will provide an interface to various Android features such as sending SMS, making a call, changing profile, etc. Callbacks are simple, autonomous Java classes that form building blocks which HMR models use to create mode complex, logic-based applications.

References

1. Bobek, S., Nalepa, G.J.: Compact representation of conditional probability for rule-based mobile context-aware systems. In: Bassiliades, N., Gottlob, G., Sadri, F., Paschke, A., Roman, D. (eds.) RuleML 2015. LNCS, vol. 9202, pp. 83–96. Springer, Cham (2015). doi:10.1007/978-3-319-21542-6_6
2. Bobek, S., Ślażyński, M., Nalepa, G.J.: Capturing dynamics of mobile context-aware systems with rules and statistical analysis of historical data. In: Rutkowski, L., Korytkowski, M., Scherer, R., Tadeusiewicz, R., Zadeh, L.A., Zurada, J.M. (eds.) ICAISC 2015. LNCS, vol. 9120, pp. 578–590. Springer, Cham (2015). doi:10.1007/978-3-319-19369-4_51
3. Di Bona, D., Lo Re, G., Aiello, G., Tamburo, A., Alessi, M.: A methodology for graphical modeling of business rules. In: 5th UKSim European Symposium on Computer Modeling and Simulation (EMS) 2011, pp. 102–106, November 2011
4. Horridge, M., Knublauch, H., Rector, A., Stevens, R., Wroe, C.: A practical guide to building OWL ontologies using the Protege-OWL plugin and CO-ODE tools edition 1.0, August 2004. http://www.co-ode.org/resources/tutorials/ProtegeOWLTutorial.pdf
5. Kirakowski, J., Corbett, M.: Sumi: the software usability measurement inventory. Br. J. Educ. Technol. **24**(3), 210–212 (1993)
6. Nalepa, G.J.: Semantic Knowledge Engineering. A Rule-Based Approach. Wydawnictwa AGH, Kraków (2011)
7. Object Management Group (OMG): Decision model and notation request for proposal. Technical report, bmi/2011-03-04, Object Management Group, 140 Kendrick Street, Building A Suite 300, Needham, MA 02494, USA, March 2011

Parallelization of Image Encryption Algorithm Based on Game of Life and Chaotic System

Dariusz Burak(✉)

Faculty of Computer Science and Information Technology,
West Pomeranian University of Technology,
49 Żołnierska St., 71-210 Szczecin, Poland
dburak@wi.zut.edu.pl

Abstract. In this paper, the results of parallelizing an image encryption algorithm based on Game of Life and chaotic system are presented. The data dependence analysis of loops is applied in order to parallelize the algorithm. The parallelism of the algorithm is demonstrated in accordance with the OpenMP standard. As a result of this study, it is stated that the most time-consuming loops of the algorithm are suitable for parallelization. The efficiency measurements of the parallel algorithm working in standard modes of operation are shown.

Keywords: Game of Life · Chaos · Image encryption · Parallelization · OpenMP

1 Introduction

One of the very important functional features of cryptographic algorithms is cipher speed. This feature is significant in case of symmetric ciphers since they usually work on large data sets. Thus even not much differences of speed may cause the choice of the faster cipher by the user. Therefore, it is all-important to parallelize such algorithms in order to achieve faster processing using multicore processors or multiprocessing systems. In 1989, British mathematician Robert Andrew Matthews firstly proposed a chaotic encryption algorithm [1]. In recent years many ciphers based on chaotic maps were proposed. On the other hand, cellular automata are introduced to design encryption algorithms considering the complicated and time-varying nature of the structures. It was the idea of Steven Wolfram who suggested for the first time the use of cellular automata for cryptography in 1985 [2]. A combination of both cellular automata and chaos for designing cryptosystems for image security have shown better performance in terms of robustness and information security due to its confusion and diffusion property. Nowadays, there are some descriptions of various ciphers based on cellular automata and chaos, for instance [3–12]. The critical issue in such ciphers is program implementation.

Unlike parallel implementations of classical symmetric ciphers, for instance [13,14] or chaotic encryption systems, for instance [15] there are no parallel

© Springer International Publishing AG 2017
L. Rutkowski et al. (Eds.): ICAISC 2017, Part II, LNAI 10246, pp. 422–431, 2017.
DOI: 10.1007/978-3-319-59060-8_38

implementations of symmetric ciphers based on cellular automata and chaotic networks. Being seemingly a research gap it is absolutely fundamental to show real functional advantages and disadvantages of the encryption algorithm using software or hardware implementation.

The main contribution of the study is developing a parallel algorithm in accordance with OpenMP standard of the cryptographic system designed by Xingyuan Wang and Canqi Jin and presented in [16] based on transformations of a source code written in the C++ language representing the sequential algorithm.

This paper is organized as follows. The next section briefly describes the image encryption algorithm based on Game of Life and chaotic system. In Sect. 3, parallelization process of the algorithm is fully characterized. In Sect. 4, the experimental results obtained for developed parallel algorithm are presented. Finally, concluding remarks are given in Sect. 5.

2 Description of the Image Encryption Algorithm Based on Game of Life and Chaotic System

The image encryption algorithm based on Game of Life and chaotic system [16] is a symmetric-key cipher composed of three separate blocks: confusion, diffusion and key generator. The Game of Life supported by the logistic map is used as confusion method. Then the higher half pixel diffusion based on the piecewise linear chaotic map (PWLCM) is applied in diffusion phase. The key generator block supports these layers by calculating some initial values and parameters.

The encryption process consists of the following steps:

Confusion phase.

1. Calculate the sum (δ) of all pixels in the image, which is used to generate the initial value (x_0) of the logistic map with (x_0') (given in advance and $x_0' > 0.1$) by using the following formula:

$$x_0 = x_0' - (\delta/10^{14} - [\delta/10^{14}])/10^2, \qquad (1)$$

where [] is meaning to take the integer part of a number. In this equation, parameter δ is used to make from the different plain-images (even one bit different) completely different cipher-images. The Number of Pixels Change Rate (NPCR) (is over %99) and the Unified Average Changing Intensity (UACI) (is over %33) randomness tests for image encryption shows that the proposed algorithm is very sensitive to tiny changes in the plain image. Even if there is only one bit difference between two plain images, the decrypted images will be different completely. Thus, the algorithm is robust against differential attack.

2. Use x_0 and α given in advance to generate a sequence ($x_1 x_1 ... x_{M \times N}$) by logistic map:

$$x_i = \Phi(x_{i-1}) = \alpha x_{i-1}(1 - x_{i-1}), \qquad (2)$$

where $x_i \in (0, 1)$, α is the logistic map parameter and the output sequence is chaotic when $\alpha \in (3.57, 4)$.

Then create two-dimensional orthogonal grid of cells $S^0_{M \times N}$ (a pixel in plain image corresponds to a cell) as the seed of the Game of Life. Each cell has exactly two possible states (dead or alive) and is represented in accordance with the following rule:

$$S^0_{M \times N} = \begin{cases} 1, (x_i \times 10^{14}) \bmod 3 > 0 \\ 0, (x_i \times 10^{14}) \bmod 3 \leq 0 \end{cases}. \tag{3}$$

3. When producing the ith generation $(S^i_{M \times N})$ by the rules of Game of Life, put the corresponding pixels of $S^i_{M \times N}$ to the scrambling image one by one, except the processed pixels.
4. After produce R generations, stop and put the rest of the pixels into the scrambling image.

Diffusion phase.

A bit can contain different amounts of information depending on its position in the pixel. The percentage of information $p(i)$ provided by the ith bit is given by:

$$p(i) = \frac{2^{i-1}}{\sum\limits_{i=1}^{8} 2^{i-1}}, i = \{1, 2, ..., 8\}. \tag{4}$$

The higher 4 bits (8th, 7th, 6th and 5th) carry 94.125 of the total information of the 8 binary image so diffusion only for the higher half pixel is executed. The diffusion method is based on piecewise linear chaotic map (PWLCM) described as:

$$x_i = F(x_{i-1}, \eta) = \begin{cases} \dfrac{x_{i-1}}{\eta}, 0 \leq x_{i-1} < \eta \\ \dfrac{x_{i-1} - \eta}{0.5 - \eta}, \eta \leq x_{i-1} < 0.5 \\ 0, x_{i-1} = 0.5 \\ F(1 - x_{i-1}, \eta), 0.5 < x_{i-1} \leq 1.0 \end{cases}, \tag{5}$$

where x_0 is the initial condition value, η is the control parameter (and can be served as a secret key), $x_{i-1} \in [0, 1]$, and $\eta \in (0, 0.5)$. PWLCM has perfect behaviour and high dynamical properties such as invariant distribution, autocorrelation function, periodicity, large positive Lyapunov exponent, and mixing property, so it can provide excellent random sequence, which is suitable for image encryption algorithm [17].

The value of each higher half pixel is altered sequentially at the pixel-level by the output of the PWLCM map and a parameter m. The operations in this phase are governed by the following equations:

$$y_i = F(y_{i-1}, \eta), \tag{6}$$

where F means PWLCM, y_0 is the PWLCM's initial value and η is the parameter.

$$d_i = (y_i \times 10^{14}) \bmod 16. \tag{7}$$

$$c_i = ((d_i + p_i) \bmod 16) \oplus ((c_{i-1} + w) \bmod 16), \tag{8}$$

where $w \in [1, 16]$ is encryption parameter (given in advance), p_i is the ith higher half pixel of the permuted image and the scanning order is from left to right and from up to down, c_i is the encrypted value of p_i and c_0 is the last p_i.

The decryption process (symmetric to encryption one) is the reverse of the encryption process. More detailed description of image encryption algorithm based on Game of Life and chaotic system is given in [16].

3   Parallelization Process of Image Encryption Algorithm

Given the fact that proposed encryption algorithm can work in block manner it is necessary to prepare a C++ source code representing the sequential algorithm working in Electronic Codebook (ECB), Cipher Block Chaining (CBC), Cipher Feedback (CFB), Output Feedback (OFB) and Counter (CTR) modes of operation. The source code of the encryption algorithm in the essential ECB mode contains twenty three *for* loops. Fourteen of them include no I/O functions. Some of these loops are time-consuming. Thus their parallelization is critical for reducing the total time of the parallel algorithm execution.

In order to find dependencies in program a research tool for analyzing array data dependencies called Petit was applied. Petit was developed at the University of Maryland under the Omega Project and is freely available for both DOS and UNIX systems [18].

The OpenMP standard was used to present parallelized loops. The OpenMP Application Program Interface (API) [19,20] supports multi-platform shared memory parallel programming in C/C++ and Fortran on all architectures including Unix and Windows platforms. OpenMP is a collection of compiler directives, library routines and environment variables which could be used to specify shared memory parallelism. OpenMP directives extend a sequential programming language with Single Program Multiple Data (SPMD) constructs, work-sharing constructs, synchronization constructs and help to operate on both shared data and private data. An OpenMP program begins execution as a single task (called a master thread). When a parallel construct is encountered, the master thread creates a team of threads. The statements within the parallel construct are executed in parallel by each thread in a team. At the end of the parallel construct, the threads of the team are synchronized. Then only the master thread continues execution until the next parallel construct will be encountered. To build a valid parallel code, it is necessary to preserve all dependencies, data conflicts and requirements regarding parallelism of a program [19,20].

The process of the encryption algorithm parallelization can be divided into the following stages:

- carrying out the dependence analysis of a sequential source code in order to detect parallelizable loops;
- selecting parallelization methods based on source code transformations;

– constructing parallel forms of program loops in accordance with the OpenMP standard.

There are the following basic types of the data dependencies that occur in *for* loops:

– A Data Flow Dependence indicates a write-before-read ordering that must be satisfied for parallel computing. This dependence cannot be avoided and limits possible parallelism. The following loop yields such dependences [21, 22]:

```
for(i=1; i<n; i++) {
    a[i] = a[i - 1];
    }
```

– A Data Anti-dependence indicates a read-before-write ordering that should not be violated when performing computations in parallel. There are techniques for eliminating such dependences. The loop below produces anti-dependences:

```
for(i=0; i<n; i++) {
    a[i] = a[i + 1];
    }
```

– An Output Dependence indicates a write-before-write ordering for parallel processing. There are techniques for eliminating such dependencies. The following loop yields output dependences:

```
for(i=0; i<n; i++) {
    a[0] = a[i];
    }
```

Additionally, control dependence determines the ordering of an instruction i, with respect to a branch instruction so that instruction i is executed in a correct program order.

To find the most time-consuming loops of the algorithm, experiments were carried out for an about 5 MB input file.

It appeared that the algorithm has two computational bottlenecks: the first is enclosed in the function *golife_enc()* and the second is enclosed in the function *golife_dec()*. The *golife_enc()* function enables enciphering of the whichever number of data blocks and the *golife_dec()* one does the same for deciphering process (analogically to similar functions of the classic block ciphers like DES- the *des_enc()*, the *des_dec()* presented in [23]). Thus the parallelization of *for* loops included in these functions is a crucial for parallelization process.

The bodies of the *golife_enc()* and the *golife_dec()* functions are as follows:

```
void golife_enc(golife_context *ctx,UINT8 *input,UINT8 *output,
            int input_length) {
    for (int i = 0; i<NUMBER_OF_BLOCKS; i++) {
        Encryption(ctx, input, output);
        input+= BLOCKSIZE;
        output+= BLOCKSIZE;
    }
};

void golife_dec(golife_context *ctx,UINT8 *input,UINT8 *output,
            int input_length){
    for (int i = 0; i<NUMBER_OF_BLOCKS; i++) {
        Decryption(ctx, input, output);
        input+= BLOCKSIZE;
        output+= BLOCKSIZE;
    }
}.
```

Taking into account the strong similarity of the above functions only the first one is examined. Subsequently this analysis is valid in the case of the second one.

In order to apply the data dependencies analysis of the loop included in *golife_enc()* function the body of the *Encryption()* function should be put in this loop.

The actual parallelization process of the loop included in the *golife_enc()* function consists of the six following stages:

- removal of the key generator operations; all these calculations have to be executed sequentially before starting the confusion phase operations;
- insertion of the following statements in the beginning of the loop body:
  $plaintext = \&input[BLOCKSIZE * i]$;
  $ciphertext = \&output[BLOCKSIZE * i]$;
- removal from the end of the loop body the following statements:
  $input+ = BLOCKSIZE$;
  $output+ = BLOCKSIZE$;
- insertion of the following statements:
  $Confusion(ciphertext, plaintext, alpha, delta, x, y)$;
  $Diffusion(ciphertext, eta, y0, w)$;
  The first statement carries out the operations specified in confusion phase (Game of Life, table S creation, logistic map), the second one accomplishes the operations included in diffusion phase (piecewise linear chaotic map, calculating the value of higher half pixel).
- suitable variables privatization
  $(i, ii, plaintext, ciphertext, alpha, delta, x, y, eta, y0, w, x0, xx0, d, c, m)$ using OpenMP (based on the results of data dependence analysis) for the loop indexing by i;

– adding appropriate OpenMP directive and clauses
(*#pragma omp parallel for private() shared()*) for the loop indexing by i.

The steps above result in the following parallel form of the loop include in the *golife_enc()* function in accordance with the OpenMP standard:

```
#pragma omp parallel private (i, ii, plaintext, ciphertext,
                              alpha, delta, x, y, eta, y0,
                              w, x0, xx0, d, c, m)
#pragma omp for
for (i=0; i<nblocks; i++) {
    plaintext=&input[BLOCKSIZE*i];
    ciphertext = &output[BLOCKSIZE*i];
    for(ii=0; ii<R; ii++) {
        Confusion(ciphertext, plaintext, alpha, delta, x, y);
        Diffusion(ciphertext, eta, y0, w);
    }
}.
```

4   Experimental Results

In order to study the efficiency of the presented encryption algorithm eight Quad-Core Intel Xeon Processors 7310 Series - 1.60 GHz and the Intel C++ Compiler (version 13.1.1 20130313 that supports the OpenMP 4.0) were used. The results received for an about 5 megabytes input file (8 bit per pixel image) using two, four, eight, sixteen and thirty-two cores versus the only one have been shown in Tables 1 and 2. The number of threads is equal to the number of processors.

The total running time of the presented encryption algorithm consists of the following operations: data receiving from an input file, data encryption, data decryption and data writing to an output file.

Thus the total speed-up of the parallel encryption algorithm depends heavily on the following five factors:

– the degree of parallelization of the loop included in the golife_enc() function;
– the degree of parallelization of the loop included in the golife_dec() function;
– the block size of the encryption algorithm.
– the method of reading data from an input file;
– the method of writing data to an output file.

The results confirm that the loops included both the golife_enc() and the golife_dec() functions are parallelizable with high speed-up (see Table 1).

During experiments the data block of encryption process (and decryption one) was set on value 16 bytes. Additional tests showed that this size of block gives a good encryption/decryption speed of the encryption algorithm.

The block method of reading data from an input file (512-bytes blocks) and writing data to an output file (64-bytes blocks)was used.

In accordance with Amdahl's Law [24] the maximum speed-up of the encryption algorithm is limited to 4.4623, because the fraction of the code that cannot be parallelized is 0.2241.

The encryption algorithm was also parallelized in the following standard modes of operation (CTR, CBC and CFB). Implementation is based on recommendation detailed described in [25]. The results are presented in Table 2.

When the encryption algorithm operates in the ECB and CTR modes of operation, both the encryption and decryption processes are parallelizable and speed-ups of the whole algorithm are similar (see details- Tables 1 and 2). For the CBC and CFB modes only the decryption process is parallelized so the values of speed-up are lower than for the ECB and CTR modes of operation (see Table 2).

Table 1. Speed-up of the parallel image encryption algorithm based on Game of Life and chaotic system working in the ECB mode of operation.

Number of threads	Speed-up of the encryption process	Speed-up of the decryption process	Speed-up of the whole algorithm
1	1.00	1.00	1.00
2	1.93	1.95	1.51
4	3.73	3.88	1.85
8	5.91	6.33	2.38
16	6.12	6.47	2.49
32	5.93	6.37	2.40

Table 2. Speed-ups of the parallel image encryption algorithm based on Game of Life and chaotic system working in the CTR, CBC and CFB mode of operation.

Number of threads	Operation	Speed-up of the CTR mode of operation	Speed-up of the CBC mode of operation	Speed-up of the CFB mode of operation
1	Encryption	1.00	1.00	1.00
1	Decryption	1.00	1.00	1.00
2	Encryption	1.90	1.00	1.00
2	Decryption	1.90	1.90	1.90
4	Encryption	3.60	1.00	1.00
4	Decryption	3.70	3.70	3.80
8	Encryption	5.80	1.00	1.00
8	Decryption	6.20	6.20	6.30
16	Encryption	6.00	1.00	1.00
16	Decryption	6.30	6.20	6.30
32	Encryption	5.80	1.00	1.00
32	Decryption	6.10	6.10	6.20

5   Conclusions

In this paper, the parallelization process of the image encryption algorithm based on Game of Life and chaotic system has been shown. The time-consuming *for* loops included in the functions responsible for the encryption and decryption processes are parallelizable. The experiments have shown that the application of the parallel encryption algorithm for multiprocessor and multicore computers would considerably boost the time of the data encryption and decryption. The speed-ups received for these operations can be admitted as satisfactory. Moreover, the developed parallel encryption algorithm can be also helpful for hardware and GPGPU implementations.

References

1. Matthews, R.: On the derivation of a chaotic encryption algorithm. Cryptologia **13**(1), 29–42 (1989)
2. Wolfram, S.: Cryptography with cellular automata. In: Williams, H.C. (ed.) CRYPTO 1985. LNCS, vol. 218, pp. 429–432. Springer, Heidelberg (1986). doi:10.1007/3-540-39799-X_32
3. Habutsu, T., Nishio, Y., Sasase, I., Mori, S.: A secret key cryptosystem by iterating a chaotic map. In: Davies, D.W. (ed.) EUROCRYPT 1991. LNCS, vol. 547, pp. 127–140. Springer, Heidelberg (1991). doi:10.1007/3-540-46416-6_11
4. Machicao, J., Marco, A., Bruno, O.: Chaotic encryption method based on life-like cellular automata. Expert Syst. Appl. **39**(16), 12626–12635 (2012)
5. Rey, A.M., Sánchez, G.R., Villa Cuenca, A.: Encrypting digital images using cellular automata. In: Corchado, E., Snášel, V., Abraham, A., Woźniak, M., Graña, M., Cho, S.-B. (eds.) HAIS 2012. LNCS, vol. 7209, pp. 78–88. Springer, Heidelberg (2012). doi:10.1007/978-3-642-28931-6_8
6. Zhang, S., Luo, H.: The research of image encryption algorithm based on chaos cellular automata. J. Multimedia **7**(1), 66–73 (2012)
7. Wang, X., Luan, D.: A novel image encryption algorithm using chaos and reversible cellular automata. Commun. Nonlinear Sci. Numer. Simul. **18**(11), 3075–3085 (2013)
8. Bakhshandeh, A., Eslami, Z.: An authenticated image encryption scheme based on chaotic maps and memory cellular automata. Opt. Lasers Eng. **51**(6), 665–673 (2013)
9. Rimsa, S., Meskauskkas, T.: Data encryption algorithm based on cellular automaton and chaotic logistic equation. Informacines technologijos, 84–88 (2014)
10. Souyah, A., Faraoun, K.: An image encryption scheme combining chaos-memory cellular automata and weighted histogram. Nonlinear Dyn. **86**(1), 639–653 (2016)
11. Zhang, X., Zhang, H., Xu, C.: Reverse iterative image encryption scheme using 8-layer cellular automata. KSII Trans. Internet Inf. Syst. **10**(7), 3397–3413 (2016)
12. Del Rey, A., Pastora, J., Sánchez, G.: 3D medical data security protection. Expert Syst. Appl. **54**, 379–386 (2016)
13. Bielecki, W., Burak, D.: Exploiting loop-level parallelism in the AES algorithm. WSEAS Trans. Comput. **1**(5), 125–133 (2006)
14. Beletskyy, V., Burak, D.: Parallelization of the IDEA algorithm. In: Bubak, M., Albada, G.D., Sloot, P.M.A., Dongarra, J. (eds.) ICCS 2004. LNCS, vol. 3036, pp. 635–638. Springer, Heidelberg (2004). doi:10.1007/978-3-540-24685-5_108

15. Burak, D., Chudzik, M.: Parallelization of the discrete chaotic block encryption algorithm. In: Wyrzykowski, R., Dongarra, J., Karczewski, K., Waśniewski, J. (eds.) PPAM 2011. LNCS, vol. 7204, pp. 323–332. Springer, Heidelberg (2012). doi:10.1007/978-3-642-31500-8_33

16. Wang, X., Jin, C.: Image encryption using game of life permutation and PWLCM chaotic system. Opt. Commun. **285**(4), 412–417 (2012)

17. Maqableh, M.: A novel triangular chaotic map (TCM) with full intensive chaotic population based on logistic map. J. Softw. Eng. Appl. **8**, 635–659 (2015)

18. Kelly, W., Maslov, V., Pugh, W., Rosser, E., Shpeisman, T., Wonnacott, D.: New user interface for Petit and other extensions. User Guide (1996)

19. Chapman, B., Jost, G., van der Pas, R.: Using OpenMP - Portable Shared Memory Parallel Programming. The MIT Press, Cambridge (2007)

20. OpenMP application program interface: Version 4.0 (2013)

21. Allen, R., Kennedy, K.: Optimizing Compilers for Modern Architectures: A Dependence based Approach. Morgan Kaufmann Publishers Inc., San Francisco (2001)

22. Aho, A., Lam, M., Sethi, R., Ullman, J.: Compilers: Principles, Techniques, and Tools, 2nd edn. Prentice Hall, New Jersey (2006)

23. Schneier, B.: Applied Cryptography: Protocols, Algorithms, and Source Code in C, 2nd edn. Wiley, New York (1995)

24. Amdahl, G.: Validity of the single-processor approach to achieving large scale computing capabilities. In: AFIPS Conference Proceedings, pp. 483–485 (1967)

25. Dworkin, M.: Recommendation for block cipher modes of operation: methods and techniques. NIST Special Publication 800–38A (2001)

Cognitive Investigation on Pilot Attention During Take-Offs and Landings Using Flight Simulator

Zbigniew Gomolka[1](✉), Boguslaw Twarog[1], and Ewa Zeslawska[2]

[1] Natural Sciences Department of Computer Engineering, Faculty of Mathematics, University of Rzeszow, ul. Pigonia 1, 35-959 Rzeszow, Poland
{zgomolka,btwarog}@ur.edu.pl
[2] Department of Applied Information, Faculty of Applied Informatics, University of Information Technology and Management in Rzeszow, ul. Sucharskiego 2, 35-225 Rzeszow, Poland
ezeslawska@wsiz.rzeszow.pl

Abstract. The paper presents cognitive studies on pilot's attention during the take-off and landing performance. The studies were conducted using SMI RED 500 eyetracker and Saitek Pro 2000 set of pilot instruments. Simulation experiments involved two groups with different flight experience, recording particular attention trajectories during respective flight phases. The NON-PILOT group comprised members who had less than 80 h of flight time and the PILOT group the ones with more than 80 h of flight time. The differences in perception of flight process in a group of people with different flight experience were presented based on the analyses of the conducted measurements. This might be a useful advice to junior pilots improving their skills and, as a result may increase passengers safety during a flight.

1 Introduction

Eyetracking is the area of modern computer science that measures eye movements, usually in response to visual, auditory and cognitive stimuli. There are various methods for recording human eye movement, starting from typical direct observation, through invasive mechanical observation, to measurement of the difference in electric potentials between both sides of eyeball [3,6,8–11]. Key parameters that describe the process of seeing are: fixations, saccades, pursuits and Gaze path. The analysis of attention trajectory allows to formulate conclusions, for example about the attention process of observed scene. It is possible to determine which areas attract observer's attention, which are omitted, and also to obtain information about the order in what an observer scans them. Currently used eyetracking systems employ Dual Purkinje, Bright Pupil, and Dark Pupil technologies or their hybrid connections that provide high measuring accuracy. The SMI Red 500, that was used in the experimental part of the paper, uses hybrid measuring system. Dynamic growth of aviation forces engineers to provide the quickest and simplest possible means of information about flight

© Springer International Publishing AG 2017
L. Rutkowski et al. (Eds.): ICAISC 2017, Part II, LNAI 10246, pp. 432–443, 2017.
DOI: 10.1007/978-3-319-59060-8_39

parameters to pilots. Due to oculographic studies it is possible to analyse patterns of saccades and fixations of pilots demonstrating different levels in flight training. The object of the research presented in the paper are the registered attention trajectories of piloting students in key procedures during take-offs and landings with the use of flight simulator [13,17,18].

2 Measuring Station and Initial Assumptions of Experiment

In order to conduct studies a few assumptions have been introduced that define the problem and allow to obtain information. Taking into consideration the above, the assumptions are as follows. The station including an eyetracker, piloting devices of flight simulator and a screen were treated as a compatible cockpit model. Each experimental session is held in conditions that reduce the impact of external environment including noise and over-illumination in a room used for conducting studies. Tests are performed in a Cessna 172 Skyhawk aircraft equipped with analogue instruments located in the front part of the plane. Piloting the aircraft is performed using only a yoke, pedals and steering with a damper. The aircraft used for doing simulation tasks is fully configured and adapted to perform a particular part of flight. The length of experiment for a person takes no longer than 60 s in order to reduce the influence of fatigue on perception and concentration. The group of people with flight experience higher than 80 h is considered to as experiences group of pilots - PILOT. The studies are conducted at the same time with the studies of brain activity and they do not introduce additional errors related to the complexity of the experiment. In the studies the SMI RED 500 eyetracker was used which was synchronised with the measuring station for the Brain Products electroencephalographic studies obtaining a joint timeline that allows to correlate the dynamics of attention trajectories with the accompanying EEG brain activity. During flight simulation, Microsoft Flight Simulator X software was used, a set of pilot instruments Saitek Pro 2000 equipped with a damper, yoke responsible for rolling and pitching of an aircraft and pedals responsible for controlling the yaw of an aircraft [1,2,7,12,16]. The measuring station is presented in Fig. 1.

During pre-selection of potential candidates for an experiment the flight experience and time of last flight performance were taken into consideration. There were two groups of different flight experience identified: NON-PILOT (five persons with zero flight time) and PILOT (five persons with 80–1200 flight time in hours). Group members had to do the following tasks:

- **Task 1.** Take-off run from the flight path 27 at the airport EPRZ to the rotation speed, next, take-off and climb up to 1200 ft AMSL (Above Medium Sea Level).
- **Task 2.** Landing on flight path 27, at the airport EPRZ, from the height of 1500 ft AMSL and decelerating an aircraft to the speed lower than 10 kts.

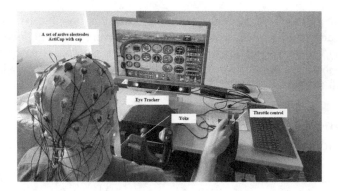

Fig. 1. Measuring station used in the studies of pilot attention with simultaneous encephalographic measurement

The tasks were executed in still air, using a configured Cessna 172 Skyhawk plane with stable three-point landing gear. The flaps swing angle of an aircraft ready to start was set in the configuration as 0, while for landing as 20 degrees. The flights were performed at the airport EPRZ – Rzeszow-Jasionka, on the runway 27 of maximum take-off length 3200 m and landing length 3192 m, width 45 m, airport elevation 693 ft and runway magnetic heading 265°.

3   Results and Analysis

The software environment SMI Experiment Center® operating with the RED 500 eyetracker was used in designing the experiment that consists of three stages. The first is the calibration stage in which, due to specific character of the study being performed, there the most precise 9-point calibration was used. It is a 9-point method that evaluates the measurement accuracy of observer's fixation coordinates for a set size of a visualisation device. Second stage is the validation, which was used for evaluating whether the calibration was sufficiently precise for current geometry of an observed scene. In the validation stage, an individual profile was assigned to each test subject in order to avoid errors during relevant measurement, which result from individual factors of a person such as positioning of the head, distance between the eyes, etc. The substantial part of the measurement concerned running the Microsoft Flight Simulator application in a previously planned place on the runway or in the air with simultaneous start of recording a scenery observed by a test subject. The following experiments have been designed:

- NON-PILOT – take-off (task 1) and landing (task 2)
- PILOT – take-off (task 1) and landing (task 2)

It has been presumed 60 s to record the attention of each person. The results were obtained in the form of video material. Videos were processed focusing on individual timeframes for each test subject. Using a dedicated component of SMI BeGaze environment, a set of areas of interest (AOI) was designed for the model plane used in the studies. 17 areas were identified having different surface areas and shapes that describe selected parts of an observed scene and particular aviation instruments (see Fig. 2). A non-standard template associated with the recorded video material allows to extract information such as fixation time on a particular device or area, order of recorded fixations or their percentage part in the observation of a selected instrument [4,5,14,15,19]. It enabled to estimate a timeframe in which a test participant did not focus on any of the instruments. The template was used in an analysis of the results for people who took part in the experiment, taking into consideration particular tasks. This delivered information about the amount of fixations in a particular area, their duration and order of observed areas. Task 1 was completed by all test subjects with different duration.

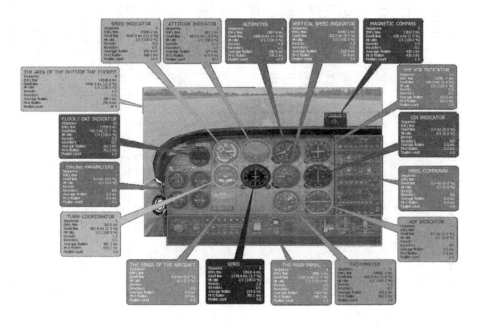

Fig. 2. Designed AOI's for the Cessna 172 Skyhawk cockpit

The obtained results in Table 1 show that during average duration of test that was about 67.5 s, almost 81% of fixations was in the areas of AOI's. Average number of fixations was 157. Therefore it may be presumed that saccades occurred in the remaining 19% of the time, batting of eyelids may have occurred which resulted in losing XY coordinates of the left and right eye or fixations were out of the areas defined as AOI. Analysis of average percentage of fixations in certain areas allowed to determine the most essential instruments (the threshold of more than 2% was heuristically assumed).

Table 1. Measurement results for test participants from the NON-PILOT and PILOT groups in task 1

Area of interest	The average percentage of all fixation [%]		The average number of fixations		The average time of all fixation [ms]		The average time a fixation [ms]	
	NON-PILOT	PILOT	NON-PILOT	PILOT	NON-PILOT	PILOT	NON-PILOT	PILOT
Altimeter (AOI_1)	27,24	7,52	35	17	22521,96	6955	645,94	275,78
The area of the outside the cockpit (AOI_2)	24,4	20,85	50	34	20624,96	10767,85	377,9	386,77
Speed indicator (AOI_3)	10,94	17,82	26	30	9086,6	16354,92	323,04	414,97
Attitude indicator (AOI_4)	6,24	8,05	17	17	5180,34	7300	346,76	345,98
The main panel (AOI_5)	2,46	0,27	7	1	2080,18	259,1	202,92	145,73
The VOR indicator (AOI_6)	2,06	0,05	3	0	1831,98	49,67	159,12	24,83
Tachometer (AOI_7)	1,66	3,28	4	10	1342,7	2993,05	354,66	198,98
Vertical speed indicator (AOI_8)	1,44	2,5	3	8	1164,56	2277,55	197,68	192,23
Gyro (AOI_9)	1,38	2,27	3	7	1144,44	2119,5	278,76	257,4
Panel COMM/NAV (AOI_10)	0,7	0	2	0	582,26	0	109,58	0
Clock/OAT indicator (AOI_11)	0,68	0,02	2	0	556,92	20,02	189,56	20,02
Turn coordinator (AOI_12)	0,62	0,83	3	3	514,72	770,68	82,5	155,5
Engine parameters (AOI_13)	0,46	0,17	1	1	380,58	163,07	319,74	29,28
Magnetic compass (AOI_14)	0,28	0	0	0	230,5	0	230,5	0
CDI indicator (AOI_15)	0,18	0,02	1	0	150,08	12,05	76,04	12,05
ADF indicator (AOI_16)	0,12	0,03	0	0	92,34	29,02	92,14	29,02
The signs of the aircraft (AOI_17)	0	0	0	0	0	0	0	0
Sum	80,86	63,68	157	128	67485,12	50071,48	3986,84	2488,54

When evaluating average percentage of the duration of fixation it can be assumed that for the NON-PILOT group, for task 1, the most essential areas of interest were the following aviation instruments and areas (Table 1): altimeter, the area of the outside the cockpit, speed indicator, attitude indicator, the main panel, the VOR indicator. In the PILOT group, in turn, for average duration of test which took 50.5 s, nearly 64% of fixations were in defined AOI areas. Average number of fixations was 128. In the PILOT group, when doing task 1, the most essential areas of interest were: the area of the outside the cockpit, speed indicator, attitude indicator, altimeter, tachometer, vertical speed indicator, gyro. For the NON-PILOT group, the variation coefficient C_v of registered times of particular fixations on the observed instruments achieves the values lower than 9.15%, and for the PILOT group, the values of 15.84%, respectively. A lower fixation percentage share of 63.68% during the task realization process indicates that the experienced pilots complete the assigned task faster than the NON-PILOT group whose share equals 80.86%. Based on the results for task 1 it can be assumed that:

- Attention in the PILOT group was mainly put on the area of the outside the cockpit and the readings of the speed indicator. It is essential to control the readings of speed indicator in training future pilots. Locating the attention of pilots having experience in this area means they have been properly trained.
- Attention in the NON-PILOT group was put mainly on the altimeter. In terms of taking off in Visual Meteorological Conditions (VMC), readings of altimeter should not be a priority. When it takes places, it may indicate the lack of training.
- The highest discrepancy of observation time occurs in case of altimeter. During take-off in VMC, an altimeter is not so important that is why more experiences pilots did not focus on it so much.
- In both test groups the area of the outside the cockpit seemed to be essential. Fixations in this area are usually related to visual determination of path direction on the ground and spatial orientation in the air, for VMC flights it is a priority area.
- For more experienced pilots, the attitude indicator was a supporting function in determination of spatial orientation of the plane after take-off, to a greater extent than it took place in case of the less experienced group. Scanning an attitude indicator is crucial that is why the more experienced pilots were expected to focus on this device.
- Readings of tachometer allow to determine whether the engine operates properly. The PILOT group subjects found this device more attractive than the NON-PILOT group subjects. Monitoring the readings of tachometer allows to identify problems with engine during take-off run on a runway. Future pilots are trained to scan this device. When pilots put their attention on this device it means that the level of training is satisfactory.
- Vertical speed indicator which determines vertical speed of the plane was used more often by subjects from the PILOT group. This suggests the intention of maintaining an optimal take-off angle from the starting airport.

- Gyro due to which running the course after the take-off was easier was found more necessary in the PILOT group. Its readings enable to take off in a straight line.
- The main panel and VOR indicator drew attention of the NON-PILOT group. Taking into consideration taking off in VMC these instruments do not seem to be crucial, that is why the more experienced participants put no focus on them.

For both test groups descriptive statistics was used and variance for each device was calculated. In task 1 both altimeter and speed indicator were the instruments with the highest data dispersion. Time intervals of attention in particular areas were presented on a timeline using the BeGaze environment. In the initial stage a pilot pays attention to tachometer. After setting maximum take-off rotation speed, plane acceleration begins and the attention of the NON-PILOT group is directed to the area of the outside the cockpit. Here the speed indicator is a supporting device, while the altimeter plays a supporting role during the take-off and that is the area to which the attention of test participants is mainly put on until the end of task 1. It is noteworthy that the test participant, during the take-off, pays attention to the instruments such as VOR, panel COMM/NAV and CDI indicator – the areas which are of little significance due to the character of conducted operation. In case of the PILOT group participants (Fig. 3 shows results for one of the test subjects) it can be observed that the pilot pays attention to the tachometer not only at the very beginning, but monitors its readings also during take-off run that did not take place in case of the less experienced pilot. In the initial phase of take-off run, the attention of the experienced pilot eyes is put on the area of the outside the cockpit and the readings of speed indicator. During the rotation, the pilot's eyes are moved in the direction of artificial horizon and speed indicator. Altimeter plays a supporting role. Unlike the NON-PILOT test subjects group, the experienced pilot, during the take-off, pays attention mainly to 4 instruments: speed indicator, attitude indicator, altimeter and tachometer. The solid line in the graph represents the moment of rotation when a pilot shifts a yoke to make an aircraft take-off. Second task was to perform landing of the plane on the path 27, at the airport EPRZ, from the height of 1500 ft AMSL and braking the plane to the speed lower than 10 kts. Table 2 presents the results on the basis of which it can be concluded that during average time of test which took about 112.5 s, almost 90% of fixations were located in defined AOI's. An average number of fixations in the NON-PILOT group was 207. In Task 2, in the NON-PILOT group, the dominant areas were: the area of the outside the cockpit, altimeter, attitude indicator, speed indicator. In the PILOT group: the area of the outside the cockpit, speed indicator. In this task, similarly as in case of Task 1, the variation coefficient of percentage share of average fixation values during the measurement process has been estimated. For the NON-PILOT group it does not exceed 8.15% and for the PILOT group it is less than 10.66% that can be assumed as a slight variation when comparing these figures. Moreover, the more experienced pilots complete the set tasks faster than the less experienced 69.34% versus 89.52%.

Table 2. Comparison of duration of instruments' observations in the NON-PILOT and PILOT groups for Task 2

Area of interest	The average percentage of all fixation [%]		The average number of fixations		The average time of all fixation [ms]		The average time a fixation [ms]	
	NON-PILOT	PILOT	NON-PILOT	PILOT	NON-PILOT	PILOT	NON-PILOT	PILOT
Altimeter (AOL1)	3,64	0,88	17	4	3555,78	1182,43	243,58	236,75
The area of the outside the cockpit (AOL2)	77,8	51,65	155	137	99363,92	67733,38	511,6	439,72
Speed indicator (AOL3)	2,1	13,48	10	51	2368,1	18141,1	233,32	279,32
Attitude indicator (AOL4)	2,78	0,73	12	4	3315,34	927,5	251,56	234,13
The main panel (AOL5)	0,4	0,32	2	2	467,86	430,62	185,1	55,73
The VOR indicator (AOL6)	0,58	0,12	2	1	652,68	145,05	150,26	87,88
Tachometer (AOL7)	0,22	0,2	1	1	288,9	271,8	227,46	72,63
Vertical speed indicator (AOL8)	0,54	1,38	2	7	655,88	1880,53	204,18	198,33
Gyro (AOL9)	0,34	0,3	2	1	405,48	208,08	214,76	144,23
Panel COMM/NAV (AOL10)	0,2	0	1	0	264,54	0	94,12	0
Clock/OAT indicator (AOL11)	0,38	0,12	2	1	468,6	116,4	116,9	38,8
Turn coordinator (AOL12)	0,16	0,03	0	0	217,28	54,07	217,28	54,07
Engine parameters (AOL13)	0,12	0,08	1	1	166,1	115,4	48,44	46,37
Magnetic compass (AOL14)	0,06	0	0	0	86,44	0	43,22	0
CDI indicator (AOL15)	0,1	0	0	0	127,64	0	60,8	0
ADF indicator (AOL16)	0,1	0,05	0	1	144,52	64,07	144,52	42,72
The signs of the aircraft (AOL17)	0	0	0	0	0	0	0	0
Sum	89,52	69,34	207	211	112549,06	91270,43	2947,1	1930,68

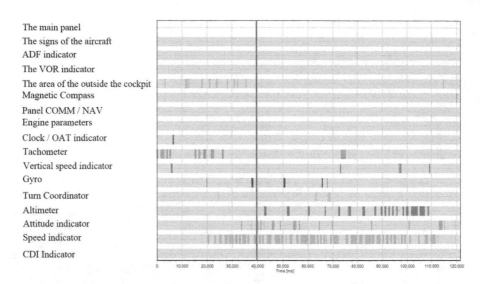

Fig. 3. Distribution of fixation in particular AOI as a function of time, PILOT group

On the basis of the obtained results the following conclusions can be formulated:

- In both test groups the area of the outside the cockpit was essential. Fixations in this area aimed at visual determination of path direction and spatial orientation.
- The NON-PILOT group directed its attention mainly towards the area of the outside the cockpit, however significant fixations can be also observed in the area of the speed indicator, altimeter and attitude indicator.
- Significant discrepancy between the two test groups can be observed in case of the speed indicator. During a difficult operation of landing it is crucial to monitor the speed, that is why the experienced pilot devoted more time to fixations in this area.
- The attitude indicator and altimeter was much more helpful for the NON-PILOT test subjects than for the group of more experienced test participants.

In task 2, data dispersion occurred beyond the area of the cockpit and on the speed indicator. Attention analysis for particular areas depending on the time of subject from the NON-PILOT group was presented in Fig. 4. The solid line represents the moment of touchdown. This subject's attention was drawn mainly to the area of the outside the cockpit for the whole period of measurement. At the beginning of the experiment the eyes of test subject are directed toward the attitude indicator in order to stabilise the position of the plane. The speed indicator, attitude indicator, altimeter, turn coordinator, gyro, vertical speed indicator, tachometer and VOR indicator are the instruments to which the less experienced pilots put little attention. During the touchdown, short scanning of

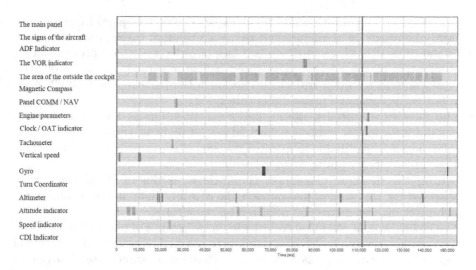

Fig. 4. Distribution of fixation in particular AOI as a function of time, NON-PILOT group

instruments occurs and after that, the eyes are directed back to the area of the outside the cockpit. Scanning of this area is maintained until the total stop of the plane on the path.

In case of the participants from the PILOT group it can be observed that at the beginning the pilot pays attention to the attitude indicator and next the fixation takes place near to the speed indicator. It results from the necessity of ensuring a safe speed of descending, which during the descent approach achieves minimum values. Next, the eyes are directed alternately between two areas: to the outside the cockpit and to speed indicator. Fixations in other areas than mentioned above are of little significance.

There is a clear difference between the representatives of the PILOT and NON-PILOT groups. Taking into consideration the safety of flight, it is unacceptable not to monitor the speed indicator. During the descent approach, the difference in monitoring the vertical speed indicator by subjects from both groups can be observed. In case of the more experienced pilot, it occurs more often and is used for maintaining optimal path of approach. The unexperienced subject skips this instrument.

4   Conclusion

The paper presented the measurement of pilot's attention trajectories during ascend and descend procedures with the use of flight simulator and SMI RED500 eyetracking system. Based on the conducted experiments, the following conclusion can be formulated:

- The degree of preparation and flight experience of test participants significantly influence the dynamics and shape of attention trajectories.
- In the NON-PILOT group there is a lack of order and chaos present during observation of the cockpit. The speed indicator turned out to be of little significance in recorded trajectories although the participants were informed about the necessity of maintaining safe speed.
- The more experienced pilots focused their attention on fewer instruments, but did it for longer period of time. In case of flight in visual meteorological conditions it is justifiable to monitor the area of the outside the cockpit. This area, in both test groups, was a priority.
- During the descend procedure, the highest results of discrepancy were observed for the area of the outside the cockpit while at the ascend operation, the highest dispersion occurred on the altimeter.
- The comparison of attention trajectories in two groups of pilots allows to evaluate the preparedness of beginner pilots to the flights performed in practical conditions. Obtained measurements and observation stands useful advice to junior pilots under the training process.
- The study of both NON-PILOT and PILOT groups showed differences in perceiving particular phases of flight depending on the flight experience of the test subjects. Some cockpit instruments could be, for example, more visible and flash to draw attention in particular flight phases crucial for passenger safety.

Acknowledgments. Simulation experiments were conducted in the scientific cooperation between Faculty of Mathematics and Natural Sciences Department of Computer Engineering at University of Rzeszow and Aviation Training Center of Rzeszow Technical University. The studies were conducted in the laboratory of Computer Graphics and Digital Image Processing at Center for Innovation and Transfer of Natural Sciences and Engineering Knowledge of Rzeszow University. Grant WMP/GD-11/2016: "Measurement and analysis of pilot attentions and the accompanying EEG brain activity during take-off and landing procedures".

References

1. Bolanowski, M., Paszkiewicz, A.: The use of statistical signatures to detect anomalies in computer network. In: Gołębiowski, L., Mazur, D. (eds.) Analysis and Simulation of Electrical and Computer Systems. LNEE, vol. 324, pp. 251–260. Springer, Cham (2015)
2. Dralus, G.: Study the quality of global neural model with regard to the local models of dynamic complex system. In: Gołębiowski, L., Mazur, D. (eds.) Analysis and Simulation of Electrical and Computer Systems. LNEE, vol. 324, pp. 35–61. Springer, Cham (2015)
3. Duchowski, A.T.: Eye Tracking Methodology: Theory and Practice. Springer, London (2007)
4. Dudek-Dyduch, E.: Algebraic logical meta-model of decision processes - new meta-heuristics. In: Rutkowski, L., Korytkowski, M., Scherer, R., Tadeusiewicz, R., Zadeh, L.A., Zurada, J.M. (eds.) ICAISC 2015. LNCS, vol. 9119, pp. 541–554. Springer, Cham (2015). doi:10.1007/978-3-319-19324-3_48

5. Dudek-Dyduch, E.: Modeling manufacturing processes with disturbances - a new method based on algebraic-logical meta-models. In: Rutkowski, L., Korytkowski, M., Scherer, R., Tadeusiewicz, R., Zadeh, L.A., Zurada, J.M. (eds.) ICAISC 2015. LNCS, vol. 9120, pp. 353–363. Springer, Cham (2015). doi:10.1007/978-3-319-19369-4_32

6. Eberz, S., Rasmussen, K.B., Lenders, V., Martinovic, I.: Preventing lunchtime attacks: fighting insider threats with eye movement biometrics. In: NDSS Symposium 2015. Internet Society, San Diego (2015). doi:10.14722/ndss.2015.23203

7. Grabowski, F., Paszkiewicz, A., Bolanowski, M.: Wireless networks environment and complex networks. In: Gołębiowski, L., Mazur, D. (eds.) Analysis and Simulation of Electrical and Computer Systems. LNEE, vol. 324, pp. 261–270. Springer, Cham (2015)

8. Hua, Y., Yang, M., Zhao, Z., Zhou, R., Cai, A.: On semantic-instructed attention: from video eye-tracking dataset to memory-guided probabilistic saliency model. Neurocomputing 168, 917–929 (2015). doi:10.1016/j.neucom.2015.05.033

9. Lu, Y., Zheng, W.-L., Li, B., Lu, B.-L.: Combining eye movements and EEG to enhance emotion recognition. In: Proceedings of the International Joint Conference on Artificial Intelligence (2015)

10. Orlosky, J., Toyama, T., Sonntag, D., Kiyokawa, K.: Using eye-gaze and visualization to augment memory. In: Streitz, N., Markopoulos, P. (eds.) DAPI 2014. LNCS, vol. 8530, pp. 282–291. Springer, Cham (2014). doi:10.1007/978-3-319-07788-8_27

11. Pfeiffer, T., Memili, C.: GPU-accelerated attention map generation for dynamic 3D scenes. In: Höllerer, T., Interrante, V., Lécuyer, A., JES II, (eds.) Proceedings of the IEEE VR 2015, pp. 257–258. IEEE (2015)

12. Rutkowski, L.: Computational Intelligence. Methods and Techniques. Springer, Heidelberg (2008)

13. Sharma, K., Hostettler, L.O., Lemaignan, S., Fink, J., Mondada, F., Dillenbourg, P.: Eye tracking with educational robots: A cautionary tale, Dissertation (2014)

14. Smith, J., Booth, T., Bailey, R.: Refresh rate modulation for perceptually optimized computer graphics. In: International Conference on Computer Graphics Theory and Application (2014)

15. Tadeusiewicz, R.: Neural networks in mining sciences-general overview and some representative examples. Arch. Min. Sci. 60(4), 971–984 (2015)

16. Teng, T.H., Tan, A.H., Zurada, J.M.: Self-organizing neural networks integrating domain knowledge and reinforcement learning. IEEE Trans. Neural Netw. Learn. Syst. 26(5), 889–902 (2015)

17. Toyama, T., Orlosky, J., Sonntag, D., Kiyokawa, K.: A natural interface for multi-focal plane head mounted displays using 3D gaze. In: Proceedings of the 2014 International Working Conference on Advanced Visual Interfaces AVI 2014, pp. 25–32 (2014). doi:10.1145/2598153.2598154

18. Xu, J., Mukherjee, L., Li, Y., Warner, J., Rehg, J.M., Singh, V.: Gaze-enabled egocentric video summarization via constrained submodular maximization. In: IEEE Computer Society Conference on Computer Vision and Pattern Recognition (CVPR) (2015)

19. Young, L.R., Sheena, D.: Behavior Research Methods and Instrumentation: Survey of eyemovement recording methods (1975)

3D Integrated Circuits Layout Optimization Game

Katarzyna Grzesiak-Kopeć[(⊠)], Leszek Nowak, and Maciej Ogorzałek

Department of Information Technologies,
Jagiellonian University in Krakow, Krakow, Poland
{katarzyna.grzesiak-kopec,leszek.nowak,maciej.ogorzalek}@uj.edu.pl

Abstract. This paper is devoted to the original approach to block-level
3D IC layout design. The circuit components are modeled as autonomous
mobile agents that explore their virtual world in order to find a globally
near-optimal layout solution. The search space is defined by geometry
features, wire connections, goals and constraints of the design task. The
approach is illustrated by the example application to one of the MCNC
benchmark circuits and implemented using Godot.

Keywords: Floorplaning · Machine learning · Steering behaviors ·
Optimization · Computer game

1 Introduction

The problem of a valid 3D layout generation can be found in many different
domains, starting from practical and scientific purposes, through virtual reality modeling, ending at urban planning or crisis management. Design solutions
that fulfill requirements and meet constraints promote minimizing the materials
and energy consumption while optimizing the functional properties. The spatial
arrangement of components also plays a crucial role in integrated circuit design.
The chip design involves myriad conditions related to chip area minimization,
thermal hot spots reduction or wire length optimization, which makes it especially challenging. An original approach to block-level 3D IC layout design has
been proposed in [7,8], where the intelligent framework architecture uses a simple shape grammar to generate topologically feasible solutions. These proposal
solutions are further optimized with a use of the extremal optimization. In this
paper, an alternative concept of a computer game like visual 3D optimal layout
design is proposed. It is inspired by swarm intelligence algorithms and steering
behaviors for autonomous agents in animation and computer games. The components are treated as autonomous agents that navigate around their world in
order to find a globally near-optimal solution. Combinations of steering behaviors
are used to achieve both goals and constraints of a specific layout design task.
The approach is illustrated by the example application to one of the MCNC
benchmark circuits [12] and implemented using Godot which is an advanced,
feature-packed, multi-platform 2D and 3D open source game engine [6].

© Springer International Publishing AG 2017
L. Rutkowski et al. (Eds.): ICAISC 2017, Part II, LNAI 10246, pp. 444–453, 2017.
DOI: 10.1007/978-3-319-59060-8_40

2   Related Work

The most critical phase in integrated circuits design is floorplanning. It is a kind
of a packing task where all circuit components have to be arranged according to
given design rules. The circuit components are rectangular modules that cannot
overlap. The minimum bounding box of a packing is called the chip [14]. The
problem have been effectively solved in 2D spaces, but the proposed algorithms
are not easily transformed to introduce the third dimension. The 3D chips allow
for the smaller footprint, higher packing density, lower interconnect power consumption and heterogeneous technology chip support [4]. However, the today's
3D IC technology has some important limitations. A truly 3D chip fabrication is
actually impossible. All the circuit components are distributed among restricted
number of device layers and the height of the inter-layers is fixed (see example
configurations in Fig. 1).

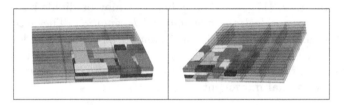

Fig. 1. Quasi-3D ICs configurations example.

The 3D integrated circuits placement problem is known to be NP-hard. Common techniques for global placements are: partitioning-based algorithms, analytic techniques and stochastic ones [10]. Recursive partitioning are constructive
techniques that recursively cut the layout into smaller parts. The most common
partitioning algorithms are the Kernighan-Lin [11] and the Fiduccia-Mattheyses
algorithm [5]. Analytic techniques use either quadratic objective functions or
sophisticated nonlinear calculations [16]. The most popular stochastic based
placement uses simulated annealing [2]. The generic approach to 3D layout design
proposed in [7] generates plausible solutions with a use of a simple shape grammar supervised by an intelligent derivation controller. Design knowledge is fed
into system in a form of predicates. The floorplan generation procedure takes
into account the current technological limitations and divides a chip into layers.
Hence, the obtained design results are quasi-3D ones.

The seminal research into steering behavior by Craig Reynolds [17] modeled
the movement patterns of flocks, and since then has been studied from many
different perspectives, like swarm robotics [9], crowd simulation [19] or artificial
life [15]. In the 3D layout design task, the components should navigate around
their world to find a globally near-optimal solution. The use of behavioral animation in generating virtual worlds is still the subject of many different research
projects. A number of them use hierarchical schemes for organizing complex
control [3].

3   Autonomous Agent

In our approach agents corresponds to physical components of the design, which should be optimally arranged in the search space. The term *autonomous agent* may be used in many different contexts. In this paper, by an agent we understand a computer system situated in a world shared by other entities, which is capable of autonomous actions that lead it to satisfy its design task [20]. It is not only reactive, perceives its environment and responds to changes that occur in it, but exhibits goal-driven behavior as well. It also interacts with other agents. Having all this features, an agent can be recognized as intelligent [21]. It is a real agent in a virtual world embodied in a physical manifestation.

Also the term of behavior has many different interpretations. Being inspired by swarm intelligence heuristics we have decided to solve a layout design task applying various motion behaviors. Combining stochastic approach and motion behaviors may provide an effective mechanism for screening large and discontinuous spaces. Thus after [17], the agent's behavior may be divide into a hierarchy of three layers: *action selection*, *steering*, and *locomotion*. The *action selection* layer involves actions strategy, goals and planning. In the *steering* level, the goal is decomposed into a series of simple subgoals that correspond to some steering behaviors and an agent path is determined. Finally, the *locomotion* layer is responsible for an actual movement.

3.1   Steering Behaviors

Steering behaviors allow autonomous agents to navigate around their environment in a life-like or any imaginative manner. They are usually defined in such a way to be largely independent of the agent's means of locomotion and have a similar structure. They take as an input the kinematic of the agent that is moving and some target information [13]. They can be divided into simple and combined behaviors presented in Table 1.

Table 1. Simple and combined steering behaviors.

Simple behaviors	Combined behaviors
Seek & flee	Pursuit & evade
Arrive	Wander, obstacle avoidance, path following, ...
Align	Flocking behavior: separation, cohesion and alignment

Simple behaviors are applicable to single agents. *Seek* steers the agent towards a specified target. It calculates the direction to the target in the global coordinate system and heads toward it as fast as possible (maximal speed). If no other behavior appears, the agent eventually pass through the target and then turn back to approach again. *Flee* is the opposite of seek. The agent turns away from the target and tries to get as far from it as possible. *Arrive* is a kind of a seek behavior that slows the agent down as it approaches the target and makes it stop there. *Align* is responsible for the agent heading. It turns the agent to reach the target orientation.

Combined behaviors are applicable not only to single agents but to groups of agents as well. *Pursuit* and *evade* derive from seek and flee behaviors respectively and used when a target is moving. *Wander* is a kind of random life-like steering that enables to move agent around the world when no target is specified. It acts as a delegated seek behavior. *Obstacle avoidance* allows an agent to maneuver in a cluttered environment by dodging around obstacles. It casts one or more rays out in its motion direction. If the collision with an obstacle occurs, a new target is calculated in such a way to avoid it. Then a moving agent simply seek on the new target. *Path following* enables an agent to steer along a predetermined path within the specified radius of the spine. It applies a seek behavior to steer toward a predicted future position. The most common group steering behavior is *flocking* [17].

In many computer games, simple steering behaviors can achieve a satisfying movement realization. Some decision making algorithms determine where the agent should move and the seek behavior is applied to perform it. However, in order to reach its goal safety and avoid collisions, an autonomous agent usually needs more than one steering behavior. There are two general methods of combining steering behaviors: *blending* and *arbitration*. Both of them, take a group of steering behaviors and generate a single overall steering output.

Blending uses a set of weights or priorities to combine the results of all the steering behaviors. There are no constraints on the blending weights. The final steering output achieved from the weighted sum may even go far beyond the moving capabilities of the agent, so it is simply trimmed according to the maximum possible value. There is no simple answer how to determine the right coefficients values. Even though there are different research projects trying to automate the tuning of model parameters using evolutionary strategies [1], as in most parametrized systems, they are greatly dependent on the system architect experience and her/his inspired lucky guess or a good trial and error. To be more efficient, weights or priorities may change over time in response to the state of the working environment.

One of the best known combined blended behavior is *flocking*. *Flocking* is a kind of coordinated motion inspired by animals groups such as bird flocks and fish schools [17]. It blends three steering behaviors, namely *separation, cohesion* and *alignment*. Separation moves an agent away from agents that are too close. Cohesion works in a quite opposite way and moves an agent toward the center of mass of the flock. Alignment lets all the agents to move in the same direction and at the same velocity. In some cases, using equal blending weights for all of these three behaviors may be sufficient. However, usually separation is more important than cohesion, which is more important than alignment. While blending, it is also possible to use priorities groups of behaviors. Each group contains behaviors with regular blending weights and is considered according to a given priority order.

Arbitration uses different schemes to select a current steering behavior. There are no restrictions imposed on the arbitration, which would enforce to return only one simple steering behavior instead of a combined one. In fact, blending and arbitration are often mixed together to get more realistic implementations.

Both blending and arbitration combine steering behaviors in an independent manner. Yet, in order to obtain more realistic model, some cooperation among different behaviors is required. Being aware of its context, a steering behavior increases its complexity and is more difficult to handle. Thus, collaborative steering behaviors implementations use more sophisticated decision making algorithms like state machines, decision trees, or a steering pipeline.

4   3D ICs Layout Design Constraints and Goals

While investigating the 3D ICs layout design problem, various constraints and goals were identified [7]. They are summarized in Tables 2 and 3 respectively. First of all, all the circuit components should be placed in a specified chip (AREA) without overlapping (NO INTERSECTION). To minimize the chip bounding box, a plausible layout has to be consistent (GLUE). The total wirelength of a chip is minimal if the connected components are as close to each other as possible (ADJACENT). Thermal management requests separating selected modules to minimize a hot spot problem (NEIGHBOR) and also may require to settle the most heating components in the outermost layers (LAYER).

Table 2. 3D ICs floorplanning constraints

Constraint	Description
AREA	A component must be placed in a specified area
NO INTERSECTION	A component does not intersect other components
GLUE	A solution is consistent
NEIGHBOR	A component is in the specified neighborhood range
ADJACENT	Neighboring faces of adjoining components are of the same type
LAYER	A component is in a boundary (intra) layer

Constraints are either true or false, while objectives can be achieved to some extent. Instead of rejecting imperfect solutions, the search procedure should change its direction toward better ones. The main goal is a chip (packing) minimization (MINIMAL SPACE). Some components are preferred to be placed as close to the boundary as possible (POSITION) and some require aligning (SPATIAL RELATION).

Table 3. 3D ICs floorplanning goals

Goal	Description
MINIMAL SPACE	Evaluates the area occupied by a current design in relation to the expected minimal area
POSITION	Evaluates whether components are generated in the expected positions (e.g. boundary)
SPATIAL RELATION	Evaluates whether components are arranged in an expected way (e.g. aligned vertically)

5 Game

In this paper, we propose an original approach to the 3D integrated circuits layout optimization problem that goes far beyond the current technological and manufacturing limitations. The circuit components are autonomous mobile agents situated in a search space defined by their geometry features, wire connections, goals and constraints. To verify the proposed method, a dedicated game has been developed using the open source Godot game engine [6]. All the examples presented in this paper are generated with a use of the original software. The approach is illustrated by the example application to one of the MCNC benchmark circuits [12], namely *apte.yal* which is composed of 9 components all connected to one another (Fig. 2).

5.1 Circuit Components

All circuit components are cuboids with specified geometry features and wire connections. They are modeled with a use of Godot *RigidType* nodes. This kind

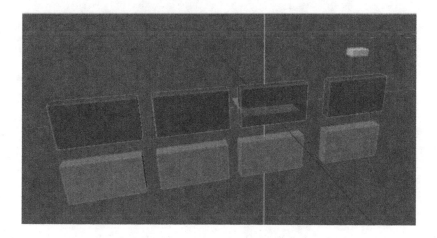

Fig. 2. Godot: *apte.yal* components.

of body has mass, friction, bounce and simulates Newtonian physics. Its motion may be affected by gravity and other entities. Its current position is generated by the simulation of linear and angular velocity from the former one. In order to meet the NO INTERSECTION constraint, each component has also appropriate *CollisionShape* assigned. It also knows his circuit connections (connected components), called *neighbors*.

5.2   Behavioral Animation

The game make use of the flocking behavior and gravity. The optimization procedure is actually driven by the physics engine implemented in Godot. The main challenge is the appropriate assignment of steering forces. Its general algorithmic scheme is very simple and proceeds as follows:

```
while(!stop){
  for each net{
    calculate the center of mass
    for each component in net{
      seek toward the center of mass
    }
  }
}
```

Before any movement begins, all the constraints and goals must be defined by the means of steering behaviors (see summary in Table 4). The whole process starts with a random positioning of chip components in a 3D search space. Then the game moving algorithm is applied where the main goal is the MINIMAL SPACE one.

Table 4. The constraints and goals mapping to steering behaviors.

Constraint/Goal	Steering behavior
AREA	Cohesion
NO INTERSECTION	Ceparation
GLUE	Cohesion and gravity
NEIGHBOR	Cohesion and separation, respectively
ADJACENT	Faces cohesion and separation, respectively
LAYER	Faces cohesion and separation, respectively
MINIMAL SPACE	Cohesion and gravity
POSITION	Cohesion and separation, respectively
SPATIAL RELATION	Alignment

The whole game may be divided in two logical stages. The aim of the first stage is to find minimal local arrangements of connected neighbors (NEIGHBOR constraint). During this stage, the gravity force is completely neglected and the main steering behavior that let this stage accomplish is *cohesion*. The components are moved toward the center of mass of the neighborhood until they collide. When no further move is possible, components try to rotate in an obtained position to minimize the neighborhood volume. Just like in the simulated annealing algorithm, better configurations are always accepted while worst are accepted with a certain probability. In the same time, while moving closer toward its connected neighbors, components are affected by repulsive forces from components that are either not directly connected to them or should not be placed close to each other (*separation steering behavior*). In order to incorporate the SPATIAL ARRANGEMENT goal, the *alignement* steering behavior is applied. The example configuration of components obtained in such a way is presented in Fig. 3. At the first glance it seems to be far away from the optimal one, but only few more actions are required to improve it.

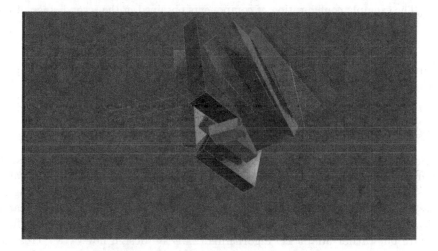

Fig. 3. The example configuration of components obtained in the first stage of the game.

The first stage is finished by the game player (designer). She/he turns on the gravity and the second stage starts. Now, the aim of the game is to squeeze the intermediate solution. To better understand the proposed approach, let's imagine the process of collecting spilled deck of cards from the table. First, we grab all the cards and try to hold them. After that, we lower the cards on the table without dropping them from the hand and they are aligned in one dimension. Then, we make a 90-degree turn and repeat the procedure to aligned them in a second dimension. The same process is applied to a components configuration. While keeping the attraction forces among neighbors, the configuration is affected by gravity. It falls down into a 90-degree V-shape (virtual table). A 90-degree V-like shape eliminates the need of turning the configuration and repeating the falling

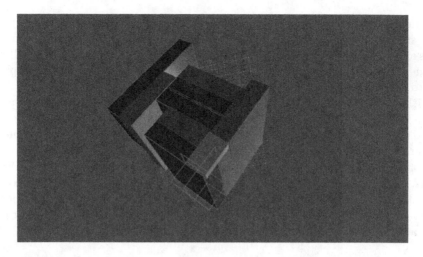

Fig. 4. The example configuration of components obtained in the second stage of the game.

procedure. After reaching the virtual ground, the chip is much more compact (see Fig. 4).

6   Conclusions and Future Prospects

The presented approach is a part of ongoing research on building a flexible software architecture framework which will enable solving the 3D integrated circuits layout problem. The task is not only up-to-date but very challenging one as well. The market electronic design automation (EDA) tools are dedicated solutions adjusted to present technology limitations. Most of them are not fully 3D aware but rather adapt 2D algorithms (2.5D IC design flow) [18]. Treating circuit components as an autonomous agents that are governed by the laws of Newtonian physics and navigate around their virtual reality, is a completely new approach to solving this problem. Both goals and constraints are described by the means of steering behaviors. Even though the final outcome of the research is still hard to predict, taking into account the preliminary results and practical applications of autonomous agents in complex and dynamic environments like a crowd simulation, we strongly belief that it is worth pursuing.

References

1. Berseth, G., Kapadia, M., Haworth, B., Faloutsos, P.: SteerFit: automated parameter fitting for steering algorithms. In: Proceedings of the ACM SIG-GRAPH/Eurographics Symposium on Computer Animation (SCA 2014), Eurographics Association, Aire-la-Ville, Switzerland, pp. 113–122 (2015)

2. Chen, T.C., Chang, Y.W.: Modern floorplanning based on B*-tree and fast simulated annealing. IEEE Trans. Comput. Aided Des. Integr. Circuits Syst. **25**(4), 637–650 (2006)
3. Donikian, S., Rutten, E.: Reactivity, concurrency, data-flow and hierarchical preemption for behavior animation. In: Veltkamp, R.C., Blake, E.H. (eds.) Programming Paradigms in Graphics. Eurographics Collection. Springer, Vienna (1995)
4. Dong, X., Xie, Y.: System-level cost analysis and design exploration for three-dimensional integrated circuits (3D ICs). In: Proceedings of the 2009 Asia and South Pacific Design Automation Conference (ASP-DAC 2009), pp. 234–241 , IEEE Press, Piscataway, NJ, USA, (2009)
5. Fiduccia, C.M., Mattheyses, R.M.: A Linear-time heuristic for improving network partitions. In: DAC, pp. 175–181 (1982)
6. Godot: An advanced, feature-packed, multi-platform 2D and 3D open source game engine (2016). https://godotengine.org/. Accessed Dec 2016
7. Grzesiak-Kopeć, K., Ogorzałek, M.: Computer-aided 3D ICs layout design. Comput. Aided Des. Appl. **11**(3), 318–325 (2014)
8. Grzesiak-Kopeć, K., Oramus, P., Ogorzałek, M.: Using shape grammars and extremal optimization in 3D IC layout design. Microelectron. Eng. **148**, 80–84 (2015)
9. Joselli, M., Passos, E.B., Zamith, M., Clua, E., Montenegro, A., Feijó, B.: A neighborhood grid data structure for massive 3D crowd simulation on GPU. In: 2009 VIII Brazilian Symposium on Games and Digital Entertainment, pp. 121–131 (2009)
10. Kahng, A.B., Lienig, J., Markov, I.L., Hu, J.: VLSI Physical Design: From Graph Partitioning to Timing Closure. Springer Publishing Company Inc., Heidelberg (2011)
11. Kernighan, B.W., Lin, S.: An efficient heuristic procedure for partitioning graphs. Bell Syst. Tech. J. **49**(2), 291–307 (1970)
12. MCNC: The MCNC set of benchmark circuits (2015). http://lyle.smu.edu/~manikas/Benchmarks/MCNC_Benchmark_Netlists.html. Accessed June 2015
13. Millington, I., Funge, J.: Artificial Intelligence for Games, 2nd edn. Morgan Kaufmann Publishers Inc., San Francisco (2009)
14. Murata, H., Fujiyoshi, K., Nakatake, S.: VLSI module placement based on rectangle-packing by the sequence pair. IEEE Trans. Comput. Aided Des. Integr. Circ. Syst. **15**(12), 1518–1524 (1996)
15. Nathan, A., Barbosa, V.C.: V-like formations in flocks of artificial birds. Artif. life **14**(2), 179–188 (2008)
16. Obermeier, B., Johannes, F.M.: Temperature-aware global placement. In: Proceedings of the 2004 Asia and South Pacific Design Automation Conference (ASP-DAC 2004), pp. 143–148, IEEE Press, Piscataway, NJ, USA, (2004)
17. Reynolds, C.: Steering behaviors for autonomous characters. In: Game Developers Conference, pp. 763–782 (1999)
18. Rhines, W.: 3D IC design challenges. In: GSA Memory Conference, San Jose, CA (2011)
19. Thalmann, D., Musse, S.R.: Crowd Simulation, 2nd edn. Springer, Heidelberg (2013)
20. Wooldridge, M.J., Jennings, N.R.: Intelligent agents: theory and practice. Knowl. Eng. Rev. **10**(2), 115–152 (1995)
21. Wooldridge, M.J.: Intelligent Agents, Multiagent Systems: A Modern Approach to Distributed Artificial Intelligence. MIT Press, Cambridge (1999)
22. Zhang, H.: The optimality of naive bayes. In: FLAIRS Conference (2004)

Multi-valued Extension of Putnam-Davis Procedure

Krystian Jobczyk[1,2](✉) and Antoni Ligeza[2]

[1] University of Basse-Normandie of Caen, Caen, France
krystian_jobczyk@op.pl
[2] AGH University of Science and Technology of Kraków, Kraków, Poland

Abstract. In 1960 M. Davis and H. Putnam introduced some logical verification procedure for propositional languages – called later Putnam-Davis procedure. It found a broad application in AI as a basis of the planning paradigm based on satisfiability of formulas. Unfortunately, this procedure refers to satisfiability in a classical two-valued logic. This paper is aimed at proposing some multi-valued extension of this procedure that may be sensitive to temporal and preferential aspects of reasoning. This method is evaluated in more practical contexts

1   Introduction

One of the most smart logical tools in planning is the so-called Putnam-Davis procedure (alternatively: Davis-Putnam procedure). Although it was initially invented by Hilary Putnam (1926–2016) and Martin Davis (1928-) already in 1960 in [12] and improved by J. Beckford, G. Logemann and D. Loveland in 1962 in [1] as a purely logical method, it waited almost 30 years to be adopted in Artificial Intelligence to planning in the satisfiability-based planning paradigm. Indeed, this paradigm – *explicitly* – was elaborated relatively late in 1992 in [10] and newly discussed in 1994 in [2] – just in the contexts of Putnam-Davis procedure. The indicators of this paradigm may be listed as follows:

– states $s_i \in \mathcal{S}$ in a planning domain may be naturally viewed as propositions of a (propositional) planning language, say $\mathcal{L}$, and $\mathcal{S} \subseteq 2^{\mathcal{L}}$,
– planning conditions, action preconditions etc. may be identified with the appropriate formulas representing them in a given planning language $\mathcal{L}$.

The Putnam-Davis procedure plays a crucial role in this satisfiability-based approach to planning. The original Putnam-Davis procedure forms a two-valued-based proof procedure and may be very briefly given by the algorithm:

Putnam-Davis procedure (Φ in CNF)

> **Input**: A set of clauses Φ
> **Output**: A Truth Value

L. Rutkowski et al. (Eds.): ICAISC 2017, Part II, LNAI 10246, pp. 454–465, 2017.
DOI: 10.1007/978-3-319-59060-8_41

The crucial role of two-valued logic as a 'basis' of this procedure also manifests itself by the fact that formulas are valid if and only if their negation are unsatisfiable.

1.1 State of the Art and the Paper Motivation

Either in Davis-Putnam's work from 1960 in [12] or in [1,2] the Putnam-Davis procedure is introduced and considered as two-valued logical method for propositional languages. It has some consequences: this procedure (in this original depiction) cannot be used in fuzzy cases; for situation of acting under uncertainty. In addition, the original Putnam-Davis procedure is a 'static' procedure, which is not sensitive to temporal aspects of acting.

Simultaneously, some promising results were recently elaborated with respect to four-valued logic by H.J. Levesque [11] and U. Straccia in [13]. Unfortunately, these meta-logical results have not incorporated to earlier researches on Putnam-Davis procedure. These all shortcomings form a main motivation factor of considerations of this paper. Some ideas of this paper – regarrding to fuzziness and multi-valency and its representation in different contexts of temporal planning – were discussed by authors in [3–9].

1.2 The Paper Objectives and Organization

According to this motivation factor – the paper is aimed at:

G1 proposing a new Multi-valued extension of the original Putnam-Davis procedure,

G2 proving some of its meta-logical features and

G3 showing how this extension allows us to propose both temporal and preferential extension of this procedure.

The paper is organized as follows. In Sect. 2 the original Putnam-Davis procedure is repeated. In Sect. 3 the new Multi-valued Putnam-Davis procedure and its meta-logical properties are presented. Section 4 presents this extension in use. Section 5 describes the same Multi-valued Putnam-Davis procedure in the preferential variant. Section 6 contains closing remarks.

2 Unit-Propagation and Putnam-Davis Procedure

Putnam-Davis procedure is essentially supported by the so-called **unit-propagation**. Its main role relies on simplifying the formulas given in a Conjunctive Normal Form (CNF). This procedure may be specified as follows.

Unit-propagation. Assume that a formula Φ of a (propositional) planning language $\mathcal{L}$ is given in a CNF, that is $\Phi = C_1 \wedge C_2 \wedge \ldots C_k$ for some k (each

$C_i \in \Phi$ is called a *clause*). On the input of the `Unit-Propagation` we have Φ and some empty model μ. In output we get a simplified formula Φ and a newly extended model μ. This model is just extended for a cost of the simplified Φ. Namely:

- we chose a unit clause $\{l\}$ in Φ (if there is),
- we throw away all clauses $C_i \in \Phi$, where l occurs,
- we throw away $\neg l$ from clauses C_i, where $\neg l$ occurs,
- the rejected unit clause is added to μ-model.

Thus, we exchange a formula Φ by $\Phi - C$ (for a unit clause $\{l\}$) and by $\Phi - C \cup \{C - \{\neg l\}\}$ for $\{\neg l\}$. Observe, that in the first case, we reject the whole clause C from Φ, in the second one – we preserve it rejecting a unit clause $\{\neg l\}$ only. This procedure in a compact form is depicted by the following `Unit-Propagate` algorithm.

`Unit-propagate`(Φ, μ)
begin
    while there is a unit clause $\{l\}$ in Φ **do**
        $\mu \leftarrow \mu \cup \{l\}$
        for every clause $C \in \Phi$
            if $l \in C$ then $\Phi \leftarrow \Phi - \{C\}$
            else if $\neg l \in C$ then $\Phi \leftarrow \Phi - C \cup \{C - \{\neg l\}\}$
end

In such a framework, Putnam-Davis procedure may be specified as based on unit-propagation, which is applied to all cases (formulas) excluding two aberrations: when $\emptyset \in \Phi$ and $\phi = \emptyset$.

    If unit-propagation is applied to Φ, then algorithm orders to select a variable P such that P or its negation occurs in Φ and to reject it from Φ – due to unit-propagation. Finally, the algorithm orders to continue the same procedure for the simplified formula Φ and for the extended model $\mu \cup \{P\}$ or $\mu \cup \{\neg P\}$ (resp.).

In terms of algorithm:
        select a variable P such that P or $\neg P$ occurs in Φ
        `Davis-Putnam` $(\Phi - \{P\}, \mu \cup \{\neg P\})$
        `Davis-Putnam` $(\Phi - \{\neg P\}, \mu \cup \{P\})$

Due to [14] – the whole Putnam-Davis procedure may be algorithmically depicted as follows.

```
Davis-Putnam(Φ, μ)
begin
    if ∅ ∈ Φ then return
    if Φ = ∅ then exit with μ
    otherwise Unit-propagate
    select a variable P such that P or ¬P occurs in Φ
    Davis-Putnam (Φ − {P}, μ ∪ {¬P})
    Davis-Putnam (Φ − {¬P}, μ ∪ {P})
end

Unit-propagate(Φ, μ)
begin
    while there is a unit clause {l} in Φ do
        μ ← μ ∪ {l}
        for every clause C ∈ Φ
            if l ∈ C then Φ ← Φ − {C}
            else if ¬l ∈ C then Φ ← Φ − C ∪ {C − {¬l}}
end
```

3 Multi-valued Extension of Putnam-Davis Procedure and Its Meta-Logical Properties

It arises a natural question of a generalization of this procedure by admissing more than two truth values. Suppose – for simplicity – that new fuzzy values are admitted in branches in, say k-level, for $k \leq n$ of Putnam-Davis procedure tree (preferable at the final step). Obviously, this new P-D procedure should radically deviate from the original one in cases of conjunctions $P \wedge \neg P$ (of literals). Since a fuzzy value of $P \wedge \neg P$ is generally different from 0 and is determined by some t-norm, it allows us to specify the new Putnam-Davis procedure as follows.

Consider a n-level tree Putnam-Davis procedure for some formulas Φ (of a first-order propositional language) given in CNF.

1. until a formula $P \wedge \neg P$[1] occurs in k-level for $k \leq n$ (if any), Putnam-Davis procedure with unit propagation works without changes,
2. when a formula $P \wedge \neg P$ occurs, we work as follows:
   - we associate fuzzy values to P and $\neg P$, i.e. $v(P), v(\neg P)$,
   - we compute $v(P \wedge \neg P) = t\text{-norm} (v(P), v(\neg P))$.

If now $v(P \wedge \neg P) = 0$, any model exists, so $\mu = \emptyset$. If $v(P \wedge \neg P) = t\text{-norm}(v(P), v(\neg P)) \neq 0$, we get a model μ with a degree $v(P \wedge \neg P)$.

It allows us to propose the following 'temporal' extension of Putnam-Davis procedure in (Łukasiewicz logic with min-norm).

---

[1] P is an atomic variable or a term of a given language built up from atomic variables.

```
Davis-Putnam(Φ, μ, v)
begin
    if ∅ ∈ Φ then return
    if Φ = ∅ then exit with μ
    otherwise Unit-propagate
    select a variable P such that P or ¬P occurs in Φ
    Davis-Putnam (Φ − {P}, μ ∪ {¬P})
    Davis-Putnam (Φ − {¬P}, μ ∪ {P})
            if P ∧ ¬P associate v(P) and v(¬P)
            compute v(P ∧ ¬P) = min(v(P), v(¬P)).
                if v = 0, then failure and μ = ∅
                if v ≠ 0, then return model μ with a degree v.
end
```

A similar 'temporal' extension of Putnam-Davis procedure may be proposed in Product Fuzzy Logic with a product norm as follows.

```
Davis-Putnam(Φ, μ, v)
begin
    if ∅ ∈ Φ then return
    if Φ = ∅ then exit with μ
    otherwise Unit-propagate
    select a variable P such that P or ¬P occurs in Φ
    Davis-Putnam (Φ − {P}, μ ∪ {¬P})
    Davis-Putnam (Φ − {¬P}, μ ∪ {P})
            if P ∧ ¬P associate v(P) and v(¬P)
            compute v(P ∧ ¬P) = v(P) • v(¬P)).
                if v = 0, then failure and μ = ∅
                if v ≠ 0, then return model μ with a degree v.
end
```

As earlier, it arises a natural question of a complexity of this procedure. Unfortunately, a solving of this problem in the whole generality seems to be problematic. Nevertheless, one can approximate a solution by considering this TP-Putnam-Davis procedure in Four-Valued Fuzzy Language – such as in [11,13].

More precisely, assume $\mathcal{L}_4$ is a propositional language with connectives $\vee, \wedge, \neg$. Assume that to each proposition of $\mathcal{L}_4$ one can associate a one of the four values: *true, false, unknown and contradiction*. If A is a formula of $\mathcal{L}_4$, its derivability from a set $\sum$ of proposition, called 'Knowledge Base' will be denoted as $\sum \models_4 A$. We can enlarge $\mathcal{L}_4$ to $\mathcal{L}_4^{fuzzy}$ by introducing fuzzy propositions of the form $[A \geq n]$, for $n \in [0,1]$ (for example: $[A \geq 0,7]$).

Whenever we refer to formulas of $\mathcal{L}_4^{fuzzy}$, we will denote $\sum$-derivability by $\sum |\approx_4 [A \geq n]$. As usual, CNF denotes in $\mathcal{L}_4^{fuzzy}$ a formula in a *Conjunctive Normal Form*. Then it holds the following

Theorem 1 *(Straccia [13]).* $\sum |\approx_4 [A \geq n]$ *is coNP-complete as* $\sum \models_4 A$ *is. Given* $\sum$ *and* $[A \geq n]$ *in CNF (i.e. both* $[A \geq n]$ *and all formulas in* $\sum$ *are formulas in CNF), then a checking* $\sum |\approx_4 [A \geq n]$ *can be done in* $O(|\sum||A|)$.

It allows us to formulate the following theorem describing a complexity problem of TP-Putnam-Davis procedure.

Theorem 2. *Assume that TP-Putnam-Davis procedure refers to formulas of* $\mathcal{L}_4^{fuzzy}$ *(given as above) with fuzzy propositions* $\sum |\approx_4 [A \geq n]$ *and a set of Knowledge Base* $\sum$*. Given* $\sum$ *and* $[A \geq n]$ *in CNF (i.e. both* $[A \geq n]$ *and all formulas in* $\sum$ *are formulas in CNF), a checking derivability for TP-Putnam-Davis procedure can be done in* $O(|\sum||A|)$.

Proof: Note that checking derivability for TP-Putnam-Davis procedure may be identified with:

1. checking $\sum |\approx_4 [A \geq n]$ for a finite set of such fuzzy formulas $[A \geq n] \in \mathcal{L}_4^{fuzzy}$ or
2. checking $\sum |\approx_4 [A \wedge B \geq nm]$ for a finite set of formulas $[A \geq n], [B \geq m] \in \mathcal{L}_4^{fuzzy}$ (since we consider fuzzy values $v(A \wedge B) = v(A) \bullet v(B)$ in TP-Putnam-Davis procedure).

(TP-Putnam-Davis procedure always contains a finite set of such formulas; in practice – a very small set.).

Assume the first case. In this case the thesis immediately follows from the theorem above, which asserts that $\sum |\approx_4 [A \geq n]$ for each such a formula $[A \geq n]$ in $\mathcal{L}_4^{fuzzy}$ can be done in $O(|\sum|, |A|)$. It is easy to see that the second case with formulas $[A \wedge B \geq nm]$ forms a unique sub-case of the earlier one.   □

4   Multi-valued Putnam-Davis Procedure in Use

One of ideas of last chapter was an introducing a new fuzzy TP-Putnam-Davis procedure. It forms a temporal-preferential extension of the original P-D procedure. TP-Putnam-Davis procedure was proposed in two variants – dependently of t-norms used to compute fuzzy values of formulas of the form $P \wedge \neg P$. In this chapter we intend to exemplify this new (fuzzy) TP-Putnam-Davis procedure. In oder to make it consider the following example.

Example 1. Consider a very basic planning situation described by a formula describing a possible situation of an agent's activity. The reservoir of the agent actions is the following: Move(P) (the agent moves from a point P), Put(A) (the agent puts a block A somewhere), Load(B) (the agent loads B somewhere). Find a possible consistent plan of agent's activity.

Solution: The solution via Putnam-Davis is illustrated in Fig. 1.

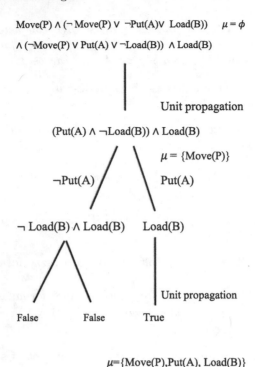

Move(P) ∧ (¬ Move(P) ∨ ¬Put(A)∨ Load(B))   μ = φ

∧ (¬Move(P) ∨ Put(A) ∨ ¬Load(B)) ∧ Load(B)

Unit propagation

(Put(A) ∧ ¬Load(B)) ∧ Load(B)

μ = {Move(P)}

¬Put(A)   Put(A)

¬ Load(B) ∧ Load(B)   Load(B)

False   False   True

Unit propagation

μ={Move(P),Put(A), Load(B)}

Fig. 1. Solution of Example 1 via Putnam-Davis procedure.

Assume now that we also have – from some *other sources* – a knowledge about temporal constrains imposed on a conjunction of actions Load(B) and ¬Load(B)[2]. For simplicity, assume that these temporal constraints – in terms of normalized values of the appropriate function – are as depicted on Fig. 2. Consider Load(B)∧¬Load(B) now. In other words, we associate to both actions a unique normalized value (as a value of a given function f – defined as earlier) such that Load(B) obtained 0.8 and ¬Load(B) obtained 0.2.

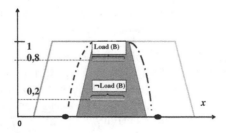

Fig. 2. The actions Load(B) and ¬Load(B) with associated values 0.8 and 0.2 (*resp.*)

---

[2] The action ¬Load(B) should be interpreted as some action different from Load(B).

As earlier mentioned, Load(B)∧¬Load(B) should not imply any contradiction in a fuzzy case as this conjunction must not necessary take 0. In order to check it, assume that we decide to consider this conjunction in Łukasiewicz logic. Recall that TP-Putnam-Davis algorithm in this fuzzy logic type is as follows.

Davis-Putnam(Φ, μ, v)
begin
    if $\emptyset \in \Phi$ then return
    if $\Phi = \emptyset$ then exit with μ
    otherwise **Unit-propagate**
    select a variable P such that P or $\neg P$ occurs in Φ
    Davis-Putnam $(\Phi - \{P\}, \mu \cup \{\neg P\})$
    Davis-Putnam $(\Phi - \{\neg P\}, \mu \cup \{P\})$
        if $P \wedge \neg P$ associate $v(P)$ and $v(\neg P)$
        compute $v(P \wedge \neg P) = \min(v(P), v(\neg P))$.
            if $v = 0$, then **failure** and $\mu = \emptyset$
            if $v \neq 0$, then return model μ with a degree v.
end

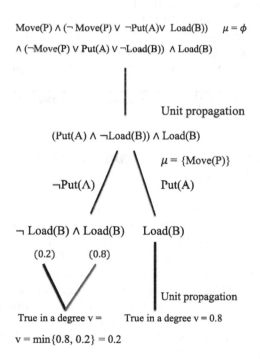

Fig. 3. Putnam-Davis procedure in a fuzzy case. In this case we do not reject Load(B)∧¬Load(B), which gives us a model μ with a true degree $v = 0.2$.

According to it, a fuzzy value $v(\texttt{Load(B)} \wedge \neg\texttt{Load(B)}) = \min\{0.8, 0.2\} = 0.2$ in Łukasiewicz fuzzy logic. In this situation (of conjunction interpreted by Łukasiewicz t-norm) the solution via modified Putnam-Davis procedure is given as visualized in Figs. 2, 3, 4 and 5. In results, we obtain the following models[3]:

1. an 'old' model $\mu_1 = \langle$ Move(P), Put(A), Load(B)$\rangle$ – with a true degree 0.8,
2. a new model $\mu_2 = \langle$ Move(P), Put(A), Load(B)$\wedge\neg Load(B)\rangle$ with a true degree 0.2.

A similar 'temporal' extension of Putnam-Davis procedure may be proposed in Product logic case as follows.

```
Davis-Putnam(Φ, μ, v)
begin
    if ∅ ∈ Φ then return
    if Φ = ∅ then exit with μ
    otherwise Unit-propagate
    select a variable P such that P or ¬P occurs in Φ
    Davis-Putnam (Φ − {P}, μ ∪ {¬P})
    Davis-Putnam (Φ − {¬P}, μ ∪ {P})
            if P ∧ ¬P associate v(P) and v(¬P)
            compute v(P ∧ ¬P) = v(P) • v(¬P)).
                if v = 0, then failure and μ = ∅
                if v ≠ 0, then return model μ with a degree v.
end
```

Due to this algorithm – the same $v(\texttt{Load(B)} \wedge \neg\texttt{Load(B)}) = 0.8 \bullet 0.2 = 0.16$ in Product Logic[4].

5   Further Extension of Putnam-Davis Procedure – Practically Motivated

Assume now that our knowledge about performing actions Load(B) and ¬Load(B) is slightly more extended. Namely, assume that the preferential line – computed as in earlier STRIPS-algorithm case – is given and it only admits points over its diagram as the admissible ones. Thus, only action Load(B) is admissible, ¬Load(B) should be rejected. In a consequence, the solution of the initial example should lead to the final situation as depicted here: In results, we obtain two times the following model as a solution: an 'old' model $\mu_1 = \langle$ Move(P), Put(A), Load(B)$\rangle$ – with a true degree 0.8.

Let us summarize the reasoning leading to this solution. Let us observe that:

---

[3] As it was signalized in 'Introduction' models μ of Putnam-Davis procedure play a role o plans.

[4] Note that the so-called *strong conjunction* Load(B)$\otimes\neg$Load(B) really gives $v = \max\{0, 0, 8 + 0.2 - 1\} = 0$.

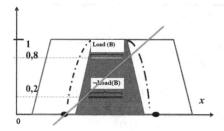

Fig. 4. The actions `Load(B)` and `¬Load(B)` with associated values 0.8 and 0.2 (*resp.*) and the preferential line, which orders to reject `¬Load(B)`. The trapezium – a common area of two fuzzy intervals – represents temporal constraints imposed on action performing in Example 1. The sloping line separates the preferred area of this trapezium (above it) from the non-preferred (under).

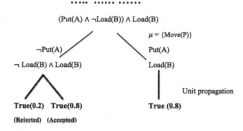

Fig. 5. Fragment of solution of the initial example with fuzzy temporal constraints and preferences. The red line leads to the rejected solution. The green ones – to the accepted ones. (Color figure online)

1. until a formula $P \wedge \neg P$ occurs, Putnam-Davis procedure with unit propagation works without changes,
2. when a formula $P \wedge \neg P$ occurs, we work as follows:
   - we associate fuzzy values to P and $\neg P$, or $v(P), v(\neg P)$,
   - we compute the preferential function `Pref`,
   - we compare $v(P), v(\neg P)$ with $|$`Pref(x)`$|$ – values of the preferential function `Pref` (in a given interval).
     - If $v(P) < \forall x |$`Pref(x)`$| < v(\neg P)$, then take $v(\neg P)$ and return model μ with $\neg P$ and degree $v(\neg P)$,
     - If $v(\neg P) < \forall x |$`Pref(x)`$| < v(P)$, then take $v(P)$ and return model μ with P and degree $v(P)$,
     - If $v(\neg P) < \forall x |$`Pref(x)`$|$ and $v(P) < \forall x |$`Pref(x)`$|$, then return model $\mu = \emptyset$.

It allows us to propose the following temporal-preferential extension of Putnam-Davis procedure.

```
Davis-Putnam(Φ, μ, v, Pref)
begin
    if ∅ ∈ Φ then return
    if Φ = ∅ then exit with μ
    otherwise Unit-propagate
    select a variable P such that P or ¬P occurs in Φ
    Davis-Putnam (Φ − {P}, μ ∪ {¬P})
    Davis-Putnam (Φ − {¬P}, μ ∪ {P})
        if P ∧ ¬P associate v(P) and v(¬P)
        compare v(P) and v(¬P) with |Pref|.
            if v(P) < ∀x|Pref(x)| < v(¬P), then return μ(¬P) and degree v(¬P),
            if v(¬P) < ∀x|Pref(x)| < v(P), then return μ(¬P) and degree v(P),
            if v(¬P) < ∀x|Pref(x)| and v(P) < ∀x|Pref(x)|, then return μ = ∅.
end
```

It is easy to see that no computing of a fuzzy value for a conjunction $v(P \wedge \neg P)$ intervene in this procedure – neither in Łukasiewicz Logic, nor in Product Logic. I makes this procedure more independent of a logical foundation of analysis.

6 Closing Remarks

It has been just shown how some Multi-valued extension of the Putnam-Davis procedure may be constructed. It has also emerged that this multi-valued Putnam-Davis procedure may be exploited in a practice twice: as a temporal extension or as a preferential extension of this procedure.

It arises a natural question of further possible extensions of this procedure. Obviously, there is still a relatively far distance between a four-valued extension of this procedure – just proposed – and its fuzzy extension. In addition, Putnam-Davis procedure might be adopted to some temporal planning contexts, where fuzziness would be introduced such as in formalism – discussed in [5,6]. Nevertheless, such a possibility seems to be conditioned by theoretic considerations.

References

1. Beckford, J., Logemann, G., Loveland, D.: A machine program for theorem proving. Commun. ACM 5(7), 394–397 (1962)
2. Dechter, R., Rish, I.: Directional resolution: the davis-putnam procedure, revisited. In: Principles of Knowledge Representation and Reasoning: Proceedings of the Fourth International Conference KR, pp. 134–145 (1994)
3. Jobczyk, K., Ligeza, A.: Fuzzy-temporal approach to the handling of temporal interval relations and preferences. In: Proceeding of INISTA, pp. 1–8 (2015)
4. Jobczyk, K., Ligeza, A.: Multi-valued halpern-shoham logic for temporal allen's relations and preferences. In: Proceedings of the Annual International Conference of Fuzzy Systems (FuzzIEEE) (2016)

5. Jobczyk, K., Ligeza, A.: Systems of temporal logic for a use of engineering. Toward a more practical approach. In: Stýskala, V., Kolosov, D., Snášel, V., Karakeyev, T., Abraham, A. (eds.) Intelligent Systems for Computer Modelling. AISC, vol. 423, pp. 147–157. Springer, Cham (2016). doi:10.1007/978-3-319-27644-1_14
6. Jobczyk, K., Ligeza, A., Kluza, K.: Selected temporal logic systems: an attempt at engineering evaluation. In: Rutkowski, L., Korytkowski, M., Scherer, R., Tadeusiewicz, R., Zadeh, L.A., Zurada, J.M. (eds.) ICAISC 2016. LNCS (LNAI), vol. 9692, pp. 219–229. Springer, Cham (2016). doi:10.1007/978-3-319-39378-0_20
7. Jobczyk, K., Bouzid, M., Ligeza, A., Karczmarczuk, J.: Fuzzy integral logic expressible by convolutions. In: Proceeding of ECAI 2014, pp. 1042–1043 (2014)
8. Jobczyk, K., Bouzid, M., Ligeza, A., Karczmarczuk, J.: Fuzzy logic for representation of temporal verbs and adverbs 'often' and 'many times'. In: Proceeding of LENSL 2011, Tokyo (2014)
9. Jobczyk, K., Ligęza, A., Bouzid, M., Karczmarczuk, J.: Comparative approach to the multi-valued logic construction for preferences. In: Rutkowski, L., Korytkowski, M., Scherer, R., Tadeusiewicz, R., Zadeh, L.A., Zurada, J.M. (eds.) ICAISC 2015. LNCS (LNAI), vol. 9119, pp. 172–183. Springer, Cham (2015). doi:10.1007/978-3-319-19324-3_16
10. Kautz, H., Selman, B.: Planning as satisfiability. In: Proceedings of the European Conference on Artificial Intelligence (ECAI) (1992)
11. Levesque, H.-J.: A logic of implicit and explicit belief. In: Proceedings of AAAI, pp. 198–202 (1984)
12. Putnam, H., Davis, M.: A computing procedure for quantification theory. J. ACM 7(3), 201–215 (1960)
13. Straccia, U.: A four valued fuzzy propositional logic. In: Proceedings of IJCAI 1997, pp. 128–133 (1997)
14. Traverso, P., Ghallab, M., Nau, D.: Automated Planning: Theory and Practice. Elsevier (1997, 2004)

Comparison of Effectiveness of Multi-objective Genetic Algorithms in Optimization of Invertible S-Boxes

Tomasz Kapuściński[1,2]([✉]), Robert K. Nowicki[1], and Christian Napoli[3]

[1] Institute of Computational Intelligence, Czestochowa University of Technology,
Al. Armii Krajowej 36, 42-200 Czestochowa, Poland
{tomasz.kapuscinski,robert.nowicki}@iisi.pcz.pl
[2] Institute of Information Technology, Radom Academy of Economics,
Domagalskiego Street 7a, 26-600 Radom, Poland
[3] Department of Mathematics and Informatics,
University of Catania, Viale A. Doria 6, 95125 Catania, Italy
napoli@dmi.unict.it
http://www.iisi.pcz.pl
http://wsh.pl/

Abstract. Strength of modern ciphers depends largely on cryptographic properties of substitution boxes, such as nonlinearity and transparency order. It is difficult to optimize all such properties because they often contradict each other. In this paper we compare two of the most popular multi-objective genetic algorithms, NSGA-II and its steady-state version, in solving the problem of optimizing invertible substitution boxes. In our research we defined objectives as cryptographic properties and observed how they change within population during experiments.

Keywords: NSGA-II · Steady state · Substitution box · Invertible S-box · Cryptography · Genetic algorithm

1 Introduction

One of the most important problems in cryptography is the construction of secure cryptographic primitives, especially block and stream ciphers. New ways of breaking existing solutions are found over time. Because of this more advanced systems need to be designed, which becomes increasingly more difficult as we need to take into account many new ways of attacking cryptographic systems. Automated methods of construction and testing based on artificial intelligence are one way to approach the subject. Various methods in artificial intelligence are very popular and have been used in fields related to cryptography. Examples include neural networks which were used in design of modern ciphers. One type of components, i.e. substitution boxes, are very important because they take major part in securing ciphers against attacks based on differential and linear

© Springer International Publishing AG 2017
L. Rutkowski et al. (Eds.): ICAISC 2017, Part II, LNAI 10246, pp. 466–476, 2017.
DOI: 10.1007/978-3-319-59060-8_42

cryptanalysis. Genetic algorithms are an effective optimization tool that was shown to work well in cryptography and various other areas [1, 6, 7, 14].

This paper shows the results of research meant to compare the effectiveness of two well-known multi-objective genetic algorithms, namely NSGA-II and its steady-state version, on the invertible S-box optimization problem. We use our previous work as a base for new experiments [8]. We introduce an additional cryptographic property that needs to be optimized.

The paper is organized as follows. In the following subsections we present the idea of invertible substitution boxes and the genetic algorithms used in the research. Section 2 shows the implementation details, the coding method and objectives. The following section describes the setup and results of our experiments. The final section covers conclusions and future plans regarding the subject.

1.1   Invertible Substitution Boxes

Substitution boxes, also knows as S-boxes, are a basic component of modern symmetric block ciphers, such as Advanced Encryption Standard and Blowfish. Substitution boxes are functions that map m-bit values to n-bit values and are commonly implemented as look-up tables. They implement Shanon's property of confusion, that is they are used to hide the relationship between the secret key and the ciphertext [12]. For this reason S-boxes must be as nonlinear as possible.

A special variant of S-boxes exists that is commonly used in substitution-permutation networks. This kind of cryptographic algorithms requires S-boxes to be invertible functions. This is because substitution needs to be reversed during the process of decryption. Invertible S-box is therefore an S-box that maps n-bit input to n-bit output and this mapping can be reversed. They effectively represent a type of permutation. And example of such S-boxes can be found in AES [3].

1.2   NSGA-II

In our previous work, we used the Nondominated Sorting Genetic Algorithm II (NSGA-II) [4]. It is a very popular multi-objective genetic algorithm that uses nondominated sorting. Unlike its predecessor, NSGA-II uses domination counts calculated for each solution to determine Pareto fronts which significantly improves the speed of the algorithm by reducing high computational complexity [13]. Domination counts are calculated based on the number of solutions that dominate given solution. Solutions with the count equal to zero are then added to the first front. The procedure is then repeated for the remaining solutions to define next Pareto fronts.

NSGA-II proved to be very effective in optimization of invertible substitution boxes because it can optimize multiple properties separately. This is especially important for substitution boxes because many cryptographic properties contradict other properties. Using a single-objective genetic algorithm to find strong

S-box may be difficult because there is no simple way to combine those properties into one value that can measure all the properties. With multi-objective algorithm, the final population will be a Pareto optimal set of solutions that can later be analyzed by a cryptographer who can choose the most appropriate solution.

1.3 Steady-State Strategy for NSGA-II

The steady-State version of NSGA-II operates on the same principles as NSGA-II. It also uses nondominated sorting based on domination counts. However, unlike NSGA-II, it does not have a concept of generation. Instead of replacing all individuals from the population by the newly generated population, it replaces pairs of solutions by their offspring if their children are better [2,9]. This ensures that population size remains constant. It is worth noting that the steady-state NSGA-II can be significantly slower than NSGA-II because it has to reevaluate Pareto fronts more often. However, if calculating objectives takes long enough, the difference becomes negligible.

2 Implementation

In this section we describe the implementation details of our research. We reused the coding method and two objectives from our previous work, modified one of the objectives and introduced a new objective that describes an important cryptographic property of substitution boxes.

2.1 Coding

The coding method used in our research is identical to the one we described in the previous work [8]. Each S-box is encoded as a selection table which values are constrained based on their position within the table. Each value has a minimum of 0 and a maximum of $n - i - 1$, where n is the length of the array and i is the index of the element. Each selection table uniquely identifies an invertible S-box. In order to decode the S-box we use an algorithm that is a modification of random shuffle. It uses values from the selection table to swap values in S-box initialized using values $[0, 1, 2, \ldots, n - 1]$. The method for decoding invertible S-boxes from selection tables is presented in Algorithm 1.

2.2 Objectives

In this section we describe the objectives for NSGA-II. For the purpose of our research, we defined four objectives in order to ensure stability of the algorithm.

Algorithm 1. Algorithm for decoding S-boxes from selection tables

function DECODESBOX(*selection*, *n*)
    Create an array *sbox* with length *n*
    for $i = 0; i < n; i \leftarrow i + 1$ **do**
      $sbox[i] \leftarrow i$
    end for
    for $i = 0; i < n; i \leftarrow i + 1$ **do**
      $index \leftarrow i + selection[i]$
      if $i \neq index$ **then**
        $temp \leftarrow sbox[i]$
        $sbox[i] \leftarrow sbox[index]$
        $sbox[index] \leftarrow temp$
      end if
    end for
      return *sbox*
end function

Nonlinearity. Similarly to our previous work, we used nonlinearity as an objective for scoring invertible S-boxes. Nonlinearity is an important property of substitution boxes that determines how difficult it is to linearly approximate them. S-boxes with high nonlinearity make the cipher highly resistant to linear cryptanalysis. As an objective we used Peak-to-Average Power Ratio (PAR) with respect to Walsh-Hadamard Transform described in [10]. Unlike nonlinearity calculated from its base definition, PAR can be used directly as NSGA-II objective because low values represent high nonlinearity. Since invertible S-boxes are balanced boolean functions, they can not be perfectly nonlinear. The lowest possible value of PAR for 8-bit invertible S-boxes is 4. PAR can be calculated using Eq. (1).

$$PAR(f) = 2^n \max_{\forall k} \left(\left| 2^{-n} \sum_{x \in Z_2^n} (-1)^{f(x) + x \cdot k} \right| \right)^2 . \qquad (1)$$

Transparency Order. Differential Power Analysis is a type of side-channel attack that is based on measurement and analysis of power consumption of a device that encrypts or decrypts data. While this type of attack can only be used with direct access to a device such as a smart card reader, it was shown that it can be much more effective than differential and linear cryptanalysis and due to the widespread use of mobile devices, it can be successfully used to break the cipher. Prouff defined in [11] the transparency order as a cryptographic property of substitution boxes that determines how resistant they are against DPA. The lower the value, the less information is leaked through power consumption and therefore the more resistant it is to DPA attack. Because of the importance of this property we decided to include it in our research as an objective. Transparency order can be calculated using Eq. (2).

$$\tau_F = \max_{\beta \in F_2^m} \left(\left| m - 2H(\beta) \right| - \frac{1}{2^{2n} - 2^n} \sum_{a \in F_2^{n*}} \left| \sum_{\substack{v \in F_2^m \\ H(v)=1}} (-1)^{v \cdot \beta} W_{D_a F}(0, v) \right| \right). \quad (2)$$

Hamming Distance Score. This objective is identical to the one we used in our previous work. It is based on the principle that the S-box should change half of the input bits on average. This can be measured using Hamming distances between all inputs and their corresponding outputs for a given S-box. Hamming distance between two binary vectors is a number of differing bits on corresponding positions. This can be calculated using exclusive OR of two vectors and counting the number of bits set to 1. We can calculate Hamming distances for all possible S-box inputs and count how many times they appear. Then we can use those numbers to define a negative score. Bit changes of 0 and 8 bits (no bits changed and all bits changed respectively) have a score $1/256$, bit changes of 1 and 7 bits have a score $1/2\,560$ and bit changes of 2 and 6 bits have a score $1/25\,600$. This way S-boxes that change too few or too many bits will have a high score and those who change 5, 6 or 7 bits will have a low score. The method for calculating this objective is presented in Algorithm 2.

Algorithm 2. Algorithm for calculating score based on Hamming distances in S-box

---

    function HAMMINGDISTANCESCORE($sbox$)
        Create an array $count$ with length 9
        for $i = 0$; $i < 9$; $i \leftarrow i + 1$ **do**
            $count[i] \leftarrow 0$
        end for
        for $x \leftarrow 0$; $x < 256$; $x \leftarrow x + 1$ **do**
            $y \leftarrow f(x)$
            $d \leftarrow x + y$
            $c \leftarrow$ number of ones in d
            $count_c \leftarrow count_c + 1$
        end for
        $sum \leftarrow count[0] + 0.1 \times count[1] + 0.01 \times count[2] + 0.01 \times count[6] + 0.1 \times count[7] + count[8]$
            return $sum \times 2^{-8}$
    end function

Acceptability Test. The diffusion score used in the previous research [8] has been replaced by validity test. To calculate this objective we check the diffusion of a modified AES-128 cipher and Hamming distances of the S-box. This objective is set to 0 when given S-box is determined acceptable in both diffusion test and Hamming distance test or 1 when the S-box is found to have at least one unacceptable property. This objective works similarly to death penalty, but unlike death penalty it does not force removal of the solution from population.

The first test is closely related to the Hamming distance score calculation. If the algorithm finds that there are any values in given S-box that represent identity transformation (Hamming distance is 0) or that it represents negation of all bits (Hamming distance is 8), then the test fails and the objective value is set to 1. This test was added because our research showed that some weak S-boxes can have high nonlinearity or low transparency order and they dominate non-weak S-boxes.

The second test calculates diffusion and compares it to a threshold value. To calculate diffusion we used a probabilistic test that randomly generates secret keys and plaintexts, and then performs simple differential analysis. For generated plaintext x_1 we compute additional plaintext x_2 with one bit inverted. Plaintexts are then encrypted using the randomly generated secret key, resulting in ciphertexts y_1 and y_2. We then perform exclusive OR operation on ciphertexts, obtaining the difference Δy. In the perfect cipher, a probability that each bit of Δy is set is 50%, which means that differences do not give us any additional information. But in a weak cipher those probabilities would likely deviate from 50%, giving us a way to break the cipher using differential cryptanalysis. Therefore, we can count how many times each bit in difference Δy was set and check how much it deviated from the perfect 50% probability. The exact algorithm for calculating diffusion score is presented below.

Algorithm 3. Algorithm for calculating diffusion score of a given S-box

```
function DIFFUSIONSCORE(sbox, tests)
    Construct cipher with sbox
    diffusion ← 0
    for b ← 0; b < 128; b ← t + 1 do
        Initialize count to 0
        for t ← 0; t < tests; t ← t + 1 do
            Generate random plaintext x₁
            y₁ ← cipher(x₁)
            Invert bit b in x₁
            y₂ = cipher(x₂)
            Δy ← y₁ + y₂
            for i ← 0; i < 128; i ← i + 1 do
                if bit i in Δy is set then
                    count[i] ← count[i] + 1
                end if
            end for
        end for
        for i = 0; i < 128; i ← i + 1 do
            diffusion ← diffusion + |count_i - ½ tests| / tests
        end for
    end for
    return 2⁻¹³ × diffusion
end function
```

In an ideal situation, this test would give 0 for the perfect cipher. However, since this is a probabilistic test we always obtain a value above 0 but within a specific margin. The average value of the diffusion test for good S-boxes using 256 as the number of tests is 0.05. After calculating diffusion score for a given S-box we compare it with threshold 0.055. If the value is higher than the threshold, we set the objective to 1. Otherwise, the S-box passes the test and the objective is set to 0.

3   Experimental Research

For the purpose of this research we used a newer version of jMetal library [5]. Version 5.1 has been redesigned and contains implementations of both NSGA-II and steady state version of NSGA-II. It contains an implementation that simplifies parallel processing using multiple threads which is appropriate for NSGA-II. Unfortunately, steady state NSGA-II could not be parallelized the same way because this algorithm does not evaluate enough solutions at once. For this reason, steady state NSGA-II was optimized in a different way which can result in varying time of computation. Additionally, implementation of algorithms was modified to analyze and save partial results to a file after each 200 evaluations.

3.1   Scenario

We performed two simulations, one for NSGA-II and one for steady-state version of NSGA-II, using an equal set of parameters. Because steady-state NSGA-II does not have a concept of generations, we used instead a common number of evaluations, which is the number of generated and evaluated solutions. We chose a population size 100, crossover operator with the probability of 0.6 and mutation operator with the probability of 0.2. The total number of evaluations was 250 000, which is an equivalent of 2 500 generations for NSGA-II. Every 200 evaluations, objectives from current the population were analyzed and stored in a file for later analysis.

3.2   Results

The experiments were performed on a computer with 6-core Intel Xeon E5-2620 processor clocked at 2.0 GHz. The results are as follows. The scenario with NSGA-II took about 16 h and 14 min to complete and the scenario with steady state version of NSGA-II took about 28 h and 42 min to complete. The total time is considerably long due to computational complexity of the algorithm that calculates transparency order. Statistics taken from the final populations after the experiment concluded are presented in Tables 1 and 2.

Figures 1 and 2 show how nonlinearity values changed during each experiment. The minimum value of 5.0625 was found immediately in the original population. We can see that the maximum value in the case of NSGA-II was fluctuating a lot around values 15–20 while for the steady-state version the

Table 1. Summary statistics of objectives in the final population for NSGA-II

Objective	Minimum	Average	Maximum
Nonlinearity	5.0625	8.2340	20.2500
Transparency order	7.7041	7.7371	7.7789
Hamming distance score	0.001719	0.005130	0.025859
Acceptability score	0.0000	0.1064	1.0000

Table 2. Summary statistics of objectives in the final population for steady state NSGA-II

Objective	Minimum	Average	Maximum
Nonlinearity	5.0625	7.5599	12.2500
Transparency order	7.7017	7.7409	7.7828
Hamming distance score	0.001484	0.005344	0.026171
Acceptability score	0.0000	0.1633	1.0000

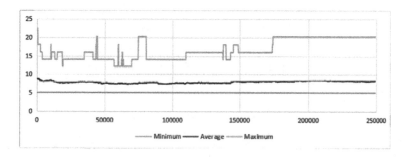

Fig. 1. Changes in nonlinearity during the experiment with NSGA-II

maximum was changing gradually towards the minimum. The steady-state NSGA-II reached slightly lower average and obtained better results.

Figures 3 and 4 describe changes in transparency order during experiments. As in the case of nonlinearity, the steady state NSGA-II provided more stable, gradual changes. It also reached the minimum value of 7.7017 after 178 000, which is considerably faster than NSGA-II. The second one reached the minimum value of 7.7041 after 239 600 evaluations.

Analysis of the Hamming distance score shows very similar results. Steady-state NSGA-II had smoother changes in the average and reached minimum values faster than standard NSGA-II. On the other hand, NSGA-II performed better in acceptability score than steady state version, which can be seen in Fig. 5.

The final population generated by NSGA-II contained 9 solutions with non-linearity of 5.0625, with the lowest transparency order of 7.7176 and the highest of 7.7672. All of them passed the acceptability test and had good Hamming distance scores. The final population generated by steady-state NSGA-II contained

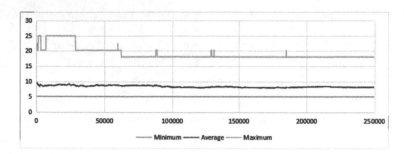

Fig. 2. Changes in nonlinearity during the experiment with steady-state NSGA-II

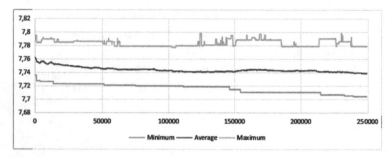

Fig. 3. Changes in transparency order during the experiment with NSGA-II

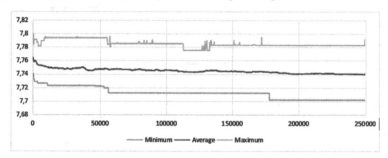

Fig. 4. Changes in transparency order during the experiment with steady-state NSGA-II

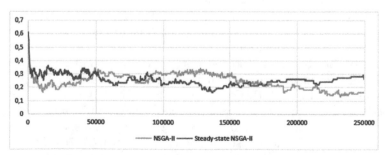

Fig. 5. Changes in acceptability order during experiment with steady-state NSGA-II

5 solutions with nonlinearity of 5.0625. Among these solutions, the minimum transparency order was 7.7340 and the maximum was 7.7644. One solution did not pass the acceptability test. The remaining solutions had good Hamming distance scores. Unlike in our previous work [8], no solution with nonlinearity 4 was created.

4   Conclusions and Future Work

In our research we showed that multi-objective genetic algorithms with steady-state strategy can be successfully used as a tool for construction of invertible substitution boxes for cryptographic purposes. Steady state version of NSGA-II performs in a similar way, although changes in population are considerably smoother and it reaches better results faster than generational NSGA-II.

Introduction of transparency order of S-boxes as an objective made S-box optimization problem more difficult because this cryptographic property conflicts with nonlinearity. Both versions proved to be very effective at solving this problem and created many Pareto optimal solutions that can be later analyzed and used in the construction of block ciphers.

Future research will adopt this method and use it to create optimal S-boxes for new block ciphers. We will also try to optimize parallel computation of steady-state NSGA-II in attempt to match the speed of parallelized version of standard NSGA-II.

References

1. Aguirre, H., Okazaki, H., Fuwa, Y.: An evolutionary multiobjective approach to design highly non-linear boolean functions. In: Proceedings of the 9th Annual Conference on Genetic and Evolutionary Computation, GECCO 2007, pp. 749–756. ACM, New York (2007)
2. Chafekar, D., Xuan, J., Rasheed, K.: Constrained multi-objective optimization using steady state genetic algorithms. In: Cantú-Paz, E., et al. (eds.) GECCO 2003. LNCS, vol. 2723, pp. 813–824. Springer, Heidelberg (2003). doi:10.1007/3-540-45105-6_95
3. Daemen, J., Rijmen, V.: AES proposal: rijndael (1999)
4. Deb, K., Pratap, A., Agarwal, S., Meyarivan, T.: A fast and elitist multiobjective genetic algorithm: NSGA-II. IEEE Trans. Evol. Comput. 6(2), 182–197 (2002)
5. Durillo, J.J., Nebro, A.J.: jmetal: a Java framework for multi-objective optimization. Adv. Eng. Softw. 42(10), 760–771 (2011)
6. El-Samak, A.F., Ashour, W.: Optimization of traveling salesman problem using affinity propagation clustering and genetic algorithm. J. Artif. Intell. Soft Comput. Res. 5(4), 239–245 (2015)
7. Ivanov, G., Nikolov, N., Nikova, S.: Reversed genetic algorithms for generation of bijective s-boxes with good cryptographic properties. Crypt. Commun. 8(2), 1–30 (2016)

8. Kapuściński, T., Nowicki, R.K., Napoli, C.: Application of genetic algorithms in the construction of invertible substitution boxes. In: Rutkowski, L., Korytkowski, M., Scherer, R., Tadeusiewicz, R., Zadeh, L.A., Zurada, J.M. (eds.) ICAISC 2016. LNCS, vol. 9692, pp. 380–391. Springer, Cham (2016). doi:10.1007/978-3-319-39378-0_33

9. Nebro, A.J., Durillo, J.J.: On the effect of applying a steady-state selection scheme in the multi-objective genetic algorithm nsga-ii. Nat.-Inspired Algorithms Optim. **193**, 435–456 (2009)

10. Parker, M.: Generalised s-box nonlinearity. NESSIE Public Document NES/DOC/UIB/WP5/020/A (2003)

11. Prouff, E.: DPA attacks and S-boxes. In: Gilbert, H., Handschuh, H. (eds.) FSE 2005. LNCS, vol. 3557, pp. 424–441. Springer, Heidelberg (2005). doi:10.1007/11502760_29

12. Shannon, C.E.: Communication theory of secrecy systems*. Bell Syst. Tech. J. **28**(4), 656–715 (1949)

13. Srinivas, N., Deb, K.: Multiobjective optimization using nondominated sorting in genetic algorithms. Evol. Comput. **2**(3), 221–248 (1994)

14. Yang, C.H., Moi, S.H., Lin, Y.D., Chuang, L.Y.: Genetic algorithm combined with a local search method for identifying susceptibility genes. J. Artif. Intell. Soft Comput. Res. **6**(3), 203–212 (2016)

The Impact of the Number of Averaged Attacker's Strategies on the Results Quality in Mixed-UCT

Jan Karwowski[1(✉)] and Jacek Mańdziuk[1,2]

[1] Faculty of Mathematics and Information Science,
Warsaw University of Technology, Koszykowa 75, 00-662 Warsaw, Poland
{j.karwowski,j.mandziuk}@mini.pw.edu.pl
[2] School of Computer Science and Engineering, Nanyang Technological University,
50 Nanyang Avenue, Singapore 639798, Singapore
j.mandziuk@ntu.edu.sg

Abstract. Mixed-UCT is a method for finding efficient defender's mixed strategy in multi-act Security Games. This paper presents experimental evaluation of the impact of the number of averaged past attackers (APA) used to define the defender's strategy on solution quality of the method. Specifically designed set of test games is proposed for evaluation of the Mixed-UCT method with different values of APA parameter. The results indicate that larger values of APA generally lead to faster convergence of the method, and in some cases also improve the results in terms of the expected defender's payoff value.

Keywords: Stackelberg Equilibrium · MCTS · UCT · Security Games

1 Introduction

Security Games (SG) is the area of research where scientifically grounded methods based on game theory are used to find the optimal behavior (activities) of security forces when protecting some relevant objects (potential targets). Recently, this research domain gained momentum due to rising terrorist threats in many places in the world. While the most noticeable applications of SG refer to homeland security, e.g. patrolling LAX airport [7] or patrolling US coast [2], SG are not limited to this type of applications and can also be applied in other scenarios, for instance, train fare controlling [5] or preventing of poaching [6].

In SG there are two players, the *defender* which represents security forces, and the *attacker*. Vast majority of SG models are built around the Stackelberg Game - an asymmetric model rooted in the game theory where one player (the attacker) is aware of the opponent's strategy, but not *vice versa*. This assumption reflects the common situation in which the attacker (e.g. an evader or a thief) is able to observe the defender's behavior before committing an attack. Such games are usually referred to as Stackelberg Security Games (SSG). SSG are generally imperfect-information games and some aspects of the current game state are

© Springer International Publishing AG 2017
L. Rutkowski et al. (Eds.): ICAISC 2017, Part II, LNAI 10246, pp. 477–488, 2017.
DOI: 10.1007/978-3-319-59060-8_43

not observable by the players. Solving SSG consists in finding the Stackelberg Equilibrium (SE) and is typically approached by solving the Mixed Integer-Linear Program (MILP) [2,6,7,11] corresponding to a given SSG instance (which must be represented in a normal (matrix) form).

Calculating SE is computationally intensive both in terms of CPU time and memory usage. Due to these limitations only very simple SSG models can be solved (exactly) in practice. For this reason, in many papers various simplifications of the MILP formulation are proposed in order to make it possible to solve larger games, at the expense of obtaining approximate (instead of exact) solutions. These approaches, however, still suffer from some of the limitations of baseline MILP formulation, e.g. the requirement of matrix-based game representation. Consequently, nearly all deployed solutions consider single-act (one-step) games with simultaneous moves.

The lack of efficient methods for solving multi-act games motivated our research on completely different approach - the Mixed-UCT [9] - which relies on massive Monte-Carlo (MC) simulations, instead of MILP. The method is suitable for multi-act games and demonstrates reasonable memory and time requirements. The approach consists in iterative computation of the defender's strategy based on the results of MC simulations of the game against a gradually improving attacker which is defined as a combination of a certain number of past attackers used in previous iterations. This paper investigates and discusses the relevance of the selection of the number of averaged past attackers (APA) used to build the current attacker. The main goal of this paper is experimental evaluation of the impact of the number of previous attacker's strategies (henceforth denoted by n) on the quality of resulting defender's strategy in the Mixed-UCT method.

The rest of the paper is organized as follows: Sect. 2 briefly introduces the Mixed-UCT method, with particular emphasis on the usage of the APA parameter (in Sect. 2.4). Section 3 describes games and test methods used to evaluate Mixed-UCT against various selections of APA. In the next section the results of experiments are presented. Finally, Sect. 5 covers discussion of results along with the final conclusions and research perspectives.

2   Mixed-UCT

Mixed-UCT algorithm, proposed in [9], is a method for finding an approximate defender strategy in multi-step SSG. The motivation behind using this method is to overcome limitations of the MILP-based approaches which make them not well suited to multi-step games due to extensive memory and time requirements. Mixed-UCT employs a variant of Monte-Carlo Tree Search (MCTS) called Upper Confidence Bounds applied to Trees (UCT) [10] to effectively address the exploration vs. exploitation dilemma while searching larger SSG trees.

The reminder of this section briefly describes the Mixed-UCT method. First, a vanilla UCT algorithm is presented. Then a variant of UCT applicable to imperfect-information games (I2-UCT) is described, followed by a presentation of Mixed-UCT, for which I2-UCT serves as one of the main building blocks.

2.1   MCTS/UCT

MCTS is a meta-heuristic method used to estimate the quality of moves in perfect information games by means of iterative move sampling in a game tree (where nodes and edges correspond to game states and game moves, respectively) from the current root state to one of the leaf nodes (terminal states of a game). Each edge in the tree is labeled with two values: the number of times the respective move was sampled in the hitherto simulations and the average payoff of these hitherto simulations that included this edge/move. MCTS builds the tree gradually, starting from a single-node tree, containing the root node (the current state of the game) only.

Each MCTS simulation consists of the following four steps:

- **Selection.** One move is selected in each of the subsequently visited nodes, starting from the root node until a leaf node of the tree (not necessarily being a terminal state of the game) is reached. The selection is made according to some *node selection policy.*
- **Expansion.** Once the process reaches a leaf node, a new successor node is added to the tree.
- **Simulation.** A game is simulated from this newly-added node until a terminal state of a game with randomly selected moves.
- **Backpropagation.** A game result is read out in the terminal state and propagated back, all the way up to the root node. Nodes lying on this path within the currently maintained game tree update their statistics (visits counters and average scores).

Various node selection policies can be used in **Selection** phase. UCT [4, 10] is a variant of MCTS that employs the UCB1 [3] heuristic function defined as follows:

$$a^* = \underset{a \in A}{\operatorname{argmax}} \left\{ Q(s,a) + C\sqrt{(\ln N(s))/N(s,a)} \right\}, \tag{1}$$

where s is the currently considered state, A is a set of possible moves in the state, $Q(s,a)$ is the average reward of playing move a in state s hitherto, $N(s,a)$ is the number of times action a was played from given state, and $N(s) = \sum_{a \in A} N(s,a)$. UCT was proven to be highly successful in board games domain, e.g. in Go [13] or General Game Playing [14,15].

Equation (1) allows for maintaining a balance between *exploitation* (the left component) of the currently best known move and *exploration* of rarely visited moves (the right term). Coefficient C is responsible for balancing these two above-mentioned contradicting tendencies.

The UCT tree generated in the above-described way is an essential component of our method for defender's mixed strategy construction. (See Sect. 2.3 for the description of the use of a UCT tree).

2.2   I2-UCT

Vanilla MCTS/UCT method described in the above section assumes that a game can be simulated from the current state to the end-of-game state. While this assumption is true in perfect-information games, where both players can fully observe the game state, is SG players commonly do not possess full knowledge about their opponent's moves and past positions. Consequently, in SG, due to imperfect-information nature of this game, vanilla UCT is inapplicable except for the initial state of the game.

In order to address this problem we have developed a method called I2-UCT [8] which samples full game states based on the information accessible to the player. In short, in I2-UCT a game is simulated from the initial state by playing the defender's moves that led to the current state of the game and simulating the optimal opponent's responses according to their strategy.

Note, that execution of I2-UCT algorithm assumes possession of the knowledge about the opponent's (attacker's) mixed strategy, which in our case is calculated as part of the Mixed-UCT framework (described below).

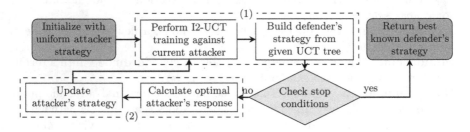

Fig. 1. The Mixed-UCT algorithm outline

2.3   Mixed-UCT

Mixed-UCT is designed for finding close-to-optimal mixed strategy in SSG for the leader player (defender in SG) by means of an iterative process. In each iteration the defender's strategy is updated using a tree obtained in the I2-UCT training against the currently optimal attacker's strategy (the method of attacker's optimal strategy calculation is presented in the next subsection).

The outline of the algorithm is presented in Fig. 1. First the attacker's strategy is initialized as uniform distribution over all possible move sequences. Then the process enters the main loop of the algorithm which consists of two parts. In the first one, labeled (1) in Fig. 1, the defender's strategy is built, and in the other one, labeled (2) in the figure, the attacker's response strategy is calculated.

A detailed description of the defender's strategy calculation method is available in [9]. In short, it consists of two steps:

– I2-UCT training is performed against the current attacker's mixed strategy.
– UCT tree generated during training is transformed into a mixed defender's strategy in the following way. Each path from the initial state to a terminal state represents a sequence of defender's moves. A probability distribution of such sequences (a mixed strategy) is generated with probabilities proportional

to the product of all visit counters of the moves (edges in the trained UCT tree) belonging to this sequence.

I2-UCT training starts-off with an empty tree. In subsequent iterations the tree generated in previous iterations is used as the initial one in the current iteration. Once the defender's strategy is calculated, the strategy of the attacker is updated with respect to this new defender's strategy (see Sect. 2.4). The training is stopped after one of the two following conditions is met:

- Currently known best defender's strategy has not been improved in the last 10 000 iterations, or
- A number of 45 000 training iterations was executed in total.

2.4   Calculating Attacker's Strategy

Game simulations require a procedure of attacker's moves generation, as indicated in part (2) of Fig. 1. To this end, first, the optimal attacker response in the sense of SE for just generated defender's strategy is calculated by exhaustive search of all possible attacker's move sequences. It can be shown that this response can be any pure strategy that maximizes attacker's and defender's payoff (in this order, due to SE requirements) or any mixture of such pure strategies (all these forms are equivalent) [11]. We have arbitrarily decided to use a uniform distribution of all feasible pure strategies, as presented in Algorithm 1.

Algorithm 1. Calculating optimal attacker against a given defender in SSG

---

1 $M \leftarrow$ AllMoves()
2 $M_1 \leftarrow \{m \in M | $AttackerPayoff$(m) = max_{m' \in M}$AttackerPayoff$(m')\}$
3 $M_2 \leftarrow \{m \in M_1 | $DefenderPayoff$(m) = max_{m' \in M_1}$DefenderPayoff$(m')\}$
4 **return** UniformStrategy(M_2) // Uniform strategy containing all pure
      strategies from M_2

The second step, i.e. the update of attacker's strategy, is motivated by the outcomes of our preliminary experiments in the early development stage of Mixed-UCT, which shown that using the average of some number of attacker's responses from previous iterations is beneficial in many test instances. Therefore, instead of using directly the attacker's mixed strategy just calculated in the current iteration of the I2-UCT training, we use an "average" mixed strategy from n most recent attacker's responses.

Let M be a set of possible move sequences in a game tree, and $f_i : M \rightarrow [0, 1]$, such that $\sum_{m \in M} f(m) = 1$ be the probability density function of the attacker's mixed strategy in the i-th iteration. The probability density function of the average strategy in the j-th iteration $(g_j(m))$ is defined as follows:

$$g_j(m) = \frac{1}{n} \sum_{i=j-n+1}^{j} f_i(m). \tag{2}$$

Algorithm 2. Drawing a sample from averaged attacker's strategy

---

Data: optimalResponses: an array of n recent mixed strategies
Result: moveSequence
1 $i \leftarrow$ DrawFrom(Uniform(1,Length(optimalResponses)))// Draw a number from
     a uniform distribution
2 moveSequence $\leftarrow$ DrawFrom(optimalResponses $[i]$)

Please note that direct implementation of Eq. (2) is not the most efficient method of calculation unless very large number of samples had been already drawn from the distribution. The method of drawing a sample from that distribution without immediate calculation of the density, which was implemented in our experiments, is presented as Algorithm 2.

3   Experimental Setup

The set of benchmark games was defined using a graph-based game model which defines the structure of the game environment. The model follows our approach used in [8] for evaluation of the I2-UCT method, with some modifications.

Each game is defined on a directed graph with unweighted edges, and involves two players: one attacker player and one defender player. Each player starts the game from a designated vertex. In all examples presented in the paper the attacker's starting vertex will be denoted by a red triangle and the defender's starting point will always be vertex number zero. Some of the vertices are highlighted as *targets* and as such are of special interest for both playing sides. More precisely, during the game the attacker attempts to reach one of the targets without being caught by the defender, and the defender tries to prevent the opponent from doing so, typically by catching them. The attacker is considered caught by the defender iff they meet in the same vertex. The attack is considered successful iff the attacker reaches the target in the absence of the defender there.

In each turn of the game players decide to move to any of the neighboring vertices or to stay in the current vertex. Players perform their moves simultaneously and *are not aware of the opponent's choice* - consequently they *do not know the current or past opponent's positions*, expect for the initial one (starting vertices). The game has a certain limit of rounds and ends immediately after any of the three possible situations takes place: a successful attack, catching the attacker, exceeding the limit of rounds.

Each vertex has two assigned values: a reward for the defender when catching the attacker in this vertex, and a penalty for the attacker for being caught there. Target vertices have two additional values: a penalty for the defender and the respective reward for the attacker, if a successful attack takes place in the target. Rewards and penalties vary among vertices and, in general, the game is not a zero-sum one. Example game graph with payoffs values is presented in Fig. 2. **Formally, the goal of the game for each of the player is to find a mixed strategy that maximizes their expected payoff.**

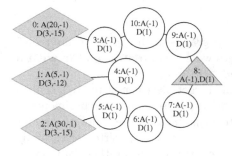

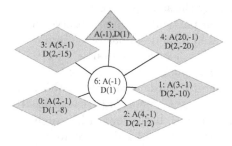

Fig. 2. *game3* graph. Red triangle vertex (number 8) denotes the attacker's starting point, vertex number 0 is the defender's starting base. Green diamond shaped vertices are targets. Labels of the vertices contain attacker's and defender's payoffs. In non-target vertices $A(\cdot)$ and $D(\cdot)$ denote the payoff of the attacker and the defender, respectively. In target vertices $A(\cdot,\cdot)$ denotes the attacker's reward for successful attack and their penalty when being caught; $D(\cdot,\cdot)$ refers to the defender's reward for catching the attacker and their penalty in the case the attack renders successful. (Color figure online)

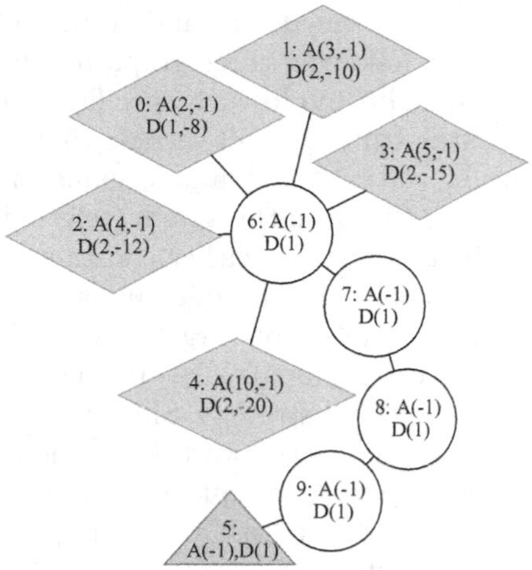

Fig. 4. *game2* graph

Variant	Target 0				Target 2			
	w_A	l_A	w_D	l_D	w_A	l_A	w_D	l_D
Base	20	−1	3	−15	30	−1	3	−15
a	20	−1	1	−15	20	−1	6	−15
b	3	−1	3	−15	5	−1	5	−1
c	20	−1	3	−15	30	−1	5	−10
d	10	−1	3	−15	30	−1	3	−5

Table 1. Reward/penalty values in five variants of *game3*. w_A, l_D, w_D, l_A denote the attacker's payoff for successful attack and the respective defender's penalty, and the defender's payoff for catching the attacker and the respective attacker's penalty.

Fig. 3. *game1* graph

Please note, that the above-described game model was chosen only because it covers a large variety of game structures, is easily configurable (by modifying the game-defining graph), and is intuitively-plausible for human analysis of possible game strategies. The discussed Mixed-UCT method is by no means limited or tailored to this particular game model.

3.1 Test Graphs

Tests were performed on hand-crafted game graphs with the game limit of 5 rounds. The following three graph topologies were used: **a star**, **a star with a long tail**, and **a cycle** with specifically located targets.

In **a star topology** all vertices (including targets) are connected to a central non-target vertex. An example game of this type used in the experiments is presented in Fig. 3. In this graph (referred to as *game1*), the defender may adopt one of the two policies: either constantly protect the central vertex or carefully balance their presence in different targets - the latter one is a rather risky policy due to high number of targets. Apparently the first option, enhanced by an attempt to catch the attacker in their base in the last move (with probability 0.5), is the optimal strategy. Since the rewards in most of the targets are higher than that in the central vertex, the defender is tempted to patrol these highly rewarding targets. Consequently, learning the optimal strategy (i.e. not to leave the central vertex at least until the penultimate round of a game) is not easy and requires extensive number of simulations.

A star topology with a long tail (referred to as *game2*) is similar to the previous one except that in this game design the attacker starting point is not connected directly to the central vertex, but is linked through a path composed of 3 vertices (see Fig. 4). Since the attacker starting node is separated from the star part of the graph by a three-node path, the defender has high chances to catch the attacker on their way to the targets, which makes games with this topology less demanding than the previous ones.

The third type of games are those with **a cycle topology** - depicted in Fig. 2. There are 10 test games of this type, with exactly the same topology and location of targets, differing by the payoff assignment in particular nodes. Observe, that in this type of game an attacker (initially located in node 8) has two alternative paths towards the targets (either $8 \rightarrow 9 \rightarrow 10 \rightarrow \ldots$ or $8 \rightarrow 7 \rightarrow 6 \rightarrow \ldots$).

In the first set of games, denoted by *game3, game3a, ..., game3d* (c.f. Fig. 2) non-target vertices are assigned small defender's rewards compared to those in the targets. In the other set, labeled *game4, game4a, ..., game4d* (not visualized) the game topology is exactly the same and in *game4*, but a reward for catching the attacker in a non-target vertex is proportional to the distance from the nearest target to that vertex. Such a payoff distribution is more intuitive as it encourages keeping the attacker at bay and catching them as early as possible.

Please observe that in all variants of *game3/game4*, due to relatively low payoff and higher distance than to the other targets, the optimal attacker would never attempt to reach target 1. This target was added to these graphs for the purpose of checking if the existence of additional, practically non-achievable, target vertex, would affect the outcomes of Mixed-UCT (mixed defender's strategies) in any way.

In order to tests the strengths and weaknesses of the proposed Mixed-UCT method, the circle graphs were tested under several hand-crafted payoff distributions designed with specific purposes in mind. Five variants of payoff distributions in targets 0 and 2 assigned to *game3,3a,3b,3c,3d*, respectively are presented in Table 1. In all of them the payoff values in the remaining vertices were assigned as presented in Fig. 2. Due to space limits we are not able to delve into detailed motivation behind each tested payoff distribution. Generally speaking, the base variant (*game3*) represents the situation with different attacker's rewards when defender's payoff values are the same in both targets; *game3a* addresses "the opposite" situation with different defender's rewards when attacker's payoffs are the same in both targets; variant *b* of *game3* presents the case with imbalanced defender's penalties in the targets; variant *c* tests the case of *skewed* payoffs, where more valuable target for the attacker has lower penalty for the defender penalty than other target; and finally in variant *d* the above-mentioned skewness is enlarged compared to the case *c*.

In *game4* graphs the payoff distributions in all five cases are exactly the same as in the respective *game3* graphs except that the defender's rewards are equal to 1 in all targets, and in ordinary vertices are proportional to the distance from that non-target vertex to the nearest target. Hence, as stated above, contrary to *game3* type games, *game4* ones are defined in a way that encourages catching the attacker far from the targets rendering the game easier for the proposed method in all tested variants of the payoff distribution.

4   Experimental Results

The results of Mixed-UCT (the defender's payoff values and computation time) on the above-described test games are presented in Table 2 together with the exact theoretical results computed by a baseline Mixed Integer Linear Programming based method [12] using SCIP [1] solver. The particular choice of MILP as a baseline method was motivated by its popularity in the SSG literature as a reference exact method.

In addition to the exact solution (which is the theoretically best possible result), a payoff resulting from playing uniform defender's strategy is also presented (serving the purpose of the "lower-end" solution). The assessment of Mixed-UCT, denoted by *sc* in the table, is the accomplished defender's reward scaled linearly to $[0, 1]$, with 0 value assigned to the reward of the uniform defender's strategy and 1 assigned to the result of the optimal (MILP calculated) value. Values of *sc* close to 1 correspond to very good results while those close to 0 are very poor. Time for calculating the uniform strategy is negligible, hence is not presented.

Mixed-UCT was tested with four values of n: 1, 10, 100, 1000. Since the method is non-deterministic (due to Monte-Carlo sampling) 20 trials for each game and each parameter setting were executed. The relatively low number of trials stems form high consistency of results with very low standard deviation. All experiments were conducted on a machine with core i7@3.40 GHz processor.

Test results indicate that the method is robust with respect to setting parameter n, as for all values of $n > 1$ only meaningless differences in defenders' payoffs can be observed and most of results ended with $sc = 1$, i.e. the highest assessment possible. The only exception is *game1* for which the highest *sc* value

Table 2. Averaged results of 20 trials of Mixed-UCT with different attackers history lengths and their comparison with baseline results computed with MILP. Six different methods are presented – four variants of Mixed-UCT with different values of n labeled respectively with numbers 1, 10, 100 and 1000, MILP exact solution calculated by the solver, and uniform defender strategy labeled with U. Columns labeled with R present the average defender's payoff value, with t – computation time in seconds, and with sc – payoff quality assessment on a linear scale from the Uniform defender's result (0) to the optimal solution (1).

game	1			10			100			1000			MILP		U
	R	t	sc	R	t	sc	R	t	sc	R	t	sc	R	t	R
1	−1.57	1448	0.81	0.34	1436	0.98	0.48	1492	0.99	0.47	1611	0.99	0.54	28052	−10.46
2	0.03	1049	0.99	0.05	1388	1	0.06	1113	1	0.07	1351	1	0.08	106	−7.21
3	−4.32	2319	0.99	−4.34	1618	0.99	−4.28	1786	1	−4.28	1135	1	−4.27	42946	−13.97
3a	−4.5	2003	1	−4.53	1452	1	−4.51	1230	1	−4.5	1487	1	−4.5	31926	−13.97
3b	2.44	1838	0.96	2.57	2020	1	2.57	1382	1	2.58	1639	1	2.58	227	−0.87
3c	−1.52	1777	0.94	-1.18	1724	0.99	−1.07	1358	1	−1.06	1169	1	−1.06	41624	−9.29
3d	0.84	2129	0.99	0.69	1595	0.96	0.9	1467	1	0.9	1218	1	0.9	37870	−4.61
4	−4.87	1777	1	−4.87	1378	1	−4.87	1995	1	−4.87	1310	1	−4.87	5312	−13.91
4a	−6	1681	1	−6	1638	1	−6	1637	1	−6	1114	1	−6	5949	−13.84
4b	0.79	1788	1	0.79	1266	1	0.79	1639	1	0.79	1061	1	0.79	5546	−0.81
4c	−2.85	1783	1	−2.85	1498	1	−2.85	1494	1	−2.85	1203	1	−2.85	5928	−9.23
4d	0.17	1948	1	0.17	1455	1	0.17	1391	1	0.16	1165	1	0.17	5155	−4.55

equals 0.99. The difference between $n = 1$ and larger n settings clearly exists, though is relatively small. For $n = 1$ half of the games yielded slightly weaker results, among which two (*game1* and *game3c*) were lower than 0.95.

Clear differences can be observed, however, in terms of computation times. Only in the case of *game2* skipping attacker's history completely led to the fastest convergence to the final result. For *game1*, *game3a* and *game3b*, when weaker results are excluded, the most time-effective setting was $n = 100$, and for all the remaining games the largest tested option, i.e. $n = 1\,000$.

Comparison of computation times between MILP solver and Mixed-UCT shows that the solver is usually slower for considered benchmark games, except for *game2* and *game3b*. It can be observed that computation times of MILP can vary significantly between games and exceed 2 orders of magnitude in extreme cases (c.f. *game2* and *game3a*, for instance). Time differences between games in the case of Mixed-UCT are much smaller and the longest execution time is slightly more than two times bigger than the shortest one (in the case of *game3b* and *game2* for $n = 100$).

In terms of memory usage (which is not presented in the table) Mixed-UCT is about 4 times more effective than MILP solver. Its memory requirements were varying between 1.0 GB and 1.2 GB depending on the graph topology compared to about 5 GB required by the solver. No significant differences in memory requirement were observed across various settings of n.

5   Discussion and Conclusions

The results show that the choice of n has an impact on performance of the Mixed-UCT method. In some cases it affects the quality of solution (as in *game1*), but most commonly influences the running time of the method. Generally speaking, in majority of the tested games increasing the number of past attacker's decreased computation time required to reach a solution. The exceptions are *game1* and *game2* which performed best (in terms of execution time and quality of solution) with $n = 100$.

Even though for some games values of n other than 1 000 actually appeared to be the best performing ones the differences compared to $n = 1\,000$ are not significant. Consequently, based on the presented results we may safely conclude that the suggested number of averaged past attacker's strategies n should be greater than 1 and setting $n = 1\,000$ seems to be an effective and reliable choice. In the context of the above-discussed shortening of the Mixed-UCT execution time it is worth be note that thanks to using Algorithm 2, the length of the attacker's history has almost no impact on the duration of a single algorithm's iteration and therefore selection of higher values of n does not introduce any additional tradeoffs.

On a general note, the results show that Mixed-UCT is able to solve the tested games effectively, leading in all cases to optimal or close-to-optimal solutions (with $sc = 1$), doing so in a reasonable time. Moreover, execution time differences between games of the same length are small which allows for reliable estimations of the amount of time required to solve a given game. This property does not hold for the reference MILP-based method whose execution time may vary significantly even for games of similar structure (e.g., time differences between *game3b* and *game3c* exceed two orders of magnitude).

Furthermore, Mixed-UCT has significantly smaller memory requirements compared to established MILP-based approach, which seems to be the real asset of this method. While there are many approximate methods built around the exact approach proposed in [11], with improved computation time, these approaches still require large amounts of memory (the same order of magnitude as the exact MILP solution).

In summary, this paper shows that Mixed-UCT method provides high quality mixed defender's strategies in artificially designed 5-step games and that the choice of the number of averaged past attacker's strategies is important, mainly in terms of the execution time and - to a smaller extent - the solution strength. Currently we investigate this issue on other types of games - defined on other graph topologies and of bigger length. We also test the Mixed-UCT time and memory scalability by applying the method to more complex games with very promising results. In a long-term perspective we believe that Mixed-UCT may become a viable alternative to the exact and approximate MILP-based SSG approaches, especially for large (complex) games, which pose real challenges to MILP methods.

References

1. Achterberg, T.: SCIP: solving constraint integer programs. Math. Program. Comput. **1**(1), 1–41 (2009)
2. An, B., Ordóñez, F., Tambe, M., Shieh, E., Yang, R., Baldwin, C., DiRenzo, J., Moretti, K., Maule, B., Meyer, G.: A deployed quantal response-based patrol planning system for the us coast guard. Interfaces **43**(5), 400–420 (2013)
3. Auer, P., Cesa-Bianchi, N., Fischer, P.: Finite-time analysis of the multiarmed bandit problem. Mach. Learn. **47**(2–3), 235–256 (2002)
4. Browne, C., Powley, E., Whitehouse, D., Lucas, S., Cowling, P., Rohlfshagen, P., Tavener, S., Perez, D., Samothrakis, S., Colton, S.: A survey of monte carlo tree search methods. IEEE Trans. Comput. Intell. AI Games **4**(1), 1–43 (2012)
5. Delle Fave, F.M., Jiang, A.X., Yin, Z., Zhang, C., Tambe, M., Kraus, S., Sullivan, J.P.: Game-theoretic patrolling with dynamic execution uncertainty and a case study on a real transit system. JAIR **50**, 321–367 (2014)
6. Fang, F., Stone, P., Tambe, M.: When Security Games go green: Designing defender strategies to prevent poaching and illegal fishing. In: IJCAI (2015)
7. Jain, M., Tsai, J., Pita, J., Kiekintveld, C., Rathi, S., Tambe, M., Ordóñez, F.: Software assistants for randomized patrol planning for the LAX airport police and the federal air marshal service. Interfaces **40**(4), 267–290 (2010)
8. Karwowski, J., Mańdziuk, J.: A new approach to security games. In: Rutkowski, L., Korytkowski, M., Scherer, R., Tadeusiewicz, R., Zadeh, L.A., Zurada, J.M. (eds.) ICAISC 2015. LNCS (LNAI), vol. 9120, pp. 402–411. Springer, Cham (2015). doi:10.1007/978-3-319-19369-4_36
9. Karwowski, J., Mańdziuk, J.: Mixed strategy extraction from UCT tree in security games. In: ECAI 2016, pp. 1746–1747. IOS Press (2016)
10. Kocsis, L., Szepesvári, C.: Bandit based monte-carlo planning. In: Fürnkranz, J., Scheffer, T., Spiliopoulou, M. (eds.) ECML 2006. LNCS (LNAI), vol. 4212, pp. 282–293. Springer, Heidelberg (2006). doi:10.1007/11871842_29
11. Paruchuri, P., Pearce, J.P., Marecki, J., Tambe, M., Ordonez, F., Kraus, S.: Efficient algorithms to solve bayesian stackelberg games for security applications. In: AAAI, pp. 1559–1562 (2008)
12. Paruchuri, P., Pearce, J.P., Marecki, J., Tambe, M., Ordonez, F., Kraus, S.: Playing games for security: an efficient exact algorithm for solving bayesian stackelberg games. In: AAMAS, pp. 895–902 (2008)
13. Silver, D., Huang, A., Maddison, C.J., Guez, A., Sifre, L., van den Driessche, G., Schrittwieser, J., Antonoglou, I., Panneershelvam, V., Lanctot, M., Dieleman, S., Grewe, D., Nham, J., Kalchbrenner, N., Sutskever, I., Lillicrap, T., Leach, M., Kavukcuoglu, K., Graepel, T., Hassabis, D.: Mastering the game of go with deep neural networks and tree search. Nature **529**, 484–503 (2016)
14. Świechowski, M., Mańdziuk, J.: Self-adaptation of playing strategies in general game playing. IEEE Trans. Comput. Intell. AI Games **6**(4), 367–381 (2014)
15. Waledzik, K., Mańdziuk, J.: An automatically-generated evaluation function in general game playing. IEEE Trans. Comput. Intell. AI Games **6**(3), 258–270 (2014)

Data-Driven Polish Poetry Generator

Marek Korzeniowski and Jacek Mazurkiewicz$^{(\boxtimes)}$

Department of Computer Engineering, Faculty of Electronics,
Wroclaw University of Science and Technology,
ul. Wybrzeze Wyspianskiego 27, 50-370 Wroclaw, Poland
borsuczek@gmail.com, Jacek.Mazurkiewicz@pwr.edu.pl

Abstract. The paper describes an attempt to create a poetry generator for Polish language. It is a data-driven approach – grammatical and
semantic structures are automatically derived from input text. The system was successfully implemented and the quality of the output "poems"
was tested in a "Poetic Turing Test": a public survey. Its participants
have been asked to distinguish between human written and computer
generated poetry.

Keywords: NLP · Natural language generation · Polish language ·
Poetry

1 Introduction

Natural language generation (NLG) is quite an old concept. In 1948 C. Shannon
proposed using Markov chains for such task [11]. In the sixties J. Weizenbaum
created the famous ELIZA program [12]. In the eighties W. Chamberlain and
T. Etter created RACTER – a program that generated a book that was good
enough to be published [1]. Although these examples demonstrate quite different approaches to the same problem, they serve a quite similar low functional
quality – amusement of people reading the generated text.

NLG also had some practical applications. In the early nineties a system
named SPOKESMAN was developed for the military – a framework for creating
natural language interfaces for various applications [6]. Another example is a system named FOG – a marine weather forecast generator [2]. These two systems
were constructed like classical expert systems – everything they could create was
limited by a hand-crafted set of rules and facts. Currently numerous corpora-
tions (i.e. Apple, Microsoft, Google, Amazon) are actively developing systems
that use NLG to provide "artificial intelligence" interfaces for their products.

This paper focuses on a rather unusual problem – poetry generation. A complex NLG system was designed and implemented to answer a simple question:
could a computer program be taught to create poems that would appear as
work of a human to an average person? Similar applications were created in the
past, but none of them could provide an output with proper rhyme and rhythmic structures. What is more, none of them were data-driven – they usually

L. Rutkowski et al. (Eds.): ICAISC 2017, Part II, LNAI 10246, pp. 489–497, 2017.
DOI: 10.1007/978-3-319-59060-8_44

depended on hand-crafted lexicons and sentence patterns. Finally, no similar work was conducted on Polish language – a highly fusional language that has been chosen to be the constructed system's output.

2   Previous Works

J. Mountstephens tried to create a system for generating mnemonic phrases – sentences with initial letters same as in a given set of words [7]. Such sentences should also be easy to memorize for humans. For example "Richard Of York Gave Battle In Vail" is a mnemonic phrases for the colours of a rainbow (Red, Orange, Yellow, Green, Blue, Violet). J. Mountstephens maintained grammatical correctness of the output by using a formal grammar description of a English language subset. The "easiness of remembering" was assessed by determining the semantic similarity of each pair of neighbouring words by looking it up in an external database.

A far more complex system was created by N. Ito and M. Hagiwara for Japanese language [3]. Their system was designed to generate sentences with a given noun and verb. Instead of using a formal grammar model of a subset of Japanese they used a database of automatically generated "case frames" taken from a large text corpus. These structures allowed creating meaningful sentences which were further refined by use of an n-gram model. A survey was conducted to determine the quality of generator's output.

S. Pudaruth and others tried to generate English song lyrics [9]. Two databases were build from an input text corpus: sentence patterns in form of sequences of parts-of-speech and a simple lexicon of words. A sentence was generated by choosing a random sentence pattern and filling it with appropriate words from the lexicon. No form of maintaining semantic correctness was proposed.

Research most closely related to this paper was conducted by H. Manurung and others. They proposed a complex approach to creating an English poetry generator [5]. Their work also included an in-depth review of countless poetry generators found on the Internet. They called them "party tricks in the mould of ELIZA" that were able to achieve quite good results thanks to ingeniously modelled heuristics. These applications could not provide output with proper rhymes or rhythm, because of their pattern based construction. To solve this problem the authors proposed a stochastic hill climbing approach. Their algorithm started from a set of input semantics which in each next iteration was appropriately transformed. Every intermediate result was scored accordingly to how well it matched the target rhythmic form. Their algorithm worked only on a small hand-crafted grammar and lexicon. It seems it was never able to produce interesting results.

3   Proposed System

Unfortunately n-grams do not work well for Polish because of the countless forms of words obtained via declension and conjugation. Describing a subset of Polish in a form of a formal grammar proves to be extremely hard because of arbitrary sentence word order. Pudaruth's approach based on deriving grammatical patterns from the input text seemed to be acceptable, but needed further enhancement. Simple sequences of parts-of-speech had to be replaced by dependency trees with full morphological information. To extract such information a dependency parser capable of parsing Polish poetry was needed. One was created explicitly for this task [4].

Randomly filling dependency trees with appropriate words and preforming a morphological synthesis would provide grammatically correct sentences but like in S. Pudaruth's approach it would also ignore semantics. This issue was solved by introducing semantic rules derived from the parsed input.

A special feature enforcement algorithm was created to ensure rhythm and rhymes in the output. Its construction was based on the following reasoning: "If we make the generator *really* fast we could simply wait until it produces results with proper rhythm and rhymes".

The complete system is shown in Fig. 1.

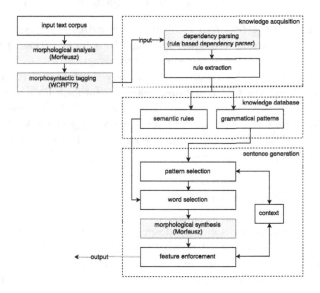

Fig. 1. The proposed poetry generation system. External modules (represented by gray fields): Morfeusz morphological analyser [13], WCRFT2 morphosyntactic tagger [10] and the rule based dependency parser [4].

4 Knowledge Acquisition

The dependency parser used in the constructed system returns a dependency tree for each sub-sentence it encounters and accepts. Semantic rules are created by simply extracting and saving all the parent-child pairs from the trees. A single rule consists of base forms and morphological tags of two words. In some cases additional information must be included, like verb reflexivity and person.

A single grammatical pattern consists of a sub-sentence, its dependency tree and information about how it is connected with neighbouring sub-sentences. This is realized by attaching to each pattern two "connectors", which can be divided into the following groups:

- initial connectors: the pattern has to be used at the beginning of a sentence,
- terminal connectors: the pattern has to end a sentence,
- intermediate connectors, which carry information about the type of connection: occurrence of a comma, the first/last word of the next/last sub-sentence and its predicate.

Two connectors are assumed equal if:

- they have the same type (terminal, initial, intermediate),
- a comma has or has not occurred in both of them,
- the predicates saved in them have the same part-of-speech,
- the first/last words saved in them have the same part-of-speech,
- if the first/last words are conjunctions they have to be equal.

The most significant problem related to acquiring knowledge was how to save it in a form that would allow fast look-up. To achieve $\mathcal{O}(\log n)$ search times multiple indexes were created using various orderings.

For testing purposes a 320k word text corpus was prepared. It was composed from poetry acquired via OCR of book scans by famous Polish poets such as Adam Mickiewicz or Bolesław Leśmian. The collected data was morphologically analyzed by Morfeusz [13] and morphosyntacticly tagged by WCRFT2 tagger [10]. A knowledge database of 75k semantic rules and 16k grammatical patterns was derived after dependency parsing.

5   Sentence Generation

The whole generation process can be described as a random search with backtracking throughout a tree of all possible word combinations. These combinations are limited by grammatical patterns and semantic rules. The search starts by choosing a grammatical pattern with an initial connector. Starting from the root of its dependency tree, the pattern gets filled with words that have compatible morphological tags. The root is chosen at random, but each next word is taken according to a semantic rule. Tags of chosen words do not need to be strictly equal. For example when choosing a noun only the case needs to match. Gender and number of depended adjectives are automatically adjusted during the morphological synthesis. When a tree gets successfully filled with words, a next pattern with an equal connector is chosen. If the connector was a terminal connector, the process ends and a complete sentence is returned.

Pronouns allow indirect referring to previously introduced nouns. This is a fact both in Polish and English. Consider an incomplete sentence "I see people. They are [adjective]". When looking for an adjective to insert, only semantic rules with "people" as their parent should be considered (instead of "they"). A special mechanism was implemented into the generator to properly support pronouns.

Polish includes constructions known as "default subjects". They allow skipping the subject of a sentence. Such sentences inherit the subject from the previous sentence. Support for these constructions has been included in the generator implementation.

A complete implementation of the generator using the test knowledge database worked with the speed of 2k words per second. The computation was held on a single core of a standard 3 GHz Intel i7 processor.

Example outputs (followed by English translations):

I była mina i pycha dokoła grobu w agonii skalnej, była, była noc z karku, zaś bogini mu na przemian kotka rzekł i myślę wrócić tysiącem serc na ślepą wieczność naszych progów.
And there was a facial expression and pride around a grave in rockery agony, it was, it was the night from one's neck, whereas the goddess alternately created a kitten from words and I think about returning into blind eternity of our doorsteps by means of a thousand hearts.

Mąż każdego cwałującego stanu śmiał ku mnie wyjść ciekawym oknem!
A husband of each galloping state had the insolence to leave through a curious window and walk in my direction!

Niech droga ziemia ciemnych nieb mnie pogrzebic.
Let the beloved land of black skies bury me.

6   Feature Enforcement

6.1   Context

A list of recently used nouns is kept in memory to provide the output text with an illusion of context. This list is updated each time a pattern is successfully filled with words. New nouns are added to its top, shifting all entries down. As the list has a constant size, the oldest entries are removed. If a given noun is already in the list, then it is moved to the top.

Subjects inserted into patterns are chosen from this list. A higher position of a noun in the list implies a higher probability of choosing it. However semantic rules still need to be respected.

The following example output is provided to demonstrate how the systems simulates context:

W wiecznym pokoju żyje?
In eternal peace I live?
Pokój ma noc, noc ma dzień strasznych bojów.
The peace has the night, the night has a day of terrifying battles.
Wszak dzień odtąd nie minie, tylko dotknięty ból jest koziołka przystrojonym dębem.

Indeed, the day from now will not pass, only touched pain is a goat's decorated oak.
Ale po wolność miski idę budować zewy?
But for freedom of a bowl I will build blood calls?
Wolność zgubiła.
The freedom lost.
Niesie ból ludzi biednych.
Pain carries poor people.

State of the recently used noun list after generating the example above: people, pain, freedom, call, bowl, oak, goat, day, battle, night, peace. Enforcing these constraints did not significantly impact the generator speed.

6.2   Rhythm

Ensuring proper rhythm in Polish is surprisingly easy when compared with other languages because of constant accent. Every word is accented in the exactly same way. When writing a poem, one has only to ensure that each verse has the same amount of syllables.

Dividing words into syllables is not an easy task. Fortunately counting syllables in Polish is trivial. It can be done by counting the non-overlapping matches of the following regular expression: [ąęaeioóuy]+.

To force the output to be divisible into proper verses a global syllable counter has been introduced. After computing each sub-sentence, the counter is updated. If its value is lesser equal to the target verse syllable count, then the sub-sentence is accepted. If the value is strictly equal, then the sub-sentence is accepted and the counter is reset. Otherwise, the sub-sentence is discarded. Sub-sentences ending with a terminal connector can only occur at the end of a verse.

Enforcing those constraints caused a drop in generator efficiency to about 152 word per second. An example eight-syllable output with an English translation:

Mgła w oku okna boleje.
A fog suffers in an eyes's window.
Znowu lawy nie wyrzucam.
Again I do not throw lava away.
Strącam w ognie, serca gryzę,
I throw into flames, I gnaw hearts,
Dla mody ganię, nie ywię!
For fashion I rebuke, I do not nourish!

6.3   Rhymes

The two most frequently occurring rhyme types in Polish language are called "male rhymes" and "female rhymes". Male rhymes apply only to monosyllabic words – they form a rhyme if their suffixes ending on the last vowel are equal. For example *"krew"* [krɛf] and *"zew"* [zɛf] both have the same rhyming suffix "ew".

Female rhymes apply to multisyllabic words – they form a rhyme if their prefixes ending on the second to last vowel are equal. For example *"rusałka"* [ru'sawka] and *"chałka"* ['xawka] both have the same rhyming suffix "ałka".

Support for simple tailed rhymes was implemented into the generator. This was achieved by forcing rhymes between last words of odd numbered verses with the last words of even numbered verses. Odd numbered verses ending with a terminal sub-sentence were disallowed to prevent creating a pause between rhyming verses.

An approach based on discarding sub-sentence without correct rhymes proved to be too inefficient to produce results in reasonable time. To force a rhyme the last word of each even verse is replaced by a random word with a strictly equal morphological tag and a correct rhyme. Semantic rules are ignored during this operation.

With these changes the efficiency of the generator dropped to about 30 words per second. An example eight-syllable output with an English translation:

Skrę wydałem, skra gorzała
A spark I gave, a spark has blazed
I gorzała jej czekała!
And it's booze has hazed! [waited]

Ja nie wyjmę, skra rozdmucha,
I will not remove, the spark will scatter,
Zboże za mną nie wybucha!
The grain behind me does not shatter! [explode]

Już setna skra się rozlewa,
Already the hundredth spark is spilling,
A skra potem się przelewa!
And then a spark starts swilling! [overfilling]

Gorzała skra, gorzała skra,
A spark has blazed, a spark has blazed,
Gorzała słońca, dała dna.
It blazed the sun, a bottom it craved [gave].

To make the translation more accurate the last words of even verses were randomly altered to form rhymes in English. The exact translation is given in square brackets.

7   Results

Finding a good metric to measure the quality of the generator's output proved difficult. A decision was made to conduct a public survey. It was composed of sixteen short poetry fragments: ten generated and six randomly chosen from classic Polish poems (by Adam Mickiewicz and Bolesław Leśmian). The task was to determine which fragments were computer generated. Survey also included questions about the surveyed person gender (male or female) and their education background (liberal arts or science).

Eighty-six people were surveyed. Each person was scored on a scale of sixteen – one point for each correctly answered question. Detailed survey results are presented in a form of a histogram (Fig. 2).

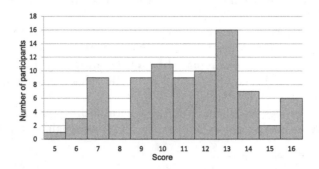

Fig. 2. A histogram illustrating detailed results of the survey.

The average score was 11.1, with no significant differences in both gender groups. The result was a bit higher in the liberal arts group (12.7) and a bit lower in the science group (10.6). Some participants with the highest scores have said to distinguish the human written fragments by simply recognizing the poems they came from (which was clearly cheating).

For each of the generated fragments a percentage of incorrect answers was calculated and averaged. The result was 34%. This value can be interpreted as the probability that a random Polish native speaker will treat the generator's output as genuine poetry written by a human being.

Only one generated fragment had the ratio of incorrect answers above 0.5. 50 out of 86 participants (58%) had classified the following fragment as work of a famous Polish poet:

Słowo pali, słowo pędzi,
The word burns, the word dashes,
Słowo się pode mną szczędzi.
The word lavishes under me.

Słowo ciska naokół, gdzie czci,
The word hurls around wherever it reveres,
Wiecznie zbywam, i słowo lśni.
Eternally I put it off and it gleams.

8   Conclusion

The survey results seem to be surprisingly good, but it needs to be mentioned that the fragments used in it were very short. Recognizing human poetry in longer portions of text would be a much easier task as the generator output has no coherent meaning.

The generator made noticeable grammatical mistakes. This was caused by two factors: errors in the input text corpus (poor quality OCR of book scans) and numerous mistakes made by the WCRFT2 tagger (which was not trained to process poetry). These errors propagated by faulty grammatical patterns into the output. Only grammatically correct output fragments were selected for the survey.

Also, the achieved output semantic correctness was lower than expected. Simple two-word rules were too weak to create meaningful and complex sentences. Perhaps the quickly growing Polish wordnet "Słowosieć" would provide means to achieve better results [8].

References

1. Chamberlain, B.: The Policeman's Beard is Half Constructed. Warner Books, New York (1984)
2. Goldberg, E., Driedger, N., Kittredge, R.I.: Using natural-language processing to produce weather forecasts. IEEE Expert **9**(2), 45–53 (1994)
3. Ito, N., Hagiwara, M.: Natural language generation using automatically constructed lexical resources. In: The 2011 International Joint Conference on Neural Networks (IJCNN), pp. 980–987, July 2011
4. Korzeniowski, M., Mazurkiewicz, J.: Rule based dependency parser for polish language. In: Artificial Intelligence and Soft Computing, ICAISC 2017 (2017, to appear)
5. Manurung, H., Ritchie, G., Thompson, H.: Towards a computational model of poetry generation. Technical report, The University of Edinburgh (2000)
6. Meteer, M.W.: Spokesman: data-driven, object-oriented natural language generation. In: Proceedings of Seventh IEEE Conference on Artificial Intelligence Applications 1991, vol. 1, pp. 435–442, February 1991
7. Mountstephens, J.: Mnemonic phrase generation using genetic algorithms and natural language processing. In: 2013 IEEE International Conference on Teaching, Assessment and Learning for Engineering (TALE), pp. 527–530, August 2013
8. Piasecki, M.: plWordNet (Slowosiec) (2008). http://hdl.handle.net/11321/43
9. Pudaruth, S., Amourdon, S., Anseline, J.: Automated generation of song lyrics using CFGs. In: 2014 Seventh International Conference on Contemporary Computing (IC3), pp. 613–616, August 2014
10. Radziszewski, A., Warzocha, R.: WCRFT2: CLARIN-PL Digital Repository (2014). http://hdl.handle.net/11321/36
11. Shannon, C.E.: A mathematical theory of communication. Bell Syst. Tech. J. **27**, 623–666 (1948)
12. Weizenbaum, J.: ELIZA–a computer program for the study of natural language communication between man and machine. Commun. ACM **9**(1), 36–45 (1966)
13. Wolinski, M.: Morfeusz-a practical tool for the morphological analysis of polish. In: Kłopotek, M.A., Wierzchoń, S.T. (eds.) Intelligent Information Processing and Web Mining. Advances in Soft Computing, vol. 35, pp. 511–520. Springer, Heidelberg (2006)

Rule Based Dependency Parser
for Polish Language

Marek Korzeniowski and Jacek Mazurkiewicz[✉]

Department of Computer Engineering, Faculty of Electronics,
Wroclaw University of Science and Technology,
ul. Wybrzeze Wyspianskiego 27, 50-370 Wroclaw, Poland
borsuczek@gmail.com, Jacek.Mazurkiewicz@pwr.edu.pl

Abstract. The paper presents a dependency parser for Polish language.
It uses a simple chain of word combining rules operating on fully mor-
phosyntactically tagged input instead of a formal grammar model or
statistical learning. The proposed approach generates robust dependency
trees and allows parsing of uncommon texts, such as poetry. This gives
a significant advantage over current state-of-the-art dependency parsers.

Keywords: NLP · Text parsing · Dependency parsing · Polish language

1 Introduction

Currently there are two leading dependency parsers available for Polish language:
Świgra and "Polish Dependency Parser" [6,8]. The first is based on a formal
grammar model of Polish language created by Marek Świdziński [4]. The second
represents a data-driven approach created using Maltparser and Składnica –
a data-driven parser generator and a data set of fully tagged and parsed Polish
sentences [2,7].

These solutions work well for multiple applications but problems arise when
one tries to analyse non-standard texts, like poetry. For complex sentences
Świgra often does not return a result in reasonable time – the computing process
often takes longer than half an hour for a single sentence. "Polish Dependency
Parser" used on texts drastically different from the ones it was trained on returns
gibberish.

Poetry-parsing capabilities are a completely unnecessary feature for a real-
world usage. Another project we worked on required a database of words used in
Polish poetry with information about semantic and syntactic relations between
them. For such analyses a dependency parser was needed.

We propose a completely different approach to dependency parsing: a method
based on a chain of simple heuristic rules, operating on words and their morphological tags. Each rule removes a word from the input and attaches it to a
different word – effectively joining them into a parent-child pair. During the final
step the last word is removed and returned as the root of the dependency tree.
If no rule can be applied to a given word the input is discarded as unrecognised.

© Springer International Publishing AG 2017
L. Rutkowski et al. (Eds.): ICAISC 2017, Part II, LNAI 10246, pp. 498–508, 2017.
DOI: 10.1007/978-3-319-59060-8_45

The proposed parser can analyse any kind of grammatically correct texts and return correctly parsed sentences only. Its operating time is insignificant when compared to the time of morphosyntactic tagging – the preliminary step of this process. For morphosyntactic tagging the WCRFT2 tagger was used [3].

The output trees are quite crude and carry much less information compared with trees returned by other parsers. Nevertheless, this method can be used to quickly find relations in complex texts, especially when other methods would fail to generate any useful results.

2   Preliminary Definitions

2.1   Parts of Speech Archetypes

The morphological tagset used by the WCRFT2 tagger defines over thirty parts of speech [5]. The following archetypes were introduced to reduce the complexity of the parser:

- noun (*subst, depr, ppron12, ppron3, siebie* and *ger*) representing parts of speech that can become the subject of a sentence,
- verb (*fin, bedzie, praet, impt, imps* and *winien*) representing parts of speech that can become the predicate,
- adjective (*num, numcol, adj, pact* and *pant*) representing parts of speech that describe nouns,
- adverb (*adja, adjp, adjc, adv, inf, pcon, pant* and *pred*) representing parts of speech that describe verbs.

2.2   Dependency Trees

The roots of the dependency trees returned by the parser are always a verb. In Polish language a simple sentence can only have a single verb and this verb is always the predicate. In most cases a compound sentence can easily be devised into sub-sentences, each contacting a single verb being the predicate. Such approach can not be used for verb-less sentences which are quite common in Polish. Such sentences are rejected by the algorithm as unrecognised.

Interjections in sentences present another problem. In most cases such sentences are rejected as unrecognised, but sometimes they can also yield incorrect results. For example, the sentence *"Znalazłem, choć nie szybko, króla."* can be word-by-word translated to "I found, but not fast, the king." with "but not fast" as the interjection informing that we looked for the king for quite some time. The parser will recognise "I found" as a correct sub-sentence and discard the rest, while the correct sub-sentence should be "I found the king".

The root (verb) of the dependency tree can have child nodes being nouns, adverbs and prepositions. Prepositions always have one child which needs to be a noun. Adverbs can have children being other adverbs or adjectives. Nouns can have children being other nouns, adjectives or prepositions. Adjectives can have

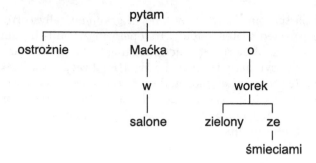

Fig. 1. Graphical representation of the dependency tree of the sentence *"Ostrożnie pytam się Maćka w salonie o zielony worek ze śmieciami."*

a single child – also being an adjective – only if they represent a compound numeral.

Parts of speech not included in the archetypes, or not being a preposition, are simply ignored and they are not included in the output.

In the paper a compact notation will be used to represent the dependency trees. A node will be represented by a word. If the node is not a leaf node, its children will be presented as a comma-separated list enclosed in square brackets. For example: the sentence *"Ostrożnie pytam się Maćka w salonie o zielony worek ze śmieciami."* which roughly translated to "Cautiously I ask Mike in the showroom about the green bag with trash" has dependency tree shown in Fig. 1 that can be written as:

```
pytam[ ostrożnie, Maćka[ w[ salonie ] ], o[ worek[ zielony,
ze[ śmieciami ] ] ] ] ]
```

English version:

```
(I) ask[ cautiously, Mike[ in[ showroom ] ], about[ bag[
green, with[ trash ] ] ] ] ]
```

3   Parsing Rule Chain

To demonstrate the state of the input text after each step of the parser's rule chain the following example sentence will be used: *"Ostrożnie zapytałem bardzo czerwoną dziewczynkę z kapturkiem równie czerwonym o zapałki, a następnie niechętnie kupiłem od niej dwadzieścia dwa pomidory i jedną marchew"*. Word-by-word translation: "(I) carefully asked (a) very red girl with (a) hood equally red about matches, and afterwards reluctantly bought from her twenty-two tomatoes and (a) single carrot". Words in parenthesis have been inserted into the translation to make it grammatically correct.

The complete parsing rule-chain is summarized in Fig. 2.

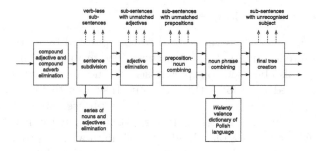

Fig. 2. The parser's rule-chain illustrated on a single diagram. The input is in form of a single sentence. The output is in form of multiple dependency trees – one for each sub-sentence that the input was composed from. While traversing the rule-chain, sub-sentences can get discarded as unrecognised for numerous reasons – shown on the diagram in form of dashed lines.

3.1 Compound Adjective and Compound Adverb Elimination

The first step of the parsing rule chain can be described as a simple iteration over all input words in search of compound numerals, adverb-adjective or adverb-adverb combinations.

If a compound numeral is found, it is replaced by a tree having the least significant numeral in its root. For example a numeral *"sto dwadzieścia dwa"* ("one hundred and twenty-two") would be replaced by "dwa[dwadzieścia[sto]]" (in English "two[twenty[one hundred]]").

If an adverb followed by a different adverb or adjective is found, the two words are replaced by a two-element tree with the second word as the root. This rule transforms phrases like *"niewiele szybciej"* ("slightly faster") or *"niesamowicie zielony"* ("amazingly green") into trees "szybciej[niewiele]" and "zielony[niesamowicie]" (in English "faster[slightly]" and "green[amazingly]").

The state of the example sentence can be written as:

Ostrożnie zapytałem czerwoną[bardzo] dziewczynkę z
kapturkiem czerwonym[równie] o zapałki, a następnie
niechętnie kupiłem od niej dwa[dwadzieścia] pomidory i
jedną marchew.

English version:

(I) carefully asked (a) red[very] girl with (a) hood red[
equally] about matches, and afterwards reluctantly bought
from her two[twenty] tomatoes and (a) single carrot.

It should be noted that all trees inserted into the text are treated as single words – equivalent with their roots. All rules described in the paper can by applied to trees as if they were words.

3.2   Sentence Subdivision

In Polish language every two verbs need to be separated by a comma or a conjunction – they create different sub-sentences. The problem is that commas and conjunctions can also be used to separate series of nouns and adjectives. A state machine was created to resolve this issue. Its task was to join series of nouns and adjectives into special trees. For example the phrase *"Znalazłem psa i kota"* would be replaced by "`Znalazłem @[ psa, kota ]`" (in English "`I found @[ cat, dog ]`") where '@' is a placeholder plural noun with appropriately derived case and gender. The machine's state diagram is shown in Fig. 3.

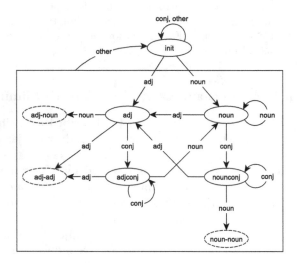

Fig. 3. State diagram of the finite-state machine responsible for finding series of nouns and adjectives. "conj" is an abbreviation of conjunction.

The state machine starts at the initial state "init" and takes whole words with their tags as input. If the machine stops in an accepting state a corresponding action is taken and the whole process is repeated. There are three accepting states:

noun-noun two nouns are joined into a tree with a '@' in its root. All words between them are removed.

adj-adj two adjectives are accumulated into a temporary list. All words between them are removed. This list needs to be joined with a noun in one of the following state machine passes.

adj-noun a noun and an adjective (or an adjective list) are joined into a tree. All words between them are removed.

The process ends when the machine is not in an accepting state after processing the last word of the input. Then the input can be safely divided into separate sub-sentences on every comma and conjunction.

There are some rare cases where the state machine will incorrectly detect series of nouns. Consider the following sentence: *"Znalazłem psa i kota tym psem poszczułem"* which can be translated to "I found a dog and I let it loose on a cat". To demonstrate the problem a word-by-word translation is more suitable: "(I) found (a) dog and cat this dog (I let) loose". The first part of the sentence "(I) found (a) dog and cat" simply means "I found a dog and a cat", but when the ending "this dog (I let) loose" is added, the meaning changes to "I found a dog and I let it loose on a cat".

The state of the example sentence can be written as:

```
Ostrożnie zapytałem dziewczynkę[ czerwoną[ bardzo ] ] z
kapturkiem czerwonym[ równie ] o zapałki,
a następnie niechętnie kupiłem od niej @[ pomidory[ dwa[
dwadzieścia ] ], marchew[ jedną ] ].
```

English version:

```
(I) carefully asked (a) girl[ red[ very ] ] with (a) hood
red[ equally ] about matches,
and afterwards reluctantly bought from her @[ tomatoes[
two[ twenty ] ], carrot[ single ] ].
```

3.3   Adjective Elimination

The previous step eliminates practically all adjectives by connecting them with nouns. In some rare cases adjectives can appear not directly before the nouns described by them. For example: *"W czarodziejskiej cię odwiedziłem wieży"* ("I visited you in a magicians tower") with a word-by-word translation "In magicians you (I) visited tower". Such constructions are not used in common spoken or written language, but they can be found in a poetry. The third step of the parsing rule chain was created to handle such cases.

For each adjective in a sub-sentence the closest noun with a compatible morphological tag (same gender, case and number) is found and the two are joined together or the sub-sentence is discarded as unrecognised.

In Polish adjectives can also be subjects, like in the sentence *"Szczęśliwi nie znają litości"* ("The happy have no mercy"). Such constructions unfortunately get discarded as unrecognised in this step.

The state of the example sentence can be written as:

```
Ostrożnie zapytałem dziewczynkę[ czerwoną[ bardzo ] ] z
kapturkiem[ czerwonym[ równie ] ] o zapałki,
a następnie niechętnie kupiłem od niej @[ pomidory[ dwa[
dwadzieścia ] ], marchew[ jedną] ].
```

English version:

```
(I) carefully asked (a) girl[ red[ very ] ] with (a) hood[
red[ equally ] ] about matches,
and afterwards reluctantly bought from her @[ tomatoes[
two[ twenty ] ], carrot[ single ] ].
```

3.4   Preposition-Noun Combining

For each preposition in the input the next word is checked. If it is a noun, the two words are joined together into a tree with the preposition in its root. If the next word is not a noun the entire sub-sentence is discarded as unrecognised.

The state of the example sentence can be written as:

```
Ostrożnie zapytałem dziewczynkę[ czerwoną[ bardzo ] ] z[
kapturkiem[ czerwonym[ równie ] ] ] o[ zapałki ],
a następnie niechętnie kupiłem od[ niej ] @[ pomidory[ dwa[
dwadzieścia ] ], marchew[ jedną] ].
```

English version:

```
(I) carefully asked (a) girl[ red[ very ] ] with[ hood[
red[ equally ] ] ] about[ matches ],
and afterwards reluctantly bought from[ her ] @[ tomatoes[
two[ twenty ] ], carrot[ single ] ].
```

3.5   Noun Phrase Combining

In Polish every noun in a sentence can to be connected with the predicate or with an adjacent noun. The connection can be direct or via a preposition.

Two adjacent nouns that are not separated by a preposition should be connected only if the second noun is in genitive case. This rule is true in most of the cases. If the two nouns are separated by a preposition there is no simple way to determine if they should be connected or not.

This problem was solved using Walenty – a valence dictionary of Polish language [1]. Two nouns separated by a preposition are allowed to be joined only if the predicate cannot join with the preposition separating the nouns. This is determined by looking up semantic frames bound with the predicate in Walenty.

Many verbs in Walenty have incomplete entries. This led to a quite high error rate of this step. To radically decrease the number of incorrectly joined nouns a special table of allowed noun-preposition-noun semantic frames was created. Two nouns separated by a preposition can be joined together only if they represent a construct listed in this table. The table has a purely heuristic nature and was created empirically from a list of most common noun phrases in Polish language. It's content is shown in Table 1. To summarise – two adjacent nouns are joined together if:

1. They are not separated by a preposition and the second one is in genitive case.
2. They are separated by a preposition and according to Walenty the predicate cannot join with the preposition. They also represent a construct listed in the table of allowed semantic frames.

Table 1. Table of allowed noun-preposition-noun semantic frames.

Prep.	Second noun case	Example phrase
do	Genitive	*maszyna do pisania* (machine for typing)
o	Locative	*książka o ptakach* (a book about birds)
o	Accusative	*walka o życie* (a fight for life)
z	Ablative	*worek z kotem* (a bag with a cat)
z	Genitive	*człowiek z miasta* (a man from a city)
na	Accusative	*worek na śmieci* (a bag for trash)

In every other case the two nouns are not joined together.

Nouns are processed from right to left. The whole noun-phrase combining can be described as applying a right associative binary operator to each adjacent noun pair in the input text.

The state of the example sentence can be written as:

```
Ostrożnie zapytałem dziewczynkę[ czerwoną[ bardzo ], z[
kapturkiem[ czerwonym[ równie ] ] ] ] o[ zapałki ],
a następnie niechętnie kupiłem od[ niej ] @[ pomidory[ dwa[
dwadzieścia ] ], marchew[ jedną] ].
```

English version:

```
(I) carefully asked (a) girl[ red[ very ], with[ hood[ red[
equally ] ] ] ] about[ matches ],
and afterwards reluctantly bought from[ her ] @[ tomatoes[
two[ twenty ] ], a carrot[ single ] ].
```

3.6 Final Tree Creation

The predicate is placed in the root of the output tree. All adverbs, preposition and nouns left in the input are added as its children. All other words are discarded.

The subject of the sentence is verified before returning the result. In Polish the subject is a noun in nominative case. If such noun is not found – the sentence is accepted as correct (is has a "default" subject). If there is exactly one such noun, the sentence is also accepted. In all other cases the sentence is discarded.

The example sentence after the complete rule-chain has the form of the following two trees:

```
zapytałem[ ostrożnie, dziewczynkę[ czerwoną[ bardzo ], z[
kapturkiem[ czerwonym[ równie ] ] ] ], o[ zapałki ] ]
kupiłem[ następnie, niechętne, od[ niej ], @[ pomidory[
dwa[ dwadzieścia ] ], marchew[ jedną] ] ]
```

English version:

```
asked[ carefully, girl[ red[ very ], with[ hood[ red[
equally ] ] ] ], about[ matches ] ]
bought[ afterwards, reluctantly, from[ her ], @[ tomatoes[
two[ twenty ] ], carrot[ single ] ] ]
```

4   Results

To test the created parser a corpus of Polish poetry was prepared. It was build
from the work of classical Polish poets such as Adam Mickiewicz and Bolesław
Leśmian. An initial preprocessing stage was applied to remove or replace all
characters not being a letter, comma, punctuation mark, question mark or an
exclamation mark. The text was tagged using the WCRFT2 tagger and supplied
as input to the parser. In 24219 input sentences 71051 sub-sentences were found
and:

- 36052 were accepted,
- 27882 were discarded as verb-less,
- 3670 were discarded in the adjective elimination step,
- 3448 were discarded for other reasons.

The parsing took 0.45 s on a single 3 GHz core of an Intel i7 processor. 4253
sentences were discarded before parsing because of words unknown to the tagger.

A comparison of results obtained from other parsers was not conducted as
currently there are no alternatives for the constructed parser. The parser's purpose was to process poetry – texts that are practically unparsable by other
parsers. Only simple texts could be used for a meaningful comparison. Such tests
were considered unnecessary as they would not provide any useful information.

A few examples of the implemented parser's output are as follows:

*"Chłopiec dorósł młodzieńca, w obce pokolenia, w dalekie zbłądził kraje i
pod wschodnim słońcem poił duszę płomieniem, on wiecznym był gońcem
na lądzie i na morzu."* — Adam Mickiewicz *Sen z Lorda Byrona*

```
dorósł[ chłopiec, młodzieńca ]
/discarded/
zbłądził[ w[ kraje[ dalekie ] ] ]
poił[ pod[ słońcem [ wschodnim ] ], duszę, płomieniem ]
był[ on, gońcem[ wiecznym ], na[ lądzie ] ]
/discarded/
```

*"Tu mnóstwo scen okropnych snuło się od razu, i ów chłopiec był cząstką
każdego obrazu."* — Adam Mickiewicz *Sen z Lorda Byrona*

```
snuło[ tu, mnóstwo[ scen[ okropnych ] ], od[ razu ] ]
był[ chłopiec[ ów ], cząstką[ obrazu [ każdego ] ] ]
```

"Te przypomnienia Kirem śmiertelnej zasłony oddzieliły na wieki
młodzieńca od żony, po cóż w takiej godzinie takie przypomnienia?" —
Adam Mickiewicz *Sen z Lorda Byrona*

```
oddzieliły[
  przypomnienia[ te ],
  Kirem[ zasłony[ śmiertelnej ] ],
  na[ wieki[ młodzieńca ] ],
  od[żony ] ]
/discarded/
```

5   Conclusions

The paper proposes a new approach to dependency parsing. A parser has been
implemented and tested on Polish poetry – texts that are mostly unparsable by
current state-of-the-art dependency parsers.

The largest drawback of the parser is the lack of support for verb-less sentences which leads to approximately 40% loss of the test corpus. This problem
should be resolved before the parser can be considered as a fully functional tool.
However, even in its current state it can be used for crude dependency analysing
of texts which so far could not have been analysed at all.

Another problem of the parser is high sensitivity to tagging errors. In most
cases a miss-tagged word leads to discarding a whole sub-sentence but sometimes
in can lead to generating incorrect output. This usually happens when the tagger
assigns a wrong part-of-speech to a word.

Even though the described solution was created strictly for Polish, it could
be applied to other languages. Especially those from the Slavic group, which
prove to be hard to formalize. The parser's rule-chain would need to be modified
to accommodate the target grammar, but the approach described in this paper
would remain the same.

References

1. Andrejewicz, J., et al.: Walenty: CLARIN-PL Digital Repository (2016). http://
   hdl.handle.net/11321/251
2. Nivre, J., Hall, J., Nilsson, J.: Maltparser: a data-driven parser-generator for dependency parsing. In: Proceedings of LREC, vol. 6, pp. 2216–2219 (2006)
3. Radziszewski, A., Warzocha, R.: WCRFT2: CLARIN-PL Digital Repository (2014).
   http://hdl.handle.net/11321/36
4. Swidzinski, M.: Formal Grammar of Polish Language. No. 349, Warsaw University
   (1992). (in Polish)
5. Wolinski, M.: System of morphosyntactic markers in IPI PAS frame. Polonica XXII-
   XXIII, pp. 39–55 (2003). (in Polish)
6. Wolinski, M.: Swigra: CLARIN-PL Digital Repository (2016). http://hdl.handle.
   net/11321/258

7. Wolinski, M., Glowinska, K., Swidzinski, M.: A preliminary version of skladnica-a treebank of Polish. In: Proceedings of the 5th Language & Technology Conference, Poznan, pp. 299–303 (2011)
8. Wroblewska, A.: Polish Dependency Parser Trained on an Automatically Induced Dependency Bank. Ph.D. dissertation, Institute of Computer Science, Polish Academy of Sciences, Warsaw (2014)

Porous Silica Templated Nanomaterials for Artificial Intelligence and IT Technologies

Magdalena Laskowska[1], Łukasz Laskowski[2(✉)], Jerzy Jelonkiewicz[2],
Henryk Piech[3], Tomasz Galkowski[2], and Arnaud Boullanger[4]

[1] Institute of Nuclear Physics Polish Academy of Sciences, 31342 Krakow, Poland
[2] Department of Microelectronics and Nanotechnology,
Czestochowa University of Technology,
Al. Armii Krajowej 36, 42-201 Czestochowa, Poland
lukasz.laskowski@kik.pcz.pl
[3] Institute of Computer Science, Czestochowa University of Technology,
Ul. Dabrowskiego 69, 42-201 Czestochowa, Poland
[4] Université Montpellier II, Chimie Moléculaire et Organisation du Solide,
Institut Charles Gerhardt, UMR 5253 CC 1701, 2 Place E. Bataillon,
34095 Montpellier Cedex 5, France

Abstract. This paper focuses on two types of novel nanomaterials based on ordered mesoporous silica designed for applications in artificial intelligence and IT technologies: molecular neural network and super dense magnetic memories. There's no doubt that electronics needs new solutions for the further development. Nanotechnology comes here with the help. Especially nanostructured functional materials can help solve the problem of miniaturization.

Keywords: Artificial intelligence · Functional materials · Hopfield Neural Network · Spin-glass · Molecular magnet

1 Introduction

Self-organization presently is one of the most important method of functional nanomaterials creation. It is possible to design the synthesis process in such a way, that atoms create assumed molecular structure by themselves. In this way the physical and chemical properties of the material can be precisely determined since they result from the molecular structure. This procedure allows obtaining materials precisely tailored for the expected applications. Particularly the IT technology requires novel materials with very small structures – nanostructures. The material that offers an ordered structure at the nanoscale is mesoporous silica which can be functionalized with various molecules [11]. The approach proposed for the first time by Kim Eric Drexler assumes that the synthesis procedure leads to the self-organization of the atoms and the formation of the assumed molecular structure is known as the Botom–Up method [9]. This process should be preceded by needs analysis - physical and chemical properties of the

© Springer International Publishing AG 2017
L. Rutkowski et al. (Eds.): ICAISC 2017, Part II, LNAI 10246, pp. 509–517, 2017.
DOI: 10.1007/978-3-319-59060-8_46

material should be precisely determined. On the base of Drexler's assumption we can establish an algorithm of functional materials fabrication. The procedure can be summarized as follows:

1. *Analysis of needs* can be written as a question: how device are we planning to "synthesize" (what kind of application is planned for the material)?
2. *Determination of material's molecular structure* in such a way, that the structure fits assumed physical properties.
3. *Synthesis of testing material* (evaluating material) in order to establish correct testing procedure and verify synthesis route.
   (a) Designing of synthesis procedure.
   (b) Realization of synthesis procedure.
   (c) Possibly complete characterization of the material obtained. If the material meets assumption, go to 4.
   (d) Modification (optimization) of synthesis procedure – go to 3.b.
4. *Synthesis of destination material*.
   (a) Working out synthesis procedure on the base of results obtained for evaluating material.
   (b) Realization of synthesis procedure.
   (c) Material's characterization. If material meets assumptions, procedure is completed.
   (d) Modification (optimization) of synthesis procedure – go to 4.b.

Above mentioned algorithm will be treated as a framework of the article. However here we focus on two first point of it: analysis of needs and determination of material's molecular structure.

2 Analysis of Needs

Computational Intelligence systems [3,5–7,16–18,24,25,27,29–34] has been a subject of rapid development during a few last decades. Especially artificial neural networks [2,4,21,22,26] revolutionized IT domain. Such a systems has been found to be promising in some specialized applications, such as image processing, speech synthesis and analysis, pattern recognition, high energy physics and so on [15,23]. Especially Hopfield's type neural networks led to a breakthrough in the Neural Networks domain, allowing for the construction of auto-associative memories [12] or systems for multi-criterion optimization [1,10,13,19]. In contrary to other artificial neural networks there was no imitation of biological structure, but projection of a spin-glass into computer science. The main feature of spin-glasses, as far as Hopfield architecture is concerned, is their complex energy landscape with numerous local minima. Such systems are subjects of Minimum Energy Law (as all physical systems) – only minimum energy configuration is stable. With a framework of Ising model, each interacts with all remaining spins and external magnetic field. Structure of Hopfield's-like networks are just a computer simulation of spin-glass. Atoms were substituted by neurons, exchange interactions by interconnections strengths.

One can think, that we have perfect tool for multicriterion optimization. Regrettably, a lot of attempts to using the Hopfield Neural Networks for multicriterion optimization brought rather disappointing results. We have found a few reasons of this situation:

- **Stacking of the network in local minima.** In the original form, introduced by J.J. Hopfield, the neural network has no possibility of local minima leaving which does not exactly represent optimal solution.
- **Errors connected with discretization.** It must be remarked, that Hopfield Neural Network only imitate spin-glass. Operation of Neural Network is not really continuous, its only some approximation of analogue neurons interactions. It involves numerous mistakes connected with problem's projection into energy landscape.
- **Long operating time.** Computer simulation works so slowly, that algorithm's using in a real-time working devices was impossible.
- **Memory needs.** During Hopfield's-like structures operating huge amount of memory is need for interconnections strengths values storing.

All above concerned one can say, that the most important disadvantage of Hopfield's like networks is the fact, that these structures only imitate spin-glass. Ideal solution would be enlarged model of spin-glass, giving possibility of atoms/nodes/neurons (next called processing units) configuration's reading and with possibility of energy landscape creation, by applying of interconnections strengths between processing units.

In this place we can therefore define precisely necessity: we need layout of small, bistable units connected in such a way, that they can interact with each other (parallel continuous operating). At the same time processing units should be large enough allowing for reading of their configurations and control of interconnections weights. Such a layout can operate much faster and more precisely as computer simulation of Hopfield network. Proposed here device works as spin-glass, not only imitates it.

At the same time it is clearly seen, that it is necessary to construct novel memory storage system, allowing for dense information writing. Such a device can have a structure similar as presented above for spin-glass model, with the exception of units interconnections.

3   Determination of Materials Molecular Structure

During designing molecular structure of materials with assumed properties it is necessary to take into account not only assumed physical properties, but also feasibility of synthesis procedure. It is worth to base assumed functional material on some proven structure giving some possibility of fitting/adjustment of molecular structure to assumed application.

The project of *enlarged model of a spin-glass*, working as ideal Hopfield network was named *molecular neural network*. In our opinion such a system should be composed of bistable magnetic units with nanometer dimension,

distributed regularly on a substrate. Regularity of distribution facilitates localization, connection and states reading. Dimension of processing units should be
adjusted in such a way to allow for their reading and obtaining large density of
their packing. Also feasibility of synthesis should be taken under consideration.
It is possible to obtain processing units with diameter of 3 nm. Such a dimension
allows for obtaining huge capacity: $13 \cdot 10^{10}$ units in mm^2, and states reading
will be possible (XMCD technique). Arrangement of processing units should be
2D hexagonal, what is feasible and allows for connections and reading. For the
reason of assumed units bistability, we cannot use ferro- or ferrimagnetic material. It seems, that the best solution would be molecular magnets, e.g. Mn_{12}
[20]. In the ground state such a magnet has spin S=10, so it is possible to read
magnetic state (nevertheless we must remark, that it will be difficult). Molecular magnets should be separated from each other, and regularly distributed
on the surface. Such a distribution can be achieved by using of porous matrix.
In the case of obtaining thin porous layer with pores oriented perpendicularly
to substrate's surface, assuming 2 nm of pores diameter, we are able to bound
individual Mn_{12} molecules at the pores bottoms. Mn_{12} molecular magnet has
diameter 1,2–1,8 nm, so only sole molecule has contact with pore's bottom. With
using of proper anchoring units and effective removal procedure, we are able to
obtain precise matrix's functionalization: each pore contains individual Mn_{12}
molecular magnet. Matrix in the form of thin films with assumed parameters
can be made with the help of self-assembly methods. As it is known, it is possible to achieve regular, hexagonal pores arrangement. Schema of proposed thin
porous layer was presented in Fig. 1.

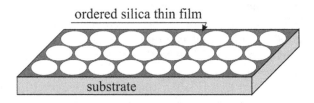

Fig. 1. Schema of thin silica layer playing the role of matrix for molecular magnets.

Schema of processing units system for molecular neural network was presented in Fig. 2.

Layer of silica containing propyl-carboxy acid groups can play a role of
anchoring layer in this case. Assuming of using derivative of Mn_{12} in the form of
acetate ($Mn_{12}ac_{16}$), we can bound molecular magnets to the surface. Additionally propyl units plays a role of spacers, minimizing possibility of losing of Mn_{12}
magnetic properties as a result of contact with surface. Control of processing
units interactions between each other can be realized through Fermi electrons,
as in the case of spin-glass. Exchange strengths can be tuned by adjusting of
electrons coherence. The problem of coherent electrons way can be overcome

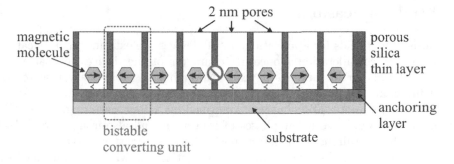

Fig. 2. Schema of the basic material for molecular neural network (system of processing units).

by conversion into a spin-wave [14] and using of Datta-Das [8] transistor for coherence adjusting. Here we must remark, that connection of processing units between each other can be an extremely difficult task. Nevertheless in our opinion scientific research should gives possibility to go beyond assumed frameworks, or actually broaden all frameworks and limitations.

Similar molecular structure can be applied for super-dense magnetic memory device. In this case however, silica pores should be filled with hard-magnetic material. Assuming pore's diameter about 2 nm, wall thickness 1 nm, it is possible to obtain $13 \cdot 10^{10}$ bits in mm^2. States reading could be possible with Magnetic Force Microscopy (MFM). Schema of such a device was presented in Fig. 3.

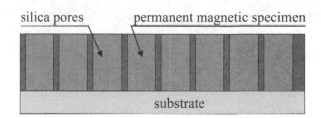

Fig. 3. Schema of molecular structure of material, that can be used as a super-dense magnetic memory storage device.

One can think, that it is not possible to obtain ferro- or ferrimagnetic material inside pores for a reason of superparamagnetic limit, but I must remark, that despite of pores diameter of 2 nm their length is from 70 to 200 nm, depending on synthesis route. Moreover our preliminary research confirmed possibility of obtaining ferromagnetic material inside silica pores.

4   Study of Feasibility

Synthesis of materials for applications in IT devices are extremely complicated. Taking under consideration pioneering character of the devices, preparing of target materials could take a few years. Nevertheless, preliminary research of target materials has brought some successes: we obtained magnetite inside silica pores (powdered material), what was confirmed by X-Ray scattering. It must be remarked that we found a few phases of iron oxide and also iron. Nevertheless we think, that obtaining a single magnetic phase is a matter of synthesis optimization. The material shown permanent magnetization, what was confirmed by magnetization measurement (SQUID). We observed hysteresis both at 2 K and at room temperature. We also obtained thin silica films with assumed structure: containing 2 nm pores perpendicular to substrate with 2D hexagonal arrangement. To this end we applied Electrochemically Assisted Self-Assembly method (EASA) [28]. As a substrate we used fluoride doped tin oxide (FTO), playing also an electrode role. Obtained thin films were shown as TEM microphotography in Fig. 4.

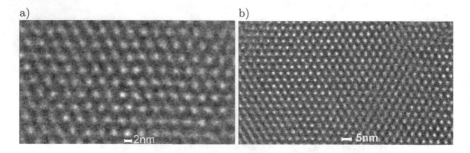

Fig. 4. Thin silica layers obtained via Electrochemically Assisted Self-Assembly method (EASA)

We also dealt with precise activation of obtained thin films with the pores on bottom side only, according to the procedure worked out by our team. Molecular structure of obtained thin layers was confirmed by Raman spectroscopy on the base of comparison with non-activated thin layers. In the Raman spectrum we found distinguished peaks taken from active, metal containing units. The results obtained by us confirmed activation's success.

All above concerned, in our opinion our research has open a way to physical realization of molecular neural network. We must admit, that for physical realization we need time, but concept's feasibility has been confirmed. Also super-dense magnetic memory storage seems to be feasible.

5   Conclusion

In the paper we have presented completely novel concepts of devices for using in IT technologies. It is molecular neural network and super-dense memory storages. Both devices are based on porous silica matrix. We shown preliminary research and confirmed feasibility of proposed devices.

Aknowledgement. Financial support for this investigation has been provided by the National Centre of Science (Grant-No: 2015/17/N/ST5/03328).

References

1. Aghdam, M.H., Heidari, S.: Feature selection using particle swarm optimization in text categorization. J. Artif. Intell. Soft Comput. Res. **5**(4), 231–238 (2015)
2. Bas, E.: The training of multiplicative neuron model based artificial neural networks with differential evolution algorithm for forecasting. J. Artif. Intell. Soft Comput. Res. **6**(1), 5–11 (2016)
3. Bello, O., Holzmann, J., Yaqoob, T., Teodoriu, C.: Application of artificial intelligence methods in drilling system design and operations: a review of the state of the art. J. Artif. Intell. Soft Comput. Res. **5**(2), 121–139 (2015)
4. Bertini Junior, J.R., Nicoletti, M.D.C.: Enhancing constructive neural network performance using functionally expanded input data. J. Artif. Intell. Soft Comput. Res. **6**(2), 119–131 (2016)
5. Cpalka, K.: A method for designing flexible neuro-fuzzy systems. In: Rutkowski, L., Tadeusiewicz, R., Zadeh, L.A., Zurada, J.M. (eds.) ICAISC 2006. LNCS, vol. 4029, pp. 212–219. Springer, Heidelberg (2006). doi:10.1007/11785231_23
6. Cpalka, K.: Design of Interpretable Fuzzy Systems, vol. 684. Springer, Heidelberg (2017)
7. Cpałka, K., Zalasiński, M., Rutkowski, L.: A new algorithm for identity verification based on the analysis of a handwritten dynamic signature. Appl. Soft Comput. **43**, 47–56 (2016)
8. Datta, S., Das, B.: Electronic analog of the electro-optic modulator. Appl. Phys. Lett. **56**(7), 665–667 (1990)
9. Drexler, K.E., Minsky, M.: Engines of Creation. Fourth Estate, London (1990)
10. El-Samak, A.F., Ashour, W.: Optimization of traveling salesman problem using affinity propagation clustering and genetic algorithm. J. Artif. Intell. Soft Comput. Res. **5**(4), 239–245 (2015)
11. Giraldo, L., López, B., Pérez, L., Urrego, S., Sierra, L., Mesa, M.: Mesoporous silica applications. In: Macromolecular symposia, vol. 258, pp. 129–141. Wiley Online Library (2007)
12. Hopfield, J.J., Feinstein, D., Palmer, R.: 'unlearning' has a stabilizing effect in collective memories (1983)
13. Hopfield, J.J., Tank, D.W., et al.: Computing with neural circuits - a model. Science **233**(4764), 625–633 (1986)
14. Hueso, L.E., Pruneda, J.M., Ferrari, V., Burnell, G., Valdés-Herrera, J.P., Simons, B.D., Littlewood, P.B., Artacho, E., Fert, A., Mathur, N.D.: Transformation of spin information into large electrical signals using carbon nanotubes. Nature **445**(7126), 410–413 (2007)

15. Lan, K., Sekiyama, K.: Autonomous viewpoint selection of robot based on aesthetic evaluation of a scene. J. Artif. Intell. Soft Comput. Res. **6**(4), 255–265 (2016)
16. Łapa, K., Cpałka, K., Wang, L.: New method for design of fuzzy systems for nonlinear modelling using different criteria of interpretability. In: Rutkowski, L., Korytkowski, M., Scherer, R., Tadeusiewicz, R., Zadeh, L.A., Zurada, J.M. (eds.) ICAISC 2014. LNCS, vol. 8467, pp. 217–232. Springer, Cham (2014). doi:10.1007/978-3-319-07173-2_20
17. Łapa, K., Przybył, A., Cpałka, K.: A new approach to designing interpretable models of dynamic systems. In: Rutkowski, L., Korytkowski, M., Scherer, R., Tadeusiewicz, R., Zadeh, L.A., Zurada, J.M. (eds.) ICAISC 2013. LNCS, vol. 7895, pp. 523–534. Springer, Heidelberg (2013). doi:10.1007/978-3-642-38610-7_48
18. Łapa, K., Szczypta, J., Venkatesan, R.: Aspects of structure and parameters selection of control systems using selected multi-population algorithms. In: Rutkowski, L., Korytkowski, M., Scherer, R., Tadeusiewicz, R., Zadeh, L.A., Zurada, J.M. (eds.) ICAISC 2015. LNCS, vol. 9120, pp. 247–260. Springer, Cham (2015). doi:10.1007/978-3-319-19369-4_23
19. Leon, M., Xiong, N.: Adapting differential evolution algorithms for continuous optimization via greedy adjustment of control parameters. J. Artif. Intell. Soft Comput. Res. **6**(2), 103–118 (2016)
20. Lis, T.: Preparation, structure, and magnetic properties of a dodecanuclear mixed-valence manganese carboxylate. Acta Crystallogr. Sect. B Struct. Crystallogr. Crystal Chem. **36**(9), 2042–2046 (1980)
21. Nobukawa, S., Nishimura, H., Yamanishi, T., Liu, J.Q.: Chaotic states induced by resetting process in izhikevich neuron model. J. Artif. Intell. Soft Comput. Res. **5**(2), 109–119 (2015)
22. Patgiri, C., Sarma, M., Sarma, K.K.: A class of neuro-computational methods for assamese fricative classification. J. Artif. Intell. Soft Comput. Res. **5**(1), 59–70 (2015)
23. Prasad, M., Liu, Y.T., Li, D.L., Lin, C.T., Shah, R.R., Kaiwartya, O.P.: A new mechanism for data visualization with tsk-type preprocessed collaborative fuzzy rule based system. J. Artif. Intell. Soft Comput. Res. **7**(1), 33–46 (2017)
24. Rutkowski, L., Cpalka, K.: Flexible weighted neuro-fuzzy systems. In: Proceedings of the 9th International Conference on Neural Information Processing, 2002. ICONIP 2002, vol. 4, pp. 1857–1861. IEEE (2002)
25. Rutkowski, L., Cpalka, K.: Neuro-fuzzy systems derived from quasi-triangular norms. In: Proceedings of the 2004 IEEE International Conference on Fuzzy Systems, vol. 2, pp. 1031–1036. IEEE (2004)
26. Sugiyama, H.: Pulsed power network based on decentralized intelligence for reliable and lowloss electrical power distribution. J. Artif. Intell. Soft Comput. Res. **5**(2), 97–108 (2015)
27. Szczypta, J., Łapa, K., Shao, Z.: Aspects of the selection of the structure and parameters of controllers using selected population based algorithms. In: Rutkowski, L., Korytkowski, M., Scherer, R., Tadeusiewicz, R., Zadeh, L.A., Zurada, J.M. (eds.) ICAISC 2014. LNCS, vol. 8467, pp. 440–454. Springer, Cham (2014). doi:10.1007/978-3-319-07173-2_38
28. Walcarius, A., Sibottier, E., Etienne, M., Ghanbaja, J.: Electrochemically assisted self-assembly of mesoporous silica thin films. Nat. Mater. **6**(8), 602–608 (2007)
29. Zalasiński, M., Cpałka, K.: A new method of on-line signature verification using a flexible fuzzy one-class classifier. In: Selected Topics in Computer Science Applications, pp. 38–53 (2011)

30. Zalasiński, M.: New algorithm for on-line signature verification using characteristic global features. In: Wilimowska, Z., Borzemski, L., Grzech, A., Świątek, J. (eds.) Information Systems Architecture and Technology: Proceedings of 36th International Conference on Information Systems Architecture and Technology – ISAT 2015 – Part IV. AISC, vol. 432, pp. 137–146. Springer, Cham (2016). doi:10.1007/978-3-319-28567-2_12

31. Zalasiński, M., Cpałka, K.: New approach for the on-line signature verification based on method of horizontal partitioning. In: Rutkowski, L., Korytkowski, M., Scherer, R., Tadeusiewicz, R., Zadeh, L.A., Zurada, J.M. (eds.) ICAISC 2013. LNCS, vol. 7895, pp. 342–350. Springer, Heidelberg (2013). doi:10.1007/978-3-642-38610-7_32

32. Zalasiński, M., Cpałka, K.: Novel algorithm for the on-line signature verification using selected discretization points groups. In: Rutkowski, L., Korytkowski, M., Scherer, R., Tadeusiewicz, R., Zadeh, L.A., Zurada, J.M. (eds.) ICAISC 2013. LNCS, vol. 7894, pp. 493–502. Springer, Heidelberg (2013). doi:10.1007/978-3-642-38658-9_44

33. Zalasiński, M., Cpałka, K., Er, M.J.: New method for dynamic signature verification using hybrid partitioning. In: Rutkowski, L., Korytkowski, M., Scherer, R., Tadeusiewicz, R., Zadeh, L.A., Zurada, J.M. (eds.) ICAISC 2014. LNCS, vol. 8468, pp. 216–230. Springer, Cham (2014). doi:10.1007/978-3-319-07176-3_20

34. Zalasiński, M., Cpałka, K., Rakus-Andersson, E.: An idea of the dynamic signature verification based on a hybrid approach. In: Rutkowski, L., Korytkowski, M., Scherer, R., Tadeusiewicz, R., Zadeh, L.A., Zurada, J.M. (eds.) ICAISC 2016. LNCS, vol. 9693, pp. 232–246. Springer, Cham (2016). doi:10.1007/978-3-319-39384-1_21

Combining SVD and Co-occurrence Matrix Information to Recognize Organic Solar Cells Defects with a Elliptical Basis Function Network Classifier

Grazia Lo Sciuto[1], Giacomo Capizzi[1], Dor Gotleyb[2], Sivan Linde[2],
Rafi Shikler[2], Marcin Woźniak[3(✉)], and Dawid Połap[3]

[1] Department of Electric, Electronic and Informatics Engineering,
University of Catania, Catania, Italy
glosciuto@dii.unict.it, gcapizzi@diees.unict.it
[2] Department of Electrical and Computer Engineering,
Ben-Gurion University of the Negev, Beersheba, Israel
rshikler@bgu.ac.il
[3] Institute of Mathematics, Silesian University of Technology,
Kaszubska 23, 44-100 Gliwice, Poland
{Marcin.Wozniak,Dawid.Polap}@polsl.pl

Abstract. This paper presents a new methodology based on elliptical basis function (EBF) networks and an innovative feature extraction technique which makes use of the co-occurrence matrices and the SVD decomposition in order to recognize organic solar cells defects. The experimental results show that our algorithm achieves an high accuracy of recognition of 96% and that the feature extraction technique proposed is very effective in the pattern recognition problems that involving the texture's analysis. The proposed methodology can be used as a tool to optimize the fabrication process of the organic solar cells. All the tests carried out for this work were made by using the organic solar cells realized in the Optoelectronic Organic Semiconductor Devices Laboratory at Ben Gurion University of the Negev.

Keywords: Organic solar cells · EBFs neural networks · Co-occurrence matrix · Singular Value Decomposition

1 Introduction

The study of organic solar cells (OSCs) has been rapidly developed in recent years. Organic solar cell technology is sought after mainly due to the ease of manufacture and their exclusive properties such as mechanical flexibility, lightweight, and transparency [1]. These properties enable OSCs to be used in unconventional applications which are not suited for conventional solar cells. Nowadays the power conversion efficiencies of OSCs are higher than 10% [2].

© Springer International Publishing AG 2017
L. Rutkowski et al. (Eds.): ICAISC 2017, Part II, LNAI 10246, pp. 518–532, 2017.
DOI: 10.1007/978-3-319-59060-8_47

This relatively high efficiency and rather low manufacturing costs allows OSCs to be used as an appropriate alternative to other power sources. Nevertheless, currently OSCs are yet to be applicable. In order to extract their full potential OSCs must be optimized. By using the power of computational calculation it can be a relatively simple task to examine, compare, and correlate between both theoretical parameters and physical properties of such device to its performance. According to how the final outcome, i.e., the output power of the device, is affected by its characteristics, e.g., electronic properties, geometry, one can adjust these characteristics to enhance the device performance. It is not yet clear what is the role of defects in the performance of organic solar cells [3]. Therefore, it is critical to examine the correlation between defects in the structure of the device and its performance. Defects can be caused by a variety of reasons, many of them are embedded within the organic materials and physics that governs those kind of devices. However, many of the defects can arise during the OSC fabrication stage. These defects can emerge for example by scratches that occurs during different stages of fabrication, or by trapped microscopic bubbles during the spin coating stage. Thus, identification and inspection of defects can lead to the improvement and preciseness in fabrication and to a more informed decision of the materials that should be used. By taking microscopic images of the morphology of the OSC and by using a state of the art mathematical models for defects detection we were able to detect, identify, and classify these defects. In most common approaches we can discuss various types of information analysis where initial data is processed by application of some fuzzy measures [5] or image processing for extraction of most important features [4, 6–8]. In this paper we would like to discuss a novel methodology based on neural network classifier.

2   Materials

Solar cells are devices that convert sunlight into electrical power. The process of light conversion takes place in the organic layer, which will be referred as the active layer. This active layer is a composition of two organic materials (blend). These two materials are similar to the semiconductors used in the inorganic industry. Just like in semiconductors, organic materials possess a band-gap between two bands, namely, the Highest Occupied Molecular Orbital (HOMO) band and the Lowest Unoccupied Molecular Orbital (LUMO) band; the occupation refers to the organic molecule electrons. The HOMO and LUMO of organic materials are analogous to the conduction and valence bands in the inorganic semiconductors respectively. Typically, the gap between HOMO and LUMO is in the range of 1 to 3eV and thus electrons can be exited, for instance, by light in the visible spectrum. Hence, it makes organic semiconductors appropriate for solar cells. Another similarity refer to the materials type. Organic materials are divided into two categories, electron donors and electron acceptors, which are analogues to the n-type (donors) and p-type (acceptors) materials that comprise the corresponding inorganic solar cells counterparts. Donors are hole transport materials and acceptors are electron transport materials, i.e., each exhibits higher mobility for its respective charge carrier.

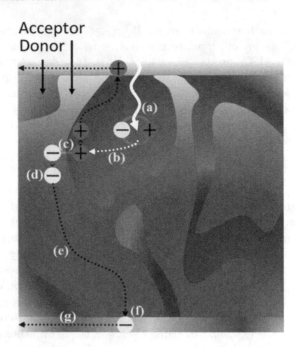

Fig. 1. The processes leading to photocurrent in the bulk heterojunction solar cell. (1) light absorption at the donor lead to the generation of excitons, (2) exciton diffuse towards donor-acceptor interface, (3) an intermediate step where exciton dissociate into $P - P$ state, (4) dissociation of $P - P$ into free two charged polarons, (5) charge transport towards the electrodes, (6) charge collection by the electrodes and charge transport within the electrodes.

The main difference between the organic and inorganic solar cells is the low dielectric constant of the former. As a result, the energy of light (photon energy) is not suffice to separate the electrons and holes into free charges (the charge carriers in a semiconductor). The physical process leading to the generation of current [9] is illustrated schematically in Fig. 1. It is a sequence of the following steps:

1. Light absorption leads to generation of excitons.
2. Exciton diffusion towards donor-acceptor interface.
3. An intermediate step where excitons dissociate into Polaron-Pair $(P - P)$ state.
4. Dissociation of $P - P$ into positive and negative polarons.
5. Charge transport towards the electrodes.
6. Charge collection by the electrodes.

As sunlight absorbs in a molecule, an electron is excited from the LUMO to the HOMO. The absorption occurs mostly in the donor material. However, instead of free charges, an exciton is generated, an electron and hole bounded

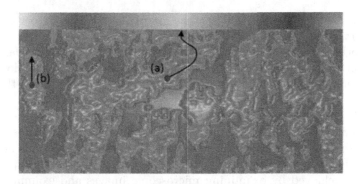

Fig. 2. An illustration of the morphology. The top gray strip stands for the ITO, the bright and dark orange signifies the acceptor and donor respectively, the purple points are the electrons and the arrows represents the path of the electrons to the ITO. (a) A path for the negatively charged polaron. (b) there is no direct path for the polaron. (Color figure online)

together by coulomb force within a molecule. It is a direct outcome of the low dielectric constant. In order to separate the exciton into free charges, the exciton must diffuse and encounter an interface between donor and acceptor materials (heterojunction). The excitons have a very low diffusion length, in the order of 10 nm. Above this length, the excitons will recombine and will not contribute to the output current. With the appropriate combination of donor and acceptor, the exciton can lower his energy by transferring one of the charges to the adjacent material. If exciton was initially created at the donor (acceptor), then electron (hole) will transfer to the acceptor (donor), namely, electron is situated at the acceptor and hole at the donor. Yet, the electron and hole are still coulomb bound. This intermediate state between excitons and free charges is the Polaron-Pair ($P-P$). Depending on the electric field, the temperature, and the materials parameters there is a finite probability for the $P-P$ to decay to the ground state or to be separated into two charged polarons (electron and hole), i.e., $P-P$ dissociation. This non-zero probability ensures the generation of free polarons. After the separation, the free polarons must diffuse and be collected at the corresponding contacts. Negatively charged polarons move in the acceptor material and are collected at the cathode and positively charged polarons move in the donor and are collected at the anode.

As was mentioned before, the active layer is made of a blend of both the donor and acceptor materials composed in a single layer. In this fashion, junctions between donors and acceptors are spatially distributed throughout the active layer. Hence, this type of OSC often called bulk heterojunction (BHJ). The obtained phase separation (separation between donor and acceptor materials) within the film is less than 30 nm [10] which is in the range of the exciton diffusion length. Thus, regardless of where an exciton generates, it will most likely encounter a junction. This formation enhances the exciton dissociation within the active layer. However, defects in the structure of the active layer can drastically reduce the output power. By inspecting the microscopic images of the active layer, one can examine whether the morphology is satisfactory or whether it needs further optimizations.

Yet another very important issue is the transportation of free polarons from the position of the specific junction to the respective contact. The downside of this blend morphology is the large extent of disorder. In the path to the desired contact, the free polarons can stumble upon different interface and recombine again to $P - P$. Although it can lead again to the dissociation of $P - P$ to free polarons, it interferes their movement and increases the chance to decay to the ground state. Therefore, knowing how the different materials and fabrication processes affects the final morphology is crucial for the process of improvement. This can be achieved by examining microscopic images and examine them by means of computational models. An illustration of the morphology is shown in Fig. 2. Once again, using such images one can detect defects that can impair the path of free carriers to the electrodes.

3 Architecture of OSC

The OSC built up as an assemblage of several layers of different materials and different functionalities. A typical cross-section of an OSC is shown in Fig. 3. It composed of the following layers, in the direction of the incoming light, transparent substrate, transparent anode, active layer, and cathode.

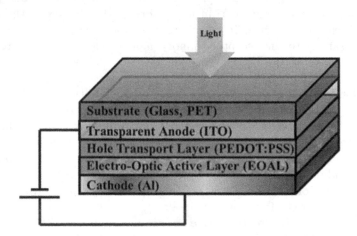

Fig. 3. A typical cross section of an organic solar cell. From the direction of the incoming light, transparent substrate (glass), transparent anode (ITO), Hole Transport Layer (PEDOT:PSS) optoelectronic active layer (P3HT:PCBM) and cathode (Aluminum).

The fabrication and measurements of the OSCs took place in the "Optoelectronic Organic Semiconductor Device Laboratory" (OOSDL) in the department of Electrical and Computer Engineering at Ben-Gurion University of the Negev in Beer-Sheva, Israel. In order to create large database, a series of samples has been made, where each sample comprised of four solar cells.

The substrate of the sample is a glass with a thickness of 0.7 mm coated with a 90 ± 10 nm layer of Indium Tin Oxide (ITO), i.e., the transparent anode. The area of the substrate is 12×12 mm^2. The ITO covers only 6×12 mm^2 of the glass and has a resistivity of 20 Ω/m^2. The active layer of the sample is a solution blend of [6, 6]-Phenyl C61 Butyric acid Methyl ester (PCBM) and Poly(3-Hexylthiophene-2, 5-diyl) (P3HT) with 1:1 ratio. The solution is dissolved in chloroform with the aid of a magnetic stirrer for one hour to make a total of 20 mg/ml. The preparation is a sequence of few procedures. First, the substrate was cleaned in an ultrasonic bath with acetone, then methanol and then isopropanol for 15 min each; next, it was treated with UV-ozone for 4 min. To facilitate the conduction of holes from the active layer to the ITO we spin coated on top of the glass+ITO a 30 nm layer of a transparent and conducting conjugated polymer Poly(3, 4-ethylenedioxythiophene)-poly(styrenesulfonate) (PEDOT:PSS). The spin coating was performed at 5000 RPM, and 1700 acceleration for 1 min. Afterwards, the sample was dried on a hot plate at 105° C for an half an hour to remove the excess water, and then for another half an hour inside a glove-box. Inside this inert atmosphere, the solution of P3HT:PCBM was than deposited by spin-casting (1000 RPM, and 600 acceleration for 1 min) at room temperature. At the top of the sample, the contact pattern was thermally evaporated to create 80 nm thickness layer of aluminum, i.e., the cathode. The sample was then annealed at 140°C for 20 min. To expose the ITO, we screeched the active layer with a toothpick. Finely, to make the contacts accessible for the external probes, we deposited on top of ITO and aluminum a conductive silver epoxy. A typical OPV sample realized is shown in Fig. 4.

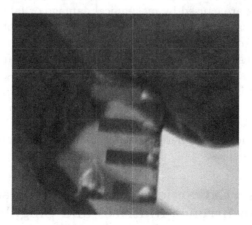

Fig. 4. A typical OPV sample realized at the "Optoelectronic Organic Semiconductor Device Laboratory".

4 Experimental Image's Dataset and Acquisition System

The images acquisition is performed with a microscope camera DeltaPix, model *DpxViewPro 1.14.8* (see Fig. 5).

Fig. 5. The microscope used to explore the fabrication defects of the organic solar cells.

Microscopy was used to observe the various defects the surface morphology on top of organic solar cells examined. We have acquired 240 images with resolution of 1280 × 1024. Some devices analyzed are good, others with various kind of defects.

A large number of defects have been observed which cracks, breaks and scratches. Scratches are caused by mechanical damage or fabrication during the handling and preparation while the shiny spots and the bubbles are due to the water infiltration and at the exposition to the air at high humidity. In addition the annealing process is responsible of the evident spots and the bubbles on surface of samples caused by different gradient temperatures.

A critical aspect is to determine properly on the ITO/aluminum interface the number and type of faults in order to understand the degradation mechanisms providing better encapsulation strategies.

In Fig. 6 we can identify different scratches due to fabrication process at the interfaces of OPV devices.

5 The Feature Extraction Methodology Based on Gray Level Co-occurrence Matrices and Singular Value Decomposition

The degradation mechanisms of the active inter-layers are fast involving the diffusion of molecular oxygen and water into the device inducing chemical reactions in polymer materials, degradation of interfaces, electrode reaction with the

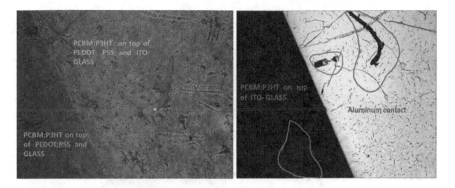

Fig. 6. Defects at the interface PSS and ITO-Glass/PSS and Glass (left), Defects at the interface area on interface PCBM:P3HT on top of ITO-Glass/Aluminum (right).

organic materials, morphological changes due to temperature, and macroscopic changes such as delamination, formation of particles, bubbles, and cracks [11].

The PEDOT:PSS is vulnerable to thermal degradation and also very sensitive to moisture and oxygen involving the irreversible structural modifications. These modifications give rise to a number of defects in the OPVs such as breaks, scratches, spot, bubbles on surface of the device etc. In the SEM images of the OPVs used in this paper, these defects manifest themselves as variation of the image texture.

One of the most popular and powerful ways to describe texture is using of color mapping co-occurrence matrix (GLCM). Since the use of the co-occurrence matrices leads to a course of dimensionality we used the singular value decomposition (SVD) to reduce the redundancy arising of description of the texture by means of the GLCM.

When a high dimensional, highly variable set of data points is taken, SVD is employed to reduce it to a lower dimensional space that exposes the substructure of the original data more clearly and orders it from most variation to the least. In this way, the region of most variation can be found and its dimensions can be reduced using the method of SVD. This implies that we can achieve a good identification of the most significant structures present in the image texture, by taking only a few largest singular values [12].

Color Mapping Co-occurrence Matrix. A GLCM is a square matrix where the number of rows and columns is equal to the number of gray levels in the image that can reveal certain properties about the spatial distribution of the gray-levels in the image texture [13,14]. The matrix gives how the pixel value l_1 of a reference pixel occurs in a specific relationship to a neighbouring pixel with pixel value l_2. So, each element (l_1, l_2) of the matrix represents the number of occurrences of the pair of pixel with pixel values l_1 and l_2 which are at a relative distance d from each other. There are many ways to specify the spatial

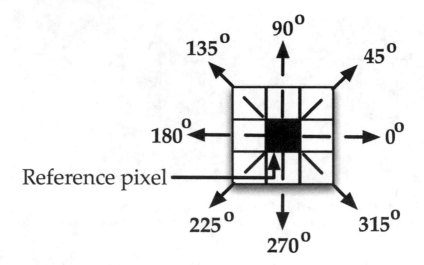

Fig. 7. Co-occurrence matrix directions for extracting texture features.

relationship between two neighbouring pixels with different offsets and angles (see Fig. 7).

Mathematically, the elements of a $L \times L$ gray-level co-occurrence matrix with displacement vector $\mathbf{d}(= d_i, d_j)$, for a given image I of size $n \times n$ is defined as:

$$M_{CO}(l_1, l_2) = \sum_{i=1}^{n} \sum_{i=1}^{n} \begin{cases} 1, & \text{if } I(i,j) = l_1 \text{ and } I(i+d_i, j+d_j) = l_2 \\ 0, & \text{otherwise} \end{cases} \quad (1)$$

Figure 7 describes how to compute the GLCM. It shows an image and its corresponding co-occurrence matrix using the default pixels spatial relationship (offset = +1 in i direction). Each element of the GLCM is the number of times that two pixels with gray tone l_1 and l_2 are in neighborhood at a distance d and direction ϕ.

Singular Value Decomposition. SVD is a potent mathematical exploration tool for matrices which gives minimum least square truncation error [15,16]. This is because the total potential degrees of freedom of the decomposed matrices is equal to the input host image. Further, SVD is a single path decomposition algorithm. Singular values represent inherent algebraic image properties and are not instable. Given an image $I(x, y)$, with dimensions $m \times n$, it can be factorized with SVD:

$$I = USV^T \quad (2)$$

Where both U and V are the orthonormal matrices, $U \in R^{m \times m}$, $V \in R^{n \times n}$, $S = [diag(\sigma_1, \sigma_2, \ldots, \sigma_q), 0]$ and $q = min(m, n)$. Besides, the singular values appear in descending order, i.e., $\sigma_1 \geq \sigma_2 \geq \ldots \geq \sigma_q \geq 0$.

The Feature Extraction. The training set must be thoroughly representative of the actual population for effective classification We have calculated the co-occurrence matrices of every image. For each channel (Red, green, Blue) we have calculated the co-occurrence matrices for $d = 1, 2$ and in the four main directions: $0°$, $45°$, $90°$ and $135°$.

The use of the co-occurrence matrices leads to a course of dimensionality because these matrices are composed of two complementary subspaces called signal subspace (the information suitable for defects classification) and noise subspace. To achieve the separation between signal and noise we have been used the SVD and for each co-occurrence matrix we took the 12 largest singular values (the criterium of truncation is $\frac{\sigma_i}{\sigma_{i+1}} \geqslant 10$), thus obtaining for each image a feature vector of 216 elements.

The magnitude of the singular values indicate the importance of the corresponding directions (vectors). The singular values reflect the amount of data variance captured by the basis elements. The first vector of the basis (the one with largest singular value) lies in the direction of the greatest data variance. The second vector captures the orthogonal direction with the second greatest variance, and so on.

This is a useful procedure because the entries of co-occurrence matrices have a large variance in correspondence of an irregular texture while a lower variance when the texture is regular.

The Figs. 8 and 9 show a marked difference between features belonging to the defective devices and the good ones. Then this feature set proves extremely suitable for the problem classification addressed in this paper.

6   The Used EBF Classifier

A PNN is predominantly a classifier: Map any input pattern to a number of classifications (in our case the neural network has to distinguish between two classes: the defective devices and the good ones). A PNN is an implementation of a statistical algorithm called kernel discriminant analysis in which the operations are organized into a multilayered feedforward network with four layers:

– Input layer
– Pattern layer
– Summation layer
– Output layer

In this paper we make use of a particular kind of PNN: the EBF network [17] that is a type of feedforward neural networks in which the hidden units evaluate the distance between the input vectors and a set of vectors called function centers or kernel centers (the centers are the data points of the training set), and the outputs are a linear combination of the hidden nodes' outputs. More specifically, the k-th network output has the form

$$y_k(x(t)) = \sum_{j=1}^{M} w_{k,j}\, \Phi_j(x(t)) \qquad (3)$$

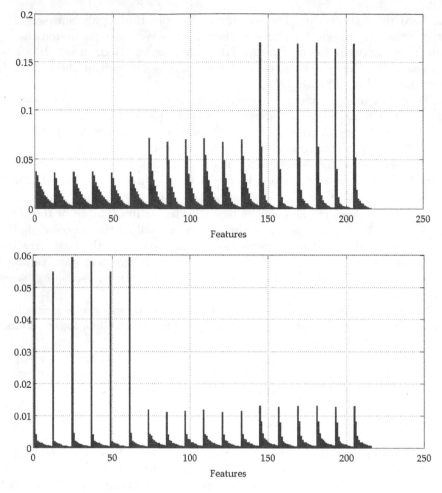

Fig. 8. Features of a defective device randomly choosen (up), Features of a device with no defect randomly choosen (down).

where

$$\Phi_j(x(t)) = \exp\left\{-\frac{1}{2\sigma_j}(x(t) - \mu_j)^T \Sigma_j^{-1}(x(t) - \mu_j)\right\} \qquad (4)$$

μ_j and Σ_j are the function center (mean vector) and covariance matrix of the j-th basis function respectively and σ_j is a smoothing parameter controlling the spread of the j-th basis function.

We restricts Σ to two global and scalar smoothing parameter, σ_1 and σ_2, where σ_1 is used in those basis functions that have centers coming from the good devices while σ_2 for the defective ones. The determination of the smoothing parameters is done by calculating the spreads of the training data set belonging to the reference classes for σ_1 and σ_2 respectively.

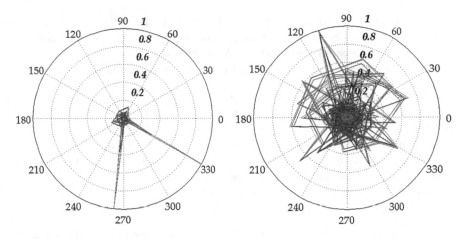

Fig. 9. Features of all defective devices present in the dataset in polar representation (left), Features of all devices with no defect present in the dataset in polar representation (right).

With these assumptions we have that the input nodes (see Fig. 10) are the samples of the feature set. The second layer (pattern layer) consists of the Gaussian functions (5) formed using the training set of data points as centers.

$$y = e^{\frac{\|X - X_{i,j}\|^2}{\sigma^2}} \tag{5}$$

The third layer (summation layer) performs an weighted average of the outputs from the second layer for each class. The fourth layer (output layer) performs a vote, selecting the largest value (the target values are: 0 for the defective devices and 1 for the good ones).

Adding and removing training samples simply involves adding or removing neurons in the pattern layer and a minimal retraining required.

For the training of the neural network simply note that the centers and spreads are predetermined then only the weights w_{ij} is required to find. The calculation can be performed by using the method of least squares.

The difference between PNN and EBF is that for EBF networks, discrimination among all the known classes is considered during the training phrase; whereas for PNNs this class discrimination is introduced during the recognition phase.

7   Results and Discussion

To evaluate the pattern recognition algorithm, dataset is randomly split into three parts: a training set consisting of 80 data points (48 data points representative of the various kind of defects and 32 representative of the devices with no defects) a validation set consisting of 80 data points and a testing set consisting of 80 data points. The training set is used to find the model parameters in the used EBF network. These parameters are the number of neuron for each class (the defective devices and the good ones) and the weights value.

To find the optimal number of neurons we proceed as follows:

1. We use the data points of the training set as the centres of the network's neurons, so obtaining a network with 80 neurons split up into two classes (48 neurons represent the defects and 32 represent the devices with no defects).
2. We calculate the network's weights by using the validation set.
3. We eliminate the neuron with a minimum weight and recalculate the network's weights by using the validation set. The procedure ends when the performance, in terms of correct classification on the validation set, falls down of the 2% with respect to the previous step.

The resulting network after the training phase is shown in Fig. 10). It consists of 21 neurons (13 of them represent the defects and 8 represent the devices with no defects).

Once the optimal parameters are found the trained algorithm is applied to classify the data points in the testing dataset into one of the two classes. A correct classification rate of 96% average has been obtained.

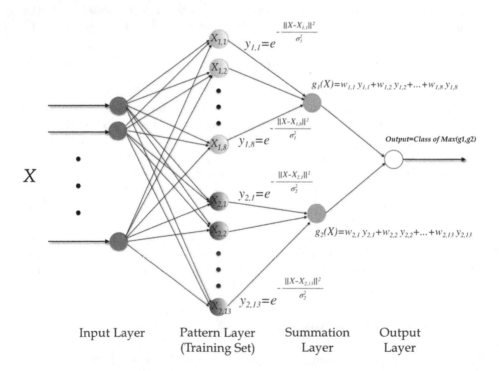

Fig. 10. Architecture of the proposed PNN classifier.

8   Conclusion

This paper presents a new methodology based on elliptical basis function (EBF) networks and an innovative feature extraction technique which makes use of the co-occurrence matrices and the SVD decomposition in order to recognize organic solar cells defects.

The organic solar cells used in this paper were realized in the Optoelectronic Organic Semiconductor Devices Laboratory at Ben Gurion University of the Negev. Microscopy was used to observe the various defects the surface morphology on top of organic solar cells examined. We have acquired 240 images with resolution of 1280×1024.

A large number of defects have been observed which cracks, breaks and scratches. Scratches are caused by mechanical damage or fabrication during the handling and preparation while the shiny spots and the bubbles are due to the water infiltration and at the exposition to the air at high humidity.

The experimental results show that our algorithm achieves an high accuracy of recognition of 96% and that the feature extraction technique proposed is very effective in the pattern recognition problems that involving the texture's analysis.

The proposed methodology can be used as a tool to optimize the fabrication process of the organic solar cells.

Acknowledgments. Authors acknowledge contribution to this project from the "Diamond Grant 2016" No. 0080/DIA/2016/45 funded by the Polish Ministry of Science and Higher Education.

References

1. Brabec, C.J., Dyakonov, V., Parisi, J., Sariciftci, N.S.: Organic Photovoltaics: Concepts and Realization, vol. 60. Springer Science & Business Media, Berlin (2013)
2. Kan, B., Zhang, Q., Li, M., Wan, X., Ni, W., Long, G., Wang, Y., Yang, X., Feng, H., Chen, Y.: Solution-processed organic solar cells based on dialkylthiol-substituted benzodithiophene unit with efficiency near 10%. J. Am. Chem. Soc. **136**(44), 15529–15532 (2014)
3. Gagorik, A.G., Mohin, J.W., Kowalewski, T., Hutchison, G.R.: Effects of delocalized charge carriers in organic solar cells: predicting nanoscale device performance from morphology. Adv. Funct. Mater. **25**(13), 1996–2003 (2015). http://dx.doi.org/10.1002/adfm.201402332
4. Korytkowski, M., Rutkowski, L., Scherer, R.: Fast image classification by boosting fuzzy classifiers. Inf. Sci. **327**, 175–182 (2016). doi:10.1016/j.ins.2015.08.030
5. Starczewski, J.T.: Centroid of triangular and gaussian type-2 fuzzy sets. Inf. Sci. **280**, 289–306 (2014). doi:10.1016/j.ins.2014.05.004
6. Cpalka, K., Zalasinski, M., Rutkowski, L.: A new algorithm for identity verification based on the analysis of a handwritten dynamic signature. Appl. Soft Comput. **43**, 47–56 (2016). doi:10.1016/j.asoc.2016.02.017

7. Grycuk, R., Gabryel, M., Nowicki, R., Scherer, R.: Content-based image retrieval optimization by differential evolution. In: IEEE Congress on Evolutionary Computation, CEC, Vancouver, BC, Canada, July 24–29 2016, pp. 86–93. IEEE (2016). doi:10.1109/CEC.2016.7743782

8. Pabiasz, S., Starczewski, J.T., Marvuglia, A.: SOM vs FCM vs PCA in 3D face recognition. In: Rutkowski, L., Korytkowski, M., Scherer, R., Tadeusiewicz, R., Zadeh, L.A., Zurada, J.M. (eds.) ICAISC 2015. LNCS, vol. 9120, pp. 120–129. Springer, Cham (2015). doi:10.1007/978-3-319-19369-4_12

9. Scarongella, M., Brauer, J.C., Douglas, J.D., Fréchet, J.M.J., Banerji, N.: Charge generation in organic solar cell materials studied by terahertz spectroscopy, pp. 95 670M-95 670M-13 (2015)

10. Heeger, A.J.: 25th anniversary article: bulk heterojunction solar cells: understanding the mechanism of operation. Adv. Mater. 26(1), 10–28 (2014)

11. Balderrama, V.S., Estrada, M., Formentin, P., Iñiguez, B., Ferré-Borrull, J., Pallarés, J., Nolasco, J.C., Palomares, E., Sánchez, A., Marsal, L.F.: Performance and degradation of organic solar cells with different p. 3ht: pcbm[70] blend composition. In: Proceedings of the 8th Spanish Conference on Electron Devices, CDE 2011, pp. 1–4, February 2011

12. Yang, J.-F., Lu, C.-L.: Combined techniques of singular value decomposition and vector quantization for image coding. IEEE Trans. Image Process. 4(8), 1141–1146 (1995)

13. Haralick, R.M., Shanmugam, K., Dinstein, I.: Textural features for image classification. IEEE Trans. Syst. Man Cybern. SMC−3(6), 610–621 (1973)

14. Capizzi, G., Sciuto, G.L., Napoli, C., Tramontana, E., Woźniak, M.: Automatic classification of fruit defects based on co-occurrence matrix and neural networks. In: 2015 Federated Conference on Computer Science and Information Systems (FedCSIS), pp. 861–867, September 2015

15. Klema, V., Laub, A.: The singular value decomposition: its computation and some applications. IEEE Trans. Autom. Control 25(2), 164–176 (1980)

16. Lange, K.: Singular Value Decomposition, pp. 129–142. Springer, New York (2010)

17. Mak, M.W., Kung, S.Y.: Estimation of elliptical basis function parameters by the em algorithms with application to speaker verification. IEEE Trans. Neural Networks 11(4), 961–969 (2000)

An Intelligent Decision Support System for Assessing the Default Risk in Small and Medium-Sized Enterprises

Diana Manjarres[1]([⊠]), Itziar Landa-Torres[1], and Imanol Andonegui[2]

[1] TECNALIA, 48160 Derio, Spain
{diana.manjarres,itziar.landa}@tecnalia.com
[2] Department of Applied Physics I.,
University of the Basque Country UPV/EHU, 48013 Bilbao, Spain
imanol.andonegui@ehu.eus

Abstract. In the last years, default prediction systems have become an important tool for a wide variety of financial institutions, such as banking systems or credit business, for which being able of detecting credit and default risks, translates to a better financial status. Nevertheless, small and medium-sized enterprises did not focus its attention on customer default prediction but in maximizing the sales rate. Consequently, many companies could not cope with the customers' debt and ended up closing the business. In order to overcome this issue, this paper presents a novel decision support system for default prediction specially tailored for small and medium-sized enterprises that retrieves the information related to the customers in an Enterprise Resource Planning (ERP) system and obtain the default risk probability of a new order or client. The resulting approach has been tested in a Graphic Arts printing company of The Basque Country allowing taking prioritized and preventive actions with regard to the default risk probability and the customer's characteristics. Simulation results verify that the proposed scheme achieves a better performance than a naïve Random Forest (RF) classification technique in real scenarios with unbalanced datasets.

Keywords: Classification · Default prediction · Clustering

1 Introduction

Along with the impact of the economic crisis worldwide, the assets structure of financial institutions has been remodelled due to the enormous bad debt acquired by many companies. Therefore, personal consumer finance loans have attracted great interest and even become a popular financial product. One of their main goals was to substantially enhance consumer finance in order to stimulate consumption and accelerate the economic recovery. However, the negative effects have gradually emerged due to the intense competition in financial market and the excessive number of financial loans that have been granted without certainly knowing whether the borrowers could repay their debt on time.

© Springer International Publishing AG 2017
L. Rutkowski et al. (Eds.): ICAISC 2017, Part II, LNAI 10246, pp. 533–542, 2017.
DOI: 10.1007/978-3-319-59060-8_48

With the emergence of new Basel requirements [1] (i.e. Basel II, Basel III) in which a global regulatory framework for more resilient banking systems is presented, financial institutions started to develop risk control strategies and analysis with the aim at ensuring a significant reduction of credit risk. In this context, many studies highlight the importance of obtaining an effective consumer loan default predicting model to make financial institutions and also the banking industry able to reduce the overdue loan without worsening their loan business [2–6]. Apart from focusing in the banking industry, related works based on credit business and peer-to-peer (P2P) lending has attracted great interest in the last years [7–12].

Specifically, authors in [2] propose the application of neural networks to credit risk assessment of Italian small businesses. For that purpose, real data coming from 76 distinct companies is utilized with the aim at detecting their economic and financial situation and obtaining a considerable accurate outcome in the default prediction model. Another related work is the one presented in [3] in which a Partial Least Squares (PLS) regression method is proposed to model the risk severity and its associated most influential variables in order to take corrective actions and prevent the loan from defaulting. In this regard, authors in [4] present a consumer loan default predicting model by means of the application of DEA-DA (Data Enveloped Analysis - Discriminant Analysis).

One of the main differential contribution of this work is the integration of non-financial related data, such as: personal information and money attitude of the borrowers, in order to predict whether they are going to pay back on-time or not. Likewise, the study proposed in [5] presents a credit-card-holder delinquency and default risk forecast model in which real customer transactions and credit bureau data is efficiently combined. More recently, authors in [6] propose a data mining approach based on Random Forest (RF) to predict the performance of the Peer-to-Peer (P2P) loan.

Regarding the prediction of probability of enterprise default, most of the related works in the literature [13–15] are mainly focused on medium and large enterprises that have detailed financial information. Less research is devised for assessing the default risk of Small and Medium Enterprises (SMEs) [16–20]. The lack of models for evaluating the risk profiles of small firms is important because SMEs represent a significant part of the economy of every country. In this regard, authors in [20] propose the application of Artificial Neural Networks (ANNs) to sample over 7000 Italian Small Enterprises (SEs). Results show that the presented system based on ANNs outperforms traditional methods when evaluating the SEs credit risk.

The presented work is then focused on providing a decision support system for small and medium-sized companies that quantitatively infers the risk associated to a new order considering the information of the different orders that have been processed by this company in the past. Specifically, the presented information is related to a Graphic Arts printing company in The Basque Country (Spain) which one of its major concerns is being able to infer whether the different orders that need to be produced would or not be paid back on time by the different customers. The proposed scheme for assessing the default risk, hereafter named as DSSDP (Decision Support System for Default Prediction), is able to predict whether a new order comes from a possible future default company and allow prioritization actions with regard to the default risk probability and the customer's characteristics.

The paper is organized as follows: Sect. 2 presents the real environment deployment. Secondly, a detailed explanation of the algorithm designed for assessing the default risk is provided in Sect. 3. Finally, Sect. 4 shows the simulation and experiments accomplished and final results and remarks are obtained in Sect. 5.

2 Real Environment Deployment and Data Characteristics

This paper proposes a novel Decision Support System for Default Prediction (DSSDP) applied in a real case study of a small and medium-sized company of The Basque Country that calculates the default risk prediction of a new order based on the historical information stored at the Enterprise Resource Planning (ERP) system. At the end of the billing period this company maintains a considerable number of orders with uncollected amounts, which means that the orders have not been paid on time. This situation leads the company to search for actions that prevent the company from having accumulated orders without being paid which severely affects its financial condition.

The available information at the ERP system consists of a set of different variables, such as: budget information, quality measures, production related data, and so on, for which the variables presented in the following list are found to be relevant features for the DSSDP:

The first indicators define in a general fashion the client and product type of the order:

1. **Client**: Identifier of the client.
2. **Product type**: Identifier of the type of product.
3. **Date**: Date of the order.
4. **Zone**: Location of the order.
5. **Market**: Market information related to the order.
6. **Internal codes of difficulty of the order**: Difficulty level of the order.

In terms of budget related information, different variables are considered for the study:

1. **Budget amount**: Amount of budget related to the order.
2. **Budget code**: Budget code of the order.
3. **Budget margin**: Margin data in percentage (%) for each budget for a specific time set period.
4. **Added value**: Added value of the budget. This quantity is obtained by subtracting to the budget the different subcontracts, consumption and materials needed for the order.

5. **Budget maturity period**: This quantity indicates the amount of time between the date of the order for which a budget is formalized and the date on which the order goes into production.
6. **Outstanding amount:** Outstanding amount in euros.

The production process has a set of magnitudes that describes distinct characteristics, such as: cost, profitability, added value, and so on.

1. **Production cost**: Cost of production of the order in euros.
2. **Production profit**: Profitability of the order in euros.
3. **Production added value**: Added value of the order.
4. **Production margin**: The margin in percentage (%) of each order for a specific period of time.
5. **Production period**: Delivery period of the order.

Regarding quality measurements, different indicators are defined for considering the number and amount value of no conformities.

1. **Number of no conformities**: Boolean value that indicates the conformity (0) or no conformity (1) of the order.
2. **Amount of no conformities**: Quantity of no conformities of the order.

Finally, there is also information with regard to commercial indicators, the communication way and the number of commercial visits are taken into account in the following indicators:

1. **Communication**: Communication way, i.e. (via web, telephone).
2. **Number of commercial visits**: Number of commercial visits to the specific client that requests the order.

Summarizing, a total of 21 features are found to be relevant for obtaining the probability of default risk of a new order. The outcome of the study is the result of a feature selection process that considers the total of variables stored at the ERP system of the company.

3   Main Procedure of the Proposed DSSDP Algorithm

The principal aim of the proposed DSSDP algorithm is to obtain the probability of default of a new order by means of considering the available information per client and order available at the ERP system.

Figure 1 depicts the general scheme of the proposed DSSDP. For generating the clusters it employs an unsupervised K-means algorithm [21], in which each order is defined by means of the 21 relevant features. In order to determine the optimum number of clusters M and test the goodness of the obtained cluster distribution a DPM (Default Prediction Metric) metric (Eq. 1) is defined. The optimum number of clusters M is obtained by means of an exhaustive search process with M in range $\{1,..., M_{max}\}$ in which each cluster distribution provided by the K-means algorithm is evaluated with the above-defined DPM metric. Note

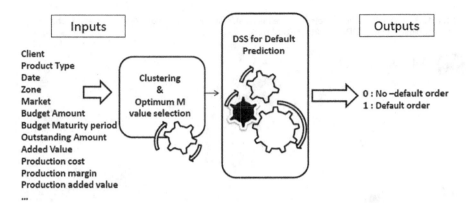

Fig. 1. General scheme of the proposed DSSDP

that the maximization of this metric focuses on creating cluster distributions in which the number of default orders is approximately 0% or 100% in order to agglutinate the no-default and default orders in different clusters and minimize the number of mixed clusters, i.e. clusters with similar number of default and no-default orders. Having cluster distributions with a high or a low number of default orders facilitates the default or no-default decision of a new order, whereas having mixed clusters worsens the decision output.

In order to apply both strategies (agglutinate the no-default and default orders in different clusters and minimize the number of mixed clusters), an objective metric DPM that merges both concepts has been defined in Eq. 1 for evaluating the obtained clustering solutions by the K-means algorithm.

$$DPM = \frac{P_{nd}(\%)}{100 - \frac{\sum_i^{M_{do}} P_{dci}(\%)}{M_{do}}} \tag{1}$$

where P_{nd} refers to the percentage of clusters in which there is no default orders, P_{dci} represents the percentage of default orders per cluster i and M_{do} the number of cluster with default orders.

More specifically, the way the proposed DSSDP detects the default and no-default orders consists on associating default orders in the same cluster by maximizing the DPM metric in an exhaustive search process; the distribution that is obtained (with K-means) maximizes the percentage of non-default clusters together with the number of default orders per cluster. Therefore, an optimised cluster distribution will be one that contains a high number of clusters without default orders and clusters with a significant percentage of default orders.

Figure 2 depicts an example of cluster distribution for which the default probability is computed for each cluster.

Note that the proposed DSSDP approach tries to agglutinate the default orders in the same cluster while maximizing the number of non-default orders in the remaining clusters.

Step 1: Based on the defined metric, create the cluster distribution and compute the Default Probability (DP) of each cluster, being : DP= [75, 20, 0, 0, 12.5] %

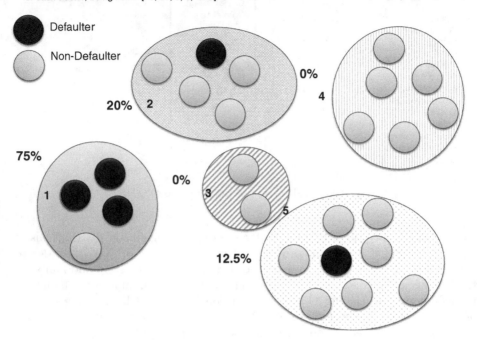

Fig. 2. Step 1 of the proposed DSSDP technique.

Step 2: Compute the default probability of new orders x_1, x_2, x_3=PD(x_1, x_2, x_3)=[75, 20,0] %

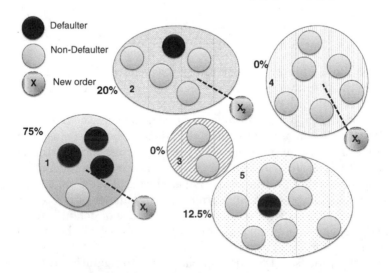

Fig. 3. Step 2 of the proposed DSSDP technique.

Step 3: Test the results of the new orders x_1, x_2, x_3 =PD(x_1, x_2, x_3)= [75, 20,0] %= [1 ,0 , 0]

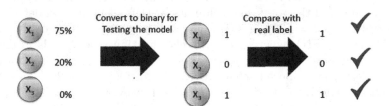

Fig. 4. Step 3 of the proposed DSSDP technique.

That being so, once the clustering distribution is settled, when a new order arrives (x_1, x_2 and x_3 in Fig. 3), the order is automatically associated to its closest centroid (in terms of Euclidean distance) from the defined clusters, and the risk of default given for the new order is the one associated to the cluster that is assigned (Fig. 4). Obviously, the accuracy of the risk of default prediction will highly depend on the homogeneity and optimal distribution of the clusters created by the proposed model.

Then, based on the risk probability of the cluster to which the new order is assigned and for comparison purposes, the proposed DSSDP approach gives as output 1 in case the new order is suggested to be a defaulter order and 0 otherwise. Next Sect. 4 presents the results obtained by applying the proposed DSSDP algorithm for Default prediction and compares the achieved results with other artificial intelligence method of the literature.

4   Simulation Results

In this section simulation results obtained by the proposed DSSDP are evaluated and compared towards another technique widely used in the literature for tackling this kind of decision problems, i.e. Random Forest (RF). In order to make a fair comparison a randomly chosen subset of test samples is separated and evaluated by means of the considered techniques along 20 Monte Carlo simulations. In both schemes the 80% of the data corresponds to training data and the 20% to testing data. Thus, among the total of 27439 orders that are kept in the ERP system that correspond to 4 years of operation of 4 small and medium-sized enterprises. 21951 orders correspond to training data and 5488 randomly chosen orders are used for testing the system. The developed DSSDP tool takes the necessary data from the ERP of the company and when a new order arrives it presents a risk value (0 (no-default order) or 1 (default order)) for the incoming order. The final results obtained by the proposed algorithm are then compared with the real label of the data (default or no-default order based on the ERP system information) in order to calculate the accuracy of the presented solution. Moreover, a Random Forest (RF) classification approach (for which balanced tests have been carried out) has been simulated and tested for comparison.

In order to measure the accuracy of both techniques (the proposed DSSDP and a RF) the confusion matrix has been computed in Eq. 2 for both approaches. As can be observed the proposed technique for Default prediction detects as no-default orders $5402 = 5399 + 3$, i.e. only 3 orders have been wrongly detected (False Negative - FN), whereas the RF classification scheme fails in detecting $FN = 15$ no-default orders. In terms of default prediction, among a total of 29 default orders, the proposed scheme is capable of identifying 26 (True Positive - TP) whereas the RF only detects $TP = 14$. However, the presented approach wrongly identifies 60 (False Positive - FP) as default orders and the RF only $FP = 5$. In this application domain, the false positives (FP) are not as important as detecting the default orders because the company is willing to know the highest number of possible default orders to take preventive actions which consist of asking for a cash advance, among others.

$$\begin{vmatrix} TN & FP \\ FN & TP \end{vmatrix} \quad DSSDP = \begin{vmatrix} 5399 & 60 \\ 3 & 26 \end{vmatrix} \quad RF = \begin{vmatrix} 5454 & 5 \\ 15 & 14 \end{vmatrix} \tag{2}$$

Table 1 depicts the comparison of the results obtained by the proposed DSSDP and RF by means of three widely known indicators: Precision, Recall and AUC (Area Under the ROC Curve), F_1 score, Specificity (SPC) and False Negative Rate (FNR). The Precision (P) in Eq. (3) is defined as the number of true positives (TP) over the number of true positives plus the number of false positives (FP). Recall (R) instead is defined as the number of true positives (TP) over the number of true positives plus the number of false negatives (FN) (see Eq. (4)). F_1 score in Eq. (5) represents a weighted average of the precision and recall with 1 as its best value and 0 the worst score. Specificity (SPC) in Eq. (6) measures the proportion of negatives that are correctly identified. It can be inferred but both techniques are able to identify a great proportion of non-default orders. However, in terms of False Negative Rate Eq. (7), i.e. the proportion of positives which yield negative outcomes with the test, is higher for the RF model 51.73% with respect to 10.35% of the proposed DSSDP approach.

Note that in this application scenario (unbalanced with 0.6% of default probability), the capability of detecting the default orders is more relevant than the prediction of no-default orders (Precision metric), since the company aims at detecting the greatest number of defaulters in order not to lose money.

$$P = \frac{TP}{TP + FP} \tag{3}$$

$$R = \frac{TP}{TP + FN} \tag{4}$$

$$F_1 = \frac{2 \cdot TP}{2 \cdot TP + FP + FN} = 2 \cdot \frac{P \cdot R}{P + R} \tag{5}$$

$$SPC = \frac{TN}{TN + FP} \tag{6}$$

$$FNR = \frac{FN}{FN + TP} = 1 - R \tag{7}$$

Table 1. Comparison of the mean results obtained by the proposed DSSDP approach and RF along 20 Monte Carlo simulations

	P (%)	R (%)	AUC [0–1]	F_1 [0–1]	SPC [0–1]	FNR (%)
RF	**73.68**	48.27	0.74	**0.58**	0.98	51.73
Proposed DSSDP	30.23	**89.65**	**0.94**	0.45	**0.99**	**10.35**

As shown in Table 1, an AUC of 0.94 is obtained by the proposed DSSDP technique in contrast to the 0.74 value achieved by the RF model. Thus, the proposed technique is capable of obtaining a higher true positive rate which translated to a highest default detection.

5   Concluding Remarks

The presented paper proposes a novel Decision Support System for assessing the default risk (named DSSDP) in small and medium-sized enterprises. Although the data set is really unbalanced, as default orders suppose just the 0.5% of the total of orders, the proposed technique is capable of ensuring a high number of default orders detection without maximizing the number of clusters with few elements. The DSSDP approach is then compared towards a naïve RF classification scheme that reduces the false negative rate at the expense of minimizing the default detection. Regarding the application domain, the main goal consists of detecting the default orders with the aim at taking some preventing actions. As this process does not translates in a greater investment of money at the expense of the company, the DSSDP approach formulated as an 'ad-hoc' solution for the considered default prediction problem achieves the best result in terms of default detection.

Acknowledgements. This work has been funded by the IG-201400315 INTEK-BERRI GAITEK Programme of the Basque Country Government (Spain).

References

1. Basel Committee on Banking Supervision. Studies on credit risk concentration. Bank for International Settlements Press & Communications, Basel, Switzerland
2. Angelinia, E., di Tollob, G., Rolic, A.: A neural network approach for credit risk evaluation. Q. Rev. Econ. Financ. **48**, 733–755 (2008)
3. Srinivasan, B.V., Gnanasambandam, N., Zhao, S., Minhas, R.: Domain-specific adaptation of a partial least squares regression model for loan defaults prediction. In: IEEE 11th International Conference on Data Mining Workshop (ICDMW), pp. 474–479 (2011)
4. Tsai, M.C., Lin, S.P., Cheng, C.C., Lin, Y.P.: The consumer loan default predicting model. An application of DEA'DA and neural network. Expert Syst. Appl. **36**, 11682–11690 (2012)

5. Khandani, A.E., Kim, A.J., Lo, A.W.: Consumer credit-risk models via machine-learning algorithms. J. Bank. Financ. **34**, 2767–2787 (2010)
6. Jin, Y., Zhu, Y.: A data-driven approach to predict default risk of loan for online peer-to-peer (P2P) lending. In: IEEE Fifth International Conference in Communication Systems and Network Technologies (CSNT), pp. 609–613 (2015)
7. llen, L., DeLong, G., Saunders, A.: Issues in the credit risk modeling of retail markets. J. Bank. Financ. **28**, 727–752 (2004)
8. Khashman, A.: Neural networks for credit risk evaluation: investigation of different neural models and learning schemes. Expert Syst. Appl. **37**, 6233–6239 (2010)
9. Yu, L., Wang, S., Lai, K.K.: Credit risk assessment with a multistage neural network ensemble learning approach. Expert Syst. Appl. **34**, 1434–1444 (2008)
10. Wang, Y., Wang, S., Lai, K.K.: A new fuzzy support vector machine to evaluate credit risk. IEEE Trans. Fuzzy Syst. **13**, 820–831 (2005)
11. Yu, L., Yue, W., Wang, S., Lai, K.K.: Support vector machine based multiagent ensemble learning for credit risk evaluation. Expert Syst. Appl. **37**, 1351–1360 (2010)
12. Shin, K.S., Lee, T.S., Kim, H.J.: An application of support vector machines in bankruptcy prediction model. Expert Syst. Appl. **28**, 127–135 (2005)
13. Altman, E.I., Saunders, A.: An analysis and critique of the BIS proposal on capital adequacy and ratings. J. Bank. Financ. **25**(1), 25–46 (2001)
14. Berger, A.N., Frame, S.W.: Small business credit scoring and credit availability. J. Small Bus. Manage. **45**(1), 5–22 (2007)
15. Berger, A.N., Udell, G.F.: A more complete conceptual framework about SME finance. J. Bank. Financ. **30**(11), 2945–2966 (2006)
16. Altman, E.I., Sabato, G.: Effects of the new basel capital accord on bank capital requirements for SMEs. J. Finan. Serv. Res. **28**(1–3), 15–42 (2005)
17. Pompe, P.P.M., Bilderbeek, J.: The prediction of bankruptcy of small- and medium-sized industrial firms. J. Bus. Ventur. **20**, 847–868 (2005)
18. Ciampi, F., Gordini, N.: Using economic financial ratios for small enterprise default prediction modeling: an empirical analysis. In: 2008 Oxford Business & Economics Conference Proceedings, pp. 1–21. Association for Business and Economics Research (ABER) (2008)
19. Ciampi, F., Gordini, N.: Default prediction modeling for small enterprises: evidence from small manufacturing firms in northern and central italy. Oxf. J. **8**(1), 13–29 (2009)
20. Ciampi, F., Gordini, N.: Small enterprise default prediction modeling through artificial neural networks: an empirical analysis of italian small enterprises. J. Small Bus. Manage. **51**(1), 23–45 (2013)
21. McQueen, J.: Some methods for classification and analysis of multivariate observations. In: Proceedings of the 5th Berkeley Symposium on Mathematics and Statistics, pp. 281–297 (1968)

Swarm Intelligence in Solving Stochastic Capacitated Vehicle Routing Problem

Jacek Mańdziuk[1,2]([⊠]) and Maciej Świechowski[3]

[1] Faculty of Mathematics and Information Science,
Warsaw University of Technology, Warsaw, Poland
mandziuk@mini.pw.edu.pl
[2] School of Computer Science and Engineering,
Nanyang Technological University, Singapore, Singapore
j.mandziuk@ntu.edu.sg
[3] Systems Research Institute, Polish Academy of Sciences, Warsaw, Poland
m.swiechowski@ibspan.waw.pl

Abstract. In this paper, the two most popular Swarm Intelligence approaches (Particle Swarm Optimization and Ant Colony Optimization) are compared in the task of solving the Capacitated Vehicle Routing Problem with Traffic Jams (CVRPwTJ). The CVRPwTJ is a highly challenging optimization problem for the following reasons: while the CVRP is already a problem of NP complexity, adding another stochastic layer to its definition (related to stochastic occurrence of traffic jams while traversing the planned vehicle routes) further increases the problem's difficulty by requiring that potential solution methods be capable of on-line adaptation of the routes, in response to changing traffic conditions. The results presented in the paper shed light on the underlying differences between ACO and PSO in terms of their suitability to solving particular instances of CVRPwTJ.

Keywords: Vehicle routing problem · Traffic jams · Particle swarm optimization · Ant colony optimization · Imperfect information

1 Introduction

Capacitated Vehicle Routing problem (CVRP) is a popular NP-complete optimization problem which consists in finding the set of routes of a minimum cumulative length (cost) for a given number of trucks that serve a given set of clients. All trucks start from and ultimately return to a pre-defined depot (with a certain 2D location). Each client is defined by its location on a plane and an amount of goods (a demand) to be delivered to them in one shot (a client cannot be served by multiple trucks). Each truck has some pre-defined capacity and all trucks (as well as goods to be delivered) are homogenous. In short, the problem combines the multiple-tour Traveling Salesman Problem with the Bin Packing Problem. For its formal definition please refer, for example, to [10].

© Springer International Publishing AG 2017
L. Rutkowski et al. (Eds.): ICAISC 2017, Part II, LNAI 10246, pp. 543–552, 2017.
DOI: 10.1007/978-3-319-59060-8_49

There are many approaches rooted in Operation Research or Computational Intelligence which can be applied to solve the CVRP (see, for instance, [10] for their overview).

In this paper, similarly to [9], the baseline problem formulation is extended by adding traffic jams which may occur on the edges (atomic parts of the routes) and therefore increase the cost of their traversal. This extension leads to the Capacitated Vehicle Routing problem with Traffic Jams (CVRPwTJ) specification. Due to highly dynamic nature of CVRPwTJ, the methods used to solving it must be able to swiftly adapt to the on-line changes in the cost function (the cost of currently planned routes) due to frequently changing traffic conditions.

The proposed idea of solving the CVRPwTJ is based on the concept of Swarm Intelligence (SI) which consists in having a population of simple objects that encode solutions to the problem in the search space. These objects iteratively communicate and influence each other, which enables them to modify the encoded solution. Each object has relatively simple rules and goals and the complexity of the system is an emergent feature resulting from maintaining a swarm as a whole. Two such metaheuristic SI methods are employed: Ant Colony Optimization [2] (which uses the notion of *an ant* as the baseline element of the swarm) and Particle Swarm Optimization (PSO) [4] (which refers to *a particle* as an atomic object).

2   ACO in CVRPwTJ

Our implementation of the ACO approach is inspired by a classical algorithm used to solve the Traveling Salesman Problem [2,3] which was adjusted to take into account the CVRP specificity [1,11] and furthermore the stochastic nature of the CVRPwTJ stemming from the existence of traffic jams in the problem definition. The approach was initially proposed and described in detail in our previous work [9] devoted to comparison of ACO and the Upper Confidence Bounds Applied to Trees (UCT) method [7]. In this section the ACO-based algorithm presentation is limited to introduction of its main components. For the full coverage and in-depth description of the method please refer to [9].

Assuming that the number of available trucks is equal to k, the initial solution is computed by the modified Clark and Wright (CW) *Savings algorithm* [13] and used to deposit the initial pheromone traits on the initial k routes. Then, in each time step of the algorithm, each ant seeks the solution for the remaining part of the problem (i.e., the complete routes for k vehicles) using the current solution as the starting point (state). Once the solution is found, its quality is evaluated based on the cumulative length of all k routes and the pheromone is deposited on the route's segments.

For each of the k routes, the ant starts its search in the current vehicle's position and looks for the next most suitable customer to be added to the route according to the current pheromone traits and considering the current (stochastic, due to the existence of TJ) cost of traversing particular edges. If the space left on the truck is not sufficient to serve any of the remaining customers or all edges from the current position to the remaining customers are jammed, the truck returns to the depot to accommodate the left customers within a new route.

A pseudo-roulette is used to select the next customer to be visited by an ant. The greedy selection (i.e., of the closest, in terms of dynamic cost, yet not visited client) takes place with probability 0.05. Otherwise, the roulette-wheel method is applied, which selects customer j while being currently at customer i with the following probability:

$$p_{ij} = \frac{\tau_{ij}^{\alpha} * \eta_{ij}^{\beta}}{\sum_{ij}(\tau_{ij}^{\alpha} * \eta_{ij}^{\beta})} \qquad \eta_{ij} = (BASE/d_{ij})^2 \qquad (1)$$

where τ_{ij} is pheromone amount deposited on edge e_{ij} and d_{ij} is the dynamic (traffic-aware) cost of traversing this edge at the moment. $BASE$ is a normalization factor equal to the length of the initial (static) solution. Coefficients α and β were set to 2 and 3, respectively, based a limited number of preliminary tests.

Once solutions are found by all ants in the current iteration, the pheromone deposit on the edge e_{ij} is incremented in the following way:

$$\Delta\tau_{ij} = \sum_a \delta_{ij}(BASE/D_a)^2 \qquad (2)$$

where D_a denotes a cost of solution $s(a)$ found by ant a, and δ_{ij} can take one of the three values: **0**, if $e_{ij} \notin s(a)$; **10**, if $e_{ij} \in s(a)$ but $s(a)$ is not the best overall solution; **20**, if $e_{ij} \in s(a)$ and $s(a)$ is the generally best current solution (among all solutions found by the ants in the current iteration).

The final step in the pheromone update procedure is the evaporation, which is defined at the level of 90% of the previous amount (due to high degree of system's dynamism) and then confined to the predefined interval $[\tau_{min}, \tau_{max}]$ by $Conf_{\tau_{min}}^{\tau_{max}}$:

$$\tau_{ij} := Conf_{\tau_{min}}^{\tau_{max}}(0.1 * \tau_{ij} + \Delta\tau_{ij}) \qquad (3)$$

Once the last iteration is completed by all ants in a given time step, the best overall solution is found and used to move the trucks one step ahead (to the next client) according to the schedule represented by this solution. At the beginning of the next step, the best solution is left out, the pheromone traits are reset, new TJ are distributed, and the system proceeds with solving the next step of the problem. Please note, that resetting pheromone traits after each main simulation step, i.e., when the new TJ are imposed on the routes, is indispensable, since the problem is highly dynamic and traces from the previous step are misleading. This necessity was fully confirmed in the preliminary simulations.

3   PSO in CVRPwTJ

In this section, the proposed approach to CVRPwTJ with the use of PSO metaheuristic is presented and discussed in detail.

3.1   Problem Encoding

We use one of the standard CVRP encodings presented in the literature [5,6], in which a population of M particles is maintained, and each particle is encoded as a vector of length N, where N is the number of yet unvisited customers. Each position in the vector is associated with a particular customer's ID. This association, i.e., $positionIndex \leftrightarrow customerID$ is maintained by means of a dictionary (see Sect. 3.2 for the details).

3.2   Initial Population

The modified CW algorithm [13] is used to obtain the initial solution to the static problem, i.e., a set of initial routes. The dictionary, which maps indices of the particle encoding vector to customers' IDs is populated in the following way:

```
int index = 0
for each route R in the initial solution
    for each customer C in R
        dictionary.Add(index, C)
        index = index+1
```

Whenever a customer is visited by a vehicle, the dictionary is updated in a way that it maintains the original order, but uses only indices that are smaller than the number of unvisited customers (i.e., $\{0, \ldots, N-1\}$). More precisely, if a given customer is visited, and consequently should be dropped from the remaining schedule, all subsequent customers are shifted to the left (assuming the schedule is sorted from left to right). The idea is depicted in Fig. 1.

In order to induce diversity within the swarm, only 20% of particles are dedicated to encoding the initial solution. The remaining 80% are initialized randomly and afterwards undergo the corrective procedure (c.f. Sect. 3.5) if needed, followed by a local optimization phase by means of 2-OPT algorithm [8].

3.3   Operational Scheme of the Proposed Method

The algorithm solves the problem iteratively, in discrete time steps. A pseudocode of one time step of the method is listed in Algorithm 1.

In each step (which corresponds to atomic movement of vehicles to their next assigned customers) the best solution from the previous step is reset and a series of MAX_{PSO} iterations is executed. For each particle, its velocity and new position are calculated (the details are presented in the next subsection) and then the vehicles' routes are decoded to a vehicle-centric representation that allows immediate analysis of the routes. Particles containing at least one route (one vehicle), which does not obey the maximum capacity constraint, are marked as invalid and undergo a corrective procedure (discussed in details in Sect. 3.5).

In the next step, for each valid particle (either initially or after correction), a 2-OPT local optimization procedure is executed. The best particle, i.e., the one

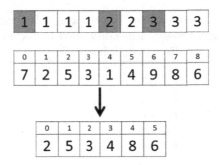

Fig. 1. The top row presents an encoding of a sample solution (3 vehicles and 9 unserved customers) stored in a particle. In the middle part a corresponding sample dictionary state is shown with the customers $(7, 2, \ldots, 8, 6)$ indexed from 0 to 8. The lowest encoding corresponds to the situation after the first customers assigned to vehicles 1, 2 and 3 (i.e., number 7, 1 and 9) have been visited and removed from the schedule. All the remaining customers are shifted to the left and their corresponding indices are renumbered to $(0, \ldots, 5)$.

having the lowest total cost of routes, is stored. Once the MAX_{PSO} iterations are over, the vehicles move to the next customers based on the encoding kept by the best particle.

3.4 Position and Velocity Update

The position of particle x at step $t + 1$ is updated according to the following equation:

$$x_{t+1}^i = (x_t^i + v_t^i) \bmod K \tag{4}$$

where x_t^i and v_t^i are the ith components of a particle's position at time step t and particle's velocity at time step t, respectively, and K is the number of available vehicles.

Particle's velocity is updated according to the following equation:

$$v_{t+1}^i = inertia * v_t^i + u_{[0;g]}^{(1)}(x_{gBest} - x_t^i) + u_{[0;l]}^{(2)}(x_{lBest} - x_t^i) \tag{5}$$

where *inertia* is a factor that specifies how much of the previous velocity is retained; x_gBest is the global best solution found so far; x_lBest is the local best solution (i.e., the one found by the current particle), g and l are global and local attractors, respectively, and $u_{[0;g]}^{(1)}$ and $u_{[0;l]}^{(2)}$ denote random variables drawn from the uniform distribution bounded by g and l, respectively.

3.5 Corrective Procedure

The *DecodeRoutes* procedure starts with an empty set of routes and iterates over the *vehicleIDs* stored in the particle encoding vector. For each *vehicleID*

Algorithm 1. A pseudocode of the PSO method. Procedures for *calculating velocity* and *updating position* are described in Sect. 3.4. *Decode routes* and *repair* procedures are discussed in Sect. 3.5.

```
1  set iteration := 0;
2  Reset BestParticle;
3  while iteration < MAX_PSO do
4      forall particle in Particles do
5          calculate velocity of particle;
6          update position of particle;
7          decode routes in particle;
8          if particle is not valid then
9              | repair particle;
10         end
11         if particle is valid then
12             run 2-OPT for particle;
13             if cost of particle < cost of BestParticle then
14                 | BestParticle := particle;
15             end
16         end
17     end
18 end
```

---

it consults the *dictionary* to identify the client associated with the current index and appends them to the route of vehicle *vehicleID*.

The above-described decoding process may create routes for which the total sum of customers' requests exceeds available capacity, in which case the corrective *Repair* procedure is executed. Please note, that the available capacity depends on two factors: the sum of all requests of customers that have already been visited and those who are planned to be visited – the corrective procedure can only influence the not yet visited customers. In the following description of the corrective procedure each vehicle with exceeded capacity will be referred to as ILV (*illegal vehicle*) and the remaining ones as LV (*legal vehicles*).

(1) First, ILVs are identified.
(2) For each ILV, a minimal number of customers, removal of whom would result in a highest capacity within the allowed limit, is identified. First, the algorithm tries to remove one customer starting from the one with the lowest requested amount. If it finds a legal situation, it stops. If it does not, it will investigate all pairs (again starting from the pair with the lowest cumulative requested amount), etc.
(3) The identified sets of customers for each ILV are candidates for transfer to other vehicles. Let us denote such sets of customers as $TSET_i$, where i is a vehicle's identifier.
(4) An *outerList* is created, which contains all vehicles, sorted in descending order, by cumulative amount of requests in $TSET_i$. The initially LV have the $TSET_i$ empty, therefore will be placed at the end of the list.

(5) An *innerList* is created, which contains all vehicles, sorted in descending order, with respect to their available space. In this step, the customers in $TSET_i$ are ignored.

(6) In a double loop possible transfers between two vehicles are investigated. The outer loop iterates over *outerList*, whereas the inner loop iterates over *innerList*. The transfer algorithm is described below. If a transfer leads to a legal situation (both vehicles' capacities are within the limit) it is applied.

(7) If there are still vehicles with exceeded capacity, the algorithm will allocate requests from the respective $TSET_i$ to new vehicles. If there are not enough available vehicles to allocate all requests, then the particle is marked as *invalid* in the current iteration of the PSO procedure. Such a particle may potentially be repaired in subsequent iterations.

The transfer algorithm in point (6) above investigates, in a fixed order, three possibilities of exchanging customers between two given vehicles (say k and l). It returns *success* as soon as it finds the first legal situation, i.e., capacities of both vehicles are below the limit.

(6a) Cross-pairing: customers from $TSET_k$ and $TSET_l$ are appended to the customers of vehicles l and k, respectively.

(6b) After performing the cross-pairing, all customers from a more loaded vehicle (by means of a sum of requests) are one-by-one tried to be transferred to the other vehicle starting from the biggest request.

(6c) After performing the cross-pairing, all pairs of customers, one per vehicle, are tried to be exchanged starting from the biggest requests.

4   Experimental Setup

Both methods are directly compared based on a set of widely-known benchmarks taken from the literature (see Sect. 4.2 for their exact selection). While the common benchmark instances are static (their definition does not include dynamic elements, such as traffic jams), they are extended to dynamic versions by adding, stochastically distributed, traffic jams. More precisely for each benchmark set, at each time step t a traffic jam of intensity I_t can be imposed on each edge with probability P, independently of other edges.

The following ranges of TJ intensity were tested: $P \in \{0.02; 0.05; 0.15\}$, $I_t = U_{INT}[10, 20]$, $L_t = U_{INT}[2, 5]$, where $U_{INT}[a, b]$ denotes random uniform selection of any integer x such that $a \leq x \leq b$ and $L_t(e)$ denotes a duration of a TJ.

For each of the three values of P and each benchmark set 50 pairwise independent distributions of TJ were samples and used in the experiments. Consequently, for both ACO and PSO we obtained 50 independent results (for pairwise the same sets of TJ distributions) which were subsequently averaged to yield the final score.

4.1 Steering Parameters

Both methods are used with the best parameterizations we were able to find. This methods' calibration was performed based on initial tests on 7 benchmarks and 30 trials per benchmark (please recall that the final experimental setup included 19 benchmarks, each tested 50 times).

ACO algorithm was run with a population of $max(100, 2n)$ ants (where n is the size of a benchmark set), for MAX_{ACO} iterations. MAX_{ACO} was set to 200 for benchmarks of size $n < 70$ and to 75 for benchmarks with $n \geq 70$.

PSO method was run for $MAX_{PSO} = 200$ iterations with the number of particles equal to 150 (for all benchmark sizes). The remaining steering parameters in Eq. (5) were set as follows: $inertia = 0.3$, $l = 0.3$, $g = 0.6$.

The above parameters, for both methods, were selected based the assumed reasonable time allotted for reaching the solution. Clearly, there is still possibility of improvement of results with bigger populations (either of ants or particles), but we believe that the current setup provides a good estimation of the general quality of both approaches, and what is more important, based on the execution times comparison it can be concluded that the selection is fair, i.e., not biased towards any of the two proposed and investigated approaches.

4.2 Benchmark Problems

A set of 19 benchmark instances for the static CVRP problem was downloaded from the webpage [12]. Dynamic traffic jams were added to these benchmarks according to the procedure described above in this section. In order to maintain diversity, those instances were chosen from five sets proposed by: Augerat et al. (3 instances of "type A" and 3 of "type P"); Christofides and Eilon (2 instances); Fisher (3 instances); Christofides, Mingozzi and Toth (2); Christofides (1), and Taillard (5). The number of customers in the selected benchmarks varies from 19 to 150 and the number of vehicles (routes) required to construct the initial solution is between 2 and 14. Moreover, the distributions of clients requests' sizes and their locations vary significantly from benchmark to benchmark.

5 Results

The results are presented in Table 1. First of all, a clear advantage of both proposed approaches over the static solution based approach can be observed. While this is an expected result, it is, nevertheless, worth noting that the improvement stemming from application of noise-adaptive methods (ACO, PSO), is quite significant, around 3–4 times, in most of the cases.

In a head-to-head comparison of both SI methods there is no clear winner, although ACO seems to be slightly more effective, in general, than PSO. In the summary of best results across all (instance, P) pairs ACO wins 31 cases compared to 26 wins of PSO (and none of Static). When it comes to stability (repeatability) of results the order is reversed: clearly the more stable method (with lower standard deviation) is PSO (46 wins out of 57 cases).

Table 1. The average values and standard deviations (in parentheses) across 50 trials. The **Static** column presents application of the initial solution (found at step 0, without any TJ imposed yet) applied to (a dynamic version of) a benchmark set. The best result for each pair (instance, P) are bolded.

P	Instance	Static (σ)	ACO (σ)	PSO (σ)	Instance	Static (σ)	ACO (σ)	PSO (σ)
0.02	P19	388.9 (214.9)	281.2 (46.9)	**251.8 (10.9)**	E76	1318.7 (295.5)	**746.0 (54.4)**	764.0 (**32.6**)
0.05	P19	612.0 (213.3)	**311.8 (93.1)**	326.1 (**23.1**)	E76	2130.0 (460.4)	**826.7** (159.9)	887.0 (**39.8**)
0.15	P19	1278.0 (358.9)	**391.2** (155.9)	541.3 (**61.3**)	E76	4536.7 (838.0)	**1037.2** (264.3)	1444.9 (**68.9**)
0.02	P45	1007.7 (326.2)	607.6 (53.9)	**590.9 (20.7)**	A80	2774.1 (625.2)	**1907.1** (146.7)	1937.9 (**58.4**)
0.05	P45	1759.6 (411.9)	**682.0** (74.4)	740.9 (**35.7**)	A80	4100.6 (830.1)	**2003.3** (442.6)	2383.4 (**121.9**)
0.15	P45	3299.5 (733.3)	**949.7** (281.7)	998.0 (**54.3**)	A80	9066.5 (1437.4)	**3161.8** (829.6)	3723.5 (**141.7**)
0.02	F45	1515.9 (703.3)	**761.5** (75.8)	771.1 (**33.7**)	Tai100a	3615.6 (849.4)	**2316.3 (227.3)**	3236.5(296.1)
0.05	F45	2078.7 (949.0)	**831.1** (159.3.)	836.4 (**77.5**)	Tai100a	5238.7 (1028.2)	**2875.4 (420.3)**	3390.6 (511.1)
0.15	F45	5060.3 (1519.1)	1138.8 (416.4.)	**1103.6 (186.4)**	Tai100a	11029.6 (1717.2)	4788.9 (**903.9**)	**3737.4** (953.4)
0.02	E51	989.2 (240.6)	**614.1 (40.0)**	637.0 (**22.2**)	Tai100b	3425.1 (938.7)	**2339.2 (337.3)**	2973.0 (388.8)
0.05	E51	1571.6 (386.3)	**650.1** (50.4)	778.9 (**40.5**)	Tai100b	5246.1 (1084.6)	3201.0 (**506.4**)	**3155.9** (586.9.)
0.15	E51	3509.7 (824.7)	**789.9** (174.8)	1215.3 (**70.2**)	Tai100b	10660.1 (1713.1)	5141.9 (**940.0**)	**3739.6** (953.4)
0.02	A54	1939.2 (542.7)	1338.7 (84.0)	**1260.5 (41.3)**	chmt100	792.3 (291.5)	469.3 (91.3)	**467.3 (27.2)**
0.05	A54	3072.4 (887.7)	1456.4 (286.0)	**1418.0 (77.9)**	chmt100	1178.7 (409.5)	**538.3** (103.9)	543.4 (**58.4**)
0.15	A54	6275.0 (1441.9)	**1829.0** (519.4)	1901.0 (**139.2**)	chmt100	2471.9 (549.1)	872.9 (319.8)	**730.0 (116.9)**
0.02	A69	2005.7 (531.8)	1395.9 (96.7)	**1297.4 (43.3)**	P101	1436.5 (262.6)	846.8 (69.2)	**809.1 (25.4)**
0.05	A69	3235.4 (644.1)	**1538.1** (294.5)	1668.3 (103.0)	P101 5	2552.0 (547.3)	**893.8** (127.3)	931.0 (**38.2**)
0.15	A69	6631.7 (1437.4)	**2096.4** (588.4)	2634.0 (**173.1**)	P101	5419.6 (801.5)	1375.0 (322.2)	**1327.7 (81.9)**
0.02	F72	445.3 (133.1)	292.3 (27.0)	**273.0 (12.2)**	F135	2062.4 (484.0)	2976.6 (125.3)	**1301.4** (261.2)
0.05	F72	712.9 (211.4)	424.2 (64.9)	**313.5 (43.3)**	F135	3390.7 (962.8)	3211.8 (**240.0**)	**1532.4** (256.6)
0.15	F72	1502.3 (279.6)	537.6 (119.3)	**455.5 (60.7)**	F135	6945.8 (1410.6)	**1936.6** (654.7)	2141.4 (**423.4**)
0.02	Tai75a	2781.6 (1015.3)	2257.1 (**276.1**)	**1819.0** (280.4)	C150D	1883.0 (368.8)	1297.0 (75.5)	**1202.2 (39.1)**
0.05	Tai75a	4036.8 (1095.0)	2447.8 (415.3)	**2213.0 (403.5)**	C150D	3099.1 (587.5)	**1392.5** (193.3)	1504.0 (**53.5**)
0.15	Tai75a	9281.3 (2008.2)	**3236.5** (873.8)	3918.8 (**830.7**)	C150D	6766.7 (840.1)	**1987.5** (492.4)	2226.6 (**110.6**)
0.02	Tai75b	2494.7 (1145.5)	2025.1 (**146.4**)	**1496.3** (293.9)	Tai150b	4994.7 (1165.5)	4367.5 (515.0)	**2790.4 (100.7)**
0.05	Tai75b	4578.9 (1283.6)	2209.2 (364.0)	**1930.1 (353.0)**	Tai150b	8751.9 (1936.7)	4834.3 (836.4)	**3201.0 (176.3)**
0.15	Tai75b	9108.3 (1752.7)	**2769.6 (777.7)**	3411.8 (804.7)	Tai150b	18104.0 (2581.2)	**7081.4** (1715.2)	7815.7 (**285.5**)
0.02	vrpnc75	817.1 (239.4)	627.6 (154.2)	**584.7 (58.6)**			-	
0.05	vrpnc75	1167.3 (349.9)	**635.4** (156.6)	741.2 (**141.2**)			-	
0.15	vrpuc75	2394.8 (653.4)	**898.2** (438.3)	1166.0 (**250.6**)			-	
Best result count		0 (0)	31 (11)	26 (46)			-	
Best P=2 count		0 (0)	6 (5)	13 (14)			-	
Best P=5 count		0 (0)	12 (3)	7 (16)			-	
Best P=15 count		0 (0)	13 (3)	6 (16)			-	

Closer examination reveals that PSO is better suited for the cases with lower amount of noise imposed by traffic jams ($P = 0.02$) with 13/19 of won cases, while ACO is superior for more noisy instances ($P = 0.15$) with exactly the same balance. For the mid-range traffic jams intensity ($P = 0.05$) the advantage is with the ACO approach, albeit, as stated above, in none of the cases is ACO stronger than PSO in terms of results' stability.

6   Conclusions

The paper compares the efficacy of two popular swarm-based methods (Particle Swarm Optimization and Ant Colony Optimization) in solving the Capacitated Vehicle Routing Problem with Traffic Jams. To this end a new approach to CVR-PwTJ relying on the PSO algorithm has been proposed and experimentally compared with the ACO-based method proposed by the authors in their previous paper [9]. Experimental results presented in this study lead to the three following conclusions: firstly, the use of swarm-based methods (either ACO or PSO) significantly improves the results compared to static (non-adaptive) approaches; secondly, ACO seem to be slightly superior than PSO (at least in the context of the particular benchmark selection), but at the same time the results yielded by PSO have much lower variance; thirdly, for the cases of relatively low amount of noise (by means of stochastic traffic jams) in the CVRPwTJ instance the preferable method is PSO, while with more dynamic situations (higher amount of noise) the ACO system manifests its upper-hand.

References

1. Bell, J.E., McMullen, P.R.: Ant colony optimization techniques for the vehicle routing problem. Adv. Eng. Inf. **18**(1), 41–48 (2004)
2. Dorigo, M.: Optimization, learning and natural algorithms. Ph.D. thesis, Politecnico di Milano (1992)
3. Dorigo, M., Gambardella, L.M.: Ant colonies for the travelling salesman problem. BioSystems **43**(2), 73–81 (1997)
4. Kennedy, J., Eberhart, R.: Particle swarm optimization. In: Proceedings of IEEE International Conference on Neural Networks, vol. 4, pp. 1942–1948 (1995)
5. Khouadjia, M.R., Talbi, E.G., Jourdan, L., Sarasola, B., Alba, E.: Multi-environmental cooperative parallel metaheuristics for solving dynamic optimization problems. J. Supercomputing **63**(3), 836–853 (2013)
6. Khouadjia, M.R., Alba, E., Jourdan, L., Talbi, E.-G.: Multi-swarm optimization for dynamic combinatorial problems: a case study on dynamic vehicle routing problem. In: Dorigo, M., et al. (eds.) ANTS 2010. LNCS, vol. 6234, pp. 227–238. Springer, Heidelberg (2010). doi:10.1007/978-3-642-15461-4_20
7. Kocsis, L., Szepesvári, C.: Bandit based monte-carlo planning. In: Fürnkranz, J., Scheffer, T., Spiliopoulou, M. (eds.) ECML 2006. LNCS, vol. 4212, pp. 282–293. Springer, Heidelberg (2006). doi:10.1007/11871842_29
8. Lin, S.: Computer solutions of the traveling salesman problem. Bell Syst. Tech. J. **44**(10), 2245–2269 (1965)
9. Mańdziuk, J., Świechowski, M.: Simulation-based approach to vehicle routing problem with traffic jams. In: 4th IEEE Symposium on Computational Intelligence for Human-Like Intelligence, pp. 1–8. IEEE, Athens (2016)
10. Mańdziuk, J., Żychowski, A.: A memetic approach to vehicle routing problem with dynamic requests. Appl. Soft Comput. **48**, 522–534 (2016)
11. Mazzeo, S., Loiseau, I.: An ant colony algorithm for the capacitated vehicle routing. Electr. Notes Discrete Math. **18**, 181–186 (2004)
12. NEO: Networking and Emerging Optmization (2013). http://neo.lcc.uma.es/vrp/vrp-instances/capacitated-vrp-instances/
13. Pichpibul, T., Kawtummachai, R.: An improved clarke and wright savings algorithm for the capacitated vehicle routing problem. Sci. Asia **38**(3), 307–318 (2012)

LSTM Recurrent Neural Networks for Short Text and Sentiment Classification

Jakub Nowak[1], Ahmet Taspinar[2], and Rafał Scherer[1(✉)]

[1] Computer Vision and Data Mining Lab,
Institute of Computational Intelligence, Częstochowa University of Technology,
Al. Armii Krajowej 36, 42-200 Częstochowa, Poland
{jakub.nowak,rafal.scherer}@iisi.pcz.pl
[2] Data Scientist at CGI Nederland, Rotterdam, Netherlands
info@ataspinar.com
http://iisi.pcz.pl, http://ataspinar.com/

Abstract. Recurrent neural networks are increasingly used to classify text data, displacing feed-forward networks. This article is a demonstration of how to classify text using Long Term Term Memory (LSTM) network and their modifications, i.e. Bidirectional LSTM network and Gated Recurrent Unit. We present the superiority of this method over other algorithms for text classification on the example of three sets: Spambase Data Set, Farm Advertisement and Amazon book reviews. The results of the first two datasets were compared with AdaBoost ensemble of feed-forward neural networks. In the case of the last database, the result is compared to the bag-of-words algorithm. In this article, we focus on classifying two groups in the first two collections, since we are only interested in whether something is classified into a SPAM or an eligible message. In the last dataset, we distinguish three classes.

Keywords: Recurrent neural networks · Long Short-Term Memory · Bidirectional LSTM · Gated Recurrent Unit · Text classification · Sentiment classification

1 Introduction

Human communication is a very complex and complicated process. One of the most important elements in verbal communication are speech sounds combined in words and sentences (phrases). We use dictionaries, i.e. collections of words describing our surroundings, feelings and thoughts. Very important is their order and context. In the paper we classify whole sentences into appropriate groups using recurrent neural networks (RNNs) [18], namely Long Short Term Memory (LSTM) [14,17], its Birecursive LSTM (BSLTM) variant and Gated Recurrent Unit (GRU). LSTM were developed by Sepp Hochreiter and Jürgen Schmidhuber. The first observations on this subject were made by Sepp Hochreiter in his 1991 thesis but the beginning of LSTM was in 1997 (see [14]). Their recurrent network performed very well in comparison with previous RNNs, mainly thanks

© Springer International Publishing AG 2017
L. Rutkowski et al. (Eds.): ICAISC 2017, Part II, LNAI 10246, pp. 553–562, 2017.
DOI: 10.1007/978-3-319-59060-8_50

to partial solving vanishing gradient problem [13] in RNNs. In the paper we use a variant with an additional modification, so called "peephole connections", proposed by Gers and Schmidhuber [8,10] with an additional link between the gate layers and the cell state. There are also other LSTM modifications with one especially interesting described in [7], where the Gated Recurrent Unit (GRU) was used [5,14]. This is a truncated LSTM version without the output gate. This translates into faster network learning but it is recommended to use the classic LSTM network for more complex datasets. According to [22], GRU in some cases performs slightly better then LSTM.

The characteristic feature of the RNN network are gradual calls of the network over time. We do not provide all the input features for a single network call, as is the case in FNNs [2,4] or convolutional networks. In the case of RNNs, we gradually modify the final result. Intuitively, we can compare it to saying consecutive words in a sentence. Each subsequent word modifies the final meaning. In the case of these networks, we can also preserve the order of words in context. If we would like to use FNNs we would have to provide a whole sentence in one go. No matter if the sentence started with one word or another, it is difficult to implement such a FNN with the possibility of marking the occurrence of words. Thus, LSTMs are better in sentence content classification. The advantages of LSTM will be further described in the next chapter. Recursive networks are very widely used. Examples can be found in modelling simple physical processes such as ball bouncing [21]. Their greatest possibilities are seen in the translation of the text [1]. Every call is analyzed and appropriate word is selected in another language. If we use them for classification we are only interested in the result of the last network call.

2   Long Term Short Term Memory Networks

LSTM networks have their origins in RNN from 1980s. Their architecture allows for the accumulation of information during operation and uses feedback to remember previous network call states. It is often possible to find information that FNN networks are like RNN networks with only one call. This statement has a lot of truth in itself since their basic principles are similar to classic neural networks. Some mechanisms used in FNN are also applicable to recurrent networks. Classical recursive networks have a number of disadvantages [3] which narrows their ability to solve more complex problems. The community started to give them a special attention again with the advent of LSTM as it turned out that this technique performs much better than classical recursive networks. Detailed description of the architecture can be found for example in [17]. When using this structure, we activate the same cell every time, modifying the state of its internal structure. The most important elements of LSTM cells are

- cell state – the state of the cell passed in sequence after to the next steps,
- forget gate – the gate that decides what information should be omitted,
- input gate – a gate that decides what should be forwarded to the next activation.

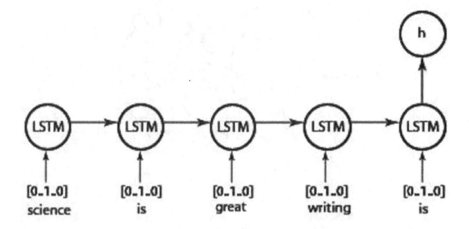

Fig. 1. Example of LSTM network call for text classification.

In short, we care about remembering some information that is crucial for the final result and it is important to have some information omitted during the operation of the network, as not everything affects positively the network performance. A general network model used in this article is shown in Fig. 1. Each circle represents an LSTM cell. The input is given as a one-hot vector representing one word. Output h will return the result we are interested in.

2.1 Bidirectional LSTM

A single LSTM layer can, however, be too small in some examples. In the experiment we have checked whether the use of two LSTM layers gives better results. This model differs from the previous one in that the content is both propagated forward and backward through the network. Unlike in FNN it is not backpropagation learning. Originally, this structure is taken from [12]. The bidirectional network example is shown in Fig. 2.

2.2 Gated Recurrent Unit

Gated Recurrent Unit (GRU) was first described by Cho et al. in [7] and has a lot in common with LSTM. Forget and input gates are merged into a single "update gate", so was the cell state and hidden state. A more accurate comparison of LSTM to GRU can be found in [6]. In both cases, they are RNNs where the most important is their last state, which has the most influence on the current state of cells. Most importantly, in the case of GRU, control is done through the update gate.

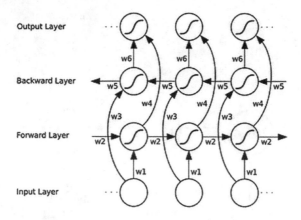

Fig. 2. Bidirectional LSTM [11]

3 Experiments

3.1 Experimental Setup

We used three datasets, i.e. Spambase Data Set (1999), Farm Advertisement (2011) and Amazon book reviews Data Set (2016). The SMS spam dataset contains a set of labeled messages that are collected from a UK forum. The data set consists of 13.4 % spam messages and 86.6 % eligible messages [19]. A second database was collected from text advertisements found on 12 websites that deal with various farm animal related topics. There are 53.3 % accepted messages and 46.7 % rejected messages [12]. The last dataset contains 213.335 book reviews for 8 different books. The experiment was limited to five books. In selected books the amount of negative and neutral opinions was higher. For this experiment, the opinions are divided to three classes and presented in 1–5 scale. Score 1 and 2 are labeled as negative, score 3 represents a neutral comment, 4 and 5 represent a positive comment. From the set we took only comment titles, farther content were not considered. The aim was to classify the content as negative, neutral or positive on the basis of the opinion title. Table 1 contains the percentage distribution of opinions of selected books. The number of negative and neutral opinions was doubled for training to reduce imbalance with the number of positive opinions.

In the this section we compare the results of RNNs with the bag-of-word algorithm which consists in choosing characteristic words for a given class. We assume that some words give more meaning to a given sentence, and if the number of occurrences of the characteristic words for the assumed class is large then we associate the sequence with this class. We can use various textual features [16]. The SPAM SMS problem is still very important as despite the age of technology we still obtain promotional offers of various products or services. In the case of short messages the spam filter should be almost faultless as it is very

Table 1. A summary of the Amazon dataset

Title	Bad	Neutral	Good	Total number of texts
The martian	2.31 %	13.34 %	84.35 %	22571
The goldfinch ·	16.22 %	23.57 %	60.22 %	22863
Fifty shades of grey	36.2 %	7.65 %	56.24 %	32977
Gone girl	22.64 %	11.94 %	65.42 %	41974
The girl on the train	19.92 %	11.2 %	68.87 %	37139

important for users to receive all legitimate messages. The last case considers a problem of book description and evaluation.

3.2 Grammar and Punctuation in Sequences

All data were normalized in the way that grammar and punctuation did not have influence on the results. It was important to obtain only a pure word stream which was better for neural network training. Namely, we took the following assumptions:

- only small letters were used,
- all special marks were deleted (e.g. . , %, ?),
- all numerical data were replaced by "SPEC-NUM" mark,
- for every dataset we created a special dictionary of unique words.

3.3 Word Representation

We created dictionaries of unique words for all three datasets without ordering them. During training, the data from the dictionaries were used to modify LSTM on the basis of only one word, (similarly to [15]) and an example is shown in Fig. 1. The number of input vector elements is equal to the number of words in the dictionary. Every word is represented by one vector value. We create as many input vectors as the number of words in one sequence. In this case the sequence is one sentence, one set of words, for which the classification result will be expected. For a FNN network it is possible to conclude the whole sequence by using one vector. All used words were represented by "1" (binary TRUE) and unused words "0" (binary FALSE). For RNN it is usually necessary to create a so-called one-hot vector where the value will be available only for one word what is the simplest possible method of representing input data.

3.4 Training

LSTM network learning is a specific process. The idea of recursive networks described earlier does not imply a fixed length of phrase and hence the number of network calls. In this case, we are required to use a special learning method called back propagation through time (BPTT) [9,23]. This algorithm was created with the intention of recursive networks. It is derived from the classical BP algorithm for FNNs except that the gradient is propagated backwards by all network calls, which entails a high memory requirement. For the first dataset (SPAM SMS) it was not so much important as sentences were not too long, so the the network was unfolded reasonable number of times. Much more demanding was the case with the Farm Advertisement dataset as the dictionary was much longer and so were the sentences. With 2 GB memory on the NVIDIA GeForce GT 740M GPU, the calculation ends with an error of insufficient memory available. To solve this problem, learning in this case was transferred to the Intel i5 2.6 GHz CPU with 12 GB RAM, however the memory was not fully utilized in this experiment. The network was implemented and trained using Microsoft CNTK [11] with the SGD (Stochastic Gradient Descent) algorithm.

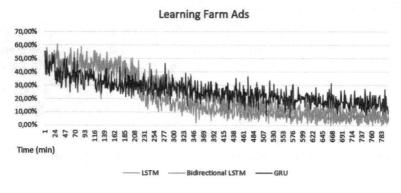

Fig. 3. Learning error.

The learning process can be compared on the example of Farm Ads. The internal parameters of the cells have been set to comparable values. Figure 3 shows that GRU achieved the worst result, but it was achieved in 346 min of the learning process. After this time, it oscillated around the constant value. Unlike in the case of GRU, LSTM and BLSTM needed a longer learning time, but the training error become lower. In this case, the graph shows that BLSTM after 277 min had smaller error than the LSTM.

4   Results

Graph in Fig. 4 shows an example of LSTM operation on one of the used datasets. For the RNN network, we can see the classification result after consecutive words in the sequence. Each line represents the result returned by the network. In this case, the sequence can be classified into three different classes. It is important to note how certain words affect the result, for example "cool" greatly increases the positive and the neutral outputs and sets low in the "bad" output.

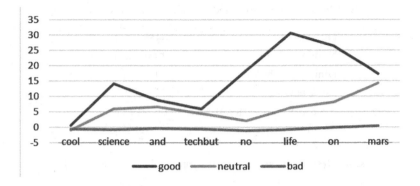

Fig. 4. An example of the incremental LSTM operation.

4.1   Results All Datasets

The results of the experiment are presented in Tables 2, 3 and 4. The result from the first dataset is very satisfactory (Table 2). The dictionary we worked on contained only 9,000 words as during our empirical tests it turned out that increasing this vocabulary did not improved the network performance and only training time was increased. In the case of the Amazon book reviews dataset (Table 4) we created one dictionary for all books. It should also be noted that GRU is not a good choice for classification based on very short phrases such as book titles and is better suited for simple datasets such as for example the SMS SPAM dataset.

Table 2. Results obtained on the SPAM Collection dataset.

Dictionary size	9054
Number of outputs	2 (spam, ham)
Dimension, hidden layer	140
Accuracy, LSTM	99.798 %
Accuracy bidirectional LSTM	99.834 %
Accuracy GRU	**99.945 %**

The results obtained on the first two datasets can be compared to ones from [19]. Table 5 presents results from the literature and our results. It was more difficult to compare the outcome of the experiments on the third dataset. At the time of writing, the only available results were based on [20] where the bag-of-words model classifies whether the opinion is negative or positive with 60 % accuracy. In our case, we tried to classify into three different classes and in the best case we obtained 86.4 % accuracy. We worked on the comment titles whereas in [20] whole comments were analyzed. Nevertheless, both LSTM and BLSTM resolved

Table 3. Results obtained on the Farm Ads dataset.

Dictionary size	54856
Number of outputs	2 (spam, ham)
Dimension, hidden layer	80
Accuracy, LSTM	94.497 %
Accuracy bidirectional LSTM	**96.017 %**
Accuracy GRU	87.521 %

Table 4. Results obtained on the Amazon book reviews dataset.

Dictionary size	16201
Number of outputs	3 (good, neutral, bad)
Dimension, hidden layer	140
Accuracy, LSTM	84.415 %
Accuracy bidirectional LSTM	**86.4 %**
Accuracy GRU	75.821 %

Table 5. Results of classification accuracy comparison of various algorithms on two datasets.

Metod	SPAM Collection	Farm Ads
LSTM	99.798 %	94.497 %
Bidirectional LSTM	99.834 %	96.017 %
GRU	99.945 %	87.521 %
FNN [19]	98.58 %	94.34 %
AdaBoosting [19]	98.98 %	84.03 %

the problem more accurately. However, LSTM training consumes more computational resources and requires more time than the bag-of-words algorithm.

5   Conclusion

Proper understanding of written texts by the reader is sometimes difficult if the author does not express them clearly. Computer text undestanding is even more challenging. Attempts that use e.g. the bag-of-word algorithm are not yet fully satisfatory. The paper shows that much better solutions are recursive neural networks, especially in both LSTM and BLSTM form. In some cases it is also possible to use GRU. Classical FNNs generally give slightly worse results and only in the case of the Farm Ads dataset they are very close to LSTM.

In the Amazon book reviews dataset there are texts that are classified as negative as well as positive. In this case, the full comment description is necessary, limiting to the review titles only does not give perfect results and achieving

100 % correct classification is not possible. It also important to take into account that RNN networks do not handle well very long phrases. Some problems appear already during the learning process, i.e. large memory requirements and a large value of error after more network calls. Moreover, in a long sentence, its first part can represent one class and the second part another one.

LSTM, BLSTM and GRU seems to be very similar constructions but as it has been shown in the experiments that GRU is well suited in simple cases where we want to obtain fast results. Advantages of GRU are faster training as they require less number of epochs to obtain the final result. Results achived by LSTM and BLSTM networks are more precise, but it is necessary to spend more time for the training process. GRU does not work with very short phrases such as in the case of third dataset. Each of the presented configurations has advantages and disadvantages and the type of dataset should be taken into account when the decision on the choice of the RNN network model is made.

References

1. Auli, M., Galley, M., Quirk, C., Zweig, G.: Joint language and translation modeling with recurrent neural networks. In: Proceedings of the 2013 Conference on Empirical Methods in Natural Language Processing, Seattle, Washington, USA, pp. 1044–1054, October 2013 Association for Computational Linguistics (2013)
2. Bas, E.: The training of multiplicative neuron model based artificial neural networks with differential evolution algorithm for forecasting. J. Artif. Intell. Soft Comput. Res. **6**(1), 5–11 (2016)
3. Bengio, Y., Simard, P., Frasconi, P.: Learning long-term dependencies with gradient descent is difficult. IEEE Trans. Neural Netw. **5**(2), 157–166 (1994)
4. Bertini Junior, J.R., Nicoletti, M.D.C.: Enhancing constructive neural network performance using functionally expanded input data. J. Artif. Intell. Soft Comput. Res. **6**(2), 119–131 (2016)
5. Britz, D.: Recurrent neural network tutorial, part 4 - implementing a GRU/LSTM RNN with python and theano. http://www.wildml.com/. Accessed 27 Oct 2015
6. Chen, C., Xia, L.: Recurrent neural network and long short-term memory. http://ace.cs.ohiou.edu/~razvan/courses/dl6890/presentations/lichen-lijie.pdf
7. Cho, K., Van Merriënboer, B., Gulcehre, C., Bahdanau, D., Bougares, F., Schwenk, H., Bengio, Y.: Learning phrase representations using RNN encoder-decoder for statistical machine translation (2014). arXiv preprint arXiv:1406.1078
8. Gers, F.A., Schmidhuber, E.: LSTM recurrent networks learn simple context-free and context-sensitive languages. IEEE Trans. Neural Netw. **12**(6), 1333–1340 (2001)
9. Gers, F.A., Schmidhuber, J., Cummins, F.: Learning to forget: continual prediction with LSTM. Neural Comput. **12**(10), 2451–2471 (2000)
10. Gers, F.A., Schraudolph, N.N., Schmidhuber, J.: Learning precise timing with LSTM recurrent networks. J. Mach. Learn. Res. **3**, 115–143 (2002)
11. Graves, A.: Supervised sequence labelling. In: Supervised Sequence Labelling with Recurrent Neural Networks, pp. 5–13. Springer (2012)
12. Graves, A., Schmidhuber, J.: Framewise phoneme classification with bidirectional lstm and other neural network architectures. Neural Netw. **18**(5), 602–610 (2005)

13. Hochreiter, S.: The vanishing gradient problem during learning recurrent neural nets and problem solutions. Int. J. Uncertainty Fuzz. Knowl. Based Syst. **6**(02), 107–116 (1998)
14. Hochreiter, S., Schmidhuber, J.: Long short-term memory. Neural Comput. **9**(8), 1735–1780 (1997)
15. Huang, Z., Xu, W., Yu, K.: Bidirectional LSTM-CRFmodels for sequence tagging (2015). arXiv preprint arXiv:1508.01991
16. Murata, M., Ito, S., Tokuhisa, M., Ma, Q.: Order estimation of Jaanese paragraphs by supervised machine learning and various textual features. J. Artif. Intell. Soft Comput. Res. **5**(4), 247–255 (2015)
17. Olah, C.: Understanding LSTM networks, August 2015. http://colah.github.io/posts/2015-08-Understanding-LSTMs
18. Patgiri, C., Sarma, M., Sarma, K.K.: A class of neuro-computational methods for assamese fricative classification. J. Artif. Intell. Soft Comput. Res. **5**(1), 59–70 (2015)
19. Shuang Bi, W.Z.: Cs294-1 final project algorithms comparison deep learning neural network — adaboost — random forest. http://bid.berkeley.edu/cs294-1-spring13/images/0/0d/ProjectReport(Shuang_and_Wenchang).pdf. Accessed 15 May 2013
20. Taspinar, A.: Sentiment analysis with bag-of-words. http://ataspinar.com/2016/01/21/sentiment-analysis-with-bag-of-words/. Accessed 21 Jan 2016
21. Trask, A.: Anyone can learn to code an LSTM-RNN in python. https://iamtrask.github.io/2015/11/15/anyone-can-code-lstm/. Accessed 15 Nov 2015
22. Trofimovich, J.: Comparison of neural network architectures for sentiment analysis of Russian tweets, 1–4 June 2016
23. Williams, R.J., Zipser, D.: Gradient-based learning algorithms for recurrent networks and their computational complexity. In: Backpropagation: Theory, Architectures, and Applications, vol. 1, pp. 433–486 (1995)

Categorization of Multilingual Scientific Documents by a Compound Classification System

Jarosław Protasiewicz[✉], Marcin Mirończuk, and Sławomir Dadas

National Information Processing Institute, Warsaw, Poland
{jaroslaw.protasiewicz,marcin.mironczuk,slawomir.dadas}@opi.org.pl

Abstract. The aim of this study was to propose a classification method for documents that include simultaneously text parts in various languages. For this purpose, we constructed a three-leveled classification system. On its first level, a data processing module prepares a suitable vector space model. Next, in the middle tier, a set of monolingual or multilingual classifiers assigns the probabilities of belonging each document or its parts to all possible categories. The models are trained by using Multinomial Naïve Bayes and Long Short-Term Memory algorithms. Finally, in the last component, a multilingual decision module assigns a target class to each document. The module is built on a logistic regression classifier, which as the inputs receives probabilities produced by the classifiers. The system has been verified experimentally. According to the reported results, it can be assumed that the proposed system can deal with textual documents which content is composed of many languages at the same time. Therefore, the system can be useful in the automatic organizing of multilingual publications or other documents.

Keywords: Multilingual text classification · Compound classification system · Multinomial Naïve Bayes · Long Short-Term Memory

1 Introduction

Nowadays, we are deluged by data coming from various sources that are spread electronically by new media, especially the Internet. Since people cannot cope manually with such variety and quantity of data, it is advisable to apply machine learning methods in modern information systems to provide them appropriately selected information. More specifically, the challenge is to organize and ensure quick access to scientific publications, which may be written in many languages. We want to classify a publication document into the multi-class topics such as Applied Sciences, Arts & Humanities, Economic & Social Sciences, General, Health Sciences, Natural Sciences.

There are many approaches to text classification. When input documents are in the same language, there is a monolingual problem. On the other hand, when entire documents are written in various languages, there is a multilingual

© Springer International Publishing AG 2017
L. Rutkowski et al. (Eds.): ICAISC 2017, Part II, LNAI 10246, pp. 563–573, 2017.
DOI: 10.1007/978-3-319-59060-8_51

classification task, which further can be divided into poly-lingual and cross-lingual issues (for more, see Sect. 2). Nonetheless, we assume that documents under examination may consist of text parts in many languages at the same time, e.g., a publication may comprise a title and keywords concurrently in English, Spanish and Polish, an abstract in English and Polish, and the rest only in Polish. Such classification problem was not discussed in the literature yet.

Our hypothesis is that a compound system containing (i) a pre-processing layer, (ii) monolingual or multilingual classifiers of whole documents or their parts, (iii) and a multilingual decision module can classify before-mentioned documents sufficiently, and in this way, may help to organize multilingual publications. We have designed and performed a series of experiments using the proposed compound system and single classifiers as a baseline. The tests have been carried out using our dataset of publications.

We have to mention that are some works exploring the concept of monolingual classifiers used to the categorization of multilingual documents. However, rather than introduce a decision module like in our approach, they use a language selection procedure to choose a suitable monolingual model [6] or select the final category as a (weighted) sum of outputs or output decisions [7]. We have to admit that our study is inspired by the idea of a system integrating various methods for better classification of multilingual resources [11].

This study is an extended version of our previous work [16]. It contains additional experiments, which are important to discuss all advantages and disadvantages of proposed methods of multilingual classification. More precisely, we compare (i) Multinomial Naïve Bayes (MNB) vs. Long Short-Term Memory (LSTM), (ii) a separate multilingual model vs. a compound system, and (iii) a decision module, which integrates multilingual models based on document parts vs. monolingual models based on entire documents. The article brings significant and additional findings in comparison to the previous work.

The paper is structured as follows: Sect. 2 covers the review of related works; Sect. 3 discusses the proposed methods; Sect. 4 contains experiments and their results; finally, conclusions and references are included.

2   Related Works

In this section, we briefly analyze works concerning the text classification issues with the particular focus on multilingual approaches.

The text classification problem is widely recognizable in the literature. According to a comprehensive literature review [10] including the period since 1997 to 2012, the most popular machine learning algorithms applied to this task are as follows: Support Vector Machines (SVM), k-Nearest Neighbors (KNN), Naïve Bayes (NB), Artificial Neural Networks (ANN), Rocchio, and Association Rule Mining (ARM). Whereas, features in input data are the most often selected or transformed by the following methods: Chi-squared test (CHI), Information Gain (IG), Mutual Information (MI), Latent Semantic Indexing (LSI) or Singular Value Decomposition (SVD), Document Frequency (FD), and others. Almost all works use Vector Space Model (VSM) to represent documents.

We have to note that the above review concerned mainly monolingual problems, whereas multilingual classification tasks are becoming more common nowadays. There are three main approaches to this issue, namely (i) cross-lingual translation, (ii) cross-lingual Esperanto, and (iii) poly-lingual [17].

The cross-lingual Esperanto approach is when all documents are translated into a chosen language. On the other hand, the cross-lingual translation (multiview) approach is when a labeled set is available only in one language, and it needs to be translated into other languages [8,17,20]. A monolingual classifier is trained on original documents and their translated version [1], however, these two steps may be merged to one by the use of the M1 statistical word alignment model [14]. Since machine translation is an error-prone process, there are proposed various strategies. It is easy to notice that a document in the original language and its translated version represent the same object. Thus, two separated classifiers, each trained on the different language, should produce similar results. Minimization by the gradient descent algorithm of train losses on both datasets may improve the performance of a classifier trained on the translated documents [8]. An objective function to minimalize may take into account misclassification of monolingual classifiers and disagreement of different views (monolingual classifiers) on the same document [1]. Instead of the minimization approach, a final decision in a multiview environment may be taken by a majority voting procedure of outputs from the Gibbs classifier [3].

Additional views added by machine translation improve the quality of classification [3]. For instance, if a classifier was trained on translated dataset from another manually labeled set, it is possible to improve its performance by using the expectation–maximization procedure, i.e., unknown labels are guessed in an expectation phase, whereas estimates of final classifier parameters are found in a maximization stage [17]. Instead of multi-view approaches, multilingual classification may utilize ontologies, e.g., the concept of region ontology mapping make use of semantic networks of ontologies, where an entire region is taken into account rather than an individual concept [5].

The poly-lingual approach is when labeled training sets are available for each considered language [2,19]. The main idea of multilingual classification is to develop a universal algorithm (not a model) that can deal with documents in various languages. The crucial issue is to use such representation of documents that is language independent [19]. A self-organizing map may automatically arrange multilingual documents into monolingual clusters producing a keyword cluster map and a document cluster map as well as associations between nodes and hierarchy [21]. Another scenario is when the same documents are available concurrently in many languages, but some of them are labeled, whereas most of them are not. In this case, for each language can be trained a monolingual model and then disagreement of the models on the unlabelled set may be minimized [2].

3 Methods

In this section, we define the problem to solve and discuss the proposed approaches to multilingual classification.

3.1   The Problem to Solve

Let us consider documents (publications) comprising text parts concurrently in various languages. A document d may be composed of several text parts f that are substantially different from each other. These parts could be for instance a title, an abstract or an introduction, a chapter, references, and so on. All parts may appear many times in many languages in one document. For instance, an introduction may be included twice, once in English and once in French, but all chapters may be published only in Chinese, whereas a title could be presented in English, French, German, and Chinese.

More precisely, let us define a set of publications D. Each publication is considered as a separate document d_i, $i = 1, 2, ..., I$, where I is the number of documents. We assume that each document d_i may contain several parts $f_i^{t,l}$, $t = 1, 2, ..., T$, where t is the type of a part, l is its language, and i is the document index. Additionally, we should be aware that there is no guarantee that each part type $f_i^{t,l}$ exists in each document. Our task is to assign the documents $d_i \in D$ into relevant to them classes (scientific domains) $c_j \in C$, where $j = 1, 2, ...J$ is the class index, and J is the number of classes [16].

3.2   Training Algorithms

In our previous work [16], we found out that the Multinomial Naïve Bayes (MNB) algorithm is suitable to train multilingual text classifiers. Thus, we decided to use it in the new experiments. The MNB approach assumes that a word position in an analyzed document does not affect classification results. Of course, this assumption is untrue in real texts, where words in sentences are logically linked (dependent). Although these classifiers work surprisingly well in practice, we decided to check experimentally an algorithm that takes into account words position in sentences. We have chosen the Long Short-Term Memory (LSTM) algorithm [9], which was invented almost twenty years ago, but nowadays is widely used in natural languages processing tasks due to its recurrent nature.

3.3   Modeling Procedures

We propose several approaches considering various representations of input documents, classifiers organization, and final prediction regarding a target class.

One Multilingual Model Without a Decision Module. It is a simple modeling approach, which is a baseline to compare succeeding more advanced systems. Only one model is trained on a whole data set without considering languages included in input documents. The model assigns a class to a document under examination without any additional decision module. The documents are represented as a bag of words in two following ways:

A: An entire document d_i is transformed to one bag of words (Fig. 1A).

A: documents $\xrightarrow{\hspace{2cm}}$ $\longrightarrow$ $\xrightarrow[preprocessing]{data}$ VSM $\xrightarrow[training]{supervised}$ a model

B: documents $\xrightarrow[transformation]{data}$ parts $\xrightarrow[preprocessing]{data}$ VSM $\xrightarrow[training]{supervised}$ a model

C: documents $\xrightarrow[transformation]{data}$ parts $\xrightarrow[preprocessing]{data}$ VSMs $\xrightarrow[training]{supervised}$ models

Fig. 1. The modeling procedures of a multilingual model (A, B) or models (C).

B: A document d_i is split into its parts f_i^t. Then, each part is transformed into a bag of words. Next, a prefix corresponding to a particular part f^t is added to each word. In this way we distinguish words occurring in different parts of the document. For example, if a word *model* appears in a title and keywords, it is represented as *t_model* and *k_model* (Fig. 1B).

Then, the documents (their bags of words) are transformed into a Vector Space Model, which can occur in two ways depending on a training algorithm: (i) TF-IDF, when a model is based on MNB; (ii) Word2vec, when LSTM is used [12].

Several Multilingual Models and a Decision Module. It is possible that each part of a document has the different impact on a target category of the document. For instance, keywords provided by an author may be more significant than an abstract. Based on this assumption, we propose a compound classification system containing (i) a preprocessing layer, (ii) a set of multilingual models, where each classifier corresponds to a particular part of a document, and (iii) a decision module combining outputs of all models and producing the final category of a processed document. The preprocessing layer splits documents into their parts, generates a bag of words for each document part, and then represents them as Vector Space Models. In this way, we can train a separate model for each document (paper) part (Fig. 1C). Since we use all data irrespective of language, the models cover multilingual information. During the classification procedure, the decision module integrates all outputs produced by multilingual classifiers (for more details, see the last paragraph of this subsection).

Several Monolingual Models and a Decision Module. Instead of modeling each document part separately, we can train monolingual classifiers on selected monolingual documents. In this way, we achieve a set of monolingual classifiers. More specifically, we propose the multilingual classification system, which is composed of (i) the preprocessing, (ii) classification, and (iii) decision layers like in the paragraph above. However, the classification layer contains monolingual models.

A Decision Module. The most interesting part of the system is a multilingual decision module, which works the same in both cases, i.e., multilingual models representing multilingual document parts and monolingual models representing

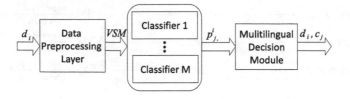

Fig. 2. The compound classification system.

whole monolingual documents (Fig. 2). The data preprocessing layer receives documents d_i and transforms them to a VSM that is suitable to a chosen training algorithm. These vectors are inputs of the classifiers $m = 1, ..., M$. Each classifier m produces the probabilities $p^i_{j,m}$ of belonging of a document d_i or its part f^t_i is to each possible category c_j. The multilingual decision module is an ensemble of one vs. others logistic regression classifiers that integrates these probabilities. It finally chooses the target category according to the following criterion:

$$
\begin{aligned}
d_i \in c_j \quad & if \ h(\varphi_i) > h_{th} \\
d_i \notin c_j \quad & if \ h(\varphi_i) \leq h_{th}
\end{aligned}, \tag{1}
$$

where h_{th} is the decision threshold and $h(\varphi_i)$ is the logistic regression function

$$
h(\varphi_i) = \frac{1}{1 + e^{-\varphi_i}}. \tag{2}
$$

and φ_i is the weighted sum of probabilities that are generated by the classifiers

$$
\varphi_i = \sum_{m=1}^{M} \left(w_{j,m} \cdot p^i_{j,m} \right), \tag{3}
$$

and $w_{j,m}$ are weights of the regression model calculated in a supervised way [16].

4 Experiments

In this section, we present the experiments and their results to verify the approaches to multilingual classification discussed in the previous section.

4.1 Setup

The experiments are performed on our dataset of publications, which is the part of a recommender system of reviewers and experts [15]. There are almost five million publications, where all of them contain a title, and almost all them contain a source, two-thirds of them contain a Polish or English abstract, and they are also described by keywords. The dataset has been labeled automatically [16] by using the Ontology of Scientific Journals (OSJ) [18] giving 2,733,991 publications in English and 119,312 publications in Polish, where some of them are multilingual.

For the purpose of the experiments, we selected monolingual and multilingual documents from this dataset randomly. They are multilingual in this way that they include sentences concurrently in two languages, i.e., Polish and English. The resulting dataset has been divided into equal subsets, where each of them contains a different distribution of languages as presented in Table 1. We have to note that the distributions were calculated as the ratio of a words number in a particular language to the total number of words in each document excluding stopwords. Since it is impossible to find documents containing strict language proportions, the numbers 25, 50, and 75 in Table 1 should be understood as the centers of a certain range, whereas proportion 0% to 100% indicates monolingual documents.

Table 1. The percentage distribution of texts in Polish (PL) and English (EN) in the equal subsets of the dataset.

Subset no.	1	2	3	4	5
PL %	0	25	50	75	100
EN %	100	75	50	25	0

Our goal is to assign each document d_i to one of six top-level domains $c_{j=1,..,6}$ of OSJ, namely Applied Sciences, Arts & Humanities, Economic & Social Sciences, General, Health Sciences, Natural Sciences. The computations are carried out with the use of the scikit-learn library [13] implementing the MNB algorithm and the Keras software [4] implementing the LSTM algorithm. The quality of each experiment is measured by F-score, precision, recall, and accuracy, which are calculated on the 10-fold cross-validation multilingual sets.

4.2   Experiments Description

We prepared four experiments regarding the various combinations of modelling approaches described in Sect. 3. More specifically, they are as follows:

Experiment 1. Only one multilingual model is trained on the entire documents (transformed to the bags of words) by using two different algorithms, i.e., (i) the MNB with the TF-IDF data representation and (ii) the LSTM with the word2vec data model.

Experiment 2. Only one multilingual model is trained on the document parts (transformed to the bags of words) by using the MNB algorithm with the TF-IDF data representation.

Experiment 3. The multilingual modules, each representing the document parts (titles, abstracts, keywords, authors) separately, are trained by using the MNB algorithm with TF-IDF data representation. Then, the decision module integrates classifiers outputs and chooses the target class.

Experiment 4. The monolingual modules, each representing different language (Polish, English) separately, are trained by using two different algorithms, namely (i) MNB with the TF-IDF data representation and (ii) LSTM with the word2vec data model. Then, the decision module integrates classifiers outputs and chooses the target class.

In experiments 3 and 4, the decision threshold h_{th} of the logistic regression function $h(\varphi_i)$ (1) is equal to 0.5.

4.3   Results and Discussion

Table 2 includes the results of all experiments achieved when only the MNB algorithm was used. We deliberately excluded the results of the models trained by the LSTM algorithm, because it helps us to compare different modeling approaches. It is clearly depicted that the compound system, which contains the set of monolingual or multilingual classifiers and the decision module (experiments 3 and 4), categorizes publications better than both single classifiers (experiments 1 and 2). Moreover, it is advisable to split a document (publication) into its parts and train a multilingual model for each part (experiment 3) rather than prepare a separate monolingual model for each language (experiment 4). If we compare the results produced by single classifiers (experiment 1 vs. experiment 2), we also observe the slight advantage of the decomposition approach (documents vs. their parts).

Table 2. The results of experiments 1–4 produced by the models trained only by the MNB algorithm.

Experiment no.	1	2	3	4
Precision	0.85	0.85	**0.92**	0.87
Recall	0.80	0.82	**0.92**	0.87
F-score	0.82	0.83	**0.92**	0.87
Accuracy	0.80	0.82	**0.92**	0.87

The detailed results of experiment 3 (Table 3) confirm our intuition that each part of a document (publication) has a different influence on its category. Unsurprisingly, the keywords provided by authors describe best the whole document. In this experiment, only the MNB algorithm was used for clarity of presentation.

It is remarkable that the decision module distinctly improves the classification quality by integrating outputs of all multilingual classifiers that separately perform worse than the whole system (Table 3). However, we do not observe such phenomenon in the case of the system with monolingual classifiers and the decision module (experiment 4). The Polish and English classifiers trained by the MNB are better than the whole system. On the contrary, we observe the opposite results, when the models are trained by the LSTM (Table 4). Thus,

Table 3. The detailed results of experiment 3, where the models were trained by the MNB algorithm. The columns T (titles), Ab (abstracts), K (keywords), and Auth (authors) show the outputs of multilingual classifiers; the column Total includes outputs of the decision module.

Classifer	T	Ab	K	Auth	Total
Precision	0.75	0.70	**0.81**	0.72	0.92
Recall	0.69	0.65	**0.77**	0.60	0.92
F-score	0.70	0.64	**0.78**	0.65	0.92
Accuracy	0.69	0.65	**0.77**	0.60	0.92

Table 4. The detailed results of experiment 4 with the use of MNB and LSTM algorithms. The columns PL (Polish) and EN (English) show the outputs of monolingual classifiers; the column Total includes outputs of the decision module.

Classifer	PL		EN		Total	
Algorithm	MNB	LSTM	MNB	LSTM	MNB	LSTM
Precision	**0.92**	0.86	**0.85**	0.82	0.87	**0.88**
Recall	**0.90**	0.86	0.81	**0.82**	0.87	**0.88**
F-score	**0.90**	0.86	**0.82**	0.82	0.87	**0.88**
Accuracy	**0.90**	0.86	0.81	**0.82**	0.87	**0.88**

Table 5. The comparison of the MNB and LSTM algorithms.

Experiment no.	1		4	
Algorithm	MNB	LSTM	MNB	LSTM
Precision	0.85	**0.87**	0.87	**0.88**
Recall	0.80	**0.87**	0.87	**0.88**
F-score	0.82	**0.87**	0.87	**0.88**
Accuracy	0.80	**0.87**	0.87	**0.88**

we can conclude that the system with monolingual classifiers is dependent on a dataset and training algorithms.

Experiments 1 and 4 compare the MNB and LSTM algorithms (Table 5). Although the single multilingual model trained by the LSTM algorithm is better than that based on the MNB algorithm (experiment 1), the multilingual classification system (experiment 4) works almost the same in the case of using both algorithms. Since the LSTM networks require pretty much more computational power than the MNB algorithm, we suggest that it is better to use the simple MNB with only slight expense on the classification quality. We have to note that the units number of the LSTM network has been selected experimentally. It was equal to 256 units (Table 6).

Table 6. The selection of the units number of the LSTM network.

Units number	256	512	128
Precision	**0.87**	0.87	0.86
Recall	**0.87**	0.87	0.86
F-score	**0.87**	0.87	0.86
Accuracy	**0.87**	0.87	0.86

5   Conclusions

The objective of this work was to propose a classification system that can categorize multilingual documents (publications). We have to underline that these documents are multilingual in this sense that they can contain text parts simultaneously in different languages.

We proposed the various combinations of the compound classification system composed of three layers as follows: (i) a preprocessing layer generating a VSM model, (ii) monolingual classifiers representing diferent languages or multilingual classifiers corresponding to different text parts, and (iii) a decision layer integrating the outputs of all classifiers and producing the final prediction regarding a target class. A baseline was a single multilingual model.

The system has been verified experimentally. According to the reported results, we found out that is advisable to decompose documents to its parts and then create a multilingual model representing each text part. What is interesting, the decision module integrates classifiers outputs in this way that the whole system is recognizable better in the categorization of documents than its each classifier working independently. We also observed that the Multinomial Naive Bayes algorithm describes documents sufficiently well and there is no need to introduce more computationally demanding algorithms like Long Short-Term Memory, which became quite popular nowadays.

Based on above findings, we believe that the proposed system can classify multilingual documents containing the text parts in various languages at the same time. The further works would consider a dataset of multilingual documents containing concurrently many languages in one document in order to conduct experiments and better evaluate the proposed approach. The dataset should be available publicly.

References

1. Amini, M.-R., Goutte, C.: A co-classification approach to learning from multilingual corpora. Mach. Learn. **79**(1–2), 105–121 (2010)
2. Amini, M.-R., Goutte, C., Usunier, N.: Combining coregularization and consensus-based self-training for multilingual text categorization. In: Proceedings of the 33rd International ACM SIGIR Conference on Research and Development in Information Retrieval, SIGIR 2010, pp. 475–482. ACM, New York (2010)

3. Amini, M.-R., Usunier, N., Goutte, C.: Learning from multiple partially observed views-an application to multilingual text categorization. In: Advances in Neural Information Processing Systems, pp. 28–36 (2009)
4. Chollet, F.: Keras (2015). https://github.com/fchollet/keras
5. Melo, G., Siersdorfer, S.: Multilingual text classification using ontologies. In: Amati, G., Carpineto, C., Romano, G. (eds.) ECIR 2007. LNCS, vol. 4425, pp. 541–548. Springer, Heidelberg (2007). doi:10.1007/978-3-540-71496-5_49
6. García-Adeva, J.-J., Calvo, R.A., de Ipiña, D.L.: Multilingual approaches to text categorisation. CEPIS promotes, p. 43 (2005)
7. Gonalves, T., Quaresma, P.: Multilingual text classification through combination of monolingual classifiers. In: Proceedings of the 4th Workshop on Legal Ontologies and Artificial Intelligence Techniques, pp. 29–38 (2010)
8. Guo, Y., Xiao, M.: Cross language text classification via subspace co-regularized multi-view learning. In: Langford, J., Pineau, J. (eds.) Proceedings of the 29th International Conference on Machine Learning (ICML 2012), pp. 1615–1622. ACM, New York (2012)
9. Hochreiter, S., Schmidhuber, J.: Long short-term memory. Neural Comput. 9(8), 1735–1780 (1997)
10. Jindal, R., Malhotra, R., Jain, A.: Techniques for text classification: literature review and current trends. Webology 12(2) (2015)
11. Lee, C.-H., Yang, H.-C.: Construction of supervised and unsupervised learning systems for multilingual text categorization. Expert Syst. Appl. 36(2), 2400–2410 (2009)
12. Mikolov, T., Chen, K., Corrado, G., Dean, J.: Efficient estimation of word representations in vector space. arXiv preprint arXiv:1301.3781 (2013)
13. Pedregosa, F., Varoquaux, G., Gramfort, A., Michel, V., Thirion, B., Grisel, O., Blondel, M., Prettenhofer, P., Weiss, R., Dubourg, V., Vanderplas, J., Passos, A., Cournapeau, D., Brucher, M., Perrot, M., Duchesnay, E.: Scikit-learn: machine learning in Python. J. Mach. Learn. Res. 12, 2825–2830 (2011)
14. Pinto, D., Civera, J., Barron-Cedeno, A., Juan, A., Rosso, P.: A statistical approach to crosslingual natural language tasks. J. Algorithms 64(1), 51–60 (2009)
15. Protasiewicz, J., Pedrycz, W., Kozłowski, M., Dadas, S., Stanisławek, T., Kopacz, A., Gałężewska, M.: A recommender system of reviewers and experts in reviewing problems. Knowl.-Based Syst. 206, 164–178 (2016)
16. Protasiewicz, J., Stanislawek, T., Dadas, S.: Multilingual and hierarchical classification of large datasets of scientific publications. In 2015 IEEE International Conference on Systems, Man, and Cybernetics, pp. 1670–1675. IEEE (2015)
17. Rigutini, L., Maggini, M., Liu, B.: An EM based training algorithm for cross-language text categorization. In: The 2005 IEEE/WIC/ACM International Conference on Web Intelligence (WI 2005), pp. 529–535 (2005)
18. Science-Metrix. Ontology of scientific journals (v1.03), September 2011
19. Suzuki, M., Yamagishi, N., Tsai, Y.-C., Hirasawa, S.: Multilingual text categorization using Character N-gram. In: IEEE Conference on Soft Computing in Industrial Applications, SMCia 2008, pp. 49–54 (2008)
20. Xiao, M., Guo, Y.: Semi-supervised representation learning for cross-lingual text classification. In: EMNLP, pp. 1465–1475. Citeseer (2013)
21. Yang, H.-C., Hsiao, H.-W., Lee, C.-H.: Multilingual document mining and navigation using self-organizing maps. Inf. Process. Manage. 47(5), 647–666 (2011)

Cognitive Content Recommendation in Digital Knowledge Repositories – A Survey of Recent Trends

Andrzej M.J. Skulimowski[1,2](✉)

[1] Chair of Automatic Control and Biomedical Engineering,
Decision Science Laboratory, AGH University of Science
and Technology, 30-050 Kraków, Poland
ams@agh.edu.pl
[2] International Centre for Decision Sciences and Forecasting,
Progress & Business Foundation, 30-048 Kraków, Poland

Abstract. This paper presents an overview of the cognitive aspects of content recommendation process in large heterogeneous knowledge repositories and their applications to design algorithms of incremental learning of users' preferences, emotions, and satisfaction. This allows the recommendation procedures to align to the present and expected cognitive states of a user, increasing the combined recommendation and repository use efficiency. The learning algorithm takes into account the results of the cognitive and neural modelling of users' decision behaviour. Inspirations from nature used in recommendation systems differ from the usual mimicking the biological neural processes. Specifically, a cognitive knowledge recommender may follow a strategy to discover emotional patterns in user behaviour and then adjust the recommendation procedure accordingly. The knowledge of cognitive decision mechanisms helps to optimize recommendation goals. Other cognitive recommendation procedures assist users in creating consistent learning or research groups. The primary application field of the above algorithms is a large knowledge repository coupled with an innovative training platform developed within an ongoing Horizon 2020 research project.

Keywords: Research recommenders · Scientific big data · Personal learning environments · Preference modelling · Mobile and ubiquitous learning

1 Introduction

Cognitive and biological inspirations are increasingly common in the design of advanced software systems. This is due to the fact that basic biological observations and the knowledge of cognitive mechanisms intervene in the background of virtually all creative processes. These include the design and implementation of web applications, where interaction with users plays a primary role.

This paper presents research on eliciting optimal functional architectures of recommendation systems supporting users of scientific and learning repositories. Such systems are expected to facilitate the use of large scientific knowledge bases and their

© Springer International Publishing AG 2017
L. Rutkowski et al. (Eds.): ICAISC 2017, Part II, LNAI 10246, pp. 574–588, 2017.
DOI: 10.1007/978-3-319-59060-8_52

future successors - global expert systems (GESs, cf. [37, 39]). The latter will ensure access to all web-based knowledge sources, including data and multimedia repositories as well as the Internet of Things (IoT)-based real-time data streams from sensors and actuators. GESs will also encompass real-time and archive information concerning human and artificial users of mutually connected social networks. The emergence of GESs and their impact on the future of scientific research has been studied recently in [37, 39, 40]. The above-cited research reveals a growing need to support the users of heterogeneous large-scale repositories while searching, classifying, analysing, managing, and further use of the retrieved content. Appropriate software agents endowed with the usual expert system functionalities to support users of large-scale scientific repositories can greatly increase the efficiency of users' interaction with such systems, cf. e.g. [17, 21, 25, 26, 49]. We will argue that the ability of such agents to align with the individual cognitive phenomena of the users determines a most promising development trend of such applications [14].

The agents will address the issues related to the following general development trends of scientific repositories that have been identified and studied in [40, Chap. 5 and 8], and investigated further in [39] and in H2020 project MOVING [19]:

- a growing number of users with an increasing diversification of individual preferences and learning goals,
- a growing number of interconnected knowledge units,
- a growing diversity of content stored in knowledge repositories,
- a growing level of integration of heterogeneous information sources,
- an increase in the amount of information and sophistication of information processing within individual units,
- a growing mean intensity of information exchange (in bauds) and the total amount of information exchanged within an individual session (in bits).
- a rapidly growing need to assist the users by informed recommendation of content and services of knowledge repositories.

The above trends are the main reason of the growing complexity of research and learning supporting applications, including recommenders. A need to acquire knowledge on the new functionalities and the corresponding users' skills is at the same time an important source of difficulty when these systems are used by the learners. This is why identifying the learner's momentary emotional state by the learning application [31, 32] can be as relevant as the elicitation of users' preferences to be used by learning and research recommender systems. This paper studies the related issues in the context of selecting the best recommendation methods.

Due to a close relationship between the decision support (DSS) and recommendation systems, cognitive research and/or learning recommenders can benefit from the existing contributions of the cognitive DSS theory and implementation experience. It turns out that the cognitive science inspirations [28, 48] are becoming common in open and distance learning environments. They often occur in the design of various advanced learning software systems. This is due to the fact that the basic biological observations and the knowledge of cognitive mechanisms intervenes in the background of virtually all creative processes, including the design and implementation of web applications, where the interaction with users plays a primary role.

The results here presented should contribute towards designing automatic decision pilots, a subclass of cognitive recommenders, providing rankings and implementing constraints, but not the final choice. They have been studied with the aim to facilitate the use of a large innovative knowledge repository storing mostly scientific papers, massive open online courses (MOOCs), and economic information (cf. [24]). Content-based recommendation for this repository, including the performance measurements and a comparison of 12 approaches focused on methods of scientific papers recommending in the area of the economy has been recently presented in the above cited paper [24]. Other methods of recommending learning- and research-related content are also discussed in [2, 8, 25, 26]. Issues related to distance learning are studied in [28, 48].

2 Cognitive Inspirations of Recommender Systems

We will start this section by presenting a formal background of recommendation systems that will be useful to defining a family of recommenders fitting best the needs of knowledge repositories.

2.1 A Formal Background of Recommendation Systems

A general single set recommendation problem can be formulated in the following way:

$$(F : U \supset U_0 \to R^n) \to min(P) \tag{1}$$

$$\left(G : 2^V \times \Pi \to R^m\right) \to min(Q), V := \{(u, f(u)) : u \in U, f(u) \in R^p\} \tag{2}$$

$$C \subset argmin\{F^*(u) : u \in U_0 \cup argmin\{G(V_r, \pi_r)\}\} \tag{3}$$

where $F = (F_1, ..., F_n)$ and $G = (G_1, ..., G_m)$ are vector performance criteria, the first one of the decision maker D who selects items from a certain subset U_0 of the set of all admissible or available items U, the second of the recommendation system owner $S(\Omega)$. Equation (1) describes the item selection problem without recommendation, where the set U_0 contains admissible items D is initially aware of or with features known to D. We assume that the function F may be represented as a composition of a selection function F^* defined on a set of characteristics of items from U_0 expressed by certain features, and a function f associating features to items, i.e. $F = F^* \circ f$. The recommender Ω is an artificial agent that recommends a subset of admissible items to the decision maker D together with an information about their p features represented numerically for data processing. It is assumed that D takes into account the same features of items from the subset U_0 known to D prior to the recommendation. It can be an essential subset of U.

When recommending admissible items to D, Ω takes in to account an estimation π_r of D's preference structure P from the set of all feasible preference structures Π. By definition, a preference structure is a partial ordering of $F(U_0)$ conforming to the natural componentwise order in R^n. In this paper we will not study in detail the course of the estimation process and the properties of Π, focussing on cognitive aspects of recommendation and on digital repository applications. The final choice C of the decision maker D depends on D's preference structure P and on the recommendation that is modelled by Eq. (3).

In the most common case, where a recommender presents an ordered sequence of items to a potential customer, the set of all k-permutations of U, for $k = 1, 2, ... K$, $\leq \#U$ is used instead of 2^U, i.e. the second equation of the recommendation problem (1–3) is replaced by

$$(G : \cup_{i=1}^{K} U^i \times \Pi \to R^m) \to \min(Q) \qquad (4)$$

yielding the ranking-based recommendation problem (1, 4, 3) with similar solution principles as in the problem (1–3).

Both recommendation processes exemplify multicriteria recommenders that have been studied e.g. in [1, 22]. While D makes decisions usually taking into account $n > 1$ criteria, the goal of a recommendation system is most often to maximize its owner's profit so that frequently $p = 1$ [7]. When recommending items from a not-for-profit or institutional subscription-based knowledge repository, it is likely that the sets of criteria $\{G_1,...,G_n\}$ and $\{F_1,...,F_n\}$ coincide or overlap considerably. This is the case when G includes indices describing the efficiency of the learning or research process that are also followed by the users. For example, the repository content recommender may be designed to optimize the recommendation criteria such as:

- the degree of representativeness of a scholarly literature necessary to complete a specific research or learning task.
- the precision of the recommended literature set (with respect to the user's learning goal).
- the goodness of fit of recommended courses to the individual learning preferences.
- increasing the creativity of learners beyond the momentary learning goal [34].

To create a common base to classifying intelligent agents, including cognitive recommenders and autonomous decision-making systems, in [36] we defined the three levels of freewill. Freedom of choice of the 1st order is the ability to choose when a set of choice criteria for a given set of admissible alternatives is specified, freedom of choice of the 2nd order allows the decision-maker to relax the constraints, finally, freedom of choice of the 3rd order is the power to select one's criteria of choice in the feature space of real-life objects selected by an intelligent artificial agent.

According to the above taxonomy, recommenders are autonomous systems of the 1st or 2nd order, depending on the fact whether implementation allows the agent to seek for information in the open web [12]. This is not allowed for the here presented knowledge repository, however. In addition, recommenders are given the capacity to learn from their past decision and efficiency experience.

2.2   Research Issues in Cognitive Recommendation and Decision Support

The predominant mechanism of cognitive inspirations in recommendation systems is somehow different than the usual mimicking living systems. As it has been evidenced by the results of a recent foresight project [40, Chap. 4], the overall development trend of recommender systems follows a strategy to discover the cognitive aspects of user behavior and adjust the recommendation procedures correspondingly, so that the recommendation goals are optimized [3, 27].

During the last two decades the theory of recommenders has become an interdisciplinary research field [4], [40, Chap. 5] situated partly within

- Decision science, specifically modelling real-life decision problems and processes,
- Computer science in terms of the implementations and computer architecture of recommenders and decision support systems, both regarded as a subclass of intelligent systems, and
- Cognitive science and mathematical psychology.

Although decision-processes related to applying recommendations involve highest-level cognitive functions of the human mind, their appurtenance to the Cognitive Science is sometimes neglected when designing recommender systems. The reasons for that are threefold:

- A partitioned character of research on elicitation and modelling of human preferences, where the psychometric research does not meet the prevailing theoretical studies on decision analysis, and the recommender systems are designed without paying enough attention to the real-life human decision-making mechanisms.
- The tendency to restrict modelling human decisions to cases, where the decision-maker(s) is either able to formulate criteria of choice or able to explicitly define a set of admissible alternatives. What follows, is a mathematical programming or a gaming problem, and the subsequent efforts are concentrated on solving it, without taking care about cognitive phenomena, such as rapidly changing preferences, an extension/contraction of the decision scope resulting from different cognitive processes.
Finally,
- The lack of adequate decision models when the information underlying decision-making has a multimedia form. In such situations autonomous learning support systems need to elicit users' preferences concerning the sequence and relevance hierarchy of learning goals. These, in turn, can be used for further recommendations after being implemented in tailored recommender systems.

Further drawbacks concerning the current state of research on decision models for recommendation systems originate from the relatively low linkage of recommendation-suitable models to the theory of decision support systems (DSS). This issue can be described as follows:

- A difficulty in applying decision-making procedures arises in situations where the decision-maker's preference structure is non-compatible with the data available or must be gradually elicited. The awareness of the relevance of using cognitive decision-making mechanisms to increase the efficiency and adequacy of recommendation in the above mentioned situations is still insufficient.
- The application of a variety of so-called interactive decision-making algorithms that are very popular in DSS can face different problems in recommenders [32] where decisions are made quickly, leaving no space for a long dialogue as it is usual in

interactive DSS. In addition, processing such a dialogue does not guarantee a final success in form of a convergence to a satisfactory compromise learning plan. The convergence conditions of interactive procedures may be non-compatible with the process of cognitive decision-making, neglecting spontaneous discoveries of potential solutions better than the next candidate for compromise solution generated by the formal procedure. Even a simple 'change of mind' by the decision-maker during the procedure can perturb the convergence.

- Unlike the human-expert advising that is capable of mitigating the results of incorrect recommendations, the existing recommenders act according to the principle that, once an item is selected, and this choice is registered, the role of the expert system is finished. When external circumstances that influenced the choice change, the decision-maker might wish further assistance to re-examine the choice, and – sometimes - change the decision. This is not straightforward in recommender systems, the decision maker must usually repeat the whole procedure.
- The crucial role of an ability to learn based on previous choices can be greatly enhanced by estimating the emotions of decision makers, cf. [14, 31].

Having said the above, one should observe that the appearance of cognitive phenomena has contributed recently to a remarkable change in the practice as well as in the philosophy of designing recommendation systems [50].

3 Principles of Cognitive Content Recommenders

When selecting recommendation principles most suitable for a learning platform, we will refer to the commonly approved taxonomy of recommendation methods, such as the breakdown into collaborative filtering and content (or item)-based recommendation [4, 5]. In addition, to respond to the needs of specialized recommenders designed for content recommendation in knowledge repositories, learning platforms and learning management systems (LMS), their functionalities should be split into content searching, filtering and presenting to the users (knowledge extraction and processing), and in recommending learning activities, following the phases of a holistic preference learning process.

Learning recommendation algorithms may recommend items (research papers, laboratories, courses, books, videos) as well as intangibles, actions to be taken, or other users or user groups as research or learning partners, cf. e.g. [10, 15, 16, 23, 29, 42]. Here, *items* denote either:

- Digital documents, such as research papers or books.
- Online courses, including MOOC, or educational games.
- Quantitative or qualitative datasets.
- Videos and graphic digests.
- Laboratories.
- Research or learning-oriented software.

Content-based recommendation is a natural way of recommending these items, but social and other recommendation modes can also be used. Collaborative filtering may be misleading as the most used (or most cited) items do not necessarily have to satisfy the needs of a particular user [5]. Similarly, a big diversification of research and learning goals makes it difficult to apply the similarity of users' activities when using knowledge repository items. Furthermore, the recommendations may be hybrid [8] or complex, pointing out objects, actions, and/or persons at one time.

For a given collection of recommended items stored in a knowledge repository, we will investigate the following content recommendation problems:

- Direct item or content based recommendation by providing a list of items (problem (1–3), in some cases also (1, 4, 3)):
- Indirect content recommendation by providing a query extension to the user.

The latter problem can be converted to the first one by considering the anticipated properties [38] of results of search with the recommended query. Based on the analysis of current approaches to learning and research recommendation, we will specify automatic decision pilots, a subclass of cognitive content-based recommenders. Decision pilots provide sets or rankings of items and implement constraints, while the final choice is performed by a user. Finally, we will propose a hybrid cognitive recommendation engine, endowed with supervised learning schemes that make it possible to achieve a high level of user satisfaction with the recommended content. The satisfaction measurements rely on the subjective user assessments [11] and on an automated evaluation of learning resources [6]. They can be estimated as an aggregation of user's interest scores assigned to the recommended query responses or to the items recommended directly. Query processing will apply knowledge fusion methods such as combinations of recommendations, ex-post assessments of retrieved content and other methods [35].

3.1   The Design of Cognitive Decision Pilots for Research Recommenders

The application of recommenders does not deny the traditional worth and the role played by the intuition and experience of the decision-maker. They remain relevant, but can be enhanced by a computational-system that ensures

- A systematic survey and automated evaluation [6] of learning and research resources available on the knowledge platform,
- A personalized learning or research items acquisition and presentation (cf. [30, 46, 47]).

The decision pilot can be regarded as a content recommender engine responsible for the automated assessment of items, formulating and solving problem (1–3) or (1, 4, 3) and gathering and representing the knowledge about the decision makers' preference structures based on cognitive behavioural analysis [33]. The degree of satisfaction of the decision-maker with the recommendations thus generated supplies the information about the model quality and creates the basis for a supervised learning scheme.

The cognitive approach applied in decision pilots is based on the following key principles:

- All information resources available are explored to a maximum extent possible, observing user-defined temporal processing constraints.

- Inconsistent or contradictory content in the repository (e.g. article duplicates with mistakes in title or other metadata) can be disambiguated or judged on their usefulness before passing them to the recommended set.
- The recommendation can be performed incrementally in an open information space, i.e. in the situation, when there is an inflow of items to the repository in real-time or if real-time processing of visual or audio information is required.
- The elicitation of users' preferences is performed in real-time as well.

Nervous, tired or irrational knowledge platform users may exhibit behaviour that leads to a chaotic choice of decisions, such that the conditions for terminating the decision-making process were never fulfilled. This may likely happen with some learners prior to exams, seeking information in a hurry etc. This is why in some situations hurrying or tired decision-makers may especially need a quick and efficient decision aid. The recommenders designed based on automatic decision pilot principles will be able either to recommend deferring the learning strategy choice to a more suitable moment or to generate a 'cautious' recommendation. By 'cautious' we mean the selection of an item or a set of items which conforms in a maximum way to the decisions made previously by this system's user based on an individual cognitive decision model.

Based on real-life experience with DSS we assume that mutually inconsistent or contradictory information found in the repository or received from the decision-maker can be treated as a result of different cognitive processes. An identification of such processes makes possible a reconstruction, re-definition or averaging of faulty resources, converting them to useful material. Thus the recommendation process avoids becoming loopy or inconsistent and the finally generated output does not depend on the subjective sequence in which the additional information was processed.

Moreover, this assumption emphasizes the need to understand human cognitive processes that accompany decision-making. Incorrect or inconsistent statements are often caused by the shortcomings of human perception of decision objects features. A recommendation system based on an intelligent decision pilot is capable of tracing the learning processes on the platform, find and indicate the possible source of inconsistencies using a cognitive perception model.

Similarly as in case of using price comparison engines and recommenders to support goods selection in e-commerce systems, the quick changes of research or learning resources available in a digital repository affect the recommendation process. In such situations optimal stopping rules should be applied. The expected rise of information inflows from the web [39] to knowledge repositories will create further needs for more adequate cognitive recommendation mechanisms.

As an example of a cognitive extension of a decision approach, the procedure which measures the user's reply time in defining aspiration levels for learning in the well-known reference set method [36], then using the observation that for a certain group of users, the faster the reply is generated, the higher the probability of getting a correct reply. Another real-life cognitive observation that can be applied when designing decision engines for recommenders is the bicriteria trade-off hypothesis. This states that irrespective of the number of criteria used to make the choice, decision makers intuitively try to group them into two aggregated criteria then solve the bicriteria problem thus formulated. This hypothesis should be a subject of further psychometric investigation to find factors that influence the sequence of aggregation, relating them to the feature perception and selection during recommendation processes.

3.2   Implementing Cognitive Content Recommenders for a Learning Platform

Successful implementation of cognitive recommenders in learning and research have been reported e.g. in [25, 26, 43, 44]. Some of the solicited principles of recommendation systems capable of facilitating the use of a learning platform [19] and making it attractive for its users can be listed as follows:

- Irrespective of how advanced mathematical methods are used to process the underlying information and to generate the recommendation, the sophisticated procedures should not be directly visible to the users (maths ignorance assumption).
- The quality of automated recommendation should be enhanced by the collaborative systematic verification of the platform content, by its administrators and involving the users.
- A trust-credibility system should be designed and implemented: trust regarding the users, credibility regarding the content items stored on the platform. Different trust and credibility models can be taken into account (cf. [9, 13, 20, 45]) focussing on those that allow for dynamic changes of trust and credibility measures in real-time.
- A user-friendly recommendation assessment coupled with supervisory learning mechanisms should be built into the recommendation system. If necessary, the user should be able to redefine the recommendation with a dedicated intelligent agent so that the user's preferences were satisfied to an optimal extent.

After a cold start and reaching a critical number of users, the recommendation algorithms will be gradually improved as well as new cognitive decision making procedures and preference elicitation methods will be added. The use of additional preference information in form of reference sets [36] and bicriteria trade-offs seems especially well suited to generating compromise recommendations and choose satisfactory items. The recommendation will be supplemented by extensive visualization and guideline procedures, providing the graphical and video object recommendations in an annotated form. Conversely, the feature space methods that originate from image processing may be adapted for use with text and multimedia files [18]. In the mid-term future, human experts will only play the role of platform supervisors taking care about management issues and ordering missing items and data.

3.3   Search Strategy Recommendation as a Cognitive Process

A relevant issue that needs to be considered when designing a content recommendation system for a digital repository is the choice of a search-and-survey strategy to process queries capable of reviewing a very large number of feasible information sources. The survey planning approach, presented e.g. in [30], cannot be used in a dynamically changing environment with a very large number of potential knowledge sources, out of which only a quotient is explicitly known *ex-ante*. Also, classical precision-and-recall assessment of responses to the query will fail for a number of reasons. In particular, the user will not be able to assess the results on his/her own and will be forced to delegate the judgment regarding the quality of the reply and corresponding decisions to autonomous agents. A heuristic search-and-survey procedure can be designed making use of the *creative decision process* notion [36], where the user defines an initial subset of information sources according to some criteria, assigns them trust coefficients and activates the procedure that runs recursively at each information source, transforming them to autonomous agents with similar capabilities as the user. The design of such a procedure can be accomplished based on the creative decision process definition provided in [36].

Further development of cognitive features of the learning platform may involve using specialized brain computer interfaces (BCI, [41]) to elicit users' preferences and identify emotions in an efficient direct way. BCI can also be helpful in adding the above mentioned creativity-support-system functionalities to the knowledge repository [34]. Ultimately, sophisticated query design, extension, and recommendation procedures will allow the innovative learning platform [19] to develop towards a genuine GES [37].

4   Discussion and Conclusions

Intelligent cognitive recommender systems constitute a new market and a social challenge. Their implementation horizon, from the current stage of development – the item search and price comparison machines, seems more or less equivalent to the expected start of implementing a new class of innovative learning platforms exemplified by [19]. Cognitive recommendation software can create a new market trend, following the research trend evidenced in [40] and shown in Fig. 1 below.

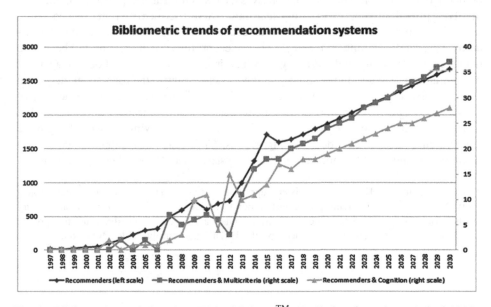

Fig. 1. Bibliometric trends based on Web of Science™ (WoS) data from the period of 1997–2015 with ARIMA(2,1,0) forecasts until 2030. Left and right scales show the number of records in WoS for the corresponding queries.

The above trend, as well as the rise of multicriteria recommenders, will be enhanced by a rapid development of natural-language-based and multimedia search engines.

New decision-making concepts and methods make it possible to design intelligent autonomous recommendation systems, able to make discoveries, anticipate the consequences of the decision made (cf. [38]) and enhance the quality of interaction with users. An anticipated application is an innovative knowledge repository coupled with a training platform developed within the ongoing EU Horizon 2020 research project MOVING [19]. Due to a growing importance of multimedia courses and other content, it is indispensable to combine visual information processing with recommendation and decision support algorithms, which is also a subject of the MOVING project.

The recommender design approach proposed in this paper extends the recommendation process to the incremental learning of users' preferences, emotions, and satisfaction while using large heterogeneous knowledge repositories. Then the recommendation mechanisms are adjusted to align to the present and expected cognitive state of a user. Thus the recommendation, user interaction with the repository and recommendation algorithms updates are combined in one anytime procedure with several levels of interaction. We claim that this principle will become the standard in future recommendation and decision support systems, and its successful implementation will be a decisive factor for the dominance on the intelligent recommender market.

The recommendation systems here described are assumed to work without any idealistic presumptions concerning the rational behaviour of users, and they are endowed with the capacity to check the consistency of a user's input and correct the choice. Together with content understanding capabilities, they will be able to propose optimal learning strategies based on multicriteria optimization algorithms when embedded in both, mobile and stationary systems. One further potential application involves providing support to group learning, where the members are matched by a recommendation mechanisms taking into account the expected compliance of users attitudes towards learning and their psychological profiles. Such systems will be able to manage the credibility and trust in group learning [37]. User reputation management mechanisms will be used to optimize the recommendations to the users to taking part in common learning or research. The recommendation will be enhanced by the activity of autonomous agents searching for statistical, patent, or bibliographic information. We expect that future recommenders will be endowed with a growing number of cognitive features and multicriteria decision algorithms. The latter will support increasingly autonomous and complex interaction of such systems with knowledge repository users.

Acknowledgement. This paper has been supported by the EU Horizon 2020 research project MOVING (http://www.moving-project.eu) under Contract No. 693092. Selected preliminary results concerning recommendation systems trends have been obtained during the project SCETIST (www.ict.foresight.pl) financed by the ERDF and contributed to MOVING.

References

1. Adomavicius, G., Kwon, Y.O.: Multicriteria recommender systems. In: Ricci, F., Rokach, L., Shapira, B., Kantor, P.B. (eds.) Recommender Systems Handbook, pp. 847–880. Springer, Heidelberg (2015)
2. Aher, S.B., Lobo, L.: Combination of machine learning algorithms for recommendation of courses in e-learning system based on historical data. Knowl. Based Syst. **51**, 1–14 (2013)
3. Bobadilla, J., Ortega, F., Hernando, A., Alcalá, J.: Improving collaborative filtering recommender system results and performance using genetic algorithms. Knowl. Based Syst. **24**(8), 1310–1316 (2011). doi:10.1016/j.knosys.2011.06.005
4. Bobadilla, J., Ortega, F., Hernando, A., Gutierrez, A.: Recommender systems survey. Knowl. Based Syst. **46**, 109–132 (2013)
5. Bobadilla, J., Serradilla, F., Hernando, A.: Collaborative filtering adapted to recommender systems of e-learning. Knowl. Based Syst. **22**, 261–265 (2009)
6. Cechinel, C., Camargo, S.D.S., Sánchez-Alonso, S., Sicilia, M.A.: Towards automated evaluation of learning resources inside repositories. In: Manouselis, N., Drachsler, H., Verbert, K., Santos, O.C. (eds.) Recommender Systems for Technology Enhanced Learning: Research Trends and Applications, pp. 25–46. Springer, New York (2014). doi:10.1007/978-1-4939-0530-0_2
7. Chen, L.S., Hsu, F.H., Chen, M.C., Hsu, Y.C.: Developing recommender systems with the consideration of product profitability for sellers. Inf. Sci. **178**, 1032–1048 (2008)
8. Chen, W., Niu, Z., Zhao, X., Li, Y.: A hybrid recommendation algorithm adapted in e-learning environments. World Wide Web **17**, 271–284 (2014)
9. Cho, J., Kwon, K., Park, Y.: Q-rater: a collaborative reputation system based on source credibility theory. Expert Syst. Appl. **36**, 3751–3760 (2009)
10. Diaz, A., Motz, R., Rohrer, E., Tansini, L.: An ontology network for educational recommender systems. In: Santos, O., Boticario, J. (eds.) Educational Recommender Systems and Technologies: Practices and Challenges, pp. 67–93. IGI Global, Hershey (2012). doi:10.4018/978-1-61350-489-5.ch004
11. Erdt, M., Fernández, A., Rensing, C.: Evaluating recommender systems for technology enhanced learning: a quantitative survey. IEEE Trans. Learn. Technol. **8**(4), 326–344 (2015). doi:10.1109/TLT.2015.2438867
12. Fernández, A., Anjorin, M., Dackiewicz, I., Rensing, C.: Recommendations from heterogeneous sources in a technology enhanced learning ecosystem. In: Manouselis, N., Drachsler, H., Verbert, K., Santos, O.C. (eds.) Recommender Systems for Technology Enhanced Learning: Research Trends and Applications, pp. 251–265. Springer, New York (2014). doi:10.1007/978-1-4939-0530-0_12
13. Gligor, V., Wing, J.M.: Towards a theory of trust in networks of humans and computers. In: Christianson, B., Crispo, B., Malcolm, J., Stajano, F. (eds.) Security Protocols 2011. LNCS, vol. 7114, pp. 223–242. Springer, Heidelberg (2011). doi:10.1007/978-3-642-25867-1_22
14. Katarya, R., Verma, O.P.: Recent developments in affective recommender systems. Phys. A **461**, 182–190 (2016)
15. Khribi, M.K., Jemni, M., Nasraoui, O.: Automatic recommendations for e-learning personalization based on web usage mining techniques and information retrieval. Educ. Technol. Soc. **12**(4), 30–42 (2009)
16. Manouselis, N., Drachsler, H., Verbert, K., Duval, E.: Recommender Systems for Learning, p. 90. Springer, Berlin (2012)
17. Lai, C.H., Liu, D.R.: Integrating knowledge flow mining and collaborative filtering to support document recommendation. J. Syst. Softw. **82**, 2023–2037 (2009)

18. Liu, H., Motoda, H.: A survey of content-based image retrieval with high-level semantics. Pattern Recogn. **40**(1), 262–282 (2007)
19. MOVING Project web site. www.moving-project.eu. Accessed 31 Mar 2017
20. Moyano, F., Fernandez-Gago, C., Lopez, J.: A conceptual framework for trust models. In: Fischer-Hübner, S., Katsikas, S., Quirchmayr, G. (eds.) TrustBus 2012. LNCS, vol. 7449, pp. 93–104. Springer, Heidelberg (2012). doi:10.1007/978-3-642-32287-7_8
21. Mangina, E., Kilbride, J.: Evaluation of keyphrase extraction algorithm and tiling process for a document/resource recommender within e-learning. Comput. Educ. **50**, 807–820 (2008)
22. Manouselis, N., Costopoulou, C.: Analysis and classification of multi-criteria recommender systems. World Wide Web Internet Web Inf. Syst. **10**(4), 415–441 (2007)
23. Moedritscher, F.: Towards a recommender strategy for personal learning environments. In: 4th ACM Conference on Recommender Systems (RecSys 2010)/5th European Conference on Technology Enhanced Learning (EC-TEL 2010), Proceedings of the 1st Workshop on Recommender Systems for Technology Enhanced Learning. Recsystel, Procedia Computer Science, Barcelona 2010, vol. 1(2), pp. 2775–2782 (2010)
24. Nishioka, C., Scherp, A.: Profiling vs. time vs. content: what does matter for top-k publication recommendation based on twitter Profiles? In: 16th ACM/IEEE-CS Joint Conference on Digital Libraries (JCDL 2016), Newark, NJ, USA, pp. 171–180, 19–23 June 2016. http://dx.doi.org/10.1145/2910896.2910898
25. Porcel, C., Lopez-Herrera, A.G., Herrera-Viedma, E.: A recommender system for research resources based on fuzzy linguistic modeling. Expert Syst. Appl. **36**, 5173–5183 (2009)
26. Porcel, C., Moreno, J.M., Herrera-Viedma, E.: A multi-disciplinar recommender system to advice research resources in University Digital Libraries. Expert Syst. Appl. **36**, 12520–12528 (2009)
27. Pu, P., Li, C., Hu, R.: Evaluating recommender systems from the user's perspective: survey of the state of the art. User Model. User Adap. Inter. **22**(4–5), 317–355 (2012)
28. Rozewski, P., Kusztina, E., Tadeusiewicz, R., Zaikin, O.: Intelligent Open Learning Systems: Concepts, Models and Algorithms. Intelligent Systems Reference Library, vol. 22, p. 257. Springer, Berlin (2011)
29. Salehi, M.: Application of implicit and explicit attribute based collaborative filtering and BIDE for learning resource recommendation. Data Knowl. Eng. **87**, 130–145 (2013). http://dx.doi.org/10.1016/j.datak.2013.07.001
30. Santos, O.C., Boticario, J.G., Pérez-Marin, D.: Extending web-based educational systems with personalised support through user centred designed recommendations along the e-learning life cycle. Sci. Comput. Program. **88**, 92–109 (2014)
31. Santos, O.C., Boticario, J.G., Manjarrés-Riesco, A.: An approach for an affective educational recommendation model. In: Manouselis, N., Drachsler, H., Verbert, K., Santos, O.C. (eds.) Recommender Systems for Technology Enhanced Learning: Research Trends and Applications, pp. 123–143. Springer, New York (2014)
32. Santos, O.C., Saneiro, M., Boticario, J., Rodriguez-Sanchez, C.: Towards interactive context-aware affective educational recommendations in computer assisted language learning. New Rev. Hypermedia Multimedia **22**(1–2), 27–57 (2015). doi:10.1080/13614568.2015.1058428
33. Shi, F., Marini, J.L., Audry, E.: Towards a psycho-cognitive recommender system. In: ERM4CT 2015: Proceedings of the International Workshop on Emotion Representations and Modelling for Companion Technologies, Seattle, pp. 25–31, 9–13 November 2015. http://dx.doi.org/10.1145/2829966.2829968

34. Sielis, G.A., Mettouris, C., Tzanavari, A., Papadopoulos, G.A.: Context-aware recommendations using topic maps technology for the enhancement of the creativity process. In: Santos, O.C., Boticario, J. (eds.) Educational Recommender Systems and Technologies: Practices and Challenges, pp. 43–66. IGI Global, Hershey (2012). doi:10.4018/978-1-61350-489-5.ch003

35. Skulimowski, A.M.J.: Optimal strategies for quantitative data retrieval in distributed database systems. In: Proceedings of the Second International Conference on Intelligent Systems Engineering, Hamburg, IEE Conference Publication No. 395, IEE, London, pp. 389–394, 5–9 September 1994. doi:10.1049/cp:19940655

36. Skulimowski, A.M.J.: Freedom of choice and creativity in multicriteria decision making. In: Theeramunkong, T., Kunifuji, S., Sornlertlamvanich, V., Nattee, C. (eds.) KICSS 2010. LNCS (LNAI), vol. 6746, pp. 190–203. Springer, Heidelberg (2011). doi:10.1007/978-3-642-24788-0_18

37. Skulimowski, A.M.J.: Universal intelligence, creativity, and trust in emerging global expert systems. In: Rutkowski, L., Korytkowski, M., Scherer, R., Tadeusiewicz, R., Zadeh, L.A., Zurada, J.M. (eds.) ICAISC 2013. LNCS (LNAI), vol. 7895, pp. 582–592. Springer, Heidelberg (2013). doi:10.1007/978-3-642-38610-7_53

38. Skulimowski, A.M.J.: Anticipatory network models of multicriteria decision-making processes. Int. J. Syst. Sci. **45**(1), 39–59 (2014). doi:10.1080/00207721.2012.670308

39. Skulimowski, A.M.J.: Impact of future intelligent information technologies on the methodology of scientific research. In: Proceedings 16th IEEE International Conference on Computer and Information Technology, Nadi, Fiji, IEEE CPS, pp. 238–247, 7–10 December 2016. doi:10.1109/CIT.2016.118

40. Skulimowski, A.M.J., Badecka, I., Czerni, M., Klamka, J., Kluz, D., Ligęza, A., Okoń-Horodyńska, E., Pukocz, P., Rotter, P., Szymlak, E., Tadeusiewicz, R., Wisła, R.: Trends and Scenarios of Selected Information Society Technologies. Advances in Decision Sciences and Futures Studies, vol. 1, p. 634. Progress & Business Publishers, Kraków (2016)

41. Skulimowski, A.M.J., Rotter, P., Tadeusiewicz, R.: Technological evolution models of neurocognitive and vision systems in medicine: prospects and scenarios for the development of brain-computer interfaces (BCI) until 2025 [in Polish]. In: Skulimowski, A.M.J. (ed.) Scenarios and Development Trends of Selected Information Society Technologies until 2025. Final Report. Progress & Business Publishers, Kraków, pp. 234–255 (2013). http://www.ict.foresight.pl

42. Tang, T.Y., Daniel, B.K., Romero, C.: Special issue on recommender systems for and in social and online learning environments. Expert Syst. **32**(2), 261–263 (2015)

43. Tejeda-Lorente, A., Porcel, C., Bernabé-Moreno, J., Herrera-Viedma, E.: REFORE: a recommender system for researchers based on bibliometrics. Appl. Soft Comput. **30**, 778–791 (2015)

44. Van Maanen, L., Van Rijn, H., van Grootel, M., Kemna, S., Klomp, M., Scholtens, E.: Personal publication assistant: abstract recommendations by a cognitive model. Cogn. Syst. Res. **11**, 120–129 (2010)

45. Victor, P., Cornelis, C., De Cock, M.: Trust Networks for Recommender Systems. Springer, Heidelberg (2011)

46. Verbert, K., Manouselis, N., Xavier, O., Wolpers, M., Drachsler, H., Bosnic, I., Duval, E.: Context-aware recommender systems for learning: a survey and future challenges. IEEE Trans. Learn. Technol. **5**(4), 318–335 (2012)

47. Vesin, B., Milicevic, A.K., Ivanovic, M., Budimac, Z.: Applying recommender systems and adaptive hypermedia for e-learning personalization. Comput. Inform. **32**(3), 629–659 (2013)
48. Zaikin, O., Tadeusiewicz, R., Różewski, P., Busk Kofoed, L., Malinowska, M., Żyławski, A.: Teachers' and students' motivation model as a strategy for open distance learning processes. Bull. Pol. Acad. Sci. Tech. Sci. **64**(4), 943–955 (2016). doi:10.1515/bpasts-2016-0103
49. Zapata, A., Menendez, V.H., Prieto, M.E., Romero, C.: A framework for recommendation in learning object repositories: an example of application in civil engineering. Adv. Eng. Softw. **56**, 1–14 (2013)
50. Zhou, M., Xu, Y.: Challenges to use recommender systems to enhance meta-cognitive functioning in online learners. In: Santos, O., Boticario, J. (eds.) Educational Recommender Systems and Technologies: Practices and Challenges, pp. 282–301. IGI Global, Hershey (2012)

Supporting BPMN Process Models with UML Sequence Diagrams for Representing Time Issues and Testing Models

Anna Suchenia (Mroczek)[1], Krzysztof Kluza[2(✉)], Krystian Jobczyk[2,3],
Piotr Wiśniewski[2], Michał Wypych[2], and Antoni Ligęza[2]

[1] Cracow University of Technology, ul. Warszawska 24, 31-155 Kraków, Poland
asuchenia@pk.edu.pl
[2] AGH University of Science and Technology,
al. A. Mickiewicza 30, 30-059 Krakow, Poland
{kluza,jobczyk,wpiotr,mwypych,ligeza}@agh.edu.pl
[3] University of Caen, Caen, France
krystian.jobczyk@unicaen.fr

Abstract. Business Process Model and Notation is a standard for process modeling. However, such models do not specify the time issues such as time of performing tasks or time of utilizing the resources. We propose the complementary UML sequence model generated from the BPMN model. Such a model can support time specification and provide direct time visualization. It is also suitable for validation in terms of time matters by domain experts as well as can be used to estimate and test methods in the systems based on random examination of the critical paths.

1 Introduction

Process models are commonly exploited for supporting communication between business domain experts, software engineers as well as other people with technical knowledge. However, crucial characteristics – i.e. time – might be represented using pure BPMN in a very limited way. Thus, other notations or descriptions must be provided for specification of such matters.

As BPMN (Business Process Model and Notation) [1] and UML (Unified Modeling Language) [2] are standardized by the OMG (Object Management Group) consortium, their integration and interoperability is not provided. The aim of our research is to examine a possibility of combining these two widely used standards: BPMN and UML. Especially, this paper is concerned with the translation from the BPMN representation to UML model, which can be interpreted by business analysts or software engineers in terms of time issues for validating the process.

The paper is supported by the AGH UST research grant.

L. Rutkowski et al. (Eds.): ICAISC 2017, Part II, LNAI 10246, pp. 589–598, 2017.
DOI: 10.1007/978-3-319-59060-8_53

1.1 Paper Motivation

A possibility of time specification is usually very limited in BPMN-models. In particular, they do not support any specification of such temporal issues as task preforming or temporal aspects of utilizing the resources. Therefore, such matters should be specified outside the original model. We also need some methods of test pattern generations. Unfortunately, the method ATPG (Automatic Test Pattern Generation) is not effective for testing object-oriented systems based on UML or BPMN models. As a unique remedy for this difficulty, an interesting alternative to the ATPG testing Object-Oriented Systems by Using a Random Sequence Diagrams of UML in [3] was proposed.

Nevertheless, the main problem – the reliability of the testing and estimation of a random sequence length (when testing the system with random sequences) – still remains unsolvable. This problem requires to achieve a desired level of quality tests.

These shortcomings form the main motivation factor – to propose a new complementary UML sequence model generated from the BPMN model. Such a UML model seems to be also suitable for validation of models in terms of time matters by domain experts. In addition, some new test methods – as naturally associated to these models – will be proposed. It seems that this method may support the estimation problem of the random sequence length. This method allows users to test the random system object using data from UML diagrams. It seems that this method constitutes a promising tool for extracting additional knowledge from UML-model based on the initial BPMN one.

The discussed method tests the system in the pseudorandom manner by computing the probability (P_k) of the every tested object and by establishing the length (L) of random path. However, the quality of the length of the test which warrants the suitable fault coverage is essential. This method will be described in detail later. We only underline that it is designed to estimate and to test the undetectable error in the object system by using UML diagrams. This task will be achieved by proving which of the tested paths give greater probability of tested objects to be selected, and which of them give smaller one. Finally, these tests will examine random critical paths.

The rest of the paper is organized as follows. In Sect. 2, the OMG standards– exploited later in the paper analysis – are presented. In addition, a short overview of the possibilities of representing time issues in these standards is also discussed. Section 3 describes the random testing approach. In Sect. 4, we provide our proposal of the translation from BPMN to UML model and in Sect. 5 the illustrative case study example for this approach is presented. Section 6 summarizes the paper results.

2 Standards for Modeling Systems

2.1 Business Process Model and Notation

BPMN [1,4] provides a metamodel for a representing business processes. These processes may be defined as collections of related tasks for providing certain services or producing specific products for customers [5]. Although BPMN provides several diagrams, mainly process diagrams are used, as the modeling process can be supported with workflow tools and such models can be directly executed in process engines [6].

2.2 Time Representation in BPMN Process Models

If it comes to the time-related issues in BPMN, some of them can be represented directly in the model (see Fig. 1). Timers are supported as *Timer start* and *Timer intermediate events* (Fig. 1(a)), specified with one of the following properties:

- *time date* – it specifies a fixed date when trigger will be fired,
- *time duration* – it specifies how long the timer should run before it is fired,
- *time cycle* – it specifies repeating interval, which can be useful for starting process periodically or for sending multiple reminders for overdue user task.

Such timers are often used with the BPMN event-based gateways. Event-based gateways with timers (Fig. 1(b)) work like exclusive gateways. However, there is no condition that must be satisfied but some events which must occur are specified in this case. Timer boundary events support following additional or alternative control flow when the timer is fired (Fig. 1(c)). They can (but do not necessary have to) interrupt the task or subprocess associated to them. Finally, event-based subprocesses are simply such subprocesses which are started by a timer event (Fig. 1(d)).

Apart from the time issues which are basically supported by the existing BPMN elements, some issues can be modeled indirectly or time related issues can be enforced in the model using very complex combination of elements [7]. As there are many equivalences in BPMN [8], some time patterns are not the only one that are possible. One can also notice that the model is quite complex and requires additional BPMN elements in some cases.

Simpler models for these relations might be found in [9, 10]. However, they require additional non-standardized elements. Thus, they extend the BPMN notation. Furthermore, time-related aspects of process models were surveyed by Cheikhrouhou et al. in [11, 12]. Their survey focused on the existing approaches to specifying and verifying temporal aspects of processes, not focusing on processes represented using the BPMN notation, but rather temporal constraints specification methods.

The detailed description how the existing time patterns from the literature can be used in BPMN as well as how they can be supported with temporal logics can be find in our previous work [7]. The objective of time patterns [9, 13] is to facilitate the analysis and comparison of PAIS (Process-Aware Information Systems) Time constraints for specific patterns may be used for generating optimized plans for processes modeled in a declarative way [14]. Temporal information along with resources can also be modeled using a modified Petri Net [15].

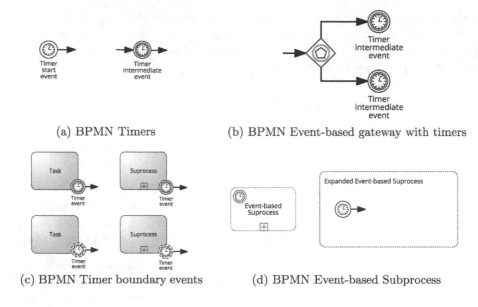

(a) BPMN Timers

(b) BPMN Event-based gateway with timers

(c) BPMN Timer boundary events

(d) BPMN Event-based Subprocess

Fig. 1. Time Representation in BPMN process models

2.3 Unified Modeling Notation

In practical software design, UML is the standard for modeling software applications [16]. This graphical modeling language, which has become the dominant notation among software engineers, provides diagrams to capture many views like user requirements (use case diagrams), collaboration between parts of software (collaboration diagrams), implementation details (class, sequence or statechart diagrams) as well as software integration (component and deployment diagrams).

2.4 Time Representation in UML Sequence Diagrams

UML Sequence diagrams support time issue specification [2], such as: duration constraints and observations or time constraints and observations (see Fig. 2).

3 Random Testing

Quality tests (Q_T) determine the quality of the tests that are recorded during an application testing. Qualitative research can be determined by two parameters: a code coverage and a case coverage [3]. In [17] test quality (quality of testing) was used for random testing methods. This method includes a set of test scenarios S which can occur, and the errors E = $\{e_1, e_2, ..., e_l\}$. Assume that a probability space $\Omega = \left(S, \mathcal{F}, p_i\right)$ is given, where $|S| = n$ for a fixed n, $\mathcal{F} \subseteq |n|^{|k|}$ and $p_i : \mathcal{F} \rightarrow [0,1]$ is a usual probability (measure) and $i \in \mathcal{A}$ – a set of agents.

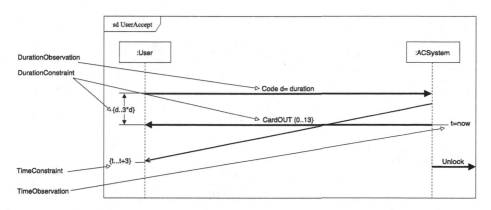

Fig. 2. A UML sequence diagram presenting ways of time specification [2]

Denote also by S_k a set of k-element sequences $\{s_k^j\}$ of elements from S which detects an error e_j, where $1 \leq j \leq l$.

This allows to observe – due to [17] – that:

– The probability $p_i(\{s_k^j\})$ $(i \in \mathcal{A})$ of the fact (that) the random sequence does not test the error e_j is equal:

$$p_i(\{s_k^j\}) = 1 - \frac{|S_k|}{|S|} \tag{1}$$

– The probability p of the fact that the random sequence does not test all errors from E satisfies the inequality:

$$max_i p_i(\{s_k^j\}) \leq p \leq \Sigma_i p_i(\{s_k^j\}) \tag{2}$$

These 'local' and 'global' computations may constitute – depending on the results to be obtained – a promising measure of a system confirmation or falsification. It corresponds to the concept of measuring the quality test (testing quality of Q_T) – introduced in [17]. Since – as stated above – $1 - Q_T = 1 - p$, the value Q_T essentially depends on the 'global' probability p and the 'local' probability p_i for each sequence $\{s_k^j\} \in S^k$ detecting errors $e_j \in E$ and for an agent $i \in \mathcal{A}$ and $1 \leq j \leq l$.

It is easy to observe that each such local and global probability encodes some piece of information about the lenght of the random sequence – as observed in [17]. In order to compute the required length of the random sequence one can use the following reasoning. Taking into account the Eq. (1) for p_i and the inequalities (2), we can put:

$$Q_D = | 1 - max_{pk} | . \tag{3}$$

Since (from Eq. (2)):

$$max_i\{p_i\} \leq \Sigma_i p_i, \tag{4}$$

thus we also obtain:

$$1 - max_i\{p_i\} \leq 1 - \Sigma_i p_i = 1 - \Sigma_i(1 - \frac{|S_k|}{|S|}) \leq 1 - p. \tag{5}$$

If we establish now p to achieve:

$$Q_D = 1 - p \tag{6}$$

and write $Q_D^L = (1 - p)^L$, where $1 \leq L \leq k$, then the length of the required random sequence:

$$L = \frac{logQ_D^L}{log(1 - p)}. \tag{7}$$

To summarize this reasoning – this method tests the system in the pseudo-random manner by computing the probabilities of every tested objects to make them us to establish the required length (L) of the test random sequence.

4   Translation Algorithm

In this section, a translation algorithm – describing how to transform a BPMN model into a UML model – is proposed. This algorithm may be depicted as follows:

1. Create a UML sequence diagram.
2. For each lane create a lifeline representing an individual participant in the interaction. If there are no tasks in the pool (collaborative public process model), create an actor representing such an empty pool.
3. According to BFS (breadth-first search) algorithm, traverse the process model (graph) performing the following actions for every visited element or flow:
   (a) for a message flow incoming from the empty pool, add a call message corresponding to the flow (a call message from the actor to the lifeline of another participant),
   (b) for a message flow outgoing to the empty pool, add a return message with the object associated with this message flow (a return message from the participant to the actor),
   (c) for a sequence flow with associated data object, add a return message passing the information (from data object) back,
   (d) for a task or subprocess:
      i. if there is no incoming message flow as well as no incoming sequence flow with associated data object, add a sequence message defining a communication between participants (previous and current participant) or a self message if the current participant is the same as the previous one.
      ii. if there is an incoming message flow or sequence flow with associated data object, add a self message in the corresponding participant.
      iii. if there is a loop marker in the task or subprocess, add a loop interaction fragment over the corresponding flows.

Table 1. Mapping of elements from BPMN to UML

BPMN process model	UML sequence diagram
Message flow	Call, self or return message
Sequence flow with data object	Return message
Task or subprocess	Sequence message or self message
Loop marker	Loop interaction fragment
Exclusive gateway	Alternative combined fragment
Parallel gateway	Parallel interaction fragment

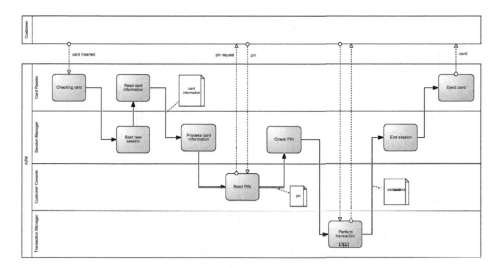

Fig. 3. A BPMN model specifying the ATM case example

To sum up, the mapping presented in Table 1 was specified.

In the following section, the case study presenting our solution using the ATM benchmark example (see Fig. 3) is described.

5   Case Study Example

Our solution will be presented using an excerpt from the ATM case study example presented in Fig. 3. ATM can handle one person at a time. Support comes down to insert the card into the reader and enter a PIN. These data will be handled by the bank for verification. After PIN authorization, the user can make multiple transactions during a single session. The card is returned only after the last transaction. We manually transformed the BPMN model presented in Fig. 3 into the UML model. The result of the translation using the algorithm provided in the previous section is presented in Fig. 4.

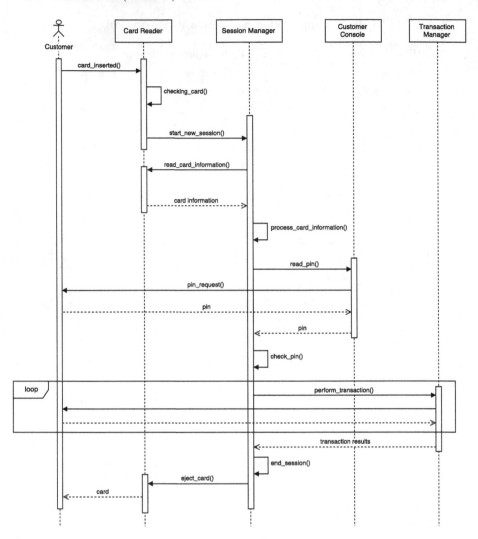

Fig. 4. A UML model specifying the ATM case example

5.1   Testing Results

Let us assume that a 3-elemental set of objects (agents) is given:

$$\mathcal{A} = \{\text{actor(A)}, \text{usecase(U)}, \text{scenario(S)}\}$$

Then, we randomly select and determine the probability for a given actor (P_A), the use case (P_U) and scenario (P_S). In our example, we discuss the actor for a use case Customer, Session Manager and its scenario. First step consists in selecting a random actor Customer. Then the use case for the actor Session Manager is selected. Next, the initial test cases can be identified in the design process. The next step is selection of the test scenario (Check PIN). The probability of drawing Customer is 0.33, the case of Session Manager is equal to 0.33, and the scenario Card Reader 0.02. Then, select the next item. The algorithm terminates when it reaches the final outcome. The results of probability and length (Eq. 7) of the random sequence. Information concerning the probability of drawing a path and the length of random sequences. Probability values of detectable error detection hardest are $P_E = 0.16$, random sequence of length $L = 37.88$, while random sequence of length is $L = 369.55$ and it is the longest path in the example.

6 Concluding Remarks

It seems that the paper objectives – to provide an interoperability solution for transforming a BPMN process model into a UML sequence model – have already been satisfied. Indeed, the algorithm which describes such translation was presented. The result of the algorithm is a UML sequence model consistent with the source BPMN model. As the UML model natively supports modeling time issues, our solution can be used for validating such issues by people aware of these notations like business analysts, software engineers, as well as other people familiar with the company processes. Moreover, the UML model can be formally described [3] for estimating and testing methods of the system objects using random sequences examination of the critical paths. In future, such description can be transformed to the algebraic logical meta-model for solving the process optimization problems [18,19].

References

1. OMG: Business Process Model and Notation (BPMN): Version 2.0 specification. Technical report formal/2011-01-03, Object Management Group, January 2011
2. OMG: Unified Modeling Language version 2.1.2 infrastructure specification. Technical report formal/2007-11-04, Object Management Group. http://www.omg.org/cgi-bin/doc?formal/2007-11-04.pdf
3. Mroczek, A.: Testing object-oriented systems by using a random sequence of UML diagrams. Int. J. Comput. Sci. Netw. Secur. (IJCSNS) 13(11), 42 (2013)
4. Chinosi, M., Trombetta, A.: BPMN: an introduction to the standard. Comput. Stand. Interfaces 34(1), 124–134 (2012)
5. Lindsay, A., Dawns, D., Lunn, K.: Business processes - attempts to find a definition. Inf. Softw. Technol. 45(15), 1015–1019 (2003). Elsevier
6. Geiger, M., Harrer, S., Lenhard, J., Wirtz, G.: BPMN 2.0: the state of support and implementation. Future Gener. Comput. Syst. (2017)
7. Kluza, K., Jobczyk, K., Wiśniewski, P., Ligęza, A.: Overview of time issues with temporal logics for business process models. In: 2016 Federated Conference on Computer Science and Information Systems (FedCSIS), pp. 1115–1123. IEEE (2016)

8. Kluza, K., Kaczor, K.: Overview of BPMN model equivalences: towards normalization of BPMN diagrams. In: Canadas, J., Nalepa, G.J., Baumeister, J. (eds.) 8th Workshop on Knowledge Engineering and Software Engineering (KESE 2012) at the Biennial European Conference on Artificial Intelligence (ECAI 2012), Montpellier, France, 28 August 2012, pp. 38–45 (2012)

9. Lanz, A., Weber, B., Reichert, M.: Time patterns for process-aware information systems. Requirements Eng. **19**(2), 113–141 (2012)

10. Gagne, D., Trudel, A.: Time-BPMN. In: 2009 IEEE Conference on Commerce and Enterprise Computing, pp. 361–367, July 2009

11. Cheikhrouhou, S., Kallel, S., Guermouche, N., Jmaiel, M.: A survey on time-aware business process modeling. In: International Conference on Enterprise Information Systems (ICEIS), 10 p., July 2013

12. Cheikhrouhou, S., Kallel, S., Guermouche, N., Jmaiel, M.: The temporal perspective in business process modeling: a survey and research challenges. SOCA **9**(1), 75–85 (2015)

13. Lanz, A., Weber, B., Reichert, M.: Workflow time patterns for process-aware information systems. In: Bider, I., Halpin, T., Krogstie, J., Nurcan, S., Proper, E., Schmidt, R., Ukor, R. (eds.) BPMDS/EMMSAD 2010. LNBIP, vol. 50, pp. 94–107. Springer, Heidelberg (2010). doi:10.1007/978-3-642-13051-9_9

14. Barba, I., Lanz, A., Weber, B., Reichert, M., Valle, C.: Optimized time management for declarative workflows. In: Bider, I., Halpin, T., Krogstie, J., Nurcan, S., Proper, E., Schmidt, R., Soffer, P., Wrycza, S. (eds.) BPMDS/EMMSAD 2012. LNBIP, vol. 113, pp. 195–210. Springer, Heidelberg (2012). doi:10.1007/978-3-642-31072-0_14

15. Xie, J., Tang, Y., He, Q., Tang, N.: Research of temporal workflow process and resource modeling. In: Proceedings of the 9th International Conference on Computer Supported Cooperative Work in Design, vol. 1, pp. 530–534. IEEE (2005)

16. Hunt, J.: Guide to the Unified Process featuring UML, Java and Design Patterns. Springer, London (2003)

17. David, R., Blanchet, G.: About random fault detection of combinational networks. IEEE Trans. Comput. **25**(6), 659–664 (1976)

18. Dudek-Dyduch, E., Kucharska, E., Dutkiewicz, L., Rączka, K.: ALMM solver - a tool for optimization problems. In: Rutkowski, L., Korytkowski, M., Scherer, R., Tadeusiewicz, R., Zadeh, L.A., Zurada, J.M. (eds.) ICAISC 2014. LNCS, vol. 8468, pp. 328–338. Springer, Cham (2014). doi:10.1007/978-3-319-07176-3_29

19. Rączka, K., Dudek-Dyduch, E., Kucharska, E., Dutkiewicz, L.: ALMM solver: the idea and the architecture. In: Rutkowski, L., Korytkowski, M., Scherer, R., Tadeusiewicz, R., Zadeh, L.A., Zurada, J.M. (eds.) ICAISC 2015. LNCS, vol. 9120, pp. 504–514. Springer, Cham (2015). doi:10.1007/978-3-319-19369-4_45

Simulation of Multi-agent Systems
with Alvis Toolkit

Marcin Szpyrka[✉], Piotr Matyasik, Łukasz Podolski, and Michał Wypych

Department of Applied Computer Science, Faculty of Electrical Engineering,
Automatics, Computer Science and Biomedical Engineering,
AGH University of Science and Technology,
al. Mickiewicza 30, 30-059 Kraków, Poland
{mszpyrka,ptm,podolski,mwypych}@agh.edu.pl

Abstract. The paper presents a method of using the Alvis formal modelling language and related software to model and simulate multi-agent systems. The approach has been illustrated with an example of a railway traffic management system for a real train station. One of the main advantages of this approach is the possibility of including artificial intelligence (AI) systems encoded in Haskell into Alvis models. Moreover, Alvis models can be developed at the level very close to the final implementation of the corresponding real system. Thus simulation logs can be treated as a virtual prototype logs.

Keywords: Alvis language · Alvis Toolkit · Multi-agent systems · Simulation · AI systems

1 Introduction

A multi-agent system is defined as a system composed of multiple interacting computing elements called agents with two important features: agents perform autonomous (to some degree) actions and are capable of interacting with other agents [17,18]. The implementation of multi-agent systems requires taking into account issues such as communication between agents, synchronization, mutual exclusion over shared resources, livelocks, deadlocks, etc. In this paper we want to show that both the Alvis language and the related software can be used for modelling, simulation, and even formal verification of multi-agent systems. Alvis has been designed as a formal language for modelling of concurrent systems. Both simulation and formal verification of models may take the passage of time into account (we define execution time of each statement). Execution of a model (simulation or generation of a labelled transition system – LTS graph) is based on the intermediate model implemented in Haskell (code generated by the Alvis Compiler). The generated source code is available for the user, which allows to modify it before the execution e.g. including user-defined Haskell functions. Such functions are used in expressions that assign new values to parameters of an agent. Evaluation of such an expression is seen as a single model step but may

© Springer International Publishing AG 2017
L. Rutkowski et al. (Eds.): ICAISC 2017, Part II, LNAI 10246, pp. 599–608, 2017.
DOI: 10.1007/978-3-319-59060-8_54

include complex operations on the data stored by the agent, e.g. one can include artificial intelligence techniques into the model, if a decision support multi-agent system is being developed [6]. The compression of a complex logic and computation into a single model step speeds up simulation and allows verification of even more complex systems that suffer from combinatorial explosion.

The paper is organized as follows. A short introduction to the tools supporting Alvis is presented in Sect. 2. The railway traffic management system case study is described in Sect. 3. Conclusions and future work are described in the final section.

2   Introduction to Alvis Toolkit

The Alvis language [14–16] and related tools (Alvis Toolkit) are being developed at AGH-UST in Kraków, Department of Applied Computer Science. The project website is located at (http://alvis.kis.agh.edu.pl). The aim of the project is to provide a formal language with engineering-like look and style.

An Alvis model is a system of *agents* that run concurrently, communicate with each other, compete for shared resources, etc. Alvis provides two types of agents – *active* that are threads of control in a concurrent system and *passive* that provide a mechanism for the mutual exclusion and data synchronization. A model is composed of two parts (levels). The graphical level is called the *communication diagram* [14]. The layer takes the form of the directed graph with nodes representing agents and edges representing communication channels between ports of agents. Agents may be grouped into subsystems represented by the so-called *hierarchical agents*, which introduce a hierarchical structure to the communication diagram. The second level contains the Alvis code that defines the behaviour of active and passive agents. Alvis provides only a few control statements and is supported by the Haskell functional programming language [12]. A short survey of the Alvis graphical components and the code statements is given in Fig. 1. For more details see the project website.

Models are developed using *Alvis Editor*. The editor is implemented in the Java language and provides the following functionality: basic hierarchy editor/viewer, visual diagrams editing, textual editor with syntax highlighting, code folding and code completion. An Alvis model is stored as an XML file and translated into its Haskell representation using *Alvis Compiler*. This Haskell model representation may be used to generate the model state space (in the form of the labelled transition system – LTS graph [16]) and to simulate the model behaviour (including step-by-step simulation).

One of the main advantages of this approach is the accessibility of the Haskell source code for the users. A developer may modify the code including for example his own verification methods implemented in Haskell. Moreover, he may modify the main function to decide what kind of operations to be performed on the model.

Let us focus on the Alvis *exec* statement, which is crucial for the presented approach. The statement is represented by the assignment operator (=) that

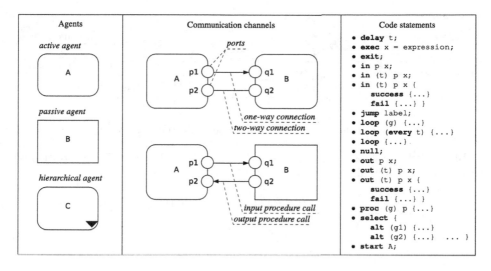

Fig. 1. Elements of Alvis language

takes an agent's parameter as its left-hand side argument and a Haskell expression as the right-hand side argument. The statement evaluates the expression and assigns its result to the parameter. This is represented as a single step in the description of the corresponding agent behaviour. The most important feature of this statement is that the Haskell expression may contain any user-defined function. For example, it enables a developer to include artificial intelligence systems into an agent's behaviour. In other words, an agent may use rule-based system, decision tree, neural network, etc. to take decisions as long as it is possible to implement such systems in Haskell. An example of including a rule-based system into an agent code is presented in Sect. 3.

3   Case Study

The model of a railway traffic management system for the Czarna Tarnowska train station was chosen to illustrate how to use the Alvis Tools to simulate multi-agent systems. The station belongs to the Polish railway line no 91 from Kraków to Medyka [13]. The system is used to ensure safe passing of trains through the station. It collects some information about current railway traffic and uses a rule-based system to choose routes for trains. The topology of the train station with original signs is shown in Fig. 2. The letters A, B, D, etc. stand for light signals, the symbols Z3, Z4, Z5, etc. stand for turnouts and JTA, JTB, JT1, etc. stand for track segments.

A train can ride through the station only if a suitable route is prepared i.e., suitable track segments are free, related turnouts and light signals are set, and exclusive rights to these elements are guaranteed for the train. There are six incoming routes and six outgoing routes in this system. For example, the symbol

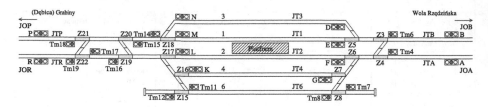

Fig. 2. Topology of the train station (Czarna Tarnowska)

Table 1. Relationships between routes: x – mutually exclusive (different position of turnouts); xx – mutually exclusive (safety reasons)

	B1	B2	B3	B4	R2	R4	F2W	G2W	K1D	L1D	M1D	N1D
B1	–	x	x	x								xx
B2	x	–	x	x	xx		x	x				
B3	x	x	–	x								
B4	x	x	x	–	xx	xx	x	x				
R2		xx		xx	–	x		xx	x	x		
R4				xx	x	–			x	x		
F2W	x		x				–	x				
G2W		x		x	xx		x	–				
K1D				x	x				–	x	x	x
L1D				x	x				x	–	x	x
M1D									x	x	–	x
N1D	xx								x	x	x	–

B4 stands for the incoming route from the light signal B to the track no. 4. The symbol F2W stands for the outgoing route from the track no. 2 (from the light signal F) to the right (to Wola Rzędzińska), etc. The route B4 can be used by a train only if: turnouts 7, 8, 15, 16 are closed, turnouts 3, 4, 6 are open, and the track segments JTB, JT4, JZ4/6 (a segment between turnouts 4 and 6), JZ7 (diagonal segment leading to the turnout 7) and JZ16 are free. Table 1 shows which routes are mutually exclusive.

The Alvis model of the railway traffic management system is composed of the following agents:

- *CentralController* – receives requests from passing trains and based on the current traffic determines the incoming and/or outgoing routes;
- *LightSignals* – manages the state of the light signals;
- *TrackSegments* – stores information about the state of the track segments (passive agent);
- *Train1–8* – simulate the train traffic in the considered system;
- *TrafficController* – manages the train traffic.

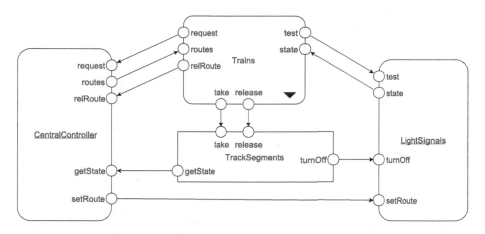

Fig. 3. Communication diagram – primary page

The communication diagram of the model is shown in Fig. 3 (the primary page) and Fig. 4. The *Trains1–8* and *TrafficController* agents are placed on a separate page, and are represented by the hierarchical agent *Trains* on the primary page. This way one can easily manage the number of trains multiplying *Train* agents on the subpage. All of these agents are connected with *Central-Controller*, *LightSignals* and *TrackSegments* agents in the same manner as the hierarchical agent (rules for the application of hierarchical diagrams in Alvis are described in [14]).

To introduce the possibility of defining the behaviour of agents in Alvis we will focus on the *CentralController* agent. The code layer for the agent is presented in Listing 1.1. The dynamics of the agent is based on an infinite loop in which it tests whether another agent initiated a communication on the port *request* (request for a new route) or *relRoute* (release of an already completed route). Determination of a new route is based on the rule-based system taken from [13] but encoded using the Haskell language (see Listing 1.1 line 11). A piece of the *findRoutes* function source code is shown in Listing 1.2.

Listing 1.1. *CentralController* – implementation

```
1   agent CentralController {
2     train :: (Int,Int) = (0,0);                    -- current train type and position
3     trainRoutes :: (Int,Int) = (0,0);                -- new routes to set
4     route :: Int = 0;                                -- route to release
5     setRoutes :: [Int] = [0,0,0,0,0,0,0,0,0,0,0,0];  -- already set routes
6     segmentsState :: [Int] = [0,0,0,0,0,0,0,0,0,0,0,0]; -- state of track segments
7
8     loop {
9       in (0) request train { success {
10        in getState segmentsState;
11        trainRoutes = findRoutes train segmentsState setRoutes;
12        out setRoute trainRoutes;
13        out routes trainRoutes;
14        setRoutes = updateSetRoutes trainRoutes setRoutes; }}
15      in (0) relRoute route { success {
16        setRoutes = decList (route-1) setRoutes;  }}
17      delay 1;
18    }}
```

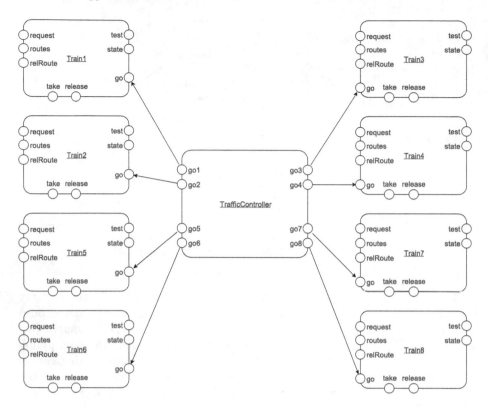

Fig. 4. Communication diagram – *Trains* subpage

Listing 1.2. *findRoutes* function implementation

```
1    findRoutes :: (Int,Int) -> [Int] -> [Int] -> (Int,Int)
2    findRoutes (tt,jt) [jta,jtb,jt1,jt2,jt3,jt4,jt6,jtp,jtr,joa,job,jop,jor]
3       [b1,b2,b3,b4,r2,r4,f2w,g2w,k1d,l1d,m1d,n1d]
4       | tt==1 && jt==11 && jt1==0 && jop==0 && b1==0 && b2==0 && b3==0
5         && b4==0 && k1d==0 && l1d==0 && m1d==0 && n1d==0 = (1,11) -- (b1, m1d)
6       | tt==1 && jt==11 && jt1 > 0 && jt3==0 && jop==0 && b1==0 && b2==0
7         && b3==0 && b4==0 && k1d==0 && l1d==0 && m1d==0 && n1d==0 = (3,12) -- b3, n1d
8    -- ...
9       | otherwise = (0,0)
```

The rule-based system contains 27 conditional and 2 decision attributes. The conditional attributes provide information about the considered train type (passing through the station (1) or stopping at the platform (0)), its position, current status of track segments (a segment is free (0) or it is taken (>0)), and already set routes (a route is already set (1) or not (0)). The decision attributes represent the incoming and outgoing routes that will be prepared for the train or 0 if the corresponding route cannot be set at the moment or is superfluous.

Depending on the *main* function included into the Haskell source code, the middle-stage model can be used to generate the complete LTS graph or to simulate the model behaviour (interactively or automatically). The LTS graph or a trace of a simulation path can be saved to a file using several formats, among others *dot* or CSV. In case of the considered model the simulation collecting traces was performed. The length of each trace was 50000 steps. The simulation was performed using 20 different scenarios of train traffic which was achieved by modifying the behaviour of the *TrafficController* agent. A small portion of the collected data is shown in Listing 1.3.

Listing 1.3. Collected traces of simulation paths (CSV format)

```
"timeout(CentralController),timeout(LightSignals),timeout(Train1)",0,0,0,0,0,...
"loop(CentralController),loop(LightSignals),loop(Train1)",0,0,0,0,0,0,0,0,0,...
"in(CentralController.request),in(LightSignals.test),out(Train1.request)",0,...
"in(CentralController.relRoute),in(LightSignals.turnOff)",0,0,0,0,0,0,0,0,0,...
"delay(CentralController),in(LightSignals.setRoute)",0,0,0,0,0,0,0,0,0,0,0,0,...
"delay(LightSignals)",0,0,0,0,0,0,0,0,0,0,0,0,0,0,0,0,0,0,0,0,0,0,0,0,0,0,0,...
"time",1,0,0,0,0,0,0,0,0,0,0,0,0,0,0,0,0,0,0,0,0,0,0,0,0,0,0,0,1,0,1,0,0,0,0,...
"timeout(CentralController),timeout(LightSignals)",0,0,0,0,0,0,0,0,0,0,0,0,0,...
"loop(CentralController),loop(LightSignals)",0,0,0,0,0,0,0,0,0,0,0,0,0,0,0,...
"in(CentralController.request),in(LightSignals.test)",0,0,0,0,0,0,0,0,0,0,0,...
"in(CentralController.getState),in(LightSignals.turnOff),in(Train1.routes)",...
"in(LightSignals.setRoute),out(TrackSegments.getState)",0,2,3,0,0,0,0,0,0,0,...
"delay(LightSignals),exit(TrackSegments)",0,2,3,0,0,0,0,0,0,0,0,0,0,0,0,0,0,...
"exec(CentralController)",0,2,3,0,0,0,0,0,0,0,0,0,0,0,0,0,0,0,1,1,0,0,0,...
"out(CentralController.setRoute)",0,2,3,0,11,0,0,0,0,0,0,0,0,0,0,0,0,0,0,1,...
"time",1,2,3,0,11,0,0,0,0,0,0,0,0,0,0,0,0,0,1,1,0,0,0,0,0,0,0,0,0,0,0,...
```

The data obtained from the simulation were processed using several Python scripts. First of all, transitions labels and values of parameters were filtered to form the raw CSV files (see Listing 1.3). Then, received records were extended with the information about the source model, the simulation number, the step number and accumulated model-time passed up to that step. Finally, the data were placed in a database.

To analyse the collected data the Python [4] and R languages [8] (with additional frameworks) were used. First, the data were used for preliminary assessment of the correctness of the model – values taken by the parameters were examined to find possible inconsistencies in the model. Then, a series of statistical analyses focused mainly on time properties was performed, e.g.

- the maximum number of trains running through any track segment at the same time;
- *throughput* – the number of trains passing through the station within a certain period of time;
- the average (maximum) time a train waits to enter the station;
- the average (maximum) number of requests sent to the *central controller* before obtaining a route;
- the average (maximum) time a train passes through the station;
- intensity of use of individual track segments.

For example, simulated time of passing through the stations for 8 types of trains used in one of the scenarios is presented in Fig. 5.

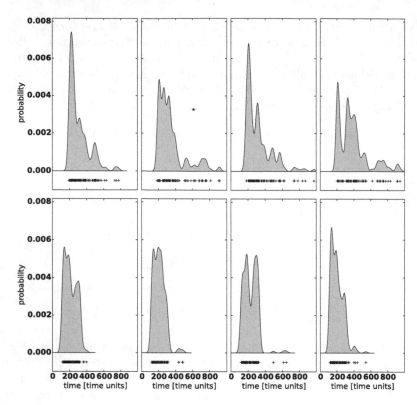

Fig. 5. Simulated time of passing through the stations for 8 types of trains used in the model. Every subplot presents single train statistics and the density plots with measurement points marked as black crosses.

4   Conclusions and Future Work

Alvis is a formal modelling language and with its syntax, similar to high level programming languages, it seems to be more convenient to use by engineers than formalisms like Petri nets, process algebras or time automata. Comparison between the model presented in this paper and the same model encoded in Petri nets [13] seems to confirm that thesis. The main advantage of formal methods is the ability to generate a state space of the given model and to use mathematical methods to check model properties. In case of more complex systems, generation of the complete state space may be impossible due to the state space explosion problem. However, even in such situations, formal models with unambiguous semantics are useful to better understand the developed system and can be used for simulation of the system behaviour. It has been shown in the paper that Alvis and related software can be used to model and simulate multi-agent systems. One of the main advantages of this approach is the possibility of modelling real systems at the level very close to their final implementation. Moreover, a developer may include AI systems in the form of rule-based systems, decision trees, neural networks, etc. into Alvis models. Alvis architecture is quite flexible. Any system that can be encoded as a Haskell function can be easily included into model. Moreover, such operation does not enlarge resulting system significantly. It will be still represented as a single model transition.

Alvis was not designed for simulating or executing multi-agent systems, however it has some small advantages over well known environments like MAGE [7], MaDKit [3], Repase [11] or Jade [1]. First of all, its communication behaviour is strictly, mathematically defined. User can apply it as a conversation framework. Moreover, multi-agent world is quite dominated by the Java language. Alvis runtime is implemented in Haskell which provides highly optimized executable code from the start. This language has however steep learning curve, but due to its type system it provides very flexible setting for experimentation. Unfortunately, current implementation of Alvis runtime lacks distribution features, thus its ability to execute very large systems might be limited. However, there are ongoing experiments to exploit Haskell's deterministic parallelism [9,10] and distribution [2].

Future work on the simulation environment based on Alvis will focus on developing export functions for direct cooperation with popular data mining tools like the R language or Weka and Pentaho tools. Another promising idea is to develop a dedicated Haskell-based query language to explore LTS-graphs or simulation logs. Moreover, we are going to check the suitability of Alvis models for modelling malware detection systems [5].

References

1. Bellifemine, F., Caire, G., Poggi, A., Rimassa, G.: Jade: a software framework for developing multi-agent applications. lessons learned. Inf. Softw. Technol. **50**(1–2), 10–21 (2008)
2. Epstein, J., Black, A., Peyton-Jones, S.: Towards haskell in the cloud. SIGPLAN Not. **46**(12), 118–129 (2011)
3. Gutknecht, O., Ferber, J.: The MADKIT agent platform architecture. In: Wagner, T., Rana, O.F. (eds.) AGENTS 2000. LNCS, vol. 1887, pp. 48–55. Springer, Heidelberg (2001). doi:10.1007/3-540-47772-1_5
4. Idris, I.: Python Data Analysis. Packt Publishing Ltd. (2014)
5. Jasiul, B., Śliwa, J., Gleba, K., Szpyrka, M.: Identification of malware activities with rules. In: Ganzha, M., Maciaszek, L., Paprzycki, M. (eds.) Proceedings of the 2014 Federated Conference on Computer Science and Information Systems. Annals of Computer Science and Information Systems, vol. 2, pp. 101–110. IEEE (2014). http://dx.doi.org/10.15439/2014F265
6. Krzywicki, D., Faber, L., Byrski, A., Kisiel-Dorohinicki, M.: Computing agents for decision support systems. Future Gener. Comput. Syst. **37**, 390–400 (2014)
7. Laleci, G.B., Kabak, Y., Dogac, A., Cingil, I., Kirbas, S., Yildiz, A., Sinir, S., Ozdikis, O., Ozturk, O.: A platform for agent behavior design and multi agent orchestration. In: Odell, J., Giorgini, P., Müller, J.P. (eds.) AOSE 2004. LNCS, vol. 3382, pp. 205–220. Springer, Heidelberg (2005). doi:10.1007/978-3-540-30578-1_14
8. Lee, A., Ihaka, R., Triggs, C.: Advanced Statistical Modelling. Course notes for University of Auckland Paper STATS 330 (2012)

9. Marlow, S.: Parallel and concurrent programming in haskell. In: Zsók, V., Horváth, Z., Plasmeijer, R. (eds.) CEFP 2011. LNCS, vol. 7241, pp. 339–401. Springer, Heidelberg (2012). doi:10.1007/978-3-642-32096-5_7

10. Marlow, S., Newton, R., Peyton Jones, S.: A monad for deterministic parallelism. In: Haskell 2011: Proceedings of the Fourth ACM SIGPLAN Symposium on Haskell. ACM (2011). http://simonmar.github.io/bib/papers/monad-par.pdf

11. North, M., Collier, N., Ozik, J., Tatara, E., Macal, C., Bragen, M., Sydelko, P.: Complex adaptive systems modeling with Repast Simphony. Complex Adapt. Syst. Model. 1(1), 3 (2013)

12. O'Sullivan, B., Goerzen, J., Stewart, D.: Real World Haskell. O'Reilly Media, Sebastopol (2008)

13. Szpyrka, M.: Modelling and analysis of real-time systems with RTCP-nets. In: Kordic, V. (ed.) Petri Net, Theory and Applications, chap. 2, pp. 17–40. I-Tech Education and Publishing, Vienna (2008)

14. Szpyrka, M., Matyasik, P., Biernacki, J., Biernacka, A., Wypych, M., Kotulski, L.: Hierarchical communication diagrams. Comput. Inf. 35(1), 55–83 (2016)

15. Szpyrka, M., Matyasik, P., Mrówka, R.: Alvis - modelling language for concurrent systems. In: Bouvry, P., Gonzalez-Velez, H., Kołodziej, J. (eds.) Intelligent Decision Systems in Large-Scale Distributed Environments. SCI, vol. 362, pp. 315–341. Springer, Heidelberg (2011). doi:10.1007/978-3-642-21271-0_15

16. Szpyrka, M., Matyasik, P., Mrówka, R., Kotulski, L.: Formal description of Alvis language with α^0 system layer. Fundamenta Informaticae 129(1–2), 161–176 (2014)

17. Weiss, G. (ed.): Multiagent Systems: A Modern Approach to Distributed Artificial Intelligence. MIT Press, Cambridge (1999)

18. Wooldridge, M.: An Introduction to MultiAgent Systems, 2nd edn. Wiley Publishing, Chichester (2009)

Tensor-Based Syntactic Feature Engineering for Ontology Instance Matching

Andrzej Szwabe[(✉)], Paweł Misiorek, Jarosław Bąk, and Michał Ciesielczyk

Institute of Control and Information Engineering, Poznan University of Technology,
ul. Piotrowo 3a, 60-965 Poznan, Poland
{andrzej.szwabe,pawel.misiorek,jaroslaw.bak,
michal.ciesielczyk}@put.poznan.pl

Abstract. We investigate a machine learning approach to ontology instance matching. We apply syntactic and lexical text analysis as well as tensor-based data representation as means for feature engineering effectively supporting supervised learning based on logistic regression. We experimentally evaluate our approach in the scenario of the SABINE Data linking subtask defined by Ontology Alignment Evaluation Initiative. We show that, as far as the prediction of non-trivial matches is concerned, the use of the proposed tensor-based modelling of lexical and syntactical properties of the ontology instances enables achieving a significant quality improvement.

Keywords: OAEI · Ontology instance matching · Machine learning · Tensor-based data modeling · Natural Language Processing · Syntactic analysis

1 Introduction

We propose a tensor-based approach to feature engineering that is applicable to a machine learning system based on a logistic regression algorithm. The purpose of the system is to estimate the degree of similarity between pairs of ontology instances described using the form of OWL Aboxes [1]. Instance matching is one of the ontology matching problems investigated by the main tracks of the Ontology Alignment Evaluation Initiative (OAEI) [12] – the major contest and research forum for ontology matching solutions. In this paper, we show the use of the dataset of the OAEI SABINE data linking subtask. SABINE is an ontology matching benchmark focused on the domain of European politics. The data linking subtask of the SABINE task is devoted to the problem of finding matches between instances of the class 'Topic' and DBpedia entities. Both the ontology topics and DBpedia entities are given as OWL files. Additionally, the subtrack provides the third file that contains the set of reference alignments of ontology instances.

What is specific for the SABINE task (as the only such task among the OAEI tasks) is that the content of the SABINE target OWL document enables syntactic feature engineering: this document contains ontology instances' properties

© Springer International Publishing AG 2017
L. Rutkowski et al. (Eds.): ICAISC 2017, Part II, LNAI 10246, pp. 609–622, 2017.
DOI: 10.1007/978-3-319-59060-8_55

definitions in natural language – they have the form of sequences of English language sentences. An original hypothesis of the research presented herein is that such sentences (i.e., sentences describing ontology instances) may be analyzed syntactically to extract features of the new kind – syntactical features – that are valuable for supervised learning.

In this paper, following the assumptions made by other researchers [9,20], we treat the ontology alignment problem as a binary classification problem. Similarly to the authors of [2], we investigate the use of a machine learning system based on supervised learning. Therefore, the presented research is not supposed to follow all the present rules of the OAEI contest [1,12] since these rules do not allow for the use of reference matches for the machine learning model training.

In the proposed system we use Logistic Regression (LR) – the method that has been already successfully used in the ontology alignment scenario [2,4]. As it has been found [3,4], LR enables to deal with the sample data described using heterogeneous features. Such an advantage is crucial from the perspective of our research since it allows us to smoothly introduce the results of syntactic text analysis into the learning model.

2   Related Work

In recent years, an increasing interest of the research community in the ontology matching field has been observed [13] and many research activities in this domain have been presented on the OAEI forum [12].

In this paper we aim at putting some light on the practical value of supervised learning solutions applied to the ontology instance alignment task, in particular in the application scenario simulated in the OAEI SABINE task [1]. Unfortunately, the use of supervised learning violates the current rules of the OAEI contest. In consequence, recently proposed ontology alignment solutions based on supervised learning techniques, e.g. [2], have been presented to the OAEI community as non-contest contributions.

Tensor-based methods are not popular among approaches to ontology alignment. So far, tensor modeling has been only used to address Subtask#4 of the OAEI Anatomy track [16,17]. The solutions presented in [16,17] are based on the model of the 3rd order tensor used to represent the Resource Definition Framework (RDF) triples describing ontology resources. In this paper, we provide completely different solution which assumes an application of tensor modeling to feature engineering, rather than as a basis for tensor decomposition.

To estimate a potential instance match or mismatch, several competing strategies are usually being applied to the same pair of entities. Subsequently, the best performing matcher is chosen [5,20] or all of the resulting matching scores are combined [14]. Such an approach requires to be automatically tuned at runtime. The AML matcher, introduced in [5], uses the string similarity, word similarity, and WordNet similarity between the properties of the instances. The system enabled to achieve the highest data linking precision in the latest OAEI contest in the Instance Matching track [11]. In [20], the authors – additionally to the label-based approach – used a structure-based approach when the literal properties of the instances are not similar. Using such a supplementary information enabled to achieve the highest recall on the SABINE data linking task [11].

While various semantic modelling methods have been successfully used by the authors of many matchmaking systems of the OAEI contest, none of these systems exploits results of any kind of syntactic analysis of natural language text. Nevertheless, an approach, in which rules and syntactic text analysis provide additional binary features for LR, was used in other applications of data mining. In [19] more than 30 rules for fraud detection were successfully used to estimate fraud likeliness by means of LR. In [7] several Natural Language Processing (NLP)-based and rule-based features provided by the experts were used in a hybrid classification system. The solution proposed in [2] (presented to the OAEI community as a non-contest contribution) applies the LR algorithm to learn an optimal combination of both lexical and structural similarity weights. The approach of [4] uses LR to customize the ontology alignment system by determining the weights assigned to different strategies based on user feedback, which may be also seen as a feature engineering solution. However, none of presented approaches exploits syntactical features nor uses tensor-based feature engineering that involves multi-linear representation of feature values.

3   Ontology Instance Matching System with Tensor-Based Syntactic Feature Engineering

The proposed *Tensor-Based Syntactic Feature Engineering System for Ontology Instance Matching* (TSFES-OIM) consists of several modules corresponding to consecutive data processing steps, as presented in Fig. 1.

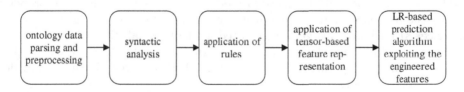

Fig. 1. Ontology instance matching system with Tensor-Based Syntactic Feature Engineering.

3.1   Application of Syntactic Analysis to Instance Property Extraction

In accordance to the above-presented data processing flow of TSFES-OIM, the syntactic analysis of ontology properties which are given as sentences, e.g., abstracts describing the ontology instances, is performed after the basic (i.e., non-syntactic) ontology parsing and pre-processing. The basic parsing and pre-processing step includes the extraction of ontology instances' properties. Then, in the case of properties which are given as text in natural language – in the current implementation of TSFES-OIM we support English only – the syntactic text analysis is performed by the execution of the following steps:

- the Stanford Parser (version 3.7.0) [18] is used to construct the syntax tree of each sentence,
- on the basis of the syntax tree, the noun phrases with adjectives are extracted separately from the sentences' subjects and from objects of verbs - in the cases of complex or compound sentences, the extraction algorithm explores all their clauses.

As a result of the execution of the above-specified steps, the additional syntactical properties of ontology instances are added to their original pre-processed representations to form the input data of the rule-based augmentation.

3.2 Rule-Based Augmentation of Matches Representation

Effective firing of the rules requires the availability of the data structures that enable the random access to all the matched instances' properties. In the context of the application scenario investigated in the research presented herein these data structures should provide the access to syntactical properties of the text representing the instances being matched.

The proposed approach to data augmentation differs from the popular approaches to data augmentation [1,4]. In particular, in our approach augmentation of the core training data involves augmentation of the core training data tuples (i.e., the pairs representing the reference matches), not just augmentation of the elements of these tuples. Specifically, the augmentation process involves the application (i.e., the firing) of rules that are manually defined for Cartesian products of the matched instances' properties, i.e., that represent some properties of the reference matches, not just the properties of the matched ontology instances. To achieve this goal we created a set of logical rules which are able to determine syntactic relations between textual parts of data representing the instances and their appropriate matches (i.e., other instances). Each logical rule is of the following form:

$$
\begin{aligned}
&if \quad set\_of\_conditional\_elements \\
&then \, return \;\; 1 \\
&else \;\; return \;\; 0
\end{aligned}
\tag{1}
$$

Logical rules (1) return only boolean values: *True* or *False*, represented herein as *1* or *0* respectively. Each set of conditional elements contains conditions that are checked when a rule is fired (executed). The conditions check if there is a relation between ontology instances. In such a way, the conditions provide a syntactic analysis of the textual parts of data.

3.3   Tensor-Based Model for Instance Matches Representation

We use a tensor-based data model to represent pre-processed and enhanced data on matches between ontology nodes. The model is based on the concept of *Multi-Tensor Hierarchy Network* (MTHN) proposed in [15]. The model is used to represent features of the 'event' of matching two nodes from two ontologies. It is worth to note that the proposed model has the most distinctive property of any tensor-based data representation in which vector space dimensions represent feature values, rather than data/training examples [10,16]. Thanks to this property, any conjunction of the features' values may be represented by its dedicated tensor entry. Moreover, the use of the multi-tensor hierarchy network provides simple means for representing the conjunction features' subset as the corresponding tensor network node – the arity of the conjunction tuple equals to the level of the tensor network.

The formal definition of the tensor-based data representation model used herein is given in [15]. In this paper we use the model in a form adapted to the ontology alignment task and extend the way the MTHN is exploited. In particular, we use the notation in which $\mathcal{A}, \mathcal{B}, \dots$ denote sets, $\mathbf{A}, \mathbf{B}, \dots$ denote tensors and $a, b, \dots$ denote scalars. We define an n-order tensor:

$$\mathbf{T} = [t_{i_1,\dots,i_n}]_{m_1 \times \cdots \times m_n},$$

in a tensor space $\mathcal{I}_1 \otimes \cdots \otimes \mathcal{I}_n$, where each $\mathcal{I}_i$, $1 \leq i \leq n$ indicates a standard basis [10] of dimension $|\mathcal{I}_i| = m_i$ used to index elements of the domain $\mathcal{F}_i$ – the domain of the feature i. In the case of the ontology alignment application scenario, each tensor entry represents the existence of the given ontology instance matching, what is formally described as the function $\psi : \mathcal{F}_1 \times \cdots \times \mathcal{F}_n \rightarrow [0,1]$ indicating the similarity between all the compared ontology instances.

The MTHN-based model enables to represent all the properties obtained as a result of data pre-processing (properties of ontology instances) and application of rules (properties of the matches) as tensor modes corresponding to basic features as well as all possible combinations of features – by means of tensors constituting the nodes of MTHN. Each tensor of the MTHN network stores the representation of feature values conjunctions. To describe MTHN we use $[n] = \{1, 2, \dots, n\}$ to denote the set enumerating features describing the matching existence. For each subset $\mathcal{S} = \{p_1, \dots, p_k\} \subset [n]$ we construct the k-th order tensor $\mathbf{T}(\mathcal{S})$ with modes corresponding to features' numbers from $\mathcal{S}$. Tensors $\mathbf{T}(\mathcal{S})$ form the hierarchical network of 2^n tensors consisting of:

- level 0 of MTHN – containing one node which is the tensor of order 0, i.e., $\mathbf{T}(\varnothing)$,
- for $k \in \{1, \dots, n-1\}$: level k of MTHN – containing $\binom{n}{k}$ kth-order tensors $\mathbf{T}(\{p_1, \dots, p_k\})$,
- level n of MTHN – containing one node – the tensor of order n, i.e., $\mathbf{T}([n]) = \mathbf{T}$.

The following modifications of the model have been introduced herein when compared to the model presented in [15]:

- we add modeling of new type of binary modes corresponding to features obtained by rule-based augmentation,
- as a consequence of adding rule-based modes, we apply our model for much bigger total number of modes (we use the total number of modes greater than 20 whereas in [15] we limit to $n = 4$),
- we apply the new method of feature selection by limiting the level up to which MTHN is used ('explored'); such an MTHN 'pruning' is necessary for the feasibility of the case of $n > 20$ (due to the problem of combinatorial explosion).

The MTHN obtained in the above-described way is used to represent the instances' feature values for the purposes of logistic regression application in the core of the recommendation engine used for the ontology instance matching. As the final result of feature engineering, each matching sample is augmented by (i) values of matched ontology instances' features, (ii) values of the alignments' features (obtained using the rule-based method), and (iii) values of the conjunctions of the feature values stored at MTHN nodes up to the given level. In particular, the use of conjunctions of features values defined by rules, i.e., conjunctions of the rules application results, gives the possibility to take into account the new features, application of which, as we shown in the experiments presented in this paper, may improve the performance of LR-based ontology instance alignment recommendation.

3.4   Supervised Learning Based on Logistic Regression

The TSFES-OIM is a machine learning system that uses reference alignments to learn the model. We apply the LR algorithm [4,6] to learn the weights of the feature values defined using the tensor-based model described in Sect. 3.3. The learned weights are used to generate the matching results in a way that follows the assumption of the OAEI SABINE task – according to which the system is expected to find all '1:n' type alignments [1]. Therefore, our system generates the matching results such as at most a single target instance is assigned to each of the SABINE source instances.

   In our implementation of the LR algorithm, we use stochastic gradient descent to learn the model weigths, cross entropy between the recommendation results and the reference results as the loss function, and L2 regularization (with regularization parameter equal to 0.000001). We set learning rate to 0.01. The loss tolerance for the stopping criterion was set to 0.00005. Following the authors of [8] we set the minimal number of stochastic gradient descent iterations to 5.

4   Evaluation Methodology

Being based on a supervised learning technique, our ontology instance alignment system (in all of its variants) requires the availability of a subset of the reference alignments set (in case of OAEI being referred to as 'expert matches'). Just like the authors of any machine learning systems applied to OAEI tasks, we assumed that the user of the ontology instance alignment system is provided with automatic 'recommendations' of the most probable 'missing' matches and that these matches are interactively 'suggested' to the user by the ontology instance alignment system. In our dataset-based experiments, we treated a random subset of the reference alignments set as the training set representing the user feedback 'observed' by the machine learning system during the process of manual ontology instance alignment.

4.1   Dataset Use

We used the dataset of the OAEI SABINE data linking subtask which is a subproblem of the OAEI SABINE task. The SABINE task is one of the three independent tasks of the OAEI Instance Matching track which is aimed at evaluating the performance of tools estimating the degree of similarity between pairs of instances – OWL Aboxes [1]. In particular, the SABINE dataset includes an ontology with about 500 topics in the domain of European politics for Social Business Intelligence. For the case of SABINE data linking subtask, the dataset consists of three parts: (i) the *sabine_source.owl* file containing descriptions of 407 instances of OWL class 'Topic', (ii) the *sabine_target.owl* file containing descriptions of 1127 DBpedia entities, and (iii) the *refalign.rdf* file containing the information of 338 matches between instances from *sabine_source.owl* and DBpedia entities from *sabine_target.owl* with the matches' weights being numbers from the range of $(0.5, 1]$.

Similarly to the authors of [4] we assumed that each pair built from an instance of *sabine_source.owl* and a DBpedia entity of *sabine_target.owl* (representing a 'potential match') that does not appear in the *refalign.rdf* document, should be explicitly represented as a non-match sample (indicating the lack of matching between the two instances). Following this assumption we complemented the set of original 338 reference matches with $1127 \times 407 - 338 = 458,689$ non-matches.

We performed our experiments (collectively presented herein) with the use of multiple random subsets of the reference alignments set given in the *refalign.rdf* document; these subsets served as the training sets. To move towards cases of using unsupervised learning techniques (i.e., a case of training ratio equal to zero), we performed experiments for training ratio equal to 0.1 and 0.05. As in [4], we generated a set of training tuples in two-step procedure. In case of the experiments presented herein, the procedure included: (i) random selection of a subset from 407 'Topic' class' instances of SABINE ontology according to the given training ratio and (ii) construction of the training set from all the reference alignments and the non-matching samples concerning selected instances. All the remaining samples (both matching and non-matching ones) were used as the test set elements. We repeated each experiment 20 times (for different random splits of the dataset into the training set and the test set) and provided the results in the form of average precision values and the corresponding standard error values.

4.2   Matching Quality Measures

In correspondence with the rules of the OAEI contest, we measured the performance of the evaluated algorithms using precision and recall [12]. Similarly to the evaluation of the SABINE subtrack of OAEI [1], when calculating precision and recall we treat all the reference alignments for which the provided similarity measure is greater than 0.5 as positives. We present the measurement results by means of precision versus recall curves, what enables a deeper analysis of compared approaches' performance – in particular as far the identification of non-trivial matches is concerned.

5   Evaluated Systems

Due to the rules of OAEI – forbidding the use of the alignments set as the training set – the results of the experimental evaluation presented herein, are only 'asymptotically comparable' to the official results reported in [5,20]; one may see the OAEI contest results as corresponding to the special case of applying machine learning to ontology alignment in which the training ratio is equal to zero. On the other hand, in our experiments we have juxtaposed the matching quality of the solutions proposed in this paper with the quality of the StringEquiv – the main baseline matcher of the OAEI contest. StringEquiv is based on string equality of lowercased local names of entities [1]. It is worth being noted that, despite its simplicity, over the years of the OAEI contest, StringEquiv outperformed many ontology alignment systems submitted by the contest competitors [1].

In our experiments we compared the following matchmaking systems:

- the StringEquiv matcher – the baseline OAEI method [12],
- TSFES-OIM(1) - TSFES-OIM variant exploiting the non-syntactical rules (i.e., the rules which do not operate on the results of syntactic analysis) and MTHN-based features representation up to the level 1 (i.e., do not exploiting the conjunctions of rule-based features' values),
- TSFES-OIM(1S) - exploiting the syntactical rules (i.e., the rules applied to the syntactic analysis results) and limited to MTHN features representation up to the level 1,
- TSFES-OIM(2) - exploiting non-syntactical rules and the MTHN-based features representation up to the level 2 (i.e., using conjunctions of rule-based features' values),
- TSFES-OIM(2S) - exploiting syntactical rules and the MTHN-based features representation up to the level 2.

The following groups of rules were used in the case of TSFES-OIM(1) and TSFES-OIM(2):

- rules comparing the label of the source instance and the label of the target instance,
- rules comparing the labels of the source instance aliases with the label of the target instance,

– rules comparing the labels of the source instance and its aliases with the noun phrases extracted from the target instance abstract, i.e., the abstract of the given DBpedia entity.

For TSFES-OIM(1S) and TSFES-OIM(2S) the set of applied rules was extended by the following groups of syntactical rules:

– rules comparing the label of the source instance and its aliases with the noun phrases extracted from the first-sentence subjects of the target instance abstract,
– rules comparing the label of the source instance and its aliases with the noun phrases from subjects of all sentences of the target instance abstract,
– rules comparing the labels of the source instance and its aliases with the noun phrases extracted from the first-sentence objects of the target instance abstract.

Each of the above-mentioned six groups of rules consists of (i) the rule checking exact phrase equality, (ii) the rule checking lowercased phrase equality, and (iii) the rule checking whether the compared phrases share the same word. In the case of the group of rules comparing the labels of source and target instances, we additionally use (iv) the rule checking whether labels share the same lowercased word, and (v) the rule checking whether the labels are equal after omitting special characters. Finally, in our experiments we used $3 \times 3 + 2$ non-syntactical rules and 3×3 syntactical rules.

6 Evaluation Results

Figures 2 and 3 present the precision-vs-recall curves for all the above-specified systems and all the training ratio values. The vertical coordinate of each point of each curve represents an average over the set of precision values corresponding to the given recall value of the point. This set has been obtained by multiple random splitting of the dataset into the training set and the test set before each individual experiment. The error bar of each point represents the standard error of the precision averaged across the individual experiments.

Both the Figs. 2 and 3 present both the whole precision-vs-recall curves (the upper subplots) as well as their parts – 'zoomed-in' to the ranges of precision and recall showing the systems' ability to identify non-trivial matches (the bottom subplots). When compared to the OAEI baseline method (StringEquiv), all the evaluated variants of the TSFES-OIM system may be seen as clearly improving the matchmaking quality: they enable achieving high precision despite exceeding the recall beyond the level achievable by means of the baseline method. Moreover, it may be seen that the best results are achievable by means of the system TSFES-OIM(2S) in case of which both the syntactical rules and the MTHN feature representation (up to the second MTHN level) are used.

There are no results in literature which may be directly compared with our results. The SABINE data-linking subtask is a new problem defined for the 2016

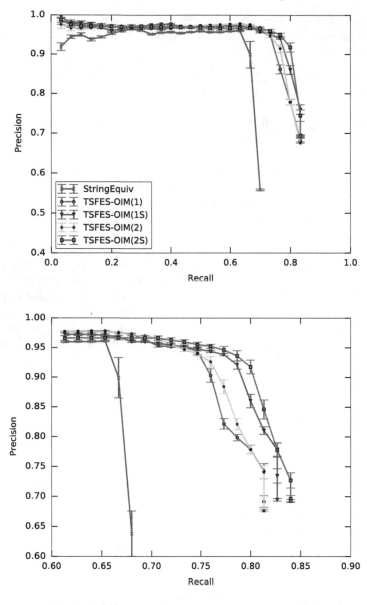

Fig. 2. P(R) curves for training ratio equal to 0.1.

OAEI contest edition. The only results in this subtask were provided by the AML system [5] (precision = 0.926, recall = 0.855) and the RiMOM system [20] (precision = 0.424, recall = 0.917). Although OAEI results of AML and RiMOM systems could not be used for a direct comparison with TSFES-OIM results (due to the application of a different evaluation methodology), one may observe that an application of these systems may lead to higher recall values. This observation

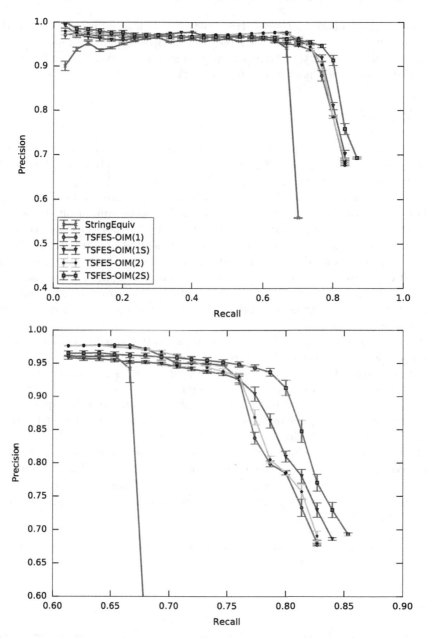

Fig. 3. P(R) curves for training ratio equal to 0.05.

is easily explainable by the fact that these systems use structural and semantic analysis in parallel to the lexical analysis of labels. In this paper we have focused on the impact of syntactic analysis results on the matching system performance. We have not investigated the analogical impact of semantic and structural similarities of ontology instances. However, due to the ability of the LR algorithm of using various types of variables, our approach may be easily extended by the use of features other than lexical and syntactic. We plan to investigate such an extension as a part of our future work.

7   Conclusions

We show that the logistic regression algorithm, supported by the proposed tensor-based feature engineering methods, may be effectively applied in the ontology instance matching scenario. As indicated by high precision values observed for recall exceeding the values achievable by means the OAEI's baseline method, following the proposed approach enables successful automatic identification of many non-trivial matches, without leading to many false positive ones.

Furthermore, the results of the evaluation presented herein indicate that, in the investigated scenario, both the tensor-based lexical properties representation and the tensor-based syntactical properties representation enable achieving a significant improvement of the matching quality. Moreover, we show that the highest matching quality is achievable when both these feature modeling methods are applied together.

One of the key finding of the research presented in the paper is that modeling of conjunctions of rule-based feature values, i.e., conjunctions of rules' firing outcomes – rather than, e.g. conjunctions of rules – gives the possibility to take a tangible advantage from the use of conjunctive features at the supervised learning stage. As we shown in the experiments presented in this paper, the use of such features may significantly improve the performance of LR-based ontology instance matching prediction.

In this paper we emphasize the practical value of syntactical features modelling. We do so because, to the best of our knowledge, until now no authors – neither those investigating ontology matching problems nor those conducting research activities in the more general field of machine learning – presented any such findings.

Acknowledgments. This work is supported by the Polish National Science Centre, grant DEC-2011/01/D/ST6/06788.

References

1. Achichi, M., et al.: Results of the ontology alignment evaluation initiative 2016. In: Proceedings of the 11th International Workshop on Ontology Matching, Kobe, Japan, 18 October 2016, pp. 73–129 (2016). http://ceur-ws.org/Vol-1766/oaei16_paper0.pdf

2. Balasubramani, B.S., Taheri, A., Cruz, I.F.: User involvement in ontology matching using an online active learning approach. In: Proceedings of the 10th International Workshop on Ontology Matching, Bethlehem, PA, USA, 12 October 2015, pp. 45–49 (2015). http://ceur-ws.org/Vol-1545/om2015_TSpaper3.pdf
3. Chapelle, O., Manavoglu, E., Rosales, R.: Simple and scalable response prediction for display advertising. ACM Trans. Intell. Syst. Technol. **5**(4), 61:1–61:34 (2014)
4. Duan, S., Fokoue, A., Srinivas, K.: One size does not fit all: customizing ontology alignment using user feedback. In: Patel-Schneider, P.F., Pan, Y., Hitzler, P., Mika, P., Zhang, L., Pan, J.Z., Horrocks, I., Glimm, B. (eds.) ISWC 2010. LNCS, vol. 6496, pp. 177–192. Springer, Heidelberg (2010). doi:10.1007/978-3-642-17746-0_12
5. Faria, D., Pesquita, C., Balasubramani, B.S., Martins, C., Cardoso, J., Curado, H., Couto, F., Cruz, I.: OAEI 2016 results of AML. In: Proceedings of the 11th International Workshop on Ontology Matching, Kobe, Japan, 18 October 2016, pp. 138–145 (2016). http://ceur-ws.org/Vol-1766/oaei16_paper2.pdf
6. Gondek, D.C., Lally, A., Kalyanpur, A., Murdock, J.W., Duboue, P.A., Zhang, L., Pan, Y., Qiu, Z.M., Welty, C.: A framework for merging and ranking of answers in DeepQA. IBM J. Res. Dev. **56**(3), 399–410 (2012)
7. Johnson, E., Baughman, W.C., Ozsoyoglu, G.: Mixing domain rules with machine learning for radiology text classification. In: Proceedings of the ACM SIGKDD Workshop on Health Informatics (HI-KDD 2014) (2014)
8. Langford, J., Li, L., Zhang, T.: Sparse online learning via truncated gradient. J. Mach. Learn. Res. **10**, 777–801 (2009)
9. Mao, M., Peng, Y., Spring, M.: Ontology mapping: as a binary classification problem. Concurr. Comput. Pract. Exp. **23**(9), 1010–1025 (2011)
10. Nickel, M., Tresp, V.: An analysis of tensor models for learning on structured data. In: Blockeel, H., Kersting, K., Nijssen, S., Železný, F. (eds.) ECML PKDD 2013. LNCS, vol. 8189, pp. 272–287. Springer, Heidelberg (2013). doi:10.1007/978-3-642-40991-2_18
11. Ontology Alignment Evaluation Initiative: OAEI 2016: Instance Matching Track (2016). http://islab.di.unimi.it/content/im_oaei/2016
12. Ontology Alignment Evaluation Initiative: Ontology Alignment Evaluation Initiative - OAEI 2016 Campaign (2016). http://oaei.ontologymatching.org/2016/
13. Otero-Cerdeira, L., Rodríguez-Martínez, F.J., Gómez-Rodríguez, A.: Ontology matching: a literature review. Expert Syst. Appl. **42**(2), 949–971 (2015)
14. Shvaiko, P., Euzenat, J.: Ontology matching: state of the art and future challenges. IEEE Trans. Knowl. Data Eng. **25**(1), 158–176 (2013)
15. Szwabe, A., Misiorek, P., Ciesielczyk, M.: Tensor-based modeling of temporal features for big data CTR estimation. In: Kozielski, S., Mrozek, D., Kasprowski, P., Małysiak-Mrozek, B., Kostrzewa, D. (eds.) BDAS 2017. CCIS, vol. 716, pp. 16–27. Springer International Publishing, Cham (2017). doi:10.1007/978-3-319-58274-0_2
16. Szwabe, A., Misiorek, P., Walkowiak, P.: Reflective relational learning for ontology alignment. In: Omatu, S., De Paz Santana, J., González, S., Molina, J., Bernardos, A., Rodríguez, J. (eds.) Distributed Computing and Artificial Intelligence. AISC, vol. 151, pp. 519–526. Springer, Heidelberg (2012). doi:10.1007/978-3-642-28765-7_62
17. Szwabe, A., Misiorek, P., Walkowiak, P.: Tensor-based relational learning for ontology matching. In: Advances in Knowledge-Based and Intelligent Information and Engineering Systems - 16th Annual KES Conference, San Sebastian, Spain, 10–12 September 2012, pp. 509–518 (2012)
18. The Stanford Natural Language Processing Group: The Stanford Parser. http://nlp.stanford.edu/software/lex-parser.shtml

19. Zhang, L., Yang, J., Chu, W., Tseng, B.: A machine-learned proactive moderation system for auction fraud detection. In: Proceedings of the 20th ACM International Conference on Information and Knowledge Management, CIKM 2011, pp. 2501–2504. ACM, New York (2011)
20. Zhang, Y., Jin, H., Pan, L., Li, J.: RiMOM results for OAEI 2016. In: Proceedings of the 11th International Workshop on Ontology Matching, Kobe, Japan, 18 October 2016, pp. 210–216 (2016). http://ceur-ws.org/Vol-1766/oaei16_paper13.pdf

Semantic Annotations for Mediation
of Complex Rule Systems

Mateusz Ślażyński, Grzegorz J. Nalepa, Szymon Bobek, and Krzysztof Kutt$^{(\boxtimes)}$

AGH University of Science and Technology,
al. A. Mickiewicza 30, 30-059 Krakow, Poland
{mslaz,gjn,sbobek,kkutt}@agh.edu.pl

Abstract. Design of Business Intelligence systems capable of effectively handling a domain knowledge is a well known, but currently not solved challenge for both Software and Knowledge Engineers. There exist several approaches to extract and model the Business Knowledge, most notably Business Processes and Business Rules. However, each of them has its own weaknesses and therefore it is often desirable to build hybrid models composed of several knowledge representations. In this paper we describe an extension to Business Rules in order to facilitate creation of such heterogeneous systems. This is achieved by introducing semantic annotations to existing rule modeling languages. We also present how the additional semantic information is leveraged in the Prosecco project.

Keywords: Semantic annotations · Rule systems · Business intelligence

1 Introduction

One of the crucial problems occurring during implementation of a complex information system is a constantly changing set of its requirements. Therefore, even if it is possible to express the problem's domain knowledge in terms of a programming language, it should be avoided, because the knowledge encoded in not dedicated and unrelated representation tends to be difficult to change. In other words, small modification of the system's definition may lead to the unpredictably big changes in its implementation. This software engineering issue is one of the main reasons of migration from waterfall models to more incremental development techniques.

Rule–based systems provide an elegant solution to this problem by an application of the knowledge engineering methods. Within this paradigm system is separated into two isolated parts: first responsible for the runtime, so called rule engine, implemented in a selected technology; and second part containing a mutable knowledge base represented as a set of rules and facts. Each rule individually encodes a certain part of the knowledge, effectively separating concerns about system's behavior.

© Springer International Publishing AG 2017
L. Rutkowski et al. (Eds.): ICAISC 2017, Part II, LNAI 10246, pp. 623–634, 2017.
DOI: 10.1007/978-3-319-59060-8_56

The *Prosecco* (Processes Semantics Collaboration for Companies) project[1] was a research and development project funded by NCBR (2012–2015), created in order to provide Small and Medium Enterprises (SME) with adequate management tools. Specifically, the main goal of the project was to simplify design and configuration of the Business Process Management (BPM) systems. Properly prepared BPM systems can significantly improve competitiveness of company, supporting not only the management, but also strategic planning within the SME sector. To achieve this goal, the domain knowledge has to be properly encoded and leveraged. The Prosecco system supports this task with a rich and heterogeneous repository of formalized knowledge. The repository contains three different types of components: (1) hierarchic Business Vocabulary represented in a form of a formal ontology, (2) BPMN (Business Process Modeling and Notation) [14] components representing the independent elements of the modeled business processes, and (3) business rules responsible for modeling constraints and effects occurring within the SME domain.

Heterogeneous character of the repository combined with its size, revealed several shortcomings of rules' storage and execution methods. Three main issues were identified during implementation and design phases:

P1 The rule systems of practical value tend to be very complex, containing rules with no clear structure and meaning. There is no standard tool to effectively query rule bases in order to find interesting parts of the system.

P2 The interoperation between different models and representations depends on the ability to recognize the same concepts occurring at different levels of abstraction. Due to the inherent heterogeneity of the repository, certain form of concept aligning has to be introduced.

P3 Despite being a declarative and unambiguous, rule systems tend to semantically differ from natural notions used in the business domain. Such semantic gap hinders the communication between user and system, preventing effective analysis of the system behavior and design.

In this paper, we propose a solution for all the P1–3 problems by augmenting the rule systems with semantic annotations. The ontology describing the vocabulary of a modeled domain is used to specify context and notions represented by rules and their components. Then, the semantic technology tools are used to query the repository of rule models in systematic and context-aware way. Furthermore, we demonstrate its practical use in the Prosecco BPM system.

The rest of the paper is composed as follows. In Sect. 2, the current state of the art is presented. Motivation and original contribution of the paper are laid out in Sect. 3. Then in Sect. 4, details of the proposed solution are discussed as well as the implementation used in the Prosecco project is described. Section 5 demonstrates the current application of the semantic annotations in the Prosecco project. The paper ends with a list of future challenges and short summary in Sect. 6.

---

[1] See http://prosecco.agh.edu.pl for the project website.

2   Related Works

There exist multiple proposals of integrating rules with ontologies. Most notably OWL–RL [15] is a subset of Web Ontology Language combining basic rules with semantic description of the domain. Semantic Web Rule Language [8] (SWRL) extends the OWL with the capabilities of Rule Markup Language in order to perform first order inference using facts and class instances stored in a triple store. SWRL Editor [16], designed to enhance the interoperability of SWRL rules with different technologies, contains a SWRL Factory component which allows storing SWRL rules within ontology and manipulate them using a Java code. For a deeper insight on issues regarding combining rule systems and ontologies reader can refer to [18]. The paper [5] presents the current state of the art of research in the domain.

Description of the Prosecco architecture, along with the detailed presentation of the SME ontology can be found in [11]. It includes an analysis of design choices and methodology of building a business domain ontology, with a special treatment on its role as a Business Vocabulary. In order to provide system with serializable and Java-compatible representation of the domain, ontology is translated to an executable model in a form of the Plain Old Java Objects (POJO). The POJO model itself is a simple object based structure responsible for communication within the system, effectively forcing concept alignment in its components.

Related problem of enhancing business modeling tools with semantic layer was examined in [9]. Applications analyzed there, use ontologies to provide user with intelligent mechanisms useful both during modeling phase, and during execution phase, when system uses semantic information to select appropriate services for current task. Extension of a popular BPMN modeling and execution environment Activiti [17] with automatic recommendations has been described in [2]. Serialization of the BPMN models into the ontology–based representation was proposed and discussed in SUPER project [19], which developed several ontologies capable of capturing the BPMN structure. In the case of combining Business Processes with Business Rules, the overview of the existing approaches can be found in [13] and the presentation of the prototype hybrid environment is described in [10].

The HeaRT rule engine [3] is one of the rule execution environments in the Prosecco project. It uses the Extended Tabular Tree (XTT2) rule representation [12], which is characterized by the clear structure, composed of interconnected tables, meant to be easy to visualize and understand. There exist tools capable of statically verifying certain aspects of rules represented in XTT. Pellet–Heart [1] discusses former attempt of connecting HeaRT with Semantic Web Stack by usage of an automatic reasoner.

3   Motivation

The primary motivation for the work presented in this paper is to provide a solution to the problems stated in Sect. 1. Despite many advantages, current systems combining ontologies with rules, presented in Sect. 2 are not applicable in the case of the heterogeneous system, such as the one used in Prosecco. Current state-of-the-art solutions aim at embedding an ontology inside the broader rule based model, therefore they depend on ontology as a knowledge base. However, in the Prosecco, ontology performs only a role of business vocabulary; other parts of the system make use of independent, business tailored solutions with underlying specific knowledge storage technologies. Usage of SWRL would require a global migration from the existing architecture, at the same time not solving issues regarding interoperation with different knowledge representations existing within the system. Furthermore, rule languages used in previous works are not entirely expressible in terms of SWRL. For the reasons stated above, proposed solution provides a semantic layer, above the rule representations, leaving the existing business logic intact and at the same time solving the three problems we identified.

4   Proposed Solution

One of the main goals of the modern web technologies is to provide users efficient and precise ways of finding and obtaining interesting knowledge from a flood of irrelevant facts and data. Semantic Web Stack [7] is a growing set of technologies aimed at enriching data with semantic meaning, used later to perform reasoning and communicate with user. This section describes how the already mature part of the Semantic Web technologies can be used to capture semantics of a rule model. Next, the current implementation of semantic annotations in selected rule execution environments is presented. Finally, applicability of the introduced techniques is shortly analyzed taking into consideration available technologies and character of the system.

4.1   Representing Rule Models in Ontology

The problem of representing rule models within an ontology consists of two parts. First, the common structure of the models has to be represented as a TBox part — namely classes corresponding to the appropriate components of the model have to be defined. Depending on the requirements, a proper granularity has to be chosen. A fine-grained view of the rule model could represent every condition and effect of rule as a separate entity, while a coarse-grained view could treat a model as an indivisible whole. Beside the pragmatical reasons, also the specification of the modeling language has to be taken into consideration. Table 1 shows a comparison of XTT2 and Drools modeling languages considered in our work. XTT2 was used as a base due to the richer structure of its models. Due to the major differences between the languages, we have decided to define two separate TBoxes within corresponding ontologies. Figure 1 presents a hierarchy of classes used to capture structure of a XTT2 model. The ontology related to Drools representation is analogous and consists of three classes: *DroolsModel*, *DroolsRule* and *DroolsFactType*; the matter of extending it is described as a future challenge

Table 1. Structure of a XTT2 model compared to the Drools equivalents.

XTT2 component	Description	Drools equivalent
Type	Defines a discrete set of admissible values (numeric or symbolic)	*Fact types* describe the structure of facts contained in a knowledge base
Attribute	Named container for a value of specified type. The contained value can change in time	There is no equivalent. Drools use a knowledge base with a dynamic amount of facts
Rule	Consists of at least one test and effect depending on the attributes values	*Rules* are dispatched based on the state of facts in the knowledge base
Table	Groups rules with similar structure and meaning. Declares use of attributes and potentially modifies inference flow	There is no independent component of this kind. Arbitrary sets of rules can be grouped using special attributes

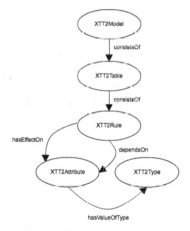

Fig. 1. TBox corresponding to the structure of a XTT2 model

in Sect. 6. The selected granularity reflects the level of abstraction sufficient to solve P1–3 problems presented in Sect. 1.

Next, the resulting rule-model ontology has to be combined with a problem domain (in the Prosecco case a SME) ontology. This can be achieved with a built-in OWL2 construct 'owl:imports'[2] which effectively imports all the axioms and classes from an external ontology. This approach has several limitations [6]. In fact, the problem domain ontology is effectively polluted with rule-specific

---

[2] For more information, please see: https://www.w3.org/TR/owl-semantics/.

knowledge. However, due to the small size of the rule-model ontology, we consider this problem as negligible. The second issue concerns inter-ontological relations between rule knowledge components and problem domain elements. In order to make our solution as domain-independent as possible, we had to define them inside the problem domain.

4.2 Semantic Annotations

Having a specified TBox we need to take care of populating the corresponding ABox with individual models, rules, tables, etc. As far as the structure of rule model can be directly inferred from its unmodified representation, the meaning of rules and their connections to the problem domain require an additional knowledge. In order to achieve this goal, both the HMR[3] and Drools Rule Language (DRL) were extended with annotation mechanism based on the Java annotation syntax[4]. There are two types of annotations allowed: one which describes data-type attributes of entities and other that defines inter-ontological relations between entities. These two types of annotations are encoded with @Attribute and @Relation keywords respectively. To define an attribute for a particular XTT2 element, it is required to include the following code before the definition of this element in HMR:

```
@Attribute(name=<attribute name>,
          value=<attribute value>)
```

where `name` denotes attribute name in the ontology, and `value` denotes its value.

Similarly, to define an inter-ontological relation for a particular XTT2 element, it is required to include the following code before its definition:

```
@Relation(name=<relation name>,
          subject=<URI of the individual>)
```

where `name` denotes a name of the relation and `subject` represents a problem domain entity with which the chosen XTT2 component should be related. A shorter, more natural version of this statement is also available in a form:

```
@<relation name>(subject=<URI of the individual>)
```

Listing 1.1 contains an example of the HMR code with semantic annotations. In the example, rule's definition (keyword 'xrule') is annotated as being not valid (a user defined data-type attribute) and is related to the individual located at URI 'user11'. The 'defines' relation is used to force an URI of the entity within the ontology.

---

[3] HMR stands for HeKaTe Meta Representation, a human readable format to represent XTT2 models. For a detailed specification of the language see: http://glados. kis.agh.edu.pl/doku.php?id=pub:software:heartdroid:documentation:hmr.

[4] For more details on the annotation system in Java see: http://docs.oracle.com/ javase/specs/jls/se8/html/jls-9.html#jls-9.7.

```
@Relation(name=concerns, subject=user11)
@Attribute(name=valid, value=false)
@Defines(subject=rule123)
xrule 'calculateFee'/2: [spentTime eq 'MoreThanDay']
 ==> [userFee set 100.0].
```

Listing 1.1. Example of HMR code with semantic annotations included

DRL is extended analogously, with a small difference in the location of the annotations. Due to the language specifics, they have to be put after rule declaration. An example of the same rule written in Drools language is presented in Listing 1.2.

```
rule "Calculate fee"
  @Relation(name=concerns, subject=user11)
  @Attribute(name=valid, value=false)
  @Defines(subject=rule123)
when
  u : User(URI == "user123",
           spentTime == "MoreThanDay")
then
  u.setFee(100.0);
end
```

Listing 1.2. Example of Drools rule with semantic annotations included

4.3 Analysis of Applicability

A simple Java API is available for the introduced annotation system giving access to the annotations in a well structured manner. Combined with Apache Jena[5], it allows to automatically generate appropriate individuals in the problem domain ontology. What is equally important, it is also possible to find the specific component based on the URI of the corresponding individual. One of the available tools allowing that is SPARQL[6]: an SQL-like language designed to express non-trivial semantic queries based on relationships and values of attributes in the ontology. It is supported by most of the semantic reasoning engines, such as HermiT, Jena, Pellet, TrOWL and much more [4].

The most notable constraint of the applicability of this solution is only one-directional communication between rule model and ontologies. Because of that, currently there is no method to validate the annotations with contents of the ontology – a minor spelling error can potentially lead to the mismatch with the rest of the system. Furthermore, every change in the ontology has to be additionally checked and reflected in annotations. This can become problematic in case of a constantly evolving or immature ontologies.

5 Application Study

The Prosecco project makes use of several independent business solutions, connected in order to design and configure BPM systems targeted at SME

---

[5] For a full listing of Jena's capabilities see: https://jena.apache.org/.

[6] For a complete specification, see: https://www.w3.org/TR/sparql11-query/.

customers. Figure 2 shows a part of Prosseco's architecture, relevant directly to the scope of this paper, leaving aside the execution environment of modeled processes and its connection with the configurable BPM system. The User Interface block represents tools available to the end user of the system; it includes hybrid modeling environment, capable of combining Business Processes with Business Rules, and a monitoring component designed to analyze traces of their execution. Both these tools have access to the vast repository of existing components, which can be modified or extracted to built new models, tailored to fit needs of a specific company.

In order to maintain consistency of knowledge distributed across different representations, there was introduced an ontology containing business vocabulary used in the domain. It contains 86 major concepts partitioned into eight types, including structure of the company, its resources, partners, documents and also more abstract ones describing the typical processes and events happening during the life of SME company.

The following paragraphs describe, respectively to the problems P1–3 introduced in Sect. 1, and how the semantic annotations were used to improve the Prosecco system.

P1 — Rule Models Filtering: As noted previously, user of the modeling environment can take an advantage of the already prepared components. This way he or she does not have to repeat work and can rely on the experience of experts responsible for the contents of the repository. The biggest issue connected with this approach is size and amount of the models contained in repository; it would be a time-consuming task to manually search the database element by element. What is more, there is no reason to assume, that user knows the contents of repository enough to recognize the components related to his or her current modeling task. Because of that the recommendation system, as shown in Fig. 2, was created. This tool has insight into the currently designed model and based on that, it proposes a suitable elements from the repository. To do that, it relies on the meaning of the elements already used in the model and looks for the components involved with the similar concepts. An example of two semantically annotated rule components is shown in Listing 1.3.

```
@Used_by(subject=Office_worker)
@Is_involved_in(subject=Shopping_list)
xschm 'calculates remaining budget':
[expenseAmount, sumOfExpenses, currentBudget] ==> [budgetExceeding].

@Used_by(subject=Office_worker)
@Confirmed_by(subject=Board)
@Is_involved_in(subject=Shopping_list)
@Affects(subject=Budget)
@Deprecated
xschm 'evaluates possibility of financing':
[expenseAmount, budgetExceeding] ==> [expense].
```

Listing 1.3. Example of simple HMR component used in the Prosecco project annotated with semantic information (concept and relation names translated to English).

In case the user was trying to model the process of shopping in a company, the recommendation mechanism recommends usage of the elements, based on shared concepts, like "Shopping List" or "Budget". This can be easily achieved with a SPARQL query similar to presented in Listing 1.4. The same mechanism can be also used in an explicit search of components within repository.

```
SELECT ?component
WHERE
   { ?component psc:Is_involved_in psc:Shopping_list}
   UNION
   { ?component psc:Affects psc:Budget}
?component rdf:type xtt:XTT2Table
```

Listing 1.4. An example of SPARQL query used by recommendation engine

P2 — Concept Alignment: The proper use of the modeled vocabulary in the resulting BPM system is ensured by its data model, which is automatically generated from the ontology. However, it would be preferred to find potential mismatches still during the modeling phase, before any deployment is attempted. This problem affects mostly rules making assumptions about structure of entities represented in the resulting system. In situation, when user models rule component that share input with different representation, e.g. an activity from BPMN process, the modeling environment should recognize semantic mismatches and propose an adequate equivalent. Having an annotated attributes in rule components, as shown in Listing 1.5, would render this problem as a simple semantic reasoning task of asserting identity between entities.

```
@Relation(name-owl:sameAs,
    value=rdf:about="http://dbpedia-live.openlinksw.com/ontology/budget")
xattr [name: projectBudget,
    abbrev: budget, class: simple,
    type: real, comm: inter ].
```

Listing 1.5. An HMR attribute annotated with semantic description.

In case of Drools fact types the same annotation is used to ensure, that its structure reflects the one modeled within the ontology.

P3 — Semantic Documentation: The last, but also important problem regarding the complexity of the repository, is so called semantic gap between the formal models and the natural meaning of concepts, as they are perceived with the business domain. It is very difficult, if not impossible, to create formalized, abstract descriptions without loss of expressibility with respect to their natural equivalents. Currently, this difficulty is solved in Software Engineering by documenting code, often in a form of in-line comments within the models. The main disadvantage of this approach is that the natural language used to explain the code is not understandable by the computer program. Semantic annotations allow to "comment" knowledge components with fixed set of relations and attributes, e.g. the @Deprecated annotation in Listing 1.3 informs the user, that related component is now discouraged to be used in new models. Furthermore, the natural language description can still be used in form of

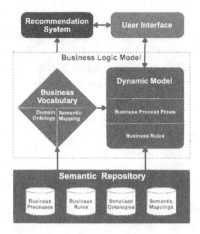

Fig. 2. Part of the Prosecco system's architecture containing a semantic repository

Fig. 3. An annotated relation from Prosecco domain ontology, seen within Protege editor

annotations connected to specific ontology concepts. The Protege[7] environment allows creation of annotations in separate languages, tracks usage of relations across different concepts and displays formal attributes of the relation. An example of annotated relation from Prosecco ontology is presented in Fig. 3.

---

[7] Protege is an ontology editor. For details see http://protege.stanford.edu/.

6 Summary and Future Works

In this paper, a technical solution to combine strengths of the Semantic Web with business rules was presented. Such a tool proved to be useful in domain of big heterogeneous systems, providing user with ability to efficiently query the rule model components based on their meaning. Furthermore, it enforces a consequent use of business vocabulary, thus aligning concepts within models on different levels of abstraction.

The further research will focus on two possible extensions. First, current representation of the Drools model lacks much information about its hidden structure. The grouping and flow dependency aspect represented in the special rules attributes should be taken into consideration. Furthermore, the structure of the fact types could be aligned with the classes contained in the Business Vocabulary. This enhancement would improve further integration of Drools models with the rest of the system.

Second, XTT2 tables could be automatically populated, based on their semantic annotations and contents of problem domain ontology. Generated tables would consist of rules capable of identifying specific instances from domain ontology. Such simple, but difficult to manually create, expert systems would extend possibilities of mediation with user in order to gather information about the current context.

References

1. Adrian, W.T., Bobek, S., Nalepa, G.J., Kaczor, K., Kluza, K.: How to reason by HeaRT in a semantic knowledge-based wiki. In: Proceedings of the 23rd IEEE International Conference on Tools with Artificial Intelligence, ICTAI 2011, Boca Raton, Florida, USA, pp. 438–441 (2011). http://ieeexplore.ieee.org/xpls/abs_all.jsp?arnumber=6103361&tag=1

2. Bobek, S., Baran, M., Kluza, K., Nalepa, G.J.: Application of bayesian networks to recommendations in business process modeling. In: Giordano, L., Montani, S., Dupre, D.T. (eds.) Proceedings of the Workshop AI Meets Business Processes 2013 co-located with the 13th Conference of the Italian Association for Artificial Intelligence (AI*IA 2013), Turin, Italy, 6 December 2013. http://ceur-ws.org/Vol-1101/

3. Bobek, S., Nalepa, G.J., Ślażyński, M.: Challenges for migration of rule-based reasoning engine to a mobile platform. In: Dziech, A., Czyżewski, A. (eds.) MCSS 2014. CCIS, vol. 429, pp. 43–57. Springer, Cham (2014). doi:10.1007/978-3-319-07569-3_4

4. Dentler, K., Cornet, R., ten Teije, A., de Keizer, N.: Comparison of reasoners for large ontologies in the OWL 2 EL profile. Semant. Web 2(2), 71–87 (2011). doi:10.3233/SW-2011-0034

5. Eiter, T., Ianni, G., Krennwallner, T., Polleres, A.: Rules and ontologies for the semantic web. In: Baroglio, C., Bonatti, P.A., Małuszyński, J., Marchiori, M., Polleres, A., Schaffert, S. (eds.) Reasoning Web 2008. LNCS, vol. 5224, pp. 1–53. Springer, Heidelberg (2008). doi:10.1007/978-3-540-85658-0_1

6. Grau, B.C., Parsia, B., Sirin, E.: Working with multiple ontologies on the semantic web. In: McIlraith, S.A., Plexousakis, D., Harmelen, F. (eds.) ISWC 2004. LNCS, vol. 3298, pp. 620–634. Springer, Heidelberg (2004). doi:10.1007/978-3-540-30475-3_43

7. Horrocks, I., Parsia, B., Patel-Schneider, P., Hendler, J.: Semantic web architecture: stack or two towers? In: Fages, F., Soliman, S. (eds.) PPSWR 2005. LNCS, vol. 3703, pp. 37–41. Springer, Heidelberg (2005). doi:10.1007/11552222_4
8. Horrocks, I., Patel-Schneider, P.F., Boley, H., Tabet, S., Grosof, B., Dean, M.: SWRL: A Semantic Web Rule Language Combining OWL and RuleML. Technical report, World Wide Web Consortium, May 2004
9. Kluza, K., Kaczor, K., Nalepa, G., Slazynski, M.: Opportunities for business process semantization in open-source process execution environments. In: 2015 Federated Conference on Computer Science and Information Systems (FedCSIS), pp. 1307–1314, September 2015
10. Kluza, K., Kaczor, K., Nalepa, G.J.: Integration of business processes with visual decision modeling. Presentation of the HaDEs toolchain. In: Fournier, F., Mendling, J. (eds.) BPM 2014. LNBIP, vol. 202, pp. 504–515. Springer, Cham (2015). doi:10.1007/978-3-319-15895-2_43
11. Nalepa, G., Slazynski, M., Kutt, K., Kucharska, E., Luszpaj, A.: Unifying business concepts for SMEs with prosecco ontology. In: 2015 Federated Conference on Computer Science and Information Systems (FedCSIS), pp. 1321–1326, September 2015
12. Nalepa, G.J., Ligęza, A., Kaczor, K.: Overview of knowledge formalization with XTT2 rules. In: Bassiliades, N., Governatori, G., Paschke, A. (eds.) RuleML 2011. LNCS, vol. 6826, pp. 329–336. Springer, Heidelberg (2011). doi:10.1007/978-3-642-22546-8_26
13. Nalepa, G.J., Kluza, K., Kaczor, K.: Proposal of an inference engine architecture for business rules and processes. In: Rutkowski, L., Korytkowski, M., Scherer, R., Tadeusiewicz, R., Zadeh, L.A., Zurada, J.M. (eds.) ICAISC 2013. LNCS, vol. 7895, pp. 453–464. Springer, Heidelberg (2013). doi:10.1007/978-3-642-38610-7_42
14. OMG: Business Process Model and Notation (BPMN): Version 2.0 specification. Technical report formal/2011-01-03, Object Management Group, January 2011
15. OWL Working Group, W3C: OWL 2 Web Ontology Language: Document Overview. W3C recommendation, W3C, October 2009
16. O'Connor, M., Knublauch, H., Tu, S., Grosof, B., Dean, M., Grosso, W., Musen, M.: Supporting rule system interoperability on the semantic web with SWRL. In: Gil, Y., Motta, E., Benjamins, V.R., Musen, M.A. (eds.) ISWC 2005. LNCS, vol. 3729, pp. 974–986. Springer, Heidelberg (2005). doi:10.1007/11574620_69
17. Rademakers, T., Baeyens, T., Barrez, J.: Activiti in Action: Executable Business Processes in BPMN 2.0. Manning Pubs Co Series, Manning Publications Company, Greenwich (2012)
18. Rosati, R.: On combining description logic ontologies and nonrecursive datalog rules. In: Calvanese, D., Lausen, G. (eds.) RR 2008. LNCS, vol. 5341, pp. 13–27. Springer, Heidelberg (2008). doi:10.1007/978-3-540-88737-9_3
19. Thomas, O., Fellmann, M.: Semantic process modeling-design and implementation of an ontology-based representation of business processes. Bus. Inf. Syst. Eng. 1(6), 438–451 (2009)

Convolutional Neural Networks for Time Series Classification

Mariusz Zębik[1], Marcin Korytkowski[2], Rafal Angryk[3], and Rafał Scherer[2(✉)]

[1] Smartvide Ltd., Moszczenica, Poland
[2] Computer Vision and Data Mining Lab,
Institute of Computational Intelligence, Częstochowa University of Technology,
Al. Armii Krajowej 36, 42-200 Częstochowa, Poland
{marcin.korytkowski,rafal.scherer}@iisi.pcz.pl
[3] Department of Computer Science,
Georgia State University, Atlanta, GA 30302-5060, USA
angryk@cs.gsu.edu
http://smartvide.com, http://iisi.pcz.pl
http://grid.cs.gsu.edu/rangryk/

Abstract. This article concerns identifying objects generating signals from various sensors. Instead of using traditional hand-made time series features we feed the signals as input channels to a convolutional neural network. The network learned low- and high-level features from data. We describe the process of data preparation, filtering, and the structure of the convolutional network. Experiment results showed that the network was able to learn to recognize objects with high accuracy.

Keywords: Convolutional neural networks · Time series classification · Accelerometer · Gyroscope

1 Introduction

Artificial intelligence in recent years is widely used in business systems. Combining artificial intelligence with the Internet of Things (IoT) concept, which deals with the collection of information by uniquely identified devices and items, enables building intelligent infrastructure that can have a wide range of applications. Time series and data streams are very important in real time classification. Examples may be medical or industrial systems that in the case of detecting an anomaly can inform the physician or a supervisor about the abnormalities that the system has detected. Accelerometers and gyroscope sensors are the basic accessories used in mobile devices. There are many systems that have a wide range of applications based on the operation of individual sensors.

Convolutional neural networks (CNNs) have been used recently extensively in the literature. They allow to depart from traditional hand-made features [4,5,7,10,16,18,19] to features that are designed automatically in the process of training from data. They are better in this task than traditional feedforward neural networks [3]. In [15] the authors predict in real time seizures from

© Springer International Publishing AG 2017
L. Rutkowski et al. (Eds.): ICAISC 2017, Part II, LNAI 10246, pp. 635–642, 2017.
DOI: 10.1007/978-3-319-59060-8_57

EEG signals. Novel machine learning techniques outperformed existing classification methods. They obtained 71 % accuracy and 0 false positives on the 21-patient Freiburg dataset. Apart from spatio-temporal data, CNNs are utilized to visual object recognition [13], speech [1], natural language processing [6], etc. According to many sources, deep neural networks are much more efficient than shallow architectures despite deep learning creates new challenges in comparison to standard gradient-based algorithms. In this paper we use convolutional neural networks [14] for classifying time series. Multichannel ability of CNNs can have many applications in e.g. processing sound or data from various sensors where one dimensions is time. An example can be health care and Congestive Heart Failure Detection [20], where authors used a multichannel CNN and a one-dimensional time series is fed to every input. In [9] authors identify smartphone users by their gait using statistical measures preceded feature dimensionality reduction by random projections. Similarly, in [8] they use also random projections but followed by the Jaccard distance computation between probability density functions of the projected data. In [11] nearest neighbors are used to classify acceleration data from a smartphone. Authors of [2] compare several classifiers, i.e. k-Nearest Neighbors, Gaussian Mixture Models, Support Vector Machines, Random Forest, k-means and Hidden Markov Models and showed that at that time k-NN achieved the best accuracy on human activity classification. Of course, k-NN drawback is the necessity of storing whole datasets. An autocorrelation matrix of fast Fourier transform features is used in [12] to analyze 3-D acceleration signals regardless of the orientations of the accelerometer obtaining 50% accuracy.

In the paper we use data from sensors to train and test convolutional neural network in CNTK (Microsoft Cognitive Toolkit) software. The training and test data preparation software was written using Microsoft Visual Studio 2015, .NET 4.5, and the C# programming language. In the next section we describe the collected data and experiments with CNN learning are presented in Sect. 3.

2   Data Collecting

We used time series from a custom-made devices with three-axis accelerometers and gyro sensors. Gyroscope is a device for measuring angular position based on the principle of angular momentum preservation. It measures the angular velocity of rotating objects relative to the X, Y, and Z axes. The unit of measurement is degrees per second. Sensor operating range is ±245, 500, and 2000°/s and examples of positions are presented in Fig. 1.

The purpose of the accelerometer is to measure linear acceleration of its own motion. In addition to determining the values of linear acceleration, it is possible to determine the spatial arrangement of the object with the sensor attached. Unit of measurement is g (earth acceleration). The accelerometer used by us can be scaled within range 2, 4, 6, 8, or 16 g. An example graph generated from the recorded data from a device with sensors attached is shown in Figs. 2 and 3.

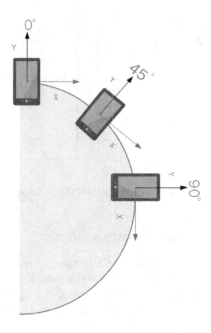

Fig. 1. Various angular positions of gyroscope sensor.

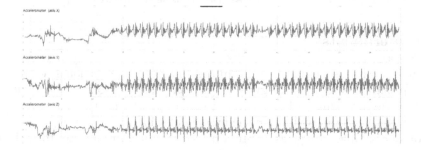

Fig. 2. Example of collected accelerometer data.

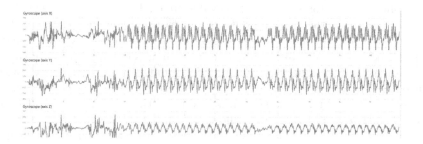

Fig. 3. Example of collected gyroscope data.

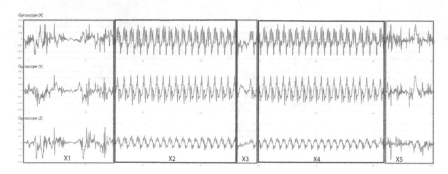

Fig. 4. Examples of data acquisition intervals for gyroscope time series.

Figure 4 shows various stages of data acquisition:

- the device starts to move - interval X1 in the figure,
- movement to the indicated destination - interval X2 in the figure,
- changing the direction of movement - interval X3 in the figure,
- return to the place where the recording was started - interval X4 in the figure,
- the device stops - interval X5 in the figure.

Data from the X2 range was used to prepare training samples for the convolutional network, whereas X4 data was used to test the network.

3   Experiments

The main problems we had to cope in designing the convolutional network for device profiles based on 3-axis gyro and accelerometer sensors were twofold. We had different data length. The recording time of profiles for the same movement length was different over time, so the number of samples was varying. The interval for one sample was of great importance for the time it took to train the network. The larger the sample, the more time was needed to train the network. Preparation of training data for 50 device profiles was performed by an algorithm that recalculated the number of samples for each of N profiles. Sample size for one channel was selected from values 0.5 s, 1 s and 2 s (Table 1). In the process of preparing the training data, the data range marked with X2, shown in Fig. 4, is the stage of the device in the direction to the target. Only these data were involved in the convolutional network learning. The preparation of the test data was done in the same way as the training data, except that the data that was used are indicated by X4 in Fig. 4 and are the return stage of the device. Convolutional neural network was simulated using the CNTK (Microsoft Cognitive Toolkit). The configuration file allows to build nearly any network structure. Table 2 shows various parameters of the network used for device classification. The structure of the entire network architecture was presented according to the operation sequence in Table 3.

Table 1. Relationship between the length of the interval and the number of samples

Interval [s]	Number of samples for one channel	Data for 6 channels
0.5 s	24	24 * 6 = 144
1 s	48	48 * 6 = 288
2 s	96	96 * 6 = 576

Table 2. CNN parameters used during experiments.

Parameter name	Value
Image shape	48 × 1 × 6
Label dimension	30
Features dimension	288
Feat scale	1/1024
Epoch size	60000
Max epochs	15
Learning rates per sample	0.001 * 10:0.0005 * 10:0.0001
Dropout rate	0.5

Table 3. CNN structure

	Parameter name	Value
1	Feat scale	1/1024
2	Convolution layer	128 feature map, (32:1) – capacity, padding
3	ReLU	Activation
4	MaxPoolingLayer	(16:1), stride = (3:1)
5	Convolution layer	256 feature map, (3:1) – capacity, padding
6	ReLU	Activation
7	MaxPoolingLayer	(8:1), stride = (2:1),
8	Dense layer	100
9	Dropout	0.5
10	ReLU	Activation
11	Linear layer	30 profiles

The network has six input channels (axis X, Y, Z for accelerometer and gyroscope). Each channel is sequentially sampled and the input data has been scaled according to the "Feat Scale" parameter listed under "1" in Table 3. The next step is convolution with 128 feature maps of size 32. After the first convolution, the ReLU activation function is applied followed by a 3:1 pooling. Filters are set by default in the CNTK software for the normal distribution. In the next step after activation, there is another convolution layer consisting of 256 feature maps of size 3. Similarly to the previous convolution, the ReLU activation function was used and pooling in this case with 2:1 stride. In step 8 in Table 3 there is a Dense Layer composed of 100 neurons. Dropout [17] rate is 0.5. In the next step, the ReLU activation function is called. The last step is a linear layer, which is designed to classify samples for 50 profiles. The learning process took 15 epochs. In the first 10 epoch the learning rate was 0.001, but in the next 5 epochs it dropped to 0.0005.

To find the optimal CNN structure we performed several test for 50 device profiles using 2 GB NVidia Geforce GT650M GPU. After many experiments the best results are presented in Table 4.

Table 4. Accuracy for optimal CNNs.

Id	Sample length	Number of epochs	Accuracy on test data	Learning time
1	0.5 s	15	70.00 %	~3 min
2	1.0 s	15	87.60 %	~6 min
3	2.0 s	15	88.10 %	~20 min

The best network performance for identifying device profiles based on sensor data has been achieved by a network in which six channels was fed with 2 s samples, however the training time was much longer than in the case of the 1 s samples.

4   Conclusion

Artificial intelligence and machine learning methods are used extensively to analyze data from many sensors. With the advent of the Internet of Things we have more and more data to process.

Especially convolutional neural networks are very successful, mainly in computer vision. In the paper we used them to classify devices by using data from accelerometers and gyroscope sensors that are one of the basic accessories used in mobile devices. We used a custom made device to collect data from devices. The CNNs utilized one-dimensional feature maps to analyze time series from the sensors. We achieved very good accuracy with relatively small CNN working on six input one-dimensional time series.

References

1. Abdel-Hamid, O., Mohamed, A., Jiang, H., Penn, G.: Applying convolutional neural networks concepts to hybrid NN-HMM model for speech recognition. In: 2012 IEEE International Conference on Acoustics, Speech and Signal Processing (ICASSP), pp. 4277–4280, March 2012
2. Attal, F., Mohammed, S., Dedabrishvili, M., Chamroukhi, F., Oukhellou, L., Amirat, Y.: Physical human activity recognition using wearable sensors. Sensors **15**(12), 31314–31338 (2015)
3. Bas, E.: The training of multiplicative neuron model based artificial neural networks with differential evolution algorithm for forecasting. J. Artif. Intell. Soft Comput. Res. **6**(1), 5–11 (2016)
4. Bertini Junior, J.R., Nicoletti, M.D.C.: Enhancing constructive neural network performance using functionally expanded input data. J. Artif. Intell. Soft Comput. Res. **6**(2), 119–131 (2016)
5. Brester, C., Semenkin, E., Sidorov, M.: Multi-objective heuristic feature selection for speech-based multilingual emotion recognition. J. Artif. Intell. Soft Comput. Res. **6**(4), 243–253 (2016)
6. Collobert, R., Weston, J.: A unified architecture for natural language processing: deep neural networks with multitask learning. In: Proceedings of the 25th International Conference on Machine Learning, ICML 2008, pp. 160–167. ACM, New York (2008)
7. Cpalka, K., Zalasinski, M., Rutkowski, L.: A new algorithm for identity verification based on the analysis of a handwritten dynamic signature. Appl. Soft Comput. **43**, 47–56 (2016)
8. Damaševičius, R., Vasiljevas, M., Šalkevičius, J., Woźniak, M.: Human activity recognition in AAL environments using random projections. Comput. Math. Methods Med. **2016** (2016)
9. Damaševičius, R., Maskeliūnas, R., Venčkauskas, A., Woźniak, M.: Smartphone user identity verification using gait characteristics. Symmetry **8**(10), 100:1–100:20 (2016). doi:10.3390/sym8100100
10. Drozda, P., Gorecki, P., Sopyla, K., Artmiejew, P.: Visual words sequence alignment for image classification. In: 2013 IEEE 12th International Conference on Cognitive Informatics and Cognitive Computing, pp. 397–402, July 2013
11. Frank, J., Mannor, S., Precup, D.: Activity and gait recognition with time-delay embeddings. In: Proceedings of the Twenty-Fourth AAAI Conference on Artificial Intelligence, AAAI 2010, pp. 1581–1586. AAAI Press (2010)
12. Kobayashi, T., Hasida, K., Otsu, N.: Rotation invariant feature extraction from 3-D acceleration signals. In: 2011 IEEE International Conference on Acoustics, Speech and Signal Processing (ICASSP), pp. 3684–3687, May 2011
13. Krizhevsky, A., Sutskever, I., Hinton, G.E.: Imagenet classification with deep convolutional neural networks. In: Pereira, F., Burges, C.J.C., Bottou, L., Weinberger, K.Q. (eds.) Advances in Neural Information Processing Systems, vol. 25, pp. 1097–1105. Curran Associates Inc. (2012)
14. Lecun, Y., Bottou, L., Bengio, Y., Haffner, P.: Gradient-based learning applied to document recognition. Proc. IEEE **86**(11), 2278–2324 (1998)
15. Mirowski, P., Madhavan, D., LeCun, Y., Kuzniecky, R.: Classification of patterns of EEG synchronization for seizure prediction. Clin. Neurophysiol. **120**(11), 1927–1940 (2009)

16. Nikulin, V.: Prediction of the shoppers loyalty with aggregated data streams. J. Artif. Intell. Soft Comput. Res. **6**(2), 69–79 (2016)
17. Srivastava, N., Hinton, G., Krizhevsky, A., Sutskever, I., Salakhutdinov, R.: Dropout: a simple way to prevent neural networks from overfitting. J. Mach. Learn. Res. **15**, 1929–1958 (2014)
18. Woźniak, M., Połap, D., Napoli, C., Tramontana, E.: Graphic object feature extraction system based on cuckoo search algorithm. Expert Syst. Appl. **66**, 20–31 (2016). doi:10.1016/j.eswa.2016.08.068
19. Woźniak, M., Połap, D., Nowicki, R.K., Napoli, C., Pappalardo, G., Tramontana, E.: Novel approach toward medical signals classifier. In: IEEE IJCNN 2015–2015 IEEE International Joint Conference on Neural Networks, Proceedings, Killarney, Ireland, 12–17 July 2015, pp. 1924–1930. IEEE (2015). doi:10.1109/IJCNN.2015.7280556
20. Zheng, Y., Liu, Q., Chen, E., Ge, Y., Zhao, J.L.: Time series classification using multi-channels deep convolutional neural networks. In: Li, F., Li, G., Hwang, S., Yao, B., Zhang, Z. (eds.) WAIM 2014. LNCS, vol. 8485, pp. 298–310. Springer, Cham (2014). doi:10.1007/978-3-319-08010-9_33

Special Session: Advances in Single-Objective Continuous Parameter Optimization with Nature-Inspired Algorithms

A DSS Based on Hybrid Ant Colony Optimization Algorithm for the TSP

Islem Kaabachi[✉], Dorra Jriji, and Saoussen Krichen

LARODEC Laboratory, Higher Institute of Management,
University of Tunis, Tunis, Tunisia
islamkaa@hotmail.com, dorrajriji@gmail.com, saoussen.krichen@isg.rnu.tn

Abstract. The traveling salesman problem (TSP) is one of the most studied problems in combinatorial optimization due to its importance and NP-hard numerous approximation methods were proposed to solve it. In this paper, we propose a new hybrid approach which combines local search with the ant colony optimization algorithm (ACO) for solving the TSP. The performance of the proposed algorithm is highlighted through the implementation of a Decision Support System (DSS). Some benchmark problems are selected to test the performance of the proposed hybrid method. We compare the ability of our algorithm with the classical ACO and against some well-known methods. The experiments show that the proposed hybrid method can efficiently improve the quality of solutions than the classical ACO algorithm, and distinctly speed up computing time. Our approach is also better than the performance of compared algorithms in most cases in terms of solution quality and robustness.

Keywords: Traveling Salesman Problem · Ant Colony Optimization · Meta-heuristic · Decision Support System

1 Introduction

The Traveling Salesman Problem (TSP) is the most well-known and extensively studied in the area of combinatorial optimization [1–3], which can be described as follows: Given a one salesman and a set of N cities within specific area. The salesman has to visit each one of these cities starting from a certain one city and returning to the same city. The TSP aims to minimize the total travel distance circuit (Hamiltonian circuit) passing through each city only once. It has important applications to real life problems, such as planning bus or taxi lines, regular distribution of goods, finding of the shortest of costumer servicing route, vehicle routing problems, etc. TSP is one of the best known NP-hard problems [4], which means that there is no exact algorithm to solve it in polynomial time. The minimal expected time to obtain optimal solution is exponential. Therefore, the utilization of the heuristic techniques are necessary to obtain a good solution. The Tabu Search by Gendreau et al. [8] and Knox [7]; Simulated Annealing

© Springer International Publishing AG 2017
L. Rutkowski et al. (Eds.): ICAISC 2017, Part II, LNAI 10246, pp. 645–654, 2017.
DOI: 10.1007/978-3-319-59060-8_58

by Geng et al. [10] and Allwright and Carpenter [9]; Particle Swarm Optimization by Pang et al. [13], Shi et al. [15], Wang et al. [14], and Zhong et al. [6]; Neural Networks by Ghaziri and Osman [11] and Leung et al. [12] have been applied to solve this problem with more or less success. Aside from the above works, several hybrid methods have also been developed to TSPs. Gunduz and Kiran [21] presented a new hierarchic method based swarm intelligence algorithms for theTSP. Masutti and Castro [22] proposed some modifications on the RABNET-TSP, an immuneinspired self-organizing neural network, for solving the TSP. On the other hand, the Genetic Algorithm and Ant Colony Optimization (ACO), can be good candidates to find a good quality solutions for the TSP within an acceptable time. In [5], the authors present a genetic algorithm operators to reduce the total distance and time for the TSP. So, a new crossover method, the sequential constructive crossover method is used. They also propose a binary matrix representation of chromosomes. The proposed method is produces the high quality solutions in reasonable time. Gambardella and Dorigo [17] conducted the first work into ACO in the TSP. Their algorithm was tested and compared to other algorithms such as Simulated Annealing, Elastic Net and Self Organizing map. The tests were performed on a set of 50 node graph problems. Subsequently, many of the proposed ACO algorithms [16,18] have initially been applied to the TSP. Despite that the TSP was the first problem to be solved by the ACO, it still inspires researchers. In [24] several ACO algorithms is used in order to solve a bi-criteria TSP to compare them and discuss their characteristics. More recently, Tavares and Pereira [25] proposed a system to evolve strategies for updating pheromone trails and the TSP was used as a test case. Although in general ACO algorithms give a high-quality solutions, there are cases where an hybridization algorithm, proves to be necessary. Therefore, in the recently years researchers have developed hybrid algorithms between ACO and other metaheuristics, such as the Local Search [26], Simulated Annealing [27], Genetic Algorithm [28].

The main research contributions of the present paper are (1) to propose a Hybrid ACO algorithm (HACO) for solving the TSPs, (2) to compare the obtained solutions with the classical ACO and our hybrid approach, (3) to develop a decision support system (DSS) that use to solve the TSP, and (4) to show the superiority of the proposed HACO against some well-known methods. The remainder of this paper is organized as follows. The HACO algorithm is proposed in Sect. 2. The main steps of the proposed DSS are outlined in Sect. 3. In Sect. 4, the computational results are presented, followed by the conclusion in Sect. 5.

2   DSS Architecture Embedding a HACO

We develop a Decision Support System (DSS) for solving the TSP in order to find to best salesman routing. This latter leads to find the best way to generate salesman's route. The DSS, that integrates the proposed HACO, has to cope with multiple system constraints while optimizing the TSP objective. The purpose of

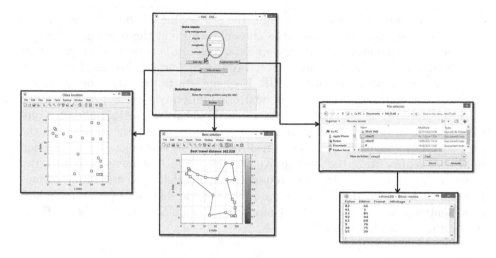

Fig. 1. A HACO based DSS for solving the TSP

using a DSS is to give users a flexible and user-friendly web based application and help them to make decisions about problems that may be rapidly changing and not easily specified. As detailed in Fig. 1, the proposed DSS consists of three parts: data inputs, resolution process and best solution display. The DSS starts by extracting the input data as: The number of cities, the geographic coordinates of the cities to find the distance matrix. The routing data is to be performed in order to generate the appropriate pathway. We then run the DSS, integrating the proposed HACO, on different benchmarks until they reach the stopping criterion. Finally the best solution that fully satisfies all users' requirements is displayed.

3 Hybrid Ant Colony Optimization Algorithm

The notations related to the proposed HACO are presented below:

- T_{ij} Trail intensity on edge $\{i, j\}$.
- I_{ij} Inverse of the distance of edge $\{i, j\}$.
- m Number of ants.
- N Number of cities.
- V_{ij} The attractive value of edge $\{i, j\}$.
- L_i The list of all feasible cities that has not been visited by the ant.
- $Lbest$ The length of the best solution found.
- Tab_s The list for first visited city by all ants.
- Tab_i The list for the cities visited by ant i.

At the beginning of HACO, we generate initial solution randomly by starting the route from a certain one city. We need to create the initial solution to initialize the trail intensities. The major part of literature of ACO used initial solutions to initialize trail intensities. Therefore, the proceeding of the algorithm is not affected by the initial trail intensities chosen, because at the beginning all edges disposed the same amounts. We use m ants, each ant generates a complete tour (ant solutions of HACO), these ants travel from one city to another city to construct routes solutions of the TSP. Each ant starts its route from a certain one city then continues by selecting next city randomly.

Each ant travels between two cities based on two attraction measures representing the probability function, the distance from current location to the proposed city (edge $\{i, j\}$) and the trail intensity on this edge. Two attraction measures representing the probability function. The first measure is the trail intensity T_{ij} which contains how frequently edge $\{ij\}$ has been used in previous iterations. The second measure is the inverse of the distance of edge I_{ij}, which represents the move desirability. Therefore, we calculate the probability P_{ij} of picking the next city j as follows:

$$P_{ij} = \begin{cases} \frac{V_{ij}}{\sum_{i \in L_i} V_{ij}} & \text{if } j \in L_i \\ 0 & \text{else where} \end{cases}$$

where the attractive value V_{ij}

$$V_{ij} = (T_{ij})^{\alpha}(I_{ij})^{\beta}$$

And L_i is the list of all feasible cities that has not been visited by the current ant. The ant continues adding customers to its route solution until there is no more cities in the list of all feasible cities (L_i is empty). Then the ant return to the first visited city from which start another trip. A history of the visited cities, during current route solution, is kept in the list for the cities visited by ant (Tab_i list). After creating routes solutions by all ants, the local search procedure is performed to enhance the solution quality. The importance of the local search is to improve the solutions generated by all ants. Therefore, the best solution and all trail intensities are updated according to the best solution found so far after applying local search. After the local search improvements in each iteration trail intensities evaporate with time on all edges and ants leave pheromone on the visited edges.

Step 1: Generate initial solution and initialize the trail intensities.
Step 2: Creating ants solutions (Construct routes for m ants).
Step 3: Applying local search scheme to improve the solution create by each ant.
Step 4: Update best solution found and trail intensities for all edges using best solution.
Step 5: Terminate the algorithm and indicate the best solution.
The process of HACO for the TSP

4   Numerical Experiments

In order to show the performance of our method, the best results of HACO were compared with the results of the classical ACO and those of other existing algorithms proposed in the literature based on some benchmark problems from TSP. They are taken from the TSPLIB website [29]. According to TSPLIB, the distance between any two cities is computed by the Euclidian distance and then rounded up. The best results are given in bold. The algorithm proposed in our solution procedure is coded in Matlab 2015a. All experiments are conducted on a personal computer with $Intel^{®}$ $Core^{TM}$ i5-5200U CPU @ 2.20 GHz 6,00 GO RAM and Windows 8.1 pro, 64-bit operating system, x64-based processor. For each problem, the algorithm is implemented for 10 runs. The reported result is the best found solution over the runs with the CPU time in seconds.

4.1   HACO Tuning

The metal tuning of HACO is very important in order to general good quality solutions. To do so, we propose to run the HACO according to numerous configurations. After a series of run, we select the following configurations: The number of the ants m controls the solution quality and affects the computational time. The computational time increases excessively without any found solution improvements after 25 ants. All ants used apply a local search to their solution. [19] found that setting pheromone factor $\alpha - 1.0$ and heuristic factor $\beta = 2.0$ gives very good solutions. The evaporation rate parameter $\rho = 0.1$ gave a good chance to update the pheromone with the new experience of ants on the account of the existing experiences. Additionally, we fixed the parameter $q_0 = 0.9$. The maximum iteration is set 1000 times for all TSPs instances to test the performance of the HACO and the classical ACO. To keep the solution time comparable with the classical ACO we used 1000 iterations and improvement is very minimum after 1000 iterations.

4.2   Computational Results

The algorithms are applied to 15 standard instances which possess a number of cities whose sizes are between 70 and 14021. These results have been presented in Table 1. We assume that:

- **Inst.:** The instance name where the linked number represents the problem dimension (i.e. the number of cities).
- **Gap%:** Percentage between the best found solution $z(x)$ obtained by the proposed HACO or the best results of the well-known methods and the optimal solution (Opt.) $z(x^{**})$ for the instance.
- **Improvement%:** Percentage between the best found solution $z(x)$ obtained by the proposed HACO and the best solution $z(x^*)$ obtained by the ACO.
- **CPU:** Average computing time needed to obtain the best solution.

Table 1. Comparison results of the algorithms for benchmark test problems

Inst.	Opt.	ACO			HACO		Gap (%)	Improv. (%)
	$z(x^{**})$	$z(x^{*)}$	CPU (s)	z(x)	CPU(s)			
berlin52	7542	7721.43	04.21	7542.15	02.68	0.20	2.32
att48	10628	11421	15.23	10710.21	13.45	0.77	6.22
eil76	538	564.68	12.54	538.36	09.78	0.07	4.66
eil51	426	442.12	04.62	426.71	03.85	0.17	3.49
ch130	6110	6162.35	16.56	6110.02	13.81	**0.00**	0.85
rd400	15281	15441.32	30.26	15282.21	38.62	**0.01**	1.03
rat99	1211	1264.57	10.47	1212.17	11.24	0.10	4.14
lin105	14379	14456.65	15.24	14382.86	14.62	0.03	0.51
d1655	62128	62541.71	124.52	62129.73	123.71	**0.00**	0.66
a280	2579	3124.62	25.47	2596.24	21.56	0.67	16.91
brd14051	469385	476312.58	184.26	470521.24	165.18	0.24	1.22
bier127	118282	123251.52	26.54	121031.20	28.25	2.32	1.80
pr144	58537	58628.84	21.36	58537.31	18.41	**0.00**	0.16
ali535	202310	223821.78	45.87	203024.68	42.72	0.35	9.29
st70	675	682.14	07.29	675.24	06.95	0.04	1.01
Average	66193.67	68565.39	37.63	66502.18	34.99	0.28	3.31

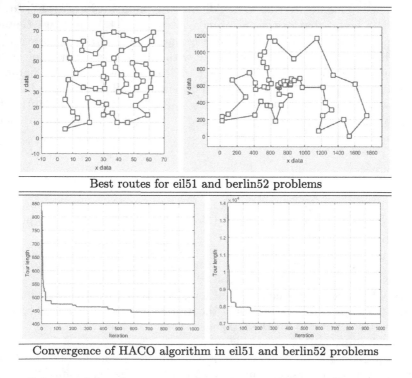

Best routes for eil51 and berlin52 problems

Convergence of HACO algorithm in eil51 and berlin52 problems

Fig. 2. Solution of the eil51 and berlin52 problems and the performance graphics

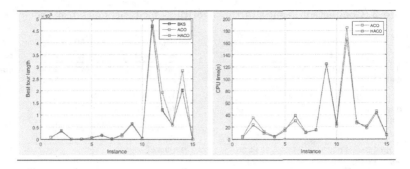

Fig. 3. (a) Pareto optimal sets comparison for TSPs instances (b) Running time representation

As can be seen from Table 1, for the 15 TSP problems, the best and average values of the proposed HACO algorithm are best than the classical ACO algorithm in the experiment. There have 4 TSP problems, which are close to the optimum. For TSP instances berlin52, eil51, eil76 the proposed HCAO algorithm can find the best known solutions. Particularly, for the other TSP instances the found best values are approaching to the best known solutions. We plot in Fig. 3(a) the Pareto optimal sets generated from both algorithms among the different TSP instances. For larger scale instances, we can notice that the proposed HACO algorithm in the best value is better than the ACO algorithm. We can found also that the Gap increases with problem complexity-that is, when the number of cities is less, ACO provides better results. However, when the number of cities and solution space is larger HACO outperforms ACO in the quality of the solutions. In order to further prove the performance of the proposed HACO algorithm, the best route for TSP instances eil51 and berlin52 found is shown in Fig. 2. Figure 2 shows the algorithm's convergence. The experiments show that the HACO algorithm converges quickly. For the berlin52 instance, our algorithm finds a solution of distance 7542.15 in 792th iteration where there is no improvement until the 1000th iteration with minor improvements of 0.24%. In addition, the comparison of the proposed method with other studies in the literature is shown in Table 2. As seen from the computational experiments (Table 2), our method has generated the closest results to optimal solution problems the eil51, eil76, berlin52, st70 and kroa100. The best results obtained for the eil51, eil76, berlin52, st70 and kroa100 are 426.71, 538.36, 7542.15, 675.24, 21283.21, respectively. These results are better than the results of studies in the literature. It is also observed that results close to optimal solution have been obtained for the lin105, ch150 and kroa200, and also reasonable results have been obtained in the eil101. As seen from the results, our method found 631.21 for the ch150 problem while it was obtained 631.28 in the literature with the least Gap 0.35. Table 2 also shows that our proposed method has produced closer results to optimal solutions. In addition, the HACO algorithm generates the optimal results in a

Table 2. Comparison results of the algorithms for benchmark test problems

Method	Inst.	eil51	eil76	berlin52	st70	kroa100	lin105	ch150	kroa200	eil101
	Opt.	426	538	7542	675	21282	14379	6528	29368	629
RABNET-TSP (2009) [22]	Best	427	541	7542	-	21333	14379	6602	29600	638
	CPU	7.95	15.07	08.91	-	15.87	20.62	39.90	15.87	21.72
	Gap	0.23	0.55	0.00	-	0.23	0.00	01.13	0.78	01.43
ACO + 2Opt (2012) [23]	Best	431.25	-	7549.52	-	22006.39	-	-	-	654.49
	CPU	18.35	-	14.69	-	23.06	-	-	-	13.58
	Gap	01.23	-	0.09	-	03.40	-	-	-	04.05
IVRS + 2Opt (2012) [23]	Best	429.11	-	7544.36	-	21309.42	-	-	-	631.28
	CPU	11.32	-	08.23	-	17.83	-	-	-	09.45
	Gap	0.73	-	0.03	-	0.12	-	-	-	0.36
ACO with Tagushi Method (2013) [20]	Best	426	544	7542	-	21382.30	14380.90	-	-	640.50
	CPU	03.32	05.92	03.15	-	11.52	21.45	-	-	20.75
	Gap	0.00	01.11	0.00	-	0.47	0.01	-	-	01.82
ACO with ABC (2014) [21]	Best	431.47	551.07	7544.37	687.24	22122.75	-	6641.69	-	672.71
	CPU	58.33	138.82	60.64	115.65	311.12	-	698.61	-	267.08
	Gap	01.28	02.42	0.03	01.81	03.95	-	01.74	-	06.94
Proposed Method HACO	Best	426.71	538.36	7542.15	675.24	21283.21	14382.86	6531.05	29372.12	631.21
	CPU	03.85	09.78	02.68	06.95	10.83	14.62	15.31	14.03	10.51'
	Gap	**0.00**	**0.00**	**0.00**	**0.00**	**0.00**	0.03	0.04	0.01	0.35

challenging CPU time. In fact, the convergence speeds of the HACO algorithm is obviously much faster than those of the ACO algorithm (see Fig. 3(b)).

5   Conclusion

In this paper, a Hybrid Ant Colony Optimization algorithm (HACO) is proposed to cope with the traveling salesman problem. We proposed a Decision Support System (DSS) to better visualize the obtained results and make it more intuitive. TSP problems are used to examine the effectiveness of the proposed algorithm. The HACO algorithm is more effective than the classical ACO algorithm in terms of convergence speed and the ability to finding better solutions. The experiment results show also that our proposed method is an alternative TSP solver and a competitive algorithm.

References

1. Lawler, E.L., Lenstra, J.K., Kan Rinnooy, A.H.G., Shmoys, D.B.: The Traveling Salesman Problem: A Guided Tour of Combinatorial Optimization, vol. 11, pp. 201–209. Wiley, Chichester (1985)
2. Johnson, D.S., McGeoch, L.A.: The Travelling Salesman Problem: A Case Study in Local Optimization. In: Aarts, E.H.L., Lenstra, J.K. (eds.) Local Search in Combinatorial Optimization, pp. 215–310. Wiley (1997)
3. Reinelt, G.: The Traveling Salesman: Computational Solutions for TSP Applications. LNCS, vol. 840. Springer, Heidelberg (1994)

4. Applegate, D.L., Bixby, R.E., Chvatal, V., Cook, W.J.: The Traveling Salesman Problem: A Computational Study. Princeton University Press, USA (2007)
5. Arananayakgi, A.: Reduce total distance and time using genetic algorithm in Traveling Salesman Problem. Int. J. Comput. Sci. Eng. Technol. **5**(08) (2014). ISSN 2229–3345
6. Zhong, W.L., Zhang, J., Chen, W.N.: A novel discrete particle swarm optimization to solve traveling salesman problem. In: IEEE Congress on Evolutionary Computation, pp. 3283–3287 (2007)
7. Knox, J.: Tabu search performance on the symmetric traveling salesman problem. Comput. Oper. Res. **21**, 867–876 (1994)
8. Gendreau, M., Laporte, G., Semet, F.: A tabu search heuristic for the undirected selective travelling salesman problem. Eur. J. Oper. Res. **106**, 539–545 (1998)
9. Allwright, J.R.A., Carpenter, D.B.: A distributed implementation of simulated annealing for the traveling salesman problem. Parallel Comput. **10**, 335–338 (1989)
10. Geng, X., Chen, Z., Yang, W., Shi, D., Zhao, K.: Solving the traveling salesman problem based on an adaptive simulated annealing algorithm with greedy search. Appl. Soft Comput. **11**, 3680–3689 (2011)
11. Ghaziri, H., Osman, I.H.: A neural network algorithm for the traveling salesman problem with backhauls. Comput. Ind. Eng. **44**, 267–281 (2003)
12. Leung, K.S., Jin, H.D., Xu, Z.B.: An expanding self-organizing neural network for the traveling salesman problem. Neurocomputing **62**, 267–292 (2004)
13. Pang, W., Wang, K.P., Zhou, C.G., Dong, L.J., Liu, M., Zhang, H.Y., Wang, J.Y.: Modified particle swarm optimization based on space transformation for solving traveling salesman problem. In: Proceedings of the 33rd International Conference on Machine Learning and Cybernetics, pp. 2342–2346 (2004)
14. Wang, C., Zhang, J., Yang, J., Hu, C., Liu, J.: A modified particle swarm optimization algorithm and its application for solving traveling salesman problem. In: International Conference on Neural Networks and Brain, vol. 2, pp. 689–694 (2005)
15. Shi, X.H., Liang, Y.C., Lee, H.P., Lu, C., Wang, Q.X.: Particle swarm optimization based algorithms for TSP and generalized TSP. Inf. Process. Lett. **103**, 169–176 (2007)
16. Stutzle, T., Hoos, H.H.: Improvements on the ant system: introducing the MAX-MIN ant system. In Albrecht, R.F., Smith, G.D., Steele, N.C. (eds.) Artificial Neural Networks and Genetic Algorithms, pp. 245–249. Springer, Wien (1998)
17. Gambardella L. and M. Dorigo. Ant-Q: A Reinforcement Learning approach to the traveling salesman problem. In: Prieditis, A., Russell, S. (eds.) Proceedings of Twelfth International Conference on Machine Learning, ML 1995, Tahoe City, CA, pp. 252–260. Morgan Kaufmann (1995)
18. Dorigo, M., Gambardella, L.M.: Ant colony system: a cooperative learning approach to the traveling salesman problem. IEEE Trans. Evol. Comput. **1**(1), 53–66 (1997)
19. Dorigo, M., Maniezzo, V., Colorni, A.: Ant system: optimization by a colony of cooperating agents. IEEE Trans. Syst. Man Cybern. Part B **26**(1), 1–13 (1996)
20. Peker, M., Sen, B., Kumru, P.Y.: An efficient solving of the traveling salesman problem: the ant colony system having parameters optimized by the Taguchi method. Turk. J. Electr. Eng. Comput. **21**, 2015–2036 (2013)
21. Gunduz, M., Kiran, M.S., Ozceylan, E.: A hierarchic approach based on swarm intelligence to solve traveling salesman problem. Turk. J. Electr. Eng. Comput. Sci. (2014). doi:10.3906/elk-1210-147

22. Masutti, T.A.S., de Castro, L.N.: A self-organizing neural network using ideas from the immune system to solve the traveling salesman problem. Inf. Sci. **179**, 1454–1468 (2009)
23. Jun-man, K., Yi, Z.: Application of an improved ant colony optimization on generalized traveling salesman problem. Energy Proc. **17**, 319–325 (2012)
24. Garcia-Martinez, C., Cordon, O., Herrera, F.: A taxonomy and an empirical analysis of multiple objective ant colony optimization algorithms for the bi-criteria TSP. Eur. J. Oper. Res. **180**(1), 116–148 (2007)
25. Tavares, J., Pereira, F.B.: Evolving strategies for updating pheromone trails: a case study with the TSP. In: Schaefer, R., Cotta, C., Kołodziej, J., Rudolph, G. (eds.) PPSN 2010. LNCS, vol. 6239, pp. 523–532. Springer, Heidelberg (2010). doi:10.1007/978-3-642-15871-1_53
26. Pour, H.D., Nosraty, M.: Solving the facility layout and location problem by ant-colony optimization-meta heuristic. Int. J. Prod. Res. **44**, 5187–5196 (2006)
27. Bouhafs, L., hajjam, A., Koukam, A.: A combination of simulated annealing and ant colony system for the capacitated location-routing problem. In: Gabrys, B., Howlett, R.J., Jain, L.C. (eds.) KES 2006. LNCS, vol. 4251, pp. 409–416. Springer, Heidelberg (2006). doi:10.1007/11892960_50
28. Altiparmak, F., Karaoglan, I.: A genetic ant colony optimization approach for concave cost transportation problems. In: IEEE Congress on Evolutionary Computation, CEC 2007, pp. 1685–1692 (2007)
29. TSPLIB. http://www.iwr.uniheidelberg.de/groups/comopt/software/TSPLIB95/

Comparing Strategies for Search Space Boundaries Violation in PSO

Tomas Kadavy$^{(\boxtimes)}$, Michal Pluhacek, Adam Viktorin, and Roman Senkerik

Faculty of Applied Informatics, Tomas Bata University in Zlin,
T.G. Masaryka 5555, 760 01 Zlin, Czech Republic
{kadavy,pluhacek,aviktorin,senkerik}@fai.utb.cz

Abstract. In this paper, we choose to compare four methods for controlling particle position when it violates the search space boundaries and the impact on the performance of Particle Swarm Optimization algorithm (PSO). The methods are: hard borders, soft borders, random position and spherical universe. The goal is to compare the performance of these methods for the classical version of PSO and popular modification – the Attractive and Repulsive Particle Swarm Optimization (ARPSO). The experiments were carried out according to CEC benchmark rules and statistically evaluated.

Keywords: Particle Swarm Optimization · PSO · ARPSO · CEC · Search space · Boundaries

1 Introduction

As rapid particle acceleration is one of the notorious problems of Particle Swarm Optimization algorithm (PSO) [1] it is common that the particles leave the feasible search space and get out of bounds.

There are various approaches for handling the out of bounds particles and all of them can have a significant impact on the performance of the original PSO algorithm [1], or its modifications like Heterogeneous Particle Swarm Optimization (HPSO) [2], Orthogonal Learning Particle Swarm Optimization (OLPSO) [3] and many others. This paper compares some of the strategies on classic PSO and popular Diversity guided PSO (ARPSO) [4].

The boundaries of the search space are defined by particular optimization problem and typically, there is definition of minimal acceptable value and maximum acceptable value of the solution in each dimension. If a particle violates search space boundaries, some position handling of a particle should be applied.

T. Kadavy—This work was supported by Grant Agency of the Czech Republic – GACR P103/15/06700S, further by the Ministry of Education, Youth and Sports of the Czech Republic within the National Sustainability Programme Project no. LO1303 (MSMT-7778/2014). Also by the European Regional Development Fund under the Project CEBIA-Tech no. CZ.1.05/2.1.00/03.0089 and by Internal Grant Agency of Tomas Bata University under the Projects no. IGA/CebiaTech/2017/004.

© Springer International Publishing AG 2017
L. Rutkowski et al. (Eds.): ICAISC 2017, Part II, LNAI 10246, pp. 655–664, 2017.
DOI: 10.1007/978-3-319-59060-8_59

In this study, four methods are selected for comparison: hard borders, soft borders, random position and spherical universe (endless space).

The selected methods are tested and compared on the CEC'15 benchmark set [5]. The research questions for this work were following:

- Is there a significant difference in performance of selected methods?
- Is there a solo best performing method?
- Is there significant difference of these methods between PSO and ARPSO?
- Does a number of dimensions have a significant impact on performance of the selected methods?

The paper is structured as follows. The classical PSO algorithm is described in Sect. 2. The ARPSO variant is described in Sect. 3. The methods for handling the out of bounds particles are described in Sect. 4. The experiment setup is detailed in Sect. 5. Section 6 contains statistical overviews of results and performance comparisons obtained during the evaluation on benchmark set. Following is the discussion and conclusion.

2   Particle Swarm Optimization

This algorithm (PSO) using natural phenomena like a movement of swarming animals and social behavior of its members. This algorithm was introduced in 1995 by Eberhart and Kennedy [1].

In PSO, the particles are moving in space of possible solution of the particular problem. Each particle, individual solution, has defined a position, velocity and remembers his best position (solution of the problem) obtained so far. This solution is tagged as the personal best solution (*pBest*). Each particle knows the global best solution (*gBest*). This solution is selected from all *pBests*. These variables set the direction of movement for each particle and then even a new position in next iteration. PSO usually stops after a number of iterations defined by the user.

The particle is represented as coordinates in n-dimensional space of solutions. These coordinates are parameters of the optimized problem. In each iteration of the algorithm, the new position of each particle is calculated based on previous position and velocity of a particle. This new position is checked if it lies in space of possible solutions. The function of the optimized problem is called Cost Function (CF). The new position of a particle is used as input parameters in CF. If the value of CF is better than the previous *pBest* of a particle, then the particle saves this new position as the new *pBest*. If the new *pBest* is better than any other *pBest* in population then this *pBest* is saved as *gBest*.

The position of a particle is calculated according to (1)

$$x_{ij}^{t+1} = x_{ij}^t + v_{ij}^{t+1} \tag{1}$$

The velocity v of a particle is calculated by the Formula (2)

$$v_{ij}^{t+1} = w \cdot v_{ij}^t + c_1 \cdot r_1 \cdot \left(pBest_{ij} - x_{ij}^t\right) + c_2 \cdot r_2 \cdot \left(gBest_j - x_{ij}^t\right) \tag{2}$$

where v_{ij}^{t+1} is a new velocity of the i particle for j dimension in iteration $t+1$. The w stands for inertia weight [6] and linearly decreases from 0.9 to 0.4. The acceleration factors c_1 and c_2 is set to value 2. Values r_1 and r_2 are random numbers drawn from uniform distribution in interval $<0,1>$.

3 Diversity-Guided Particle Swarm Optimizer

This algorithm (ARPSO) was proposed by Riget and Vesterstrøm in 2002 [4]. It was created to avoid premature convergence, which is one of the well-known weaknesses of classical PSO. The mechanic is very similar to PSO described in previous Sect. 2. In each iteration, the divergence of a population is calculated using Formula (3). Where NP is population size, $|L|$ is longest diagonal in search space, dim is dimension size and $\overline{x_j}$ stands for average value (position) of particles in j dimension.

$$divergence = \frac{1}{NP \cdot |L|} \cdot \sum_{i=1}^{NP} \sqrt{\sum_{j=1}^{dim} (x_{ij} - \overline{x_j})^2} \tag{3}$$

If this value drops under defined threshold value $(5 \cdot 10^{-6})$, then a repulsive phase is activated. If the divergence value is higher than an upper threshold value (0.25), then an attractive phase is activated. The velocity calculation is then adjusted to (4)

$$v_{ij}^{t+1} = w \cdot v_{ij}^t + dir \cdot (c_1 \cdot r_1 \cdot (pBest_{ij} - x_{ij}^t) + c_2 \cdot r_2 \cdot (gBest_j - x_{ij}^t)) \tag{4}$$

where dir is direction and can be only 1 (attractive phase) or -1 (repulsive phase).

4 Strategies for Search Space Boundaries Violation

In each iteration of an algorithm (PSO or ARPSO) the position of every particle is updated. A particle has to be checked if its new position is in the appropriate boundaries (inside a space of possible solutions). If a particle is not in this space, then its position has to be updated. There are several possible strategies.

4.1 Hard Borders

In this method, a particle that is outside of the search space boundary is moved at the boundary itself in the corresponding dimension. A velocity of a particle remains the same. For computational complexity, the method is approximately linear $(O(n))$ and depends only on dimension size.

4.2 Soft Borders

If a particle is outside of the acceptable search space, then no restriction is applied on that particle. However, its cost function is not calculated and therefore *pBest* is not updated. A particle will return inside the search space by itself if required conditions are met [7]. This method requires no additional computation.

4.3 Random Position

This method generates a new value for the position in a dimension where a particle violated a search space boundary. New value is generated from a uniform distribution between the lower and upper boundaries of the violated dimension. The velocity of the corresponding particle is set to zero. The computational complexity is similar to Hard border method and its approximately linear $O(n)$.

4.4 Spherical Universe

Under this method, the space of available solutions looks like an endless spherical surface. The upper boundary is neighboring the lower boundary of corresponding dimensions and vice versa. If a particle violates a search space boundary, it comes out on the "opposite" side. This method has also approximately linear complexity due to the fact, that only depends on dimension size like hard border method.

5 Experimental Setup

The experiments were performed for dimension sizes (dim) 10, 30 and 50. The maximal number of cost function evaluations is set to $10\,000 \cdot dim$ according to the definition in CEC'15 benchmark set. The population size (NP) is set to 40 for all dimensions. Every test function is repeated for 51 independent runs also according to CEC'15. This benchmark includes 15 functions separated into four

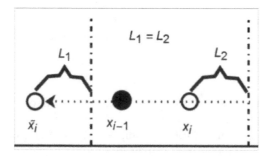

Fig. 1. Explanation of spherical method. x_{i-1} is the particle position in last iteration, $\bar{x}_i$ is uncorrected position and x_i is the final correct position.

categories: unimodal, multimodal, hybrid and composite. In the next section, the functions are denoted with a letter which indicates the category (u is unimodal, m is multimodal, h is hybrid and c is composite).

6   Results

The Friedman test [8] is used for statistical comparison of used methods performance. For each test function, defined by CEC'15, and for both algorithms the Friedman tests are independent. The critical distance CD is calculated by a Formula (5) from [9]. Where $q_{0.05} = 2.69$, k is a number of compared methods (in this case $k = 4$) and N is a number of samples ($N = 204$ that is 51 independent runs times k tested methods).

$$CD = q_{0.05} \cdot \sqrt{\frac{k \cdot (k+1)}{6 \cdot N}} \tag{5}$$

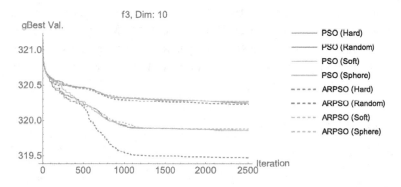

Fig. 2. Comparison of $gBests$ mean history over 51 runs.

The best method from Friedman test serves as referencing method (method with highest rank). If any other method is differing by at least the CD value computed by (5), then this method is significantly worse than referencing method. The results of this Friedman test are presented in Tables 1 and 2. The methods which are not differencing at least the value CD (cannot be decided if the method is significantly worse), including the referencing method, are given in bold numbers. The methods with bold labels in tables (for same dimension) are best performing ones. For example, in Table 1 for f_5 the Sphere method is solo best performing in $dim = 10$ due to fact that results from Friedman test of others methods differ more than CD value. Next example is also for Table 1, but for test function f_{10}. In this example, only one method differs more than CD value, so the other three methods (random, soft and sphere) performances are equivalent on average. Furthermore, some examples of mean $gBest$ value history are shown in Figs. 2, 3, 4, 5, 6, 7 and 8.

Table 1. Friedman test, PSO.

f_x	Method											
	Hard			Random			Soft			Sphere		
	10d	30d	50d	10d	30d	50d	10d	30d	50d	10d	30d	50d
f_1^u	2.33	1.39	1.16	**2.63**	**3**	**3.24**	**2.84**	**2.82**	**3.12**	2.20	**2.78**	2.49
f_2^u	2.14	1.16	1	**2.57**	**3**	**3.14**	2.45	2.69	2.77	**2.84**	**3.16**	**3.10**
f_3^m	**2.51**	**2.77**	**2.73**	**2.53**	2.41	2.35	**2.57**	**2.67**	**2.49**	**2.39**	2.16	**2.43**
f_4^m	2.47	2.20	2	**2.84**	2.45	**2.86**	2.49	2.46	**2.76**	2.20	**2.89**	2.37
f_5^m	2.29	**2.61**	2.16	2.39	**2.53**	**2.90**	2.43	**2.51**	2.33	**2.88**	2.35	**2.61**
f_6^h	1.86	1.59	1.65	2.51	**3.02**	**2.82**	2.22	2.39	**2.71**	**3.41**	**3**	**2.82**
f_7^h	1.39	1.14	1.71	**3.11**	**3.18**	2.53	2.30	2.69	**3.37**	**3.20**	**3**	2.39
f_8^h	**2.33**	1.19	1.94	**2.61**	2.57	**2.65**	**2.53**	2.27	**2.67**	2.53	**3.22**	**2.75**
f_9^c	1.39	1.43	1.35	**3.12**	**2.82**	**3.14**	**3.24**	**2.77**	**2.92**	2.26	**2.98**	2.59
f_{10}^c	2.12	2.26	2.26	**2.75**	2.37	2.08	**2.41**	**3.10**	**3.82**	**2.73**	2.28	1.84
f_{11}^c	1.98	1.45	1.35	**2.59**	**2.75**	**3.08**	**2.71**	2.73	**3**	**2.73**	**3.08**	2.57
f_{12}^c	1.90	1.06	1	**2.61**	**3.16**	**3.22**	**2.63**	**3.20**	**3.55**	**2.86**	2.59	2.24
f_{13}^c	1.80	2.27	2.73	**3.39**	**3.20**	2.43	2.78	2.77	**3.28**	2.02	1.77	1.57
f_{14}^c	2.04	1.37	1.08	2.35	2.86	3.18	**2.78**	**4**	**3.75**	**2.83**	1.76	2
f_{15}^c	1.68	1	1	**2.78**	**3**	**3.01**	**2.78**	**3**	**3.01**	**2.78**	**3**	**2.98**

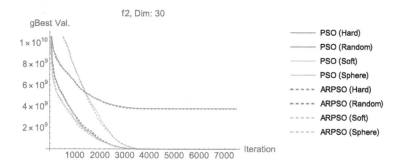

Fig. 3. Comparison of *gBests* mean history over 51 runs.

7 Results Discussion

In this section, the obtained results are discussed. By brief look at the Tables, it seems that first method, hard borders, perform significantly worse on almost every test function in all dimensions.

The ratio between a successful and unsuccessful performance of methods remains similar in all dimension for PSO algorithm. For ARPSO the ratio slightly differs across dimension size. For some cases, if a method is best performing on

Table 2. Friedman test, ARPSO.

f_x	Method											
	Hard			Random			Soft			Sphere		
	10d	30d	50d	10d	30d	50d	10d	30d	50d	10d	30d	50d
f_1^u	2.28	1.28	1.12	**2.80**	**3.16**	**3.06**	2.43	2.65	**3**	**2.49**	**2.92**	**2.82**
f_2^u	2.04	1.02	1	**2.84**	**3.10**	**3.02**	**2.55**	2.63	2.78	**2.57**	**3.26**	**3.20**
f_3^m	**2.88**	**2.51**	**2.73**	**2.55**	2.49	**2.55**	2.22	**2.53**	**2.39**	2.35	**2.47**	2.33
f_4^m	**2.46**	1.90	1.88	**2.38**	**2.61**	**2.84**	**2.66**	**2.73**	**2.71**	**2.50**	**2.76**	**2.57**
f_5^m	2.02	2.24	2.08	**2.69**	2.31	**3.02**	2.47	**2.73**	2.12	**2.82**	**2.73**	**2.78**
f_6^h	1.78	1.86	1.82	2.84	**2.90**	2.71	1.98	2.06	2.39	**3.39**	**3.18**	**3.08**
f_7^h	1.67	1.06	1.82	**2.86**	**3.28**	2.41	2.55	2.84	**3.47**	**2.92**	2.82	2.29
f_8^h	2.14	1.90	2.10	**2.55**	**2.92**	**2.63**	2.53	2.43	**2.63**	**2.78**	**2.75**	**2.65**
f_9^c	1.35	1.33	1.35	**3.16**	**3.04**	2.69	**3**	**2.78**	**3.20**	2.49	**2.84**	2.77
f_{10}^c	2.26	1.90	2.22	**2.45**	2.59	2.16	**2.77**	**3.02**	**3.80**	**2.53**	2.49	1.82
f_{11}^c	2.39	1.59	1.12	2.35	**3**	**3**	2.45	2.53	**3.39**	**2.80**	**2.88**	2.49
f_{12}^c	1.67	1	1	2.59	**3.35**	3.08	2.67	**3.37**	**3.69**	**3.01**	2.28	2.24
f_{13}^c	1.73	2.29	2.61	**3.31**	**3.29**	**2.84**	2.82	2.67	**3.16**	2.14	1.75	1.39
f_{14}^c	1.80	1.06	1.02	2.41	2.92	3.12	**2.90**	**4**	**3.78**	**2.88**	2.02	2.08
f_{15}^c	1.94	1	1	2.24	2.34	2.36	2.67	2.66	2.67	**3.10**	**4**	**3.97**

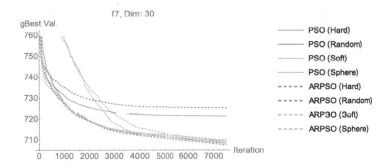

Fig. 4. Comparison of *gBests* mean history over 51 runs.

one test function (e.g. f_{14}), then this method remains, at least, one of the best methods for all dimensions and for both used algorithms.

In Fig. 6 it is shown that methods have similar performance and convergence speed regardless of used algorithm. The same results are seen on others Figures. Except for Fig. 2 where the random method has better convergence than others compared methods. The graphs also show that hard border method usually reaches stagnation first. This might explain the poor performance according to Friedman test results.

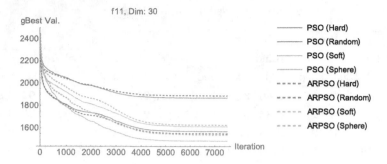

Fig. 5. Comparison of *gBests* mean history over 51 runs.

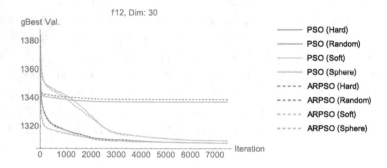

Fig. 6. Comparison of *gBests* mean history over 51 runs.

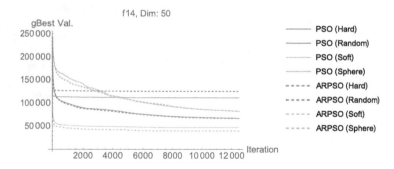

Fig. 7. Comparison of *gBests* mean history over 51 runs.

As for convergence speed, the sphere method seems to be the slowest from all compared methods. The fastest method seems to be in most cases the soft borders method followed by random position method. These results are confirmed by Friedman test as shown in the corresponding tables.

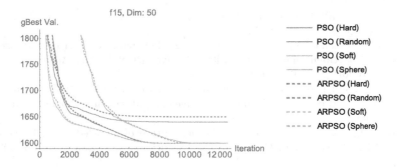

Fig. 8. Comparison of *gBests* mean history over 51 runs.

8   Conclusion

In this paper was presented results of four possible methods for handling particle position in cases where it violates border restriction. The methods were compared with two algorithms, classical PSO and its modification ARPSO. For comparison the benchmark set CEC'15 was used. The results were presented and tested for statistical significance. Based on these results the research questions can be answered as follows:

- There is a statistically significant difference in the performance of these selected methods. The best performing methods are random position and soft borders methods. Following is the sphere method with slow convergence speed, but reliable results. And the last of the methods with overall worse results is the hard border method.
- There is not a single one best performing method.
- There is no significant difference of used methods between PSO and ARPSO algorithms.
- The number of dimensions has no significant impact on these chosen methods.

The goal of this study was to show differences in performance of four possible methods for a particle position handling on two algorithms, PSO and ARPSO. The results of this study will be further used in future studies to suggest possible improvements for controlling the position of particles that violates search space boundaries.

References

1. Kennedy, J., Eberhart, R.: Particle swarm optimization. In: Proceedings of the IEEE International Conference on Neural Networks, pp. 1942–1948 (1995)
2. Nepomuceno, F.V., Engelbrecht, A.P.: A self-adaptive heterogeneous PSO for real-parameter optimization. In: 2013 IEEE Congress on Evolutionary Computation (CEC), pp. 361–368. IEEE (2013)

3. Zhan, Z.-H., et al.: Orthogonal learning particle swarm optimization. IEEE Trans. Evol. Comput. **15**(6), 832–847 (2011)
4. Riget, J., Vesterstrom, J.S.: A diversity-guided particle swarm optimizer-the ARPSO. Department of Computer Science, University of Aarhus, Aarhus, Denmark, Technical report 2002–02 (2002)
5. Chen, Q., et al.: Problem definitions and evaluation criteria for CEC 2015 special session on bound constrained single-objective computationally expensive numerical optimization
6. Kennedy, J.: The particle swarm: social adaptation of knowledge. In: Proceedings of the IEEE International Conference on Evolutionary Computation, pp. 303–308 (1997)
7. Eberhart, R.C., Shi, Y.: Comparing inertia weights and constriction factors in particle swarm optimization. In: Proceedings of the 2000 Congress on Evolutionary Computation, CEC 2000, pp. 84–88. IEEE (2000)
8. Friedman, M.: The use of ranks to avoid the assumption of normality implicit in the analysis of variance. J. Am. Stat. Assoc. **32**, 675–701 (1937)
9. Demsar, J.: Statistical comparisons of classifiers over multiple data sets. J. Mach. Learn. Res. **7**, 1–30 (2006)

PSO with Attractive Search Space Border Points

Michal Pluhacek[(✉)], Roman Senkerik, Adam Viktorin, and Tomas Kadavy

Faculty of Applied Informatics, Tomas Bata University in Zlin,
Nam T.G. Masaryka 5555, 760 01 Zlin, Czech Republic
{pluhacek,senkerik,aviktorin,kadavy}@fai.utb.cz

Abstract. One of the biggest drawbacks of the original Particle Swarm Optimization is the premature convergence and fast loss of diversity in the population. In this paper, we propose and discuss a simple yet effective modification to help the PSO maintain diversity and avoid premature convergence. The particles are randomly attracted towards the border points of the search space. We use the CEC13 Benchmark function set to test the performance of proposed method and compare it to original PSO.

Keywords: Particle swarm optimization · PSO · Diversity

1 Introduction

The Particle Swarm Optimization (PSO) [1–4] belongs among the most prominent representatives of evolutionary computational techniques (ECTs) that are used with great results in many areas of global optimization.

One of the biggest drawbacks of the original Particle Swarm Optimization is the fast loss of diversity in the population leading to premature convergence into local sub-optima. This issue has been recognized very soon [2] and since then there have been many modifications proposed to tackle it with mixed results. One of the excellent examples is the Diversity guided PSO (ARPSO) proposed in 2002 by Riget and Vestterstrom [5]. This work is partially inspired by the principles proposed in the ARPSO algorithm with respect to its limitations. Usually, the modifications that work with population diversity use the Euclidian distance to compute the diversity of the swarm. This operation is very computationally expensive, especially in higher dimensions.

The aim of this study is to propose a computational inexpensive modification that would help avoid premature convergence of the PSO, show promising

M. Pluhacek—This work was supported by Grant Agency of the Czech Repub-lic – GACR P103/15/06700S, further by the Ministry of Education, Youth and Sports of the Czech Republic within the National Sustainability Programme Project no. LO1303 (MSMT-7778/2014). Also by the European Regional Development Fund under the Project CEBIA-Tech no. CZ.1.05/2.1.00/03.0089 and by Internal Grant Agency of Tomas Bata University under the Projects no. IGA/CebiaTech/2017/004.

© Springer International Publishing AG 2017
L. Rutkowski et al. (Eds.): ICAISC 2017, Part II, LNAI 10246, pp. 665–675, 2017.
DOI: 10.1007/978-3-319-59060-8_60

performance and would be simple enough for easy utilization in advanced PSO variants.

The rest of the paper is structured as follows: The original (canonical) PSO is described in the next section; the following is the description of proposed modification. The experiment is designed in the next section followed by results and discussion. The paper concludes with proposals for future research.

2   Particle Swarm Optimization Algorithm

Original PSO [1] takes the inspiration from the behavior of fish and birds. The knowledge of global best solution, (typically noted $gBest$) is shared among the individuals (particles) in the swarm. Furthermore, each particle has the knowledge of its own (personal) best solution (noted $pBest$). Last important part of the algorithm is the velocity of each particle that is taken into account during the calculation of the particle movement. The new position of each particle is then given by (1), where x_i^{t+1} is the new particle position; x_i refers to current particle position and v_i^{t+1} is the new velocity of the particle.

$$x_i^{t+1} = x_i^t + v_i^{t+1} \tag{1}$$

To calculate the new velocity, the distances from $pBest$ and $gBest$ are taken into account, alongside with current velocity that is multiplied by inertia weight value (2)

$$v_{ij}^{t+1} = w \cdot v_{ij}^t + c_1 \cdot Rand \cdot (pBest_{ij} - x_{ij}^t) + c_2 \cdot Rand \cdot (gBest_j - x_{ij}^t) \tag{2}$$

Where:

v_{ij}^{t+1} - New velocity of the ith particle in iteration $t+1$. (component j of the dimension D).
w - Inertia weight.
v_{ij}^t - Current velocity of the ith particle in iteration t. (component j of the dimension D).
c_1, c_2 = 2- Acceleration constants.
$pBest_{ij}$ - Local (personal) best solution found by the ith particle. (component j of the dimension D).
$gBest_j$ - Best solution found in a population. (component j of the dimension D).
x_{ij}^t - Current position of the ith particle (component j of the dimension D) in iteration t.
$Rand$ - Pseudo random number, interval (0, 1).

3   Proposed Modification

The issue of premature convergence is caused mainly by the existence of single dominant attraction point ($gBest$) and the notorious "one step forward, two steps back" problem where in some dimension components the solution is getting closer to the global optima but moves away in another dimension. Often the diversity of population drops suddenly to zero or a near-zero value in some (or all) dimension components and original design given by (1) and (2) leads to stagnation.

We choose to address this issue by partial substitution of gBest for border points (lower or upper bounds of the search space). The goal is to make the particles move rapidly from the converged swarm and enrich the overall diversity of parameter values in the population.

In the proposed modification, each particle has a 10% chance (value based on tuning experiment) that it will use the lower or upper bound of search space instead of the *gBest* position during the velocity calculation. This helps to maintain the population diverse and improves the ability to avoid premature convergence.

The algorithm can be summarized in the following steps. During the new velocity calculation a random number r from uniform distribution $<0, 1>$ is generated:

If $r \leq 0.9$ the standard PSO velocity formula (1) is used.
If $r > 0.9$, another random number r_2 is generated and:
If $r_2 \leq 0.5$ (3)

$$v_{ij}^{t+1} = w \cdot v_{ij}^t + c_1 \cdot Rand \cdot (pBest_{ij} - x_{ij}^t) + c_2 \cdot Rand \cdot (low_j - x_{ij}^t) \qquad (3)$$

If $r_2 > 0.5$ (4)

$$v_{ij}^{t+1} = w \cdot v_{ij}^t + c_1 \cdot Rand \cdot (pBest_{ij} - x_{ij}^t) + c_2 \cdot Rand \cdot (high_j - x_{ij}^t) \qquad (4)$$

Where

low_j – is the search space lower bound of the dimension j,
$high_j$ – is the search space upper bound of the dimension j.

4   Experiment

In this study, the performance of the proposed modification was tested on the IEEE CEC 2013 benchmark set [6] for dimension setting (dim) = 10 and 30. According to the benchmark rules, 51 separate runs were performed for each algorithm and the maximum number of cost function evaluations (CFE) was set to 10000·dim. The population size was set to 20. According to the literature [7,8], the values of control parameters were set as follows:

c_1, c_2 = 1.49618;
w = 0.7298;

In the following results overview the performance of original PSO (noted further PSO) is compared to the proposed method with attractive border points (noted PSO_B).

The following Table 1 contains the statistical evaluation of the performance of the proposed PSO_B on the CEC13 benchmark in dim = 10. The results for canonical PSO are presented in Table 2. Tables 3 and 4 contain the results for PSO_B and PSO in dim = 30. All results in tables are given in the form of error values from the function optima.

The Wilcoxon Rank sum test (also known as Mann–Whitney U test) with the level of significance $\alpha = 0.05$ was used to determine the statistical significance of the differences in performance. If the p-value is lower than α, the results are

Table 1. Error values – PSO_B, $dim = 10$, max. CFE $= 100000$

$f(x)$	f_{min}	Mean	Std. Dev.	Median	Max	Min
f_1^u	−1400	2.30E−02	1.85E−02	1.79E−02	1.01E−01	3.72E−03
f_2^u	−1300	5.62E+05	4.20E+05	4.51E+05	2.27E+06	5.18E+04
f_3^u	−1200	1.43E+08	2.63E+08	4.36E+07	1.62E+09	7.93E+04
f_4^u	−1100	1.02E+03	5.81E+02	8.09E+02	3.22E+03	2.44E+02
f_5^u	−1000	1.92E+00	1.24E+01	1.37E−01	8.84E+01	4.14E−02
f_6^m	−900	1.96E+01	2.56E+01	1.01E+01	1.00E+02	1.95E−01
f_7^m	−800	2.06E+01	1.83E+01	1.67E+01	7.87E+01	1.18E+00
f_8^m	−700	2.04E+01	7.03E−02	2.04E+01	2.05E+01	2.02E+01
f_9^m	−600	4.39E+00	1.33E+00	4.38E+00	7.98E+00	1.31E+00
f_{10}^m	−500	8.74E+00	1.44E+01	1.60E+00	6.45E+01	8.04E−01
f_{11}^m	−400	1.24E+01	5.88E+00	1.13E+01	2.80E+01	3.31E+00
f_{12}^m	−300	1.55E+01	6.44E+00	1.36E+01	3.04E+01	4.38E+00
f_{13}^m	−200	2.33E+01	9.87E+00	2.40E+01	4.22E+01	3.38E+00
f_{14}^m	−100	4.48E+02	1.78E+02	4.02E+02	1.03E+03	9.33E+01
f_{15}^m	100	8.26E+02	2.56E+02	8.20E+02	1.38E+03	2.47E+02
f_{16}^m	200	1.12E+00	2.10E−01	1.14E+00	1.52E+00	6.47E−01
f_{17}^m	300	3.75E+01	5.46E+00	3.82E+01	5.15E+01	2.76E+01
f_{18}^m	400	4.28E+01	5.36E+00	4.40E+01	5.44E+01	3.04E+01
f_{19}^m	500	2.50E+00	6.81E−01	2.57E+00	3.70E+00	9.51E−01
f_{20}^m	600	3.06E+00	6.87E−01	3.23E+00	4.03E+00	1.60E+00
f_{21}^c	700	3.91E+02	4.04E+01	4.00E+02	4.00E+02	2.04E+02
f_{22}^c	800	5.51E+02	2.38E+02	4.92E+02	9.91E+02	1.48E+02
f_{23}^c	900	1.02E+03	3.45E+02	1.01E+03	1.93E+03	3.29E+02
f_{24}^c	1000	2.19E+02	5.52E+00	2.19E+02	2.32E+02	2.08E+02
f_{25}^c	1100	2.15E+02	6.20E+00	2.16E+02	2.24E+02	2.03E+02
f_{26}^c	1200	2.38E+02	8.65E+01	3.11E+02	3.28E+02	1.08E+02
f_{27}^c	1300	4.71E+02	8.98E+01	4.83E+02	6.25E+02	3.19E+02
f_{28}^c	1400	3.84E+02	1.42E+02	3.11E+02	7.05E+02	1.02E+02

Table 2. Error values – PSO, $dim = 10$, max. CFE $= 100000$

$f(x)$	f_{min}	Mean	Std. Dev.	Median	Max	Min
f_1^u	−1400	8.02E−14	1.19E−13	0.00E+00	4.55E−13	0.00E+00
f_2^u	−1300	1.66E+06	2.35E+06	4.59E+05	1.10E+07	6.05E+03
f_3^u	−1200	6.05E+08	1.71E+09	4.44E+07	9.70E+09	3.27E+00
f_4^u	−1100	1.26E+04	8.33E+03	1.09E+04	4.68E+04	1.23E+03
f_5^u	−1000	8.94E−13	2.01E−12	3.41E−13	1.23E−11	1.14E−13
f_6^m	−900	4.21E+01	3.99E+01	3.24E+01	1.94E+02	9.00E−04
f_7^m	−800	6.36E+01	2.91E+01	6.46E+01	1.23E+02	1.63E+01
f_8^m	−700	2.03E+01	7.08E−02	2.03E+01	2.05E+01	2.02E+01
f_9^m	−600	6.33E+00	1.52E+00	6.31E+00	9.35E+00	3.00E+00
f_{10}^m	−500	2.30E+01	7.15E+01	1.60E+00	4.80E+02	5.42E−02
f_{11}^m	−400	2.11E+01	1.05E+01	1.99E+01	4.97E+01	3.98E+00
f_{12}^m	−300	3.46E+01	1.87E+01	3.38E+01	8.80E+01	4.97E+00
f_{13}^m	−200	5.42E+01	2.26E+01	5.16E+01	1.27E+02	1.84E+01
f_{14}^m	−100	6.73E+02	2.32E+02	6.81E+02	1.08E+03	1.94E+02
f_{15}^m	100	9.74E+02	2.88E+02	9.51E+02	1.73E+03	3.29E+02
f_{16}^m	200	5.37E−01	2.41E−01	4.89E−01	1.28E+00	1.61E−01
f_{17}^m	300	3.02E+01	8.22E+00	2.92E+01	5.70E+01	1.43E+01
f_{18}^m	400	3.46E+01	1.12E+01	3.33E+01	6.34E+01	1.70E+01
f_{19}^m	500	1.52E+00	7.88E−01	1.39E+00	3.35E+00	3.85E−01
f_{20}^m	600	3.58E+00	3.85E−01	3.53E+00	4.51E+00	2.62E+00
f_{21}^c	700	3.90E+02	4.13E+01	4.00E+02	4.00E+02	2.00E+02
f_{22}^c	800	9.59E+02	3.17E+02	9.17E+02	1.70E+03	2.23E+02
f_{23}^c	900	1.39E+03	4.42E+02	1.45E+03	2.48E+03	5.15E+02
f_{24}^c	1000	2.23E+02	1.50E+01	2.25E+02	2.38E+02	1.24E+02
f_{25}^c	1100	2.22E+02	6.91E+00	2.24E+02	2.34E+02	2.08E+02
f_{26}^c	1200	1.95E+02	7.48E+01	1.80E+02	3.29E+02	1.07E+02
f_{27}^c	1300	5.46E+02	9.24E+01	5.72E+02	7.36E+02	4.00E+02
f_{28}^c	1400	4.55E+02	2.15E+02	3.00E+02	8.58E+02	1.00E+02

different with statistical significance. The corresponding p-values are presented in Table 5 alongside with the scoring (1 – PSO_B win, 0 – draw, 2 – PSO win) and total ratio of wins and losses (PSO_B: PSO).

In Tables 1, 2, 3, 4, 5 the benchmark functions are divided by notation into unimodal (noted with ^{u}), basic multimodal (noted with ^{m}) and composite functions (noted with ^{c}).

Table 3. Error values – PSO_B, $dim = 30$, max. CFE = 300000

$f(x)$	f_{min}	Mean	Std. Dev.	Median	Max	Min
f_1^u	−1400	6.94E + 02	8.07E + 02	5.11E + 02	2.69E + 03	2.04E + 00
f_2^u	−1300	2.26E + 07	1.41E + 07	2.14E + 07	9.08E + 07	7.92E + 06
f_3^u	−1200	8.45E + 09	8.41E + 09	6.79E + 09	5.17E + 10	2.73E + 08
f_4^u	−1100	3.91E + 03	1.43E + 03	3.61E + 03	1.07E + 04	2.39E + 03
f_5^u	−1000	1.78E + 02	1.10E + 02	1.69E + 02	5.95E + 02	2.65E + 00
f_6^m	−900	1.39E + 02	7.17E + 01	1.26E + 02	3.25E + 02	2.53E + 01
f_7^m	−800	1.18E + 02	4.77E + 01	1.09E + 02	2.58E + 02	4.26E + 01
f_8^m	−700	2.09E + 01	5.50E − 02	2.09E + 01	2.11E + 01	2.07E + 01
f_9^m	−600	2.37E + 01	3.71E + 00	2.32E + 01	3.23E + 01	1.60E + 01
f_{10}^m	−500	2.52E + 02	1.86E + 02	2.36E + 02	1.04E + 03	1.90E + 01
f_{11}^m	−400	1.13E + 02	2.89E + 01	1.10E + 02	1.95E + 02	6.42E + 01
f_{12}^m	−300	1.78E + 02	3.19E + 01	1.77E + 02	2.67E + 02	1.03E + 02
f_{13}^m	−200	1.95E + 02	2.52E + 01	1.92E + 02	2.83E + 02	1.48E + 02
f_{14}^m	−100	2.57E + 03	6.18E + 02	2.57E + 03	4.16E + 03	9.37E + 02
f_{15}^m	100	5.81E + 03	5.57E + 02	5.73E + 03	7.30E + 03	4.80E + 03
f_{16}^m	200	2.38E + 00	3.88E − 01	2.46E + 00	3.05E + 00	1.42E + 00
f_{17}^m	300	2.31E + 02	2.06E + 01	2.33E + 02	2.80E + 02	1.76E + 02
f_{18}^m	400	2.50E + 02	2.58E + 01	2.47E + 02	3.56E + 02	1.99E + 02
f_{19}^m	500	3.18E + 01	2.19E + 01	2.26E + 01	1.06E + 02	1.20E + 01
f_{20}^m	600	1.27E + 01	8.97E − 01	1.26E + 01	1.50E + 01	1.14E + 01
f_{21}^c	700	3.92E + 02	1.36E + 02	3.42E + 02	9.03E + 02	1.55E + 02
f_{22}^c	800	3.28E + 03	7.12E + 02	3.20E + 03	4.71E + 03	1.84E + 03
f_{23}^c	900	6.09E + 03	7.63E + 02	5.98E + 03	7.96E + 03	4.40E + 03
f_{24}^c	1000	2.89E + 02	1.14E + 01	2.88E + 02	3.14E + 02	2.66E + 02
f_{25}^c	1100	3.01E + 02	1.10E + 01	2.99E + 02	3.34E + 02	2.83E + 02
f_{26}^c	1200	3.30E + 02	6.52E + 01	3.59E + 02	3.87E + 02	2.01E + 02
f_{27}^c	1300	9.85E + 02	1.05E + 02	9.94E + 02	1.19E + 03	7.53E + 02
f_{28}^c	1400	1.10E + 03	4.92E + 02	1.21E + 03	1.80E + 03	1.63E + 02

For a better understanding of the performance differences between the two compared methods, several examples of mean *gBest* history are presented in Figs. 1, 2, 3 and 4. The results discussion follows.

According to Tables 1, 2, 3, 4, 5 the proposed method PSO_B managed to outperform the original PSO on 15 functions from the benchmark set in dim = 10 and on 21 functions in dim = 30. This proves the superiority and good scalability of the proposed method.

It seems that the performance of the proposed method PSO_B is mostly favorable on complex problems (multimodal and composite functions) with some

Table 4. Error values – PSO, $dim = 30$, max. CFE = 300000

$f(x)$	f_{min}	Mean	Std. Dev.	Median	Max	Min
f_1^u	−1400	8.62E+01	3.77E+02	1.99E−09	2.43E+03	2.96E−12
f_2^u	−1300	1.63E+08	1.10E+08	1.51E+08	4.65E+08	5.64E+05
f_3^u	−1200	3.15E+13	1.17E+14	8.61E+10	6.30E+14	7.11E+09
f_4^u	−1100	6.19E+04	1.64E+04	6.27E+04	9.85E+04	2.40E+04
f_5^u	−1000	4.88E−01	3.48E+00	1.88E−10	2.49E+01	5.23E−12
f_6^m	−900	1.25E+03	8.17E+02	1.01E+03	4.36E+03	8.76E+01
f_7^m	−800	5.50E+03	2.56E+04	3.78E+02	1.81E+05	1.02E+02
f_8^m	−700	2.09E+01	4.86E−02	2.09E+01	2.10E+01	2.08E+01
f_9^m	−600	3.53E+01	2.88E+00	3.53E+01	4.04E+01	2.89E+01
f_{10}^m	−500	2.01E+03	9.71E+02	2.18E+03	4.02E+03	8.61E−02
f_{11}^m	−400	2.43E+02	5.86E+01	2.32E+02	4.02E+02	1.17E+02
f_{12}^m	−300	2.74E+02	8.05E+01	2.82E+02	4.79E+02	1.35E+02
f_{13}^m	−200	4.10E+02	8.67E+01	4.10E+02	5.68E+02	2.00E+02
f_{14}^m	−100	3.38E+03	5.91E+02	3.43E+03	4.76E+03	2.01E+03
f_{15}^m	100	4.78E+03	6.55E+02	4.79E+03	6.37E+03	3.68E+03
f_{16}^m	200	1.39E+00	3.65E−01	1.42E+00	2.15E+00	7.44E−01
f_{17}^m	300	3.06E+02	7.37E+01	3.01E+02	4.68E+02	1.46E+02
f_{18}^m	400	3.30E+02	7.82E+01	3.08E+02	4.81E+02	1.85E+02
f_{19}^m	500	2.88E+03	5.38E+03	4.87E+02	2.64E+04	3.34E+01
f_{20}^m	600	1.43E+01	5.51E−01	1.45E+01	1.50E+01	1.13E+01
f_{21}^c	700	3.12E+02	1.24E+02	3.00E+02	7.81E+02	1.00E+02
f_{22}^c	800	4.59E+03	1.07E+03	4.60E+03	6.78E+03	2.92E+03
f_{23}^c	900	5.79E+03	9.71E+02	5.64E+03	7.89E+03	3.85E+03
f_{24}^c	1000	3.24E+02	1.59E+01	3.23E+02	3.64E+02	2.95E+02
f_{25}^c	1100	3.56E+02	1.73E+01	3.53E+02	4.06E+02	3.13E+02
f_{26}^c	1200	3.39E+02	8.04E+01	3.86E+02	4.09E+02	2.00E+02
f_{27}^c	1300	1.29E+03	9.39E+01	1.29E+03	1.47E+03	1.03E+03
f_{28}^c	1400	3.02E+03	6.94E+02	3.06E+03	4.93E+03	1.59E+03

exceptions and not exclusively favorable on simpler (e.g. unimodal) problems. We analyzed this issue also from the provided mean history of *gBest* (Figs. 1, 2). The Figs. 3, 4 highlight the change in performance of compared methods with increasing dimensionality of the problem. For dim = 10 the canonical PSO outperformed the proposed PSO_B, however for dim = 30 the proposed PSO_B managed to outperform the canonical PSO despite its initial slower convergence.

Based on the evidence presented in Figs. 1, 2, 3, 4 the slight drawback of the proposed method is slower convergence speed. This issue is highlighted in Fig. 1.

Table 5. p-values of Wilcoxon rank sum test; $\alpha = 0.05$;

$f(x)$	$dim = 10$	1 0 2	$dim = 30$	1 0 2
f_1^u	0.00	2	0.00	2
f_2^u	0.99	0	0.00	1
f_3^u	0.40	0	0.00	1
f_4^u	0.00	1	0.00	1
f_5^u	0.00	2	0.00	2
f_6^m	0.00	1	0.00	1
f_7^m	0.00	1	0.00	1
f_8^m	0.02	2	0.32	0
f_9^m	0.00	1	0.00	1
f_{10}^m	0.30	0	0.00	1
f_{11}^m	0.00	1	0.00	1
f_{12}^m	0.00	1	0.00	1
f_{13}^m	0.00	1	0.00	1
f_{14}^m	0.00	1	0.00	1
f_{15}^m	0.02	1	0.00	2
f_{16}^m	0.00	2	0.00	2
f_{17}^m	0.00	2	0.00	1
f_{18}^m	0.00	2	0.00	1
f_{19}^m	0.00	2	0.00	1
f_{20}^m	0.00	1	0.00	1
f_{21}^c	0.00	2	0.00	2
f_{22}^c	0.00	1	0.00	1
f_{23}^c	0.00	1	0.10	0
f_{24}^c	0.00	1	0.00	1
f_{25}^c	0.00	1	0.00	1
f_{26}^c	0.11	0	0.00	1
f_{27}^c	0.00	1	0.00	1
f_{28}^c	0.07	0	0.00	1
Score:		15:8		21:5

However, there is no hint of premature convergence and the algorithm continuous to improve during the whole CFE limit in strong contrast with the canonical PSO algorithm that usually falls into stagnation in the first half of provided CFE limit and is unable to escape the stagnation when it occurs. It seems that the proposed very simple modification could improve the performance of PSO on many types of fitness landscapes. In higher dimensions, the performance improves significantly in comparison with the canonical PSO.

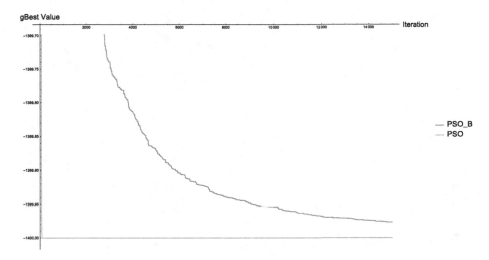

Fig. 1. Mean best value history comparison $- f_1$; $dim = 10$

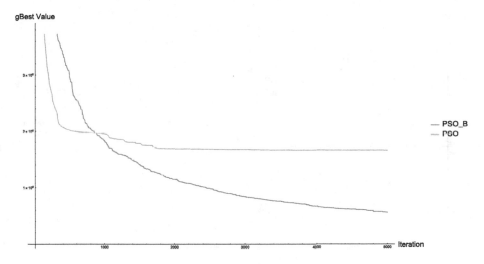

Fig. 2. Mean best value history comparison $- f_2$; $dim = 10$

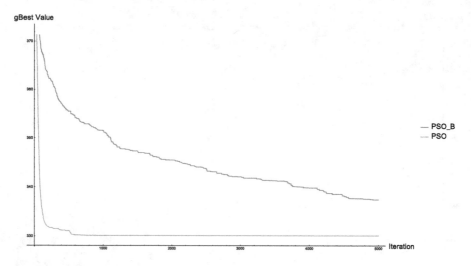

Fig. 3. Mean best value history comparison – f_{17}; $dim = 10$

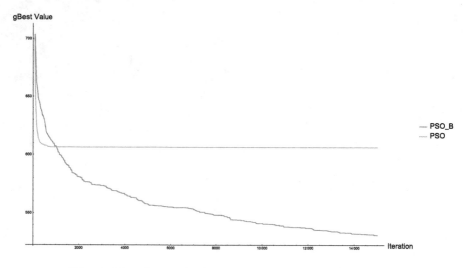

Fig. 4. Mean best value history comparison – f_{17}; $dim = 30$

5   Conclusion

In this study, we choose to address one of the biggest issues of PSO algorithm. It has been shown that the premature convergence, which is usually caused by the loss of diversity in the swarm, can be mostly prevented by a very simple modification presented in this paper. The proposed modification outperformed the original algorithm on 16 (dim = 10) and 21 functions (dim = 30) out of 28 in the benchmark set. As encouraging as the initial results are a drawback of slower convergence speed has been identified instantly and caused poor performance on smooth and simple fitness landscapes. Addressing this problem will be the main direction of future research. In addition, the future research will explore the possibilities of implementation of this approach into more advanced PSO variants.

References

1. Kennedy, J., Eberhart, R.: Particle swarm optimization. In: Proceedings of the IEEE International Conference on Neural Networks, pp. 1942–1948 (1995)
2. Shi, Y., Eberhart, R.: A modified particle swarm optimizer. In: Proceedings of the IEEE International Conference on Evolutionary Computation (IEEE World Congress on Computational Intelligence), pp. 69–73. I. S (1998)
3. Kennedy, J.: The particle swarm: social adaptation of knowledge. In: Proceedings of the IEEE International Conference on Evolutionary Computation, pp. 303–308 (1997)
4. Nickabadi, A., Ebadzadeh, M.M., Safabakhsh, R.: A novel particle swarm optimization algorithm with adaptive inertia weight. Appl. Soft Comput. **11**(4), 3658–3670 (2011). ISSN 1568-4946
5. Riget, J., Vesterstrøm, J.S.: A diversity-guided particle swarm optimizer-the ARPSO. Dept. Comput. Sci., Univ. of Aarhus, Aarhus, Denmark, Technical report 2 (2002)
6. Liang, J.J., Qu, B.-Y., Suganthan, P.N., Hernández-Díaz, A.G.: Problem Definitions and Evaluation Criteria for the CEC 2013 Special Session and Competition on Real-Parameter Optimization. Technical report 201212, Computational Intelligence Laboratory, Zhengzhou University, Zhengzhou China and Technical report, Nanyang Technological University, Singapore (2013)
7. Van Den Bergh, F., Engelbrecht, A.P.: A study of particle swarm optimization particle trajectories. Inf. Sci. **176**(8), 937–971 (2006)
8. Eberhart, R.C., Shi, Y.: Comparing inertia weights and constriction factors in particle swarm optimization. In: Proceedings of the IEEE Congress on Evolutionary Computation, San Diego, USA, pp. 84–88 (2000)

Differential Evolution Driven Analytic Programming for Prediction

Roman Senkerik[1]([⊠]), Adam Viktorin[1], Michal Pluhacek[1], Tomas Kadavy[1], and Ivan Zelinka[2]

[1] Faculty of Applied Informatics, Tomas Bata University in Zlin,
Nam T.G. Masaryka 5555, 760 01 Zlin, Czech Republic
{senkerik,aviktorin,pluhacek,kadavy}@fai.utb.cz
[2] Faculty of Electrical Engineering and Computer Science,
Technical University of Ostrava,
17. listopadu 15, 708 33 Ostrava-Poruba, Czech Republic
ivan.zelinka@vsb.cz

Abstract. This research deals with the hybridization of symbolic regression open framework, which is Analytical Programming (AP) and Differential Evolution (DE) algorithm in the task of time series prediction. This paper provides a closer insight into applicability and performance of connection between AP and different strategies of DE. AP can be considered as powerful open framework for symbolic regression thanks to its applicability in any programming language with arbitrary driving evolutionary/swarm based algorithm. Thus, the motivation behind this research, is to explore and investigate the differences in performance of AP driven by basic canonical strategies of DE as well as by the state of the art strategy, which is Success-History based Adaptive Differential Evolution (SHADE). Simple experiment has been carried out here with the time series consisting of 300 data-points of GBP/USD exchange rate, where the first 2/3 of data were used for regression process and the last 1/3 of the data were used as a verification for prediction process. The differences between regression/prediction models synthesized by means of AP as a direct consequences of different DE strategies performances are briefly discussed within conclusion section of this paper.

Keywords: Analytic programming · Differential evolution · SHADE · Time series prediction

R. Senkerik—This work was supported by Grant Agency of the Czech Republic - GACR P103/15/06700S, further by This work was supported by the Ministry of Education, Youth and Sports of the Czech Republic within the National Sustainability Programme project no. LO1303 (MSMT-7778/2014) and also by the European Regional Development Fund under the project CEBIA-Tech no. CZ.1.05/2.1.00/03.0089., partially supported by Grant SGS 2017/134 of VSB-Technical University of Ostrava; and by Internal Grant Agency of Tomas Bata University under the project no. IGA/CebiaTech/2017/004.

© Springer International Publishing AG 2017
L. Rutkowski et al. (Eds.): ICAISC 2017, Part II, LNAI 10246, pp. 676–687, 2017.
DOI: 10.1007/978-3-319-59060-8_61

1   Introduction

This paper provides an insight into hybridization of symbolic regression open framework, which is Analytical Programming (AP) [1], and Differential Evolution (DE) [2] algorithm in the task of time series prediction.

The most current intelligent methods are mostly based on soft computing, representing a set of methods of special algorithms, and belonging to the artificial intelligence paradigm. The most popular of these methods are fuzzy logic, neural networks, evolutionary algorithms (EA's) and symbolic regression approaches like genetic programming (GP). Currently, EA's together with symbolic regression techniques are known as a powerful set of tools for almost any difficult and complex optimization problems. One of such a challenging problem is naturally the regression/prediction of data/time series. In recent years, it attracts the researches' attention, and it has been solved by GP or hybrid mutual connection of EA's, GP, fuzzy systems, neural networks and more complex models [3–5].

The organization of this paper is following: Firstly, the related works and motivation for this research is proposed. The next sections are focused on the description of the concept of AP and used DE strategies. Experiment design, results and conclusion with discussion follow afterwards.

2   Related Works and Motivation

Analytical Programming (AP) is a novel approach to symbolic structure synthesis which uses EA for its computation. Since it can utilize arbitrary evolutionary/swarm based algorithm and it can be easily applied in any programming language, it can be considered as powerful open framework for symbolic regression. AP was introduced by I. Zelinka in 2001 and since its introduction; it has been proven on numerous problems to be as suitable for symbolic structure synthesis as Genetic Programming (GP) [6–10]. AP is based on the set of functions and terminals called General Functional Set. The individual of an EA is trans lated from individual domain to program domain using this set (more in the next section).

Currently, DE [11–14] is a well-known evolutionary computation technique for continuous optimization purposes solving many difficult and complex optimization problems. A number of DE variants have been recently developed with the emphasis on adaptivity/selfadaptivity. DE has been modified and extended several times by means of new proposals of versions; and the performances of different DE variant instance algorithms have been widely studied and compared with other evolutionary algorithms. Over recent decades, DE has won most of the evolutionary algorithm competitions in major scientific conferences [15–22], as well as being applied to several applications.

This research is an extension and continuation of the previous successful experiment with connection of state of the art Success-History based Adaptive Differential Evolution (SHADE) [22] algorithm and AP on regression of simple functions.

Since the open AP framework has been used recently only with basic and canonical versions of SOMA algorithm, Simulated Annealing, and many other algorithms as well as mostly with one basic strategy of DE, our motivation was to provide a closer insight into applicability and performance of the connection between AP and both canonical and state of the art powerful strategies of DE. The motivation can be summarized in following points:

– To show the results of DE driven AP for time series regression/prediction problem.
– To investigate the differences in performances of AP driven by basic canonical strategies of DE and state of the art SHADE.

3   Analytic Programming

The basic functionality of AP is formed by three parts – General Functional Set (GFS), Discrete Set Handling (DSH) and Security Procedures (SPs). GFS contains all elementary objects which can be used to form a program, DSH carries out the mapping of individuals to programs and SPs are implemented into mapping process to avoid mapping to pathological programs and into cost function to avoid critical situations.

3.1   General Function Set

AP uses sets of functions and terminals. The synthesized program is branched by functions requiring two and more arguments and the length of it is extended by functions which require one argument. Terminals do not contribute to the complexity of the synthesized program (length) but are needed in order to synthesize a non-pathological program (program that can be evaluated by cost function). Therefore, each non-pathological program must contain at least one terminal.

Combined set of functions and terminals forms GFS which is used for mapping from individual domain to program domain. The content of GFS is dependent on user choice. GFS is nested and can be divided into subsets according to the number of arguments that the subset requires. GFS_{0arg} is a subset which requires zero arguments, thus contains only terminals. GFS_{1arg} contains all terminals and functions requiring one argument, GFS_{2arg} contains all objects from GFS_{1arg} and functions requiring two arguments and so on, GFS_{all} is a complete set of all elementary objects. For the purpose of mapping from individual to the program, it is important to note that objects in GFS are ordered by a number of arguments they require in descending order.

3.2   Discrete Set Handling

DSH is used for mapping the individual to the synthesized program. Most of the EAs use individuals with real number encoded individuals. The first important step in order for DSH to work is to get individual with integer components which are done by rounding real number values. The integer values of an individual are indexes into the discrete set, in this case, GFS and its subsets. If the index value is greater than the size of used GFS, modulo operation with the size of the discrete set is performed. An illustrative example of mapping is given in (1).

$$Individual = \left\{ \begin{array}{l} 0.12, \ 4.29, \ 6.92, \ 6.12, \ 2.45, \\ 6.33, \ 5.78, \ 0.22, \ 1.94, \ 7.32 \end{array} \right\}$$
$$Rounded \ individual = \{0, \ 4, \ 7, \ 6, \ 2, \ 6, \ 6, \ 0, \ 2, \ 7\} \qquad (1)$$
$$GFS_{all} = \{+, \ -, \ *, /, \ sin, \ cos, \ x, \ k\}$$
$$\textbf{Program: } \sin x + k$$

The objects in GFS$_{all}$ are indexed from 0 and the mapping is as follows: The first rounded individual feature is 0 which represents + function in GFS$_{all}$. This function requires two arguments and those are represented by next two indexes: 4 and 7, which are mapped to function sin and constant k. The sin function requires one argument which is given by next index (rounded feature) – 6 and it is mapped to variable x. Since there is no possible way of branching the program further, other features are ignored and synthesized program is: $\sin x + k$.

3.3   Security Procedures

SPs are used in AP to avoid critical situations. Some of the SPs are implemented into the AP itself and some have to be implemented into the cost function evaluation. The typical representatives of the later are checking synthesized programs for loops, infinity and imaginary numbers if not expected (dividing by 0, square root of negative numbers, etc.).

The most significant SP implemented in AP is checking for pathological programs. Pathological programs are programs which cannot be evaluated due to the absence of arguments in the synthesized function. For example, individual with rounded features of $\{5, 5, 5, 5, 5\}$ would be mapped to program $cos(cos(cos(cos(cos\_))))$ which lacks constant or variable at the empty position denoted by _ and thus represent a pathological program. Such situation can be avoided by a simple procedure which checks remaining positions (parameters) of the individual during mapping and according to that maps rounded individual features to subsets of GFS$_{all}$ which do not require too many arguments.

3.4   Constants Handling

The constant values in synthesized programs are usually estimated by second EA (meta-heuristic) or by non-linear fitting, which can be very time-demanding. In this work, we have utilized a novel approach, which uses the extended part of the individual in EA for the evolution of constant values. The important task was to determine, what is the correct size of an extension (2).

$$k = l - floor \left((l - 1) / max\_arg \right) \qquad (2)$$

Where k is the maximum number of constants that can appear in the synthesized program (extension) of length l and max_arg is the maximum number of arguments needed by functions in GFS. Also the $floor()$ is a common floor round function. The final individual dimensionality (length) will be $k + l$ and the example might be:

1. Program length $l = 10$
2. GFS: $\{+, -, *, /, sin, cos, x, k\}$
3. GFS maximum argument $max\_arg = 2$
4. Extension size $k = 10 - floor((10-1)/2) = 6$
5. Dimensionality of the extended individual $k + l = 16$

This means, that the EA will work with individuals of length 16, but only the first 10 features will be used for indexing into the GFS and the rest will be used as constant values.

While mapping the individual into a program, the constants are indexed and later replaced by the value from individual. Simple example can be seen in (3). Individual features in bold are the constant values. It is worthwhile to note that only features which are going to be mapped to GFS are rounded and the rest is omitted.

$$Individual = \left\{ \begin{array}{l} 5.08, \ 1.64, \ 6.72, \ 1.09, \ 6.20, \\ 1.28, \ \mathbf{0.07}, \ \mathbf{3.99}, \ \mathbf{5.27}, \ \mathbf{2.64} \end{array} \right\}$$
$$Rounded \ individual = \{5, 2, 7, 1, 6, 1\}$$
$$GFS_{all} = \{+, -, *, /, sin, cos, x, k\}$$
$$\textbf{Program: } \cos(k1 * (x - k2))$$
$$\textbf{Replaced: } \cos(0.07 * (x - 3.99))$$

$$(3)$$

The first index to GFS_{all} is 5, which represents cos function, its argument is chosen by the next index – 2 representing function $*$ which needs two arguments. Arguments are indexed 7 and 1: constant $k1$ and function (operator) $-$. After this step, two arguments are needed and only two features are left in the program part of the individual. Therefore, the security procedure takes place and those last two features are indexed into GFS_{0arg}. Thus indexes 6 and 1 are mapped to variable x (6 mod size(GFS_{0arg}) = 0) and constant $k2$. The synthesized program is therefore $\cos(k1 * (x - k2))$. The constants are replaced by the remaining features 0.07 and 3.99 respectively.

4 Differential Evolution

This section describes the basics of canonical DE strategies and SHADE strategy. The original [1] has four static control parameters – number of generations G, population size NP, scaling factor F and crossover rate CR. In the evolutionary process of DE, these four parameters remain unchanged and depend on the user initial setting. SHADE algorithm, on the other hand, adapts the F and CR parameters during the evolution. The values that brought improvement to the optimization task are stored into according historical memories M_F

and M_{CR}. SHADE algorithm thus uses only three control parameters – number of generations G, population size NP and size of historical memories H. Also, the mutation strategy is different than that of canonical DE. The concept of basic operations in DE and SHADE algorithms is shown in following sections, for a detailed description on either canonical DE refer to [1] or for feature constraint correction, update of historical memories and external archive handling in SHADE see [22].

4.1 Canonical DE

In this research, we have used canonical DE "rand/1/bin" (4) and "best/1/bin" (5) mutation strategies and binomial crossover (6).

Mutation Strategies and Parent Selection. In canonical forms of DE, parent indices (vectors) are selected by classic pseudo-random generator (PRNG) with uniform distribution. Mutation strategy "rand/1/bin" uses three random parent vectors with indexes $r1$, $r2$ and $r3$, where $r1 = U[1, NP]$, $r2 = U[1, NP]$, $r3 = U[1, NP]$ and $r1 \neq r2 \neq r3$. Mutated vector $v_{i,G}$ is obtained from three different vectors x_{r1}, x_{r2}, x_{r3} from current generation G with help of static scaling factor F as follows:

$$v_{i,G} = x_{r1,G} + F\left(x_{r2,G} - x_{r3,G}\right) \tag{4}$$

Whereas mutation strategy "best/1/bin" uses only two random parent vectors with indexes $r1$, and $r2$, and best individual solution in current generation. The selection respects the very same rules and features as in the previous case. Mutated vector $v_{i,G}$ is obtained as follows:

$$v_{i,G} = x_{best,G} + F\left(x_{r1,G} - x_{r2,G}\right) \tag{5}$$

Crossover and Elitism. The trial vector $u_{i,G}$ which is compared with original vector $x_{i,G}$ is completed by crossover operation (6). CR_i value in canonical DE algorithm is static, i.e. $CR_i = CR$.

$$u_{j,i,G} = \begin{cases} v_{j,i,G} & \text{if } U[0,1] \leq CR_i \text{ or } j = j_{rand} \\ x_{j,i,G} & \text{otherwise} \end{cases} \tag{6}$$

Where j_{rand} is randomly selected index of a feature, which has to be updated ($j_{rand} = U[1, D]$), D is the dimensionality of the problem.

The vector which will be placed into the next generation $G+1$ is selected by elitism. When the objective function value of the trial vector $u_{i,G}$ is better than that of the original vector $x_{i,G}$, the trial vector will be selected for the next population. Otherwise, the original will survive (7).

$$x_{i,G+1} = \begin{cases} u_{i,G} & \text{if } f\left(u_{i,G}\right) \leq f\left(x_{i,G}\right) \\ x_{i,G} & \text{otherwise} \end{cases} \tag{7}$$

4.2   Success-History Based Adaptive Differential Evolution

The differences between canonical DE and SHADE strategy are given in following subsections.

Mutation Strategies and Parent Selection. In the original version of SHADE algorithm [22], parent selection for mutation strategy is carried out by the PRNG with uniform distribution. The mutation strategy used in SHADE is "current-to-$p$best/1" and uses four parent vectors – current i-th vector $\boldsymbol{x}_{i,G}$, vector $\boldsymbol{x}_{pbest,G}$ randomly selected from the $NP \times p$ best vectors (in terms of objective function value) from current generation G. The p value is randomly generated by uniform PRNG $U[p_{min}, 0.2]$, where $p_{min} = 2/NP$. Third parent vector $\boldsymbol{x}_{r1,G}$ is randomly selected from the current generation and last parent vector $\boldsymbol{x}_{r2,G}$ is also randomly selected, but from the union of current generation G and external archive A. Also, indices of vectors $\boldsymbol{x}_{i,G}$, $\boldsymbol{x}_{r1,G}$ and $\boldsymbol{x}_{r2,G}$ have to differ. The mutated vector $\boldsymbol{v}_{i,G}$ is generated by (8).

$$\mathbf{v}_{i,G} = \mathbf{x}_{i,G} + F_i \left(\mathbf{x}_{pbest,G} - \mathbf{x}_{i,G}\right) + F_i \left(\mathbf{x}_{r1,G} - \mathbf{x}_{r2,G}\right) \tag{8}$$

The i-th scaling factor F_i is generated from a Cauchy distribution with the location parameter $M_{F,r}$ (selected randomly from the scaling factor historical memory M_F) and scale parameter value of 0.1 (9). If $F_i > 1$, it is truncated to 1 also if $F_i \leq 0$, Eq. (9) is repeated.

$$F_i = C\left[M_{F,r}, 0.1\right] \tag{9}$$

Crossover and Elitism. SHADE algorithm uses the very same crossover (6) and elitism (7) schemes as canonical DE with following differences. CR value is not static, CR_i is generated from a normal distribution with a mean parameter value $M_{CR,r}$(selected randomly from the crossover rate historical memory M_{CR}) and standard deviation value of 0.1 (10). If the CR_i value is outside of the interval [0, 1], the closer limit value (0 or 1) is used. The crossover compare rule is given in (10).

$$CR_i = N\left[M_{CR,r}, 0.1\right] \tag{10}$$

Also the elitism process is almost identical to that described in (7), with the addition of historical archive. If the objective function value of the trial vector $\boldsymbol{u}_{i,G}$ is better than that of the current vector $\boldsymbol{x}_{i,G}$, the trial vector will become the new individual in new generation $\boldsymbol{x}_{i,G+1}$ and the original vector $\boldsymbol{x}_{i,G}$ will be moved to the external archive of inferior solutions A. Otherwise, the original vector remains in the population in next generation and external archive remains unchanged.

Due to the limited space here, for the detailed information about historical memory update processes for F and CR parameters, please refer to [22].

5   Experiment Design

For the purpose of performance comparison of AP driven by three different DE strategies, an experiment with time series prediction has been carried out. Time series consisting of 300 data-points of GBP/USD exchange rate has been utilized. The first 2/3 of data (200 points – noted as black) were used for regression process and the last 1/3 of the data (100 points – noted as red) were used as a verification for prediction process.

The parameter settings for both canonical DE and SHADE were following: Population size of 75, internal canonical DE parameters $F = 0.5$ and $Cr = 0.8$; internal SHADE parameter $H = 20$. The maximum number of generations was fixed at 2000 generations. The cost function (CF) was defined as a simple difference between given time series and the synthesized model by means of AP (11).

$$CF = \sum_{i=1}^{nreg} |dataTS_i - dataAP_i| \qquad (11)$$

Where $nreg$ represents the length of the time series regression part ($nreg$ data points), $dataTS$ given time series data; and $dataAP$ synthesized model given by AP.

Setting for AP was following:

- Max length of the individual (max D) = 150, where 100 positions were used for functions and 50 positions for constants.
- GFS$_{All}$={+, −, ∗, /, abs, cos, x^3, exp, ln, log10, mod, x^2, sin, sigmoid, sqrt, tan, a'b, x}

Experiments were performed in the environment of *Java* and *Wolfram Mathematica*. All experiments used different initialization, i.e. different initial population was generated within the each run of Canonical DE or SHADE. Overall, 30 independent runs for each DE strategy were performed.

6   Results

Statistical results of the experiments are shown in comprehensive Table 1 for all 30 repeated runs of DE/SHADE. This table contains basic statistical characteristics for cost function values like: *minimum, maximum, mean, median* and *standard deviation*. The last presented attribute is noted as *"Avg. CFE Best Sol."* which stands for the average cost function evaluations required for finding the best solution for all 30 independent run of particular DE strategy. The bold values depict the best obtained results (except the last attribute).

For the graphical comparisons, it have been selected the best (Fig. 1) as well as the second best obtained results (Fig. 2). Synthesized prediction models given by means of AP are depicted in (12) as the illustrative example of the AP outputs for the best results of particular DE strategy.

Table 1. Simple statistical comparisons for all DE strategies and 30 runs

DE strategy	Min	Max	Mean	Median	Std. Dev.	Avg. CFE best Sol.
DE "rand/1/bin"	0.8713	**1.3922**	**1.2113**	1.3045	**0.1972**	28953
DE "best/1/bin"	0.9171	15.4077	2.971	1.3922	4.4243	7442
SHADE	**0.8234**	3.4694	1.3245	**1.065**	0.7807	33987

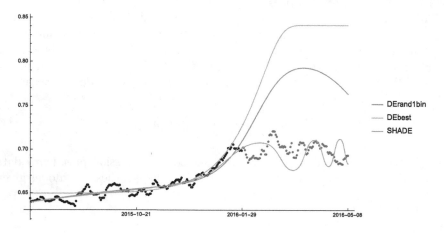

Fig. 1. Comparison of the best results given by three different DE strategies and AP framework for time series prediction problem of GBP/USD exchange rate. Black points (200) used for regression, red points (100) as reference for prediction. (Color figure online)

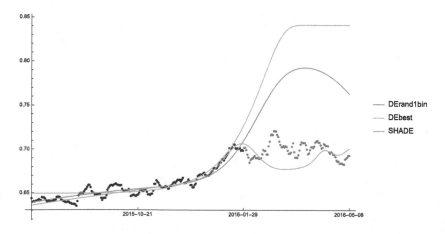

Fig. 2. Comparison of the second best results given by three different DE strategies and AP framework for time series prediction problem of GBP/USD exchange rate. Black points (200) used for regression, red points (100) as reference for prediction. (Color figure online)

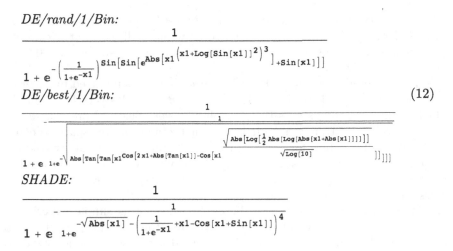

DE/rand/1/Bin:

DE/best/1/Bin: (12)

SHADE:

7 Conclusion and Results Analysis

This paper presented an insight into performance of connection between AP and different strategies of DE. Since AP can be considered as powerful open framework for symbolic regression thanks to its applicability in arbitrary programming language with arbitrary driving evolutionary/swarm based algorithm, the motivation behind this research, was to explore and investigate the differences in performance of AP driven by basic canonical strategies of DE as well as by the powerful state of the art SHADE strategy. The findings can be summarized as follows:

- Average required time per one run of any DE strategy was around 3 min (for maximum of 150 000 cost function evaluations). Considering the "*Avg. CFE Best Sol.*" values in Table 1, we can roughly estimate that SHADE required approx. 40 s for good solution, and canonical DE "best/1/bin" approx. 10 s.
- Obtained graphical comparisons depicted in Figs. 1 and 2 together with statistical data in Table 1 support the claim that there are significant performance differences between particular DE strategies in the task of synthesizing time series regression/prediction models by means of AP.
- The primary logical assumption, that state of the art SHADE strategy will outperform the canonical DE strategies has been confirmed, nevertheless based mostly on the graphical outputs of synthesized models in Figs. 1 and 2. The statistical data presented in the Table 1 shows the mixed results between SHADE and basic canonical DE/rand/1/bin strategy. The Wilcoxon sum rank test (MannWhitney test in SW *Wolfram Mathematics*) for the aforementioned two strategies returns the value $p = 0.38$ with significance level of 0.05. Therefore, we can accept the null hypothesis, that there is no significant difference between SHADE and DE/rand/1/bin strategy based on the statistical non-parametric test.

- The secondary logical assumption that DE/best/1/bin strategy will outperform DE/rand/1/bin strategy turned to be wrong. All, the statistical data,
  graphical comparisons and paired Wilcoxon sum rank tests (p-value 0.0062)
  reject such an assumption.
- The possible confirmation of aforementioned claims is supported by recorded
  parameter noted in Table 1 as *"Avg. CFE Best Sol."*. Its very low value for
  "best/1/bin" strategy shows the possibility of very fast and premature convergence to local extremes in high dimensional complex search space. Whereas
  markedly higher values for SHADE and the "rand/1/bin" strategy confirms
  the longer searching process before stagnation/not-updating of the best result.
- Complexity of the best results (synthesized models) was higher than of the
  worse results.
- An interesting phenomenon was discovered within this simple experiment.
  It seems to be a very good choice to hybridize the AP and powerful state
  of the art DE strategy, to obtain a very good synthesized model structure
  fitting and predicting the data with higher accuracy. Even though from the
  statistical point of view, there are mixed/similar characteristics with much
  simpler strategy (mostly given also by the very simple definition of used cost
  function). The mutual connection of AP and SHADE was able to search in
  very complex high dimensional space for fine individual (solution) structure
  for discrete set handling process inside AP resulting in good synthesized model
  structure securing not only the regression phase, but also with the tendencies
  for approximate prediction of the time series.

References

1. Zelinka, I., et al.: Analytical programming - a novel approach for evolutionary
   synthesis of symbolic structures. In: Kita, E. (ed.) Evolutionary Algorithms, InTech
   2011, ISBN 978-953-307-171-8
2. Storn, R., Price, K.: Differential evolution - a simple and efficient heuristic for
   global optimization over continuous spaces'. J. Global Optim. **11**, 341–359 (1997)
3. Wang, W.C., Chau, K.W., Cheng, C.T., Qiu, L.: A comparison of performance of
   several artificial intelligence methods for forecasting monthly discharge time series.
   J. Hydrol. **374**(3), 294–306 (2009)
4. Santini, M., Tettamanzi, A.: Genetic programming for financial time series prediction. In: Miller, J., Tomassini, M., Lanzi, P.L., Ryan, C., Tettamanzi, A.G.B.,
   Langdon, W.B. (eds.) EuroGP 2001. LNCS, vol. 2038, pp. 361–370. Springer, Heidelberg (2001). doi:10.1007/3-540-45355-5_29
5. Palit, A.K., Popovic, D.: Computational Intelligence in Time Series Forecasting:
   Theory and Engineering Applications. Springer Science & Business Media, London
   (2006)
6. Koza, J.R.: Genetic Programming. MIT Press, New York (1998)
7. Zelinka, I., Oplatkova, Z., Nolle, L.: Boolean symmetry function synthesis by means
   of arbitrary evolutionary algorithms-comparative study. Int. J. Simul. Syst. Sci.
   Technol. **6**(9), 44–56 (2005). ISSN 1473-8031
8. Oplatkova, Z., Zelinka, I.: Investigation on artificial ant using analytic programming. In: Proceedings of the 8th Annual Conference on Genetic and Evolutionary
   Computation, pp. 949–950. ACM (2006)

9. Zelinka, I., Chen, G., Celikovsky, S.: Chaos synthesis by means of evolutionary algorithms. Int. J. Bifurcat. Chaos **18**(4), 911–942 (2008)

10. Senkerik, R., Oplatkova, Z., Zelinka, I., Davendra, D.: Synthesis of feedback controller for three selected chaotic systems by means of evolutionary techniques: Analytic programming. Math. Comput. Modell. **57**(1), 57–67 (2013)

11. Price, K.V., Storn, R.M., Lampinen, J.A., Evolution, D.: Differential Evolution: A Practical Approach to Global Optimization. Natural Computing Series. Springer, Berlin (2005)

12. Neri, F., Tirronen, V.: Recent advances in differential evolution: a survey and experimental analysis. Artif. Intell. Rev. **33**(1–2), 61–106 (2010)

13. Das, S., Suganthan, P.N.: Differential evolution: a survey of the state-of-the-art. IEEE Trans. Evol. Comput. **15**(1), 4–31 (2011)

14. Das, S., Mullick, S.S., Suganthan, P.N.: Recent advances in differential evolution - an updated survey. Swarm Evol. Comput. **27**, 1–30 (2016)

15. Brest, J., Greiner, S., Boskovic, B., Mernik, M., Zumer, V.: Self-adapting control parameters in differential evolution: a comparative study on numerical benchmark problems. IEEE Trans. Evol. Comput. **10**(6), 646–657 (2006)

16. Qin, A.K., Huang, V.L., Suganthan, P.N.: Differential evolution algorithm with strategy adaptation for global numerical optimization. IEEE Trans. Evol. Comput. **13**(2), 398–417 (2009)

17. Zhang, J., Sanderson, A.C.: JADE: adaptive differential evolution with optional external archive. IEEE Trans. Evol. Comput. **13**(5), 945–958 (2009)

18. Das, S., Abraham, A., Chakraborty, U., Konar, A.: Differential evolution using a neighborhood-based mutation operator. IEEE Trans. Evol. Comput. **13**(3), 526–553 (2009)

19. Mininno, E., Neri, F., Cupertino, F., Naso, D.: Compact differential evolution. IEEE Trans. Evol. Comput. **15**(1), 32–54 (2011)

20. Mallipeddi, R., Suganthan, P.N., Pan, Q.K., Tasgetiren, M.F.: Differential evolution algorithm with ensemble of parameters and mutation strategies. Appl. Soft Comput. **11**(2), 1679–1696 (2011)

21. Brest, J., Korosec, P., Silc, J., Zamuda, A., Boskovic, B., Maucec, M.S.: Differential evolution and differential ant-stigmergy on dynamic opti- misation problems. Int. J. Syst. Sci. **44**(4), 663–679 (2013)

22. Tanabe, R., Fukunaga, A.S., Improving the search performance of SHADE using linear population size reduction. In: 2014 IEEE Congress on Evolutionary Computation (CEC), pp. 1658–1665. IEEE (2014)

Archive Analysis in SHADE

Adam Viktorin$^{(\boxtimes)}$, Roman Senkerik, Michal Pluhacek, and Tomas Kadavy

Faculty of Applied Informatics, Tomas Bata University in Zlin,
T.G. Masaryka 5555, 760 01 Zlin, Czech Republic
{aviktorin,senkerik,pluhacek,kadavy}@fai.utb.cz

Abstract. The aim of this research paper is to analyze the current optional archive in Success-History based Adaptive Differential Evolution (SHADE) which is used during mutation. The usefulness of the archive is analyzed on CEC 2015 benchmark set of test functions where the impact of successful archive use on final test function value is studied. This paper also proposes a new version of optional archive named Enhanced Archive (EA), which is also tested on CEC 2015 benchmark set and the results are compared with the canonical version. Two research questions are discussed: Whether SHADE with EA has better performance than canonical SHADE and whether it makes a better use of the archive.

Keywords: Differential evolution · SHADE · Archive

1 Introduction

Differential Evolution (DE) was introduced to the world in 1995 [1] and since its introduction, it has been recognized as one of the best performing optimization algorithms for continuous optimization. Thus, a lot of research has been done in this field in order to further improve the performance and robustness of the algorithm. The variety of proposed variants of DE is enormous, but was neatly summarized in [2,3] and most recently in [4].

The canonical version of DE works with three control parameters – scaling factor F, crossover rate CR and population size NP. These three control parameters along with the stopping criterion are set by the user and the performance on given problem depends on this initial setting. This leads to tedious fine-tuning of the control parameters in pursuance of the best possible performance for given optimization task. Thus, a branch of adaptive DE algorithms emerged in an effort to avoid the fine-tuning and leave the setting of the control parameters to the algorithm. This group of algorithms mostly adapts during the

A. Viktorin—This work was supported by Grant Agency of the Czech Republic – GACR P103/15/06700S, further by the Ministry of Education, Youth and Sports of the Czech Republic within the National Sustainability Programme Project no. LO1303 (MSMT-7778/2014). Also by the European Regional Development Fund under the Project CEBIA-Tech no. CZ.1.05/2.1.00/03.0089 and by Internal Grant Agency of Tomas Bata University under the Projects no. IGA/CebiaTech/2017/004.

© Springer International Publishing AG 2017
L. Rutkowski et al. (Eds.): ICAISC 2017, Part II, LNAI 10246, pp. 688–699, 2017.
DOI: 10.1007/978-3-319-59060-8_62

optimization process and does not require user intervention. Examples of such algorithms are jDE [5], SDE [6], SaDE [7], MDE_pBX [8] and JADE [9]. The last listed along with parameter adaptation proposed a novel mutation strategy "current-to-$p$best/1", which makes use of an optional archive of inferior solutions A. According to the original paper [9], the archive provides information about the progress direction and is also capable of improving the diversity of the population.

JADE created a foundation for currently one of the best DE variants named Success-History based Adaptive DE (SHADE) [10], which placed 3rd on CEC 2013 competition on real parameter single-objective optimization [11] and SHADE with linear decrease in population size titled L-SHADE [12] placed 1st the next year on CEC 2014 competition [13]. SHADE preserved the same mutation strategy and optional archive but introduced new memories for historical values of F and CR and used these memories in the adaptation of these parameters. As a current state of art method, SHADE was studied in this research.

The purpose of this paper was to analyze the influence of the archive on the overall performance of the algorithm and in order to do that, it was tested on CEC 2015 benchmark set of test functions [14]. After the preliminary results, Enhanced Archive (EA) was proposed for SHADE. The new variant EA-SHADE was compared with the canonical SHADE and two research questions were proposed and answered:

1. Is the performance of EA-SHADE better on CEC 2015 benchmark set, than that of canonical SHADE?
2. Does EA-SHADE use optional archive better than canonical SHADE?

The remainder of the paper is structured as follows: Next section describes DE algorithm. Section three presents SHADE algorithm along with the EA modification. Sections four and five are dedicated to the experimental setting and results respectively and sections that follow are results discussion and conclusion.

2   Differential Evolution

The DE algorithm is initialized with a random population of individuals P, that represent solutions of the optimization problem. The population size NP is set by the user along with other control parameters – scaling factor F and crossover rate CR. In continuous optimization, each individual is composed of a vector x of length D, which is a dimensionality (number of optimized attributes) of the problem, and each vector component represents a value of the corresponding attribute, and of objective function value $f(x)$. For each individual in a population, three mutually different individuals are selected for mutation of vectors and resulting mutated vector v is combined with the original vector x in crossover step. The objective function value $f(u)$ of the resulting trial vector u is evaluated and compared to that of the original individual. When the quality (objective function value) of the trial individual is better, it is placed into the next generation, otherwise the original individual is placed there. This step is called selection. The process is repeated until the stopping criterion is met (e.g. the maximum number of objective function evaluations, the maximum number of generations, the low bound for diversity between objective function values in population). The following sections describe four steps of DE: Initialization, mutation, crossover and selection.

2.1   Initialization

As aforementioned, the initial population P, of size NP, of individuals is randomly generated. For this purpose, the individual vector x_i components are generated by Random Number Generator (RNG) with uniform distribution from the range which is specified for the problem by *lower* and *upper* bound (1).

$$x_{j,i} = U\left[lower_j,\ upper_j\right] \text{ for } j = 1,\ \ldots,\ D \tag{1}$$

where i is the index of a current individual, j is the index of current attribute and D is the dimensionality of the problem.

In the initialization phase, a scaling factor value F and crossover value CR has to be assigned as well. The typical range for F value is $[0, 2]$ and for CR, it is $[0, 1]$.

2.2   Mutation

In the mutation step, three mutually different individuals x_{r1}, x_{r2}, x_{r3} from a population are randomly selected and combined in mutation according to the mutation strategy. The original mutation strategy of canonical DE is "rand/1" and is depicted in (2).

$$v_i = x_{r1} + F\left(x_{r2} - x_{r3}\right) \tag{2}$$

where $r1 \neq r2 \neq r3 \neq i, F$ is the scaling factor value and v_i is the resulting mutated vector.

2.3   Crossover

In the crossover step, mutated vector v_i is combined with the original vector x_i and produces trial vector u_i. The binary crossover (3) is used in canonical DE.

$$u_{j,i} = \begin{cases} v_{j,i} \text{ if } U\left[0,1\right] \leq CR \text{ or } j = j_{rand} \\ x_{j,i} \qquad\qquad \text{otherwise} \end{cases} \tag{3}$$

where CR is the used crossover rate value and j_{rand} is an index of an attribute that has to be from the mutated vector u_i (ensures generation of a vector with at least one new component).

2.4 Selection

The selection step ensures, that the optimization progress will lead to better solutions, because it allows only individuals of better or at least equal objective function value to proceed into next generation $G+1$ (4).

$$x_{i,G+1} = \begin{cases} u_{i,G} \text{ if } f(u_{i,G}) \leq f(x_{i,G}) \\ x_{i,G} \qquad \text{otherwise} \end{cases} \qquad (4)$$

where G is the index of current generation.

The whole DE algorithm is depicted in pseudo-code below.

Algorithm 1. DE

1: Set NP, CR, F and stopping criterion;
2: $G = 0$, $x_{best} = \{\}$;
3: Randomly initialize (1) population $P = (x_{1,G}, \ldots, x_{NP,G})$;
4: $P_{new} = \{\}$, $x_{best} =$ best from population P;
5: **while** stopping criterion not met **do**
6:     **for** $i = 1$ to NP **do**
7:         $x_{i,G} = P[i]$;
8:         $v_{i,G}$ by mutation (2);
9:         $u_{i,G}$ by crossover (3);
10:       **if** $f(u_{i,G}) < f(x_{i,G})$ **then**
11:           $x_{i,G+1} = u_{i,G}$;
12:       **else**
13:           $x_{i,G+1} = x_{i,G}$;
14:       **end if**
15:       $x_{i,G+1} \rightarrow P_{new}$;
16:     **end for**
17:     $P = P_{new}$, $P_{new} = \{\}$, $x_{best} =$ best from population P;
18: **end while**
19: **return** x_{best} as the best found solution

---

3 Success-History Based Adaptive Differential Evolution and Enhanced Archive

In SHADE, the only control parameter that can be set by the user is population size NP, other two (F, CR) are adapted to the given optimization task, a new parameter H is introduced, which determines the size of F and CR value memories. The initialization step of the SHADE is, therefore, similar to DE. Mutation, however, is completely different because of the used strategy "current-to-$p$best/1" and the fact, that it uses different scaling factor value F_i for each individual. Crossover is still binary, but similarly to the mutation and scaling factor values, crossover rate value CR_i is also different for each individual. The selection step is the same and therefore following sections describe only different aspects of initialization, mutation and crossover steps. The last section is devoted to the proposed EA and its built into the SHADE algorithm.

3.1   Initialization

As aforementioned, initial population P is randomly generated as in DE, but additional memories for F and CR values are initialized as well. Both memories have the same size H and are equally initialized, the memory for CR values is titled M_{CR} and the memory for F is titled M_F. Their initialization is depicted in (5).

$$M_{CR,i} = M_{F,i} = 0.5 \text{ for } i = 1, \dots, H \tag{5}$$

Also, the external archive of inferior solutions A is initialized. Since there are no solutions so far, it is initialized empty $A = \emptyset$ and its maximum size is set to NP.

3.2   Mutation

Mutation strategy "current-to-$p$best/1" was introduced in [9] and unlike "rand/1", it combines four mutually different vectors, where $pbest \neq r1 \neq r2 \neq i$ (6).

$$v_i = x_i + F_i \left(x_{pbest} - x_i \right) + F_i \left(x_{r1} - x_{r2} \right) \tag{6}$$

where x_{pbest} is randomly selected from the best $NP \times p$ best individuals in the current population. The p value is randomly generated for each mutation by RNG with uniform distribution from the range $[p_{min}, 0.2]$. Where $p_{min} = 2/NP$. Vector x_{r1} is randomly selected from the current population and vector x_{r2} is randomly selected from the union of current population P and archive A. The scaling factor value F_i is given by (7).

$$F_i = C \left[M_{F,r}, 0.1 \right] \tag{7}$$

where $M_{F,r}$ is a randomly selected value (by index r) from M_F memory and C stands for Cauchy distribution, therefore the F_i value is generated from the Cauchy distribution with location parameter value $M_{F,r}$ and scale parameter value 0.1. If the generated value $F_i > 1$, it is truncated to 1 and if it is $F_i \leq 0$, it is generated again by (7).

3.3   Crossover

Crossover is the same as in (3), but the CR value is changed to CR_i, which is generated separately for each individual (8). The value is generated from the Gaussian distribution with mean parameter value of $M_{CR,r}$, which is randomly selected (by the same index r as in mutation) from M_{CR} memory and standard deviation value of 0.1.

$$CR_i = N \left[M_{CR,r}, 0.1 \right] \tag{8}$$

3.4   Historical Memory Updates

Historical memories M_F and M_{CR} are initialized according to (5), but its components change during the evolution. These memories serve to hold successful values of F and CR used in mutation and crossover steps. Successful in terms of producing trial individual better than the original individual. During one generation, these successful values are stored in corresponding arrays S_F and S_{CR}. After each generation, one cell of M_F and M_{CR} memories is updated. This cell is given by the index k, which starts at 1 and increases by 1 after each generation. When it overflows the size limit of memories H, it is again set to 1. The new value of k-th cell for M_F is calculated by (9) and for M_{CR} by (10).

$$M_{F,k} = \begin{cases} \mathrm{mean}_{WL}\left(S_F\right) & \text{if } S_F \neq \emptyset \\ M_{F,k} & \text{otherwise} \end{cases} \tag{9}$$

$$M_{CR,k} = \begin{cases} \mathrm{mean}_{WA}\left(S_{CR}\right) & \text{if } S_{CR} \neq \emptyset \\ M_{CR,k} & \text{otherwise} \end{cases} \tag{10}$$

where mean $_{WL}()$ and mean $_{WA}()$ are weighted Lehmer (11) and weighted arithmetic (12) means correspondingly.

$$\mathrm{mean}_{WL}\left(S_F\right) = \frac{\sum_{k=1}^{|S_F|} w_k \bullet S_{F,k}^2}{\sum_{k-1}^{|S_F|} w_k \bullet S_{F,k}} \tag{11}$$

$$\mathrm{mean}_{WA}\left(S_{CR}\right) = \sum_{k=1}^{|S_{CR}|} w_k \bullet S_{CR,k} \tag{12}$$

where the weight vector w is given by (13) and is based on the improvement in objective function value between trial and original individuals.

$$w_k - \frac{\mathrm{abs}\left(f\left(u_{k,G}\right) - f\left(x_{k,G}\right)\right)}{\sum_{m=1}^{|S_{CR}|} \mathrm{abs}\left(f\left(u_{m,G}\right) - f\left(x_{m,G}\right)\right)} \tag{13}$$

And since both arrays S_F and S_{CR} have the same size, it is arbitrary which size will be used for the upper boundary for m in (13). Complete SHADE algorithm is depicted in pseudo-code below.

1. (a) Enhanced Archive

The original archive of inferior solutions A in SHADE is filled during the selection step and contains the original individuals, that had worse objective function value than trial individuals produced from them. The maximum size of the archive is NP and wherever it overflows, a random individual is removed from the archive. Because individuals from the archive are used in at least 50% (depends on the implementation) of the mutations, it is interesting to analyze whether or not these individuals help to improve the performance of the algorithm and whether this particular implementation of the archive is optimal.

Algorithm 2. SHADE

1: Set NP, H and stopping criterion;
2: $G = 0$, $\boldsymbol{x}_{best} = \{\}$, $k = 1$, $p_{min} = 2/NP$, $\boldsymbol{A} = \emptyset$;
3: Randomly initialize (1) population $\boldsymbol{P} = (\boldsymbol{x}_{1,G}, \ldots, \boldsymbol{x}_{NP,G})$;
4: Set $\boldsymbol{M}_F$ and $\boldsymbol{M}_{CR}$ according to (5);
5: $\boldsymbol{P}_{new} = \{\}$, $\boldsymbol{x}_{best}$ = best from population $\boldsymbol{P}$;
6: **while** stopping criterion not met **do**
7:     $\boldsymbol{S}_F = \emptyset$, $\boldsymbol{S}_{CR} = \emptyset$;
8:     **for** $i = 1$ to NP **do**
9:         $\boldsymbol{x}_{i,G} = \boldsymbol{P}[i]$;
10:        $r = U[1, H]$, $p_i = U[p_{min}, 0.2]$;
11:        Set F_i by (7) and CR_i by (8);
12:        $\boldsymbol{v}_{i,G}$ by mutation (6);
13:        $\boldsymbol{u}_{i,G}$ by crossover (3);
14:        **if** $f(\boldsymbol{u}_{i,G}) < f(\boldsymbol{x}_{i,G})$ **then**
15:            $\boldsymbol{x}_{i,G+1} = \boldsymbol{u}_{i,G}$;
16:            $\boldsymbol{x}_{i,G} \rightarrow \boldsymbol{A}$;
17:            $F_i \rightarrow \boldsymbol{S}_F$, $CR_i \rightarrow \boldsymbol{S}_{CR}$;
18:        **else**
19:            $\boldsymbol{x}_{i,G+1} = \boldsymbol{x}_{i,G}$;
20:        **end if**
21:        **if** $|\boldsymbol{A}| > NP$ **then**
22:            Randomly delete an ind. from $\boldsymbol{A}$;
23:        **end if**
24:        $\boldsymbol{x}_{i,G+1} \rightarrow \boldsymbol{P}_{new}$;
25:    **end for**
26:    **if** $\boldsymbol{S}_F \neq \emptyset$ and $\boldsymbol{S}_{CR} \neq \emptyset$ **then**
27:        Update $\boldsymbol{M}_{F,k}$ (9) and $\boldsymbol{M}_{CR,k}$ (10), k++;
28:        **if** $k > H$ **then**
29:            $k = 1$;
30:        **end if**
31:    **end if**
32:    $\boldsymbol{P} = \boldsymbol{P}_{new}$, $\boldsymbol{P}_{new} = \{\}$, $\boldsymbol{x}_{best}$ = best from population $\boldsymbol{P}$;
33: **end while**
34: **return** $\boldsymbol{x}_{best}$ as the best found solution

---

During the preliminary testing, results shown, that the frequency of successful uses of individuals from the archive is quite low and that there might be a possibility of implementing more beneficial archive solution. Therefore, the EA was proposed as an alternative which was inspired from the field of discrete optimization. The basic idea is, that trial individuals unsuccessful in selection should not be discarded as bad solutions, but the best of them should be stored in the archive in order to provide promising search directions. This is done by placing only unsuccessful trial individuals in the archive and when the archive overflows the maximum size, the worst individual (in terms of objective function value) is discarded. Thus, the original SHADE algorithm was partially changed

Algorithm 3. SHADE changes to accommodate EA
14: **if** $f(\boldsymbol{u}_{i,G}) < f(\boldsymbol{x}_{i,G})$ **then**
15:     $\boldsymbol{x}_{i,G+1} = \boldsymbol{u}_{i,G}$;
16:     ~~$\boldsymbol{x}_{i,G} \to A$;~~
17:     $F_i \to \boldsymbol{S}_F$, $CR_i \to \boldsymbol{S}_{CR}$;
18: **else**
19:     $\boldsymbol{x}_{i,G+1} = \boldsymbol{x}_{i,G}$, $u_{i,G} \to A$;
20: **end if**
21: **if** $|A| > NP$ **then**
22:     Delete the worst individual from A;
23: **end if**
24: $\boldsymbol{x}_{i,G+1} \to \boldsymbol{P}_{new}$;

to implement this novel archive and the changes in pseudo-code are depicted below, the new algorithm was titled EA-SHADE.

4   Experimental Setting

Archive analysis was done on CEC2015 benchmark set, where the results (final error values for each test function, the total number of archive uses and the total number of successful archive uses) were recorded for both SHADE and EA-SHADE algorithms in $10D$ space with the population of 100 individuals and the stopping criterion was the number of function evaluations set to 10,000 × D. A total of 51 independent runs was carried out.

5   Results

In order to answer the first research question, whether the performance of EA-SHADE is better than that of canonical SHADE, median and mean values for both algorithms are reported in Table 1 alongside the Wilcoxon signed rank test p-value. The statistical test was done with the alternative hypothesis that the algorithm with the worse mean value produces overall worse results. The significance level was set to 5% and statistically significant p-values are highlighted by the bold font as well as better median and mean values.

For the purpose of answering the second research question, whether EA-SHADE uses archive better than SHADE, following Tables 2 and 3 are reported. In these tables, there are basic statistical characteristics of archive use, good archive use (when the use of archive produced trial individual that succeeded in selection), best and worst obtained values are reported with corresponding percentages of good archive uses and also the Pearson's correlation coefficient is evaluated for the pair of good archive hits and objective function value. Statistics are done again on CEC 2015 benchmark set.

Values in columns Min good, Max good, Mean good are compared between Tables 2 and 3 and the higher ones are highlighted by bold font, but only if

Table 1. Median and mean values of two compared algorithms SHADE and EA-SHADE on CEC 2015 benchmark set. The last column contains p-values of Wilcoxon signed rank test.

f	SHADE		EA-SHADE		p-value
	Median	Mean	Median	Mean	
1	**0.00**	**0.00**	**0.00**	**0.00**	-
2	**0.00**	**0.00**	**0.00**	**0.00**	-
3	**20.04**	**17.16**	**20.04**	18.42	0.107
4	**3.11**	3.24	3.16	**3.22**	0.483
5	33.49	60.20	**26.85**	**54.67**	0.228
6	**0.42**	3.30	**0.42**	**0.46**	**0.002**
7	0.22	0.25	**0.20**	**0.21**	**0.041**
8	**0.32**	0.29	**0.32**	**0.29**	0.406
9	**100.19**	**100.19**	100.20	100.20	0.064
10	**216.54**	216.66	**216.54**	**216.59**	0.172
11	**2.95**	**107.49**	3.95	125.38	0.064
12	101.59	**101.55**	**101.54**	101.56	0.321
13	**27.71**	**27.49**	28.07	28.10	**0.009**
14	6682.97	4919.47	**2935.54**	**4496.48**	0.161
15	**100.00**	**100.00**	**100.00**	**100.00**	-

the correlation is negative (also highlighted). This indicates that higher success rate in the use of archive is beneficial – negative correlation shows that higher successful rate translates into lower objective function value, which is the goal of minimization task.

6 Results Discussion

The answer to the first research question, whether the performance of EA-SHADE is better than that of canonical SHADE, can be found in Table 1, where out of 15 test functions, EA-SHADE obtains significantly better results on two of them ($f6$, $f7$) and significantly worse result is obtained on one function ($f13$). Results on other test functions are not significantly different, therefore the answer to the first research question is that both archive implementations are comparable in terms of performance.

More interesting results can be seen in other two Tables 2 and 3. These tables serve to answer the second research question, whether EA-SHADE uses archive better than SHADE. The most important part is the correlation between successful use of an archive and obtained objective function value. While the percentage of archive uses is predictably the same for all functions (around 50,7%), which is due to the implementation and is not influenced by the objective function, the

Table 2. SHADE results for archive use. The columns depict mean use of archive in %, minimum of good uses in %, maximum of good uses in %, mean value of good uses in %, best obtained value with corresponding good use in %, worst obtained value with corresponding good use in % and correlation between good use in % and obtained value.

f	Mean use [%]	Min good [%]	Max good [%]	Mean good [%]	Best (good [%])	Worst (good [%])	Corr.
1	50.72	9.36	10.13	9.68	0.00 (9.66)	0.00 (9.66)	-
2	50.74	10.71	11.80	11.17	0.00 (11.65)	0.00 (11.65)	-
3	50.71	1.36	1.69	1.54	2.72 (1.55)	20.08 (1.59)	**−0.09**
4	50.73	4.45	4.96	4.79	1.30 (4.45)	5.03 (4.76)	0.19
5	50.72	**3.67**	**4.23**	**3.93**	7.79 (3.80)	168.94 (3.84)	**−0.35**
6	50.73	3.50	9.60	6.04	0.00 (7.32)	119.95 (9.55)	0.20
7	50.73	4.21	**7.36**	**6.26**	0.10 (6.62)	0.52 (5.20)	**−0.48**
8	50.72	4.15	10.74	5.75	0.00 (6.03)	1.11 (10.74)	0.65
9	50.75	2.22	2.64	2.43	100.11 (2.41)	100.27 (2.50)	**−0.01**
10	50.69	8.92	9.54	9.22	216.54 (9.39)	219.00 (9.38)	0.23
11	50.71	2.56	13.15	6.08	0.67 (3.33)	300.08 (12.12)	1.00
12	50.75	2.11	2.41	2.26	100.81 (2.20)	102.02 (2.37)	0.01
13	50.73	3.83	4.50	4.17	23.30 (4.20)	29.60 (4.13)	**−0.13**
14	50.73	10.97	20.55	14.75	2935.54 (12.16)	6682.97 (20.55)	0.86
15	50.71	14.76	16.47	15.47	100.00 (15.16)	100.00 (15.16)	-

Table 3. EA-SHADE results for archive use. The columns depict mean use of archive in %, minimum of good uses in %, maximum of good uses in %, mean value of good uses in %, best obtained value with corresponding good use in %, worst obtained value with corresponding good use in % and correlation between good use in % and obtained value.

f	Mean use [%]	Min good [%]	Max good [%]	Mean good [%]	Best (good [%])	Worst (good [%])	Corr.
1	50.72	10.19	11.03	10.70	0.00 (10.71)	0.00 (10.71)	-
2	50.72	11.73	12.56	12.17	0.00 (12.16)	0.00 (12.16)	-
3	50.70	1.40	1.70	1.56	3.43 (1.69)	20.08 (1.55)	**−0.16**
4	50.68	**4.67**	**5.26**	**4.94**	1.03 (5.26)	5.09 (4.93)	**−0.31**
5	50.73	3.72	4.30	4.03	6.36 (3.98)	262.07 (4.02)	0.03
6	50.69	**4.35**	**10.79**	**9.27**	0.00 (10.75)	3.61 (10.30)	**−0.02**
7	50.68	**5.13**	7.23	6.18	0.05 (7.23)	0.46 (5.44)	**−0.35**
8	50.69	5.04	12.84	8.08	0.00 (6.80)	1.12 (12.80)	0.70
9	50.69	**2.29**	**2.67**	**2.49**	100.12 (2.59)	100.27 (2.48)	**−0.07**
10	50.72	10.24	11.08	10.71	216.54 (10.83)	218.99 (10.71)	0.00
11	50.68	2.65	14.00	7.34	0.77 (3.32)	300.08 (13.65)	1.00
12	50.70	**2.12**	**2.52**	**2.32**	101.09 (2.34)	102.01 (2.12)	**−0.27**
13	50.71	**3.95**	**4.55**	**4.26**	23.09 (4.46)	31.80 (4.20)	**−0.03**
14	50.70	12.51	19.02	15.34	100.00 (17.82)	6682.97 (16.11)	0.65
15	50.72	16.19	18.97	17.11	100.00 (16.84)	100.00 (16.84)	-

percentage of successful archive uses differs from function to function and ranges from 1.36% to 20.55%. The benchmark set contains four types of functions – unimodal, simple multimodal, hybrid and composition. Unimodal functions tend to have higher percentage of successful uses of the archive in both SHADE and EA-SHADE algorithms, but for the other types, results are inconclusive.

In order to compare SHADEs and EA-SHADEs use of the archive, the better one should have not only higher successful use percentage on given function, but also the correlation has to be negative. Negative correlation mirrors that higher the successful use, lower the obtained objective function value. In Table 2, there are 5 negative and 7 positive correlations. The correlation for three test functions ($f1$, $f2$ and $f15$) could not be established, because there is no standard deviation in the obtained objective function value in 51 independent runs. This suggests, that higher archive use was in SHADE beneficial for only 5 test functions, while in Table 3, there are 7 negative and 5 positive correlations. Thus, the archive was more beneficial in EA-SHADE case. Nevertheless, the combination of higher percentage of successful archive use and higher negative value of correlation is the key to interpreting the findings. In SHADE, there are two cases ($f5$ and $f7$), in EA-SHADE, there are six cases ($f3$, $f4$, $f6$, $f8$, $f12$ and $f13$). Therefore, the score is 6:2 in favor of EA-SHADE and the answer to the second research question is positive – EA-SHADE algorithm uses optional archive better than SHADE algorithm.

The paradoxical results occur in two cases ($f7$ and $f13$). While the use of archive on $f7$ function is better in SHADE algorithm, significantly better results are achieved with EA-SHADE algorithm. In the case of $f13$, it is vice-versa, EA-SHADE uses archive better, but SHADE produces significantly better results.

7   Conclusion

This research paper presented an analysis of optional archive in SHADE algorithm and proposed a novel implementation of the archive EA. The results of canonical SHADE and EA-SHADE algorithms were compared on the CEC 2015 benchmark set of test functions and two research questions were answered:

1. Is the performance of EA-SHADE better on CEC 2015 benchmark set, than that of canonical SHADE?
2. Does EA-SHADE use optional archive better than canonical SHADE?

The first answer is that both algorithms are comparable in terms of performance. The second question is answered positive, the EA-SHADE algorithm uses archive better in 6 cases opposed to 2 cases, where SHADE uses the archive better. Two paradoxical states also occurred on functions ($f7$ and $f13$), where the better use of archive was not beneficial for the overall performance of the algorithm.

The future research will deal with more thorough analysis of the archive contents and the influence on the performance. Another interesting future direction is in using networks created during the optimization process and their characteristics to produce archive management independent on the objective function value, which would not require distance evaluation and would sustain the diversity in archive contents.

References

1. Storn, R., Price, K.: Differential Evolution-a Simple and Efficient Adaptive Scheme for Global Optimization Over Continuous Spaces, vol. 3. ICSI, Berkeley (1995)
2. Neri, F., Tirronen, V.: Recent advances in differential evolution: a survey and experimental analysis. Artif. Intell. Rev. **33**(1–2), 61–106 (2010)
3. Das, S., Suganthan, P.N.: Differential evolution: a survey of the state-of-the-art. IEEE Trans. Evol. Comput. **15**(1), 4–31 (2011)
4. Das, S., Mullick, S.S., Suganthan, P.N.: Recent advances in differential evolution-an updated survey. Swarm Evol. Comput. **27**, 1–30 (2016)
5. Brest, J., Greiner, S., Bošković, B., Mernik, M., Zumer, V.: Self-adapting control parameters in differential evolution: a comparative study on numerical Benchmark problems. IEEE Trans. Evol. Comput. **10**(6), 646–657 (2006)
6. Omran, M.G.H., Salman, A., Engelbrecht, A.P.: Self-adaptive differential evolution. In: Hao, Y., Liu, J., Wang, Y., Cheung, Y., Yin, H., Jiao, L., Ma, J., Jiao, Y.-C. (eds.) CIS 2005. LNCS, vol. 3801, pp. 192–199. Springer, Heidelberg (2005). doi:10.1007/11596448_28
7. Qin, A.K., Huang, V.L., Suganthan, P.N.: Differential evolution algorithm with strategy adaptation for global numerical optimization. IEEE Trans. Evol. Comput. **13**(2), 398–417 (2009)
8. Islam, S.M., Das, S., Ghosh, S., Roy, S., Suganthan, P.N.: An adaptive differential evolution algorithm with novel mutation and crossover strategies for global numerical optimization. IEEE Trans. Syst. Man Cybern. Part B (Cybern.) **42**(2), 482–500 (2012)
9. Zhang, J., Sanderson, A.C.: JADE: adaptive differential evolution with optional external archive. IEEE Trans. Evol. Comput. **13**(5), 945–958 (2009)
10. Tanabe, R., Fukunaga, A.: Success-history based parameter adaptation for differential evolution. In: 2013 IEEE Congress on Evolutionary Computation (CEC), pp. 71–78. IEEE, June 2013
11. Liang, J.J., Qu, B.Y., Suganthan, P.N., Hernández-Díaz, A.G.: Problem definitions and evaluation criteria for the CEC 2013 special session on real-parameter optimization. Computational Intelligence Laboratory, Zhengzhou University, Zhengzhou, China and Nanyang Technological University, Singapore, Technical report, 201212 (2013)
12. Tanabe, R., Fukunaga, A.S.: Improving the search performance of SHADE using linear population size reduction. In: 2014 IEEE Congress on Evolutionary Computation (CEC), pp. 1658–1665. IEEE, July 2014
13. Liang, J.J., Qu, B.Y., Suganthan, P.N.: Problem definitions and evaluation criteria for the CEC 2014 special session and competition on single objective real-parameter numerical optimization. Computational Intelligence Laboratory, Zhengzhou University, Zhengzhou China and Technical report, Nanyang Technological University, Singapore (2013)
14. Liang, J.J., Qu, B.Y., Suganthan, P.N., Chen, Q.: Problem definitions and evaluation criteria for the CEC 2015 competition on learning-based real-parameter single objective optimization. Technical report 201411A, Computational Intelligence Laboratory, Zhengzhou University, Zhengzhou China and Technical report, Nanyang Technological University, Singapore (2014)

Special Session: Stream Data Mining

Learning in Nonstationary Environments: A Hybrid Approach

Cesare Alippi[1,2], Wen Qi[1], and Manuel Roveri[1(✉)]

[1] Politecnico di Milano, Milan, Italy
{cesare.alippi,wen.qi,manuel.roveri}@polimi.it
[2] Università della Svizzera Italiana, Lugano, Switzerland

Abstract. Solutions present in the literature to learn in nonstationary environments can be grouped into two main families: passive and active. Passive solutions rely on a continuous adaptation of the envisaged learning system, while the active ones trigger the adaptation only when needed. Passive and active solutions are somehow complementary and one should be preferred than the other depending on the nonstationarity rate and the tolerable computational complexity. The aim of this paper is to introduce a novel hybrid approach that jointly uses an adaptation mechanism (as in passive solutions) and a change detection triggering the need to retrain the learning system (as in active solutions).

1 Introduction

Recent years have been characterized by an increasing interest in the research on learning in nonstationary environments [1,2]. Dealing with changes in stationarity is crucial in many real-world scenarios where data, typically available in a streaming manner, are characterized by a time varying behaviour (also called "concept drift").

Learning algorithms designed to operate in nonstationary environments typically rely on adaptation strategies to track the data generating process and adapt the learning system to new operating conditions over time to maintain the accuracy [2,3]. Two approaches for addressing the nonstationary environments are typically available in the literature: active and passive [4,5]. These approaches differ in the adaptation mechanism used to cope with time variance in the data-generating process. Solutions following the active approach rely on an explicit detection of a change to activate the adaptation mechanism, while passive ones continuously adapt the learning-system parameters over time regardless possible time variance. These two approaches are described and commented in Sect. 2. We emphasize that, in passive solutions, false positive and negative detections do not represent an issue since the learning-system parameters are continuously updated as soon as new data become available. However, if the parameter update is computationally expensive, e.g., as in case of large neural networks, passive solutions might become inappropriate for embedded applications, which might end

This work was supported by the Polish National Science Center under Grant No. 2014/15/B/ST7/05264.

L. Rutkowski et al. (Eds.): ICAISC 2017, Part II, LNAI 10246, pp. 703–714, 2017.
DOI: 10.1007/978-3-319-59060-8_63

up in a prohibitive energy consumption and processing time[1]. Differently, active solutions might suffer from false positive/negative detections induced by the change-detection mechanism. However, active approaches update the learning-based system only when it is needed. It should be noted that a false positive detection will only introduce an unnecessary update of the application parameters whereas a false negative one is likely to be perceived with a higher latency (provided that the change increases in time). It emerges that, depending on the evolution rate, an active approach might be preferable than a passive one if we need to trade-off accuracy with computational complexity. Unfortunately, this trade-off depends on the level of time variance characterizing the environment the system interacts with, an information that is rarely available at design time.

The aim of this paper is to introduce a novel generation of adaptation mechanisms based on a hybrid approach for learning in nonstationary environments. The proposed hybrid solution jointly considers a passive learning mechanism and a change detection mechanism as in active ones. The effectiveness and efficiency of the proposed hybrid solution is tested on nonlinear regression applications.

The paper is organized as follows. Sect. 2 describes the passive, active and hybrid approaches, while Sect. 3 introduces the specific hybrid solution for regression in nonstationary environments. Experimental results are detailed in Sect. 4.

2 Learning in Nonstationary Environments: Passive, Active and Hybrid

2.1 Problem Formulation

Let $\mathcal{P}$ be a data-generating process generating a sequence of couples $(\mathbf{x}_t, y_t), t \in \mathcal{N}$ from the unknown possibly time-variant probability distribution $p_t(\mathbf{x}, y)$. In particular, $\mathbf{x}_t \in \mathbb{R}^d$ represents the feature vector sampled at time t from the unknown probability distribution $p_t(\mathbf{x})$ and $y \in \Lambda$ is the output value[2] drawn from the unknown conditional distribution $p_t(y|\mathbf{x})$.

The goal of learning is to build a model $\hat{y}_t = f_t(\mathbf{x}_t, \hat{\theta}_t)$ parametrized in $\theta \in \Theta$, being Θ the parameter space. Since in time-variant scenarios $p_t(\mathbf{x}, y)$ may change over time, $f_t(\mathbf{x}_t, \hat{\theta}_t)$ needs to be adapted following the changes. How and when to adapt $f_t(\mathbf{x}_t, \hat{\theta}_t)$ depends on the particular approach considered to address the learning in nonstationary environments as described in the rest of this section.

2.2 The Passive Approach

Learning solutions following the passive approach continuously update model $f_t(\mathbf{x}_t, \hat{\theta}_t)$ every time a new couple $(\mathbf{x}_t, y_t)$ is provided. A schematic diagram showing the learning methods is given in Fig. 1.

---

[1] This problem becomes even more relevant when large-scale network-of-networks systems are considered, producing high-dimension/high-velocity data streams and units are battery powered.

[2] Λ is a discrete class label in classification problems and a subset of $\mathbb{R}^p$ in regression ones.

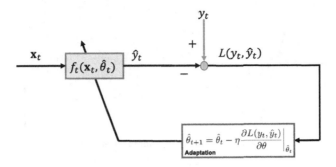

Fig. 1. The passive approach: the model $f_t(\mathbf{x}_t, \hat{\theta}_t)$ is updated every time a new couple $(\mathbf{x}_t, y_t)$ is available. $L(y_t, \hat{y}_t)$ measures the discrepancy between y_t, and $\hat{y}_t$. Adaptation is carried out as described in Eq. (1).

Adaptation, which is carried out by modifying the parameters $\hat{\theta}_t$, is continuous. In more detail, at time t parameter vector $\hat{\theta}_t$ is updated as

$$\hat{\theta}_{t+1} = \hat{\theta}_t - \eta \frac{\partial L(y_t, f_t(\mathbf{x}_t, \theta_t))}{\partial \theta}\bigg|_{\hat{\theta}_t} \tag{1}$$

where η is the learning rate and L is the loss function, e.g., the squared error. This adaptation scheme is typical for online passive solutions that follow a gradient descent approach [6]. Other adaptation mechanisms can be considered as well.

Passive solutions have shown to be particularly effective in dealing with gradual changes (e.g., see [4]). Many single- and ensemble-model learning algorithms following the passive approach have been presented in the literature (e.g., see [1,6] and references therein). Vary-fast decision trees (VFDT) [7] represents the most popular single-model learning algorithm following the passive approach. A passive fuzzy-based single-model learner for nonstationary classification problems has been introduced in [8], while [9] proposes a passive mechanism combining an online extreme learning machine with a time-variant neural network for nonstationary datastreams. Ensemble models are quite popular in passive approaches, e.g., [4,10–12]. There, the passive mechanism is integrated with the addition, removal/modification of models composing the ensemble.

2.3 The Active Approach

Solutions following the active approach rely on change detection mechanisms to trigger the adaptation of model $f_t(\mathbf{x}_t, \hat{\theta}_t)$ [1]. More specifically, the change detection phase aims at identifying changes in $p_t(\mathbf{x}, y)$ and activating the adaptation phase, whose goal is to adapt $f_t(\mathbf{x}_t, \hat{\theta}_t)$ to the new working conditions. A schematic for the active approach is shown in Fig. 2. There, the change detection mechanism inspects changes in the loss function $L(y_t, \hat{y}_t)$, e.g., in the residual $y_t - \hat{y}_t$. Other active solutions can inspect changes in other features associated with process $\mathcal{P}$, such as $\mathbf{x}_t$s or other information as done in [5,13].

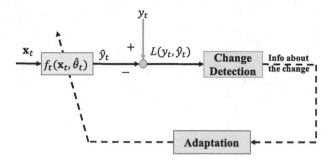

Fig. 2. The active approach: the Adaptation phase of model $f_t(\mathbf{x}_t, \hat{\theta}_t)$ is triggered by the Change Detection module inspecting e.g., the loss function $L(y_t, \hat{y}_t)$. Other active approaches can inspect $\mathbf{x}_t$ or features extracted from it.

Popular change-detection mechanisms are hypothesis tests, change-point methods and sequential change detection tests [1,6]. Among these mechanisms, sequential change detection tests (CDTs) are particularly suitable for active solutions since they are capable to sequentially investigate changes in datastreams in an online manner, as discussed in [5,14,15].

Once a change has been detected, the adaptation phase is triggered to adapt $f_t(\mathbf{x}_t, \hat{\theta}_t)$ to the new working conditions. Active-based adaptation mechanisms can be grouped in three families: windowing, weighting and random sampling [16]. In windowing, the goal is to identify a window over the last acquired samples to retrain the model. Following in this framework we find the adaptive-window solutions [17,18] and Just-In-Time Adaptive Classifiers [5,13,19]. Solutions following the weighting approach [16] do not identify a window but all the previously-acquired samples, suitably weighted (e.g., according to their acquisition time), are considered in the adaptation phase. Finally, random sampling [20] considers a window of samples randomly extracted from the available ones.

2.4   The Proposed Hybrid Approach

Passive and active approaches are characterized by advantages and disadvantages that are somehow complementary: passive approaches work well with gradual changes but are typically computationally intensive if either complex or ensemble-based models are considered, while active ones are more effective with abrupt changes and tend to be computationally lighter if the adaptation is triggered with low probability. Clearly, a trade-off has to be identified, which is based on model complexity, the type of change and its occurrence probability.

Here, we introduce a family of hybrid approaches as shown in Fig. 3. In more detail, at each new incoming couple $(\mathbf{x}_t, y_t)$, model $f_t(\mathbf{x}_t, \hat{\theta}_t)$ is updated through a suitable-defined hybrid-passive adaptation (similar to passive solutions). Here, adaptation is meant to support (to the best extent possible) the adaptation to gradual changes without requiring any explicit detections of changes. In the meanwhile, the discrepancy between y_t and $\hat{y}_t$ measured through the loss function

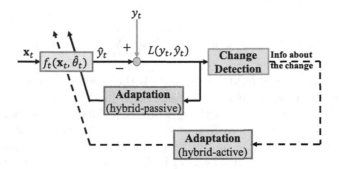

Fig. 3. The proposed hybrid approach: a low complexity *hybrid-passive adaptatation* phase is carried out at each time instant to update $f_t(\mathbf{x}_t, \hat{\theta}_t)$ (as in passive approaches). In the meanwhile, the loss function $L(y_t, \hat{y}_t)$ measuring the discrepancy between y_t and $\hat{y}_t$ is inspected by the Change Detection module. When a change is detected, the *hybrid-active adaptatation* phase is activated and retrains $f_t(\mathbf{x}_t, \hat{\theta}_t)$ to the new working conditions.

$L(y_t, \hat{y}_t)$ is inspected by the Change Detection module. When a change is detected, the hybrid-active adaptation module is triggered to adapt $f_t(\mathbf{x}_t, \hat{\theta}_t)$ to new working conditions (as in active solutions). Here, the change detection inspects the loss function $L(y_t, \hat{y}_t)$ but other solutions comprising the monitoring of $\mathbf{x}_t$s or features extracted from it could be considered as well.

3   Regression in Nonstationary Environments

3.1   The Regression Problem

In this section we focus on the nonlinear regression problem in nonstationary environments. Here, $\mathcal{P}$ follows the system model

$$y_t = g_t(\mathbf{x}_t) + \epsilon_t \tag{2}$$

and $y_t \in \mathbb{R}$. ϵ_t is a (possibly time-varying) random variable accounting for the noise affecting function $g_t(\mathbf{x}_t)$ that might change over time as well. In this work we assume $\mathcal{P}$ to be initially stationary, with time variance developing only after the configuration phase at time τ

$$g_t(\mathbf{x}_t) = \begin{cases} g(\mathbf{x}_t) & t < \tau \\ g'_t(\mathbf{x}_t), & t \geq \tau. \end{cases} \tag{3}$$

Both $g_t(\mathbf{x}_t)$ and $g'_t(\mathbf{x}_t)$ are unknown.

3.2   A Hybrid Solution for Regression in Nonstationary Environments

In the proposed hybrid solution, the approximating model family $f_t(\mathbf{x}_t, \theta_t)$ for $g_t(\mathbf{x}_t)$ and $g'_t(\mathbf{x}_t)$ is the single hidden-layer feedforward neural network

$$\hat{y}_t = f_t(\mathbf{x}_t, \theta_t) = b_t^0 + \sum_{j=1}^{h} w_t^{j,0} \Psi \left(\sum_{i=1}^{d} \omega_t^{i,j} x_t^i + b_t^j \right) \tag{4}$$

where h is the number of hidden neurons, d is the number of input variables, Ψ is the (hyperbolic tangent) activation function of hidden neurons, x_t^i is the i-th component of $\mathbf{x}_t$ with $i = \{1, \ldots, d\}$, $\omega_t^{i,j}$s are the hidden weights connecting the i-th input with the j hidden neuron at time t, b_t^j is the bias of the j-th hidden neuron at time t, $w_t^{j,0}$s are the output weights connecting the j-th hidden neuron with the output neuron[3] at time t and b_t^0 is the bias of the output neuron at time t. Parameter vector θ_t accounts for all the parameters of the network. We opted for single hidden-layer feedforward neural networks containing a finite number of neurons since they approximate any continuous function defined on compact subsets [21]. Other model family could be considered.

The parameters in (4) can be grouped into the *hidden* θ_t^h and the *output* parameter θ_t^o vectors as follows

$$\hat{\theta}_t^h = [\omega_t^{1,1}, \ldots, \omega_t^{n,h}, b_t^1, \ldots, b_t^h]$$
$$\hat{\theta}_t^o = [w_t^{1,0}, \ldots, w_t^{h,0}, b_t^0] \tag{5}$$

As such, $\theta_t = [\theta_t^h : \theta_t^o]$. After training on an initial training sequence TS that is assumed to be acquired in stationary conditions, a parameter vector $\hat{\theta}_t$ is provided and associated with model $f_t(\mathbf{x}_t, \hat{\theta}_t)$.

Figure 4 shows the proposed hybrid solution for regression in nonstationary environments. In more detail, the *hybrid-passive adaptation* is based on a *recursive least square* (RLS) estimation that updates the output parameters $\hat{\theta}_t^o$ in correspondence with each new couple $(\mathbf{x}_t, y_t)$ as:

$$\begin{cases} \psi_{t+1}^j = \Psi \left(\sum_{i=1}^{d} \omega_t^{i,j} x_{t+1}^i + b_t^j \right) \\ \mathbf{z}_{t+1} = [\psi_{t+1}^1, \ldots, \psi_{t+1}^h] \\ S_{t+1} = S_t + \mathbf{z}_{t+1}^T \mathbf{z}_{t+1} \\ \mathbf{q}_{t+1} = S_{t+1}^{-1} \mathbf{z}_{t+1}^T \\ \hat{\theta}_{t+1}^o = \hat{\theta}_t^o + (y_t - \mathbf{z}_{t+1} \hat{\theta}_t^o) \mathbf{q}_{t+1} \end{cases}$$

where ψ_t^j is the activation of the j-th hidden neuron at time t, and $\mathbf{z}_t$ the vector storing the activations of all the hidden neurons at time t; T is the transpose operator and $S_t = ([\mathbf{z}_1, \ldots, \mathbf{z}_t]^T [\mathbf{z}_1, \ldots, \mathbf{z}_t])$. S_0 and $\hat{\theta}_t^o$ are initialized on TS.

---

[3] The activation function of the output layer is linear.

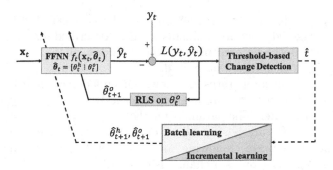

Fig. 4. The specific hybrid solution for regression in nonstationary environments: the model is a single hidden-layer feedforward neural network (FFNN), whose parameters are updated in a passive way through Recursive Least Square (RLS) estimation. The loss function $L(y_t, \hat{y}_t)$ measuring the squared error $(y_t - \hat{y}_t)^2$ is inspected by the Change Detection module. When a change is detected, one of the two possible adaptation phases (i.e., batch learning or incremental learning) is activated by exploiting the information $\hat{t}$ related to the estimated time of change.

Following the active approach, the discrepancy $L(y_t, \hat{y}_t)$ between y_t and $\hat{y}_t$ is monitored over time through a change detection test (CDT). In the proposed solution $L(y_t, \hat{y}_t)$ is the squared error $e_t = (y_t - \hat{y}_t)^2$ and the considered CDT relies on two automatically-defined thresholds:

$$T_{up} = \mu_e + \gamma_{up}\sigma_e$$
$$T_{dw} = \mu_e + \gamma_{dw}\sigma_e \tag{6}$$

with μ_e and σ_e being the sample mean and standard deviation of the squared error e_t computed on TS, respectively and $\gamma_{up} \in \mathbb{R}^+$ and $\gamma_{dw} \in \mathbb{R}^+$ (with $\gamma_{up} \geq \gamma_{dw}$) two user-defined parameters.

A change is detected at time $\bar{t}$ when $c_{\bar{t}} > T_{up}$. Afterwards, the threshold based CDT is also able to provide an estimate $\hat{t}$ of the time instant τ the change started in $\mathcal{P}$ by relying on threshold T_{dw} as follows:

$$\hat{t} = min\left(t | e_t \geq T_{dw}\right). \tag{7}$$

Intuitively, once a change has been detected, the threshold-based CDT re-processes the previously computed squared errors to identify[4] the first time instant in which e_t overcomes T_{dw} and this is used as a better estimate for τ. All samples between $\hat{t}$ and $\bar{t}$ can be associated to the new working conditions and used for retraining model (4).

---

[4] Change-point methods, i.e., statistical techniques able to identify the presence of a change and localize it within a fixed sequence of data, would have been the statistically-grounded solution to provide $\hat{t}$. Unfortunately, they are characterized by a high computational complexity; hence, their use would have significantly increase the overall complexity of the proposed solution.

When a change is detected by the threshold-based CDT, the hybrid-active adaptation is triggered. Two mechanisms are here described for this step: batch and incremental. Both mechanisms rely on a window of W training data but they differ in the way data are processed.

The batch mechanism operates as follows. Given a detection at time $\bar{t}$ and an estimate $\hat{t}$, the hybrid-active adaptation stores the W couples $(\mathbf{x}_t, y_t)$ between $\hat{t}$ and $\hat{t}+W-1$ into a buffer. The parameter update of Eq. (4) and the configuration of the thresholds in (6) are performed on once W couples have been acquired.

The second mechanism, following an incremental approach, activates the adaptation strategy as soon as a change is detected.

4   Experimental Results

The effectiveness of the proposed hybrid solution has been compared with both a passive and an active solution on two synthetic experiments and a real dataset.

Description of the experiments. The first synthetic experiment $(S1)$ refers to the function with mono-dimensional input $(d = 1)$ proposed in [6]:

$$y_t = g_t^{S1}(\mathbf{x}_t) = -x_t sin\,(x_t)^2 + \frac{e^{-0.23x_t}}{1 + (x_t)^4} + \epsilon_t$$

with $x_t \sim U(-2, 2)$, where $U(a, b)$ is a continuous uniform distribution in the interval $[a, b]$, and $\epsilon_t \sim \mathcal{N}(0, 0.05^2)$.

Differently, the second synthetic experiment $(S2)$ refers to the function with multi-dimensional input $(d = 4)$:

$$y_t = g_t^{S2}(\mathbf{x}_t) = -x_t^1 sin(x_t^4)^2 + \frac{e^{-0.23x_t^2}}{1 + (x_t^3)^4} + cos(2\pi x_t^4) + \epsilon_t$$

with $x_t^1 \sim U(-2, 2)$, $x_t^2 \sim U(-3, 0)$, $x_t^3 \sim U(-3, -1)$, $x_t^4 \sim U(-2, -1)$ and $\epsilon_t \sim \mathcal{N}(0, 0.05^2)$. Both $S1$ and $S2$ last 1500 samples.

The real dataset (R) refers to the "Communities and Crime Data Set", publicly available at the UCI Machine Learning Repository [22]. More specifically, we considered the first 40 features of this dataset (i.e., $d = 40$) and limited to the first 1500 samples as per the synthetic experiments. For both experiments $S1$ and $S2$ and dataset R, the first 900 samples have been used for training, the next 100 for validation and the last 500 for testing. Changes have been artificially injected at time $\tau = 1201$ in $S1$, $S2$, and R according to Eq. (3). Four different kinds of changes, i.e., abrupt additive (AA), abrupt multiplicative (AM), drift (DR) and function change (FC), have been considered for the synthetic experiments, while only three of them have been considered for the real dataset. All the changes for $S1$, $S2$ and R are detailed in Table 1 together with the different configurations of the change parameters.

The configuration of the proposed and alternative solutions. The specific hybrid solution presented in Sect. 3.2 has been configured with $\gamma_{dw} = 3$,

Table 1. The different changes for experiments $S1$ and $S2$ and the real dataset R. The first column details the considered experiment/dataset; the second column refers to the type of change; the third one shows the effect of the change on the original function $g_t(\mathbf{x}_t)$ as modelled in (3); the forth one specifies the identifier of the specific kind of change used in Tables 2 and 3; the last one shows the value of the corresponding change parameter.

Exp.	Change type	$g'_t(\mathbf{x}_t)$	Change ID	Parameter
$S1$	Abrupt additive (AA)	$g_t^{S1}(\mathbf{x}_t) + \Delta$	$S1-AA_1$	$\Delta = 0.5$
			$S1-AA_2$	$\Delta = 1.5$
	Abrupt multiplicative (AM)	$g_t^{S1}(\mathbf{x}_t)\Delta$	$S1-AM_1$	$\Delta = 0.5$
			$S1-AM_2$	$\Delta = 1.5$
	Drift (DR)	$g_t^{S1}(\mathbf{x}_t) + \Delta(t - \tau + 1)$	$S1-DR_1$	$\Delta = 0.005$
			$S1-DR_2$	$\Delta = 0.01$
			$S1-DR_3$	$\Delta = 0.015$
	Function change (FC)	$-\sin(4x_t^1 + 7) + e^{x_t^1} + \cos(x_t^1) + x_t^1$	$S1-FC$	
$S2$	Abrupt additive (AA)	$g_t^{S2}(\mathbf{x}_t) + \Delta$	$S2-AA_1$	$\Delta = 0.5$
			$S2-AA_2$	$\Delta = 1.5$
	Abrupt multiplicative (AM)	$g_t^{S2}(\mathbf{x}_t)\Delta$	$S2-AM_1$	$\Delta = 0.5$
			$S2-AM_2$	$\Delta = 1.5$
	Drift (DR)	$g_t^{S2}(\mathbf{x}_t) + \Delta(t - \tau + 1)$	$S2-DR_1$	$\Delta = 0.005$
			$S2-DR_2$	$\Delta = 0.01$
			$S2-DR_3$	$\Delta = 0.015$
	Function change (FC)	$-\sin(4x_t^1 + 7) + e^{x_t^2} + \cos(x_t^3) + x_t^4$	$S2-FC$	
R	Abrupt additive (AA)	$y_t + \Delta$	$R-AA_1$	$\Delta = 0.5$
			$R-AA_2$	$\Delta = 1.5$
	Abrupt multiplicative (AM)	$y_t\Delta$	$R-AM_1$	$\Delta = 0.5$
			$R-AM_2$	$\Delta = 1.5$
	Drift (DR)	$y_t + \Delta(t - \tau + 1)$	$R-DR_1$	$\Delta = 0.005$
			$R-DR_2$	$\Delta = 0.01$
			$R-DR_3$	$\Delta = 0.015$

$\gamma_{up} = 5$, and $W = 200$; h has been experimentally set to 4 for $S1$ and 25 for $S2$ and R. Both the learning mechanisms described in Sect. 3.2, i.e., batch and incremental learning, have been considered in this experimental analysis. As a comparison we considered a *passive solution* (described in Sect. 2.2) where the model $f_t(\mathbf{x}_t, \hat{\theta}_t)$ is the same feedforward neural network of the hybrid one. We also considered an *active solution* sharing the same model (4) of passive and hybrid and, to ease the comparison, the change detection mechanism is the threshold-based CDT used in the hybrid solution, while the adaptation mechanism is the *batch* one described in Sect. 3.2.

To evaluate the effectiveness of the proposed solution we considered:

- The mean square error (*mse*) computed on the test set;
- The computational time (*ct*) of the execution of one algorithm run (in seconds). All the considered solutions have been implemented in Matlab; the hardware platform was based on an Intel i5 Core 2.5 GHz with 8 GB of RAM.

Comments. The experimental results for experiments $S1$ and $S2$ and the real dataset R are shown in Tables 2 and 3, respectively. Results in the former table (i.e., the one referring to the synthetic experiments) are averaged over 25 runs.

As expected, the hybrid solution based on incremental learning provides a lower $mses$ than the batch one, at the expense of an increased ct. More specifically, the incremental learning of the hybrid solution allows to quickly react to the change by activating the adaptation as soon as a change is detected. This is underlined by lower values of $mses$ in all the different kinds of change. The drawback is a significant increase in ct since the adaptation mechanism in (1) is activated for all the W samples in the adaptation window. On the contrary, the batch solution activates the adaptation only when the window of W samples is filled up (as explained in Sect. 3.2). This reduces the computational time but forces the hybrid solution to operate with an obsolete model up to time $t = \bar{t} + W$. Interestingly, the hybrid solution based on incremental learning is able to provide $mses$ that are in line or smaller than the ones provided by the passive one (expect for the drift case in $S1$ in R) but with a significantly smaller ct. This is not surprising since passive solutions typically provide good performance in case of drift changes but might suffer from a high computational complexity as commented in Sect. 2.

Table 2. Experimental results on the dataset $S1$ and $S2$.

	Hybrid-batch update		Hybrid-incremental update		Active		Passive	
	mse	ct	mse	ct	mse	ct	mse	ct
$S1$ - No change	0.0026 ± 0.0002	3.66	0.0027 ± 0.0003	6.04	0.0029 ± 0.0014	1.76	0.0039 ± 0.0003	89.99
$S1 - AA_1$	0.0886 ± 0.0146	2.48	0.0049 ± 0.0003	37.70	0.0921 ± 0.0044	1.38	0.0050 ± 0.0002	89.01
$S1 - AA_2$	0.7938 ± 0.0924	2.41	0.0180 ± 0.0014	36.90	0.8139 ± 0.0191	1.35	0.0147 ± 0.0015	90.12
$S1 - AM_1$	0.0965 ± 0.0139	2.42	0.0267 ± 0.0022	36.56	0.0982 ± 0.0049	1.38	0.0275 ± 0.0016	89.27
$S1 - AM_2$	0.0992 ± 0.0149	2.48	0.0317 ± 0.0029	36.76	0.1012 ± 0.0086	1.33	0.0343 ± 0.0019	90.70
$S1 - DR_1$	0.1996 ± 0.0194	2.43	0.0445 ± 0.0225	37.20	0.2396 ± 0.0101	1.34	0.0051 ± 0.0008	90.35
$S1 - DR_2$	0.7262 ± 0.0230	2.75	0.1948 ± 0.0570	37.68	0.9060 ± 0.0282	1.37	0.0072 ± 0.0009	90.65
$S1 - DR_3$	1.5475 ± 0.1256	2.47	0.4511 ± 0.1393	37.70	2.0881 ± 0.2902	1.42	0.0110 ± 0.0023	90.13
$S1 - FC$	5.3306 ± 0.6507	2.55	0.7298 ± 0.2620	38.45	5.5843 ± 0.6445	1.44	0.8633 ± 0.2889	90.45
$S2$ - No change	0.0032 ± 0.0003	3.77	0.0034 ± 0.0004	6.13	0.0032 ± 0.0003	1.82	0.0054 ± 0.0004	95.75
$S2 - AA_1$	0.0880 ± 0.0134	3.13	0.0470 ± 0.0451	37.13	0.0936 ± 0.0030	1.52	0.0433 ± 0.1149	95.95
$S2 - AA_2$	0.7846 ± 0.0959	2.76	0.1504 ± 0.1527	37.37	0.8124 ± 0.0273	1.53	0.1362 ± 0.1485	96.71
$S2 - AM_1$	0.1142 ± 0.0163	2.72	0.0666 ± 0.0179	37.11	0.1186 ± 0.0111	1.50	0.0721 ± 0.0154	94.41
$S2 - AM_2$	0.1153 ± 0.0223	2.80	0.0843 ± 0.0422	37.48	0.1213 ± 0.0124	1.57	0.0839 ± 0.0320	95.78
$S2 - DR_1$	0.2111 ± 0.0291	2.76	0.0961 ± 0.1377	37.42	0.2448 ± 0.0162	1.52	0.0802 ± 0.1773	96.73
$S2 - DR_2$	0.7247 ± 0.1064	2.95	0.3298 ± 0.1976	37.07	0.9951 ± 0.1504	1.59	0.2495 ± 0.3033	95.24
$S2 - DR_3$	1.5657 ± 0.2583	2.74	0.6817 ± 0.3364	37.07	2.2294 ± 0.4979	1.49	0.3442 ± 0.4707	95.11
$S2 - FC$	1.7379 ± 0.2319	2.81	0.8897 ± 0.2267	37.22	1.8109 ± 0.1425	1.54	0.9202 ± 0.1143	95.91

Table 3. Experimental results on the real dataset R.

	Hybrid-batch retrain		Hybrid-incremental retrain		Active		Passive	
	mse	*ct*	*mse*	*ct*	*mse*	*ct*	*mse*	*ct*
No change	0.0217	4.90	0.0226	17.08	0.0221	2.88	0.0352	90.66
$R - AA_1$	0.1052	3.99	0.0384	44.96	0.1170	2.86	0.0386	90.75
$R - AA_2$	0.7689	4.29	0.0578	44.54	0.8111	2.83	0.0443	89.15
$R - AM_1$	0.0171	4.61	0.0155	22.41	0.0290	2.81	0.0297	90.56
$R - AM_2$	0.0434	3.74	0.0374	44.82	0.0443	2.76	0.0542	88.64
$R - DR_1$	0.2272	4.33	0.0679	36.77	0.3068	2.94	0.0395	90.07
$R - DR_2$	0.9881	4.02	0.1439	51.37	1.1135	3.00	0.0416	92.01
$R - DR_3$	2.0021	4.18	0.3160	50.14	2.2252	3.12	0.0593	92.16

5 Conclusion

This paper introduced a novel hybrid approach to learn in nonstationary environments. This approach that is inspired by passive and active solutions present in the literature, jointly uses an incremental adaptation mechanism (as in passive solutions) and a change detection triggering the need to retrain the learning system (as in active solutions). A specific hybrid solution for regression problems in nonstationary environments has been introduced. The effectiveness and the efficiency of the proposed solution have been experimentally evaluated on both synthetic and real data and contrasted with a passive and a active solution.

References

1. Ditzler, G., Roveri, M., Alippi, C., Polikar, R.: Learning in nonstationary environments: a survey. IEEE Comput. Intell. Mag. **10**(4), 12–25 (2015)
2. Gama, J., Žliobaitė, I., Bifet, A., Pechenizkiy, M., Bouchachia, A.: A survey on concept drift adaptation. ACM Comput. Surv. **46**(4), 44 (2014)
3. Lindstrom, P., Mac Namee, B., Delany, S.J.: Drift detection using uncertainty distribution divergence. Evolving Syst. **4**(1), 13–25 (2013)
4. Elwell, R., Polikar, R.: Incremental learning of concept drift in nonstationary environments. IEEE Trans. Neural Netw. **22**(10), 1517–1531 (2011)
5. Alippi, C., Roveri, M.: Just-in-time adaptive classifiers–part I: detecting nonstationary changes. IEEE Trans. Neural Netw. **19**(7), 1145–1153 (2008)
6. Alippi, C.: Intelligence for Embedded Systems. Springer, Switzerland (2014)
7. Domingos, P., Hulten, G.: Mining high-speed data streams. In: ACM SIGKDD, pp. 71–80. ACM (2000)
8. Cohen, L., Avrahami, G., Last, M., Kandel, A.: Info-fuzzy algorithms for mining dynamic data streams. Appl. Soft Comput. **8**(4), 1283–1294 (2008)
9. Ye, Y., Squartini, S., Piazza, F.: Online sequential extreme learning machine in nonstationary environments. Neurocomputing **116**, 94–101 (2013)
10. Street, W.N., Kim, Y.: A streaming ensemble algorithm (SEA) for large-scale classification. In: ACM SIGKDD, pp. 377–382. ACM (2001)

11. Kolter, J.Z., Maloof, M.A.: Dynamic weighted majority: a new ensemble method for tracking concept drift. In: ICDM, pp. 123–130. IEEE (2003)

12. Brzezinski, D., Stefanowski, J.: Reacting to different types of concept drift: the accuracy updated ensemble algorithm. IEEE Trans. Neural Netw. Learn. Syst. **25**(1), 81–94 (2014)

13. Alippi, C., Boracchi, G., Roveri, M.: Just-in-time classifiers for recurrent concepts. IEEE Trans. Neural Netw. Learn. Syst. **24**(4), 620–634 (2013)

14. Basseville, M., Nikiforov, I.V., et al.: Detection of Abrupt Changes: Theory and Application, vol. 104. Prentice Hall, Englewood Cliffs (1993)

15. Alippi, C., Boracchi, G., Roveri, M.: Hierarchical change-detection tests. IEEE Trans. Neural Netw. Learn. Syst. **28**, 246–258 (2016)

16. Klinkenberg, R.: Learning drifting concepts: example selection vs. example weighting. Intell. Data Anal. **8**(3), 281–300 (2004)

17. Gama, J., Medas, P., Castillo, G., Rodrigues, P.: Learning with drift detection. In: Bazzan, A.L.C., Labidi, S. (eds.) SBIA 2004. LNCS, vol. 3171, pp. 286–295. Springer, Heidelberg (2004). doi:10.1007/978-3-540-28645-5_29

18. Bifet, A., Gavalda, R.: Learning from time-changing data with adaptive windowing. In: SIAM Conference on Data Mining, pp. 443–448. SIAM (2007)

19. Alippi, C., Boracchi, G., Roveri, M.: An effective just-in-time adaptive classifier for gradual concept drifts. In: IJCNN, pp. 1675–1682. IEEE (2011)

20. Aggarwal, C.C.: On biased reservoir sampling in the presence of stream evolution. In: VLDB, pp. 607–618. VLDB Endowment (2006)

21. Hornik, K.: Approximation capabilities of multilayer feedforward networks. Neural Netw. **4**(2), 251–257 (1991)

22. Redmond, M., Baveja, A.: A data-driven software tool for enabling cooperative information sharing among police departments. Eur. J. Oper. Res. **141**(3), 660–678 (2002)

Classifier Concept Drift Detection and the Illusion of Progress

Albert Bifet[(⊠)]

LTCI, Télécom ParisTech, Université Paris-Saclay, 75013 Paris, France
`albert.bifet@telecom-paristech.fr`

Abstract. When a new concept drift detection method is proposed, a common way to show the benefits of the new method, is to use a classifier to perform an evaluation where each time the new algorithm detects change, the current classifier is replaced by a new one. Accuracy in this setting is considered a good measure of the quality of the change detector. In this paper we claim that this is not a good evaluation methodology and we show how a non-change detector can improve the accuracy of the classifier in this setting. We claim that this is due to the existence of a temporal dependence on the data and we propose not to evaluate concept drift detectors using only classifiers.

Keywords: Concept drift · Data streams · Incremental · Classification · Evolving · Online

1   Introduction

IoT Analytics is a term used to identify machine learning done using data streams from the Internet of Things (IoT). Dealing with IoT data streams, or in data streams in general, drift detection is a very important component in adaptive modeling, since detecting a change gives a signal about when to adapt models [16,20,21]. Typically, the streaming error of predictive models is monitored and when the detector raises a change alarm, then the model is updated or replaced by a new one.

We start by discussing an example of how researchers evaluate a concept drift detector using two real datasets representing a data stream. The Electricity dataset due to [11] is a popular benchmark for testing adaptive classifiers. It has been used in over 50 concept drift experiments, for instance, [6,9,14,17]. The Electricity Dataset was collected from the Australian New South Wales Electricity Market. In this market, prices are not fixed and are affected by the demand and supply of the market. Prices are set every five minutes. The dataset contains 45,312 instances which record electricity prices at 30 min intervals. The class label identifies the change of the price (UP or DOWN) related to a moving average of the last 24 h. The data is subject to concept drift due to changing consumption habits, unexpected events and seasonality.

© Springer International Publishing AG 2017
L. Rutkowski et al. (Eds.): ICAISC 2017, Part II, LNAI 10246, pp. 715–725, 2017.
DOI: 10.1007/978-3-319-59060-8_64

The second, *Forest Covertype*, contains the forest cover type for 30×30 meter cells obtained from US Forest Service (USFS) Region 2 Resource Information System (RIS) data. It contains $581,012$ instances and 54 attributes, and has been used in several papers on data stream classification

Let us test two state-of-the-art data stream classifiers on this dataset. We test an incremental Naive Bayes classifier, and an incremental (streaming) decision tree learner. As a streaming decision tree, we use the Hoeffding Tree [12] with functional leaves, using Naive Bayes classifiers at the leaves. The Hoeffding Tree employs a strategy based on the Hoeffding bound to incrementally grow a decision tree. A node is expanded by splitting as soon as there is sufficient statistical evidence, based on the data seen so far, to support the split and this is decision is based on the distribution-independent Hoeffding bound.

Tables 1 and 2 show the performance of a Naive Bayes classifier and a Hoeffding Tree Classifier that uses a change detector to start a new classifier when a change is detected. As we can see in these tables, the best performance is due to the No-Change Detector. This detector outputs change every 60 instances; it is a no-change detector in the sense that it is not detecting change in the stream. Surprisingly, the

Table 1. Evaluation results of an adaptive Naive Bayes classifier on electricity and covertype datasets.

Change detector	Forest covertype		Electricity	
	Accuracy	κ	Accuracy	κ
ADWIN	83.24	73.25	81.03	60.79
CUSUM	81.55	70.66	79.21	56.83
DDM	88.03	80.78	81.18	61.14
Page-Hinckley	80.06	68.40	78.04	54.43
EDDM	86.08	77.67	84.83	68.96
No-change	**88.79**	**81.97**	**86.16**	**71.65**

Table 2. Evaluation results of an adaptive Hoeffding Tree classifier on electricity and covertype datasets.

Change detector	Forest covertype		Electricity	
	Accuracy	κ	Accuracy	κ
ADWIN	83.36	73.37	83.23	65.41
CUSUM	83.01	72.91	81.71	62.05
DDM	87.35	79.71	85.41	70.05
Page-Hinckley	81.65	70.75	81.95	62.60
EDDM	86.00	77.48	84.91	69.08
No-change	**88.04**	**80.71**	**85.54**	**70.27**

classifiers using this no-change detector are getting better results than using the standard change detectors.

This experiment shows us that it is not enough to show the performance of a change detector working with a classifier. There is need to use other evaluation techniques.

In Sect. 2, we present the state of the art of change detector algorithms. In Sect. 3, we perform an experimental evaluation of concept drift detectors not using classifiers. In Sect. 4, we discuss temporal dependence in data streams, and Sect. 5 concludes the paper.

2   Change Detectors

A *change detector* or *drift detector* is an algorithm that takes a stream of instances as input and outputs an alarm if it detects a change in the distribution of the data. A detector may often be combined with a predictive model to output a prediction of the next instance to come. In general, the input to a change detection algorithm is a sequence $x_1, x_2, \ldots, x_t, \ldots$ of data points whose distribution varies over time in an unknown way. At each time step the algorithm outputs:

1. an estimate of the parameters of the input distribution, and
2. an alarm signal indicating whether a change in this distribution has recently occurred, or not.

We consider a specific, but very frequent case, of this setting with all x_t being real values. The desired estimate is usually the current expected value of x_t, and sometimes other statistics of the distribution such as, for instance, variance. The only assumption about the distribution of x is that each x_t is drawn independently from each other. This assumption may be not satisfied if x_t is an error produced by a classifier that updates itself incrementally, because the update depends on the performance, and the next performance depends on whether we updated it correctly. In practice, however, this effect is negligible, so treating them independently is a reasonable approach.

The most general structure of a change detection algorithm contains three components:

1. *Memory* is the component where the algorithm stores the sample data or data summaries that are considered to be relevant at the current time, i.e., the ones that describe the current data distribution.
2. *Estimator* is an algorithm that estimates the desired statistics on the input data, which may change over time. The algorithm may or may not use the data contained in Memory. One of the simplest Estimator algorithms is the *linear estimator,* which simply returns the average of the data items contained in Memory. Other examples of run-time efficient estimators are Auto-Regressive, Auto Regressive Moving Average, and Kalman filters [13].

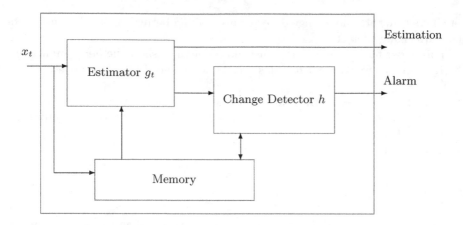

Fig. 1. General framework

3. *Change detector* (hypothesis testing) outputs an alarm signal when it detects a change in the input data distribution. It uses the output of the Estimator, and may or may not in addition use the contents of Memory.

There are many different algorithms to detect change in time series. We discuss the classical ones used in statistical quality control [2], time series analysis [18], statistical methods and more recent ones such as ADWIN[4].

2.1   Statistical Tests with Stopping Rules

These tests decide between the hypothesis that there is change and the hypothesis that there is no change, using a stopping rule. When this stopping rule is achieved, then the change detector method signals a change. The following methods differ in their stopping rule.

The CUSUM Test. The cumulative sum (CUSUM algorithm), which was first proposed in [15], is a change detection algorithm that raises an alarm when the mean of the input data is significantly different from zero. The CUSUM input ϵ_t can be any filter residual, for instance the prediction error from a Kalman filter.

The stopping rule of the CUSUM test is as follows:

$$g_0 = 0$$

$$g_t = \max\left(0, g_{t-1} + \epsilon_t - v\right)$$

if $g_t > h$ then alarm and $g_t = 0$

The CUSUM test is memoryless, and its accuracy depends on the choice of parameters v and h. Note that CUSUM is a one sided, or *asymmetric* test. It assumes that changes can happen only in one direction of the statistics, detecting only increases.

The Page Hinckley Test. The Page Hinckley Test [15] stopping rule is as follows, when the signal is increasing:

$$g_0 = 0, \qquad g_t = g_{t-1} + (\epsilon_t - v)$$

$$G_t = \min(g_t, G_{t-1})$$

$$\text{if } g_t - G_t > h \text{ then alarm and } g_t = 0$$

When the signal is decreasing, instead of $G_t = \min(g_t, G_{t-1})$, we should use $G_t = \max(g_t, G_{t-1})$ and $G_t - g_t > h$ as the stopping rule. Like the CUSUM test, the Page Hinckley test is memoryless, and its accuracy depends on the choice of parameters v and h.

2.2 Drift Detection Method

The drift detection method (DDM) proposed by Gama et al. [10] controls the number of errors produced by the learning model during prediction. It compares the statistics of two windows: the first contains all the data, and the second contains only the data from the beginning until the number of errors increases. Their method doesn't store these windows in memory. It keeps only statistics and a window of recent errors data.

The number of errors in a sample of n examples is modelled by a binomial distribution. For each point t in the sequence that is being sampled, the error rate is the probability of misclassifying (p_t), with standard deviation given by $s_t = \sqrt{p_t(1 - p_t)/t}$. They assume that the error rate of the learning algorithm (p_t) will decrease while the number of examples increases if the distribution of the examples is stationary. A significant increase in the error of the algorithm, suggests that the class distribution is changing and, hence, the actual decision model is supposed to be inappropriate. Thus, they store the values of p_t and s_t when $p_t + s_t$ reaches its minimum value during the process (obtaining p_{min} and s_{min}). DDM then checks if the following conditions trigger:

- $p_t + s_t \geq p_{min} + 2 \cdot s_{min}$ for the warning level. Beyond this level, the examples are stored in anticipation of a possible change of context.
- $p_t + s_t \geq p_{min} + 3 \cdot s_{min}$ for the drift level. Beyond this level the concept drift is supposed to be true, the model induced by the learning method is reset and a new model is learnt using the examples stored since the warning level triggered. The values for p_{min} and s_{min} are reset.

In the standard notation, they have two hypothesis tests h_w for warning and h_d for detection.

$$g_t = p_t + s_t$$

$$\text{if } g_t > h_w, \text{ then alarm warning,}$$

$$\text{if } g_t > h_d, \text{ then alarm detection,}$$

where $h_w = p_{min} + 2s_{min}$ and $h_d = p_{min} + 3s_{min}$.

The test is nearly memoryless, it only needs to store the statistics p_t and s_t, as well as switch on some memory to store an extra model of data from the time of warning until the time of detection.

This approach works well for detecting abrupt changes and reasonably fast changes, but it has difficulties detecting slow gradual changes. In the latter case, examples will be stored for long periods of time, the drift level can take too much time to trigger and the examples in memory may overflow.

Baena-García et al. proposed a new method EDDM [1] in order to improve DDM. It is based on the estimated distribution of the distances between classification errors. The window resize procedure is governed by the same heuristics.

2.3   ADWIN: ADaptive sliding WINdow algorithm

ADWIN[3] is a change detector and estimator that solves in a well-specified way the problem of tracking the average of a stream of bits or real-valued numbers. ADWIN keeps a variable-length window of recently seen items, with the property that the window has the maximal length statistically consistent with the hypothesis "there has been no change in the average value inside the window".

More precisely, an older fragment of the window is dropped if and only if there is enough evidence that its average value differs from that of the rest of the window. This has two consequences: one, that change can reliably be declared whenever the window shrinks; and two, that at any time the average over the existing window can be reliably taken as an estimate of the current average in the stream (barring a very small or very recent change that is still not statistically visible). These two points appears in [3] in a formal theorem.

ADWIN is data parameter- and assumption-free in the sense that it automatically detects and adapts to the current rate of change. Its only parameter is a confidence bound δ, indicating how confident we want to be in the algorithm's output, inherent to all algorithms dealing with random processes.

ADWIN does not maintain the window explicitly, but compresses it using a variant of the exponential histogram technique. This means that it keeps a window of length W using only $O(\log W)$ memory and $O(\log W)$ processing time per item.

3   Concept Drift Evaluation

Change detection is a challenging task due to a fundamental limitation [10]: the design of a change detector is a compromise between detecting true changes and avoiding false alarms.

When designing a change detection algorithm one needs to balance false and true alarms and minimize the time from the change actually happening to detection. The following existing criteria [2,10] formally capture these properties for evaluating change detection methods.

Mean Time between False Alarms (MTFA) characterizes how often we get false alarms when there is no change. The false alarm rate FAR is defined as 1/MTFA. A good change detector would have high MTFA.

Mean Time to Detection (MTD) characterizes the reactivity of the system to changes after they occur. A good change detector would have small MTD.

Missed Detection Rate (MDR) gives the probability of not receiving an alarm when there has been drift. It is the fraction of non-detected changes in all the changes that happened. A good detector would have small or zero MDR.

Average Run Length (ARL(θ)) generalizes over MTFA and MTD. It quantifies how long we have to wait before we detect a change of size θ in the variable that we are monitoring.

$$ARL(\theta = 0) = MTFA, \quad ARL(\theta \neq 0) = MTD$$

To do a fair comparison of change detectors, the evaluation framework needs to know ground truth changes in the data for evaluation of change detection algorithms. Thus, we need to use synthetic datasets with ground truth. Before a true change happens, all the alarms are considered as false alarms. After a true change occurs, the first detection that is flagged is considered as the true alarm. After that and before a new true change occurs, the consequent detections are considered as false alarms. If no detection is flagged between two true changes, then it is considered a missed detection.

In [7] a new quality evaluation measure was proposed that monitors the compromise between fast detection and false alarms:

$$MTR(\theta) = \frac{MTFA}{MTD} \times (1 - MDR) = \frac{ARL(0)}{ARL(\theta)} \times (1 - MDR). \tag{1}$$

This measure MTR (Mean Time Ratio) is the ratio between the mean time between false alarms and the mean time to detection, multiplied by the probability of detecting an alarm. An ideal change detection algorithm would have a low false positive rate (which means a high mean time between false alarms), a low mean time to detection, and a low missed detection rate.

Comparing two change detectors for a specific change θ is easy with this new measure: the algorithm that has the highest $MTR(\theta)$ value is to be preferred.

3.1 Comparative Experimental Evaluation

We performed a comparison using MOA [5] with the following methods: DDM, ADWIN, EDDM, Page-Hinckley Test, and CUSUM Test. The two last methods were used with $v = 0.005$ and $h = 50$ by default.

The experiments were performed simulating the error of a classifier system with a binary output 0 or 1. The probability of having an error is maintained as 0.2 during the first t_c instances, and then it changes gradually, linearly increasing by a value of α for each instance. The results were averaged over 100 runs.

Tables 3 and 4 show the results. Every single row represents an experiment where four different drifts occur at different times in Table 3, and four different drifts with different incremental values in Table 4. Note that MTFA values come from the no-change scenario. We observe the tradeoff between faster detection

Table 3. Evaluation results with change of $\alpha = 0.0001$.

Method	Measure	No change	$t_c = 1,000$	$t_c = 10,000$	$t_c = 100,000$	$t_c = 1,000,000$
ADWIN	1−MDR		0.13	1.00	1.00	1.00
	MTD		111.26	1,062.54	1,044.96	1,044.96
	MTFA	5,315,789				
	MTR		6,150	5,003	5,087	5,087
CUSUM(h=50)	1−MDR		0.41	1.00	1.00	1.00
	MTD		344.50	902.04	915.71	917.34
	MTFA	59,133				
	MTR		70	66	65	64
DDM	1−MDR		0.44	1.00	1.00	1.00
	MTD		297.60	2,557.43	7,124.65	42,150.39
	MTFA	1,905,660				
	MTR		2,790	745	267	45
Page-Hinckley(h=50)	1−MDR		0.17	1.00	1.00	1.00
	MTD		137.10	1,320.46	1,403.49	1,431.88
	MTFA	3,884,615				
	MTR		4,769	2,942	2,768	2,713
EDDM	1−MDR		0.95	1.00	1.00	1.00
	MTD		216.95	1,317.68	6,964.75	43,409.92
	MTFA	37,146				
	MTR		163	28	5	1

Table 4. Evaluation results with change at $t_c = 10,000$.

Method	Measure	No change	$\alpha = 0.00001$	$\alpha = 0.0001$	$\alpha = 0.001$
ADWIN	1−MDR		1.00	1.00	1.00
	MTD		4,919.34	1,062.54	261.59
	MTFA	5,315,789.47			
	MTR		1,080.59	5,002.89	20,320.76
CUSUM	1−MDR		1.00	1.00	1.00
	MTD		3,018.62	902.04	277.76
	MTFA	59,133.49			
	MTR		19.59	65.56	212.89
DDM	1−MDR		0.55	1.00	1.00
	MTD		3,055.48	2,557.43	779.20
	MTFA	1,905,660.38			
	MTR		345.81	745.15	2,445.67
Page-Hinckley	1−MDR		1.00	1.00	1.00
	MTD		4,659.20	1,320.46	405.50
	MTFA	3,884,615.38			
	MTR		833.75	2,941.88	9,579.70
EDDM	1−MDR		0.99	1.00	1.00
	MTD		4,608.01	1,317.68	472.47
	MTFA	37,146.01			
	MTR		7.98	28.19	78.62

and smaller number of false alarms. Page Hinckley with $h = 50$ and ADWIN are the methods with fewer false positives, however CUSUM is faster at detecting change for some change values. Using the new measure MTR, ADWIN seems to be the algorithm with the best results.

In a second experiment, we apply three different Page-Hinckley tests with three different values for $h = 25, 50, 75$. Table 5 contains the results. We observe that for $h = 25$, the test is the fastest detecting change, but it has more false positives. On the other hand, with $h = 50$, there are fewer false positives, but the detection is slower. Looking at the new MTR measure, the test with h = 75, is performing better than using other values.

This type of test, has the property that by increasing h we can reduce the number of false positives, at the expense of increasing the detection delay.

Table 5. Evaluation results of different PageHinckley tests with change of $\alpha = 0.0001$.

Method	Measure	No change	$t_c = 1,000$	$t_c = 10,000$	$t_c = 100,000$	$t_c = 1,000,000$
PageHinckley (h=25)	1−MDR		0.17	1.00	1.00	1.00
	MTD		137.10	1,315.35	1,396.56	1,386.92
	MTFA	1,346,666.67				
	MTR		1,653.31	1,023.81	964.27	970.98
PageHinckley (h=50)	1−MDR		0.17	1.00	1.00	1.00
	MTD		137.10	1,320.46	1,403.49	1,431.88
	MTFA	3,884,615.38				
	MTR		4,769.15	2,941.88	2,767.84	2,712.95
PageHinckley (h=75)	1−MDR		0.17	1.00	1.00	1.00
	MTD		137.10	1,326.01	1,410.96	1,473.90
	MTFA	4,208,333.33				
	MTR		5,166.58	3,173.68	2,982.60	2,855.23

4   Temporal Dependence in Data Streams

The excellent results of a No-Change Detector that outputs change every 60 instances, with the Electricity dataset is surprising. The reason of this good performance, could be due to the temporal dependence in the Electricity dataset [8,19,22]. If the price goes UP now, it is more likely than by chance to go UP again, and vice versa. Secondly, the prior distribution of classes in this data stream is evolving. Figure 2 plots the class distribution of this dataset over a sliding window of 1000 instances and the autocorrelation function of the target label. We can see that data is heavily autocorrelated with very clear cyclical peaks at every 48 instances (24 h), due to electricity consumption habits.

Let us consider a *No-Change* classifier that uses temporal dependence information by predicting that the next class label will be the same as the last seen class label. It can be compared to a naive weather forecasting rule: the weather tomorrow will be the same as today. The performance of this classifier is shown in Table 6. We see that this classifier is getting better results than most of the classifiers using state-of-the-art concept drift techniques. We observe that in the

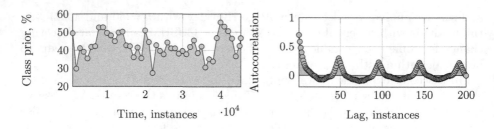

Fig. 2. Characteristics of the electricity dataset

Table 6. Evaluation results of a No-change classifier on electricity and covertype datasets.

Classifier	Forest covertype		Electricity	
	Accuracy	κ	Accuracy	κ
No-change classifier	**95.06**	**92.07**	**85.33**	**69.98**

case of the Forest Covertype dataset the performance of this No-Change classifier is much better than the methods using concept drift detection techniques.

5   Conclusions

Change detection is an important component of systems that need to adapt to changes in their input data. We discussed the surprising result that non-change detectors can outperform change-detectors when used in a classification streaming evaluation. We argued that this may be due to the temporal dependence on data, and we argued that evaluation of change detectors should not be done using only classifiers. We wish that this paper will open several directions for future research.

Acknowledgments. This work was supported by the Polish National Science Center under Grant No. 2014/15/B/ST7/05264.

References

1. Baena-García, M., del Campo-Ávila, J., Fidalgo, R., Bifet, A., Gavaldá, R., Morales-Bueno, R.: Early drift detection method. In: Fourth International Workshop on Knowledge Discovery from Data Streams (2006)
2. Basseville, M., Nikiforov, I.V.: Detection of Abrupt Changes: Theory and Application. Prentice-Hall Inc., Upper Saddle River (1993)
3. Bifet, A., Gavaldà, R.: Learning from time-changing data with adaptive windowing. In: SIAM International Conference on Data Mining (2007)
4. Bifet, A., Gavaldà, R.: Adaptive Learning from Evolving Data Streams. In: Adams, N.M., Robardet, C., Siebes, A., Boulicaut, J.-F. (eds.) IDA 2009. LNCS, vol. 5772, pp. 249–260. Springer, Heidelberg (2009). doi:10.1007/978-3-642-03915-7_22

5. Bifet, A., Holmes, G., Kirkby, R., Pfahringer, B.: MOA: massive online analysis. J. Mach. Learn. Res. **11**, 1601–1604 (2010)
6. Bifet, A., Holmes, G., Pfahringer, B., Kirkby, R., Gavaldà, R.: New ensemble methods for evolving data streams. In: Proceedings of the 15th ACM SIGKDD International Conference on Knowledge Discovery and Data Mining, KDD, pp. 139–148 (2009)
7. Bifet, A., Read, J., Pfahringer, B., Holmes, G., Žliobaitė, I.: CD-MOA: change detection framework for massive online analysis. In: Tucker, A., Höppner, F., Siebes, A., Swift, S. (eds.) IDA 2013. LNCS, vol. 8207, pp. 92–103. Springer, Heidelberg (2013). doi:10.1007/978-3-642-41398-8_9
8. Bifet, A., Read, J., Žliobaitė, I., Pfahringer, B., Holmes, G.: Pitfalls in benchmarking data stream classification and how to avoid them. In: Blockeel, H., Kersting, K., Nijssen, S., Železný, F. (eds.) ECML PKDD 2013. LNCS, vol. 8188, pp. 465–479. Springer, Heidelberg (2013). doi:10.1007/978-3-642-40988-2_30
9. Gama, J., Medas, P., Castillo, G., Rodrigues, P.: Learning with drift detection. In: Bazzan, A.L.C., Labidi, S. (eds.) SBIA 2004. LNCS (LNAI), vol. 3171, pp. 286–295. Springer, Heidelberg (2004). doi:10.1007/978-3-540-28645-5_29
10. Gustafsson, F.: Adaptive Filtering and Change Detection. Wiley, Chichester (2000)
11. Harries, M.: SPLICE-2 comparative evaluation: Electricity pricing. Technical report, University of New South Wales (1999)
12. Hulten, G., Spencer, L., Domingos, P.: Mining time-changing data streams. In: Proceedings of the 7th ACM SIGKDD International Conference on Knowledge Discovery and Data Mining, KDD, pp. 97–106 (2001)
13. Kobayashi, H., Mark, B.L., Turin, W.: Probability, Random Processes, and Statistical Analysis. Cambridge University Press, Cambridge (2011)
14. Kolter, J., Maloof, M.: Dynamic weighted majority: an ensemble method for drifting concepts. J. Mach. Learn. Res. **8**, 2755–2790 (2007)
15. Page, E.S.: Continuous inspection schemes. Biometrika **41**(1/2), 100–115 (1954)
16. Read, J., Bifet, A., Pfahringer, B., Holmes, G.: Batch-incremental versus instance-incremental learning in dynamic and evolving data. In: Hollmén, J., Klawonn, F., Tucker, A. (eds.) IDA 2012. LNCS, vol. 7619, pp. 313–323. Springer, Heidelberg (2012). doi:10.1007/978-3-642-34156-4_29
17. Ross, G., Adams, N., Tasoulis, D., Hand, D.: Exponentially weighted moving average charts for detecting concept drift. Pattern Recogn. Lett. **33**, 191–198 (2012)
18. Takeuchi, J., Yamanishi, K.: A unifying framework for detecting outliers and change points from time series. IEEE Trans. Knowl. Data Eng. **18**(4), 482–492 (2006)
19. Zliobaite, I.: How good is the electricity benchmark for evaluating concept drift adaptation. CoRR, abs/1301.3524 (2013)
20. Zliobaite, I., Bifet, A., Gaber, M.M., Gabrys, B., Gama, J., Minku, L.L., Musial, K.: Next challenges for adaptive learning systems. SIGKDD Explor. **14**(1), 48–55 (2012)
21. Zliobaite, I., Bifet, A., Holmes, G., Pfahringer, B.: MOA concept drift active learning strategies for streaming data. In: WAPA 2011, pp. 48–55 (2011)
22. Zliobaite, I., Bifet, A., Read, J., Pfahringer, B., Holmes, G.: Evaluation methods and decision theory for classification of streaming data with temporal dependence. Mach. Learn. **98**(3), 455–482 (2015)

Heuristic Regression Function Estimation Methods for Data Streams with Concept Drift

Maciej Jaworski[1(✉)], Piotr Duda[1], Leszek Rutkowski[1,2], Patryk Najgebauer[1], and Miroslaw Pawlak[2,3]

[1] Institute of Computational Intelligence, Czestochowa University of Technology, Armii Krajowej 36, 42-200 Czestochowa, Poland
{maciej.jaworski,piotr.duda,leszek.rutkowski, patryk.najgebauer}@iisi.pcz.pl
[2] Information Technology Institute, University of Social Sciences, 90-113 Łódź, Poland
[3] Department of Electrical and Computer Engineering, University of Manitoba, Winnipeg, Canada
pawlak@ee.umanitoba.ca

Abstract. In this paper the regression function methods based on Parzen kernels are investigated. Both the modeled function and the variance of noise are assumed to be time-varying. The commonly known kernel estimator is extended by adopting two popular tools often applied in concept drifting data stream scenario. The first tool is a sliding window, in which only a constant number of recently received data elements affects the estimator. The second one is the forgetting factor. In this case at each time step past data become less and less important. These heuristic approaches are experimentally compared with the basic mathematically justified estimator and demonstrate similar accuracy.

Keywords: Data streams · Concept drift · Regression estimation · Parzen kernels

1 Introduction

In recent years the amount of data processed in various fields of human activity increases constantly. Therefore data stream mining became a very important research area [1–4,10,27,30,42]. Usually the data stream is understood as a potentially infinite sequence of data elements. Because of limited available memory and computational power the traditional data mining algorithms designed for static data cannot be applied in this case. Another important issue connected with data streams is the concept drift [5,7,21,47]. It means that the distribution of data can change in time. Developed algorithms should be able to react to these changes. Among the existing in literature data stream mining methods the

M. Pawlak carried out this research at USS during his sabbatical leave from University of Manitoba.

most popular seem to be those based on decision trees [8, 33–35, 38–41]. However, they are mainly designed for data classification. In this paper a data regression problem is considered in which the target variable is a numerical value. The considered stream is a sequence of pairs (X_n, Y_n), $n = 1, 2, \ldots$, where variables X_n come from some unknown probability distribution described by a density function $f(x)$. Variables Y_i are assumed to be obtained with the following model

$$Y_n = \phi_n(X_n) + Z_n, \quad n = 1, 2, \ldots, \tag{1}$$

where $\phi_n(x)$ is a sequence of unknown functions. Noise variables Z_n come from the Gaussian distribution with zero mean

$$Z_n \sim \mathbb{N}(0, s_n), \quad s_n = z_0 n^\alpha, \quad z_0, \alpha \geq 0. \tag{2}$$

It should be noted that the variance of noise variables can change in time. Examples of problems with time-varying variance of noise can be found in [11] or [46].

In this paper the nonparametric regression methods based on Parzen kernels for estimating functions $\phi_n(x)$ are considered [9, 13–20, 22, 24–26, 28, 29, 36]. These kind of methods were also presented in literature under the name of general regression neural networks [45], first to deal with the stationary model

$$Y_n = \phi(X_n) + Z_n, \quad n = 1, 2, \ldots. \tag{3}$$

Let $R(x) = \phi(x) f(x)$, then the function $\phi(x)$ can be expressed in the following way if $f(x) \neq 0$

$$\phi(x) = \frac{\phi(x) f(x)}{f(x)} = \frac{R(x)}{f(x)}. \tag{4}$$

In the mentioned regression methods functions $R(x)$ and $f(x)$ are estimated separately and the estimator of $\phi_n(x)$ is expressed as their ratio. The simplest general form of density function estimator using Parzen kernels $K(u)$ [31] and based on sequence of observations $X_1, \ldots, X_n$ is given by (see e.g. [6])

$$\hat{f}_n(x) = \frac{1}{n} \sum_{i=1}^n \frac{1}{h_i} K\left(\frac{x - X_i}{h_i}\right), \tag{5}$$

where the bandwidth sequence h_i may or not change with i. The Parzen kernel function should satisfy some basic properties such as boundedness or integrability to 1. There are many possible Parzen kernels like uniform, Gaussian or triangular. In this paper the Epanechnikov's kernel [12] will be taken into account

$$K(u) = \begin{cases} 0.75 \left(1 - u^2\right) & |x| \leq 1 \\ 0, & |x| > 1 \end{cases} \tag{6}$$

This kind of methods for density function estimation was also presented in literature under the name probabilistic neural networks [43, 44].

The estimator for function $R(x)$ is analogous and can be expressed by

$$\hat{R}_n(x) = \frac{1}{n} \sum_{i=1}^{n} \frac{Y_i}{h_i} K\left(\frac{x - X_i}{h_i}\right),\tag{7}$$

Generally estimators (5) and (7) can use different sequences of bandwidths, however, in our paper to reduce the number of considered parameters the same sequence h_n is assumed for both $\hat{R}_n(x)$ and $\hat{f}_n(x)$. The estimator of function $\phi(x)$ can then be expressed as follows [23]

$$\hat{\phi}_n(x) = \frac{\hat{R}_n(x)}{\hat{f}_n(x)} = \frac{\sum_{i=1}^{n} Y_i \frac{1}{h_i} K\left(\frac{x - X_i}{h_i}\right)}{\sum_{i=1}^{n} \frac{1}{h_i} K\left(\frac{x - X_i}{h_i}\right)}.\tag{8}$$

In this paper we will adopt procedure (8) to deal with stream data generated by model (1).

The remainder of this paper is organized as follows. In Sect. 2 a recurrent estimator able to track time-changing functions $\phi_n(x)$ is recalled. In Sect. 3 the heuristic modifications of considered methods are presented. Two standard tools used in data stream mining approaches are adopted to existing methods - the sliding window and the forgetting factor. In Sect. 4 the experimental comparison of the considered estimators is shown. Section 5 concludes the paper.

2   Recurrent Regression with Parzen Kernels

The estimator $\hat{\phi}_n(x)$ of regression function $\phi_n(x)$ is defined as a ratio of estimator $\hat{R}_n(x)$ of function $R_n(x)$ and estimator $\hat{f}_n(x)$ of density function $f(x)$

$$\hat{\phi}_n(x) = \frac{\hat{R}_n(x)}{\hat{f}_n(x)}.\tag{9}$$

The estimators $\hat{R}_n(x)$ and $\hat{f}_n(x)$ can be expressed respectively in the following recurrent forms

$$\hat{R}_n(x) = \frac{n-1}{n}\hat{R}_{n-1}(x) + \frac{Y_n}{nh_n} K\left(\frac{x - X_n}{h_n}\right),\tag{10}$$

$$\hat{f}_n(x) = \frac{n-1}{n}\hat{f}_{n-1}(x) + \frac{1}{nh_n} K\left(\frac{x - X_n}{h_n}\right).\tag{11}$$

The sequence h_n is a sequence of Parzen kernel bandwidths and is usually given in the following general form

$$h_n = Dn^{-H}.\tag{12}$$

Since the parameters $D, H > 0$, the sequence h_n decreases with n.

The estimator (9) with estimator for $R_n(x)$ given by (10) works satisfactorily well for stationary functions ($\phi_n(x) \equiv \phi(x)$) with potentially increasing variance of noise ($\alpha \geq 0$). However, if the considered function is time-varying (e.g. $\phi_n(x) = n^\beta \phi(x)$ or $\phi_n(x) = \phi(x - n^\beta)$) it is better to apply a different estimator for $R_n(x)$. In [32,37] the authors, drawing the idea from stochastic approximation, replaced the factor $\frac{1}{n}$ by a general sequence a_n. In [32] the following estimator was proposed

$$\tilde{R}_n(x;\gamma) = (1 - a_{n,\gamma})\tilde{R}_{n-1}(x;\gamma) + a_{n,\gamma}\frac{Y_n}{h_n}K\left(\frac{x - X_n}{h_n}\right), \tag{13}$$

where the sequence $a_{n,\gamma}$ is given in the following form

$$a_{n,\gamma} = n^{-\gamma}. \tag{14}$$

It should be noticed that the stochastic approximation concerns only the estimator of $R_n(x)$. The estimator of density function remains the same. Finally, the estimator of function $\phi_n(x)$ is given by

$$\tilde{\phi}_n(x;\gamma) = \frac{\tilde{R}_n(x;\gamma)}{\hat{f}_n(x)}. \tag{15}$$

In [32] a set of inequalities is given (see conditions (38)–(43) there) which relates together the values of parameters α, β, H and γ. It is proved that if the inequalities are held then estimator (15) converges in probability to the true function $\phi_n(x)$.

3   Heuristic Modifications

Estimator (15) is mathematically justified for sufficiently low values of β, i.e. for sufficiently small changes of estimated function $\phi_n(x)$. However, for high values of β which do not fit into the mentioned inequalities, the behavior of estimator (15) is theoretically unpredictable - it becomes an heuristic approach. In this paper we present two heuristic regression methods based on Parzen kernels for any value of β. Estimator (9) is combined with two common tools often applied in data stream mining scenario, i.e. sliding windows and forgetting factor. The application of such tools let the estimator be more able to deal with various types of concept drifts in data.

Based on formula (8), one can introduce a following interpretation of estimator $\hat{\phi}_n(x)$ in point x: it is a weighted average of values of Y_i of incoming data elements, where each Y_i is included to the average with weight $\frac{1}{h_i}K\left(\frac{x-X_i}{h_i}\right)$. In the case of concept-drifting data the most recent data contain more information about the current data distribution than data from the past. Therefore, it is worth adding additional weights in formula (8) which depend increasingly on time of reading the data element (or on the index of data element). The general form of estimator with time-dependent weights is given by

$$\Psi_n(x;\theta) = \frac{\sum_{i=1}^n w_i(\theta)Y_i\frac{1}{h_i(\theta)}K\left(\frac{x-X_i}{h_i(\theta)}\right)}{\sum_{i=1}^n w_i(\theta)\frac{1}{h_i(\theta)}K\left(\frac{x-X_i}{h_i(\theta)}\right)}. \tag{16}$$

The weights $w_i(\theta)$ are parametrized by θ. We also assume that the applied parameters should affect the formula of sequence h_i (hence $h_i(\theta)$). Various forms of $w_i(\theta)$ lead to various types of estimators of function $\phi_n(x)$.

3.1   Sliding Window

In the case of sliding window at each moment of time only W recent data elements is taken into account for estimation, where W is the size of the sliding window. Therefore, the weights are given as follows

$$w_i(W) = \begin{cases} 1, & i > n - W, \\ 0, & i \le n - W. \end{cases} \tag{17}$$

Combining (17) with (16) the following estimator $\tilde{\Psi}_n(x; W)$ of function $\phi_n(x)$ is obtained

$$\tilde{\Psi}_n(x; W) = \frac{\tilde{r}_n(x; W)}{\tilde{F}_n(x, W)} = \frac{\sum_{i=n-W+1}^{n} Y_i \frac{1}{h_i(W)} K\left(\frac{x - X_i}{h_i(W)}\right)}{\sum_{i=n-W+1}^{n} \frac{1}{h_i(W)} K\left(\frac{x - X_i}{h_i(W)}\right)}, \tag{18}$$

where estimators $\tilde{r}_n(x; W)$ and $\tilde{F}_n(x; W)$ are respectively given by the following recurrent formulas

$$\tilde{r}_n(x; W) = \begin{cases} \tilde{r}_{n-1}(x; W) + \frac{Y_n K\left(\frac{x - X_n}{h_n(W)}\right)}{h_n(W)} - \frac{Y_{n-W} K\left(\frac{x - X_{n-W}}{h_{n-W}(W)}\right)}{h_{n-W}(W)}, & n > W \\ \tilde{r}_{n-1}(x; W) + \frac{Y_n K\left(\frac{x - X_n}{h_n(W)}\right)}{h_n(W)}, & n \le W \end{cases} , \tag{19}$$

$$\tilde{F}_n(x; W) = \begin{cases} \tilde{F}_{n-1}(x; W) + \frac{K\left(\frac{x - X_n}{h_n(W)}\right)}{h_n(W)} - \frac{K\left(\frac{x - X_{n-W}}{h_{n-W}(W)}\right)}{h_{n-W}(W)}, & n > W \\ \tilde{F}_{n-1}(x; W) + \frac{K\left(\frac{x - X_n}{h_n(W)}\right)}{h_n(W)}, & n \le W \end{cases} . \tag{20}$$

If the current number n of data elements is lower than the size of the sliding window, estimators (19) and (20) are the same as estimators (10) and (11), respectively. However, if the window is full, then additionally the 'oldest' (i.e. the $(n - W) - th$) data element is removed from it as well as its influence on estimators.

To keep the form of sequence h_n analogous to (12) a quantity $M(n, W)$ called an 'effective number of data' is introduced, which expresses how many data elements actually contributes to the current value of the estimator. In the case of sliding window the formula for $M(n, W)$ is very simple

$$M(n, W) = \min\{n, W\}. \tag{21}$$

The form of sequence $h_n(W)$ is therefore given by

$$h_n(W) = D\left[M(n, W)\right]^{-H} = D\left(\min\{n, W\}\right)^{-H}. \tag{22}$$

3.2   Forgetting Factor

In the following heuristic method for regression in is assumed that the significance of data elements decreases geometrically as new data are taken into account. The i-th data element is always λ times less significant that the subsequent $(i+1)$-th data element, where $\lambda < 1$. The formula for weights in this case is given by

$$w_i(\lambda) = \lambda^{n-i}, \tag{23}$$

Combining (23) with (16) the following estimator $\hat{\Psi}_n(x; \lambda)$ of function $\phi_n(x)$ is obtained

$$\hat{\Psi}_n(x; \lambda) = \frac{\hat{r}_n(x; \lambda)}{\hat{F}_n(x, \lambda)} = \frac{\sum_{i=1}^n \lambda^{n-i} Y_i \frac{1}{h_i(\lambda)} K\left(\frac{x-X_i}{h_i(\lambda)}\right)}{\sum_{i=1}^n \lambda^{n-i} \frac{1}{h_i(\lambda)} K\left(\frac{x-X_i}{h_i(\lambda)}\right)}, \tag{24}$$

where estimators $\hat{r}_n(x; \lambda)$ and $\hat{F}_n(x; \lambda)$ are recurrently expressed as follows

$$\hat{r}_n(x; \lambda) = \lambda \hat{r}_{n-1}(x; \lambda) + Y_n \frac{1}{h_n(\lambda)} K\left(\frac{x - X_n}{h_n(\lambda)}\right), \tag{25}$$

$$\hat{F}_n(x; \lambda) = \lambda \hat{F}_{n-1}(x; \lambda) + \frac{1}{h_n(\lambda)} K\left(\frac{x - X_n}{h_n(\lambda)}\right). \tag{26}$$

In estimators (25) and (26) at each time step n the previous value is multiplied by $\lambda < 1$, hence the influence of past data on the value of estimator is continuously forgotten. Therefore λ is called the 'forgetting factor'.

The effective number of data $M(n, \lambda)$ is slightly more abstract than in the case of sliding window and requires additional explanation. Since the i-th data element contributes to the value of estimator with weight λ^{n-i} it can be considered as an incomplete data element, i.e. a fraction λ^{n-i} of an element. Therefore the effective number of data can be expressed as the partial sum of geometric series

$$M(n, \lambda) = 1 + \cdots + \lambda^{n-1} = \sum_{i=0}^{n-1} \lambda^i = \frac{1 - \lambda^n}{1 - \lambda}. \tag{27}$$

Analogously to formulas (12) and (22) the form of sequence $h_n(\lambda)$ is given by

$$h_n(\lambda) = D\left[M(n, \lambda)\right]^{-H} = D\left(\frac{1 - \lambda^n}{1 - \lambda}\right)^{-H}. \tag{28}$$

4   Simulation Results

In this section all the presented estimators were compared experimentally. To evaluate the performance of an estimator the Mean Square Error was used. In some cases a quantity called the average Mean Square Error (aMSE) was applied.

If $MSE(\hat{\phi}_n)$ denotes the MSE of estimator $\hat{\phi}_n(x)$ then the aMSE is computed as follows

$$aMSE(\hat{\phi}) = \frac{1}{n-100} \sum_{i=100}^{n} MSE\left(\hat{\phi}_i\right), \tag{29}$$

where the summation starts from 100 to reject the influence of the unreliable estimator values at the very beginning stages of dataset processing.

4.1   Regression Function Estimation

To compare the performance of regression function estimators (15), (18) and (24) a synthetic data generator was used. Data were generated using model (1), where the family of functions $\phi_n(x)$ was set to be in the following form

$$\phi_n(x) = 10 \frac{2x^3 - x}{\cosh(2x)} n^\beta .. \tag{30}$$

Random variables X_n were generated from the uniform distribution in the interval $[-3{:}3]$. Noise variables Z_n were generated using (2) with coefficients $z_0 = 1$ and $\alpha = 0.1$. In all experiments the parameters of sequence h_n were set to $D = 2$ and $H = 2$.

At the beginning the case of relatively slowly changing function was considered, i.e. with $\beta = 0.15$. Firstly optimal values of parameters W, λ and γ for considered estimators were found. The values of parameters were investigated in intervals $W \in [1000{:}40000]$, $\lambda \in [0.999{:}1]$ and $\gamma \in [0.5; 0.95]$. For each parameter 20 different values were examined. For each value of each parameter five dataset with 100000 elements was generated randomly and the average aMSE of corresponding estimators over five runs of algorithm were calculated. The dependences of the aMSE on the value of W for $\tilde{\Psi}_n(x; W)$, λ for $\hat{\Psi}_n(x; \lambda)$ and γ for $\hat{\phi}_n(x; \gamma)$ are presented in Figs. 1, 2 and 3, respectively.

As can be seen, for each of the considered estimators an optimal value of appropriate parameter can be determined. Approximately these are: $W \approx 7000$, $\lambda \approx 0.99982$ and $\gamma \approx 0.81$. Therefore, in next experiment the performance of the following estimators were examined: $\tilde{\Psi}_n(x; 7000)$, $\hat{\Psi}_n(x; 0.99982)$ and $\hat{\phi}_n(x; 0.81)$. In Fig. 4 the MSE values in a function of number of data elements n are compared.

The estimator $\hat{\phi}_n(x; 0.82)$ seems to be the best in almost all range of numbers of data elements. The difference in MSE values is particularly visible at the beginning stages of dataset processing. It is worth noticing that estimator $\tilde{\Psi}_n(x; 7000)$ keeps approximately constant MSE value until the sliding window is not fulfilled. Then, after 7000 elements old data are removed and the MSE value starts decreasing drastically. In Fig. 5 the three considered estimators are compared with the true function (30) for $n = 100000$. Assessing it visually one can say that all estimators are fitted to the true function satisfactorily well. However, for some regions of x there are noticeable discrepancies for some estimators.

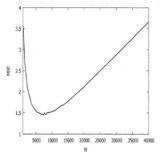

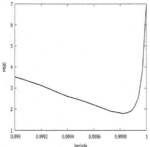

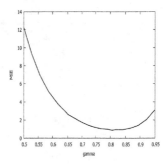

Fig. 1. The aMSE of estimator $\tilde{\Psi}_n(x; W)$ as a function of sliding window size W.

Fig. 2. The aMSE Squared Error of estimator $\hat{\Psi}_n(x; \lambda)$ as a function of forgetting factor λ.

Fig. 3. The aMSE of estimator $\tilde{\phi}_n(x; \gamma)$ as a function of parameter γ.

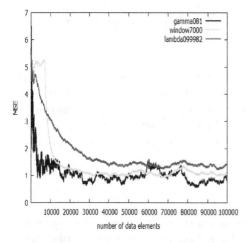

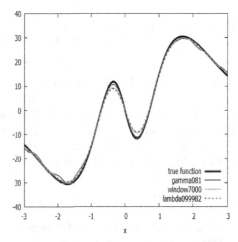

Fig. 4. The MSE as a function of the number of processed data elements n for $\beta = 0.15$ for three various estimators: $\tilde{\Psi}_n(x; 7000)$, $\hat{\Psi}_n(x; 0.99982)$ and $\tilde{\phi}_n(x; 0.82)$.

Fig. 5. Comparison of three various estimators: $\tilde{\Psi}_n(x; 7000)$, $\hat{\Psi}_n(x; 0.99982)$ and $\tilde{\phi}_n(x; 0.82)$ with the true function $\phi_n(x)$ for $\beta = 0.15$ and $n = 100000$.

In the previous experiment considered function was relatively slowly-changing, since β was set to 0.15. In the next experiment the behavior of estimators for a wide range of β values was investigated. For each value of $\beta \in [0; 0.9]$ with step $\Delta\beta = 0.02$ (i.e. for 41 different values of β) 20 datasets of size $n = 100000$ were generated randomly and the aMSE value was calculated as an average over 20 runs. The dependence between the aMSE in logarithmic scale and the value of β is presented in Fig. 6.

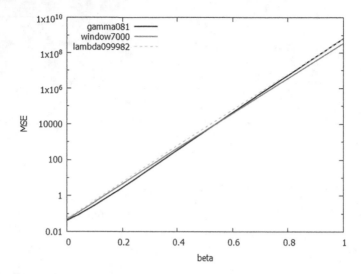

Fig. 6. The dependence between the average MSE and the value of parameter β for three various estimators: $\tilde{\Psi}_n(x; 7000)$, $\hat{\Psi}_n(x; 0.99982)$ and $\tilde{\phi}_n(x; 0.81)$.

For all considered estimators the value of aMSE increases approximately exponentially with increasing β. In the considered range of values of β all estimators seem to be comparable (i.e. they have similar orders of magnitude). However, it can be observed that for lower values of β estimator $\tilde{\phi}_n(x; 0.81)$ outperforms the other two. On the other hand for larger values of β the estimator with sliding window dominates.

5   Conclusions

In this paper the problem of regression in non-stationary environment was considered. The changes of both the modeled function and variance of noise were taken into account. The commonly known regression function estimator based on Parzen kernels was taken as a basis for proposed methods. This estimator was modified by adopting two popular tools often applied in various data stream mining algorithms - the sliding window and the forgetting mechanism. In the sliding window only a constant number of recently received data elements is taken into account. In the case of forgetting mechanism at each time step the past data become less and less important. These heuristic approaches are experimentally compared with the mathematically justified estimator based on stochastic approximation and demonstrate similar accuracy. For high non-stationary degrees, when the theoretical assumptions of the reference estimator are not satisfied, the estimator with sliding window slightly outperformed the other two estimators.

Acknowledgments. This work was supported by the Polish National Science Center under Grant No. 2014/15/B/ST7/05264.

References

1. Aggarwal, C.C.: Data Streams: Models and Algorithms. Advances in Database Systems. Springer, New York (2006)
2. Alippi, C., Boracchi, G., Roveri, M.: Hierarchical change-detection tests. IEEE Trans. Neural Netw. Learn. Syst. **28**(2), 246–258 (2017)
3. Bifet, A., Holmes, G., Kirkby, R., Pfahringer, B.: MOA: massive online analysis. J. Mach. Learn. Res. **11**, 1601–1604 (2010)
4. Bilski, J., Smolag, J.: Parallel architectures for learning the RTRN and Elman dynamic neural networks. IEEE Trans. Parallel Distrib. Syst. **26**(9), 2561–2570 (2015)
5. Brzezinski, D., Stefanowski, J.: Reacting to different types of concept drift: the accuracy updated ensemble algorithm. IEEE Trans. Neural Netw. Learn. Syst. **25**(1), 81–94 (2014)
6. Devroye, L.P.: On the pointwise and the integral convergence of recursive kernel estimates of probability densities. Utilitas Math. (Canada) **15**, 113–128 (1979)
7. Ditzler, G., Roveri, M., Alippi, C., Polikar, R.: Learning in nonstationary environments: a survey. IEEE Comput. Intell. Mag. **10**(4), 12–25 (2015)
8. Domingos, P., Hulten, G.: Mining high-speed data streams. In: Proceedings of the 6th ACM SIGKDD International Conference on Knowledge Discovery and Data Mining, pp. 71–80 (2000)
9. Duda, P., Hayashi, Y., Jaworski, M.: On the strong convergence of the orthogonal series-type kernel regression neural networks in a non-stationary environment. In: Rutkowski, L., Korytkowski, M., Scherer, R., Tadeusiewicz, R., Zadeh, L.A., Zurada, J.M. (eds.) ICAISC 2012. LNCS, vol. 7267, pp. 47 54. Springer, Heidelberg (2012). doi:10.1007/978-3-642-29347-4_6
10. Duda, P., Jaworski, M., Pietruczuk, L.: On pre-processing algorithms for data stream. In: Rutkowski, L., Korytkowski, M., Scherer, R., Tadeusiewicz, R., Zadeh, L.A., Zurada, J.M. (eds.) ICAISC 2012. LNCS, vol. 7268, pp. 56–63. Springer, Heidelberg (2012). doi:10.1007/978-3-642-29350-4_7
11. Ellis, P.: The time-dependent mean and variance of the non-stationary Markovian infinite server system. J. Math. Stat. **6**, 68–71 (2010)
12. Epanechnikov, V.A.: Non-parametric estimation of a multivariate probability density. Theory Probab. Appl. **14**(1), 153–158 (1969)
13. Er, M.J., Duda, P.: On the weak convergence of the orthogonal series-type kernel regresion neural networks in a non-stationary environment. In: Wyrzykowski, R., Dongarra, J., Karczewski, K., Waśniewski, J. (eds.) PPAM 2011. LNCS, vol. 7203, pp. 443–450. Springer, Heidelberg (2012). doi:10.1007/978-3-642-31464-3_45
14. Galkowski, T., Rutkowski, L.: Nonparametric recovery of multivariate functions with applications to system identification. Proc. IEEE **73**(5), 942–943 (1985)
15. Galkowski, T., Rutkowski, L.: Nonparametric fitting of multivariate functions. IEEE Trans. Autom. Control **31**(8), 785–787 (1986)
16. Gałkowski, T.: Kernel estimation of regression functions in the boundary regions. In: Rutkowski, L., Korytkowski, M., Scherer, R., Tadeusiewicz, R., Zadeh, L.A., Zurada, J.M. (eds.) ICAISC 2013. LNCS, vol. 7895, pp. 158–166. Springer, Heidelberg (2013). doi:10.1007/978-3-642-38610-7_15
17. Galkowski, T., Pawlak, M.: Nonparametric extension of regression functions outside domain. In: Rutkowski, L., Korytkowski, M., Scherer, R., Tadeusiewicz, R., Zadeh, L.A., Zurada, J.M. (eds.) ICAISC 2014. LNCS, vol. 8467, pp. 518–530. Springer, Cham (2014). doi:10.1007/978-3-319-07173-2_44

18. Galkowski, T., Pawlak, M.: Nonparametric function fitting in the presence of non-stationary noise. In: Rutkowski, L., Korytkowski, M., Scherer, R., Tadeusiewicz, R., Zadeh, L.A., Zurada, J.M. (eds.) ICAISC 2014. LNCS, vol. 8467, pp. 531–538. Springer, Cham (2014). doi:10.1007/978-3-319-07173-2_45

19. Galkowski, T., Pawlak, M.: Orthogonal series estimation of regression functions in nonstationary conditions. In: Rutkowski, L., Korytkowski, M., Scherer, R., Tadeusiewicz, R., Zadeh, L.A., Zurada, J.M. (eds.) ICAISC 2015. LNCS, vol. 9119, pp. 427–435. Springer, Cham (2015). doi:10.1007/978-3-319-19324-3_39

20. Galkowski, T., Pawlak, M.: Nonparametric estimation of edge values of regression functions. In: Rutkowski, L., Korytkowski, M., Scherer, R., Tadeusiewicz, R., Zadeh, L.A., Zurada, J.M. (eds.) ICAISC 2016. LNCS, vol. 9693, pp. 49–59. Springer, Cham (2016). doi:10.1007/978-3-319-39384-1_5

21. Gama, J., Žliobaitė, I., Bifet, A., Pechenizkiy, M., Bouchachia, A.: A survey on concept drift adaptation. ACM Comput. Surv. (CSUR) **46**(4), 44 (2014)

22. Greblicki, W., Krzyzak, A., Pawlak, M.: Distribution-free pointwise consistency of kernel regression estimate. Ann. Stat. **12**, 1570–1575 (1984)

23. Greblicki, W., Pawlak, M.: Necessary and sufficient consistency conditions for a recursive kernel regression estimate. J. Multivar. Anal. **23**(1), 67–76 (1987)

24. Greblicki, W., Pawlak, M.: Nonparametric System Identification. Cambridge University Press, Cambridge (2008)

25. Györfi, L., Kohler, M., Krzyzak, A., Walk, H.: A Distribution-free Theory of Nonparametric Regression. Springer Science & Business Media, New York (2006)

26. Jaworski, M., Er, M.J., Pietruczuk, L.: On the application of the parzen-type kernel regression neural network and order statistics for learning in a non-stationary environment. In: Rutkowski, L., Korytkowski, M., Scherer, R., Tadeusiewicz, R., Zadeh, L.A., Zurada, J.M. (eds.) ICAISC 2012. LNCS, vol. 7267, pp. 90–98. Springer, Heidelberg (2012). doi:10.1007/978-3-642-29347-4_11

27. Jaworski, M., Pietruczuk, L., Duda, P.: On resources optimization in fuzzy clustering of data streams. In: Rutkowski, L., Korytkowski, M., Scherer, R., Tadeusiewicz, R., Zadeh, L.A., Zurada, J.M. (eds.) ICAISC 2012. LNCS, vol. 7268, pp. 92–99. Springer, Heidelberg (2012). doi:10.1007/978-3-642-29350-4_11

28. Krzyzak, A., Pawlak, M.: Almost everywhere convergence of a recursive regression function estimate and classification. IEEE Trans. Inf. Theory **30**(1), 91–93 (1984)

29. Krzyzak, A., Pawlak, M.: The pointwise rate of convergence of the kernel regression estimate. J. Stat. Plann. Infer. **16**, 159–166 (1987)

30. Nikulin, V.: Prediction of the shoppers loyalty with aggregated data streams. J. Artif. Intell. Soft Comput. Res. **6**(2), 69–79 (2016)

31. Parzen, E.: On estimation of probability density function and mode. Ann. Math. Stat. **33**, 1065–1076 (1962)

32. Pietruczuk, L., Rutkowski, L., Jaworski, M., Duda, P.: The Parzen kernel approach to learning in non-stationary environment. In: 2014 International Joint Conference on Neural Networks (IJCNN), pp. 3319–3323 (2014)

33. Pietruczuk, L., Rutkowski, L., Jaworski, M., Duda, P.: A method for automatic adjustment of ensemble size in stream data mining. In: 2016 International Joint Conference on Neural Networks (IJCNN), pp. 9–15 (2016)

34. Pietruczuk, L., Duda, P., Jaworski, M.: Adaptation of decision trees for handling concept drift. In: Rutkowski, L., Korytkowski, M., Scherer, R., Tadeusiewicz, R., Zadeh, L.A., Zurada, J.M. (eds.) ICAISC 2013. LNCS, vol. 7894, pp. 459–473. Springer, Heidelberg (2013). doi:10.1007/978-3-642-38658-9_41

35. Pietruczuk, L., Rutkowski, L., Jaworski, M., Duda, P.: How to adjust an ensemble size in stream data mining? Inf. Sci. **381**, 46–54 (2017)

36. Rao, B.P.: Nonparametric Functional Estimation. Academic Press, Orlando (2014)
37. Rutkowski, L.: Generalized regression neural networks in time-varying environment. IEEE Trans. Neural Netw. **15**, 576–596 (2004)
38. Rutkowski, L., Jaworski, M., Pietruczuk, L., Duda, P.: The CART decision tree for mining data streams. Inf. Sci. **266**, 1–15 (2014)
39. Rutkowski, L., Jaworski, M., Pietruczuk, L., Duda, P.: Decision trees for mining data streams based on the Gaussian approximation. IEEE Trans. Knowl. Data Eng. **26**(1), 108–119 (2014)
40. Rutkowski, L., Jaworski, M., Pietruczuk, L., Duda, P.: A new method for data stream mining based on the misclassification error. IEEE Trans. Neural Netw. Learn. Syst. **26**(5), 1048–1059 (2015)
41. Rutkowski, L., Pietruczuk, L., Duda, P., Jaworski, M.: Decision trees for mining data streams based on the McDiarmid's bound. IEEE Trans. Knowl. Data Eng. **25**(6), 1272–1279 (2013)
42. Serdah, A.M., Ashour, W.M.: Clustering large-scale data based on modified affinity propagation algorithm. J. Artif. Intell. Soft Comput. Res. **6**(1), 23–33 (2016)
43. Specht, D.F.: Probabilistic neural networks. Neural Netw. **3**(1), 109–118 (1990)
44. Specht, D.F.: Probabilistic neural networks and the polynomial adaline as complementary techniques for classification. IEEE Trans. Neural Netw. **1**(1), 111–121 (1990)
45. Specht, D.F.: A general regression neural network. IEEE Trans. Neural Netw. **2**(6), 568–576 (1991)
46. Wong, K.F.K., Galka, A., Yamashita, O., Ozaki, T.: Modelling non-stationary variance in eeg time series by state space garch model. Comput. Biol. Med. **36**(12), 1327–1335 (2006)
47. Zliobaite, I., Bifet, A., Pfahringer, B., Holmes, G.: Active learning with drifting streaming data. IEEE Trans. Neural Netw. Learn. Syst. **25**(1), 27–39 (2014)

Author Index

Printed in the United States
By Bookmasters